# Molecular Vaccines

Matthias Giese
Editor

# Molecular Vaccines

From Prophylaxis to Therapy
Volume 2

*Editor*
Matthias Giese, PhD
Institute for Molecular Vaccines
Heidelberg
Germany

ISBN 978-3-319-00977-3     ISBN 978-3-319-00978-0   (eBook)
DOI 10.1007/978-3-319-00978-0
Springer Cham Heidelberg New York Dordrecht London

Library of Congress Control Number: 2013953240

© Springer International Publishing Switzerland 2014

This work is subject to copyright. All rights are reserved by the Publisher, whether the whole or part of the material is concerned, specifically the rights of translation, reprinting, reuse of illustrations, recitation, broadcasting, reproduction on microfilms or in any other physical way, and transmission or information storage and retrieval, electronic adaptation, computer software, or by similar or dissimilar methodology now known or hereafter developed. Exempted from this legal reservation are brief excerpts in connection with reviews or scholarly analysis or material supplied specifically for the purpose of being entered and executed on a computer system, for exclusive use by the purchaser of the work. Duplication of this publication or parts thereof is permitted only under the provisions of the Copyright Law of the Publisher's location, in its current version, and permission for use must always be obtained from Springer. Permissions for use may be obtained through RightsLink at the Copyright Clearance Center. Violations are liable to prosecution under the respective Copyright Law.

The use of general descriptive names, registered names, trademarks, service marks, etc. in this publication does not imply, even in the absence of a specific statement, that such names are exempt from the relevant protective laws and regulations and therefore free for general use.

While the advice and information in this book are believed to be true and accurate at the date of publication, neither the authors nor the editors nor the publisher can accept any legal responsibility for any errors or omissions that may be made. The publisher makes no warranty, express or implied, with respect to the material contained herein.

Printed on acid-free paper

Springer is part of Springer Science+Business Media (www.springer.com)

**Mens agitat molem.**

*To my children Geraldine, Sebastian and Laura, who make everything worthwhile.*

# Preface

The idea of this book was born out of rage.

I am developing vaccines for more than 20 years – a troublesome and laborious business. It can be troublesome because pure theory and successful application of a theory in lab models, and later its translation into clinical practice, are poles apart. And it can be laborious because until now there is no complete scientific compendium covering all essential aspects of modern vaccine development in one book, what makes the busy normal lab day more difficult for a developer.

A developer is obliged to read several textbooks: You will need one book to understand all immunological key aspects; to learn the way of presentation of chemically different antigens by professional cells embedded in various tissues and organs, triggered by different cytokines; to understand the interactions between pathogen recognition receptors and their PAMP or DAMP ligands; to use bioinformatics for prediction of epitopes; to follow the pathways of gene activation and their regulations; and to understand how B- and T-cell memory work or what the consequences of immunosenescence are on vaccines for elderly or how malnutrition strongly influences the immune system in very young people with consequences on vaccine efficacy. Another book would be needed to understand all aspects of modern adjuvant developments, to get a feeling for the different classes and origins of immunostimulants, and to see the multiple immune reactions caused by various adjuvants. Other books would be required to understand the different vaccine types, the different delivery technologies, the right use of nanoparticles, or the helpful assistance of biomarkers.

The here presented two volumes of *Molecular Vaccines: From Prophylaxis to Therapy* cover most of all essential aspects of modern vaccine development in different fields such as infectious, non-infectious or cancer diseases. Moreover, patent claiming strategies will be discussed and also requirements for international licensing. These are two books that will satisfy a great need that up to now has been unfulfilled.

150 authors, from more than 20 nations, from five continents, Asia, Australia, Africa, America, and Europe, contributed to this magnificent book. I am deeply impressed by the enormous responses I got upon my invitation to join our international author team. So I trust that readers of this book, academic and industrial researchers, professors, physicians and graduate students in biochemistry, molecular biology, biotechnology, and (vet) medicine,

will benefit from the comprehensive expertise and will be enabled to provide successful innovative research and development in modern vaccines.

I would like to take this opportunity to thank all authors who generously contributed their knowledge and insights to this book. Special thanks go to Raphael Lekscha, Heidelberg, for his excellent technical preparations of my illustrations. I am grateful to Springer Publishing, particularly to Claudia Panuschka, Vienna, who made my book idea possible, and Wilma McHugh, Heidelberg, for her active support.

Heidelberg, Germany                                     Matthias Giese, PhD

# Contents

**Volume II: From Prophylaxis to Therapy**

**Part V  Noninfectious and Noncancer (NINC) Vaccines**

27  **Vaccines for Hypertension and Atherosclerosis** .............  451
    Hiroyuki Sasamura, Tasuhiko Azegami, and Hiroshi Itoh

28  **Anti-ghrelin Therapeutic Vaccine: A Novel Approach
    for Obesity Treatment** ................................  463
    Sara Andrade, Marcos Carreira, Felipe F. Casanueva,
    Polly Roy, and Mariana P. Monteiro

29  **Vaccines for Type 1 Diabetes**............................  477
    Sandeep Kumar Gupta

30  **Novel Vaccines for Type I Allergy** ......................  489
    Sandra Scheiblhofer, Josef Thalhamer, and Richard Weiss

31  **Rice Seed-Based Allergy Vaccines: Induction
    of Allergen-Specific Oral Tolerance Against
    Cedar Pollen and House Dust Mite Allergies** ..............  503
    Fumio Takaiwa and Takachika Hiroi

**Part VI  Adjuvants and Nanotechnology**

32  **Laser for Skin Vaccine Delivery and Adjuvantation** ........  519
    Xinyuan Chen and Mei X. Wu

33  **Bacterial Lipopolysaccharide as Adjuvants** ...............  527
    Jesús Arenas

34  **Bacterial Toxins Are Successful Immunotherapeutic
    Adjuvants and Immunotoxins** ..........................  537
    Irena Adkins

35  **Plant Heat-Shock Protein-Based
    Self-Adjuvanted Immunogens** ..........................  551
    Selene Baschieri

36  **Functionalised Nanoliposomes for Construction of Recombinant Vaccines: Lyme Disease as an Example** ..... 561
Jaroslav Turánek, Josef Mašek, Michal Křupka, and Milan Raška

37  **Emerging Nanotechnology Approaches for Pulmonary Delivery of Vaccines** ..... 579
Amit K. Goyal, Goutam Rath, and Basant Malik

38  **Antigen Delivery Systems as Oral Adjuvants** ..... 603
Carlos Gamazo and Juan M. Irache

39  **Chitosan-Based Adjuvants** ..... 623
Guro Gafvelin and Hans Grönlund

40  **Mechanism of Adjuvanticity of Aluminum-Containing Formulas** ..... 633
Mirjam Kool and Bart N. Lambrecht

41  **From Polymers to Nanomedicines: New Materials for Future Vaccines** ..... 643
Philipp Heller, David Huesmann, Martin Scherer, and Matthias Barz

## Part VII  In Silico and Delivery Systems

42  **Considerations for Vaccine Design in the Postgenomic Era** .. 677
Christine Maritz-Olivier and Sabine Richards

43  **Vaccine Delivery Using Microneedles** ..... 697
Ryan F. Donnelly, Sharifa Al-Zahrani, Marija Zaric, Cian M. McCrudden, Cristopher J. Scott, and Adrien Kissenpfenning

44  **Nasal Dry Powder Vaccine Delivery Technology** ..... 717
Anthony J. Hickey, Herman Staats, Chad J. Roy, Kenneth G. Powell, Vince Sullivan, Ginger Rothrock, and Christie M. Sayes

45  **Nanotechnology in Vaccine Delivery** ..... 727
Martin J. D'Souza, Suprita A. Tawde, Archana Akalkotkar, Lipika Chablani, Marissa D'Souza, and Maurizio Chiriva-Internati

46  **Vaccine Delivery Systems: Roles, Challenges and Recent Advances** ..... 743
Aditya Pattani, Prem N. Gupta, Rhonda M. Curran, and R. Karl Malcolm

47  **APC-Targeted (DNA) Vaccine Delivery Platforms: Nanoparticle Aided** ..... 753
Pirouz Daftarian, Paolo Serafini, Victor Perez, and Vance Lemmon

| | | |
|---|---|---|
| 48 | **Lactic Acid Bacteria Vector Vaccines**.................... | 767 |
| | Maria Gomes-Solecki | |
| 49 | **Electroporation-Based Gene Transfer**.................... | 781 |
| | Mattia Ronchetti, Michela Battista, Claudio Bertacchini, and Ruggero Cadossi | |
| 50 | **Why Does an I.M. Immunization Work?**................. | 793 |
| | Emanuela Bartoccioni | |

## Part VIII  Patenting, Manufacturing, Registration

| | | |
|---|---|---|
| 51 | **Patentability of Vaccines: A Practical Perspective**.......... | 807 |
| | Stacey J. Farmer and Martin Grund | |
| 52 | **Influenza Cell-Culture Vaccine Production**................ | 823 |
| | Markus Hilleringmann, Björn Jobst, and Barbara C. Baudner | |
| 53 | **United States Food and Drug Administration: Regulation of Vaccines**............................... | 839 |
| | Valerie Marshall | |
| 54 | **Vaccines: EU Regulatory Requirements**................. | 845 |
| | Bettina Klug, Patrick Celis, Robin Ruepp, and James S. Robertson | |
| 55 | **Licensing and Permitting of Veterinary Vaccines in the USA: US Regulatory Requirements**................ | 851 |
| | Louise M. Henderson and AdaMae Lewis | |
| 56 | **Veterinary Vaccines: EU Regulatory Requirements**......... | 859 |
| | Rhona Banks | |

**Index** ................................................. 867

## Volume I: From Prophylaxis to Therapy

### Part I  Molecular Vaccines – From Prophylaxis to Therapry

1 **From Pasteur to Personalized Vaccines**
  Matthias Giese

### Part II  Vaccine Immunology

2 **Basic Vaccine Immunology**
  Matthias Giese

3 **Gut Immunology and Oral Vaccination**
  Sharon M. Tennant, Khitam Muhsen,
  and Marcela F. Pasetti

4 **Pediatric Immunology and Vaccinology**
  Sofia Ygberg and Anna Nilsson

### Part III  Vaccines For Infectious Diseases

5 **Subunit Vaccine Candidates Engineered from
  the Central Conserved Region of the RSV G Protein
  Aimed for Parenteral or Mucosal Delivery**
  Thien N. Nguyen, Christine Libon, and Stefan Ståhl

6 **Ebolavirus Vaccines**
  Thomas Hoenen

7 **Experimental Dengue Vaccines**
  Sathyamangalam Swaminathan and Navin Khanna

8 **Viruslike Particle Vaccines for Norovirus Gastroenteritis**
  Qiang Chen

9 **Toward a New Vaccine Against Measles**
  Alexander N. Zakhartchouk and George K. Mutwiri

10 **Development of Subunit Vaccines Against
   Shigellosis: An Update**
   Francisco J. Martinez-Becerra, Olivia Arizmendi,
   Jamie C. Greenwood II, and Wendy L. Picking

11 **Development of Subunit Vaccines for Group A
   Streptococcus**
   Colleen Olive

12 **Vaccination Against Malaria Parasites: Paradigms,
   Perils, and Progress**
   Noah S. Butler

13  **TB Vaccines: State of the Art and Progresses**
    Rodrigo Ferracine Rodrigues, Rogério Silva Rosada,
    Fabiani Gai Frantz, Frederico Gonzalez Colombo Arnoldi,
    Lucimara Gaziola de la Torre, and Celio Lopes Silva

14  **Paracoccidioidomycosis: Advance Towards
    a Molecular Vaccine**
    Luiz R. Travassos, Glauce M.G. Rittner,
    and Carlos P. Taborda

15  **Oral Vaccination of Honeybees Against
    Varroa Destructor**
    Sebastian Giese and Matthias Giese

16  **Lyme Disease: Reservoir-Targeted Vaccines**
    Maria Gomes-Solecki

17  **Anti-tick Vaccines for the Control of Ticks
    Affecting Livestock**
    Cassandra Olds, Richard Bishop, and Claudia Daubenberger

18  **Development of Safe and Efficacious
    Bluetongue Virus Vaccines**
    Polly Roy and Meredith Stewart

19  **Non-typhoidal Salmonellosis**
    Beatriz San Román, Victoria Garrido, and María-Jesús Grilló

**Part IV   Cancer Vaccines**

20  **Cancer Vaccines and the Potential Benefit
    of Combination with Standard Cancer Therapies**
    Eva Ellebæk, Mads Hald Andersen, and Inge Marie Svane

21  **Personalized Peptide Vaccine as a Novel Immunotherapy
    Against Advanced Cancer**
    Nobukazu Komatsu, Satoko Matsueda, Masanori Noguchi,
    Akira Yamada, Kyogo Itoh, and Tetsuro Sasada

22  **Molecular Immunotherapeutics and Vaccines for Renal
    Cell Carcinoma and Its Vasculature**
    Nina Chi Sabins, Jennifer L. Taylor, Devin B. Lowe,
    and Walter J. Storkus

23  **Lung Cancer Immunotherapy: Programmatic
    Development, Progress, and Perspectives**
    Edward A. Hirschowitz, Terry H. Foody, and John R. Yannelli

24  **Melanoma: Perspectives
    of a Vaccine Based on Peptides**
    Mariana H. Massaoka, Alisson L. Matsuo, Jorge A.B. Scutti,
    Denise C. Arruda, Aline N. Rabaça, Carlos R. Figueiredo,
    Camyla F. Farias, Natalia Girola, and Luiz R. Travassos

25  **Parvoviruses: The Friendly Anticancer
    Immunomodulator**
    Zahari Raykov, Svitlana P. Grekova, Assia L. Angelova,
    and Jean Rommelaere

26  **Dendritic Cells Pulsed with Viral Oncolysate**
    Philippe Fournier and Volker Schirrmacher

**Index**

# Contributors

**Irena Adkins, PhD** Institute of Microbiology of the ASCR, v.v.i., Prague, Czech Republic

**Archana Akalkotkar** Vaccine Nanotechnology Laboratory, Department of Pharmaceutical Sciences, College of Pharmacy and Health Sciences, Mercer University, Atlanta, GA, USA

**Sharifa Al-Zahrani, BSc** School of Pharmacy, Queen's University Belfast, Belfast, UK

**Mads Hald Andersen, PhD** Center for Cancer Immune Therapy (CCIT), Department of Haematology, Copenhagen University Hospital, Herlev, Denmark

**Sara Andrade, MBSc** Department of Anatomy and UMIB (Unit for Multidisciplinary Biomedical Research) of ICBAS, University of Porto, Porto, Portugal

**Assia L. Angelova, PhD** Programme Infection and Cancer, Tumor Virology Division, German Cancer Research Center (DKFZ), Heidelberg, Germany

**Jesús A. Arenas, PhD** Department of Microbiology, Utrecht University, Utrecht, The Netherlands

**Olivia Arizmendi** Department of Microbiology and Molecular Genetics, Oklahoma State University, Stillwater, OK, USA

**Frederico Gonzalez Colombo Arnoldi, PhD** Department of Biochemistry and Immunology, Center for Tuberculosis Research, School of Medicine of Ribeirão Preto, University of São Paulo, São Paulo, Brazil

**Denise C. Arruda, PhD** Experimental Oncology Unit, Department of Microbiology, Immunology and Parasitology, Federal University of São Paulo, São Paulo, Brazil

**Tasuhiko Azegami, MD, PhD** Department of Internal Medicine, School of Medicine, Keio University, Tokyo, Japan

**Rhona Banks, MIBiol, PhD** Triveritas Ltd., Bank Barn, How Mill, Brampton, UK

**Emanuela Bartoccioni, PhD**  General Pathology Institute, Università Cattolica S. Cuore, Rome, Italy

**Matthias Barz, PhD**  Institute of Organic Chemistry, Johannes Gutenberg-University Mainz, Mainz, Germany

**Selene Baschieri, PhD**  Biotechnology Laboratory ENEA, Casaccia Research Center, Rome, Italy

**Michela Battista, PhD**  IGEA S.p.A., Carpi, Italy

**Barbara Baudner, PhD**  Vaccine Research, Novartis Vaccines and Diagnostics Srl, Siena, Italy

**Claudio Bertacchini, MSc**  IGEA S.p.A, Carpi, Italy

**Richard Bishop, PhD**  Department of Biotechnology, International Livestock Research Institute, Nairobi, Kenya

**Noah S. Butler, PhD**  Department of Microbiology and Immunology, University of Oklahoma Health Sciences Center, Oklahoma City, OK, USA

**Ruggero Cadossi, MD**  IGEA S.p.A., Carpi, Italy

**Marcos Carreira, PhD**  CIBER de Fisiopatologia Obesidad y Nutricion (CB06/03), Instituto Salud Carlos III, Santiago de Compostela, Spain

**Felipe F. Casanueva, MD, PhD**  CIBER de Fisiopatologia Obesidad y Nutricion (CB06/03), Instituto Salud Carlos III, Santiago de Compostela, Spain

Department of Medicine, USC University Hospital Complex, University of Santiago de Compostela, Santiago de Compostela, Spain

**Patrick Celis, PhD**  European Medicines Agency (EMA), London, UK

**Lipika Chablani**  Vaccine Nanotechnology Laboratory, Department of Pharmaceutical Sciences, College of Pharmacy and Health Sciences, Mercer University, Atlanta, GA, USA

**Qiang Chen, PhD**  Center for Infectious Diseases and Vaccinology, Biodesign Institute at Arizona State University, Tempe, AZ, USA

College of Technology and Innovation, Arizona State University, Mesa, AZ, USA

**Xinyuan Chen, PhD**  Department of Dermatology, Harvard Medical School (HMS), Boston, MA, USA

Wellman Center for Photomedicine, Massachusetts General Hospital (MGH), Boston, MA, USA

Harvard-MIT Division of Health Sciences and Technology (HST), Cambridge, MA, USA

**Maurizio Chiriva-Internati, PhD**  Texas Tech University Health Science Center, School of Medicine, Lubbock, TX, USA

**Rhonda M. Curran, PhD** University of Ulster, Jordanstown, Northern Ireland, UK

**Pirouz Daftarian, PhD** Department of Ophthalmology, University of Miami, Miami, USA

Department of Biochemistry and Molecular Biology, University of Miami, Miami, FL, USA

**Claudia Daubenberger, PhD** Medical Parasitology and Infection Biology, Swiss Tropical and Public Health Institute, Basel, Switzerland

University of Basel, Basel, Switzerland

**Lucimara Gaziola de la Torre, PhD** School of Chemical Engineering, University of Campinas, UNICAMP, Campinas, Brazil

**Ryan F. Donnelly, PhD** School of Pharmacy, Queen's University Belfast, Belfast, UK

**Martin J. D'Souza, PhD** Vaccine Nanotechnology Laboratory, Department of Pharmaceutical Sciences, College of Pharmacy and Health Sciences, Mercer University, Atlanta, GA, USA

**Eva Ellebæk, MD** Centre for Cancer Immune Therapy (CCIT), Department of Haematology, Copenhagen University Hospital, Herlev, Denmark

Department of Oncology, Copenhagen University Hospital, Herlev, Denmark

**Camyla F. Farias** Experimental Oncology Unit, Department of Microbiology, Immunology and Parasitology, Federal University of São Paulo, São Paulo, Brazil

**Stacey J. Farmer, PhD** Grund Intellectual Property Group, Munich, Germany

**Carlos R. Figueiredo, PhD** Experimental Oncology Unit, Department of Microbiology, Immunology and Parasitology, Federal University of São Paulo, São Paulo, Brazil

**Terry H. Foody, RN** Division of Pulmonary and Critical Care Medicine, Department of Internal Medicine, University of Kentucky, Chandler Medical Center, Lexington, KY, USA

**Philippe Fournier, PhD** German Cancer Research Center (DKFZ), Heidelberg, Germany

**Fabiani Gai Frantz, PhD** DACTB, School of Pharmaceutical Sciences of Ribeirão Preto, University of São Paulo, São Paulo, Brazil

**Guro Gafvelin, PhD** Department of Medicine, Clinical Immunology and Allergy Unit, Karolinska Institutet, Clin. Immunol, Karolinska University Hospital, Stockholm, Sweden

Viscogel AB, Solna, Sweden

**Carlos Gamazo, PhD**  Department of Microbiology, University of Navarra, Pamplona, Spain

**Victoria Garrido, PhD**  Institute of Agrobiotechnology (CSIC-UPNA), Animal Health Research Group, Consejo Superoir de Investigaciones Científicas – Universidad Pública de Navarra, Pamplona, Spain

**Matthias Giese, PhD**  Institute for Molecular Vaccines, IMV, Heidelberg, Germany

**Sebastian Giese, MSc**  Institute for Molecular Vaccines, IMV, Heidelberg, Germany

**Natalia Girola**  Recepta Biopharma, São Paulo, Brazil

**Maria Gomes-Solecki, DVM**  Department of Microbiology, Immunology and Biochemistry, University of Tennessee Health Science Center, Memphis, TN, USA

**Amit Kumar Goyal, PhD**  Department of Pharmaceutics, ISF College of Pharmacy, Moga, Punjab, India

**Jamie C. Greenwood II**  Department of Microbiology and Molecular Genetics, Oklahoma State University, Stillwater, OK, USA

**Svitlana P. Grekova, PhD**  Programme Infection and Cancer, Tumor Virology Division, German Cancer Research Center (DKFZ), Heidelberg, Germany

**María-Jesús Grilló, PhD**  Institute of Agrobiotechnology (CSIC-UPNA), Animal Health Research Group, Consejo Superior de Investigaciones Científicas – Universidad Pública de Navarra, Pamplona, Spain

**Hans Grönlund, PhD**  Therapeutic Immune Design Unit, Department of Clinical Neuroscience, Center for Molecular Medicine, Karolinska Institutet, Karolinska University Hospital, Stockholm, Sweden

**Martin Grund, PhD**  Grund Intellectual Property Group, Munich, Germany

**Prem N. Gupta, PhD**  Formulation & Drug Delivery Division, Indian Institute of Integrative Medicine, Jammu, India

**Sandeep Kumar Gupta, MD**  Department of Pharmacology, Dhanlaxmi Srinivasan Medical College and Hospital, Siruvachur, Perambalur, Tamilnadu, India

**Philipp Heller, Dipl. Chemist**  Institute of Organic Chemistry, Johannes Gutenberg-University Mainz, Mainz, Germany

**Louise M. Henderson, PhD**  Henderson Consulting, LLC, Consultants for Veterinary Biologics, LLC, ND, USA

**Anthony J. Hickey, PhD**  Center for Aerosols & Nanomaterials Engineering, RTI International, Research Triangle Park, NC, USA

**Markus Hilleringmann, PhD**  Department of Applied Sciences and Mechatronics, University of Applied Sciences Munich, Munich, Germany

**Takachika Hiroi, PhD** Department of Allergy and Immunology, The Tokyo Metropolitan Institute of Medical Science, Tokyo, Japan

**Edward A. Hirschowitz, MD** Division of Pulmonary and Critical Care Medicine, Department of Internal Medicine, University of Kentucky, Chandler Medical Center, Lexington, KY, USA

Lexington Veteran's Administration Medical Center, Lexington, KY, USA

**Thomas Hoenen, PhD** Laboratory of Virology, Division of Intramural Research, National Institute of Allergy and Infectious Diseases, National Institutes of Health, Rocky Mountain Laboratories, Hamilton, MT, USA

**David Huesmann, Dipl. Chemist** Institute of Organic Chemistry, Johannes Gutenberg-University Mainz, Mainz, Germany

**Juan M. Irache, PhD** Department of Pharmacy and Pharmaceutical Technology, University of Navarra, Pamplona, Spain

**Hiroshi Itoh, MD, PhD** Department of Internal Medicine, School of Medicine, Keio University, Tokyo, Japan

**Kyogo Itoh, MD, PhD** Department of Immunology and Immunotherapy, Kurume University School of Medicine, Kurume, Japan

**Björn Jobst, PhD** Manufacturing Science & Technology (MS&T), Novartis Vaccines and Diagnostics GmbH, Marburg, Germany

**Navin Khanna, PhD** Department of Biological Sciences, Birla Institute of Technology and Science-Pilani, Hyderabad, Andhra Pradesh, India

**Adrien Kissenpfenning, PhD** School of Pharmacy, Centre for Infection and Immunology, Queen's University Belfast, Belfast, UK

**Bettina Klug, MD** Paul-Ehrlich Institut, Langen, Germany

**Mirjam Kool, PhD** Department Pulmonary Medicine, Erasmus MC, Rotterdam, The Netherlands

Laboratory of Immunoregulation and Mucosal Immunology, Ghent University, Ghent, Belgium

**Nobukazu Komatsu, PhD** Department of Immunology and Immunotherapy, Kurume University School of Medicine, Kurume, Japan

**Michal Křupka, PhD** Department of Immunology, Faculty of Medicine and Dentistry, Palacky University, Olomouc, Czech Republic

**Bart N. Lambrecht, MD, PhD** Department Pulmonary Medicine, Erasmus MC, Rotterdam, The Netherlands

Department for Molecular Biomedical Research, VIB, Ghent, Belgium and Laboratory of Immunoregulation and Mucosal Immunology, Ghent University, Ghent, Belgium

**Vance Lemmon, PhD** Miami Project to Cure Paralysis, University of Miami, Miami, FL, USA

**AdaMae Lewis, PhD** Lewis Biologics, Inc., Consultants for Veterinary Biologics, LLC, Ames, IA, USA

**Christine Libon, PhD** Department of Microbiotechnology, Institut de Recherche Pierre Fabre, Toulouse, France

**Devin B. Lowe, PhD** Department of Dermatology, University of Pittsburgh School of Medicine, Pittsburgh, PA, USA

**R. Karl Malcolm, PhD** Queen's University of Belfast, Belfast, UK

**Basant Malik** Department of Pharmaceutics, ISF College of Pharmacy, Moga, Punjab, India

**Christine Maritz-Olivier, PhD** Faculty of Natural and Agricultural Sciences, Department of Genetics, University of Pretoria, Pretoria, South Africa

Department of Genetics, School of Biological Sciences, University of Pretoria, Pretoria, South Africa

**Josef Mašek, PharmDr** Department of Pharmacology and Immunotherapy Brno, Veterinary Research Institute, Brno, Czech Republic

**Mariana H. Massaoka, PhD** Experimental Oncology Unit, Department of Microbiology, Immunology and Parasitology, Federal University of São Paulo, São Paulo, Brazil

**Valerie Marshall, MPH** Lieutenant Commander, United States Public Health Service Commissioned Corps, Food and Drug Administration, Center for Biologics Evaluation and Research, Office of Vaccines Research and Review, Rockville, MD, USA

**Francisco J. Martinez-Becerra, PhD** Department of Microbiology and Molecular Genetics, Oklahoma State University, Stillwater, OK, USA

**Satoko Matsueda, PhD** Cancer Vaccine Development Division, Research Center for Innovative Cancer Therapy, Kurume University, Fukuoka, Japan

**Alisson L. Matsuo, PhD** Recepta Biopharma, São Paulo, Brazil

**Cian M. McCrudden, PhD** School of Pharmacy, Queen's University Belfast, Belfast, UK

**Mariana P. Monteiro, MD, PhD** Department of Anatomy and UMIB (Unit for Multidisciplinary Biomedical Research) of ICBAS, University of Porto, Porto, Portugal

**Khitam Muhsen, PhD** Department of Medicine, Center for Vaccine Development, University of Maryland School of Medicine, Baltimore, MD, USA

**George K. Mutwiri, DVM, PhD** Vaccine and Infectious Disease Organization - International Vaccine Center (VIDO-InterVac), University of Saskatchewan, Saskatoon, SK, Canada

School of Public Health, University of Saskatchewan, Saskatchewan, SK, Canada

**Anna Nilsson, MD** Department of Women's and Children's Health, Karolinska Institute, Astrid Lindgrens Childrens Hospital, Stockholm, Sweden

**Thien N. Nguyen, PhD** Department of Microbiotechnology, Institut de Recherche Pierre Fabre, Toulouse, France

**Masanori Noguchi, MD, PhD** Clinical Research Division, Research Center for Innovative Cancer Therapy, Kurume University, Kurume, Japan

**Cassandra Olds, PhD** Department of Biotechnology, International Livestock Research Institute, Nairobi, Kenya

Department Medical Parasitology and Infection Biology, Swiss Tropical and Public Health Institute, Basel, Switzerland

University of Basel, Basel, Switzerland

Parasites, Vectors and Vector Borne Diseases, ARC Onderstepoort Veterinary Institute, Onderstepoort, South Africa

**Colleen Olive, Dr** Department of Immunology, Queensland Institute of Medical Research, Herston, QLD, Australia

**Marcela F. Pasetti, PhD** Department of Pediatrics, Center for Vaccine Development, University of Maryland School of Medicine, Baltimore, MD, USA

**Aditya Pattani, PhD** Kairav Chemofarbe Industries Ltd & NanoXpert Technologies, Mumbai, India

**Victor Perez, PhD** Department of Ophthalmology, University of Miami, Miami, FL, USA

Department of Microbiology and Immunology, University of Miami, Miller School of Medicine, Miami, FL, USA

**Wendy L. Picking, PhD** Department of Microbiology and Molecular Genetics, Oklahoma State University, Stillwater, OK, USA

**Kenneth G. Powell** BD Technologies, Research Triangle Park, NC, USA

**Aline N. Rabaça** Experimental Oncology Unit, Department of Microbiology, Immunology and Parasitology, Federal University of São Paulo, São Paulo, Brazil

**Milan Raška, MD, PhD** Department of Immunology, Faculty of Medicine and Dentistry, Palacky University, Olomouc, Czech Republic

**Goutam Rath** Department of Pharmaceutics, ISF College of Pharmacy, Moga, Punjab, India

**Zahari Raykov, MD** Programme Infection and Cancer, Tumor Virology Division, German Cancer Research Center (DKFZ), Heidelberg, Germany

**Sabine Richards, MSc** Faculty of Natural and Agricultural Sciences, Department of Genetics, Lynnwood road, University of Pretoria, Pretoria, South Africa

**Glauce M.G. Rittner, PhD** Department of Microbiology, Institute of Biomedical Sciences, University of São Paulo, São Paulo, SP, Brazil

**Rodrigo Ferracine Rodrigues, PhD** Department of Biochemistry and Immunology, Center for Tuberculosis Research, School of Medicine of Ribeirão Preto, University of São Paulo, São Paulo, Brazil

**James S. Robertson, PhD** National Institute for Biological Standards and Control, Blanche Lane, South Mimms, Potters Bar, UK

**Jean Rommelaere, PhD** Programme Infection and Cancer, Tumor Virology Division, German Cancer Research Center (DKFZ), Heidelberg, Germany

**Mattia Ronchetti, BSc** IGEA S.p.A., Carpi, Italy

**Rogério Silva Rosada, PhD** Department of Biochemistry and Immunology, Center for Tuberculosis Research, School of Medicine of Ribeirão Preto, University of São Paulo, São Paulo, Brazil

**Ginger Rothrock** Center for Aerosols and Nanomaterials Engineering, RTI International, Research Triangle Park, NC, USA

**Chad J. Roy, PhD** Department of Microbiology and Immunology, Tulane University School of Medicine, New Orleans, LA, USA

**Polly Roy, PhD** Department of Infectious and Tropical Diseases, London School of Hygiene and Tropical Medicine, London, UK

**Robin Ruepp, PhD** European Medicines Agency (EMA), London, UK

**Nina Chi Sabins, PhD** Department of Immunology, University of Pittsburgh School of Medicine, Pittsburgh, PA, USA

**Hiroyuki Sasamura, MD, PhD** Department of Internal Medicine, School of Medicine, Keio University, Tokyo, Japan

**Tetsuro Sasada, MD, PhD** Department of Immunology and Immunotherapy, Kurume University School of Medicine, Kurume, Japan

**Beatriz San Román, PhD** Institute of Agrobiotechnology (CSIC-UPNA), Animal Health Research Group, Consejo Superior de Investigaciones Científicas – Universidad Pública de Navarra, Pamplona, Spain

**Christie M. Sayes** Center for Aerosols & Nanomaterials Engineering, RTI International, Research Triangle Park, NC, USA

**Sandra Scheiblhofer, PhD** Department of Molecular Biology, University of Salzburg, Salzburg, Austria

**Martin Scherer, Dipl. Chemist** Institute of Organic Chemistry, Johannes Gutenberg-University Mainz, Mainz, Germany

**Volker Schirrmacher, PhD** German Cancer Research Center (DKFZ), Heidelberg, Germany

IOZK Cologne, Cologne, Germany

**Cristopher J. Scott, PhD** School of Pharmacy, Queen's University Belfast, Belfast, UK

**Jorge A.B. Scutti** Experimental Oncology Unit, Department of Microbiology, Immunology and Parasitology, Federal University of São Paulo, São Paulo, Brazil

**Paolo Serafini, PhD** Department of Microbiology and Immunology, Miller School of Medicine, University of Miami, Miami, FL, USA

**Celio Lopes Silva, PhD** Department of Biochemistry and Immunology, Center for Tuberculosis Research, School of Medicine of Ribeirão Preto, University of São Paulo, São Paulo, Brazil

**Herman Staats, PhD** Departments of Pathology and Immunology, Duke University Medical Center, Durham, NC, USA

**Stefan Ståhl, PhD** Dean, Division of Molecular Biotechnology, School of Biotechnology, Alba Nova University Center, KTH Royal Institute of Technology, Stockholm, Sweden

**Meredith Stewart, PhD** Department of Infectious and Tropical Diseases, London School of Hygiene and Tropical Medicine, London, UK

**Walter J. Storkus, PhD** Department of Immunology, University of Pittsburgh School of Medicine, Pittsburgh, PA, USA

Department of Dermatology, University of Pittsburgh School of Medicine, Pittsburgh, PA, USA

Departments of Dermatology and Immunology, University of Pittsburgh Cancer Institute, Pittsburgh, PA, USA

**Vince Sullivan** BD Technologies, Research Triangle Park, NC, USA

**Inge Marie Svane, MD** Centre for Cancer Immune Therapy (CCIT), Department of Haematology, Copenhagen University Hospital, Herlev, Denmark

Department of Oncology, Copenhagen University Hospital, Herlev, Denmark

**Sathyamangalam Swaminathan, PhD** Department of Biological Sciences, Birla Institute of Technology and Science-Pilani, Hyderabad, Andhra Pradesh, India

**Carlos P. Taborda, PhD** Laboratory of Medical Mycology, Institute of Tropical Medicine of São Paulo, University of São Paulo, São Paulo, SP, Brazil

Department of Microbiology, Institute of Biomedical Sciences, University of São Paulo, São Paulo, SP, Brazil

**Fumio Takaiwa, PhD** National Institute of Agrobiological Sciences, Tsukuba Ibaraki, Japan

**Suprita A. Tawde** Vaccine Nanotechnology Laboratory, Department of Pharmaceutical Sciences, College of Pharmacy and Health Sciences, Mercer University, Atlanta, GA, USA

**Jennifer L. Taylor, PhD** Department of Dermatology, University of Pittsburgh School of Medicine, Pittsburgh, PA, USA

**Sharon M. Tennant, PhD** Department of Medicine, Center for Vaccine Development, University of Maryland School of Medicine, Baltimore, MD, USA

**Josef Thalhamer, PhD** Department of Molecular Biology, University of Salzburg, Salzburg, Austria

**Luiz R. Travassos, MD, PhD** Cell Biology Division, Department of Microbiology, Immunology and Parasitology, Federal University of São Paulo, São Paulo, SP, Brazil

**Jaroslav Turánek, Dr. Sc.** Department of Pharmacology and Immunotherapy Brno, Veterinary Research Institute, Brno, Czech Republic

**Richard Weiss, PhD** Department of Molecular Biology, University of Salzburg, Salzburg, Austria

**Mei X. Wu, MD, PhD** Department of Dermatology, Harvard Medical School (HMS), Boston, MA, USA

Wellman Center for Photomedicine, Massachusetts General Hospital (MGH), Boston, MA, USA

Harvard-MIT Division of Health Sciences and Technology (HST), Cambridge, MA, USA

**Akira Yamada, PhD** Cancer Vaccine Development Division, Research Center for Innovative Cancer Therapy, Kurume University, Kurume, Japan

**John R. Yannelli, PhD** Department of Microbiology, Immunology and Human Genetics, University of Kentucky, Lexington, KY, USA

**Sofia Ygberg, MD, PhD** Department of Women's and Children's Health, Karolinska Institute, Astrid Lindgrens Childrens Hospital, Stockholm, Sweden

**Alexander N. Zakhartchouk, DVM, PhD** Vaccine and Infectious Disease Organization - International Vaccine Center (VIDO-InterVac), University of Saskatchewan, Saskatoon, SK, Canada

**Marija Zaric, MB** Centre for Infection and Immunology, Queen's University Belfast, Belfast, UK

# Part V
# Noninfectious and Noncancer (NINC) Vaccines

## Overview of Part V

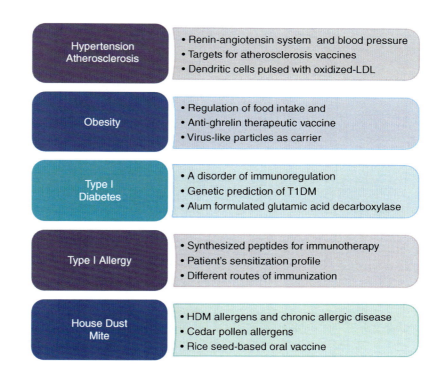

Priorities for vaccine developments are traditionally viral and bacterial infectious diseases. Meanwhile therapeutical anticancer vaccines are pushing to market authorization. Consistent with the molecular insights into various noninfectious and noncancer (NINC) diseases and the analyses of their pathways and networks present different targets for innovative NINC vaccination strategies. The most prominent example of a NINC vaccine is Alzheimer's disease. Dementia is the fastest-growing global health epidemic (WHO). A vaccine targeting the amyloid β peptide combined with monophosphoryl lipid A (see Chaps. 33 and 34) as adjuvant is able to stimulate the brain's natural defense mechanisms in people with Alzheimer's disease. In this part you will get perspectives from scientific leaders on some most pressing issues and promising approaches on vaccine development. NINC vaccines will dramatically help to improve public health.

*Hypertension and atherosclerosis.* Systemic blood pressure is determined by the product of cardiac output and total peripheral resistance, finely controlled by multiple mechanisms involving the kidney, the endocrine system, the central nervous system, and the vasculature. One of the important endocrine regulators of blood pressure is the renin-angiotensin system (RAS). Although vaccines for hypertension have focused predominantly on the RAS, in the case of vaccines for atherosclerosis, multiple targets in different cells have become candidates for atherosclerosis vaccine and immunomodulatory treatments.

*Obesity treatment.* Ghrelin is a gastrointestinal hormone that promotes food intake and decreases energy expenditure. Ghrelin is produced predominantly in the gastric fundus and conveys orexigenic signals to the hypothalamus. The suppression of endogenous ghrelin bioactivity with anti-ghrelin vaccines using keyhole limpet hemocyanin as carrier protein or bovine serum albumin was also tested in mice and pigs, respectively. These vaccines were able to decrease body weight gain and fat mass. These anti-ghrelin vaccinations present several limitations when applied to humans, e.g., the risk of immune response.

*Type I diabetes.* The pathogenic immune response is believed to be mediated by T lymphocytes that are reactive to islet β cell self-antigen(s) (autoreactive T cells), whereas a protective immune response may be mediated by T cells that suppress the autoreactive T cells (regulatory T cells). Glutamic acid decarboxylase (GAD) is found in islet cells. GAD is a major autoantigen in autoimmune diabetes and a target for vaccine development. The administration of recombinant human GAD with or without adjuvants did not induce adverse side effects or exacerbate T1DM in man and mice in preclinical studies and a phase I clinical trial.

*Type I allergy.* The next generation of allergy diagnostics and therapeutics will be based on recombinant proteins allowing tailor-made treatment according to the patient's sensitization profile. Besides the benefits of a highly standardized product, production of recombinant allergens also allows modification of the allergen of interest. Modified allergens with low-IgE binding potential are called hypoallergens. As genetic vaccination represents a highly versatile platform to design advanced types of vaccines, various innovative approaches have been tested in animal models of allergy.

**Rice against allergies.** Japanese cedar pollinosis is the most predominant seasonal allergic disease in Japan and is caused by pollen spread over most areas of Japan in early spring from February to April. House dust mite (HDM) is also a major source of inhalant allergens which cause chronic allergic disease such as bronchial asthma. About 45–80 % of patients with allergic asthma are sensitized to allergens in HDM. As a new form of oral allergy vaccine, rice seeds that accumulate hypoallergenic cedar pollen allergens and HDM allergen derivatives can be used as a vehicle to deliver to GALT.

# Vaccines for Hypertension and Atherosclerosis

## 27

Hiroyuki Sasamura, Tasuhiko Azegami, and Hiroshi Itoh

## Contents

| | | |
|---|---|---|
| 27.1 | **Vaccines for Hypertension** | 451 |
| 27.1.1 | Disease | 451 |
| 27.1.2 | Pathophysiology and Classical Therapy | 452 |
| 27.1.3 | Vaccines | 452 |
| 27.1.4 | Strengths and Weaknesses | 455 |
| 27.2 | **Vaccines for Atherosclerosis** | 456 |
| 27.2.1 | Disease | 456 |
| 27.2.2 | Pathophysiology and Classical Therapy | 456 |
| 27.2.3 | Vaccines | 457 |
| 27.2.4 | Strengths and Weaknesses | 459 |
| **References** | | 459 |

H. Sasamura, MD, PhD (✉) • T. Azegami, MD, PhD
H. Itoh, MD, PhD
Department of Internal Medicine,
School of Medicine, Keio University, Tokyo, Japan
e-mail: sasamura@a8.keio.jp

### Abstract

Both hypertension and atherosclerosis are chronic medical conditions which require continuous therapy. Recent basic and clinical studies have shown that vaccination against the renin-angiotensin system is effective in reducing blood pressure. In the case of atherosclerosis, vaccines targeting endothelial, macrophage, immune system, and lipid metabolism targets have been shown to reduce atherosclerosis in animal models. The development of vaccines for hypertension and atherosclerosis may be an important strategy for the prevention and treatment of these diseases.

## 27.1 Vaccines for Hypertension

### 27.1.1 Disease

Hypertension is a chronic medical condition characterized by elevation of blood pressure in the arterial system. Although hypertension is usually asymptomatic, it is a potent risk factor for cardiovascular diseases, such as stroke, myocardial infarction, heart failure, and end-stage renal disease [1].

It has been estimated that approximately 26.4 % of the adult population in the world had hypertension in 2000 [2]. The prevalence is projected to reach a value of 29.2 % by the year 2025. This is surprising, when we consider that a large number of new and effective antihypertensive agents have become widely available in recent years, and the choice of drug classes available has increased dramatically during the past 30 years [3].

The fact that prevalence of hypertension and its complications continues to increase in spite of many important advances in our knowledge of hypertension pathophysiology and treatment has been called the "hypertension paradox" [4]. One of the causes of this paradox is the problem of low rates of treatment and control of high blood pressure, which is often caused by low adherence to drug therapy. One of the countries with the greatest success rates for hypertension control is the United States; even so, recent data indicate that approximately 20 % of Americans with hypertension are unaware of their condition, 28 % are not receiving therapy, and approximately half do not have their blood pressure controlled to recommended levels [5]. The data for other countries are generally much lower [3]. For these reasons, new strategies to deal with the growing epidemic of hypertension are required.

## 27.1.2 Pathophysiology and Classical Therapy

Systemic blood pressure is determined by the product of cardiac output and total peripheral resistance. Both cardiac output and total peripheral resistance are finely controlled by multiple mechanisms involving the kidney, the endocrine system, the central nervous system, and the vasculature. One of the important endocrine regulators of blood pressure is the renin-angiotensin system (RAS) [6].

Renin is a proteolytic enzyme that is synthesized in the juxtaglomerular apparatus (JGA) of the kidney and stored in intracellular granules. During states of low blood pressure, renin is secreted from the JGA cells and enters the circulation. Angiotensinogen, which is the physiological substrate of renin, is a glycoprotein that is synthesized in the liver and enters the bloodstream. After proteolytic cleavage by renin, a 10-amino acid peptide (angiotensin I) is formed. This is then converted to the 8-peptide angiotensin II after interaction with angiotensin-converting enzyme (ACE) which is located primarily on endothelial cells.

Angiotensin II acts on multiple target organs, including the vasculature, heart, kidney, adrenal, and central nervous system. The effects to increase blood pressure are mediated almost exclusively by the G-protein-coupled angiotensin type 1 (AT1) receptor. In contrast, the AT2 receptor has predominantly antagonistic effects to the AT1 receptor and may act to decrease blood pressure [7].

The importance of the RAS in blood pressure control has been confirmed by the efficacy of RAS inhibitors not only for the control of blood pressure but also for the reduction of hypertensive complications, including cardiovascular disease and stroke [8]. Although many other regulatory systems other than the RAS contribute importantly to the pathophysiology of hypertension [1], the importance of RAS derives from the fact that inhibition of RAS is a safe and effective method to decrease hypertension in humans. It is therefore easy to understand why the RAS is the main target for attempts to produce effective vaccines for hypertension (Fig. 27.1).

## 27.1.3 Vaccines

The first clinical study to attempt immunization against the RAS for the treatment of hypertension was reported by Goldblatt et al. in 1951 [9]. In this study, the authors injected heterologous (hog) renin into patients with severe hypertension and examined the effects on antirenin titers and blood pressures. Their paper describes the results in 8 patients, who were given between 7 and 66 injections (28,000–142,000 units) of hog renin over a period of 3.5–33 weeks. Almost all patients developed detectable increases in serum antirenin antibodies. However, no significant decrease in blood pressure was detectable in these patients, and the authors concluded that this could be because the renin administered was of hog origin and not human. Since there is species specificity in the actions of renin, the production of antibodies to heterologous renin may not affect the actions of native human renin.

**Fig. 27.1** Overview of the renin-angiotensin system and major targets for hypertension vaccines

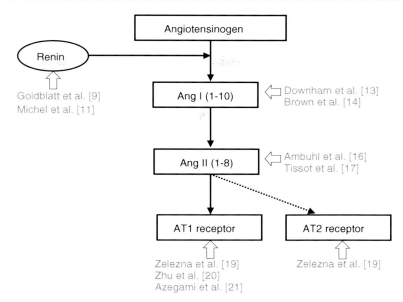

Further studies were performed using animal models. In 1987, Michel et al. reported the results of using pure human renin to immunize marmosets. Three subcutaneous injections of 30 μg, each of pure renin protein, resulted in a high titer of renin antibodies.

Furthermore, blood pressure was significantly reduced from 125 ± 13 to 87 ± 8 mmHg, together with significant decreases in plasma renin activity and plasma aldosterone. However, the authors found that the marmosets also developed immunological renal disease, characterized by the presence of immunoglobulin and macrophage infiltration colocalizing with renin in the kidney [10]. Similarly, when spontaneously hypertensive rats (SHRs) were immunized with purified mouse submandibular renin, the systolic blood pressure was significantly decreased; however, the authors again found that the kidneys of renin-immunized SHR showed a chronic autoimmune interstitial nephritis associated with the renin-producing JGA cells of the kidney [11].

These experiments on renin immunization were performed using the whole renin protein as the immunogen. It has been suggested that the large size of renin may have facilitated the development of autoimmune disease, and the fact that renin production occurs at only one anatomic site may have resulted in localized accumulation of antibodies [12].

Consequently, the target for vaccination against the RAS shifted to the angiotensin peptide itself. After performing preliminary studies in rats and humans [13], Brown et al. performed a clinical randomized double-blind placebo-controlled study of a vaccine against angiotensin I (PMD3117) in patients with essential hypertension. In their protocol, patients were given either 3 doses of 100 μg of peptide equivalent vaccine at 21-day intervals or 4 doses at 14-day intervals, and blood pressure changes were estimated as the differences in the pre- and post-vaccination rise in 24 h ambulatory blood pressure after 2 weeks of withdrawal of ACE inhibitors or ARB between the active vaccine and placebo. Although the authors found significant changes in plasma renin and urine aldosterone, they did not detect a significant difference in blood pressure between the active vaccine- and placebo-treated groups [14].

The next important breakthrough for angiotensin II vaccines was the development of new viruslike particle (VLP) technology by the Zurich-based company Cytos [15]. The VLP are macromolecular assemblies formed from bacteriophage coat protein monomers, which spontaneously assemble in a capsid structure. The

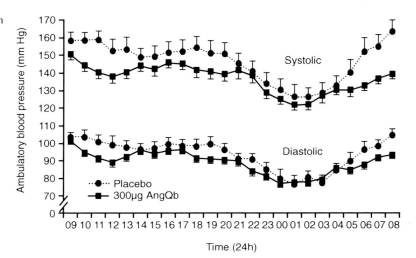

**Fig. 27.2** Effects of injection with angiotensin II vaccine (AngQb) or placebo on 24 h ambulatory blood pressure profiles in patients with mild-to-moderate hypertension (From Tissot et al. [17] with permission)

angiotensin peptide was fused at its N-terminus to a CGG spacer sequence and conjugated to VLP. Using this vaccine (AngQb), Ambuhl et al. examined the effects of subcutaneous administration of angiotensin II vaccine first in a rat model (SHR) and then in healthy human volunteers [16].

Next, Tissot et al. performed a multicenter, double-blind, placebo-controlled phase IIa trial of the efficacy of the AngQb vaccine in 72 patients with mild-to-moderate hypertension. Patients were randomized to receive either 100 μg, 300 μg vaccine, or placebo, at weeks 0, 4, and 12. 24 h ambulatory blood pressure was measured before treatment and at week 14. When the mean ambulatory daytime blood pressures were compared, the authors found a significant decrease of 9 mmHg in the systolic blood pressure ($p=0.015$), with a similar trend for diastolic blood pressure (4 mmHg reduction, $p=0.064$) in the 300 μg group. Furthermore, the 300 μg vaccine reduced the early morning blood pressure surge compared with placebo (25 mmHg reduction in systolic blood pressure ($p<0.0001$) and 13 mmHg reduction in diastolic blood pressure ($p=0.0035$)) [17] (Fig. 27.2). A significant increase in mild, transient reactions at the injection site was found, and a total of ten participants reported mild, transient influenza-like symptoms. The study by Tissot et al. can be considered a landmark study, because it was the first study to show that vaccination can reduce blood pressure in humans [18].

Several groups, including our own, have focused on the angiotensin type 1 (AT1 receptor) as a potential future target for the development of vaccines for hypertension and hypertensive complications [19–21]. One of the theoretical advantages of the AT1 receptor vaccine over the angiotensin II vaccine is that direct inhibition at the receptor level may result in more complete suppression of the RAS and enhanced effectiveness of the vaccine. Furthermore, unlike the angiotensin II vaccine which blocks the action of angiotensin II at both AT1 and AT2 receptors, specific blockade of the AT1 receptor does not affect the actions of angiotensin II at the AT2 receptor, and this may be an advantage for organ protection [7].

In our laboratory, we synthesized a seven-amino-acid peptide sequence corresponding to amino acids 181–187 in the second extracellular loop of rat AT1a receptor (AFHYESR) first described by Zhu et al. [20] and conjugated it to the carrier protein keyhole limpet hemocyanin (KLH) (Fig. 27.3). In preliminary experiments, we compared the effects of 1, 3, and 6 injections of vaccine on blood pressure and AT1 receptor antibody titers in the spontaneously hypertensive rat, a rodent model of essential hypertension [21]. We found that a single injection of the AT1 vaccine caused increased antibody titers, but these levels were significantly lower than the titers

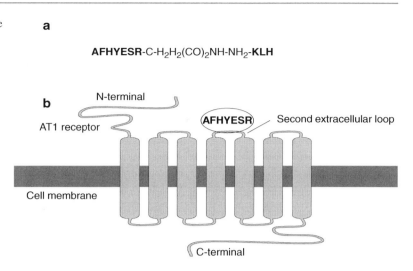

**Fig. 27.3** (**a**) Structure of the KLH-conjugated AT1 receptor peptide sequence used in our studies on hypertension and atherosclerosis. (**b**) Structure of angiotensin type 1 (*AT1*) receptor and localization of the peptide sequence in the second extracellular loop of the receptor

produced by 3 or 6 injections. In contrast, both 3 and 6 injections of AT1 vaccine produced comparable increases in AT1 antibody titers and significant decreases in the blood pressure.

We also examined the effects of the AT1 receptor vaccine on the development of renal injury, using the NO synthase inhibitor L-NAME. Previous studies from our and other laboratories have shown that the use of L-NAME results in renal injury, including proteinuria and glomerular and renal vascular injury [22]. We found that rats given 3 injections of the AT1 receptor vaccines had a significant suppression of proteinuria after L-NAME administration. Moreover, histological indicators of renal injury were significantly suppressed in the AT1 receptor vaccine group, similar to the results found in rats treated continuously with the ARB candesartan, suggesting that vaccination against the RAS may be effective not only for the attenuation of hypertension but also for the prevention of hypertensive renal injury [21].

### 27.1.4 Strengths and Weaknesses

The study by Tissot et al. demonstrated the proof of concept that patients with hypertension could be treated by vaccination against a vasoactive substance. Three injections of angiotensin vaccine resulted in an increase of antibody titers over several months with an average half-life of about 4 months after the third injection. This half-life could be compatible with a treatment regimen of several injections per year [17]. Such treatment could be very attractive for increasing adherence to hypertension therapy, which currently requires daily and lifelong medication.

Although these results are encouraging, there are many important caveats when considering the vaccine approach as a strategy for the treatment of hypertension. First, multiple antihypertensive agents are already available; therefore, it is possible to question the importance of the development of hypertension vaccines [23]. On the other hand, the fact that the prevalence of hypertension and hypertensive complications is increasing despite the availability of these medications suggests that new approaches to hypertension treatment should be considered and the risk/benefit of the vaccine approach compared to other strategies [12, 24–26].

An important concern is that the effects of vaccine therapy may be irreversible and result in hypotension, particularly during states of salt and volume depletion, where an active RAS contributes to survival. Other important safety issues include the potential for developing stimulatory as well as inhibitory antibodies, as well as the risk of developing autoimmune disease, as was reported in the renin-immunized animal models [18, 25].

Although no major adverse effects related to vaccination were reported in the study by Tissot et al., local injection site-related erythema was found in all groups including the placebo group. It is possible that such local reactions could be avoided by using alternative vaccination routes (e.g., by oral or nasal vaccines). Because hypertension itself is asymptomatic, it is important that vaccines for hypertension should have minimal side effects, if they are to be developed for routine clinical use.

## 27.2 Vaccines for Atherosclerosis

### 27.2.1 Disease

Atherosclerosis is predominantly a disease of the medium to large arteries and is characterized by the presence of lipid-containing plaques in the vessel wall, eventually resulting in luminal narrowing and susceptibility to plaque rupture and lumen thrombosis [27]. Atherosclerosis is directly responsible for the major syndromes of cardiovascular disease, including stroke, angina pectoris, myocardial infarction, aneurysmal disease, and peripheral artery disease, and is therefore a major cause of morbidity and mortality throughout the world [27].

Multiple risk factors are known to increase the susceptibility to atherosclerosis. One of the most important is age, and atherosclerosis in some form is known to occur as part of the aging process. However, increased risk is also associated with hypercholesterolemia, hypertension, diabetes, obesity, and male gender

Atherosclerosis may be considered a "final common pathway" for the deleterious effects not only of hypertension but also of other lifestyle-related diseases, including diabetes, metabolic syndrome, and dyslipidemia, and is therefore an important target for therapeutic strategies to reduce the incidence of stroke, myocardial infarction, and other cardiovascular diseases.

### 27.2.2 Pathophysiology and Classical Therapy

Although the association of atherosclerosis with high cholesterol levels has been well recognized, recent studies have shown that atherosclerosis is essentially a chronic inflammatory disease, mediated by a complex interaction between the endothelial cells and smooth muscle cells of the vessel wall with multiple immunomodulatory cells [28, 29].

Cholesterol is transported in the bloodstream in association with a family of apolipoproteins to form lipoproteins. An important lipoprotein for the pathogenesis of atherosclerosis is the low-density lipoprotein (LDL), which can accumulate in the arterial intima, where it is susceptible to oxidative modification into oxidized LDL. Oxidized LDL is known to cause damage to arterial walls and triggers a complex series of reactions resulting in inflammation and immune cell activation in the vessel wall.

Important components of the immune response to oxidized-LDL-associated vascular wall injury include macrophages, monocytes, leukocytes, dendritic cells, and T cells. In particular, the macrophages ingest the oxidized LDL and turn into foam cells, enlarging in size until they eventually rupture, and cause further deposition of oxidized LDL, resulting in a vicious cycle with further recruitment of immunomodulatory cells.

The inflammatory process occurring in the arteries results in the activation of multiple cytokines, growth factors, and other vasoactive compounds. These contribute to smooth muscle cell proliferation and invasion, deposition of collagen and other extracellular matrix proteins, and formation of the fibrous cap which can rupture and cause thrombosis.

Current therapy of atherosclerosis starts with non-pharmaceutical interventions aimed at changing lifestyles. These are aimed at reducing the risk factors for atherosclerosis development and include changes such as dietary modifications, cessation of smoking, weight control, and increased exercise.

The pharmaceutical intervention for which there is the greatest clinical evidence for both primary and secondary prevention is the use of HMG-CoA reductase inhibitors (statins), which reduce serum cholesterol levels by inhibiting the pathway of cholesterol production in the liver.

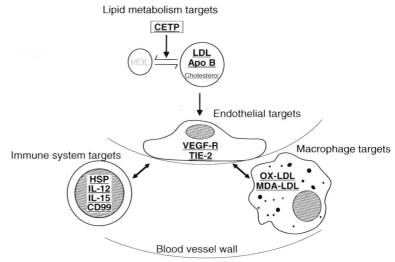

**Fig. 27.4** Simplified overview of major targets for atherosclerosis vaccines/immune therapy (vaccine targets are *underlined*)

Although it is probable that most of the beneficial actions of statins are mediated through their effects on lipid metabolism, it has been suggested that their anti-inflammatory actions could also be involved in the beneficial effects of these agents [27].

Other agents for which there is evidence for secondary prevention of cardiovascular events include nicotinic acid (niacin) and antiplatelet agents. Furthermore, treatment with RAS inhibitors may have a beneficial effect beyond their antihypertensive effect.

### 27.2.3 Vaccines

Similarly to vaccines for hypertension, the development of vaccines for atherosclerosis has a long history. Although vaccines for hypertension have focused predominantly on the RAS, in the case of vaccines for atherosclerosis, multiple targets in different cells have become candidates for atherosclerosis vaccine and immunomodulatory treatments (Fig. 27.4).

One of the earliest approaches for inhibition of atherosclerosis was the use of immunization with beta-lipoprotein reported by Gero et al. in 1959 [30]. Subsequent studies examined the effects of immunization with LDL and its derivatives. Palinski et al. examined the effects of immunization of hypercholesterolemic rabbits with malondialdehyde-modified LDL and found a significant reduction in the extent of atherosclerotic lesions [31]. Ameli et al. reported a similar effect using immunization with homologous LDL, with a smaller (nonsignificant) effect using oxidized LDL [32]. These results have been replicated in other studies [33, 34].

ApoB-100 is the main protein component of LDL, and oxidation of LDL results in degradation of apoB into smaller fragments which are further modified by reactive aldehydes. It has been suggested by Nilsson et al. that vaccines based on apoB peptide antigens may be an important candidate for the development of vaccines for atherosclerosis [35]. Fredrikson et al. examined the effects of immunization with apoB-100 peptide sequences on the development of atherosclerotic lesions in hypercholesterolemic mice and found a 60 % reduction in atherosclerosis [36]. Interestingly, it was later shown that an antiatherogenic effect could be found in the absence of an increase in peptide-specific IgG, suggesting the contribution of cellular immune responses in the atheroprotective effect [37].

Another approach has been the development of vaccines against cholesteryl ester transfer protein, which is involved in the regulation of the balance of HDL and LDL [38]. Rittershaus et al. immunized rabbits with a peptide containing a region of CETP known to be required for neutral lipid transfer function. They found that HDL levels were increased, and LDL levels decreased in the vaccinated rabbits fed with a high-cholesterol diet. They also found a 39.6 % reduction in the aortic atherosclerotic area [39]. In a phase I human trial, the same group found that 8 out of 15 subjects (53 %) developed anti-CETP antibodies after two injections, without major side effects [40]. A subsequent phase 2 clinical trial on patients with low HDL levels confirmed that the vaccine was well tolerated and produced anti-CETP antibodies in greater than 90 % of patients and that the % increase in HDL was correlated with the peak antibody titer [38]. It should be pointed out that these results were reported before the data of increased morbidity and mortality with the CETP inhibitor torcetrapib were published [41].

Newer approaches for the development of vaccines for atherosclerosis include the use of DNA vaccines and dendritic cell-based strategies. In the study by Mao et al., hypercholesterolemic rabbits were immunized intramuscularly with plasmids containing an epitope of CETP C-terminal fragment, as well as immunomodulatory CpG sequences, and a marked (80 %) reduction in atherosclerotic plaque lesions was found [42]. Beneficial results have also been reported for DNA vaccination strategies targeting cytokines and growth factors such as IL-15 [43], CD99 [44], TIE2 [45], and VEGFR-2 [46, 47]. Another experimental approach is the use of dendritic cells to induce a specific humoral immune response. Favorable results have been reported for dendritic cells pulsed with oxidized LDL [48] or apoB-100 [49]. (Further details of studies on other lipid targets, inflammatory cytokines, and endothelial cell markers may be found in several recent reviews [33, 34, 50].)

In our laboratory, we have recently examined the effects of vaccination against the AT1 receptor on the development of atherosclerosis in ApoE-deficient mice fed with a high-fat and high-salt diet. Our preliminary results suggest that the use of this vaccine targeting the AT1 receptor results not only a decrease in the Sudan red-positive atherosclerotic area in the aorta in this model but also an attenuation of albuminuria (Azegami et al. unpublished observations). Previous studies from our and other laboratories have suggested that RAS inhibition with drugs can attenuate atherosclerosis by multiple mechanisms, including decreased vascular proinflammatory response, decreased retention of atherosclerotic lipoproteins, and increased lipid release [51], and we are planning further studies to confirm whether AT1 receptor vaccines may have an antiatherogenic as well as an antihypertensive effect.

Other groups have examined the use of oral and nasal mucosal immunization strategies, as a method to induce tolerance and immune unresponsiveness to proatherogenic antigens. Among the most studied are heat shock proteins (HSPs), which are thought to be involved in proatherogenic immune-mediated responses. Harats et al. and Maron et al. examined the effects of oral or nasal immunization with HSP 65 in mouse models, and both groups found an attenuation of atherosclerosis [52, 53]. Similarly, van Puijvelde et al. reported that oral administration of HSP60 or a HSP60 peptide to atherosclerotic mice resulted in induction of oral tolerance and a significant reduction in the atherosclerotic plaque size [54]. On the other hand, Yuan et al. reported that intranasal immunization of a DNA vaccine directed towards CETP resulted in a significant anti-CETP IgG response which lasted for 28 weeks and a significant decrease in atherogenesis [55, 56]. Further studies are required to clarify the advantages of different approaches to atherosclerosis reduction.

In a small-scale clinical study, Bourinbaiar et al. examined the effects of administration of a tablet containing pooled antigens derived from pig adipose tissue to 13 volunteers over 3 months [57]. They reported an increase in HDL-C levels in 12 out of 13 patients and a significant reduction in waist, mid-arm, and thigh circumferences. Because of the small experiment design, it is difficult to characterize the mechanisms and clinical implications of these preliminary observations.

## 27.2.4 Strengths and Weaknesses

The major challenge in developing a vaccine against atherosclerosis for widespread clinical use is the difficulty of performing clinical studies of sufficient length and in sufficient numbers of patients to enable conclusions about the safety and efficacy of the vaccine treatment.

In the case of hypertension, it is possible to design a clinical protocol with a time scale of weeks or months. In contrast, the development of atherosclerosis occurs over years or even decades, and accurate assessment of atherosclerosis may require expensive and invasive methods such as angiography or intravascular ultrasound. The use of biomarkers as surrogate markers is one way to circumvent this problem, but their predictive ability for hard endpoint markers is still limited at present [27].

Although the difficulty of designing and performing clinical studies to test the efficacy of atherosclerosis vaccines in humans should not be underestimated, it should also be recognized that encouraging clinical results have already emerged that another form of vaccination, i.e., vaccination against influenza, may be associated with a reduction in cardiovascular events.

In 2005, Nichol et al. assessed the influence of influenza vaccination on the risk of hospitalization for heart disease, stroke, pneumonia, influenza, and death for all causes in two cohorts of community-dwelling members who were at least 65 years old [58]. In addition to the expected 29–32 % decrease in risk for hospitalization for influenza or pneumonia, the authors also noted a 19 % decrease in the risk of hospitalization for cardiac disease, a 16–23 % decrease in the risk of hospitalization for cerebrovascular disease, and a 48–50 % decrease in the risk of death from all causes.

In the FLUVACS prospective study, Gurfinkel et al. enrolled patients with myocardial infarction or planned coronary interventions and randomly allocated them to receive flu vaccination or remain unvaccinated. The authors found a reduction in cardiovascular death and ischemic events in the vaccine-treated group at 1-year follow-up [59]. Similarly, Phrommintikul et al. examined the effects of influenza vaccine in 439 patients who were admitted for acute coronary syndrome. They found a significant reduction in the major cardiovascular events compared to the no vaccine group [60].

Taken together, the results of these prospective clinical studies provide encouragement for the notion that vaccine therapy may become a viable strategy for management of the complications of atherosclerotic disease in the future. However, it is also clear that translating the findings of animal experiments into clinical trials poses many new challenges, which need to be overcome in order to clarify the efficacy and safety of vaccines for atherosclerosis prevention in humans.

## References

1. Kaplan, N.M., Victor, R.G.: Clinical Hypertension, 10th edn. Lippincott Williams and Wilkins, Philadelphia (2010)
2. Kearney, P.M., et al.: Global burden of hypertension: analysis of worldwide data. Lancet **365**, 217–223 (2005)
3. Israili, Z.H., Hernandez-Hernandez, R., Valasco, M.: The future of antihypertensive treatment. Am. J. Ther. **14**, 121–134 (2007)
4. Chobanian, A., Shattuck Lecture, V.: The hypertension paradox – more uncontrolled disease despite improved therapy. N. Engl. J. Med. **361**, 878–887 (2009)
5. Egan, B.M., Zhao, Y., Axon, R.N.: US trends in prevalence, awareness, treatment, and control of hypertension, 1988–2008. JAMA **303**, 2043–2050 (2010)
6. Nguyen, D.C.A., Touyz, R.: M. A new look at the renin-angiotensin system – focusing on the vascular system. Peptides **32**, 2141–2150 (2011)
7. Wright, J.W., Yamamoto, B.J., Harding, J.W.: Angiotensin receptor subtype mediated physiologies and behaviors: new discoveries and clinical targets. Prog. Neurobiol. **84**, 157–181 (2008)
8. Weber, M.: Achieving blood pressure goals: should angiotensin II receptor blockers become first-line treatment in hypertension? J. Hypertens. Suppl. **27**, S9–S14 (2009)
9. Goldblatt, H., Haas, E., Lamfrom, H.: Antirenin in man and animals. Trans. Assoc. Am. Physicians **64**, 122–125 (1951)
10. Michel, J.B., et al.: Active immunization against renin in normotensive marmoset. Proc. Natl. Acad. Sci. U. S. A. **84**, 4346–4350 (1987)

11. Michel, J.B., et al.: Physiological and immunopathological consequences of active immunization of spontaneously hypertensive and normotensive rats against murine renin. Circulation **81**, 1899–1910 (1990)
12. Gradman, A.H., Pinto, R.: Vaccination: a novel strategy for inhibiting the renin-angiotensin-aldosterone system. Curr. Hypertens. Rep. **10**, 473–479 (2008)
13. Downham, M.R., et al.: Evaluation of two carrier protein-angiotensin I conjugate vaccines to assess their future potential to control high blood pressure (hypertension) in man. Br. J. Clin. Pharmacol. **56**, 505–512 (2003)
14. Brown, M.J., et al.: Randomized double-blind placebo-controlled study of an angiotensin immunotherapeutic vaccine (PMD3117) in hypertensive subjects. Clin. Sci. (Lond.) **107**, 167–173 (2004)
15. Jegerlehner, A., et al.: A molecular assembly system that renders antigens of choice highly repetitive for induction of protective B cell responses. Vaccine **20**, 3104–3112 (2002)
16. Ambuhl, P.M., et al.: A vaccine for hypertension based on virus-like particles: preclinical efficacy and phase I safety and immunogenicity. J. Hypertens. **25**, 63–72 (2007)
17. Tissot, A.C., et al.: Effect of immunisation against angiotensin II with CYT006-AngQb on ambulatory blood pressure: a double-blind, randomised, placebo-controlled phase IIa study. Lancet **371**, 821–827 (2008)
18. Samuelsson, O., Herlitz, H.: Vaccination against high blood pressure: a new strategy. Lancet **371**, 788–789 (2008)
19. Zelezna, B., et al.: Influence of active immunization against angiotensin AT1 or AT2 receptor on hypertension development in young and adult SHR. Physiol. Res. **48**, 259–265 (1999)
20. Zhu, F., et al.: Target organ protection from a novel angiotensin II receptor (AT1) vaccine ATR12181 in spontaneously hypertensive rats. Cell. Mol. Immunol. **3**, 107–114 (2006)
21. Azegami, T., Sasamura, H., Hayashi, K., Itoh, H.: Vaccination against the angiotensin type 1 receptor for the prevention of L-NAME-induced nephropathy. Hypertens. Res. **35**, 492–499 (2012)
22. Ishiguro, K., Sasamura, H., Sakamaki, Y., Itoh, H., Saruta, T.: Developmental activity of the renin-angiotensin system during the "critical period" modulates later L-NAME-induced hypertension and renal injury. Hypertens. Res. **30**, 63–75 (2007)
23. Menard, J.: A vaccine for hypertension. J. Hypertens. **25**, 41–46 (2007)
24. Brown, M.J.: Therapeutic potential of vaccines in the management of hypertension. Drugs **68**, 2557–2560 (2008)
25. Campbell, D.J.: Angiotensin vaccination: what is the prospect of success? Curr. Hypertens. Rep. **11**, 63–68 (2009)
26. Pandey, R., Quan, W.Y., Hong, F., Jie, S.L.: Vaccine for hypertension: modulating the renin-angiotensin system. Int. J. Cardiol. **134**, 160–168 (2009)
27. Weber, C., Noels, H.: Atherosclerosis: current pathogenesis and therapeutic options. Nat. Med. **17**, 1410–1422 (2011)
28. Ross, R.: Atherosclerosis – an inflammatory disease. N. Engl. J. Med. **340**, 115–126 (1999)
29. Libby, P.: Inflammation in atherosclerosis. Nature **420**, 868–874 (2002)
30. Gero, S., et al.: Inhibition of cholesterol atherosclerosis by immunisation with beta-lipoprotein. Lancet **2**, 6–7 (1959)
31. Palinski, W., Miller, E., Witztum, J.L.: Immunization of low density lipoprotein (LDL) receptor-deficient rabbits with homologous malondialdehyde-modified LDL reduces atherogenesis. Proc. Natl. Acad. Sci. U. S. A. **92**, 821–825 (1995)
32. Ameli, S., et al.: Effect of immunization with homologous LDL and oxidized LDL on early atherosclerosis in hypercholesterolemic rabbits. Arterioscler. Thromb. Vasc. Biol. **16**, 1074–1079 (1996)
33. de Carvalho, J.F., Pereira, R.M., Shoenfeld, Y.: Vaccination for atherosclerosis. Clin. Rev. Allergy Immunol. **38**, 135–140 (2010)
34. de Jager, S.C., Kuiper, J.: Vaccination strategies in atherosclerosis. Thromb. Haemost. **106**, 796–803 (2011)
35. Nilsson, J., Fredrikson, G.N., Bjorkbacka, H., Chyu, K.Y., Shah, P.K.: Vaccines modulating lipoprotein autoimmunity as a possible future therapy for cardiovascular disease. J. Intern. Med. **266**, 221–231 (2009)
36. Fredrikson, G.N., et al.: Inhibition of atherosclerosis in apoE-null mice by immunization with apoB-100 peptide sequences. Arterioscler. Thromb. Vasc. Biol. **23**, 879–884 (2003)
37. Fredrikson, G.N., Bjorkbacka, H., Soderberg, I., Ljungcrantz, I., Nilsson, J.: Treatment with apo B peptide vaccines inhibits atherosclerosis in human apo B-100 transgenic mice without inducing an increase in peptide-specific antibodies. J. Intern. Med. **264**, 563–570 (2008)
38. Ryan, U.S., Rittershaus, C.W.: Vaccines for the prevention of cardiovascular disease. Vascul. Pharmacol. **45**, 253–257 (2006)
39. Rittershaus, C.W., et al.: Vaccine-induced antibodies inhibit CETP activity in vivo and reduce aortic lesions in a rabbit model of atherosclerosis. Arterioscler. Thromb. Vasc. Biol. **20**, 2106–2112 (2000)
40. Davidson, M.H., et al.: The safety and immunogenicity of a CETP vaccine in healthy adults. Atherosclerosis **169**, 113–120 (2003)
41. Barter, P.J., et al.: Effects of torcetrapib in patients at high risk for coronary events. N. Engl. J. Med. **357**, 2109–2122 (2007)
42. Mao, D., et al.: Intramuscular immunization with a DNA vaccine encoding a 26-amino acid CETP epitope displayed by HBc protein and containing CpG DNA inhibits atherosclerosis in a rabbit model of atherosclerosis. Vaccine **24**, 4942–4950 (2006)
43. van Es, T., et al.: IL-15 aggravates atherosclerotic lesion development in LDL receptor deficient mice. Vaccine **29**, 976–983 (2011)

44. van Wanrooij, E.J., et al.: Vaccination against CD99 inhibits atherogenesis in low-density lipoprotein receptor-deficient mice. Cardiovasc. Res. **78**, 590–596 (2008)
45. Hauer, A.D., et al.: Vaccination against TIE2 reduces atherosclerosis. Atherosclerosis **204**, 365–371 (2009)
46. Hauer, A.D., et al.: Vaccination against VEGFR2 attenuates initiation and progression of atherosclerosis. Arterioscler. Thromb. Vasc. Biol. **27**, 2050–2057 (2007)
47. Petrovan, R.J., Kaplan, C.D., Reisfeld, R.A., Curtiss, L.K.: DNA vaccination against VEGF receptor 2 reduces atherosclerosis in LDL receptor-deficient mice. Arterioscler. Thromb. Vasc. Biol. **27**, 1095–1100 (2007)
48. Habets, K.L., et al.: Vaccination using oxidized low-density lipoprotein-pulsed dendritic cells reduces atherosclerosis in LDL receptor-deficient mice. Cardiovasc. Res. **85**, 622–630 (2010)
49. Hermansson, A., et al.: Immunotherapy with tolerogenic apolipoprotein B-100-loaded dendritic cells attenuates atherosclerosis in hypercholesterolemic mice. Circulation **123**, 1083–1091 (2011)
50. Chyu, K.Y., Nilsson, J., Shah, P.K.: Immune mechanisms in atherosclerosis and potential for an atherosclerosis vaccine. Discov. Med. **11**, 403–412 (2011)
51. Hayashi, K., Sasamura, H., Azegami, T., Itoh, H.: Regression of atherosclerosis in apolipoprotein E-deficient mice is feasible using high-dose angiotensin receptor blocker, candesartan. J. Atheroscler. Thromb. **19**(8), 736–746 (2012)
52. Harats, D., Yacov, N., Gilburd, B., Shoenfeld, Y., George, J.: Oral tolerance with heat shock protein 65 attenuates Mycobacterium tuberculosis-induced and high-fat-diet-driven atherosclerotic lesions. J. Am. Coll. Cardiol. **40**, 1333–1338 (2002)
53. Maron, R., et al.: Mucosal administration of heat shock protein-65 decreases atherosclerosis and inflammation in aortic arch of low-density lipoprotein receptor-deficient mice. Circulation **106**, 1708–1715 (2002)
54. van Puijvelde, G.H., et al.: Induction of oral tolerance to HSP60 or an HSP60-peptide activates T cell regulation and reduces atherosclerosis. Arterioscler. Thromb. Vasc. Biol. **27**, 2677–2683 (2007)
55. Yuan, X., et al.: Intranasal immunization with chitosan/pCETP nanoparticles inhibits atherosclerosis in a rabbit model of atherosclerosis. Vaccine **26**, 3727–3734 (2008)
56. Jun, L., et al.: Effects of nasal immunization of multi-target preventive vaccines on atherosclerosis. Vaccine **30**, 1029–1037 (2012)
57. Bourinbaiar, A.S., Jirathitikal, V.: Effect of oral immunization with pooled antigens derived from adipose tissue on atherosclerosis and obesity indices. Vaccine **28**, 2763–2768 (2010)
58. Nichol, K.L., et al.: Influenza vaccination and reduction in hospitalizations for cardiac disease and stroke among the elderly. N. Engl. J. Med. **348**, 1322–1332 (2003)
59. Gurfinkel, E.P., Leon de la Fuente, R., Mendiz, O., Mautner, B.: Flu vaccination in acute coronary syndromes and planned percutaneous coronary interventions (FLUVACS) study. Eur. Heart J. **25**, 25–31 (2004)
60. Phrommintikul, A., et al.: Influenza vaccination reduces cardiovascular events in patients with acute coronary syndrome. Eur. Heart J. **32**, 1730–1735 (2011)

# Anti-ghrelin Therapeutic Vaccine: A Novel Approach for Obesity Treatment

## 28

Sara Andrade, Marcos Carreira, Felipe F. Casanueva, Polly Roy, and Mariana P. Monteiro

## Contents

| | | |
|---|---|---|
| 28.1 | **Introduction** | 464 |
| 28.2 | **Obesity: Diagnosis and Classical Therapy** | 464 |
| 28.3 | **Anti-ghrelin Therapeutic Vaccine** | 466 |
| 28.3.1 | Regulation of Food Intake and Energy Homeostasis | 466 |
| 28.3.2 | Ghrelin | 467 |
| 28.3.3 | Rational for the Use of an Anti-ghrelin Vaccine | 468 |
| 28.3.4 | Anti-ghrelin Vaccine Using Virus-Like Particles | 469 |
| 28.3.5 | Animal Models and Feasibility Study | 469 |
| 28.3.6 | Safety and Efficacy of the Vaccine | 469 |
| 28.4 | **Strengths and Weaknesses** | 472 |
| 28.5 | **Concluding Remarks** | 474 |
| **References** | | 474 |

S. Andrade, MBSc • M.P. Monteiro, MD, PhD (✉)
Department of Anatomy and UMIB (Unit for Multidisciplinary Biomedical Research) of ICBAS, University of Porto, Porto, Portugal
e-mail: mpmonteiro@icbas.up.pt

M. Carreira, PhD
CIBER de Fisiopatologia Obesidad y Nutricion (CB06/03), Instituto Salud Carlos III, Santiago de Compostela, Spain

F.F. Casanueva, MD, PhD
CIBER de Fisiopatologia Obesidad y Nutricion (CB06/03), Instituto Salud Carlos III, Santiago de Compostela, Spain

Department of Medicine, USC University Hospital Complex, University of Santiago de Compostela, Santiago de Compostela, Spain

P. Roy, PhD
Department of Pathogen Molecular Biology, London School of Hygiene and Tropical Medicine, London, UK

## Abstract

Obesity is currently a major public health problem, due to the worldwide increasing rates of the disease and burden of the associated co-morbidities, such as, type 2 diabetes, cardiovascular disease and cancer. Despite its increasing clinical relevance, there are still very few tools to treat obesity. The cornerstones for obesity treatment are still diet and exercise; anti-obesity drugs, which cause anorexia or malabsorption of nutrients, can be used as adjuvant therapy, however achieve only a modest weight loss and often short-term due to weight regain. For severe obesity the only proven effective therapy is bariatric surgery, an invasive procedure that carries inherent risks and is only recommended for selected patients.

Ghrelin is the only known hormone that stimulates food intake. In physiological conditions, ghrelin levels rise with fasting and decrease after meals. Most obese individuals have low fasting ghrelin levels that rise after food restriction and weigh loss, an explanation for the difficulty of weight loss maintenance. In contrast, in spite of major weight loss, the increase in ghrelin levels is prevented by some bariatric surgery techniques, which could contribute to sustain weight loss.

As ghrelin is the only known orexigenic hormone, it has been hypothesized that blocking reactive ghrelin increase could induce a sustained weight control.

Previous attempts to neutralize ghrelin orexigenic effects included passive immunizations by the inoculation of monoclonal anti-ghrelin antibodies and mixtures of monoclonal antibodies targeting different ghrelin haptens, which were able to decrease ghrelin-mediated and deprivation-induced food intake, while promoted an increase in energy expenditure, but had the limitation of having only acute effects; use of ghrelin receptor antagonists that demonstrated to improve glucose tolerance, suppress appetite and promote weight loss; and active immunization against ghrelin using keyhole limpet hemocyanin and bovine serum albumin as carrier proteins, which required the use of adjuvants that may be responsible for inflammatory responses and have limited use in humans.

A novel molecular approach is the use of an anti-ghrelin vaccine using Virus-like Particles as immunogenic carrier, which appears to be well-tolerated, decrease food intake and increase energy expenditure in both normal weight and diet-induced obese (DIO) mice. Vaccinated DIO mice also display a significant decrease of NPY gene expression in the basal hypothalamus reflecting a decrease in central orexigenic drive. All together, data suggests that this novel therapeutic anti-ghrelin vaccine is safe, has a positive impact on energy homeostasis and may be a useful tool for obesity treatment.

## 28.1 Introduction

The prevalence of overweight, obesity, and extreme obesity has been increasing worldwide in the last decades, affecting not only adults but starting as early as in childhood and adolescence [1, 2]. More recent estimates suggest that body weight gain will continue to increase, particularly in the younger people [3].

Obesity is a well-known risk factor for many chronic conditions including type 2 diabetes mellitus, hypertension, metabolic syndrome, cardiovascular diseases, and cancer [4].

The adverse health consequences occur not only in overweight individuals but start to increase at the upper limit of the normal body mass index (BMI 22–24.9 kg/m$^2$), and weight loss improves or resolves several comorbid conditions associated with the disorder [4].

The estimated economic impact of obesity in health-care systems due to both direct (personal health care, hospital care, physician services, allied health services and medications) and indirect costs (lost output as a result of a reduction or cessation of productivity due to morbidity or mortality) is enormous [5, 6]. These can be attributed not only to obesity per se, as excess physician visits, lost of workdays, restricted activity, and in-bed days, but particularly to the associated comorbidities such as type 2 diabetes mellitus, coronary heart disease, breast, endometrial and colon cancer, and osteoarthritis [7]. Furthermore, obesity is also associated with denial of employment, restriction of career advancement and higher insurance premiums [7].

Obesity and overweight are associated with large decreases in life expectancy, independently of comorbidities such as hypertension and diabetes that are major, potentially preventable, causes of premature morbidity and death [8]. The decrease in life expectancy and increase in early mortality associated with obesity is similar to those seen of smokers [8].

In view of the fact that obesity is a leading cause of preventable death worldwide, authorities have now considered the disease as one of the most serious public health problems of the century [9].

## 28.2 Obesity: Diagnosis and Classical Therapy

Obesity is defined as a medical condition characterized by accumulation of excess body fat to the extent that it may have adverse effects on health [2].

The body mass index (BMI), calculated by the ratio of weight (kg) for the square of height (m$^2$), is a measurement tool routinely used in the clinic to diagnose overweight and obesity, which in spite of providing no information concerning body fat distribution, with few exceptions, correlates well with the percentage of body fat. Body mass index defines people as normal weight if their BMI is between 18.5 and 24.9 kg/m$^2$, overweight if their BMI is between 25 and 29.9 kg/m$^2$, and obese when it is greater than 30 kg/m$^2$ [10, 11].

Obesity is most often the result of a combination of excessive food energy intake and a lack of physical activity in genetically predisposed individuals. Only a limited number of cases are due primarily to monogenetic causes; endocrine disorders, such as Cushing syndrome and hypothyroidism; or previous use of drugs that cause weight gain [4].

Obesity is a chronic disease, as evidenced by the high likelihood of weight regain after weight loss attained by medical therapies, and therefore, there is a need for a long-term approach to the disease [12]. However, clinicians have few tools to fight obesity. Diet and exercise are still the cornerstones for obesity treatment, and current antiobesity drugs achieve only relative short-term weight loss and are often followed by weight regain [10, 12].

The available weight loss treatments include different combinations of diet, exercise, behavioral modification and pharmacotherapy. Many diets with different macronutrient compositions have demonstrated efficacy in weight loss; even though there is currently no evidence that clearly supports a superiority of a single dietary approach above the other diets used for weight loss. The degree of adherence to the prescribed calorie reduction appears to be the most important determinant of success [12]. Physical activity is also a valuable aid for weight loss; however, it is even more important for weight maintenance once weight loss is achieved [12]. Pharmacotherapy for obesity includes drugs that can either suppress appetite or alter nutrient absorption, with the purpose of inducing weight loss.

These drugs generally are capable to induce 5–10 % weight loss, the minimum requirement for a drug to be approved for weight loss by the regulatory authorities such as the Food and Drug Administration (FDA) and the European Medicines Agency (EMEA). In addition to weight reduction, these drugs should provide a good safety profile and beneficial actions on several cardiovascular risk factors [13]. Even when there is only a modest weight loss and the patient does not reach a normal body weight, these are able to confer health benefits to the patient and improvement of obesity comorbidities [13]. Weight loss in people suffering from obesity is associated with a reduction in low-density lipoprotein cholesterol, total cholesterol, and blood pressure, with decreased risk of development of type 2 diabetes and may be beneficial for cardiovascular disease in the long term [14].

At the present, there are two antiobesity drug classes available in the market approved by the FDA, appetite suppressors and the lipase inhibitor orlistat, of which the former is the only drug authorized by EMEA and marketed in Europe. Appetite suppressant drugs include the central nervous system stimulants, such as phentermine, phendimetrazine, and diethylpropion. Phentermine is the most widely used drug for weight loss, owing to the fact of being in the US market for decades and to its low cost. Phentermine hydrochloride is a noradrenergic sympathetic amine approved for the short-term treatment of obesity. Phentermine belongs to a class of drugs that stimulate the central nervous system and conveys a response of "fight or flight" which blocks the feeding drive and induces anorexia. The most common side effects associated with phentermine include insomnia, irritability, and increase in blood pressure [12]. Sibutramine and rimonabant are another two appetite suppressant drugs that were recently withdrawn from the market due to safety concerns, after suspension being recommended due to cardiovascular and psychiatric side effects, respectively.

Orlistat is a lipase inhibitor that prevents hydrolysis of dietary triglycerides and consequently prevents the absorption of dietary fat that is excreted unaltered by the gastrointestinal tract. The drug is widely available and approved for long-term use; nevertheless, it produces only a modest weight loss and is associated with high rates of gastrointestinal side effects, such as steatorrhea, fecal incontinence and flatulence [15].

The FDA has recently approved two new drugs that will be available in the market in the near future, which are lorcaserin, a selective 5-hydroxytryptamine receptor 2c agonist [16], and a combination of phentermine and topiramate [17]. Topiramate is a sulfamate-substituted monosaccharide marketed since 1996 and formerly approved by the FDA for seizure disorders and prevention of migraine headaches. Topiramate, among other drugs for which previous clinical trials data suggested that could promote weight loss as a side effect of their therapeutic usage, was used off-label as adjuvant therapy in obesity treatment, along with the antidepressants fluoxetine, sertraline, and bupropion and the antidiabetic drug metformin [18].

For morbid obesity, the surgical approach, termed bariatric surgery, is the only therapy that provides sustainable weight reduction [19]. In 1992, the National Institutes of Health Consensus Development Conference, in a Position Statement, affirmed the superiority of surgical over nonsurgical approaches in this condition [20, 21]. Bariatric surgery is reserved for patients in whom medical weight loss treatments are known to fail and have not shown long-term effectiveness, namely, in patients with $BMI > 40$ kg/m$^2$ or $>35$ kg/m$^2$ associated with high-risk comorbid conditions [22].

In severe obesity, surgical treatments, by decreasing the risk for development of new obesity associated comorbidities and improving the existing ones, are also more cost-effective at producing and maintaining weight loss [23]. Obesity surgery provides a significant risk reduction for the development of new health-related conditions, namely, cardiovascular, cancer, endocrine (including diabetes mellitus and hypertension), respiratory, musculoskeletal, infectious, psychiatric, and mental disorders [24]. Weight loss, after bariatric surgery, usually results in improvement or resolution of multiple medical conditions that eliminates the use of medications and absenteeism from work in patients who were previously morbidly obese. When compared to matched controls, the patients submitted to surgery display a significant reduction in health-care use rates and total direct health-care costs. Bariatric surgery also significantly decreases overall 5-year mortality rate, with a reduction in the relative risk of death of 89 % compared to controls [24].

Understandably, in response to the relative ineffectiveness of medical therapy in severe obesity, the demand for bariatric surgery has greatly increased in recent years [25].

## 28.3 Anti-ghrelin Therapeutic Vaccine

### 28.3.1 Regulation of Food Intake and Energy Homeostasis

The physiological systems that regulate energy homeostasis include brain centers, such as the hypothalamus, brainstem, and reward centers in the limbic system, which regulate food intake and energy expenditure through the secretion of neuropeptides. These centers are modulated by neural and hormonal signals coming from the periphery. Hormones synthesized by the adipose tissue, like leptin, reflect the long-term nutritional status of the body and are able to influence long-term body weight regulation, while gastrointestinal hormones, like ghrelin, peptide tyrosine-tyrosine (PYY) and glucagon-like peptide 1 (GLP-1), among several other hormones, modulate these pathways acutely and are able to regulate food intake and energy expenditure [26] (Fig. 28.1). In the hypothalamus are located the most important food intake-regulating nuclei, the arcuate (ARC) and the paraventricular nuclei (PVN). The ARC of the basal hypothalamus receives signals from the periphery and plays an integrative role in appetite regulation. The ARC projects second-order neurons to the PVN, which is involved in the regulation of visceral efferent activity. In the ARC, there are two well-characterized neuronal populations involved, one appetite stimulating that co-expresses neuropeptide Y and Agouti-related protein (NPY/AgRP) and another appetite inhibiting that co-expresses proopiomelanocortin and cocaine- and amphetamine-regulated transcript (POMC/CART) [27] (Fig. 28.2).

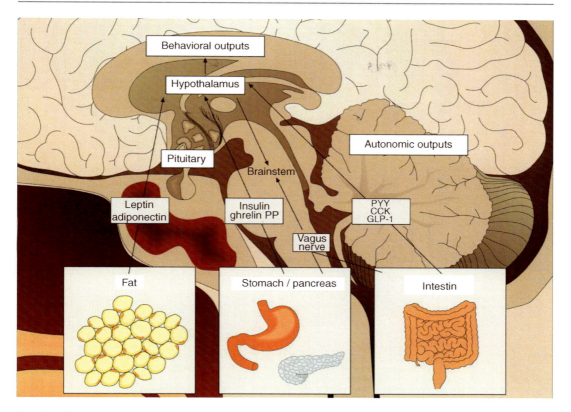

**Fig. 28.1** The physiological systems that regulate energy homeostasis. Brain centers, such as the hypothalamus, brainstem and reward centers in the limbic system, which regulate food intake and energy expenditure through the secretion of neuropeptides. These centers are modulated by neural and hormonal signals coming from the periphery. Hormones synthesized by the adipose tissue, like leptin, and gastrointestinal hormones, like ghrelin, peptide tyrosine-tyrosine (*PYY*) and glucagon-like peptide 1 (*GLP-1*), modulate these pathways and are able to regulate food intake and energy expenditure

## 28.3.2 Ghrelin

Ghrelin is a gastrointestinal hormone that promotes food intake and decreases energy expenditure [28]. Ghrelin is produced predominantly in the gastric fundus [29] and conveys orexigenic signals to the hypothalamus [30].

Ghrelin acts in the ARC of the basal hypothalamus, stimulating the production and release of NPY and suppressing POMC [31]. NPY is the most potent signal in the central nervous system that stimulates food intake and decreases energy expenditure, while POMC is a precursor protein that through proteolytic cleavage originates various peptides, among which α-MSH that decreases appetite and increases energy expenditure [32, 33].

Ghrelin plasma levels rise before meals and are suppressed after food intake [34] in lean but not in obese patients [35]. There is a negative correlation between fasting ghrelin levels and body mass index. Fasting serum ghrelin levels are usually lower in obese subjects compared with controls [36], and fasting plasma levels rise after diet-induced weight loss [34, 37] and in patients with nervous anorexia [38]. Therefore, with the exception of patients with Prader-Willi syndrome [39], ghrelin does not seem to play a causative role in obesity in general, and its decreased concentrations are believed to represent a physiological adaptation to the positive energy balance.

Recent studies suggest that weight loss attained after bariatric surgery is also due to

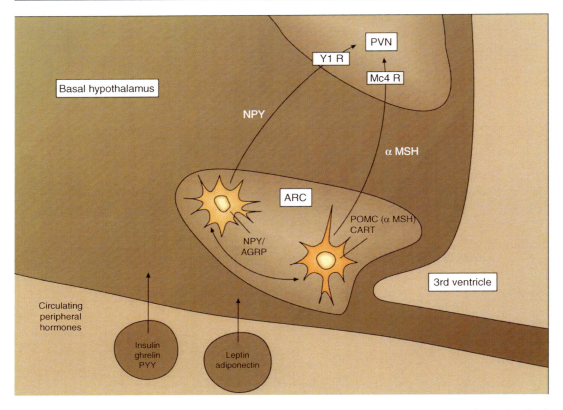

**Fig. 28.2** The arcuate (ARC) and the paraventricular nuclei (PVN) are located in the basal hypothalamus. The ARC receives signals from the periphery and plays an integrative role in appetite regulation. The ARC projects second-order neurons to the PVN, which is involved in the regulation of visceral efferent activity. In the ARC, there are two well-characterized neuronal populations involved, one appetite stimulating that co-expresses neuropeptide Y and Agouti-related protein (NPY/AgRP) and another appetite inhibiting that co-expresses proopiomelanocortin and cocaine- and amphetamine-regulated transcript (POMC/CART)

endocrine effects of the surgery, which are able to interfere with appetite pathways by suppressing the rise in ghrelin levels that is usually observed after caloric deprivation [34, 40].

### 28.3.3 Rational for the Use of an Anti-ghrelin Vaccine

In view of the fact that ghrelin is the only orexigenic hormone identified so far, it has been considered a promising target in the development of new treatments for obesity.

As a proof of this concept, it was demonstrated that inoculation of monoclonal anti-ghrelin antibodies in mice inhibited acute ghrelin-mediated orexigenic effects, but it was unable to change long-term food intake [41]. More recently, another study suggested that the use of a mixture of monoclonal antibodies targeting different haptens, but not the antibodies individually, promotes not only an increase in energy expenditure but also reduced deprivation-induced food intake [42]. Ghrelin receptor antagonists, GSH-R1, demonstrated improved glucose tolerance, suppressed appetite and promoted weight loss [43], thus confirming the potential of ghrelin blocking as a potential treatment target for obesity. The suppression of endogenous ghrelin bioactivity with anti-ghrelin vaccines using keyhole limpet hemocyanin (KLH) as carrier protein [44] or bovine serum albumin (BSA) [45] was also tested in mice and pigs, respectively. These vaccines were able to induce the development of antibodies against the active form of ghrelin [44] and also to decrease body weight gain and fat mass [45]. However, these vaccines required the use of adjuvants, such as alum and Freund incomplete adjuvant, which may be associated of inflammatory responses or have limited use in humans.

### 28.3.4 Anti-ghrelin Vaccine Using Virus-Like Particles

Virus-Like particles (VLPs) have been used as immunogenic molecules in several recombinant vaccines in the last few years in order to induce the production of specific antibodies against endogenous molecules with a preponderant role in chronic diseases [46], such as the anti-angiotensin vaccine developed for arterial hypertension treatment [47].

The main goal of our research work was to develop an effective anti-ghrelin vaccine using a chemical conjugate of active ghrelin with protein tubules of NS1 of the bluetongue virus (BTV) [48]. Although this protein is not part of the viral capsid, NS1 tubules possess the same immunogenic characteristics as classical VLPs [49].

### 28.3.5 Animal Models and Feasibility Study

Male adult C57BL6/J mice (Charles River, Barcelona, Spain), normal weight mice, and diet-induced obesity mice (DIO) ($n=18$/group) were randomized into three weight-matched groups ($n=6$/group). Normal weight mice had unrestricted access to tap water and regular rat chow and DIO mice to a hypercaloric diet with 60 % of fat (Charles River, Barcelona, Spain), after weaning and until a week before the first immunization study when food was switched to regular rat chow.

Mice received three intraperitoneal (i.p.) injections with 2-week intervals, containing 500 µl of 75 µg of immunoconjugate, 75 µg of NS1 protein alone, or PBS. A dose was chosen after performing a dose-finding study in which the 75 µg dose of the immunoconjugate has demonstrated to be adequate in inducing the development of anti-ghrelin antibodies and reducing food intake.

After the immunizations, energy expenditure was accessed by indirect calorimetry. For that mice were individually placed in a small grid cage to limit locomotor activity, which was placed into a sealed chamber containing a sodium hydroxide recipient to adsorb carbon dioxide. The lid of the chamber was sealed and pierced by a volumetric pipette to measure the volume of oxygen consumed. The time elapsed until 1 ml was consumed was registered and repeated until five concordant values were obtained. The energy expenditure was then calculated considering that 4.82 kcal is the average energy released per liter of $O_2$ consumed.

Anti-ghrelin antibodies titer was determined 2 weeks after each immunization. Plasma levels of active ghrelin (EZRGRA-90K, Linco Research, St. Charles, Mo, USA, range 25–2,000 pg/ml), leptin (EZML-82K, Linco Research, St. Charles, Mo, USA, range 0.2–30 ng/ml), insulin (EZRMI-13K, Linco Research, St. Charles, Mo, USA, range 0.2–10 ng/ml), growth hormone (EZRMGH-45K, Linco Research, St. Charles, Mo, USA, range 0.07–50 ng/ml), IGF-1 (E25, Mediagnost, Reutlingen, Germany, range 0.5–18 ng/ml), and TNF-α (Quantikine, R&D Systems, Abingdon, United Kingdom, range 15.6–1,000 pg/ml) were determined by ELISA using specific commercial kits according to the manufacturer instructions.

Two weeks after the third immunization, mice were sacrificed and the stomach fundus and the hypothalamus were recovered and immediately frozen by immersion in liquid nitrogen to evaluate ghrelin, neuropeptide Y (NPY) and proopiomelanocortin (POMC) expression.

### 28.3.6 Safety and Efficacy of the Vaccine

Normal weight mice treated with the immunoconjugate displayed a significant decrease in daily food intake (0.44 g NS1-Ghr vs. 0.14 g PBS vs. 0.16 g NS1, $p<0.001$) (Fig. 28.3). In addition, after the first two inoculations, there was also an acute decrease in food intake in the group of mice that received the Immunoconjugate when compared to

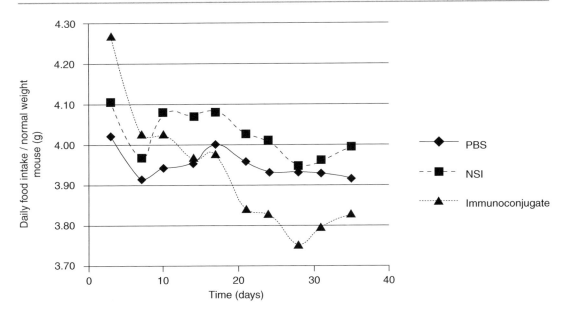

**Fig. 28.3** Normal weight mice treated with the immunoconjugate displayed a significant decrease in daily food intake when compared to control mice (0.44 g NS1-Ghr vs. 0.14 g PBS vs. 0.16 g NS1, $p<0.001$)

the PBS control, corresponding to 95.3 and 94.8 % of the PBS control, respectively, although without reaching statistical significance. There were no significant differences in body weight gain between the different groups of mice during the study span (3.83 g±0.40 g NS1-Ghr vs. 5.00 g±0.26 g PBS vs. 5.33 g±0.49 g NS1, $p=$NS) [50].

In DIO mice, after changing from the hypercaloric to the standard diet, there was an increase in daily food intake followed by rapid stabilization. DIO mice inoculated with the immunoconjugate did not display a significant decrease in cumulative food intake when compared to controls (147.64±2.46 g NS1-Ghr, 147.80±5.89 g NS1, 150.37±3.65 g PBS, $p=$NS), although there was a significant decrease of food intake in the 24 h immediately after each inoculation of the immunoconjugate, corresponding to 66.16 % ($p=0.036$), 82.22 % ($p=0.008$) and 50.09 % ($p=0.039$) of the food intake of the PBS group, after the three inoculations, respectively. DIO mice body weight decreased in response to the change from the hypercaloric to the standard diet (13.84 % compared to baseline), although after the inoculations, there were no significant differences in body weight among the different experimental groups (32.17±0.872 g NS1-Ghr vs. 31.33±1.282 g NS1 vs. 31.83±0.833 g PBS, $p=$NS) [50].

Normal weight mice inoculated with the immunoconjugate developed specific anti-ghrelin antibodies, with increasing titers after each inoculation, reaching a maximum of 1,265±492 2 weeks after the last inoculation. The control groups that received either NS1 protein alone or PBS presented basal titers of 332±114 and 324±143, $p=0.035$, respectively, which were maintained throughout the study and were not altered by the immunizations, which suggests nonspecific bindings related to the detection method. DIO mice inoculated with the immunoconjugate also developed specific anti-ghrelin antibodies in increasing titers until reaching a maximum after the third inoculation in contrast with control groups that maintained their basal titers (2,680±1,197 NS1-Ghr, 458±31 NS1 and 257±78 PBS group, respectively, $p=0.03$) [50].

Energy expenditure was significantly higher in normal weight mice inoculated with immunoconjugate when compared with controls (0.0146±0.001 kcal/h/kg NS1-Ghr,

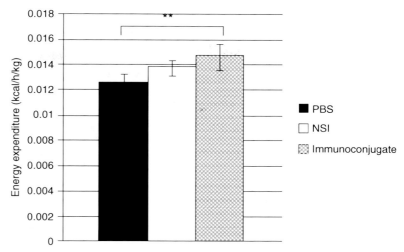

**Fig. 28.4** Energy expenditure was significantly higher in normal weight mice inoculated with immunoconjugate when compared with controls (0.0146±0.001 kcal/h/kg NS1-Ghr, 0.0138±0.001 kcal/h/kg NS1, 0.0129±0.001 kcal/h/kg PBS, p=0.038)

0.0138±0.001 kcal/h/kg NS1, 0.0129±0.001 kcal/h/kg PBS, p=0.038) (Fig. 28.4). DIO mice inoculated with the immunoconjugate also showed higher energy expenditure when compared to the control groups (0.0207±0.01 kcal/h/kg NS1-Ghr, 0.0140±0.002 kcal/h/kg NS1, 0.0159±0.002 kcal/h/kg PBS; p=0,044, NS1-Ghr vs. PBS and p=0,008, NS1-Ghr vs. NS1) [50].

Fasting plasma levels of active ghrelin were significantly higher in normal weight mice that received the immunoconjugate (361.3±79.9 pg/ml) when compared to control groups (186.9±14.8 pg/ml NS1 and 114.1±27.9 pg/ml PBS, p=0.009). DIO mice inoculated with the immunoconjugate also presented higher levels of fasting plasma ghrelin than the controls (429.63±179.27 pg/ml NS1-Ghr, 147.29±53.17 pg/ml NS1, 105.88±27.76 pg/ml PBS, p=NS) although not statistically significant. There were no significant differences in plasma levels of leptin, insulin, glucose, growth hormone, IGF-1 or TNF-α between the groups. ELISA confirmed the presence of circulating immune complexes of ghrelin-antighrelin antibodies in the plasma of normal weight mice inoculated with the immune conjugate. There was also a positive correlation between ghrelin plasma levels and the titer of circulating immune complexes (r=0.846) (Fig. 28.5). Search for immunoglobulins deposits in the kidney by immunohistochemistry failed to reveal any evidence of deposited immune complexes on the glomerular basement membranes [50].

In normal weight mice, there was no significant difference in ghrelin expression in the gastric fundus between the three experimental groups of mice (0.94±0.17 NS1-Ghr, 1.79±0.35 NS1, 1.00±0.30 PBS, p=NS). There was also no significant difference in NPY expression in the basal hypothalamus between the study groups (1.32±0.17 NS1-Ghr, 0.94±0.10 NS1, 1.00±0.20 PBS, p=NS). In contrast, POMC mRNA expression was significantly lower in mice inoculated with the immunoconjugate when compared to controls (0.20±0.17 NS1-Ghr, 0.93±0.17 NS1, 1.00±0.10 PBS, p<0.05). In DIO mice the expression of ghrelin after normalization for GAPDH expression in stomach cells was also not significantly different among the different study groups (1.27±0.30 NS1-Ghr, 0.38±0.13 NS1, 1.00±0.12 PBS, p=NS). However, DIO mice inoculated with the Immunoconjugate had a lower expression of NPY in the basal hypothalamus when compared to control groups (0.59±0.09 NS1-Ghr, 1.03±0.12 NS1, 1.00±0.13 for PBS, p<0.05). The expression of POMC in the basal hypothalamus was not significantly different between the different groups in study (1.04±0.14 NS1-Ghr, 1.32±0.25 NS1, 1.00±0.12 PBS, p=NS) [50].

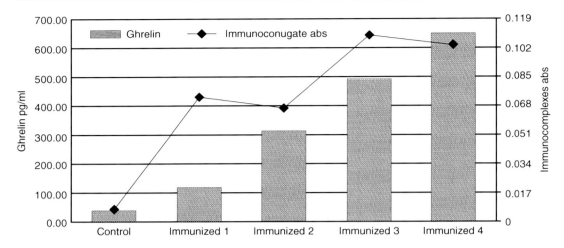

**Fig. 28.5** Circulating ghrelin-anti-ghrelin antibodies immune complexes titers and plasma ghrelin levels. The presence of circulating immune complexes of ghrelin-anti-ghrelin antibodies in the plasma of normal weight mice inoculated with the immune conjugate was confirmed by ELISA. There was also a positive correlation between ghrelin plasma levels and the titer of circulating immune complexes ($r=0.846$)

## 28.4 Strengths and Weaknesses

Obesity is nowadays a major public health problem [11, 51] for which there is a lack of medical therapeutic resources [12, 18]. Since ghrelin is the only orexigenic hormone identified so far, it has been pointed as a promising treatment target for obesity [52]. Several research groups have previously attempted ghrelin neutralization. Passive transfer of monoclonal anti-ghrelin antibodies was unable to change long-term food intake in mice [41]. Antibodies targeted to hydrolyze the octanoyl moiety of ghrelin to form desacyl ghrelin, which has no biological activity, resulted in increased metabolic rate and suppressed 6 h refeeding after 24 h of food deprivation in mice, but this approach would imply the need of periodic antibodies administration [53].

More recently, another study concluded that an oligoclonal response is required to maintain increased energy expenditure during fasting and deprivation-induced food intake as well as to reduce overall food intake upon refeeding [42]. Ghrelin receptor antagonists have also been tested, and GSH-R1a decreased food intake and body weight and improved glucose tolerance due to increased glucose-dependent insulin secretion [43]. Anti-ghrelin vaccines using KLH or BSA as immunogenic substances decreased body weight gain by decreasing feed efficiency in rats [44] and food intake and body weight in pigs [45].

However, these anti-ghrelin vaccination and neutralization strategies present several limitations when applied to humans because of the need to use adjuvants, the risk of exacerbated immune response against an endogenous substance, and, in the case of passive immunization, acquired tolerance and lack of long-term effectiveness. When compared with classic immunization techniques, VLPs are safe due to the lack of genetic material, since VLPs consist only of viral proteins and induce an efficient B cell activation. The highly repetitive nature of these structures has the advantage of allowing B cell receptor cross-linking due to the ordered presentation of epitopes in molecule surface and a high immunogenicity regardless of the route of the immunization, which allows the use of a low number of immunizations and a lower quantity of vaccine, making this type of vaccination protocol more efficient and cost-effective [54].

The main goal of the current vaccine approach was to develop a safer and more effective anti-ghrelin vaccine that could be used for human treatment. For that we developed an immunoconjugate composed of ghrelin and NS1 protein of

BTV. The choice of NS1 tubules as VLP-like carrier protein was driven by its previous use as a distribution system for molecules of prophylactic vaccines against common human infectious diseases, such as proteins of the foot-and-mouth disease and influenza A virus [49, 55].

The ability of the vaccine to trigger an immune response was tested in normal weight and DIO male mice that developed increasing titers of specific anti-ghrelin antibodies, confirming the hypothesis that a vaccine consisting of immunoconjugate only is able to trigger an immune response without the need adjuvants. Furthermore, antibody titers attained after the immunization protocol were not very high, when compared to antibodies titers after common infectious diseases, which is also reassuring in safety concerns, since complete neutralization of ghrelin was not the purpose of an anti-ghrelin vaccination strategy for obesity treatment as ghrelin also intervenes in several key biological processes besides appetite regulation, such as growth hormone secretion and gastrointestinal and cardiovascular functions [52].

Vaccinated mice showed significantly higher energy expenditure than the animals of the groups that received either NS1 protein alone or PBS. Higher energy expenditure usually translates into greater ease of weight loss and maintenance. Ghrelin is known to suppress energy metabolism, and ghrelin replacement partially reverses the reduction in body weight and body fat in gastrectomized mice [56]. Ghrelin has been shown to have a long-term effect on energy homeostasis by increasing the respiratory quotient, through decreasing utilization of fat as energy [30]. In addition, ghrelin knockout mice compared to wild-type mice present no change in food intake but have a decreased respiratory quotient when fed with high-fat diet, suggesting that endogenous ghrelin plays a more prominent role in determining the type of metabolic substrate that is used for maintenance of energy balance than in the regulation of food intake [57]. Although vaccinated animals gained less weight when compared with control animals, this difference failed to reach statistical significance, which may be explained by the short follow-up time or the activation of compensatory mechanisms of energy homeostasis pathways.

Paradoxically, vaccinated mice had higher ghrelin levels compared to controls. Given that these increased levels of ghrelin did not appear to have a biological effect, we hypothesize that circulating ghrelin could be in the form of immune complexes of ghrelin-anti-ghrelin antibodies, which was confirmed. Previous reports on anti-ghrelin vaccines have also documented an increase of ghrelin in immunized animals, although the presence of circulating immunocomplexes has not been documented [58]. The presence of circulating immunocomplexes, which could be due to a lower rate of elimination, raised the concern of renal toxicity due to the deposition in the glomerular basement membrane that has been excluded. Since there was no difference in ghrelin expression in the stomach, ghrelin appears to be synthetized in immunized animals as in controls, and after neutralization of ghrelin biological activity, there is no upregulation of ghrelin expression in order to maintain the homeostasis.

In vaccinated normal weight mice, there were no significant differences in the genetic expression of NPY gene in the basal hypothalamus in comparison to control mice. However, in vaccinated DIO mice, there was a significant decrease of NPY gene expression in the basal hypothalamus compared with controls reflecting a decrease in central orexigenic signals [59]. The expression of POMC in the basal hypothalamus was significantly lower in vaccinated normal weight animals compared to controls that could represent a compensatory mechanism to the decreased peripheral orexigenic signals in order to prevent the reduction in feeding threshold of normal weight mice, which could also explain why these findings only occurred in the normal weight mice but not in DIO mice.

Ghrelin is a growth hormone secretagogue [52] and ghrelin neutralization could induce alterations in GH/IGF-1 axis. Since this vaccine appears to have no effect in GH and IGF-1 levels, this suggests that our vaccine is unlikely to cause endocrine adverse effects on the growth hormone axis.

The regulatory mechanisms of energy homeostasis and appetite control are very complex processes that include highly redundant signalling pathways [31]. Therefore, it is possible that the lack of significant differences in some biological parameters, such as food intake and body weight, may be due to activation of compensatory mechanisms for the decrease in available active ghrelin similar to that which occurs in ghrelin knockout mice [57, 60].

## 28.5 Concluding Remarks

This anti-ghrelin vaccine appears to be well tolerated by the animals, and there were no signs of inflammatory reaction or toxicity. The production of anti-ghrelin antibodies was effective in decreasing acute food intake and increasing energy expenditure in the vaccinated animals compared to control animals, which are important contributions to establish a negative energy balance and thus promote weight loss.

Most obese patients have low ghrelin levels; therefore, it is not expected for the vaccine to be effective in the absence of diet-induced ghrelin rise, so an anti-ghrelin vaccine would be beneficial for patients enrolling a diet and exercise program as adjuvant therapy for weight loss and prevention of weight regain [34]. Additionally, obese patients with high ghrelin levels could benefit from ghrelin blockade through this anti-ghrelin vaccine, such as individuals with Prader-Willi syndrome [39].

In conclusion, these results suggest that this anti-ghrelin vaccine has a positive impact on energy homeostasis and may be a useful tool for obesity treatment.

**Acknowledgements** UMIB is funded by grants from FCT POCTI/FEDER Fcomp-01-0124-FEDER-015893 and Project Grant EXPL/BIM-MET/0618/2012. Portugal.

## References

1. do Carmo, I., et al.: Overweight and obesity in Portugal: national prevalence in 2003–2005. Obes. Rev. **9**, 11–19 (2008). doi:10.1111/j.1467-789X.2007.00422.x. OBR422 [pii]
2. WHO: Obesity and overweight fact sheet N°311. http://www.who.int/mediacentre/factsheets/fs311/en/print.html (2006). Accessed 3 July 2009
3. Ogden, C.L., Carroll, M.D., Kit, B.K., Flegal, K.M.: Prevalence of obesity and trends in body mass index among US children and adolescents, 1999–2010. JAMA **307**, 483–490 (2012). doi:10.1001/jama.2012.40. jama.2012.40 [pii]
4. Pi-Sunyer, F.X.: The obesity epidemic: pathophysiology and consequences of obesity. Obes. Res. **10**(Suppl 2), 97S–104S (2002). doi:10.1038/oby.2002.202
5. Finkelstein, E.A., Trogdon, J.G., Cohen, J.W., Dietz, W.: Annual medical spending attributable to obesity: payer-and service-specific estimates. Health Aff. **28**, w822–w831 (2009). doi:10.1377/hlthaff.28.5.w822
6. Rappange, D.R., Brouwer, W.B., Brouwer, W.B., Hoogenveen, R.T., Van Baal, P.H.: Healthcare costs and obesity prevention: drug costs and other sector-specific consequences. Pharmacoeconomics **27**, 1031–1044 (2009). doi:10.2165/11319900-000000000-00000
7. Wolf, A.M., Colditz, G.A.: Current estimates of the economic cost of obesity in the United States. Obes. Res. **6**, 97–106 (1998)
8. Peeters, A., et al.: Obesity in adulthood and its consequences for life expectancy: a life-table analysis. Ann. Intern. Med. **138**, 24–32 (2003). doi:200301070-00008 [pii]
9. WHO: Diet, nutrition and the prevention of chronic diseases. http://www.who.int/mediacentre/news/releases/2003/pr20/en/ (2003). Accessed 3 July 2009
10. NHLBI Obesity Education Initiative Expert Panel on the Identification, Evaluation, and Treatment of Obesity in Adults (US). Clinical Guidelines on the Identification, Evaluation, and Treatment of Overweight and Obesity in Adults: The Evidence Report. Bethesda (MD): National Heart, Lung, and Blood Institute; Sep 1998 Available from: http://www.ncbi.nlm.nih.gov/books/NBK2003/
11. Puska, P., Nishida, C., Porter, D., World Health Organization: Obesity and overweight. http://www.who.int/dietphysicalactivity/publications/facts/obesity/en/print.html (2006). Accessed 3 July 2009
12. Bray, G.A.: Lifestyle and pharmacological approaches to weight loss: efficacy and safety. J. Clin. Endocrinol. Metab. **93**, S81–S88 (2008). doi:10.1210/jc.2008-1294. jc.2008-1294 [pii]
13. Fujioka, K.: Management of obesity as a chronic disease: nonpharmacologic, pharmacologic, and surgical options. Obes. Res. **10**(Suppl 2), 116S–123S (2002). doi:10.1038/oby.2002.204
14. Avenell, A., et al.: Systematic review of the long-term effects and economic consequences of treatments for obesity and implications for health improvement. Health Technol. Assess. **8**(iii–iv), 1–182 (2004). doi:99-02-02 [pii]
15. Rucker, D., Padwal, R., Li, S.K., Curioni, C., Lau, D.C.: Long term pharmacotherapy for obesity and overweight: updated meta-analysis. BMJ **335**, 1194–1199 (2007). doi:10.1136/bmj.39385.413113.25. bmj.39385.413113.25 [pii]

16. Bays, H.E.: Lorcaserin: drug profile and illustrative model of the regulatory challenges of weight-loss drug development. Expert Rev. Cardiovasc. Ther. **9**, 265–277 (2011). doi:10.1586/erc.10.22
17. Bays, H.E., Gadde, K.M.: Phentermine/topiramate for weight reduction and treatment of adverse metabolic consequences in obesity. Drugs Today (Barc.) **47**, 903–914(2011).doi:10.1358/dot.2011.47.12.1718738. 1718738 [pii]
18. Snow, V., Barry, P., Fitterman, N., Qaseem, A., Weiss, K.: Pharmacologic and surgical management of obesity in primary care: a clinical practice guideline from the American College of Physicians. Ann. Intern. Med. **142**, 525–531 (2005). doi:142/7/525 [pii]
19. Bult, M.J., van Dalen, T., Muller, A.F.: Surgical treatment of obesity. Eur. J. Endocrinol. **158**, 135–145 (2008). doi:10.1530/EJE-07-0145. 158/2/135 [pii]
20. NIH: The Practical Guide: Identification, Evaluation, and Treatment of Overweight and Obesity in Adults. National Institutes of Health, National Heart, Lung, and Blood Institute, and North American Association for the Study of Obesity, Bethesda (2000) (NIH Publication Number 00-4084)
21. Fisher, B.L., Schauer, P.: Medical and surgical options in the treatment of severe obesity. Am. J. Surg. **184**, 9S–16S (2002). doi:S000296100201173X [pii]
22. Schneider, B.E., Mun, E.C.: Surgical management of morbid obesity. Diabetes Care **28**, 475–480 (2005). doi:28/2/475 [pii]
23. Picot, J., et al.: The clinical effectiveness and cost-effectiveness of bariatric (weight loss) surgery for obesity: a systematic review and economic evaluation. Health Technol. Assess. **13**, 1–190, 215–357, iii–iv (2009). doi:10.3310/hta13410
24. Christou, N. V. et al.: Surgery decreases long-term mortality, morbidity, and health care use in morbidly obese patients. Ann. Surg. **240**, 416–423; discussion 423–414, (2004). doi:00000658-200409000-00003 [pii]
25. Buchwald, H., Williams, S.E.: Bariatric surgery worldwide. Obes. Surg. **14**, 1157–1164 (2003). doi:10.1381/0960892042387057 (2004)
26. Field, B.C., Chaudhri, O.B., Bloom, S.R.: Obesity treatment: novel peripheral targets. Br. J. Clin. Pharmacol. **68**, 830–843 (2009). doi:10.1111/j.1365-2125.2009.03522.x. BCP3522 [pii]
27. Schwartz, M.W., Woods, S.C., Porte Jr., D., Seeley, R.J., Baskin, D.G.: Central nervous system control of food intake. Nature **404**, 661–671 (2000). doi:10.1038/35007534
28. De Vriese, C., Delporte, C.: Ghrelin: a new peptide regulating growth hormone release and food intake. Int. J. Biochem.Cell Biol.**40**,1420–1424(2008).doi:10.1016/j.biocel.2007.04.020. S1357-2725(07)00138-0 [pii]
29. Kojima, M., et al.: Ghrelin is a growth-hormone-releasing acylated peptide from stomach. Nature **402**, 656–660 (1999). doi:10.1038/45230
30. Tschop, M., Smiley, D.L., Heiman, M.L.: Ghrelin induces adiposity in rodents. Nature **407**, 908–913 (2000). doi:10.1038/35038090
31. Cone, R.D., et al.: The arcuate nucleus as a conduit for diverse signals relevant to energy homeostasis. Int. J. Obes. Relat. Metab. Disord. **25**(Suppl 5), S63–S67 (2001). doi:10.1038/sj.ijo.0801913
32. Millington, G.W.: The role of proopiomelanocortin (POMC) neurones in feeding behaviour. Nutr. Metab. (Lond.) **4**, 18 (2007). doi:10.1186/1743-7075-4-18. 1743-7075-4-18 [pii]
33. Williams, G., Harrold, J.A., Cutler, D.J.: The hypothalamus and the regulation of energy homeostasis: lifting the lid on a black box. Proc. Nutr. Soc. **59**, 385–396 (2000). doi:S0029665100000434 [pii]
34. Cummings, D.E., et al.: Plasma ghrelin levels after diet-induced weight loss or gastric bypass surgery. N. Engl. J. Med. **346**, 1623–1630 (2002). doi:10.1056/NEJMoa012908346/21/1623 [pii]
35. English, P.J., Ghatei, M.A., Malik, I.A., Bloom, S.R., Wilding, J.P.: Food fails to suppress ghrelin levels in obese humans. J. Clin. Endocrinol. Metab. **87**, 2984 (2002)
36. Stock, S., et al.: Ghrelin, peptide YY, glucose-dependent insulinotropic polypeptide, and hunger responses to a mixed meal in anorexic, obese, and control female adolescents. J. Clin. Endocrinol. Metab. **90**, 2161–2168 (2005). doi:10.1210/jc.2004-1251. jc.2004-1251 [pii]
37. Hansen, T.K., et al.: Weight loss increases circulating levels of ghrelin in human obesity. Clin. Endocrinol. (Oxf) **56**, 203–206 (2002). doi:1456 [pii]
38. Janas-Kozik, M., Krupka-Matuszczyk, I., Malinowska-Kolodziej, I., Lewin-Kowalik, J.: Total ghrelin plasma level in patients with the restrictive type of anorexia nervosa. Regul. Pept. **140**, 43–46 (2007). doi:10.1016/j.regpep.2006.11.005. S0167-0115(06)00223-0 [pii]
39. Goldstone, A.P., et al.: Elevated fasting plasma ghrelin in prader-willi syndrome adults is not solely explained by their reduced visceral adiposity and insulin resistance. J. Clin. Endocrinol. Metab. **89**, 1718–1726 (2004)
40. Monteiro, M.P., et al.: Increase in ghrelin levels after weight loss in obese Zucker rats is prevented by gastric banding. Obes. Surg. **17**, 1599–1607 (2007). doi:10.1007/s11695-007-9324-7
41. Lu, S.C., et al.: An acyl-ghrelin-specific neutralizing antibody inhibits the acute ghrelin-mediated orexigenic effects in mice. Mol. Pharmacol. **75**, 901–907 (2009). doi:10.1124/mol.108.052852. mol.108.052852 [pii]
42. Zakhari, J.S., Zorrilla, E.P., Zhou, B., Mayorov, A.V., Janda, K.D.: Oligoclonal antibody targeting ghrelin increases energy expenditure and reduces food intake in fasted mice. Mol. Pharm. **9**, 281–289 (2012). doi:10.1021/mp200376c
43. Esler, W.P., et al.: Small-molecule ghrelin receptor antagonists improve glucose tolerance, suppress appetite, and promote weight loss. Endocrinology **148**, 5175–5185 (2007). doi:10.1210/en.2007-0239. en.2007-0239 [pii]
44. Zorrilla, E.P., et al.: Vaccination against weight gain. Proc. Natl. Acad. Sci. U. S. A. **103**, 13226–13231 (2006). doi:10.1073/pnas.0605376103. 0605376103 [pii]
45. Vizcarra, J.A., Kirby, J.D., Kim, S.K., Galyean, M.L.: Active immunization against ghrelin decreases weight

46. Jennings, G.T., Bachmann, M.F.: Immunodrugs: therapeutic VLP-based vaccines for chronic diseases. Annu. Rev. Pharmacol. Toxicol. **49**, 303–326 (2009). doi:10.1146/annurev-pharmtox-061008-103129
47. Brown, M.J.: Therapeutic potential of vaccines in the management of hypertension. Drugs **68**, 2557–2560 (2008). doi:68182 [pii]
48. Hewat, E.A., Booth, T.F., Wade, R.H., Roy, P.: 3-D reconstruction of bluetongue virus tubules using cryo-electron microscopy. J. Struct. Biol. **108**, 35–48 (1992). doi:1047-8477(92)90005-U [pii]
49. Ghosh, M.K., Borca, M.V., Roy, P.: Virus-derived tubular structure displaying foreign sequences on the surface elicit CD4+ Th cell and protective humoral responses. Virology **302**, 383–392 (2002). doi:S004268220291648X [pii]
50. Andrade, S., Carreira, M., Ribeiro, A.: Development of an Anti-Ghrelin Vaccine for Obesity Treatment Endocr. Rev. 32, P2-305 (2011). (The Endocrine Society).
51. Moayyedi, P.: The epidemiology of obesity and gastrointestinal and other diseases: an overview. Dig. Dis. Sci. **53**, 2293–2299 (2008). doi:10.1007/s10620-008-0410-z
52. Kojima, M., Kangawa, K.: Ghrelin: structure and function. Physiol. Rev. **85**, 495–522 (2005). doi:10.1152/physrev.00012.2004. 85/2/495 [pii]
53. Mayorov, A.V., et al.: Catalytic antibody degradation of ghrelin increases whole-body metabolic rate and reduces refeeding in fasting mice. Proc. Natl. Acad. Sci. U. S. A. **105**, 17487–17492 (2008). doi:10.1073/pnas.0711808105
54. Lechner, F., et al.: Virus-like particles as a modular system for novel vaccines. Intervirology **45**, 212–217 (2002). doi:10.1159/000067912int45212 [pii]
55. Mikhailov, M., Monastyrskaya, K., Bakker, T., Roy, P.: A new form of particulate single and multiple immunogen delivery system based on recombinant bluetongue virus-derived tubules. Virology **217**, 323–331 (1996). doi:10.1006/viro.1996.0119. S0042-6822(96)90119-1 [pii]
56. Dornonville de la Cour, C., et al.: Ghrelin treatment reverses the reduction in weight gain and body fat in gastrectomised mice. Gut **54**, 907–913 (2005). doi:10.1136/gut.2004.058578
57. Wortley, K.E., et al.: Genetic deletion of ghrelin does not decrease food intake but influences metabolic fuel preference. Proc. Natl. Acad. Sci. U. S. A. **101**, 8227–8232(2004).doi:10.1073/pnas.0402763101 0402763101 [pii]
58. Kellokoski, E., et al.: Ghrelin vaccination decreases plasma MCP-1 level in LDLR(−/−)-mice. Peptides **30**, 2292–2300 (2009). doi:10.1016/j.peptides.2009.09.008. S0196-9781(09)00366-0 [pii]
59. Kalra, S.P., Dube, M.G., Sahu, A., Phelps, C.P., Kalra, P.S.: Neuropeptide Y secretion increases in the paraventricular nucleus in association with increased appetite for food. Proc. Natl. Acad. Sci. U. S. A. **88**, 10931–10935 (1991)
60. De Smet, B., et al.: Energy homeostasis and gastric emptying in ghrelin knockout mice. J. Pharmacol. Exp. Ther. **316**, 431–439 (2006). doi:10.1124/jpet.105.091504. jpet.105.091504 [pii]

# Vaccines for Type 1 Diabetes

## 29

Sandeep Kumar Gupta

## Contents

| | | |
|---|---|---|
| 29.1 | **Introduction** | 478 |
| 29.2 | **Type 1 Diabetes: A Disorder of Immunoregulation** | 478 |
| 29.2.1 | Role of Cytokine in Immune Response | 479 |
| 29.3 | **Prediction of T1DM** | 479 |
| 29.3.1 | Genetic Prediction of T1DM | 479 |
| 29.3.2 | Autoantibodies for Prediction of T1DM | 480 |
| 29.4 | **Rationale of Vaccine for T1DM** | 480 |
| 29.5 | **Whom to Vaccinate?** | 482 |
| 29.6 | **Various Interventions of Immune Prevention** | 482 |
| 29.6.1 | Antibody-Based Immunotherapy | 482 |
| 29.6.2 | Antigen-Based Immunotherapy | 483 |
| **References** | | 488 |

S.K. Gupta, MD (✉)
Department of Pharmacology,
Dhanalakshmi Srinivasan Medical College and Hospital,
Siruvachur, Perambalur, Tamil Nadu, India
e-mail: drsandeep_gupta@rediffmail.com

## Abstract

By primary prevention aimed at avoiding or averting environmental factors thought to promote disease in genetically at-risk individuals T1D would be eradicated ideally, but these factors have not been identified and may be ubiquitous. Since the early 1980s, secondary prevention, after the disease process has started, has been the focus of considerable attention and with many candidate agents, mainly immunosuppressive drugs have been trialled, usually after the onset of clinical diabetes. Prevention is however more applicable to early, preclinical disease rather than to recent onset clinical disease, when beta cell destruction is more advanced. Prevention of the infectious disease by exposing the immune system to the weakened or dead infectious agent has been a traditional vaccination method. Alternate method called "inverse vaccination" (the inhibition of immune response) arrests autoimmunity through manipulation of the innate and adaptive arms of the immune system. Inverse vaccination specifically reduces a pathological adaptive autoimmune response. Targeted reduction of unwanted antibody and T-cell responses to autoantigens is allowed by inverse vaccination, while leaving the remainder of the immune system intact. Varying degree of success in suppression of β-cell autoimmunity in NOD mice have been shown by current options for treatment of autoimmunity such as immunosuppressive drugs (e.g. cyclosporine) and anti

T-cell antibodies (e.g. anti-CD3 antibodies). But the drawback with these methods is that it requires repeated administration and may lead to non-specific harmful effects such as interference with normal immune system functions. Whereas, antigen specific immunotherapy (ASI) uses inverse vaccination for a specific auto-antigen. The advantage with the ASI is selective inactivation of auto-reactive T cells without interference in normal immune function. Major examples of antigen specific immunotherapy agent at various stage of clinical trials are alum formulated glutamic acid decarboxylase and heat shock protein and peptide 277.

## 29.1 Introduction

Type 1 diabetes mellitus (T1DM) is an autoimmune disease. There has been an overall increase in incidence of about 3 % yearly over the last decades, and it is estimated that there are approximately 65,000 new cases/year in children less than 15 years of age [1]. The innate and adaptive immune cells progressively destroy insulin-producing β-cell in the islets of pancreas in T1DM [2].

Not only T1DM but also latent autoimmune diabetes in adults (LADA) is included in autoimmune diabetes. Latent autoimmune diabetes in adults is similar to and is frequently confused with type 2 diabetes. But LADA is distinguished by the presence of autoantibodies, mainly glutamic acid decarboxylase (GAD) autoantibodies (GADA). LADA patients usually become insulin dependent at much faster rates in contrast to classic type 2 diabetes. LADA has been put as a distinct entity and type of diabetes by some; others believe that LADA is just a mild variant of T1DM and should be treated as such [3]. Table 29.1 summarizes characteristic features of type 1 DM, type 2 DM, and LADA.

## 29.2 Type 1 Diabetes: A Disorder of Immunoregulation

The selective destruction of the insulin-producing β-cells in the pancreatic islets of Langerhans leads to T1DM. An autoimmune response mediated by T lymphocytes (T cells) that reacts specifically to one or more β-cell proteins (autoantigens) results in destruction of pancreatic islet β-cells. Genetic and environmental factors interact and confer either susceptibility or resistance to disease, depending on the gene/allele possessed by the individual and the environmental agent to which that individual is exposed. Disease susceptibility leads to a pathogenic immune response whereas disease resistance leads to a protective immune response.

**Table 29.1** Characteristic features of type 1 DM, type 2 DM, and LADA [4]

| | Type 1 DM | Type 2 DM | LADA |
|---|---|---|---|
| Age of onset | Youth or adult (<35 years) | Adult (>35 years) | Adult (>35 years) |
| Progression to insulin dependence | Rapid (days/weeks) | Slow (years) | Latent (months/years) |
| Presence of autoantibodies | Yes | No | Yes |
| Insulin dependence | At diagnosis | Over time, if at all | Within 6 years |
| Insulin resistance | No | Yes | Some |
| Response to lifestyle modification or oral agents | Poor | Good | Initial mixed then worsening |
| Frequency of DKA | High | Low | Low |
| Family history of DM | Uncommon | Common | Uncommon |
| Body habitus | Fit or lean | Overweight to obese | Normal to overweight |
| Acanthosis nigricans | No | Yes | No |
| Metabolic syndrome | No | Yes | No |
| C-peptide level | Undetectable | Normal | Low/normal |

The pathogenic immune response is believed to be mediated by T lymphocytes (T cells) that are reactive to islet β-cell self-antigen(s) (autoreactive T cells), whereas a protective immune response may be mediated by T cells that suppress the autoreactive T cells (regulatory T cells). Dominance of the pathogenic immune response would lead to islet inflammation (insulitis). This is characterized by infiltration of the islet by macrophages and T cells that are cytotoxic, both directly and indirectly by producing cytokines (e.g., IL-1, TNF-α, TNF-β, and IFN) and free radicals that damage β-cells. A major part of the islets is deficient in β-cells at the time of clinical symptoms, and compared with control individuals, the total pancreatic volume is significantly reduced. Genetic and environmental factors may also directly increase or decrease the ability of β-cells to repair damage and prevent irreversible β-cell death, insulinopenia, and diabetes [2].

### 29.2.1 Role of Cytokine in Immune Response

During the process of positive and negative selection, the precursors of T cells mature in the thymus. Cells with T cell receptors (TCR) that recognize and bind antigen presented by HLA molecules will survive positive selection, while the others will die. The T cells are supposed to recognize only foreign peptides presented by HLA [5].

Surviving the selection process, T cells leave the thymus and circulate continually from the blood to peripheral lymphoid tissues (lymph nodes, spleen, and mucosal tissues). When antigen-presenting cells (APC) recognize antigen in the periphery, they transport it to a lymph node where the antigen is presented to T cells. The T cells with a receptor specific for the antigen will bind and then proliferate and differentiate into an effector T cell. Cytotoxic T cells recognizing peptides from intracellular pathogens, presented by HLA class I molecules, kill the pathogen-infected cells. Apoptosis is induced by interaction between Fas receptors on target cells and Fas ligand on infiltrating cells, but also by the cytotoxic effects of perforin and granzyme [5].

The T cells can be further classified into different subpopulations based on their expression of CD4 and CD8, which defines T helper (Th) and cytotoxic T cells (Tc), respectively. The Th cells can be divided into Th1 and Th2 cells based on their cytokine profile. An imbalance between Th1- and Th2-associated cytokines has been suggested to be of importance in mediating the β-cell destruction, seen in T1DM [5].

Th1-secreted cytokines (e.g., interferon (IFN)) are thought to initiate and propagate the inflammatory process in early diabetes and Th2-secreted anti-inflammatory cytokines (e.g., interleukin (IL)-4, IL-10) to suppress it. This has led to the hypothesis that diabetes can be prevented by using Th2-secreted cytokines. Another approach that can skew the cytokine cascade from a Th1 to a Th2 response is the use of a non-depleting anti-CD3 antibody [6].

## 29.3 Prediction of T1DM

The development of T1DM can be predicted with the combined use of genetic, islet autoantibody and metabolic testing.

### 29.3.1 Genetic Prediction of T1DM

Human leucocyte antigen (HLA) genes in the major histocompatibility complex (MHC), specifically alleles at the HLA DR and DQ loci, are the single most important genetic determinants of T1DM. Risk is highest with the HLA DR3,4-DQ 2,8 haplotype, whereas the HLA DR2-DQ6 haplotype gives protection against T1DM. HLA molecule shapes good or bad immune responses by binding specific antigenic peptides for recognition by T-cell receptors. HLA genetics accounts in large measures for why most of us don't develop T1DM. The non-HLA genes, environmental factors such as microbes, and dietary components may as well promote or retard the development of T1DM [2, 7].

## 29.3.2 Autoantibodies for Prediction of T1DM

We can detect autoantibodies against β-cell antigens well in advance before the clinical onset of T1DM, but their role in human disease is not clear. By immunofluorescence technique, islet cell autoantibodies (ICA) were the first T1DM-associated autoantibodies detected in human pancreas sections. Today, the major autoantibodies used for prediction of T1DM are those directed against insulin (i.e., anti-insulin antibodies (IAA)), the tyrosine phosphatases insulinoma antigen (IA)-2 and IA-2 β, and glutamic acid decarboxylase (GAD). More recently, antibodies against the zinc transporter (ZnT8) were discovered and are now used for the prediction and diagnosis of diabetes. We can identify individuals at high risk of developing T1DM with the presence of multiple antibodies [7].

### 29.3.2.1 Insulin
The first β-cell antigen detected in newly diagnosed T1DM patients was insulin. The first autoantibodies to appear in individuals developing T1DM are insulin autoantibodies (IAA). 40–70 % of newly diagnosed patients have IAA [5].

### 29.3.2.2 Glutamic Acid Decarboxylase
GAD is found in islet cells, the central nervous system, and the testes. GAD was originally detected as a 64 kDa protein in plasma from T1DM patients. Further studies showed that antibodies in sera from newly diagnosed T1DM patients were directed against this pancreatic islet cell protein, which later was identified as the enzyme GAD65. Early after the initiation of autoimmune insulitis, antibodies to GAD appear and are found in 70 % of patients at diagnosis. GAD is an enzyme involved in the conversion of glutamic acid to the inhibitory neurotransmitter gamma-aminobutyric acid (GABA). The physiological role of GAD in the pancreatic islets is unknown, but it has been suggested that GAD may function as a negative regulator of insulin secretion in response to glucose [5, 8].

**Table 29.2** Autoantibodies for prediction of T1DM

| |
|---|
| Insulin |
|   The first autoantibodies to appear in individuals developing T1D are insulin autoantibodies (IAA) |
|   40–70 % of newly diagnosed patients have IAA |
| Glutamic acid decarboxylase |
|   GAD is found in islet cells, the central nervous system, and the testes |
|   Early after the initiation of autoimmune insulitis, antibodies to GAD appear and are found in 70 % of patients at diagnosis |
|   GAD is an enzyme involved in the conversion of glutamic acid to the inhibitory neurotransmitter gamma-aminobutyric acid (GABA) |
|   The physiological role of GAD in the pancreatic islets is unknown, but it has been suggested that GAD may function as a negative regulator of insulin secretion in response to glucose |
| Tyrosine phosphatase-like protein |
|   IA-2 is localized in the secretory granule membranes of islets and other neuroendocrine cells |
|   Autoantibodies against IA-2 (IA-2A) are directed to the intracellular part of the protein |

### 29.3.2.3 Tyrosine Phosphatase-Like Protein
Two other antigenic targets, the 40 and 37 kDa proteins that bound antibodies strongly associated with progression to T1DM, were revealed in additional analyses of the 64 kDa protein, identified as GAD65. The 37 kDa antigen has been suggested to be a different protein with structural similarity to IA-2, while the 40 kDa antigen was identified as the tyrosine phosphatase-like protein IA-2 (ICA512). IA-2 is localized in the secretory granule membranes of islets and other neuroendocrine cells. Autoantibodies against IA-2 (IA-2A) are directed to the intracellular part of the protein. Table 29.2 summarizes autoantibodies for prediction of T1DM.

## 29.4 Rationale of Vaccine for T1DM

By primary prevention aimed at avoiding or averting environmental factors thought to promote disease in genetically at-risk individuals, T1DM would be eradicated ideally, but these factors have not been identified and may be ubiquitous. The prospects for primary prevention remain uncertain without knowing what they are and without being able to modify genetic susceptibility. Since the early 1980s, secondary prevention, after the disease process has started, has been the focus of considerable attention, and with many candidate agents, mainly immunosuppressive drugs have been trialled, usually after the onset of clinical diabetes. Prevention is however more applicable to early, preclinical disease rather than to recent-onset clinical disease, when beta-cell destruction is more advanced [2].

Individual in the preclinical phase of T1DM can be identified by the presence of circulating autoantibodies to specific islet antigens: (pro) insulin, the molecular weight 65,000 isoform of glutamic acid decarboxylase (GAD65), and tyrosine phosphatase-like insulinoma antigen 2 (IA2). The decision to intervene early in these asymptomatic, at-risk individual not only depends on the likelihood of greater efficacy but requires careful consideration of safety. In the preclinical phase, vaccines to promote protective immune homeostasis should be relatively safe and efficacious, whereas potentially toxic immunosuppressive drug would require strong justification in asymptomatic individual, especially children. In individuals with recent-onset disease, immunosuppressive drugs could be used to reduce the burden of pathogenic immunity and allow the emergence or active induction of pathogenetic mechanisms [2, 9].

Prevention of the infectious disease by exposing the immune system to the weakened or dead infectious agent has been a traditional vaccination method. Alternate method called "inverse vaccination" (the inhibition of immune response) arrests autoimmunity through manipulation of the innate and adaptive arms of the immune system. Inverse vaccination specifically reduces a pathological adaptive autoimmune response. Targeted reduction of unwanted antibody and T-cell responses to autoantigens is allowed by inverse vaccination while leaving the remainder of the immune system intact [10, 11].

Varying degrees of success in suppression of β-cell autoimmunity in NOD mice have been shown by current options for treatment of autoimmunity such as immunosuppressive drugs (e.g., cyclosporine) and anti-T-cell antibodies (e.g., anti-CD3 antibodies). But the drawback with these methods is that it requires repeated administration and may lead to nonspecific harmful effects such as interference with normal immune system functions. Whereas, antigen-specific immunotherapy (ASI) uses inverse vaccination for a specific autoantigen. The advantage with the ASI is selective inactivation of autoreactive T cells without interference in normal immune function [10, 11].

Diabetes vaccines may work through various mechanisms: (1) changing the immune response from a destructive (e.g., Th1) to a more benign (e.g., Th2) response, (2) inducing antigen-specific regulatory T cells, (3) deleting autoreactive T cells, or (4) preventing immune cell interaction [12]. The Th1–Th2 shift occurs via a change in the type of cytokine signaling molecules being released by regulatory T cells. Instead of pro-inflammatory cytokines, the regulatory T cells begin to release cytokines that inhibit inflammation [13].

A basic requirement in the drug development is the demonstration of efficacy and safety in animal models. The NOD mouse is the most commonly used animal models of T1DM, and it has contributed immensely in our knowledge of disease mechanisms and prevention theory of T1DM. Human T1DM and autoimmune diabetes in the NOD mouse share features such as polygenic inheritance dominated by genes for antigen-presenting molecule in the MHC autoimmune responses to (pro) insulin and GAD65, transfer of disease by bone marrow, and protracted preclinical phase. But in contrast to humans, the NOD mouse is inbred and responds to many immune and other interventions. However, most interventions prevent disease in only a proportion of NOD mice [2].

However, in recent studies, it has been reported that the immune pathogenesis of NOD mice and susceptible humans might be rather dissimilar, i.e., the degree of insulitis in humans is much milder and involve maximally about 15–35 % of all pancreatic islets at any given time. Moreover, in contrast to NOD mice, it is hard to imagine that inflammation can be driven autonomously by neo-formation of lymphoid structures in human islets. Lastly, in human insulitis, CD8 lymphocytes predominate, whereas in NOD mice, CD4 T cells are mainly found. The lymphopenic BB rat and several antigen-driven mouse models have been suggested as alternative models, which can be useful in highlighting aspects of diabetogenesis that are not accurately reflected in the NOD mouse. An attractive alternative can be "humanized" mice that are partly reconstituted with components of the human immune system [9].

## 29.5 Whom to Vaccinate?

The neonatal screening for high-risk HLA class II susceptibility genes can identify most of those destined to develop T1DM. Young people with first-degree relative positive for autoantibodies positive for one or more islet autoantigen, i.e., (pro) insulin, GAD, and IA2, are at high risk for T1DM. In such people, the 5-year risk of developing T1DM is of the order of ≤25 %, 26–50 %, and >50 % if they have autoantibodies to one, two, and three autoantigens, respectively. In addition, the first-phase insulin response (FPIR) to intravenous glucose below the 10th percentile indicates a poor prognosis. In autoantibody-positive relatives with a normal FPIR, the insulin resistance indicates the highest risk [2].

## 29.6 Various Interventions of Immune Prevention

### 29.6.1 Antibody-Based Immunotherapy

#### 29.6.1.1 Anti-CD3 Monoclonal Antibody

The monoclonal antibody OKT3, directed against CD3, inhibits Tc cell-mediated lysis of target cells. But, OKT3 has a strong mitogenic activity and induces massive amounts of cytokines leading to adverse events [14]. A majority of patients experienced some degree of cytokine release syndrome. A number of side effects are seen in most patients, such as chills, nausea, hypotension, breathing difficulties, fever, muscle pain, thrombocytopenia with risk of bleedings, leukocytopenia with increasing frequency of infections, and anemia [3].

Modified anti-human CD3 monoclonal antibody was thought to be the next alternative to previously used OKT3 antibody for the prevention of T1DM in NOD mouse. Modified non-Fc receptor (FcR)-binding CD3 antibodies have been tested in clinical trials. They have been found to be less mitogenic and were equally tolerogenic compared to the function of Fc CD3 antibody [14].

The preservation of endogenous insulin secretion assessed by C-peptide response with concomitant reduction in hemoglobin A1c levels and insulin requirement in the treated group over 2 years in new-onset T1DM have been reported in two phase II trials using the two different humanized anti-CD3 (teplizumab and otelixizumab). The beneficial effects that extended over a period of up to 5 years after one dose have been shown in recent follow-up studies. The phase III trials are currently under way for humanized anti-CD3 monoclonal antibody [1].

#### 29.6.1.2 Anti-CD20 Monoclonal Antibody (Rituximab)

Initially, B cells were considered to play an important role in priming T cells. But, recently it has been shown that B cells promote the survival of CD8+T cells in the islets and thus promote the disease. All mature B cells express a cell surface marker CD20. Rituximab is a humanized anti-CD20 monoclonal antibody. Rituximab was originally introduced for therapy of B-cell lymphomas, and subsequently it has been shown to be an effective rheumatoid arthritis treatment. Rituximab has been shown to successfully deplete human B cells from peripheral circulation by mechanisms involving Fc- and complement-mediated cytotoxicity and via pro-apoptotic signals. The efficacy and safety of Rituximab is being tested in clinical trial in patients with new-onset T1DM [14].

A four-dose course of rituximab (an anti-CD20) partially preserved beta-cell function over a period of 1 year in a phase II trial in patients with newly diagnosed T1DM [1].

### 29.6.2 Antigen-Based Immunotherapy

#### 29.6.2.1 Vaccination by Exogenous Antigens

The diabetes development can be suppressed by vaccination with "nonspecific" immunostimulatory agents such as BCG (Bacillus Calmette-Guérin) vaccination in NOD mice. These agents stimulate innate immune pathway and reset immune homeostasis. But initial human trial of BCG vaccination did not show benefit to demonstrate residual β-cell function [2].

Recently, Faustman DL et al. [15] conducted a small, proof-of-concept study using the BCG vaccine in six subjects (mean age, 35 years) with long-standing type 1 diabetes (mean: 15.3 years). The BCG vaccine was chosen because it stimulates innate immunity by inducing the host to produce tumor necrosis factor (TNF), which, in turn, kills disease-causing autoimmune cells and restores pancreatic beta-cell function through regeneration. The six subjects were randomly assigned to either injections of BCG or a placebo and compared to self, healthy paired controls ($n=6$) or reference subjects with ($n=57$) or without ($n=16$) type 1 diabetes, depending upon the outcome measure. The vaccine and placebo injections were given on 2 occasions, 4 weeks apart. Blood samples were monitored weekly for 20 weeks for insulin-autoreactive T cells, regulatory T cells (Tregs), glutamic acid decarboxylase (GAD) and other autoantibodies, and C-peptide, a marker of insulin secretion. It was discovered that subjects who received the BCG vaccine showed an increase in insulin-autoreactive T cells, increases in the number of Tregs compared with paired healthy controls, and an improvement in insulin sensitivity as demonstrated by transient but significant increases in C-peptide levels in two of the three vaccine recipients [15].

Virus could lead to T1DM and if a specific virus is clearly indicated as a causative factor, vaccination should be done early in the life of children provided vaccine has got good safety profile. Children suffering from congenital rubella born to mothers who contracted rubella early in pregnancy had evidence of infection in the brain, pancreas, and other tissues and approximately 20 % developed T1DM. Approximately double proportion of such population of children with congenital rubella has been reported to develop islet cell antibodies. Moreover, children with congenital rubella who subsequently developed diabetes were reported to have higher frequency of T1DM susceptibility haplotype HLA-A1-B8-(DR3-DQ2) [2].

Children born to mothers who were infected during pregnancy with some enteroviruses such as Coxsackievirus B (CVB) and echoviruses had more chances of suffering from diabetes early in life but this has not been confirmed in recent studies [2, 9].

Mumps vaccination has not been found to be of value for preventing T1DM. The evidence for cytomegalovirus (CMV) infection in T1DM is weak. Rotavirus infection has been found to be temporally associated with increase in islet autoantibodies. It has been shown that rotavirus can infect β-cell in islets from mice, pigs, and monkeys [2].

#### 29.6.2.2 Vaccination by Endogenous Antigens

**Alum-Formulated Glutamic Acid Decarboxylase (GAD)**

GAD is a 65 kDa protein that is found in islet cells, the central nervous system, and the testes. GAD is a major autoantigen in autoimmune diabetes. The reason why GAD is a major autoantigen in autoimmune diabetes is not known. Antibodies to GAD appear early after the initiation of autoimmune insulitis and are found in 70 % of patients at diagnosis. Autoantibodies to GAD (GADA) may be an early sign of the autoimmune process of diabetes, and GADA has become one of the most important predictive markers of T1DM risk. Vaccination by glutamic acid decarboxylase modulates the immune system and thus prevents the destruction of beta cells. GAD65 isoform has been shown to prevent autoimmune destruction of beta cells in studies of nonobese mice with diabetes [3, 6].

GAD vaccine with aluminum hydroxide (alum) as adjuvant to enhance the presentation of antigens to antigen-presenting cells has been produced. Antigen-presenting cells process the injected GAD65 to provide peptide fragments recognized by T cells. This leads to Th1/Th2 shift consisting of induction and proliferation of a subset of GAD65-specific regulatory T cells. These specific T cells downregulate antigen-specific killer T cells that would otherwise attack the beta cells [3].

The administration of recombinant human GAD with or without adjuvants did not induce adverse side effects or exacerbate T1DM in man and mice in preclinical studies and a phase I clinical trial. A subsequent phase II trial in LADA (latent autoimmune diabetes in adults) patients further supported clinical development of alum-formulated GAD vaccine [8, 16].

Subjects received placebo or GAD/alum (4, 20, 100, or 500 µg) subcutaneously, twice in clinical trials of GAD/alum vaccination, first conducted in individuals with LADA (LADA describes adults with a slowly progressive form of type 1 diabetes). The diagnosis of LADA is based on (1) adult onset of diabetes, (2) circulating islet autoantibodies, and (3) insulin independence at diagnosis. About 10 % of adults with non-insulin-requiring diabetes have LADA. The $CD4^+CD25^+/CD4^+CD25^-$ cell ratio, as well as serum C-peptide levels, increased from baseline only in the 20 µg dose group after 6 months, and only the 500 µg dose boosted GAD autoantibody levels. Evidently, antigen-based therapy (ABT) can have a beneficial immunomodulatory effect without changing humoral responses to the administered antigen, at least in LADA patients. No significant study-related adverse effects were reported in a 5-year follow-up and that C-peptide levels were significantly higher only in the 20 µg dose group [3, 8].

A subsequent larger clinical trial with newly diabetic children has further supported the beneficial effect of the 20 µg GAD/alum dose. GAD/alum (20 µg) vaccination preserved β-cell function in patients treated within 6 months of type 1 diabetes onset but not in those treated >6 months after type 1 diabetes onset. The effectiveness of treatment in more recently diagnosed patients is likely to reflect greater remaining β-cell mass. This parallels findings with anti-CD3 treatment in which the treatment was most effective in those with the highest residual β-cell function at the time of treatment. The treatment induced higher levels of IL-5, IL-10, and IL-13 GAD-specific responses accompanied by higher frequency of $Foxp3^+$ and TGF-β secreting T cells, even after 15 months [8].

The only antigen-based vaccine candidate which has been shown to be effective in both T1DM and LADA is alum-formulated GAD [13]. However, in two recent phase II/III studies performed independently by Diamyd and TrialNet, the alum-formulated GAD vaccine failed to meet primary efficacy endpoint in preserving insulin production [16].

## Heat Shock Protein and Peptide 277

Heat shock proteins (Hsp) are highly conserved in all prokaryotes and eukaryotes and have an intracellular role as chaperone molecules. Human Hsp60 also has a role in the regulation of the innate immune system. This effect is transmitted through stimulation of Toll-like receptor 4 (TLR-4) on macrophages (a stimulatory inflammatory effect) and through TLR-2 on T cells (an immunomodulatory anti-inflammatory effect) [6].

Heat shock protein (Hsp) is another important self-antigen in the pathogenesis of diabetes. In NOD mice that were developing insulitis, antibodies directed against Mycobacterial Hsp65 and its human variant Hsp60 were found. These autoantibodies disappeared as diabetes developed and were not present in NOD mice that did not develop diabetes. Anti-Hsp60 T-cell clones transplanted into healthy mice induced insulitis. The

Hsp60 epitope recognized by T cells was identified as a 24 amino acid peptide termed peptide 277 (p277) [6].

Both insulitis and diabetes were prevented in NOD mice or mice exposed to low-dose streptozotocin injected with p277 in incomplete Freund's adjuvant. The splenic population of Th1 cells shifted to the Th2 immune-modulating phenotype accompanied by a decrease in leukocyte numbers and Th1-produced cytokines in the islets following vaccination. T cells recovered from islets of the p277-treated mice had reduced capacity to transfer diabetes into NOD-recipient mice. Even after the onset of the disease, the vaccine prevented deterioration of diabetes in NOD mice [6].

A human p277 vaccine, DiaPep277, has been developed based on the protective and therapeutic effects of p277 in animal models as well as the findings in the human disease. In a double-blind study in patients with recently diagnosed type 1 diabetes, the efficacy of this vaccine was tested. Thirty-five patients received a subcutaneous injection of either DiaPep277 or placebo at 0, 1, 6, and 12 months. The preservation of beta-cell function detected by a halt in the loss of C-peptide production was the primary endpoint of the study. A decreased need for exogenous insulin, a reduced hemoglobin A1c (HbA1c) level, and a shift in the T-cell cytokine phenotype were secondary endpoints. C-peptide levels rapidly declined in the control group but were preserved in the DiaPep277 group ($n=15$) at 10 months' follow-up. At 7 and 10 months of follow-up and also after 18 months, this difference was statistically significant. Moreover, significantly less exogenous insulin to achieve the same level of HbA1c (7 %) at the end of the trial was required in DiaPep277 intervention group. The cytokine response of T cells exposed to Hsp60 significantly shifted from a Th1-IFN response to a Th2, IL-10, and IL-13 response by the end of the trial [6]. The results of this vaccine from clinical trials performed in LADA patients were inconclusive [16].

## Insulin

Insulin was the major β-cell autoantigen selected for its therapeutic potential. The biologically active form of insulin is processed from its precursor, preproinsulin (PPIns), by sequential enzymatic cleavages that release the leader peptide (to make proinsulin; PIns) and the C-peptide [16].

Following oral administration of porcine insulin, protection from diabetes was first reported in NOD mice by Zhang et al. [17]. Many subsequent studies reported that proinsulin/insulin or epitopes from insulin could partially protect NOD mice from developing diabetes when administered orally. The proliferation of CD4+ T regulatory cells that protect pre-diabetic mice from onset of diabetes can be induced by oral administration of human insulin to NOD mice. The ability of insulin to promote an anti-inflammatory state in dendritic cells (DCs) ultimately leading to immunological suppression of T cell function was thought be the reason behind expansion of Th2 cell population. Also, insulin can inhibit diabetes onset when administered by different routes of entry into the body, i.e., proinsulin inoculated intranasally or insulin B chain peptide B: 9–23 delivered subcutaneously was reported to be effective in partially suppressing diabetes onset in NOD mice [10, 17].

Till now, the results have been disappointing in human clinical trials conducted using insulin as a therapeutic or prophylactic immunotherapy. The individuals at risk of developing T1DM were treated with 1.6 mg of aerosolized insulin daily for 10 consecutive days, followed by 2 days/week of 1.6 mg for 6 months in a safety assessment trial. The adverse effects or accelerated β-cell loss was not seen, but decreased insulin-specific T-cell responses were seen. The intranasal insulin decreased serum insulin autoantibodies (IAAs) and, in a subset of patients, reduced insulin-specific T-cell (IFN-γ) responses in a study in recent-onset diabetic individuals; however, β-cell destruction was not delayed.

The intranasal insulin (1 unit/kg daily) did not delay or prevent T1DM when given to infants carrying high-risk HLA haplotypes or to siblings positive for two or more T1DM-associated autoantibodies in the Finnish T1DM Prediction and Prevention Study (DIPP). The result of T1DM Prevention Trial (DPT-1) conducted to study the efficacy of ultralente insulin (0.25 unit/kg/day subcutaneously and one annual 4-day continuous intravenous infusion) as a prophylactic vaccine in at-risk individuals presenting with a 5-year projected risk of >50 % showed that this protocol did not prevent or delay T1DM development over a median 3.7-year follow-up period. The T1DM was neither delayed nor prevented in another DPT-1 trial in which first- or second-degree relatives with a 5-year projected risk of 26–50 % (determined by metabolic, immunological, and genetic staging) received oral insulin (7.5 mg/day) or placebo and followed-up for median period of 4.3 years. It was however revealed in subgroup analyses that oral insulin had beneficial effects in patients with high IAA levels. TrialNet is conducting an

Oral Insulin Prevention Trial based on these results and enrolling subjects with similar characteristics to the subgroup mentioned above—relatives with normal glucose tolerance carrying at least two autoantibody specificities in serum, one of which must be anti-insulin, and presenting with 35 % risk of T1DM within 5 years [16].

The oral insulin (5 mg/day) was given in conjunction with intensive subcutaneous insulin therapy to recent-onset diabetic patients within 4 weeks of diagnosis for 12 months in the immunotherapy T1DM (IMDIAB) trial. The oral insulin failed to preserve C-peptide levels and reduce insulin requirement at 12 months of follow-up. Moreover, accelerated β-cell loss was evident in patients younger than 15 years. Daily oral insulin administration (2.5 mg or 7.5 mg/day) also failed to blunt established T1DM in recent-onset diabetic patients in the T1DM Insuline Orale study. In a phase I clinical trial, a single intramuscular injection of insulin B chain/IFA to recent-onset diabetic subjects showed that this approach elicits robust antibody and T-cell responses against insulin without causing adverse events. However, no significant differences in C-peptide level could be detected [16]. Table 29.3 summarizes various interventions of immune prevention of autoimmune diabetes.

Table 29.3 Summary of various interventions of immune prevention of autoimmune diabetes

| Agent | Target/mechanism | Methods of delivery | Animal experience | Human experience | Safety |
|---|---|---|---|---|---|
| Anti-CD3 monoclonal antibody | Direct effects on pathogenic T cells, the induction of populations of regulatory cells, or both [18] | Parenteral | Induces long-term remission of overt autoimmunity in NOD mice [19] | Better maintenance of C-peptide levels and reduced insulin requirement [1] | Modified non-Fc receptor (FcR) binding CD3 antibodies are less mitogenic and equally tolerogenic [14] |
| Anti-CD20 monoclonal antibody (Rituximab) | Deplete human B cells from peripheral circulation by mechanisms involving Fc- and complement-mediated cytotoxicity and via proapoptotic signals [14] | Parenteral | Limited by the absence of anti-CD20 reagents that can induce B-cell depletion in mice [20] | A four-dose course partially preserved beta-cell function over a period of 1 year [21] | Mostly grade 1 or grade 2 adverse events, after the first infusion. The reactions appeared to be minimal with subsequent infusions [21] |
| BCG vaccination | Stimulate innate immune pathway and reset immune homeostasis [2] | Intradermal | The diabetes development can be suppressed by vaccination with BCG in NOD mice [2] | Increase in insulin-autoreactive T cells and the number of Tregs and improved insulin sensitivity [15, 22] | Repeated BCG vaccination at low doses was safe and well tolerated [22] |
| Alum-formulated glutamic acid decarboxylase | Th1/Th2 shift consisting of induction and proliferation of a subset of GAD65-specific regulatory T cells [3] | Injection of GAD65 | Did not induce adverse effects or exacerbated T1D in NOD mice [8, 16] | The only antigen-based vaccine candidate which has been shown to be effective in both T1D and LADA [14] | Clinical findings support the safety of immunomodulation by GAD65 immunization [3] |
| Heat shock protein and peptide 277 | Human Hsp60 has a role in the regulation of the innate immune system [6] | Parenteral | Both insulitis and diabetes was prevented in NOD mice [6] | C-peptide levels preserved in the DiaPep277 group at 10 months' follow-up [6] | Clinical findings support the safety of immunomodulation by DiaPep277 immunization [23] |
| Insulin | Induce autoantigen-specific tolerance by induction of regulatory T cells [24] | Subcutaneous, oral, intranasal | Partially protected NOD mice from developing diabetes when administered by different routes of entry [10, 17] | Till now, the results have been disappointing in human clinical trials [16] | In safety assessment trials, insulin did not generate significant adverse effects [16] |

# References

1. Cernea, S., Dobreanu, M., Raz, I.: Prevention of type 1 diabetes: today and tomorrow. Diabetes Metab. Res. Rev. **26**, 602–605 (2010)
2. Harrison, L.C.: The prospect of vaccination to prevent type 1 diabetes. Hum. Vaccin. **1**, 143–150 (2005)
3. Ludvigsson, J.: The role of immunomodulation therapy in autoimmune diabetes. J. Diabetes Sci. Technol. **3**, 320–330 (2009)
4. Appel, S.J., et al.: Latent autoimmune diabetes of adulthood (LADA): an often misdiagnosed type of diabetes mellitus. J. Am. Acad. Nurse Pract. **21**, 156–159 (2009)
5. Hjorth, M. Immunological profile and aspects of immunotherapy in type 1 diabetes. Linköping University Medical dissertations, vol. 1161 (2010). liu.diva-portal.org/smash/get/diva2:279876/FULLTEXT01. Accessed 26 Aug 2012
6. Raz, I., Eldor, R., Naparstek, Y.: Immune modulation for prevention of type 1 diabetes mellitus. Trends Biotechnol. **23**, 128–134 (2005)
7. Zhang, L., Eisenbarth, G.S.: Prediction and prevention of type 1 diabetes mellitus. J. Diabetes **3**, 48–57 (2011)
8. Tian, J., Kaufman, D.L.: Antigen-based therapy for the treatment of type 1 diabetes. Diabetes **58**, 1939–1946 (2009)
9. Boettler, T., von Herrath, M.: Type 1 diabetes vaccine development: animal models vs. humans. Hum. Vaccin. **7**, 19–26 (2011)
10. Nicholas, D., Odumosu, O., Langridge, W.H.: Autoantigen based vaccines for type 1 diabetes. Discov. Med. **11**, 293–301 (2011)
11. Steinman, L.: Inverse vaccination, the opposite of Jenner's concept, for therapy of autoimmunity. J. Intern. Med. **267**, 441–451 (2010)
12. Petrovsky, N., Silva, D., Schatz, D.A.: Vaccine therapies for the prevention of type 1 diabetes mellitus. Paediatr. Drugs **5**, 575–582 (2003)
13. Gupta, S.K.: Vaccines for type 1 diabetes in the late stage of clinical development. Indian J. Pharmacol. **43**, 485 (2011)
14. Sanjeevi, C.B.: Type 1 diabetes research: Newer approaches and exciting developments. Int. J. Diabetes Dev Ctries **29**, 49–51 (2009)
15. Faustman, D.L., et al.: Proof-of-concept, randomized, controlled clinical trial of Bacillus-Calmette-Guerin for treatment of long-term type 1 diabetes. PLoS One **7**, e41756 (2012)
16. Clemente-Casares, X., Tsai, S., Huang, C., Santamaria, P.: Antigen-specific therapeutic approaches in type 1 diabetes. Cold Spring Harb. Perspect. Med. **2**, a007773 (2012)
17. Zhang, Z.J., Davidson, L., Eisenbarth, G., Weiner, H.L.: Suppression of diabetes in nonobese diabetic mice by oral administration of porcine insulin. Proc. Natl. Acad. Sci. U. S. A. **88**, 10252–10256 (1991)
18. Herold, K.C., et al.: Anti-CD3 monoclonal antibody in new-onset type 1 diabetes mellitus. N. Engl. J. Med. **346**, 1692–1698 (2002)
19. Chatenoud, L., Thervet, E., Primo, J., Bach, J.F.: Anti-CD3 antibody induces long-term remission of overt autoimmunity in nonobese diabetic mice. Proc. Natl. Acad. Sci. U. S. A. **91**, 123–127 (1994)
20. Bour-Jordan, H., Bluestone, J.A.: B cell depletion: a novel therapy for autoimmune diabetes? J. Clin. Invest. **117**, 3642–3645 (2007)
21. Pescovitz, M.D., et al.: Rituximab, B-lymphocyte depletion, and preservation of beta-cell function. N. Engl. J. Med. **361**, 2143–2152 (2009)
22. Huppmann, M., Baumgarten, A., Ziegler, A.G., Bonifacio, E.: Neonatal Bacille Calmette-Guerin vaccination and type 1 diabetes. Diabetes Care **28**, 1204–1206 (2005)
23. Fischer, B., Elias, D., Bretzel, R.G., Linn, T.: Immunomodulation with heat shock protein DiaPep277 to preserve beta cell function in type 1 diabetes - an update. Expert Opin. Biol. Ther. **10**, 265–272 (2010)
24. Rewers, M., Gottlieb, P.: Immunotherapy for the prevention and treatment of type 1 diabetes. Diabetes Care **32**, 1769–1782 (2009)

# Novel Vaccines for Type I Allergy

## 30

Sandra Scheiblhofer, Josef Thalhamer, and Richard Weiss

## Contents

30.1    **Disease** ............................................. 490
30.2    **Allergens** .......................................... 490
30.3    **Risk Factors for Allergic Sensitization** ................................. 492
30.4    **Diagnostic and Classical Therapy (Clinical "Golden Standard")** ........ 493
30.4.1   Diagnosis ........................................... 493
30.4.2   Medication ......................................... 494
30.4.3   Specific Immunotherapy ................. 495
30.5    **Novel Vaccines Against Allergy** ......... 495
30.5.1   Recombinant (Hypo)-Allergens ...... 495
30.5.2   Peptide Immunotherapy .................. 496
30.5.3   Viral and Bacterial Vector Vaccines .. 497
30.5.4   Gene Vaccines .................................. 497
30.5.5   Novel Routes for Allergen-Specific Immunotherapy ................................ 497

**References** ................................................. 499

## Abstract

Today, type I allergies affect more than 30% of the population in western industrialized countries posing an increasing burden on public health systems. In this book chapter we provide an overview on the molecular characteristics of allergens and the mechanisms of allergic sensitization. Risk factors for sensitization such as genetic predisposition or environmental factors ("hygiene hypothesis") are discussed and the current standards of diagnosis and medication are summarized. As classical allergen-specific immunotherapy suffers from unwanted side-effects, low patient compliance as well as insufficient efficacy, this chapter focuses on novel therapeutic approaches to overcome these limitations. These include new molecules, such as recombinant (hypo-) allergens or peptides but also advanced vector vaccines, and genetic vaccines. Such vaccine types address specific receptors of the innate immune system, resulting in increased immunogenicity and modulation of unwanted TH2 type responses. Finally, alternative routes to the standard subcutaneous injection or sublingual application are presented, which target highly immunocompetent tissues such as the skin or the lymph nodes.

S. Scheiblhofer, PhD • J. Thalhamer, PhD
R. Weiss, PhD (✉)
Department of Molecular Biology,
University of Salzburg, Salzburg, Austria
e-mail: richard.weiss@sbg.ac.at

## 30.1 Disease

Meanwhile, in highly developed countries up to one third of the population is affected by type I allergic diseases, placing an enormous economic burden on public health systems. This development has promoted intense research on elucidation of the underlying mechanisms, identification of risk factors, and development of novel therapeutic interventions.

Under certain conditions, otherwise harmless ubiquitous proteins derived from sources such as tree and grass pollen, dust mites, cockroaches, animal dander, food, venom of stinging insects, molds, or latex can act as allergens, which elicit exaggerated T helper 2 (TH2)-driven immune responses termed type I (or immediate) allergic reactions or hypersensitivity. The term atopy, which is sometimes also used to denote this disease, should only be used to designate a hereditary predisposition for the development of allergic disorders. Characteristically, elevated levels of serum IgE lead to activation of mast cells and basophils in the blood. The resulting inflammatory response causes clinical symptoms including eczema (atopic dermatitis), rhinoconjunctivitis ("hay fever"), gastrointestinal symptoms, asthma, and systemic anaphylaxis.

Frequently, atopic patients first suffer from gastrointestinal symptoms and atopic dermatitis predominantly caused by food allergens from milk and wheat in early childhood. Later in life, allergic disease can either resolve or progress, the latter leading to a series of conditions denoted the "atopic march" in adulthood [1].

The three major components for clinical manifestation of allergy were first described in 1921 by Prausnitz and Küstner: a disease-eliciting antigen (allergen), a transferable serum factor (IgE), which allows for discrimination between allergic and healthy individuals, and a tissue component (mast cells).

Allergic sensitization takes place at first encounter with an allergen entering via epithelial barriers (Fig. 30.1). Once activated, professional antigen-presenting cells migrate to draining lymph nodes and promote the development of naïve T lymphocytes into allergen-specific TH2 cells involving the key cytokine Interleukin-4 (IL-4) [2]. TH2 cells either migrate to B-cell zones in lymph nodes inducing affinity maturation as well as class switch and production of allergen-specific immunoglobulin E (IgE) by B cells or enter the site of allergen encounter to act as TH2 effector cells.

IgE antibodies bind to the high-affinity receptor FcεRI on basophils and mast cells. Upon repeated contact with the same allergen, IgE receptors on the surface of these cells can become cross-linked by binding of allergen to bound IgE molecules [3]. This cross-linking initiates a cascade of intracellular signals ultimately activating basophils, eosinophils, and mast cells, which release inflammatory mediators including histamine, leukotrienes, and prostaglandins from their granules causing localized or systemic symptoms of acute disease. These include contraction of smooth muscles, vasodilation, mucus secretion, edema, or even life-threatening anaphylactic shock.

Late-phase reactions, which usually occur up to 24 h after the acute reactions, originate from neutrophils, lymphocytes, eosinophils, and macrophages, but also TH2 lymphocytes, migrating to the initial site of allergen encounter causing allergic inflammation.

## 30.2 Allergens

Though humans are exposed to a multitude of proteins and compounds, only a minute fraction of them have the property of allergenicity, i.e., the capacity to induce inappropriate TH2-biased immune responses leading to the production of specific IgE – a process termed allergic sensitization. Intrinsic as well as extrinsic factors have been proposed to contribute to allergenicity by influencing the entry of allergens through epithelial barriers, the uptake, processing and presentation of allergens to T cells, and processes regarding antigen-presenting cells such as maturation and induction of signaling pathways.

Extrinsic factors, concomitantly taken up with the allergen, which have impact on sensitization, include substances derived from allergen sources such as pollen-associated lipid

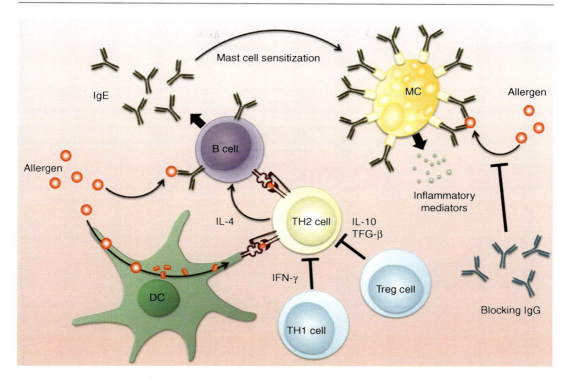

**Fig. 30.1** Allergic sensitization. Allergen is taken up via epithelial surfaces such as the skin or the mucosa of airways and gut. Dendritic cells (DC) phagocytose and process allergens into peptides which in turn are presented on MHC class II and induce generation of TH2 cells. This process can be inhibited by IFN-γ secreting TH1 cells or regulatory T cells (Treg) secreting IL-10 and/or TGF-β. TH2 cells produce IL-4 which promotes class switch of allergen-specific B cells to IgE. IgE antibodies bind to FcεRI receptors on mast cells and upon allergen re-encounter induce release of inflammatory mediators such as histamine. Presence of high levels of blocking IgG can inhibit allergen binding to mast cell surface IgE

mediators [4], chitin [5], but also environmental pollutants [6]. Air pollution, cigarette smoke, or ozone can influence the release of pollen allergens [7], can lead to nitration of inhalant allergens, thereby generating novel epitopes [8], or can directly act on antigen-presenting cells [9]. Lipopolysaccharide (LPS) derived from the outer cell membrane of Gram-negative bacteria represents an agent with the exceptional property to either promote or diminish allergic sensitizations, depending on its concentration upon allergen encounter. Whereas extremely low dose LPS together with the model allergen ovalbumin can lead to tolerance, low dose can induce TH2 responses, and high dose even can promote TH1 immune reactions [10, 11]. LPS acts via binding to the CD14/toll-like receptor 4/MD-2 complex leading to secretion of pro-inflammatory cytokines particularly from B cells and macrophages.

The LPS sensor toll-like receptor 4 (TLR4) belongs to the family of toll-like receptors of the innate immune system, which specifically recognize highly conserved microbial structures [12]. MD-2, also designated lymphocyte antigen 96, functions as an adapter molecule by associating with TLR4 on the cell surface enabling LPS-induced signaling [13].

Intrinsic properties of allergens contributing to sensitization include fold stability, chemical modifications, oligomerization, molecular dynamics, and ligand binding. One of the protein allergen functions identified to facilitate sensitization is protease activity. It has been shown that the house dust mite allergen Der p 1 acts as a cysteine protease, cleaving tight junction proteins at epithelial barriers leading to enhanced penetration of allergens. Additionally, this allergen proteolytically degrades several receptors on T and B cells and dendritic cells (DC), which are associated with the initiation or augmentation of TH2 responses [14]. The allergen papain, also acting as a cysteine protease, obviously can induce TH2 responses with IgE production via activation of basophils [15].

Serine protease activity in some house dust mite and mold allergens as well as cockroach extracts induces molecular changes in epithelial cells, thereby driving dendritic cells to polarize T cells towards a TH2 phenotype [16]. The capability of lipid binding and subsequent activation of TLRs represents another intrinsic factor responsible for allergic sensitization. It has been demonstrated that the house dust mite allergen Der p 2 substitutes MD-2 by binding LPS, thereby promoting TH2 responses via TLR4 signaling [17]. C-type lectin receptors on dendritic cells bind carbohydrates and have been identified to tailor immune responses to pathogens [18]. Uptake of allergens from house dust mite, dog dander, cockroach, and peanut by DCs has been shown to be mediated by the mannose receptor, leading to TH2 polarization [19]. Hence, the authors suggest a major role for the mannose receptor in recognition of glycoallergens and in shaping TH2 responses by DCs.

## 30.3 Risk Factors for Allergic Sensitization

Atopic individuals have a genetically mediated predisposition to mount excessive IgE-mediated reactions. It has been found from hereditary and twin studies that with both parents being atopic, the risk for a progeny to develop IgE-mediated disease increases up to 60 % compared to 5–10 % for the offspring of healthy individuals [20]. Some genetic polymorphisms have already been identified, which either promote or prevent the development of atopy. Among them, mutations in the gene encoding filaggrin, leading to a weakened barrier function of the skin, have been linked with a facilitated transition from eczema to asthma. These polymorphisms can be utilized as markers for prediction of asthma development prior to onset of disease symptoms [21]. In contrast, a reduction of the prevalence to develop atopy has been found to be associated with certain polymorphisms in genes encoding receptors of the innate immune system such as TLR4 [22].

A multitude of factors associated with lifestyle in highly developed countries has been closely investigated for correlation with type I allergic diseases, including dietary habits, air pollution, passive smoking, and a reduction of the number of infectious diseases during childhood. The latter, which is achieved by increased hygiene standards and mass vaccinations, has been found to be an important risk factor for atopy [23].

In the late 1990s, the first report about significantly reduced risk for allergic sensitization among farmer's children compared with children from the same areas, but growing up without close contact with stables and livestock was published. Designated as the "hygiene hypothesis," an explanation for the increasing prevalence of atopy in the last decades was established (Fig. 30.2). Accordingly, a reduction of exposure to microbial compounds during early childhood may account for a decreased stimulation of the innate immune system as well as a concomitant shift towards a TH2-biased adaptive immune response against allergens. Meanwhile, some key factors contributing to the reduced risk of allergies and asthma in farm children have been identified.

These include contact with livestock (cattle, pigs, poultry), contact with animal feed (hay, grain, straw, silage), and the consumption of unprocessed cow's milk [24]. The main routes of uptake are inhalation and ingestion. Best protection is conferred by exposure already in utero and during the first years of life [25, 26]. On the molecular level, the compounds responsible for the protective effect are derived from bacteria and fungi. Among them are muramic acid, a component of the cell wall of Gram-positive bacteria and extracellular polysaccharides from fungal species belonging to the genera Penicillium and Aspergillus [27, 28]. Interestingly, LPS seems to prevent from allergic sensitization, but not from childhood asthma and wheeze [29].

Peripheral blood cells from farm children have been found to express significantly higher levels of TLR2, TLR4, and CD14 mRNA com-

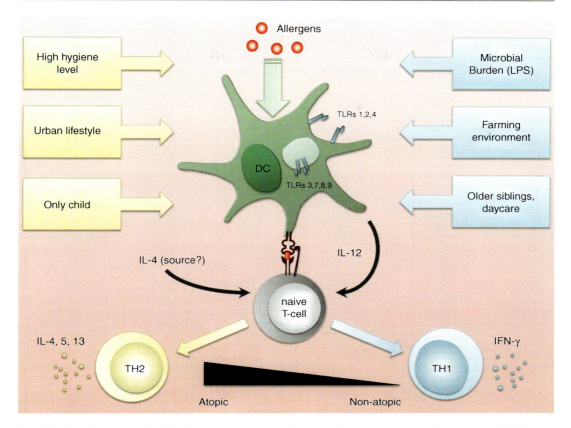

**Fig. 30.2** Hygiene hypothesis. Extrinsic factors leading to the generation of a protective TH1 immunity via stimulation of innate immune receptors promoting secretion of IL-12. High hygiene levels resulting in insufficient immune stimulation promote the generation of TH2 reactions and atopy. The role of IL-4 and its cellular source in this process is still highly debated

pared to nonfarm children at school age [25, 30]. Also, significantly increased seasonal allergen-specific IgE antibody levels were observed in cord blood of newborn babies whose mothers had no contact to animal sheds and fodder. IgE responses correlated with reduced production of the TH1-associated cytokines IFN-γ and TNF [31]. These findings support the assumption that microbial compounds in the context of a farming environment are sensed by the innate immune system, which subsequently shapes the adaptive immune system.

It remains to be elucidated whether the "missing immune deviation" towards TH1 represents the only mechanism accounting for the constant rise in prevalence of type I allergies. Studies in animal models and epidemiological evidence suggest that also a lack of immune suppression caused by decreased activity of T regulatory cells could be involved. Most likely, a combination of both (and possibly other) mechanisms might be responsible [32].

## 30.4 Diagnostic and Classical Therapy (Clinical "Golden Standard")

### 30.4.1 Diagnosis

Two different methods with similar specificity and sensitivity are available for assessment of allergen-specific IgE, i.e., the skin prick test and the allergy blood test.

For skin testing, also designated prick or puncture testing, a small metal or plastic device is used to produce a series of small pricks or punctures in the skin at the inside forearm or the back of the patient. Before, suspected allergens or allergen extracts are applied to areas on the skin, which are marked with a dye or pen. By pricking, the respective allergen(s) or extracts are introduced into the skin. Alternatively, they can be intradermally injected using needle and syringe. In case the patient is allergic against one or more of the tested allergens or extracts, an inflammatory reaction will usually become visible within 30 min. The results of a skin prick test are scored from borderline reactivity (±) to a strong reaction (4+), reflecting the skin reaction ranging from slight reddening to a so-called wheal and flare reaction reminiscent of a mosquito bite.

Additionally, the diameter of the wheal and flare can be determined. If the patient suffers from widespread skin disease, has recently taken antihistamines, or has already been afflicted by a life-threatening anaphylactic reaction, usually a blood test will be the preferred method. Compared to skin prick testing, this test harbors several advantages: The patient's age, skin condition, medication, symptoms, or disease severity is no exclusion criteria. This method is also more acceptable for very young children and babies because a single needle stick for collection of one blood sample is sufficient for testing a broad panel of allergens. The patient has no contact with potentially sensitizing material, providing an additional safety feature. By blood testing, the concentration of allergen-specific IgE is determined. Originally developed in the 1970s to work with anti-IgE antibodies labeled with radioactive isotopes and marketed as RAST (radioallergosorbent test) by the Swedish company Pharmacia Diagnostics AB, today the so-called ImmunoCap test utilizing fluorescence-labeled IgE-binding antibodies has rendered the term RAST (as a colloquialism for several in vitro allergy tests) obsolete.

Besides skin prick testing and the allergy blood test, also other methods are available. Challenge testing is especially useful in case of food and medication allergies. Small amounts of the suspected allergen are introduced orally or by inhalation and the patient is closely supervised by an allergist. For patch testing, a method employed to identify the cause of skin contact allergy or contact dermatitis, allergic chemicals or skin sensitizers are applied to the back of the patient using adhesive patches. Subsequently, the skin is examined for local reactions.

### 30.4.2 Medication

Whereas avoidance of allergens may be a simple and effective management in case of food allergies, which may reduce symptoms or even prevent life-threatening anaphylaxis, this measurement is difficult to accomplish for patients suffering from airborne allergies.

For standard pharmacotherapy, several immunosuppressive and anti-inflammatory drugs are available, including antihistamines, glucocorticoids, and β-agonists. Antihistamines block the effects of mediators released by mast cells and basophils during degranulation. Intranasally or systemically applied corticosteroids including prednisone act as anti-inflammatory drugs to treat allergic rhinitis. β-2-adrenoreceptor agonists such as Salbutamol (short-acting) or Salmeterol (long-acting) are effective bronchodilators for rapid relief of asthma symptoms.

Whereas all these agents alleviate the symptoms of allergy, they do not target the underlying immunological disorder.

Another treatment, which has been approved for moderate to severe asthma, is subcutaneous injection of humanized monoclonal anti-IgE antibodies [33], such as omalizumab (marketed as Xolair). Due to its high costs, omalizumab is mainly prescribed for patients with severe persistent asthma. The monoclonal antibody selectively binds free IgE in the blood and interstitial fluid and membrane-bound IgE on the surface of B lymphocytes, but not IgE bound by the high-affinity IgE receptor on the surface of mast cells, basophils, and antigen-presenting DCs. Omalizumab blocks the binding of IgE to its high-affinity receptor on mast cells and basophils by partly masking the site on IgE to which the receptor binds. As soon as IgE has bound to this receptor, omalizumab can no longer bind due to steric hindrance, avoiding anaphylactic effects,

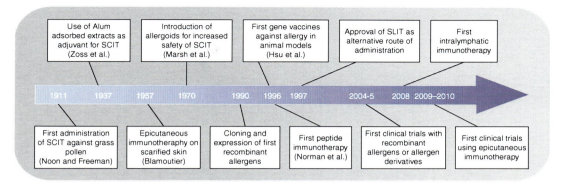

**Fig. 30.3** Milestones in allergen-specific immunotherapy (SIT)

which would occur by cross-linking of IgE molecules followed by mediator release from mast cells and basophils. Most importantly, omalizumab also depletes free IgE in patients leading to gradual downregulation of IgE receptors on basophils, mast cells, and DCs. Thereby, these cells become less sensitive to the stimulation by allergens [34].

### 30.4.3 Specific Immunotherapy

Allergen-specific immunotherapy (SIT) has been introduced by Noon and Freeman more than 100 years ago [35] and still represents the only causal treatment for allergic patients, which redirects inappropriate and exaggerated TH2-driven immune responses against allergens (Fig. 30.3). The hallmarks of these immunological changes are the promotion of TH1 cytokines such as IFN-γ and the induction of IL-10/TGF-β secretion by T regulatory cells in blood and inflamed airways. SIT is also associated with suppression of allergen-specific IgE and induction of IgG4, and suppression of mast cells, basophils, and eosinophils. Clinical practice of SIT has not been substantially changed or improved since its first use. The therapy is mostly performed by 50–80 subcutaneous injections (SCIT) of gradually increasing allergen doses over 3–5 years [36]. Local or systemic side effects caused by this therapy are reported [37]. Because barely characterized allergen extracts are used, there is the potential risk for therapy-induced new sensitizations.

As a more patient-friendly alternative, sublingual immunotherapy (SLIT) with drops or tablets has been approved [38]. This approach avoids the use of needle and syringe and offers the possibility of self-administration. SLIT requires daily intake of large amounts of allergen over a time interval comparable with SCIT and is frequently accompanied by oral as well as gastrointestinal side effects [39]. Also, SLIT is discussed to be less effective than SCIT due to poor allergen uptake caused by the relatively short contact with the oral mucosa [40].

## 30.5 Novel Vaccines Against Allergy

The side effects associated with classical SCIT as well as the low compliance and lack of efficacy of SLIT have fueled efforts to develop novel therapeutic and – more recently – also prophylactic vaccination approaches (Fig. 30.4).

### 30.5.1 Recombinant (Hypo)-Allergens

Today, SIT is based on allergen extracts from natural sources, which are often poorly defined and difficult to standardize. Additionally, the large number of different proteins contained within such an extract always poses the risk of new sensitizations against components the patient had not been sensitized before therapy [41]. Therefore,

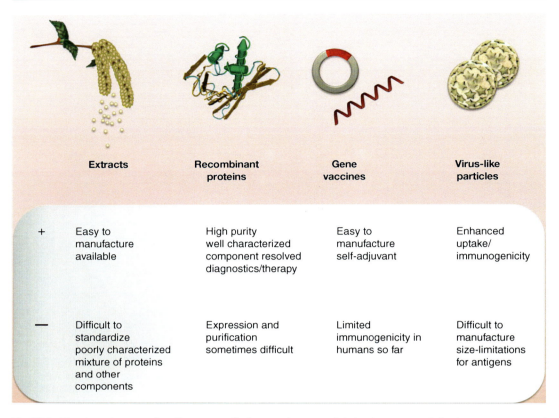

**Fig. 30.4** Novel vaccine types for allergen-specific immunotherapy and their advantages and disadvantages compared to classical allergen extracts

the next generation of allergy diagnostics and therapeutics will be based on recombinant proteins allowing tailor-made treatment according to the patient's sensitization profile [42]. Besides the benefits of a highly standardized product, production of recombinant allergens also allows modification of the allergen of interest. By changing the three-dimensional structure of the protein B-cell epitopes can be destroyed, reducing the binding of the allergen to preexisting IgE in the patient, thereby avoiding release of mast cell mediators, which are the major source of side effects during conventional SIT. Such modified allergens with low-IgE-binding potential are called hypoallergens [43–45]. Hypoallergens can also be systematically generated via so-called in silico mutation. Consequences of mutations on fold stability of an encoded molecule can be predicted by z-score calculation. We have recently demonstrated that novel hypoallergens based on decreased fold stability, which display reduced IgE-binding capacity, can be created by z-score calculation [45].

### 30.5.2 Peptide Immunotherapy

Peptide immunotherapy takes this concept one step further, as it only applies synthesized peptides. Because B-cell epitopes are usually discontinuous (i.e., they consist of sequentially separated amino acid stretches brought together by folding of the protein), these vaccines can no longer bind to patients' IgE but retain their ability to stimulate T-cell responses. Although this is a persuading concept, peptide vaccines often suffer from low immunogenicity, and clinical trials showed that although immediate reactions could be avoided, in the early phases of therapy IgE-independent late asthmatic reactions have been described [46].

### 30.5.3 Viral and Bacterial Vector Vaccines

Besides administration of purified recombinant allergen or allergen peptides, there is also the possibility to administer therapeutic protein by recombinant viral or bacterial vectors. These systems, which have only been tested in preclinical studies in animals so far, offer the benefit of improved delivery as well as co-stimulation originating from components of the microbial vector. For example, Modified Vaccinia Virus Ankara (MVA)-based [47] as well as recombinant Bacillus Calmette–Guérin (BCG) [48] or Lactobacillus plantarum [49] expressing (model-) allergens protect from allergic sensitization by establishing TH1-biased immune responses through stimulation of innate immune receptors. Due to the safety concerns when using live or live-attenuated vectors and the problems associated with anti-vector immunogenicity, a reductive approach is the use of virus-like particles displaying an antigen of interest, which has been recently demonstrated for Fel d 1, the major cat allergen. Blocking IgG antibodies were identified as major effector mechanism in this study [50].

### 30.5.4 Gene Vaccines

An elegant approach that became popular in the 1990s, which combines the immune stimulatory properties of viral infections with the safety of recombinant proteins, is the use of genetic vaccination. Genetic vaccination applies only the genetic information of an antigen, either as plasmid DNA containing a eukaryotic promoter (DNA vaccination) [51] or as synthesized messenger RNA (mRNA vaccination) [52]. The genetic material is taken up by the host cells, such as keratinocytes or fibroblasts (when administered intradermally), and translated into the respective protein. Besides using somatic cells as protein factories, the introduced DNA or mRNA can stimulate the immune system via innate immune receptors, such as TLR9 (DNA) and TLRs 3, 7, and 8 (RNA). mRNA vaccines have an advantage over DNA vaccines, as they cannot integrate into the genome and are only expressed transiently [53]. As genetic vaccination represents a highly versatile platform to design advanced types of vaccines, various innovative approaches have been tested in animal models of allergy. By targeting the expression of a plasmid-encoded allergen towards the proteasome via covalent linkage of ubiquitin to its N-terminus, we have rendered the gene product hypoallergenic, i.e., whereas T-cell responses remain essentially unaffected, complete degradation of the native allergen prevents formation of antibodies [43].

Self-replicating DNA and RNA vaccines represent a further class of genetic vaccines, which we have already employed for prevention and therapy of allergic diseases in mice. Encoding an alphavirus-derived replicase molecule, which drives its own amplification as well as transcription and translation of the encoded allergen, self-replicating DNA and RNA vaccines have demonstrated equal efficacy at doses 100-fold lower than conventional vectors [52, 54, 55]. Their superior immunogenicity is based on providing danger signals via double-stranded RNA intermediates, which are recognized by several innate antiviral sensors [56].

### 30.5.5 Novel Routes for Allergen-Specific Immunotherapy

Although the clinical efficacy of SCIT and SLIT has been proven [57], only a small percentage of allergic patients decide to undergo this treatment instead of taking medication for symptom relief. The main reasons for low acceptance and high drop-out rates of classical allergen-specific immunotherapy are the long treatment duration and therapy-induced side effects. Hence, approaches for innovative treatment of type I allergies (Fig. 30.5) aim at increased efficacy to shorten schedules and at increased safety by avoiding contact of the allergen with the general circulation.

The skin represents a promising target tissue, which is easily accessible, rich in antigen-presenting cells, and non-vascularized at its

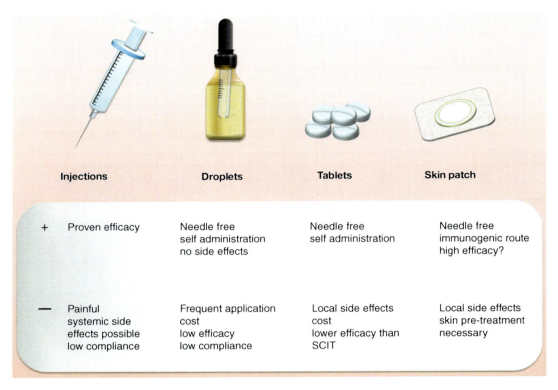

**Fig. 30.5** Advantages and disadvantages of different routes of immunization for allergen-specific immunotherapy

superficial layers [58]. The dermal and epidermal layers of the skin harbor highly immunocompetent cell types such as dermal dendritic cells and Langerhans cells, but also other cell types such as mast cells, natural killer cells, and keratinocytes contribute to the immunological functions of the skin [59]. Additionally, the skin is efficiently drained by local lymph nodes, another advantage over subcutaneous tissue. Already introduced in the 1950s [60], cutaneous immunotherapy has been recently rediscovered and termed transcutaneous (TCIT) or epicutaneous immunotherapy (EPIT).

Clinical studies have been performed with essentially two different methods to break the outermost layer of the skin, the stratum corneum, for increased uptake of allergen. One approach uses tape stripping with a simple adhesive tape for several times, followed by application of the allergen in solution included in a patch [61, 62]. The other method employs a patch containing the allergen as dry powder put on intact skin. By leaving the patch attached for a prolonged period of time, a kind of humid chamber is created, weakening the barrier function of the skin, leading to hydration of the powder and facilitated uptake [63]. An innovative approach to circumvent the skin barrier in a highly reproducible but also adaptable manner is laser-based microporation.

We have demonstrated that by using a device emitting far-infrared laser beams, aqueous micropores of variable number, density, and depth can be generated and that transcutaneous immunization via these pores represents a suitable technique for induction of specific immune responses [64]. In a mouse model of allergic asthma, we compared transcutaneous immunotherapy via laser-generated micropores to standard subcutaneous immunotherapy and found these therapeutic approaches equally effective in reducing airway hyperresponsiveness and leukocyte infiltration into the lungs. Transcutaneous application avoided the therapy-associated systemic increase in TH2 cytokines observed with subcutaneous injection [65].

Another interesting route for allergen-specific immunotherapy is intralymphatic application, termed ILIT, directly delivering the allergen into a subcutaneous lymph node by needle. Clinical studies with grass pollen allergic patients and patients reactive to cat dander have been performed with promising results [66, 67].

## References

1. Locksley, R.M.: Asthma and allergic inflammation. Cell **140**, 777–783 (2010). doi:10.1016/j.cell.2010.03.004
2. Mowen, K.A., Glimcher, L.H.: Signaling pathways in Th2 development. Immunol. Rev. **202**, 203–222 (2004). doi:10.1111/j.0105-2896.2004.00209.x
3. Turner, H., Kinet, J.P.: Signalling through the high-affinity IgE receptor Fc epsilonRI. Nature **402**, B24–B30 (1999)
4. Gilles, S., et al.: Pollen allergens do not come alone: pollen associated lipid mediators (PALMS) shift the human immune systems towards a T(H)2-dominated response. Allergy Asthma Clin. Immunol. **5**, 3 (2009). doi:10.1186/1710-1492-5-3
5. Burton, O.T., Zaccone, P.: The potential role of chitin in allergic reactions. Trends Immunol. **28**, 419–422 (2007). doi:10.1016/j.it.2007.08.005
6. Morgenstern, V., et al.: Atopic diseases, allergic sensitization, and exposure to traffic-related air pollution in children. Am. J. Respir. Crit. Care Med. **177**, 1331–1337 (2008). doi:10.1164/rccm.200701-036OC
7. Behrendt, H., Becker, W.M.: Localization, release and bioavailability of pollen allergens: the influence of environmental factors. Curr. Opin. Immunol. **13**, 709–715 (2001). doi:S0952-7915(01)00283-7 [pii]
8. Gruijthuijsen, Y.K., et al.: Nitration enhances the allergenic potential of proteins. Int. Arch. Allergy Immunol. **141**, 265–275 (2006). doi:10.1159/000095296
9. Williams, M.A., et al.: Disruption of the transcription factor Nrf2 promotes pro-oxidative dendritic cells that stimulate Th2-like immunoresponsiveness upon activation by ambient particulate matter. J. Immunol. **181**, 4545–4559 (2008)
10. Eisenbarth, S.C., et al.: Lipopolysaccharide-enhanced, toll-like receptor 4-dependent T helper cell type 2 responses to inhaled antigen. J. Exp. Med. **196**, 1645–1651 (2002)
11. Herrick, C.A., Bottomly, K.: To respond or not to respond: T cells in allergic asthma. Nat. Rev. Immunol. **3**, 405–412 (2003). doi:10.1038/nri1084
12. Poltorak, A., et al.: Defective LPS signaling in C3H/HeJ and C57BL/10ScCr mice: mutations in Tlr4 gene. Science **282**, 2085–2088 (1998)
13. Shimazu, R., et al.: MD-2, a molecule that confers lipopolysaccharide responsiveness on Toll-like receptor 4. J. Exp. Med. **189**, 1777–1782 (1999)
14. Furmonaviciene, R., et al.: The protease allergen Der p 1 cleaves cell surface DC-SIGN and DC-SIGNR: experimental analysis of in silico substrate identification and implications in allergic responses. Clin. Exp. Allergy **37**, 231–242 (2007). doi:10.1111/j.1365-2222.2007.02651.x
15. Sokol, C.L., Barton, G.M., Farr, A.G., Medzhitov, R.: A mechanism for the initiation of allergen-induced T helper type 2 responses. Nat. Immunol. **9**, 310–318 (2008). doi:10.1038/ni1558
16. Comeau, M.R., Ziegler, S.F.: The influence of TSLP on the allergic response. Mucosal Immunol. **3**, 138–147 (2010). doi:10.1038/mi.2009.134
17. Trompette, A., et al.: Allergenicity resulting from functional mimicry of a Toll-like receptor complex protein. Nature **457**, 585–588 (2009). doi:10.1038/nature07548
18. van Kooyk, Y.: C-type lectins on dendritic cells: key modulators for the induction of immune responses. Biochem. Soc. Trans. **36**, 1478–1481 (2008). doi:10.1042/BST0361478
19. Royer, P.J., et al.: The mannose receptor mediates the uptake of diverse native allergens by dendritic cells and determines allergen-induced T cell polarization through modulation of IDO activity. J. Immunol. **185**, 1522–1531 (2010). doi:10.4049/jimmunol.1000774
20. Kjellman, N.I.: Atopic disease in seven-year-old children. Incidence in relation to family history. Acta Paediatr. Scand. **66**, 465–471 (1977)
21. Marenholz, I., et al.: An interaction between filaggrin mutations and early food sensitization improves the prediction of childhood asthma. J. Allergy Clin. Immunol. **123**, 911–916 (2009). doi:10.1016/j.jaci.2009.01.051
22. Senthilselvan, A., et al.: Association of polymorphisms of toll-like receptor 4 with a reduced prevalence of hay fever and atopy. Ann. Allergy Asthma Immunol. **100**, 463–468 (2008). doi:10.1016/S1081-1206(10)60472-3
23. Floistrup, H., et al.: Allergic disease and sensitization in Steiner school children. J. Allergy Clin. Immunol. **117**, 59–66 (2006). doi:10.1016/j.jaci.2005.09.039
24. von Mutius, E., Vercelli, D.: Farm living: effects on childhood asthma and allergy. Nat. Rev. Immunol. **10**, 861–868 (2010). doi:10.1038/nri2871
25. Ege, M.J., et al.: Prenatal farm exposure is related to the expression of receptors of the innate immunity and to atopic sensitization in school-age children. J. Allergy Clin. Immunol. **117**, 817–823 (2006). doi:10.1016/j.jaci.2005.12.1307
26. Riedler, J., et al.: Exposure to farming in early life and development of asthma and allergy: a cross-sectional survey. Lancet **358**, 1129–1133 (2001). doi:10.1016/S0140-6736(01)06252-3
27. Ege, M.J., et al.: Not all farming environments protect against the development of asthma and wheeze in children. J. Allergy Clin. Immunol. **119**, 1140–1147 (2007). doi:10.1016/j.jaci.2007.01.037
28. van Strien, R.T., et al.: Microbial exposure of rural school children, as assessed by levels of

29. Vogel, K., et al.: Animal shed Bacillus licheniformis spores possess allergy-protective as well as inflammatory properties. J. Allergy Clin. Immunol. **122**, 307–312, 312 e301–e308 (2008). doi:10.1016/j.jaci.2008.05.016
30. Lauener, R.P., et al.: Expression of CD14 and Toll-like receptor 2 in farmers' and non-farmers' children. Lancet **360**, 465–466 (2002). doi:10.1016/S0140-6736(02)09641-1
31. Pfefferle, P.I., et al.: Cord blood allergen-specific IgE is associated with reduced IFN-gamma production by cord blood cells: the Protection against Allergy-Study in Rural Environments (PASTURE) Study. J. Allergy Clin. Immunol. **122**, 711–716 (2008). doi:10.1016/j.jaci.2008.06.035
32. Romagnani, S.: The increased prevalence of allergy and the hygiene hypothesis: missing immune deviation, reduced immune suppression, or both? Immunology **112**, 352–363 (2004). doi:10.1111/j.1365-2567.2004.01925.x
33. Chang, T.W., Wu, P.C., Hsu, C.L., Hung, A.F.: Anti-IgE antibodies for the treatment of IgE-mediated allergic diseases. Adv. Immunol. **93**, 63–119 (2007). doi:10.1016/S0065-2776(06)93002-8
34. Scheinfeld, N.: Omalizumab: a recombinant humanized monoclonal IgE-blocking antibody. Dermatol. Online J. **11**, 2 (2005)
35. Calderon, M., Cardona, V., Demoly, P.: One hundred years of allergen immunotherapy European Academy of Allergy and Clinical Immunology celebration: review of unanswered questions. Allergy **67**, 462–476 (2012). doi:10.1111/j.1398-9995.2012.02785.x
36. Cox, L., Calderon, M.A.: Subcutaneous specific immunotherapy for seasonal allergic rhinitis: a review of treatment practices in the US and Europe. Curr. Med. Res. Opin. **26**, 2723–2733 (2010). doi:10.1185/03007995.2010.528647
37. Frew, A.J.: Allergen immunotherapy. J. Allergy Clin. Immunol. **125**, S306–S313 (2010). doi:10.1016/j.jaci.2009.10.064
38. Canonica, G.W., et al.: Sub-lingual immunotherapy: World Allergy Organization Position Paper 2009. Allergy **64**(Suppl 91), 1–59 (2009). doi:10.1111/j.1398-9995.2009.02309.x
39. Frew, A.J.: Sublingual immunotherapy. N. Engl. J. Med. **358**, 2259–2264 (2008). doi:10.1056/NEJMct0708337
40. Razafindratsita, A., et al.: Improvement of sublingual immunotherapy efficacy with a mucoadhesive allergen formulation. J. Allergy Clin. Immunol. **120**, 278–285 (2007). doi:10.1016/j.jaci.2007.04.009
41. Ball, T., et al.: Induction of antibody responses to new B cell epitopes indicates vaccination character of allergen immunotherapy. Eur. J. Immunol. **29**, 2026–2036 (1999). doi:10.1002/(SICI)1521-4141(199906)29:06&#60;2026::AID-IMMU2026&#62;3.0.CO;2-2
42. Pauli, G., Malling, H.J.: The current state of recombinant allergens for immunotherapy. Curr. Opin. Allergy Clin. Immunol. **10**, 575–581 (2010). doi:10.1097/ACI.0b013e32833fd6c5
43. Bauer, R., et al.: Generation of hypoallergenic DNA vaccines by forced ubiquitination: preventive and therapeutic effects in a mouse model of allergy. J. Allergy Clin. Immunol. **118**, 269–276 (2006). doi:10.1016/j.jaci.2006.03.033
44. Purohit, A., et al.: Clinical effects of immunotherapy with genetically modified recombinant birch pollen Bet v 1 derivatives. Clin. Exp. Allergy **38**, 1514–1525 (2008). doi:10.1111/j.1365-2222.2008.03042.x
45. Thalhamer, T., et al.: Designing hypoallergenic derivatives for allergy treatment by means of in silico mutation and screening. J. Allergy Clin. Immunol. **125**, 926–934, e910 (2010). doi:10.1016/j.jaci.2010.01.031
46. Haselden, B.M., Kay, A.B., Larche, M.: Immunoglobulin E-independent major histocompatibility complex-restricted T cell peptide epitope-induced late asthmatic reactions. J. Exp. Med. **189**, 1885–1894 (1999)
47. Albrecht, M., et al.: Vaccination with a Modified Vaccinia Virus Ankara-based vaccine protects mice from allergic sensitization. J. Gene Med. **10**, 1324–1333 (2008). doi:10.1002/jgm.1256
48. Kumar, M., Behera, A.K., Matsuse, H., Lockey, R.F., Mohapatra, S.S.: A recombinant BCG vaccine generates a Th1-like response and inhibits IgE synthesis in BALB/c mice. Immunology **97**, 515–521 (1999)
49. Rigaux, P., et al.: Immunomodulatory properties of Lactobacillus plantarum and its use as a recombinant vaccine against mite allergy. Allergy **64**, 406–414 (2009). doi:10.1111/j.1398-9995.2008.01825.x
50. Schmitz, N., et al.: Displaying Fel d1 on virus-like particles prevents reactogenicity despite greatly enhanced immunogenicity: a novel therapy for cat allergy. J. Exp. Med. **206**, 1941–1955 (2009). doi:10.1084/jem.20090199
51. Weiss, R., et al.: Is genetic vaccination against allergy possible? Int. Arch. Allergy Immunol. **139**, 332–345 (2006). doi:10.1159/000091946
52. Roesler, E., et al.: Immunize and disappear-safety-optimized mRNA vaccination with a panel of 29 allergens. J. Allergy Clin. Immunol. **124**, 1070–1077, e1071-1011, doi:10.1016/j.jaci.2009.06.036 (2009).
53. Weiss, R., Scheiblhofer, S., Roesler, E., Weinberger, E., Thalhamer, J.: mRNA vaccination as a safe approach for specific protection from type I allergy. Expert Rev. Vaccines **11**, 55–67 (2012). doi:10.1586/erv.11.168
54. Gabler, M., et al.: Immunization with a low-dose replicon DNA vaccine encoding Phl p 5 effectively prevents allergic sensitization. J. Allergy Clin. Immunol. **118**, 734–741 (2006). doi:10.1016/j.jaci.2006.04.048. S0091-6749(06)00943-2 [pii]

55. Scheiblhofer, S., et al.: Inhibition of type I allergic responses with nanogram doses of replicon-based DNA vaccines. Allergy **61**, 828–835 (2006). doi:10.1111/j.1398-9995.2006.01142.x. ALL1142 [pii]
56. Leitner, W.W., Bergmann-Leitner, E.S., Hwang, L.N., Restifo, N.P.: Type I Interferons are essential for the efficacy of replicase-based DNA vaccines. Vaccine **24**, 5110–5118 (2006). doi:10.1016/j.vaccine.2006.04.059
57. Eifan, A.O., Shamji, M.H., Durham, S.R.: Long-term clinical and immunological effects of allergen immunotherapy. Curr. Opin. Allergy Clin. Immunol. **11**, 586–593(2011).doi:10.1097/ACI.0b013e32834cb994
58. Bal, S.M., Ding, Z., Jiskoot, W., Bouwstra, J.A.: Advances in transcutaneous vaccine delivery: do all ways lead to Rome? J. Control. Release **148**, 266–282 (2010). doi:10.1016/j.jconrel.2010.09.018
59. Gutowska-Owsiak, D., Ogg, G.S.: The epidermis as an adjuvant. J. Invest. Dermatol. **132**, 940–948 (2012). doi:10.1038/jid.2011.398
60. Blamoutier, P., Blamoutier, J., Guibert, L.: Treatment of pollinosis with pollen extracts by the method of cutaneous quadrille ruling. Presse Med. **67**, 2299–2301 (1959)
61. Senti, G., et al.: Epicutaneous allergen administration as a novel method of allergen-specific immunotherapy. J. Allergy Clin. Immunol. **124**, 997–1002 (2009). doi:10.1016/j.jaci.2009.07.019
62. Senti, G., et al.: Epicutaneous allergen-specific immunotherapy ameliorates grass pollen-induced rhinoconjunctivitis: a double-blind, placebo-controlled dose escalation study. J. Allergy Clin. Immunol. **129**, 128–135 (2012). doi:10.1016/j.jaci.2011.08.036
63. Dupont, C., et al.: Cow's milk epicutaneous immunotherapy in children: a pilot trial of safety, acceptability, and impact on allergic reactivity. J. Allergy Clin. Immunol. **125**, 1165–1167 (2010). doi:10.1016/j.jaci.2010.02.029
64. Weiss, R., et al.: Transcutaneous vaccination via laser microporation. J. Control. Release (2012). doi:10.1016/j.jconrel.2012.06.031
65. Bach, D., et al.: Transcutaneous immunotherapy via laser-generated micropores efficiently alleviates allergic asthma in Phl p 5-sensitized mice. Allergy (2012). doi:10.1111/all.12005
66. Senti, G., et al.: Intralymphatic immunotherapy for cat allergy induces tolerance after only 3 injections. J. Allergy Clin. Immunol. **129**, 1290–1296 (2012). doi:10.1016/j.jaci.2012.02.026
67. Senti, G., et al.: Intralymphatic allergen administration renders specific immunotherapy faster and safer: a randomized controlled trial. Proc. Natl. Acad. Sci. U. S. A. **105**, 17908–17912 (2008). doi:10.1073/pnas.0803725105

# Rice Seed-Based Allergy Vaccines: Induction of Allergen-Specific Oral Tolerance Against Cedar Pollen and House Dust Mite Allergies

## 31

Fumio Takaiwa and Takachika Hiroi

## Contents

| | | |
|---|---|---|
| 31.1 | **Major Allergic Diseases in Japan** | 503 |
| 31.2 | **Allergic Pathogens** | 504 |
| 31.3 | **Conventional Therapy** | 504 |
| 31.4 | **Development of Novel Vaccine for Allergic Disease** | 505 |
| 31.4.1 | Expression of Recombinant Hypoallergenic Tolerogens with Reduced Allergenicity | 505 |
| 31.4.2 | T-Cell Epitope Peptide as a Minimum Tolerogen | 506 |
| 31.4.3 | Seed-Based Oral Vaccine | 506 |
| 31.4.4 | Delivery to GALT as Oral Seed-Based Allergy Vaccine | 507 |
| 31.4.5 | Induction of Immune Tolerance in GALT | 508 |
| 31.4.6 | Efficacy of Rice-Based Allergy Vaccines | 510 |
| 31.5 | **Practical Application of Rice-Based Allergy Vaccines** | 512 |
| **References** | | 514 |

F. Takaiwa, PhD (✉)
Functional Transgenic Crop Research Unit,
National Institute of Agrobiological Sciences,
Tsukuba, Ibaraki, Japan
e-mail: takaiwa@nias.affrc.go.jp

T. Hiroi, DDS, PhD
Department of Allergy and Immunology,
Tokyo Metropolitan Institute of Medical Science,
Setagaya-ku, Tokyo, Japan
e-mail: hiroi-tk@igakuken.or.jp

## Abstract

Japanese cedar pollen and house dust mite allergens are major causes of immunoglobulin (Ig)E-mediated type I allergy. For these allergies allergen-specific immunotherapy using systemic immunization with crude allergen extracts has been achieved as an only curative treatment. Here, we introduce a novel allergen-specific immunotherapy using rice seed-based oral vaccines comprising genetically modified hypoallergenic tolerogen or T-cell epitope peptides derived from allergens as a desirable alternative.

## 31.1 Major Allergic Diseases in Japan

In industrial countries, 30–40 % of the population suffers from some kind of IgE-mediated type I allergic disease such as asthma, allergic rhinitis, conjunctivitis, and atopic dermatitis [1, 2]. Japanese cedar (*Cryptomeria japonica:* Cry j) pollinosis is the most predominant seasonal allergic disease in Japan and is caused by pollen spread over most areas of Japan in early spring from February to April. Epidemiological studies indicate that about 27 % of the Japanese population is afflicted with this allergic disease [3].

Furthermore, more than half of the general population has circulating IgE specific for cedar pollen allergens. The number of patients with *Cryptomeria* pollinosis and the economic cost associated with it is expected to increase steadily, resulting in a strong social demand for the development of a reliable and effective way of controlling this pollinosis.

House dust mite (HDM) is also a major source of inhalant allergens which cause chronic allergic disease such as bronchial asthma in as much as 10 % of the population in many parts of the world. About 45–80 % of patients with allergic asthma are sensitized to allergens in HDM, especially *Dermatophagoides farinae* (Der f) and *D. pteronyssinus* (Der p) [4]. Thus, these patients suffer from allergic symptoms or reactions in response to HDM and/or have elevated allergen-specific serum IgE, suggesting that HDM allergens are crucial for the development of bronchial asthma [5, 6].

## 31.2 Allergic Pathogens

Two major allergens of Japanese cedar pollen, Cry j 1 and Cry j 2, have been isolated and characterized. More than 90 % of patients with Japanese cedar pollinosis have IgE specific to both [6]. Cry j 1 is a basic glycoprotein pectate lyase with an apparent molecular mass of 41–45 kDa and a pI of 8.9–9.2. Cry j 2 is also a basic protein with polygalacturonate activity and a molecular mass of 37 kDa and pI of 8.6–8.8; they are specifically localized in the cell walls of papilla and amyloplasts in cedar pollen, respectively. Both Cry j 1 and Cry j 2 have several isoforms that differ in primary structure and result from posttranslational modifications. The nucleotide sequences of Cry j 1 and Cry j 2 cDNAs have been determined, and their amino acid sequences deduced [7–10].

Those derived from HDM are some of the most common indoor allergens associated with bronchial asthma, rhinitis, and atopic dermatitis. HDM is responsible for more than 70 % of childhood bronchial asthma cases and, to date, more than 20 HDM allergens have been identified and characterized [5, 11]. The major HDM allergens are classified into group 1 (Der f 1 and Der p 1, molecular mass: 25 kDa) and group 2 (Der f 2 and Der p 2, molecular mass: 14 kDa). Sera from 50 to 70 % of mite-sensitive patients are reactive to either or both of these allergens. Group 1 allergen is a heat labile acidic glycoprotein and is mainly found in HDM feces. This protein has papain-like cysteine protease activity that is involved in the pathogenesis of allergy through cleavage of CD23 and CD25 from the surface of immune cells. Group 1 allergen is synthesized as an enzymatically inactive pro-group 1 protein which is then processed into the more allergenic mature and enzymatically active proteins of 222 (Der p 1) and 223 (Der f 1) amino acids which share 82 % sequence homology [12, 13]. IgE binding to group I allergens is highly dependent on their tertiary structure. Group 2 allergens comprise 129 amino acid residues and share 87 % homology [14–17]. More than 80 % of mite-allergic patients are sensitized against group 2 allergens, which are present at high concentrations in mite feces [18]. Although the biological function of group 2 allergens is not known, they show similarities in sequence, size, and distribution of cysteine residues to a family of epididymal proteins [19].

## 31.3 Conventional Therapy

Allergic diseases are characterized by allergen-specific IgE production and the activation of effector cells including eosinophils, mast cells, and basophils. These events are regulated by Th2 cells, which preferentially produce IL-4, IL-5, and IL-13. Therefore, allergic diseases have been defined as inadequate peripheral regulation of allergen-specific T cells.

Treatment strategies for these allergic diseases generally involve pharmacotherapy using antihistamines, corticosteroids, etc. However, although these approaches reduce clinical symptoms by blocking the release of critical mediators of allergic reactions or by inhibiting allergic inflammation, they are not curative and sometimes induce impaired performance as a result of side effects. It is well established that allergen-specific immunotherapy is the only way to modulate these immune reactions, bringing about effects that persist for many years without the need for further treatment and which reduce the risk of development of sensitization to further allergens or development of asthma in children with rhinitis [20, 21]. Conventional allergen-specific immunotherapy has been practiced for almost a century [22]. Success can be achieved by repeated subcutaneous injections of increasing doses of native allergen extracts over a period of at least 3–5 years for induction of immune tolerance (desensitization).

This treatment is sometimes accompanied by severe side effects, such as anaphylaxis caused by capture of the allergen together with specific anti-allergen IgE on the surface of mast cells and basophils. Recently, sublingual immunotherapy has been developed and has provided a safer and more beneficial route [23, 24]. The basic principle of allergen-specific immunotherapy is to induce immune tolerance to allergens through multiple cellular and molecular mechanisms [25] leading to reduction in inflammatory cell recruitment and activation, and in mediator secretion. The induction of a tolerant state in peripheral T cells represents an essential step in allergen-specific immunotherapy [26–28]. Peripheral T-cell tolerance is characterized mainly by the generation of allergen-specific regulatory T cells (Tregs) (Foxp3+ CD4+CD25+, induced Foxp3+CD4+, Tr1, and Th3), leading to suppressed T-cell proliferation and Th1 and Th2 cytokine responses against allergen. This is accompanied by a significant increase in allergen-specific IgG4 and IgG1 subclasses of IgG antibody and in IgA and a decrease in IgE in the late stage of disease.

## 31.4 Development of Novel Vaccine for Allergic Disease

Safer and noninvasive oral immunotherapy has long been desired as an alternative to the conventional subcutaneous application. Modified allergens with reduced IgE-binding activity (low allergenicity) represent the most promising and elegant approach to circumvent the risk of anaphylactic reactions in allergen-specific immunotherapy. To address these requirements, rice seed-based oral allergy vaccines have been recently developed, containing modified recombinant allergens with reduced or no allergenicity, or allergen-derived T-cell epitope peptides. These oral vaccines are a simple, safe, and cost-effective form of immunotherapy.

### 31.4.1 Expression of Recombinant Hypoallergenic Tolerogens with Reduced Allergenicity

To create hypoallergenic tolerogens, natural allergen extracts used for induction of immune tolerance (desensitization), which can bind to specific IgE on mast cells and basophils and lead to anaphylactic side effects, need to be replaced by recombinant proteins with reduced allergenicity [29]. New approaches to allergen-specific immunotherapy using such modified hypoallergenic tolerogens are therefore required [30–33]. It has been reported that binding of the allergen to specific IgE is determined by a continuous stretch of amino acids or by conformational structures; hence the quest for ideal hypoallergenic tolerogens with low-IgE-binding activity has involved investigating the reassembly of allergen fragments to form various mosaic structures, achieved by means such as deletion, site-directed mutagenesis, fragmentation, oligomeric formation, or molecular shuffling [30–33]. As a result, amino acids and peptides involved in IgE binding would be removed and allergen folding would be changed by sequence alteration.

In particular, tertiary structure can be significantly altered by mutation of cysteine residue(s) disrupting disulphide bond-dependent conformation. Recombinant hypoallergenic allergen derivatives exhibit reduced IgE reactivity but retain T-cell reactivity and immunogenicity, hence they do not induce IgE-mediated side effects even when relatively high doses of tolerogen are administered to allergic patients. Reduced allergenicity enables higher doses to be administered, which results in greater efficacy of the allergen-specific immunotherapy. Interestingly, recombinant oligomers preserve IgE reactivity but lose allergenic activity because of altered presentation of IgE epitopes. The efficacy of recombinant hypoallergenic allergen derivatives of the major birch pollen allergen Bet v 1 has been confirmed in clinical trials (currently at Phase III) with allergic patients [34, 35]. The production of allergen-specific IgG antibodies (especially IgG4) with blocking activities that inhibit not only allergen-induced release of inflammatory mediators from mast cells and basophils, but also IgE-facilitated antigen presentation to T-cells, has been shown to be induced by subcutaneous and sublingual immunotherapy.

### 31.4.2 T-Cell Epitope Peptide as a Minimum Tolerogen

Allergens are recognized by T-cell receptors (TCR) and major histocompatibility complex (MHC)-II through interaction with their T-cell epitope. Immunotherapy using T-cell epitope peptide is an attractive approach for safe allergen-specific immunotherapy because it does not contain an IgE cross-linking epitope that could induce anaphylaxis [36, 37]. However, T-cell epitopes derived from allergens, which are recognized by specific T cells, vary according to MHC-II class haplotypes. This is a serious obstacle in the development of peptide immunotherapy that takes advantage of T-cell epitope peptides. To overcome this problem, artificial hybrid epitope peptides composed of the complete T-cell epitope repertoires derived from one or a few allergens have been created.

Several major T-cell epitopes from many allergy patients were identified and chosen for inclusion in the artificial hybrid polypeptide. In the case of Japanese cedar pollen allergy, five or seven candidate T-cell epitopes derived from the two major allergen molecules (Cry j 1 and Cry j 2) were linked together to create hybrid epitope peptides [38, 39]. These hybrids have several benefits when used as immunotherapeutic tolerogens, since they exhibit little or no IgE reactivity due to a lack of B-cell epitopes and conformational changes. For example, the hybrid 7Crp peptide (seven linked epitopes) showed no binding to specific IgE in the sera of 48 patients with Japanese cedar pollinosis symptoms [39]. It is interesting to note that the 7Crp peptide induced T-cell proliferation with 100-fold higher immunogenicity than a mixture of the T-cell epitopes used for construction of the hybrid peptide. Clinical trials have been mainly performed with peptides from the major cat and bee venom allergens Fel d 1 and Api m 1, respectively. It is known that peptide immunotherapy performed with peptides from Fel d 1 or the phospholipase A2 Api m 1 induced a Tr1-type allergen-specific immune response with decreased Th2-type cytokines and increased IL-10 and allergen-specific IgG4 production [40]. Low doses of peptides have also been shown to improve clinical symptoms without inducing adverse events.

### 31.4.3 Seed-Based Oral Vaccine

Seed provides a suitable production platform for recombinant allergy vaccines and, when orally administered, acts as a delivery vehicle to gut-associated lymphoid tissue (GALT), including Peyer's patches (PPs) and a large cluster of lymphoid follicles of the mucosal immune tissues. The production of vaccines in seed offers several advantages, including no requirement for cold transportation and storage, needle-free administration, no risk of contamination with human pathogens, low cost, and a production process that can easily be scaled up as required [41, 42].

For practical use of plant-made oral allergy vaccines, the continuous administration of high-dose tolerogen required for induction of immune tolerance (desensitization) means that high and consistent expression is critical for efficacy. To maximize yield in this system, the transferred genes encoding these tolerogens will need to be synthesized using codons optimized for expression in seed [43, 44]. Furthermore, because accumulation of recombinant proteins is also largely dependent on intracellular localization, it is important to target them stably and at high concentration to suitable locations by means of intracellular targeting signals [45, 46]. In the case of seeds, a signal peptide for targeting the secretory pathway and a sorting signal to target protein bodies (PBs) are mandatory for stable accumulation of product. Strong tissue-specific promoters rather than constitutive promoters are preferable for high-level expression of antigens in the desired tissue [47].

We have reported that rice seed endosperm is a good production platform for recombinant proteins, because there is ample and stable storage space for foreign products [48]. For expression of antigens in transgenic rice seed strong endosperm-specific promoters, such as those for major storage protein glutelin (GluB), 26 kDa globulin, or 10 kDa and 16 kDa prolamin, have been used for expression in rice endosperm [49]. Using these promoters, target gene products were highly accumulated as distinctly detectable proteins in CBB-stained SDS-PAGE gels, accounting for 4–6 % of total seed proteins (3–5 mg/g dry seeds). Thus far, using this system, hybrid T-cell epitope peptides (7Crp) from Japanese cedar pollen allergens Cry j 1 and Cry j 2 and their structurally disrupted full length forms, subunit antigen from major mite allergen Der p 1 and a modified mite allergen Der f 2, in which Cys residues implicated in disulfide bonds were mutated to Ser (Ala), were all found to be highly accumulated in the endosperm of transgenic rice [50, 51]. They were deposited predominantly in ER-derived PBs or in the unique ER-derived compartment, the Der f body. These recombinant proteins are consistently inherited over several subsequent generations without silencing. Furthermore, these antigens remain stable without loss of immunogenicity even when the seed is stored at ambient temperature for several years.

### 31.4.4 Delivery to GALT as Oral Seed-Based Allergy Vaccine

Oral delivery of plant-based antigens induces not only mucosal but also systemic immune responses. Repeated oral administration is inclined to induce immune tolerance to the fed antigen rather than sensitization, especially when the antigen is deposited in particulate PBs as the storage organelle of the seed. PBs of <10 nm are principally taken up by M cells in the follicle-associated epithelium of PPs in GALT of the small intestine and then presented to antigen-presenting cells (APCs) such as dendritic cells (DCs) in adjacent mucosal T-cell areas [52, 53]. Moreover, there is a possibility that DCs in the lamina propria (LP) extend their dendrites into the lumen and sample PBs. The GALT contains an organized macro-architecture of zones of B and T lymphocytes that respond to antigens presented by DCs, which can induce memory B and T cells. There are at least three subsets of DCs with distinct tissue distribution in PPs [54, 55]. $CD11b^+$ DCs are present in the sub-epithelial domain (SED) regions, $CD8^+$ DCs in the T-cell-rich interfollicular region (IFR), and double-negative $CD4^-$ $CD8\alpha^-$ DCs in both SED and IFRs (Fig. 31.1).

IL-10-producing immature DCs contribute to immunologic tolerance through inducing Tr1-regulatory cells. It should be noted that GALT is the preferential site for peripheral induction of $Foxp3^+$ Tregs (Fig. 31.2). $CD103^+$ DCs of the LP of small intestine (mucosal DCs) are particularly competent at inducing $Foxp3^+$ T cells via TGF-$\beta$ and retinoic acid (RA) (Fig. 31.2). Plasmacytoid DCs (pDCs) presenting dietary antigens are responsible for induction of oral tolerance and immune suppression (Fig. 31.2).

It has been demonstrated that to achieve the same level of efficacy, orally delivered mucosal immunotherapy requires >100–1,000-fold as much antigen as subcutaneous injection [56]. This is because the purified antigen is degraded by digestive enzymes in the gastrointestinal tract and by exposure to the harsh environment in the stomach before arrival at the mucosal immune cells in the GALT. By contrast, when orally delivered via cereal seeds such as rice grains, bio-encapsulation of antigen by the double barrier of PBs and cell wall characteristic of plant cells has the advantage of protecting the antigen from proteolysis [57]. This is not the case when the antigen is delivered by other plant vegetative cells. It should be noted that PBs derived from the ER are more resistant to gastrointestinal enzyme digestion than those in protein storage vacuoles (PSVs), as shown by comparing digestibility and immune tolerance-inducing capacity between

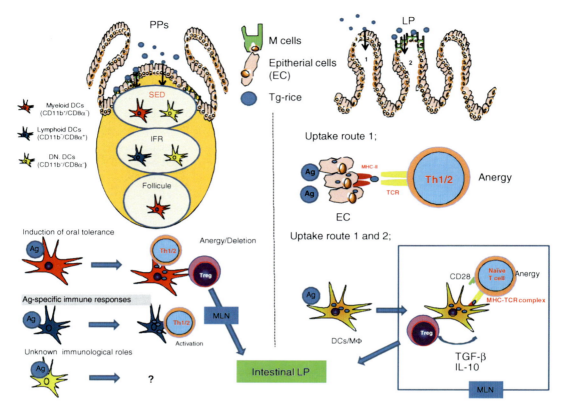

**Fig. 31.1** Fed allergens are taken to the mucosal immune system in Peyer's patches and intestinal lamina propria. Antigens fed by oral vaccination cross M cells of both Peyer's patches (*PPs*) and intestinal epithelial cells in the lamina propria (*LP*). Antigens are processed and presented on MHC-II molecules by CD11b⁺ CD8α⁻ myeloid DCs in PPs to induce oral tolerance, but not CD11b⁻ CD8α⁺ lymphoid DCs. DCs/macrophage (Mφ)-captured antigens in intestinal LP are also processed and presented on MHC-II molecules, and thereafter carried to the mesenteric lymph node (*MLN*) as an attending lymph node in the intestinal mucosa. Moreover, antigens are also processed and directly presented on MHC molecules by intestinal epithelial cells to induce oral tolerance in intestinal LP. *SED* subepithelial domain, *IFR* interfollicular region

ER-derived PB (PB-I) and PSV (PB-II)-localized antigens [58]. This physical difference may be related to the polymerized or aggregated arrangement of antigen formed with cysteine-rich prolamins observed in ER-derived PBs. In fact, when transgenic rice seeds containing several allergen derivatives deposited in ER-derived PB or PSV are fed to mice, PB-I-localized allergen derivatives manifest greater resistance to enzymatic digestion than PB-II-localized or synthesized naked forms. The dose of PB-I-localized T-cell epitope required for suppression of allergen-specific IgE levels in mice was about 20-fold and threefold lower than the amounts required of synthetic peptide and PB-II-localized form, respectively [58].

The rice-based allergy vaccine is heat-stable and can thus tolerate storage at room temperature for several years and boiling in the process of rice cooking. It has been confirmed that steamed or cooked transgenic rice retains its efficacy for inducing oral immune tolerance. In fact, extracts from the treated seeds can still elicit T-cell proliferative activity even after autoclaving for 20 min.

### 31.4.5 Induction of Immune Tolerance in GALT

The immunological mechanism implicated in tolerance induction is dependent on the frequency, duration, and dose of antigen administered.

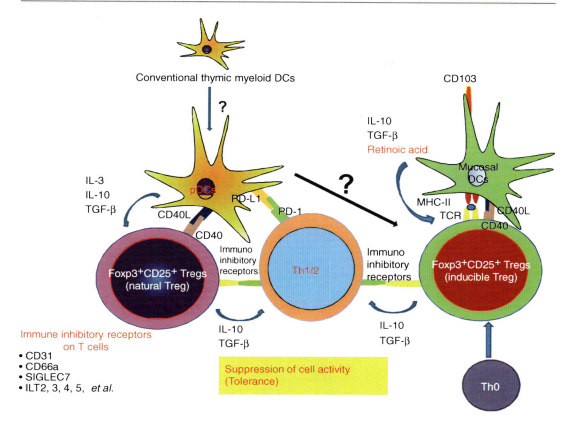

**Fig. 31.2** Two subsets of Foxp3+ CD25+ CD4+ Tregs for the induction of antigen-specific tolerance. Plasmacytoid dendritic cells (pDCs) derived from conventional thymic myeloid DCs produce IL-10 and TGF-β and induce more IL-10-producing natural regulatory T cells (*Tregs*), contributing to the suppression of immune responses. The environment of the gut mucosa is particularly permissive for the generation of inducible Tregs. CD103+ DCs excel in converting CD4+ naive T cells into Foxp3+ inducible Tregs in the absence of TGF-β, under the production of the vitamin A metabolite retinoic acid in small intestinal lamina propria

Repeated low-dose administration of antigen favors tolerance driven by Tregs, whereas high doses favor anergy or cell deletion-driven tolerance, of the kind thought to be responsible for the induction of tolerance as a quiescent immune state [59–61]. It has been shown that for oral tolerance, Fas (CD95)-dependent apoptosis is responsible for the deletion of effector T cells [62] (Fig. 31.3a). T lymphocyte anergy occurs through TCR ligation in the absence of co-stimulatory signals and is prevented by cognate interaction between CD80/86 (B7) on APCs and CD28 (CTLA-4) on effector T cells (Fig. 31.3b). CTLA-4 is known to play a pivotal role in controlling the suppressive functions of Tregs. The interactions of co-stimulatory molecules play a key role in the regulation of T-cell activation and tolerance [63]. Participation of other B7 and CD28 family members as co-stimulation factors involved in immune regulation has been also been discovered (Fig. 31.3b).

By contrast, orally administered low-dose antigen can induce active suppression by induction of inhibitory immune responses mediated by Treg cells. These cells include naturally occurring Foxp3+CD4+ CD25+ Treg cells and antigen-inducible Tregs comprising Foxp3+ inducible CD4+ cells, IL-10 producing Tr1 cells, TGF-β producing Th3 cells, and regulatory CD8+ T cells [64]. Naturally occurring Tregs are generated in the thymus, whereas induced Treg cells are generated in the periphery from naïve T cells.

It has been shown that peripheral conversion of CD4+ T cells to Tregs occurs primarily in

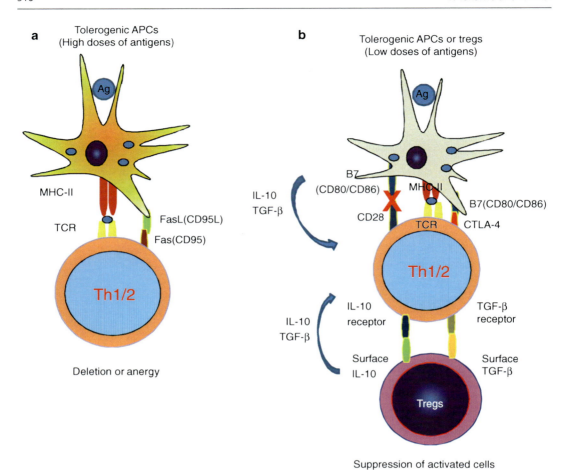

**Fig. 31.3** Immunological mechanism implicated in immune tolerance is dependent on the dose of antigen. The generation of an immune response requires ligation of the T-cell receptor (*TCR*) with peptide-MHC-II complexes whether with high dose of antigen or low dose without regulatory T cells (*Tregs*). (**a**) In oral tolerance induction by high-dose antigen, inhibitory ligations (Fas-Fas-L) lead to immune regulatory responses such as deletion and anergy of allergen-specific CD4⁺ Th2 cells. (**b**) In oral tolerance induction by repeated low dose of antigen, Tregs and/or tolerogenic antigen-presenting cells (*APCs*) are elicited. Ligation of cell surface T-cell-suppressive cytokines IL-10 and TGF-β by Tregs suppresses immune responses, which are maintained by production of soluble forms of these cytokines. Tolerogenic APCs lead to suppression of antigen-specific immune responses in the absence of B7-CD28 ligation

GALT, because the gut is an entry site for numerous antigens such as foods and commensal bacteria. Differentiation of naïve CD4⁺ T cells to IL-10-producing Tr1 cells is mediated by IL-27 produced by the Treg-modified DCs in the presence of TGF-β and IL-6 [65]. Generation of Foxp3⁺ inducible Treg cells is induced by TGF-β and RA produced by intestinal CD103⁺ DCs in gut LP [65]. RA has been characterized as a key regulator of the TGF-β-dependent immune response. IL-6 inhibits TGF-β-induced Foxp3 induction, resulting in Th17 cells, while RA can counteract this inhibition [66, 67]. Suppression of Th1- and Th17-cell response by Tregs is involved in prevention of inflammation.

### 31.4.6 Efficacy of Rice-Based Allergy Vaccines

In a mouse model of Japanese cedar pollinosis, oral administration of the dominant T-cell epitope

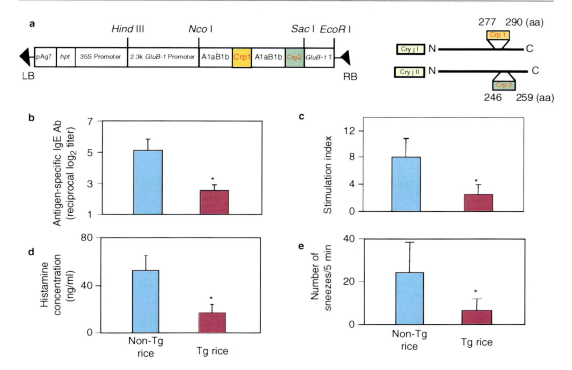

**Fig. 31.4** Transgenic rice expressing T-cell epitopes for Cry j 1 and Cry j 2 induces oral tolerance by inhibiting Th2-mediated IgE responses. Schematic representation of the transformation plasmid and the two selected major T-cell epitopes of Cry j 1 and Cry j 2 (Crp 1 and Crp 2, respectively) (**a**). Inhibition of allergen-specific serum IgE levels (**b**) and CD4+ T-cell responses (**c**) by oral administration of transgenic (*Tg*) rice seeds. Serum histamine levels (**d**) and the number of sneezes (**e**) were inhibited in the group of mice fed with Tg rice seeds. Data are expressed as mean ± SD. *, $p<0.01$ for the group of mice fed with Tg rice seeds (*blue*) in comparison with the group of mice fed with non-Tg rice seeds (*magenta*)

of Cry j 2 inhibited specific T-cell responses in Cry j 2 sensitized mice [68]. Sneezing frequency, as a measurable clinical symptom, was decreased not only by systemic injection, but also by oral administration of the dominant T-cell epitope of Cry j 2 [69, 70]. The efficacy of oral transgenic rice seed that accumulated the major mouse T-cell epitopes derived from Cry j 1 and Cry j 2 as fusion proteins with soybean seed protein glycinin was then examined by oral feeding [71] (Fig. 31.4a). When systemically challenged with total protein of cedar pollens after daily administration for 4 weeks, the production of allergen-specific serum IgE and IgG antibodies was inhibited, when compared with mice fed with non-transgenic rice seeds (Fig. 31.4b). Both allergen-specific CD4+ T-cell proliferative responses and levels of Th2 cytokines such as IL-4, IL-5, and IL-13 were inhibited (Fig. 31.4c). Histamine release from mast cells and sneezing, as measures of clinical symptoms of allergy, were also suppressed (Fig. 31.4d and e). These results showed that a rice-based oral vaccine expressing T-cell epitope peptides can induce oral tolerance against cedar pollen allergens and alleviate clinical symptoms prophylactically.

Based on this evidence, transgenic rice seed accumulating 7Crp peptide composed of human major T-cell epitopes from Cry j 1 and Cry j 2 has been developed as a rice-based peptide vaccine for cedar pollen allergy in humans [72]. The 7Crp peptide was highly accumulated in transgenic rice seed (about 3 mg/g dry seed) and was deposited in ER-derived PBs suitable for delivery to GALT. To examine its safety, chronic toxicity and reproductive toxicity were examined by feeding transgenic rice seeds to monkeys and rats; no adverse effects were detected [73].

However, despite its safety, because humans with different genetic backgrounds respond differently to various T-cell epitopes, peptide immunotherapy using T-cell epitopes is not applicable to all patients. To encompass a broader array of allergy patients an alternative strategy has recently been tested. Recombinant whole allergen derivatives with reduced IgE binding have been generated for Cry j 1 and Cry j 2 by creating mosaic structures by molecular shuffling or fragmentation to affect the tertiary structure required to recognize allergen-specific IgE [74]. When transgenic rice seeds were fed daily to mice for 3 weeks and then challenged with crude cedar pollen allergen, allergen-specific CD4+ T-cell proliferation and IgE and IgG levels were markedly suppressed as compared with mice fed non-transgenic rice seeds. Sneezing frequency, as a clinical symptom of pollinosis, and infiltration of inflammatory cells, such as eosinophils and neutrophils in the nasal tissue, were also significantly reduced. These results imply that oral administration of transgenic rice seeds containing the structurally disrupted Cry j 1 and Cry j 2 antigens is a promising therapeutic allergy vaccine applicable for all patients suffering with Japanese cedar pollinosis.

For HDM allergy, transgenic rice grains accumulating mature Der p 1, Der p 1 fragment (p45-145) which covers most human and mice major T-cell epitopes, and Der f 2 derivatives with reduced IgE reactivity resulting from disrupted intramolecular disulfide bonds have been generated [51, 75, 76]. The plant-derived Der p 1 was posttranslationally modified with a high-mannose type glycan structure. Glycosylated Der p 1 displayed reduced IgE-binding capacity in comparison with its unglycosylated counterpart [75] (Fig. 31.5a).

When mite allergen transgenic rice seeds containing the Der p 1 fragment (p45-145) were fed to mice that were subsequently immunized with recombinant Der p 1, allergen-specific CD4+ T cell proliferation (Fig. 31.5b) and production of allergen-specific IgE and IgG (Fig. 31.5c) were suppressed compared to mice fed non-transgenic rice seeds [76]. Production of Th2 cytokines IL-4, IL-5, and IL-13 were also significantly diminished by oral vaccination of transgenic rice seeds (Fig. 31.5d). As measureable clinical symptoms, development of allergic airway inflammation by infiltration of eosinophils and lymphocytes into the airway (Fig. 31.5e) and bronchial hyperresponsiveness (Fig. 31.5f) were also inhibited by prophylactic vaccination with transgenic rice seeds.

When transgenic rice seeds containing Der f 2 derivatives were fed to mice, immune tolerance against Der f 2 antigen was induced in a similar way to other allergens [51]. Interestingly, the inhibition of Der f 2-specific IgE and IgG production indicated that the efficacy of these derivatives was related to the resistance of PBs containing Der f 2 to digestive enzymes in the gastrointestinal tract.

## 31.5 Practical Application of Rice-Based Allergy Vaccines

As a vaccine antigen production system, rice seed has several advantages over other crops, including established processing, greater yield, and low risk of transgene escape due to self-pollination. In addition, the rice transformation system has been established and the genetic information, regarding the full genome sequence and expression, is available. As a new form of oral allergy vaccine, rice seeds that accumulate hypoallergenic cedar pollen allergens and HDM allergen derivatives can be used as a vehicle to deliver to GALT. When these rice-based vaccines were orally administered to mice, immunological and biochemical responses associated with these aeroallergens including the production of antigen-specific IgE antibodies, CD4+ T cell proliferation, and histamine release, were all inhibited. Clinical symptoms of the allergy, including sneezing, allergic airway inflammation and bronchial hyperresponsiveness were also alleviated in mice vaccinated with the transgenic rice after exposure to cedar pollen and HDM allergens.

Most vaccinations commonly administered to humans are formulated for parenteral delivery, requiring full purification [77]. Production of vaccine antigens must conform with the specific

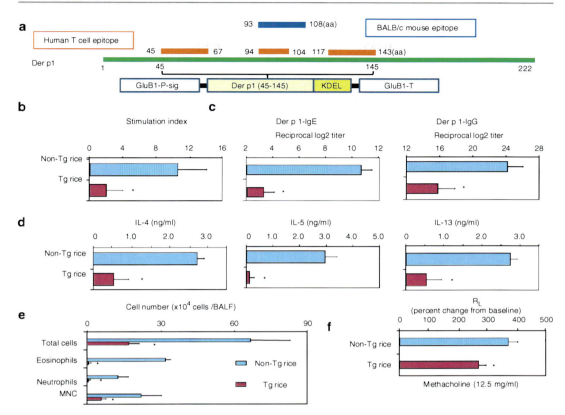

**Fig. 31.5** Transgenic rice expressing T-cell epitopes for Der p 1 induces oral tolerance by inhibiting Der p 1 specific immune responses. Schematic representation of the transformation plasmid and the selected major T-cell epitope of Der p 1 (**a**). Splenic CD4+ T cells prepared from negative control mice (data not shown), mice fed non-transgenic (Tg) rice (*magenta*), or mice fed Tg rice (*blue*) were cultured with irradiated normal splenocytes and allergen for 7 days. Oral administration of Tg rice seeds led to inhibition of allergen-specific CD4+ T cell proliferative responses (**b**), and levels of serum IgE and IgG (**c**). Concentration of cytokines (IL-4, IL-5, and IL-13) released into the culture supernatant were assayed (**d**). Inhibition of allergen-induced inflammatory cell-infiltration by prophylactic oral vaccination with Der p 1-Tg rice. The number of total cells, eosinophils, neutrophils, and mononuclear cells (*MNC*) in the bronchoalveolar lavage fluid (*BALF*) was measured (**e**). Bronchial hyperresponsiveness to aerosolized methacholine (MCh) was measured 24 h after the second allergen challenge (**f**). Data are expressed as mean ± SD. *, $p < 0.01$ for the group of mice fed with Tg rice seeds in comparison with the group of mice fed with non-Tg rice seeds

guidelines and regulations for pharmaceutical products under conditions of good manufacturing practice (GMP) [78]. It is, therefore, necessary to develop reliable processing steps (purification) for each product. To date, plant-made vaccine candidates for parenteral delivery entering clinical trials have been mainly developed using transient expression systems by viral and Agrobacterium vectors. The downstream processing stage of production is generally expensive, representing up to 80 % of overall production costs [77]. It is important to note that direct oral vaccination using products of transgenic crops without any further processing is highly advantageous in terms of stability and efficient delivery to GALT, and, as crop seeds containing antigen can be considered as oral formulations with a natural coating, they are appropriate vehicles for mucosal delivery. One possible obstacle will be validation of plant production systems for making a commercial pharmaceutical reagent. In addition, there are likely to be variations in accumulation of product due to physiological differences in biological systems.

Furthermore, in the case of field cultivation of pharmaceutical transgenic plants, there are special concerns about outcrossing with wild-type crops. Zero tolerance regarding the risk of food chain contamination is desired from consumers and environmental organizations. Thus, it has been proposed that pharmaceutical crops should be produced under the control of specific regulations and guidelines [78]. Closed molecular farming facilities (closed glasshouses) with strictly controlled environmental conditions would be ideal and are recommended to reduce variation of product concentration in targeted tissue as well as to prevent outcrossing. However, in the case of allergy vaccines, very large quantities of seeds will be required for commercialization. For example, daily oral consumption of cooked transgenic rice (about 70 g) for a minimum of 3 months, and continuing for a number of years may be required for desensitization against cedar pollen allergens. Such transgenic rice will have to be produced for about 20 million patients afflicted with Japanese cedar pollinosis.

An alternative strategy must therefore be developed, as cultivating the required amount of transgenic rice in closed glasshouses, as is done for most plant-made pharmaceuticals, will not be possible. Different production systems such as cultivation in physically isolated fields or greenhouses comprising fine-meshed nets already in use in the EU for commercial crops and biological containment will have to be considered, and specific guidelines and regulations will have to be developed [79]. Additionally, entry into the food chain during transportation and handling will need to be strictly avoided. Issues of GMP regarding product consistency and the control of pharmaceutical crops and produce still present serious challenges for the field of rice-based allergy vaccine cultivation [78].

Immunotherapy by immune tolerance induction for IgE-mediated type I allergens can be controlled more easily than immune induction therapies with vaccines against infectious diseases, because medically trained persons are not required to administer the therapy to avoid immune tolerance. However, it will be very important to check whether oral administration of rice seed-based vaccines containing hypoallergenic derivatives comes with any risk of causing new allergies (sensitization).

## References

1. Gerth van Wijk, R.: Allergy: a global problem. Allergy **57**, 1097–1110 (2002)
2. Bauchau, V., Durham, S.R.: Prevalence and rate of diagnosis of allergic rhinitis in Europe. Eur. Respir. J. **24**, 758–764 (2004)
3. Okamoto, Y., et al.: Present situation of cedar pollinosis in Japan and its immune responses. Allergol. Int. **58**, 155–162 (2009)
4. Tovey, E.R., Chapman, M.D., Platts-Mills, T.A.: Mite faces are a major source of house dust allergens. Nature **289**, 592–593 (1981)
5. Thomas, W.R., et al.: Characterization and immunobiology of house dust mite allergens. Int. Arch. Allergy Immunol. **129**, 1–18 (2002)
6. Hashimoto, M., et al.: Sensitivity to two major allergens (Cry j 1 and Cry j 2) in patients with Japanese cedar (*Cryptomeria japonica*) pollinosis. Clin. Exp. Allergy **25**, 848–852 (1995)
7. Yasueda, H., et al.: Isolation and partial characterization of the major allergen from Japanese cedar (*Cryptomeria japonica*) pollen. J. Allergy Clin. Immunol. **71**, 77–86 (1993)
8. Sone, T., et al.: Cloning and sequencing of cDNA coding for Cry j 1, a major allergen of Japanese cedar pollen. Biochem. Biophys. Res. Commun. **199**, 619–625 (1994)
9. Sakaguchi, M., et al.: Identification of the second major allergen of Japanese cedar pollen. Allergy **45**, 309–312 (1990)
10. Namba, M., et al.: Molecular cloning of the second allergen, Cry j II, from Japanese cedar pollen. FEBS Lett. **353**, 124–128 (1994)
11. Kawamoto, S., et al.: Toward elucidating the full spectrum of mite allergens-state of the art. J. Biosci. Bioeng. **94**, 285–298 (2002)
12. Chua, K.Y., et al.: Sequence analysis of cDNA coding for a major house dust mite allergen, Der p 1. Homology with cysteine proteases. J. Exp. Med. **167**, 175–182 (1988)
13. Dilworth, R.J., Chua, K.Y., Thomas, W.R.: Sequence analysis of cDNA coding for a major house dust mite allergen, Der f I. Clin. Exp. Allergy **21**, 25–32 (1991)
14. Thomas, W.R., Smith, W.: House-dust-mite allergens. Allergy **53**, 821–832 (1998)
15. Haide, M., et al.: Allergens of the house dust mite *Dermatophagoides farinae*-immunochemical studies of four allergenic fractions. J. Allergy Clin. Immunol. **75**, 686–692 (1985)
16. Chua, K.Y., et al.: Analysis of sequence polymorphism of a major allergen, Der p 2. Clin. Exp. Allergy **26**, 829–837 (1996)

17. Yuuki, T., et al.: Cloning and sequencing of cDNA corresponding to mite major allergen Der f II. Jpn. J. Allergy **39**, 557–561 (1990)
18. Park, G.M., et al.: Localization of a major allergen, Der p 2, in the gut and faecal pellets of *Dermatophagoides pteronyssinus*. Clin. Exp. Allergy **30**, 1293–1297 (2000)
19. Thomas, W.R., Chua, K.Y.: The major mite allergen Der p 2 – a secretion of the male mite reproductive tract? Clin. Exp. Allergy **25**, 667–669 (1995)
20. Bousquet, J., Lockey, R.F., Malling, H.U.: Allergen immunotherapy: therapeutic vaccines for allergic diseases. WHO position paper. Allergy **53**, 1–42 (1998)
21. Frew, A.J.: Immunotherapy of allergic diseases. J. Allergy Clin. Immunol. **111**, S712–S719 (2003)
22. Noon, L.: Prophylactic inoculation against hay fever. Lancet **2**, 1572–1573 (1911)
23. Cox, L.S., et al.: Sublingual immunotherapy: a comprehensive review. J. Allergy Clin. Immunol. **117**, 1021–1035 (2006)
24. Moingeon, P., et al.: Immune mechanisms of allergen-specific sublingual immunotherapy. Allergy **61**, 151–165 (2006)
25. Larche, M., Akids, C.A., Valenta, R.: Immunological mechanisms of allergen-specific immunotherapy. Nat. Rev. Immunol. **6**, 761–766 (2006)
26. Verhagen, J., et al.: Mechanisms of allergen-specific immunotherapy: T-regulatory cells and more. Immunol. Allergy Clin. N. Am. **26**, 207–231 (2006)
27. Till, S.J., et al.: Mechanisms of immunotherapy. J. Allergy Clin. Immunol. **113**, 1024–1034 (2004)
28. Akdis, M., Akids, C.A.: Mechanisms of allergen-specific immunotherapy. J. Allergy Clin. Immunol. **119**, 780–791 (2007)
29. Linhart, B., Valenta, R.: Molecular design of allergy vaccines. Curr. Opin. Immunol. **17**, 646–655 (2005)
30. Valenta, R., Kraft, D.: From allergen structure to new forms of allergen-specific immunotherapy. Curr. Opin. Immunol. **14**, 718–727 (2002)
31. Cromwell, O., Hafner, D., Nandy, A.: Recombinant allergens for specific immunotherapy. J. Allergy Clin. Immunol. **127**, 865–872 (2011)
32. Valenta R. *et al.* From allergen genes to allergy vaccines. *Ann Rev Immunol* **28**, 211-241 (2010).
33. Valenta, R., et al.: Recombinant allergens: what does the future hold? J. Allergy Clin. Immunol. **127**, 860–864 (2011)
34. Purohit, A., et al.: Clinical effects of immunotherapy with genetically modified recombinant birth pollen Bet v 1 derivatives. Clin. Exp. Allergy **38**, 1514–1525 (2008)
35. Gronlund, H., Gafvelin, G.: Recombinant Bet v 1 vaccine for treatment of allergy to birch pollen. Hum. Vaccin. **6**, 970–977 (2010)
36. Larche, M.: Peptide immunotherapy for allergic diseases. Allergy **62**, 325–331 (2007)
37. Ali, F.R., Larche, M.: Peptide-based immunotherapy: a novel strategy for allergic disease. Exp. Rev. Vaccine **4**, 881–889 (2005)
38. Sone, T., et al.: T cell epitopes in Japanese cedar (*Cryptomeria japonica*) pollen allergens: choice of major T cell epitopes in Cry j 1 and Cry j 2 toward design of the peptide-based immunotherapeutics for management of Japanese cedar pollinosis. J. Immunol. **161**, 448–457 (1998)
39. Hirahara, K., et al.: Preclinical evaluation of an immunotherapeutic peptide comprising 7 T-cell determinants of Cry j 1 and Cry j 2, the major Japanese cedar pollen allergens. J. Allergy Clin. Immunol. **108**, 94–100 (2001)
40. Larche, M.: Update on the current status of peptide immunotherapy. J. Allergy Clin. Immunol. **119**, 906–909 (2007)
41. Streatfield, S.J., et al.: Plant-based vaccines: unique advantages. Vaccine **19**, 2742–2748 (2001)
42. Nochi, T., et al.: Rice-based mucosal vaccine as a global strategy for cold-chain- and needle-free vaccination. Proc. Natl. Acad. Sci. U. S. A. **104**, 10986–10991 (2007)
43. Stoger, E., et al.: Sowing the seeds of success: pharmaceutical proteins from proteins. Curr. Opin. Biotech. **16**, 167–173 (2005)
44. Lau, O.S., Sun, S.S.M.: Plant seeds as bioreactors for recombinant protein production. Biotechnol. Adv. **27**, 1015–1022 (2009)
45. Vitale, A., Hinz, G.: Sorting of proteins to storage vacuoles: how many mechanisms? Trends Plant Sci. **10**, 315–323 (2005)
46. Benchabane, M., et al.: Preventing unintended proteolysis in protein biofactories. Plant Biotech. J. **6**, 633–648 (2008)
47. Kawakatsu, T., Takaiwa, F.: Cereal seed storage protein synthesis: fundamental processes for recombinant protein production in cereal grains. Plant Biotech. J. **8**, 939–953 (2010)
48. Takaiwa, F., et al.: Endosperm tissue is good production platform for artificial recombinant proteins in transgenic rice. Plant Biotech. J. **5**, 84–92 (2007)
49. Qu, L.Q., Takaiwa, F.: Tissue specific expression and quantitative potential evaluate of seed storage component gene promoters in transgenic rice. Plant Biotech. J. **2**, 113–125 (2004)
50. Takaiwa, F., et al.: Deposition of a recombinant peptide in ER-derived protein bodies by retention with cysteine-rich prolamins in transgenic rice seed. Planta **229**, 1147–1158 (2009)
51. Yang, L., et al.: Expression of hypoallergenic Der f 2 derivatives with altered intramolecular disulphide bonds induces the formation of novel ER-derived protein bodies in transgenic rice seeds. J. Exp. Bot. **63**, 2947–2959 (2012)
52. Kunisawa, J., Kurashima, Y., Kiyono, H.: Gut-associated lymphoid tissues for the development of oral vaccines. Adv. Drug Deliv. Rev. **64**, 523–530 (2012)
53. Neutra, M.R., Kozlowski, P.A.: Mucosal vaccines: the promise and the challenge. Nat. Rev. Immunol. **6**, 148–158 (2006)

54. Kelsall, B.L., Leon, F.: Involvement of intestinal dendritic cells in oral tolerance, immunity to pathogens, and inflammatory bowel disease. Immunol. Rev. **206**, 132–148 (2005)
55. Tsuji, N.M., Kosaka, A.: Oral tolerance: intestinal homeostasis and antigen-specific regulatory T cells. Trends Immunol. **29**, 532–540 (2008)
56. Streatfiled, S.J.: Mucosal immunization using recombinant plant-based oral vaccines. Methods **38**, 150–157 (2006)
57. Takaiwa, F.: Seed-based oral vaccines as allergen-specific immunotherapy. Hum. Vaccin. **7**, 357–366 (2011)
58. Takagi, H., et al.: Rice seed ER-derived protein body as an efficient delivery vehicle for oral tolerogenic peptides. Peptides **31**, 421–425 (2010)
59. Weiner, H.L.: Oral tolerance: immune mechanisms and treatment of autoimmune diseases. Immunol. Today **18**, 335–343 (1997)
60. Mayer, L., Shao, L.: Therapeutic potential of oral tolerance. Nat. Rev. Immunol. **4**, 407–419 (2004)
61. Burks, A.W., Laubach, S., Jones, S.M.: Oral tolerance, food allergy, and immunotherapy: Implications for future treatment. J. Allergy Clin. Immunol. **121**, 1344–1350 (2008)
62. Appeman, L.J., Boussiotis, V.A.: T cell anergy and co-stimulation. Immunol. Rev. **192**, 161–180 (2003)
63. Greenwald, R.J., Freeman, G.J., Sharpe, A.H.: B7 family revised. Annu. Rev. Immunol. **23**, 515–548 (2005)
64. Ozdemir, C., Akids, M., Akids, C.A.: T regulatory cells and their counterparts: masters of immune regulation. Clin. Exp. Allergy **39**, 626–639 (2009)
65. Awasthi, A., et al.: A dominant function for interleukin 27 in generating interleukin 10-producing anti-inflammatory T cells. Nat. Immunol. **12**, 1380–1390 (2007)
66. Bettelli, E., et al.: Reciprocal developmental pathways for generation of pathogenic effector Th17 and regulatory T cells. Nature **441**, 235–238 (2006)
67. Mucida, D., Salek-Ardakani, S.: Regulation of Th17 cells in the mucosal surfaces. J. Allergy Clin. Immunol. **123**, 997–1003 (2009)
68. Hirahara, K., et al.: Oral administration of a dominant T-cell determinant peptide inhibits allergen-specific TH1 and TH2 cell response in Cry j 2-primed mice. J. Allergy Clin. Immunol. **102**, 961–967 (1998)
69. Murasugi, T., et al.: Oral administration of a T cell epitope inhibits symptoms and reactions of allergic rhinitis in Japanese cedar pollen allergen-sensitized mice. Eur. J. Pharm. **510**, 143–148 (2005)
70. Tsunematsu, M., et al.: Effect of Cry-consensus peptide, a novel recombinant peptide for immunotherapy of Japanese cedar pollinosis, on an experimental allergic rhinitis model in B10.S mice. Allegol. Int. **56**, 465–472 (2007)
71. Takagi, H., et al.: A rice-based edible vaccine expressing multiple epitopes induces oral tolerance for inhibition of Th2-mediated IgE response. Proc. Natl. Acad. Sci. U. S. A. **102**, 17525–17530 (2005)
72. Takagi, H., et al.: Oral immunotherapy against a pollen allergy using a seed-based peptide vaccine. Plant Biotech. J. **3**, 521–533 (2005)
73. Domon, E., et al.: 26-week oral safety study in macaques for transgenic rice containing major human T-cell epitope peptides from Japanese cedar pollen allergens. Agric. Food Chem. **57**, 5633–5638 (2009)
74. Wakasa, Y., et al.: Oral immunotherapy with transgenic rice seed containing destructed Japanese cedar pollen allergens, Cry j 1 and Cry j 2, against Japanese cedar pollinosis. Plant Biotech. J. **11**, 66–76 (2013)
75. Yang, L., et al.: Generation of a transgenic rice seed-based edible vaccine against house dust mite allergy. Biochem. Biophys. Res. Commun. **365**, 334–338 (2008)
76. Suzuki, K., et al.: Prevention of allergic asthma by vaccination with transgenic rice seed expressing mite allergen: induction of allergen-specific oral tolerance without bystander suppression. Plant Biotech. J. **9**, 982–990 (2011)
77. Wilken, L.R., Nikolow, Z.L.: Recovery and purification of plant-made recombinant proteins. Biotechnol. Adv. **30**, 419–433 (2012)
78. Fisher, R., et al.: GMP issues for recombinant plant-derived pharmaceutical proteins. Biotechnol. Adv. **30**, 434–439 (2012)
79. Spok, A., et al.: Evolution of a regulatory framework for pharmaceuticals derived from genetically modified plants. Trends Biotechnol. **26**, 506–517 (2008)

# Part VI
# Adjuvants and Nanotechnology

## Overview of Part VI

**Laser**
- Laser vaccine adjuvant prior to vaccination
- Laser enhances migration of dermal APCs
- Patch vaccine delivery

**Bacterial adjuvants**
- From exo- and endotoxins
- LPS-based adjuvants and lipid A analogons
- Bacterial toxins for antigen delivery

**Heat shock proteins**
- Plants as biofactories
- Self-adjuvanted antigens
- Plant HSPs and their immune properties

**Nanoparticles**
- Liposome-based vaccines
- Nanotechnology for pulmonary vaccine delivery
- Nanoparticles as oral adjuvants

**Alum and chitosan**
- Formulation of aluminum-containing adjuvants
- Molecular pathways and working mechanism
- Chitosan, a carbohydrate polymer as adjuvant

**Polymers**
- From colloids to dendrimers
- Favorable properties of nanoparticles
- Controlled release strategies

A complete vaccine development does not stop at the antigen. Most antigens are poorly immunogenic and needs the support of strong adjuvants. This counts for recombinant vaccines based on protein antigens but also for DNA vaccines. Neither pure protein nor naked DNA does really work. The essential pro-inflammatory response is triggered by adjuvants, activator and enhancer of the innate immune response. Adjuvant and antigen must form a perfect mix. Until now only two adjuvants are globally licensed for human vaccines. Unfortunately, research on vaccine adjuvants has so far received little attention. More efforts are necessary to develop innovative nontoxic adjuvants, especially for mucosal vaccine delivery.

*Laser adjuvantation.* Laser vaccine adjuvant (LVA) is based on a brief (2 min) laser illumination of the injection site prior to ID vaccination, which primes our body for a better response to the vaccine.

*LPS adjuvants.* The generation of new detoxified LPS species with higher adjuvant characteristics than alum and acceptable toxicity has opened new perspectives to the vaccination.

*Bacterial toxins.* Endotoxins as well as exotoxins have been exploited as adjuvants, mainly mutated toxins with ablated or reduced enzymatic activity or modified chemical structure to reduce their cytotoxic effects.

*Plant heat shock proteins.* The data evidence a general ability of plant HSPs of stimulating immune responses and shed light also on a more specific use of HSP-polypeptide complexes derived from plant tissues expressing recombinant antigens.

*Functionalized nanoliposomes.* It has become clear that the physicochemical properties of liposomal vaccines – method of antigen attachment, lipid composition, bilayer fluidity, particle charge, and other properties – exert strong effects on the resulting immune response.

*Emerging nanotechnology.* Development of dry powder-based vaccines can potentially reduce or may eliminate cold-chain requirements, promote sterility, and increase the overall stability of antigens and thus reduce the overall cost of the product.

*Oral adjuvants.* Concerning oral vaccination, a big challenge for the immune system is to reach the right equilibrium between tolerance and inflammatory response at mucosal level. The intestine is the largest lymphoid organ in mammals and contains more immune cells and the largest concentration of antibodies than any other organ including the spleen and liver.

*Chitosan.* In the search for new types of adjuvants, the carbohydrate polymer chitosan has gained increasing interest, due to its demonstrated immunostimulatory effect. Chitosan derives from the natural product chitin.

*Aluminum-containing formulas.* Currently, alum is still the only licensed vaccine adjuvant in the USA. In Europe, however, since the 1990s of the last century, MF59, an oil-in-water emulsion, MPL (monophospholipid A, an LPS analog) + alum, and AS03 are also approved for use.

**New polymers.** Nano-sized systems can be inorganic colloids, organic colloids (synthesized by emulsion polymerization or mini/nano-emulsion techniques), polymeric aggregates (micelles or polymersomes), core cross-linked aggregates (nanohydrogels, cross-linked micelles, or polyplexes), multifunctional polymer coils, dendritic polymers, or perfect dendrimers.

# Laser for Skin Vaccine Delivery and Adjuvantation

## 32

Xinyuan Chen and Mei X. Wu

## Contents

| | | |
|---|---|---|
| 32.1 | **Background** | 520 |
| 32.2 | **Laser-Mediated Adjuvant and Delivery of Vaccines** | 520 |
| 32.2.1 | Laser Vaccine Adjuvant | 520 |
| 32.2.2 | Laser-Facilitated Patch Vaccine Delivery | 521 |
| 32.3 | **Preclinical Development and Efficacy** | 523 |
| 32.4 | **Strengths and Weakness** | 524 |
| **References** | | 525 |

X. Chen, PhD • M.X. Wu, MD, PhD (✉)
Wellman Center for Photomedicine,
Massachusetts General Hospital (MGH),
Boston, MA, USA

Department of Dermatology,
Harvard Medical School (HMS),
Boston, MA, USA

Harvard-MIT Division of Health Sciences and
Technology (HST), Cambridge, MA, USA
e-mail: mwu2@partners.org

## Abstract

Skin has received considerable attention for vaccination in the past decade owing to its rich in antigen-presenting cells (APCs) and better immune responses as compared to the muscle. Vaccination via the skin instead of the muscle may also make it possible to develop safer adjuvants and needle-free delivery systems. In the past 5 years we have developed a novel laser-based vaccine adjuvant (LVA) in which the inoculation site is briefly illuminated by a safe laser that enhances the motility and antigen-uptake of APCs. The brief illumination also facilitates emigration of APCs from the skin to draining lymph nodes. While greatly enhancing vaccination efficacies, LVA exhibits minimal local side effects and has no long-term side effects since no foreign or self materials are administered into the body apart from antigen itself. This novel strategy primes the inoculation site for better immune responses, in contrast to conventional vaccine adjuvants that all augment the immunogenicity by altering a form of the antigens. As such, LVA doesn't require any formation with the antigen and can potentially act as a universal vaccine adjuvant. Indeed, this physical type LVA can sufficiently boost immune responses induced by ovalbumin (OVA), influenza vaccine, nicotine vaccine, and so on. Besides LVA, we have also explored a laser-facilitated patch delivery system for needle-free, painless transcutaneous immunization. The system is based on ablative fractional laser (AFL) treatment to generate an array of self-renewable microchannels (MCs) in the skin, through which topically applied vaccines can enter the skin readily. The laser treatment also results in active recruitment of APCs to the vicinity of each MC where APCs can more effectively take up antigens around or within the MCs, leading to ~100 times higher immune response against OVA than tape stripping-based patch delivery. In this chapter, we introduce these technologies and their preclinical studies and discuss their strengths and weaknesses.

## 32.1 Background

Most vaccines are administered intramuscularly to date, because of relative convenience and readily management, but the muscular tissue contains few antigen-presenting cells (APCs) and it is not an effective target for vaccination. In contrast, skin is rich in resident APCs and in lymphatic vessel networks. Accumulating evidence suggests that delivery of vaccines directly into the skin induces more potent immune responses than delivery of the vaccine into the muscle [1–4]. The effectiveness of cutaneous immunization is best highlighted by the tremendous success of smallpox vaccination that was delivered via scarification about two centuries ago, eventually eradicating this highly contagious disease in 1979 [5]. Scarification is no longer used in modern era since it damages the skin, produces scars, and poses risks of infection [6, 7]. Mantoux method is an intradermal (ID) injection technique used in today's skin test of tuberculosis, but it is not for routine immunizations due to the need of special training personnel and the technical difficulty for reliable skin target [7]. To improve patient compliance and vaccine immunogenicity, we are developing a laser-facilitated patch delivery system for convenient, needle-free, painless skin immunization.

Apart from reliable skin delivery of vaccines, incorporation of adjuvants to further enhance skin immunization faces additional challenges owing to a small injection volume allowed for ID injection and a high sensitivity of the skin to inflammation [2–4]. Many chemicals and biomolecules are able to enhance vaccine-induced immune responses, but they also induce a high level of inflammation in the skin and thus are not suitable for cutaneous immunization [8]. Here, we introduce a novel, safe approach to enhance immune responses against vaccines skin delivered.

## 32.2 Laser-Mediated Adjuvant and Delivery of Vaccines

### 32.2.1 Laser Vaccine Adjuvant

Laser vaccine adjuvant (LVA) is based on a brief (2 min) laser illumination of the injection site prior to ID vaccination, which primes our body for a better response to the vaccine [9]. The illumination is conducted by a Q-switched 532-nm Nd:YAG laser with a pulse width 5–7 ns, a beam diameter 7 mm, and a frequency 10 Hz (Spectra-Physics Inc., Mountain View, CA) [9]. This noninvasive laser illumination causes little microscopic or visible skin damage, while increasing the motility and function of local skin APCs [9]. When injected into the site of laser exposure, the antigen can be sufficiently taken up by the highly mobile APCs and transported from the skin to the draining lymph nodes, augmenting immune responses [9].

In comparison with traditional vaccine adjuvants, LVA is safe and easy to use for boosting cutaneous vaccination [8]. As can be seen in Fig. 32.1, no inflammation, skin damage, or skin redness occurred visibly and histologically in laser-treated skin [8]. In contrast, aluminum salt-based (alum) adjuvants, the most widely used adjuvant for intramuscular vaccination, remained in the skin for months and provoked strong and persistent skin reaction following ID inoculation (upper panel), with a great number of inflammatory cells infiltrated into the adjuvant-treated site (low panel) [8]. Likewise, adjuvants such as the water-in-oil emulsion montanide ISA 720 and the toll-like receptor-7 (TLR7) agonist imiquimod (R837) each induced severe local skin irritation, concurrent with skin ulceration and inflammation for weeks [8]. The severe and persistent skin irritation was also observed with a combination of the TLR4 agonist monophosphoryl lipid A

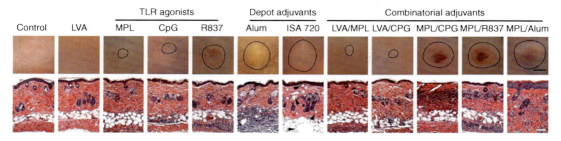

**Fig. 32.1** Reactogenicity of various adjuvants following ID injection. BALB/c mice were illuminated with laser or ID injected with indicated adjuvants in a volume of 20 μl each with or without laser illumination. Local skin reactions were analyzed 5 days later. *Upper panel*, skin photos (scale bar, 2 mm) and lower panel, histological examination (scale bar, 100 μm). The lesion size is outlined by a circle in *the upper panel* and *arrows* in the lower panel of montanide ISA 720-injected skin point to void bulbs preoccupied by the water-in-oil adjuvant

(MPL) with the TLR9 agonist unmethylated CpG oligonucleotides (CpG), R837, or alum [8]. These adjuvants, in particular, emulsive adjuvants, are precluded from clinic use for cutaneous vaccination as persistent skin inflammation can potentially breach the integrity of the skin, rendering it susceptible to various infections [8]. Among the adjuvants tested, soluble MPL or CpG gave rise to only mild local reaction that was completely resolved within 2 weeks and could thus be used for skin immunization [8]. Interestingly, a combination of LVA with MPL or CpG did not exaggerate local reactions, allowing us to explore the synergistic effects of LVA and MPL or CpG on skin immunization.

The mechanism underlying the laser-mediated immune enhancement is not completely understood. Laser illumination appears to significantly augment the motility of APCs: their pseudopods moved much faster and their distribution was more dispersed in laser-treated skin than in control skin [9]. Apart from increased motility of skin APCs, transmission electron microscopy revealed that dermal collagen fibers were disorganized and the interaction between dermal cells and the surrounding tissue scaffolds was disrupted after laser illumination, which could presumably reduce tissue resistance to APC migration [8, 10]. This loose skin connective tissue contrasts the untreated dermal tissue that is filled with dense and well-aligned collagen fibers in close association with dermal cells [8, 10]. Increased motility of APCs and diminished tissue resistance both facilitate trafficking of APCs from the skin to the draining lymph nodes, as reflected by formation of many dermal cords right after laser treatment, owing to migration of large amounts of APCs into the lymphatic vessels (Fig. 32.2) [10]. No dermal cords were seen in nontreated skin. Laser illumination also enhanced antigen uptake in both skin and draining lymph nodes by APCs as a result of an increase in their motility [9]. While increasing the motility of APCs, laser illumination did not induce inflammation or significantly increase inflammatory cytokine production [8, 9]. It also had no effects on the expression of CD86, CD80, and CD40 on local dermal dendritic cells (DCs) [8–10]. As such, LVA is defined as a noninflammatory vaccine adjuvant. In brief, laser treatment of the skin with a specific setting disorganizes the dense dermal connective tissue and increases the motility of skin APCs and antigen uptake, boosting vaccine-induced immune responses.

### 32.2.2 Laser-Facilitated Patch Vaccine Delivery

Superficial stratum corneum (SC) layer comprises the major barrier for efficient transcutaneous delivery of vaccines [11]. Various physical (e.g., thermal, ultrasound), mechanical (e.g., tape stripping, microneedles), and chemical methods (e.g., dimethyl sulfoxide, alcohol) have been explored to disrupt SC to facilitate transcutaneous vaccine delivery [12]. Yet, these methods are either inefficient or have a high risk of evoking skin irritation or other local or systemic reactions [12].

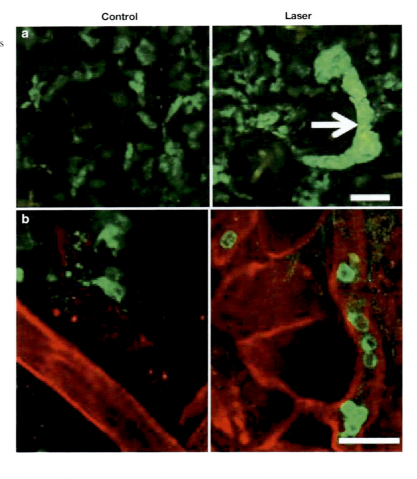

**Fig. 32.2** Laser enhances migration of dermal APCs. (**a**) Formation of dermal cords (*arrow*) within 1 h after cutaneous laser illumination of MHC II-EGFP transgenic mice. Scale bar, 50 μm. (**b**) Representative whole-mount ear images showing that GFP + cells enter lymphatic vessels of laser-illuminated ear within 30 min. *Red* collagen IV, *green* MHC II + cells. Scale bar, 25 μm

Laser-mediated ablation of the SC was also evaluated for its potential in facilitation of transcutaneous vaccine delivery, but traditional full surface ablation requires a relatively long period (weeks) of recovery time and is not safe for vaccine delivery [13]. In the past few years, ablative fractional laser (AFL) has been used to replace traditional skin resurfacing with a significantly shortened downtime [14]. AFL emits a microlaser beam and generates an array of self-renewable microchannels in the skin surface with the diameter and depth about 100–300 μm in each microchannel depending on the energy of the micro-laser beam [12]. These microchannels prove to be transmissible to vaccines topically applied to [12]. AFL treatment followed by topical application of vaccine-coated gauze patch can deliver vaccines directly into the skin through these microchannels, from which the vaccine is further dissipated to the surrounding tissue in cylindrical dimensions and taken up by APCs recruited around the microchannels [12].

It was found that topical application of ovalbumin (OVA)-coated gauze patch to AFL-treated skin increased immune responses by 100-fold compared to the same OVA-coated gauze patch applied to tape-stripped skin [12]. Moreover, laser-treated skin gained quick reepithelialization within 24 h, whereas it took at least 2 days for tape-stripped skin to recover [12]. Quick reepithelialization ensures the integrity of the skin affected and the safety of this needle-free technology for transcutaneous vaccine delivery.

To directly visualize vaccine delivery and APC recruitment by AFL, fluorescently labeled OVA was topically applied to AFL-treated skin in mice engineered to express GFP in-frame infused with MHC-II molecule on most APCs.

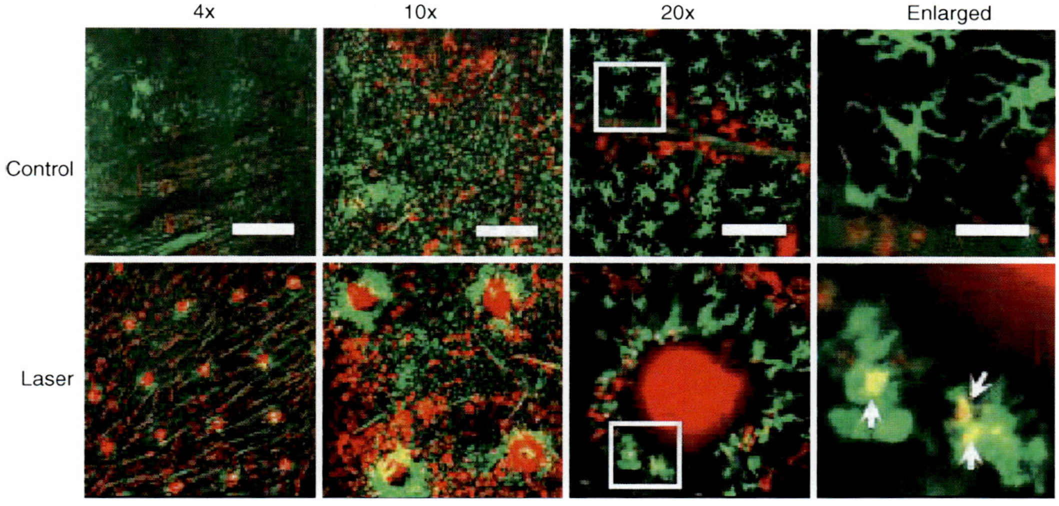

**Fig. 32.3** Efficient antigen delivery and recruitment of APCs. Ears of MHC II-EGFP mice were treated with AFL or left untreated followed by topical application of Alexa Fluor 647 (AF647)-conjugated OVA-coated gauze patch for 30 min. Epidermal layer was subjected to intravital confocal imaging 24 h later. Representative images of a low (*1st panel*, scale: 750 μm), medium (*2nd panel*, scale: 300 μm), and high magnification (*3rd panel*, scale: 75 μm) are shown. Antigen uptake by individual APCs (*arrow*) is shown in *4th panel*, scale: 25 μm, which is highlighted in the rectangle in the *3rd panel*. No antigen uptake occurred in untreated control ear. *Green* GFP-labeled APCs, *red* AF647-OVA, and *yellow* AF647-OVA uptake by APCs

AFL-generated microchannels were found not only to provide free paths for OVA entry into the skin (Fig. 32.3, red spots, laser) but also to attract large amounts of GFP-labeled APCs to the vicinity of microchannels (Fig. 32.3, 1st and 2nd panels), permitting direct and efficient antigen uptake occurring around each microchannel (Fig. 32.3, 3rd and 4th panels). Enhanced antigen uptake by recruited APCs functions as "built-in" adjuvant and is expected to enhance vaccine-induced immune responses. Unlike laser used for LVA above, which only stresses but not kills skin cells, AFL introduces photothermal effects, causing irreversible tissue damages to form microchannels and microthermal zones around each microchannel. Material release from apoptotic or necrotic cells in the microthermal zones is likely to send "danger signals" to the immune system and induce secretion of chemokines by local macrophages or infiltrated neutrophils [15], which attract a great deal of APCs. APC recruitment is a unique feature of AFL technology and cannot be recapitulated by other SC-disruption methods, to the best of our knowledge.

## 32.3 Preclinical Development and Efficacy

LVA can potentially enhance immune responses against any existing or new, protein-based vaccines, as laser illumination enhances general function of APCs. We have attained similar vaccine adjuvanicity by this technology with a model antigen OVA [9], nicotine vaccines, seasonal and pandemic influenza vaccines, or hepatitis B surface antigen (HBsAg). Anti-nicotine immunotherapy emerges as a novel and promising therapy to treat nicotine addiction by inducing anti-nicotine antibodies (NicAb) to bind and block entry of nicotine into the brain [16–18]. Yet, only 30 % smokers developed a relatively high NicAb titer, and it was among this subgroup who had a significantly increased abstinence rate as compared to placebo after 5–7 intramuscular (IM) immunizations in the presence of alum adjuvant [16, 17]. There was no significant increase in the abstinence rate if serum NicAb titer was in a low or median level [16, 17]. The positive correlation between a NicAb titer and an abstinence rate

**Table 32.1** LVA/ID immunization is superior to IM immunization with alum adjuvant

|          | Immunization times | Peak NicAb titer[a] ($\times 10^3$) | Percent increase | $p$ value |
|----------|--------------------|---------------------------------------|------------------|-----------|
| IM       | 7                  | 10.1 ± 2.7                            | –                | –         |
| Alum + IM | 4                 | 20.3 ± 5.0                            | 100              | **0.0352** |
| ID       | 3                  | 15.7 ± 3.1                            | 60               | 0.283     |
| LVA + ID | 3                  | 38.2 ± 5.1                            | 280              | **0.0028** |

Values of NicAb titer are expressed as means ± SEM (standard error of the mean), and student $t$-test is used to calculate the $p$ value in comparison with 7-IM vaccination regimen, $p$ value less than 0.05 is regarded in bold
[a]Nicotine antibody titer peaked generally 2 weeks after the last immunization

raises the necessity of improving nicotine vaccine immunogenicity, which is furthered by the recent failure of a phase III clinical trial of NicVAX, a nicotine vaccine conjugated to recombinant exoprotein A [19]. We thus evaluated whether a combination of ID immunization with LVA could significantly enhance nicotine vaccine immunogenicity in mice.

Nicotine conjugated to keyhole limpet hemocyanin (Nic-KLH) was used to evaluate laser adjuvant effects. Mice were immunized every 2 weeks for various times as shown in Table 32.1. We found three-ID immunization induced a 60 % higher NicAb titer than seven-IM immunization and a comparable NicAb titer to four-IM immunization with alum adjuvant at peaking levels (Table 32.1), confirming that ID immunization is superior to IM for nicotine vaccines. Incorporation of LVA into the primary ID immunization further increased NicAb titer by 140 % over ID vaccination or 280 % over IM vaccination (Table 32.1). Moreover, LVA/ID immunization sustained the peak NicAb titer for more than 4.5 months or beyond, while the peak NicAb production dwindled gradually but significantly ($p < 0.05$) after ~3 months with IM or ID immunization in the absence of adjuvant. In comparison with non-immunized controls, seven-IM immunization or three-ID immunization significantly inhibited entry of nicotine into the brain by 37 and 40 %, respectively, after nicotine challenge, while incorporation of LVA in the primary immunization of the three-ID immunization regimen reduced brain nicotine entry by 60 %.

Unlike the rather weak immunogenicity of nicotine vaccines, influenza vaccines display strong immunogenicity, which can be also boosted similarly with LVA. Our investigation showed that a combination of LVA with ID immunization enhanced hemagglutination inhibition (HAI) antibody titer considerably, offering a 5–10-fold dose sparing effect against mouse-adapted PR8 vaccine in adult mice over ID immunization. In addition, a combination of LVA with ID immunization but not ID immunization alone significantly increased HAI antibody titer and the survival in old mice following lethal viral challenge. These preclinical studies demonstrate the ability of LVA to boost immune responses against vaccines with either a weak or a strong immunogenicity in both adult and old mice.

## 32.4 Strengths and Weakness

Vaccine adjuvanicity of this noninflammatory laser treatment is relatively weak compared to the majority of the chemical or biological adjuvants. Yet, LVA has many advantages over traditional adjuvants. First, LVA doesn't induce local reactions or long-term side effects because no foreign materials are injected into the body besides antigen itself. On the contrary, with few exceptions, almost all traditional adjuvants are foreign and can potentially cause long-term side effects by molecular mimicry inducing self-destructive immune cross-reactions. Secondly, LVA doesn't need to modify the vaccine manufacture procedure or to develop specific vaccine/adjuvant formulations and can be used immediately and repeatedly at any time. This confers a great advantage over traditional adjuvants when we encounter vaccine shortages in an event of influenza pandemic, an outbreak of a new viral strain, or a bioterrorist attack. Thirdly, LVA can be conveniently combined with newly developed ID or cutaneous immunization strategies to further enhance vaccine-induced immune responses.

LVA can be also approved as a stand-alone adjuvant for various clinical vaccines.

To deliver vaccines into the skin, several strategies have been developed in the past decade, including jet injectors, microneedles, and microprojective array patches [20–23]. AFL has the advantages over these strategies. Firstly, it is needle-free and painless, and it doesn't induce any wheals or skin shape change, and thus more patient-compliant. Secondly, AFL guides vaccines delivered directly to individual microchannels, minimizing local reaction while increasing the efficiency of vaccine delivery as compared to delivery of vaccines to a single big spot. Thirdly, AFL not only provides sufficient vaccine delivery into the skin but also potentially enhances vaccine-induced immune response owing to the "built-in" adjuvanicity as shown by APC recruitment occurring around each microchannel and presumably release of danger signals from microthermal zones to the immune system. A small, handheld apparatus that integrates AFL and microchannel-guided vaccine delivery is being fabricated, which shall accelerate the broad application of this novel technology for vaccine delivery and adjuvantation.

## References

1. Barraclough, K.A., et al.: Intradermal versus intramuscular hepatitis B vaccination in hemodialysis patients: a prospective open-label randomized controlled trial in nonresponders to primary vaccination. Am. J. Kidney Dis. **54**, 95–103 (2009)
2. Belshe, R.B., et al.: Serum antibody responses after intradermal vaccination against influenza. N. Engl. J. Med. **351**, 2286–2294 (2004)
3. Beran, J., et al.: Intradermal influenza vaccination of healthy adults using a new microinjection system: a 3-year randomised controlled safety and immunogenicity trial. BMC Med. **7**, 13 (2009)
4. Kenney, R.T., Frech, S.A., Muenz, L.R., Villar, C.P., Glenn, G.M.: Dose sparing with intradermal injection of influenza vaccine. N. Engl. J. Med. **351**, 2295–2301 (2004)
5. Lofquist, J.M., Weimert, N.A., Hayney, M.S.: Smallpox: a review of clinical disease and vaccination. Am. J. Health Syst. Pharm. **60**, 749–756 (2003)
6. Kim, Y.C., Jarrahian, C., Zehrung, D., Mitragotri, S., Prausnitz, M.R.: Delivery systems for intradermal vaccination. Curr. Top. Microbiol. Immunol. **351**, 77–112 (2012)
7. Liu, L., et al.: Epidermal injury and infection during poxvirus immunization is crucial for the generation of highly protective T cell-mediated immunity. Nat. Med. **16**, 224–227 (2010)
8. Chen, X., Wu, M.X.: Laser vaccine adjuvant for cutaneous immunization. Expert Rev. Vaccines **10**, 1397–1403 (2011)
9. Chen, X., et al.: A novel laser vaccine adjuvant increases the motility of antigen presenting cells. PLoS One **5**, e13776 (2010)
10. Chen, X., Zeng, Q., Wu, M.X.: Improved efficacy of dendritic cell-based immunotherapy by cutaneous laser illumination. Clin. Cancer Res. **18**, 2240–2249 (2012)
11. Li, N., Peng, L.H., Chen, X., Nakagawa, S., Gao, J.Q.: Transcutaneous vaccines: novel advances in technology and delivery for overcoming the barriers. Vaccine **29**, 6179–6190 (2011)
12. Chen, X., et al.: Facilitation of transcutaneous drug delivery and vaccine immunization by a safe laser technology. J. Control. Release **159**(1), 43–51 (2012)
13. Lee, W.R., et al.: Erbium:YAG laser enhances transdermal peptide delivery and skin vaccination. J. Control. Release **128**, 200–208 (2008)
14. Manstein, D., Herron, G.S., Sink, R.K., Tanner, H., Anderson, R.R.: Fractional photothermolysis: a new concept for cutaneous remodeling using microscopic patterns of thermal injury. Lasers Surg. Med. **34**, 426–438 (2004)
15. Chen, G.Y., Nunez, G.: Sterile inflammation: sensing and reacting to damage. Nat. Rev. Immunol. **10**, 826–837 (2010)
16. Cornuz, J., et al.: A vaccine against nicotine for smoking cessation: a randomized controlled trial. PLoS One **3**, e2547 (2008)
17. Hatsukami, D.K., et al.: Immunogenicity and smoking-cessation outcomes for a novel nicotine immunotherapeutic. Clin. Pharmacol. Ther. **89**, 392–399 (2011)
18. Pentel, P.R., et al.: A nicotine conjugate vaccine reduces nicotine distribution to brain and attenuates its behavioral and cardiovascular effects in rats. Pharmacol. Biochem. Behav. **65**, 191–198 (2000)
19. Fahim, R.E., Kessler, P.D., Fuller, S.A., Kalnik, M.W.: Nicotine vaccines. CNS Neurol. Disord. Drug Targets **10**, 905–915 (2011)
20. Prausnitz, M.R., Mikszta, J.A., Cormier, M., Andrianov, A.K.: Microneedle-based vaccines. Curr. Top. Microbiol. Immunol. **333**, 369–393 (2009)
21. Matriano, J.A., et al.: Macroflux microprojection array patch technology: a new and efficient approach for intracutaneous immunization. Pharm. Res. **19**, 63–70 (2002)
22. Taberner, A.J., Ball, N.B., Hogan, N.C., Hunter, I.W.: A portable needle-free jet injector based on a custom high power-density voice-coil actuator. Conf. Proc. IEEE Eng. Med. Biol. Soc. **1**, 5001–5004 (2006)
23. Hingson, R.A., Davis, H.S., Bloomfield, R.A., Brailey, R.F.: Mass inoculation of the Salk polio vaccine with the multiple dose jet injector. GP **15**, 94–96 (1957)

# Bacterial Lipopolysaccharide as Adjuvants

## 33

Jesús Arenas

## Contents

33.1 LPS Structure and Biological Activity ..... 528
33.2 Lipid A Analogous Structures and Its Role as Adjuvants ..... 529
Conclusions ..... 533
References ..... 533

## Abstract

Lipopolysaccharide (LPS, endotoxin) is a major component of the outer membrane of gram negative bacteria. It is an activator of humoral and cellular responses in humans with potential use as adjuvant in vaccine technology. Importantly, LPS has a large capacity to induce Th1-type responses and stimulate cytotoxic T lymphocytes, which are poorly obtained by standard adjuvants but required for specific immune stimulatory therapies. In contrast, LPS possess an extreme toxicity that limit its clinical use in humans. Alteration of its chemical structure led the generation of LPS-based derivatives with reduced toxicity but retaining adjuvant properties. Monophosphoryl lipid A (MPLA) has been the most successful LPS-based adjuvant, currently incorporated in approved vaccine preparations and extensively used in vaccine trials and preclinical studies. Novel designed structures, analogous to LPS and generated by chemical synthesis, can offer lower production cost and lesser heterogenic formulations than MPLA and, in addition, be even most suitable for specific immune therapies. Thus, LPS-based structures are valuable contributions as adjuvants in human vaccinology and open new possibilities to existing demands for specific therapies.

J. Arenas, PhD
Section Molecular Microbiology,
Department of Biology,
Utrecht University, Padualaan 8, 3584CH,
Utrecht, The Netherlands
e-mail: jesusarenasbust@yahoo.es, j.a.arenasbusto@uu.nl

The development of subunit vaccines has improved the safety of human vaccine prophylaxis but requires strong adjuvants. Alum (diverse aluminum salts) is the most used adjuvant with an acceptable profile of side effects and induction of optimal protection against many human pathogens. However, alum is not suitable against certain pathogens or for new vaccine therapies like cancer or allergy. Lipopolysaccharide is a component of the outer membrane of gram-negative bacteria largely studied as adjuvants by their inherent ability to stimulate immune responses. In contrast, LPS induces inacceptable toxic effects in humans.

The generation of new detoxified LPS species with higher adjuvant characteristics than alum and acceptable toxicity has opened new perspectives to the vaccination. In this book chapter, the basic aspects of LPS structure, toxicity, and activation of the immune system are first introduced. Next, adjuvant characteristics of alum and corresponding drawbacks are briefly cited. Followed, the most promising detoxified LPS molecules and adjuvant characteristics are further discussed with special emphasis in current advances. A final overview section summarized the most relevant points.

## 33.1 LPS Structure and Biological Activity

Lipopolysaccharide is a component of the external leaflet of the outer membrane of gram-negative bacteria. It is a complex glycolipid formed by three domains, a fatty acid-rich domain (lipid A), an oligosaccharide domain (core), and a repeating oligosaccharide domain (O-antigen) [1]. Figure 33.1 represents the typical LPS organization.

The lipid A domain is a β-1, 6-linked D-glucosamine disaccharide linked to variable number of ester- and amide-linked 3-hydroxy fatty acids and phosphate groups (Fig. 33.1). Its architecture is highly conserved, but different microorganisms may present variations in the number and length of the fatty acid side chains, the presence of terminal phosphate residues, and associated

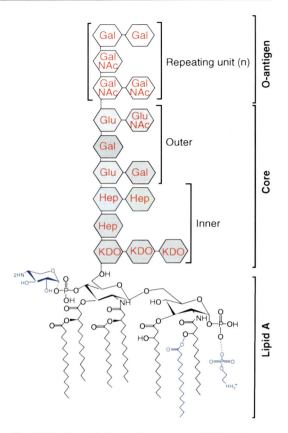

**Fig. 33.1** General chemical structure of LPS of gram-negative bacteria. The LPS structure of *Escherichia coli* is depicted. LPS consists of three regions: lipid A, core, and O-antigen. The chemical structure of the lipid A of *E. coli* that displays the maximal immunostimulatory or endotoxic activities in humans (the topic of this work) is further detailed (*black colored*). Additionally, those substituents that possess the lipid A of *Salmonella minnesota* used to obtain MPL are blue colored. Residues of core and O-antigen region are schematized and abbreviated as *KDO* 2-keto-3-deoxyoctonoic acid, *Hep* D-GLYCERO-D-manno-heptose, *Glu* D-glucose, *Gal* D-galactose, *GluNAc* N-acetyl-glucosamine, and *GalNAc* N-acetyl-galactosamine. The length of the O-antigen depends on the number of repeating units. Additional substituents or modifications can be found in nature, but they are not shown for clarity

modifications. The core domain is a branched oligosaccharide region formed by nine or ten sugars, and its composition is more variable between species than lipid A. Finally, the O-antigen, if present, is the most variable of the tree domains and consists of up to 50 repeating oligosaccharide units formed of 2–8 monosaccharide moieties.

**Table 33.1** Relevant effects of LPS in humans

| Beneficial | | Harmful | |
|---|---|---|---|
| Effect | Mechanism | Effect | Mechanism |
| Antimicrobial immunity | Activation of DC, priming B and T cells, opsonization, stimulation of natural killer cells, activation of macrophages | High fever | IL-1 release |
| Antiviral immunity | Stimulating CD8(+) T-cell immunity | Strong inflammatory responses | Large secretion of various inflammatory mediators |
| Antitumor immunity | Stimulating CD8(+) T-cell immunity | Disseminated intravascular coagulation | Reduction of elements involved in blood coagulation |
| Mitigation of Th2 response to Th1 | Enhancing Th1 phenotype | Shock, hypotension, lymphopenia | Reduced blood flow |

In addition, certain modifying enzymes can alter the composition of LPS contributing to increase the LPS heterogeneity [2–5].

Toll-like receptors (TLR)s belong to a family of receptors that recognize a broad diversity of specific but conserved structures of pathogen microorganisms [6]. Immediately after stimulation, TLRs initiated the activation of immune defense mechanisms. Toll-like receptor 4 (TLR4), in complex with the glycoprotein MD-2, constitutes the LPS receptor [7, 8]. TLR4 is a membrane-spanning protein present on antigen-presenting cells (APC) (macrophages and dendritic cells) and epithelial cells of humans. Its stimulation requires the cooperation of associated molecules, like the LPS-binding protein and CD14 that facilitate LPS transfer to the receptor. TLR4 stimulation induces the formation of intracellular protein complexes that leads to the activation of intracellular signaling cascades [9, 10]. These reactions trigger the biosynthesis and secretion of diverse proinflammatory cytokines (IL-1, IL-8, IL-12, TNFα, and IFNγ) and the production of co-stimulatory molecules [11] that finally activates humoral and cellular responses including activation of the complement system [12, 13], activation of macrophages [14, 15] B and T cells, and enhancement of cellular cooperation [11]. Consequently, this response is beneficial for the control of local infections. In contrast, high LPS dose, specially released to the blood system during sepsis, leads to large secretion of cytokines and inflammatory mediators with severe [16, 17] and/or fatal consequences [18, 19]. Table 33.1 summarizes the beneficial and harmful effects of LPS in humans, but for further details see revision [20, 21]. In summary, LPS is a strong activator of the immune system (adjuvant) but also a highly toxic substance (endotoxin).

The lipid A region is the major responsible of the TLR4 stimulation. Variations in its structure, mainly regarding the number and length of acyl acid chains, and the charge are crucial in this regard [22]. The hexa-acylated *E. coli* lipid A (canonical LPS structure and depicted in Fig. 33.1) with fatty acids of 12–14 carbons and two phosphate residues is the maximal stimulator of human TLR4 (hTLR4) [23, 24]. In contrast, the tetra-acylated lipid IVa with fatty acids of 18–16 carbons and a phosphate residue, an intermediate in the biosynthetic pathway of lipid A, does not stimulate hTLR4 (canonical hTLR4 antagonist) [25].

## 33.2 Lipid A Analogous Structures and Its Role as Adjuvants

Vaccine based on infectious attenuated or inactivated whole pathogens contains a large variety of target structures for TLRs and, subsequently, promotes strong and long protection. However, they generate a large variety of side effects even with fatal consequences [26, 27]. Vaccines based on one or certain purified components (subunit vaccines) show acceptable safely but a poor immunogenicity and require additional immune stimulators (adjuvants). Alum refers to several aluminum salts and is the most used adjuvant. It is safe and elicits predominantly a Th2-type antibody response that shows to be effective in a large variety of vaccines [28]. However, alum hardly promotes Th1-type antibody responses [29]. Adjuvants that favor Th1 or more balanced Th1/Th2 responses are required to induce optimal immune protection against certain pathogens [30] or diseases as cancer [31] or allergy [32]. Apart from that, alum poorly stimulates mucosal immunity. Mucosa tissues are the first line of defense against many pathogens and the ecological niche of commensal and opportunistic microorganisms, for example, *Neisseria meningitidis*.

Therefore, mucosal immunity is considered the gold therapy to evade pathogen colonization and confer herd immunity against certain particular pathogens. Vaccine adjuvants that target mucosal immunization must promote a large series of biological and complex activities as Th17 cell development, APC proliferation, and IgA production [33, 34]. In this regard, several substances have been extensively studied as bacterial toxins or CpG, among others [34], but, till date, no available approved adjuvant exists (with the exception of MPL, to be discussed next). Alternatively to alum, three additional adjuvants were licensed: MF59, AS03, and RC-529. MF59 is an oil in water emulsion with low oil content, and it is included in an approved influenza vaccine [35]. Although MF59 induces a more balanced Th1/Th2 response than alum, it shows partial efficacy, often requiring the coadministration of Th1 enhancers. ASO3 and RC-529 contain LPS-based substances to be discussed later.

LPS has attracted large attention as adjuvant by its high capacity to induce Th1-type responses against coadministrated antigens. One of the most relevant factors involved in the development of this response is IL-12. Note that LPS is a stimulator of this interleukin. Interestingly, TLR4 receptors are tactically present at mucosa surfaces; therefore, it would be expected that TLR4 agonists can promote immune responses at local and distal mucosal sites. In the past decades, several strategies were followed to reduce its extreme toxicity without altering this inherent capacity; variation of the LPS composition, in particular lipid A, by chemical treatments and chemical synthesis of lipid A analogues is a good example (see detailed revision [36]). As a result, a diverse lipid A species was generated, although only few exhibited the desirable properties. Next, the most relevant substances and clinical applications are further discussed.

The chemical hydrolysis of the LPS of *S. minnesota*, which contains a lipid A with seven acyl chains and three phosphate groups as depicted in Fig. 33.1 (with blue-colored substituent), generated one of the most successful LPS-based adjuvants, the monophosphoryl lipid A (MPL). This derivate structure is a six-acyl side chain lipid A with one phosphoryl group [37] (see Fig. 33.2). MPL demonstrated to be less toxic than the parent (0.1% of toxicity) [38] with a toxic side effect profile comparable to alum [39, 40] while retaining the stimulatory properties of LPS. At present, MPL is adjuvant of approved vaccine preparations for humans in Europe (human papillomavirus (Cervarix) [33, 34] and pollen allergy (Pollinex Quattro) [41, 42]) and Australia (hepatitis B virus (Fendrix) [43]), and it has been used extensively in human vaccine trials for several infectious diseases like malaria [44], tuberculosis, [45, 46] or tumor growth [47].

Because MPL is highly hydrophobic and generates aggregates in aqueous solution that may considerably affect the TLR4 activation, it is often formulated in combination with alum or other delivery systems [48]. These combinations, together with other factors (accompanying antigen or administration route), can alter its adjuvant action. For example, in aqueous formulations MPL promotes antibody production, while in oil in water emulsions, it better stimulates T-cell responses. In contrast, MPL combined with other delivery systems is a strong stimulator of cytotoxic T lymphocyte proliferation. Delivery systems can also modify their biological properties. Liposomes are spherical vesicles formed by phospholipid bilayers extensively used to deliver antigens in its native conformation. Incorporation of MPL into liposomes reduced the residual

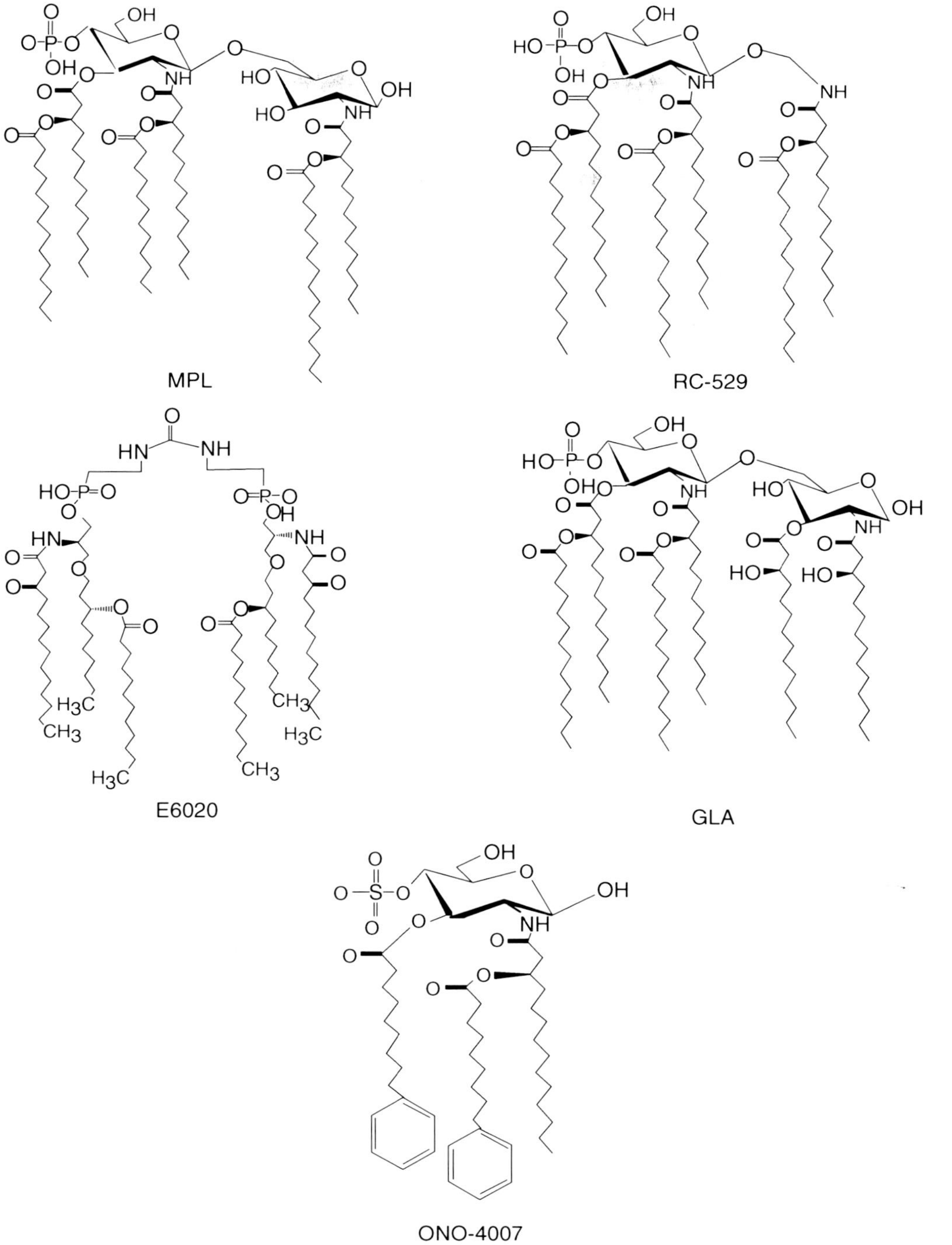

**Fig. 33.2** Chemical structure of relevant LPS-based adjuvants

toxicity of MPL but retained intact its adjuvant potential [49]; this effect was also observed with other detoxified LPS species [50]. Therefore, liposomal MPL formulations were extensively studied in human trials for different indications as malaria [49], pneumococcal disease, [51] or genital herpes type 2 [52] and in experimental animals for *Streptococcus pyogenes* infections [53] or toxin neutralization of *E. coli* [54]. Finally, it is significant to notice MPL's success in providing mucosal immunity after mucosal [55] and intramuscular [56, 57] administration; MPL formulations for mucosal vaccines have been extensively explored for the treatment of different diseases; genital herpes [58] and mucosal leishmaniasis [59] are good examples.

One of the main drawbacks attributed to MPL is the heterogeneity of lipid A congeners generated during its production with subsequent purification cost. A solution to this problem could be the generation of synthetic lipid A analogues. Chemical synthesis generates pure and defined structures that reduce production cost. Like natural LPS species, these structures interact with hTLR4. Therefore, numerous analogues with variation in acyl chain length and positions, phosphate groups, or the backbone unit were generated and their biological activity further analyzed. Till date, the most suitable molecules in vaccine development are RC-529, E6020, GLA, and ONO-4007 (see chemical structure in Fig. 33.2).

RC-529 is a synthetic analogue of MPL composed by a monosaccharide backbone with six fatty acyl chains [38]. It is a very attractive adjuvant. Like MPL, it is well tolerated and effective during clinical trials [38] but with a lower production cost. In fact, its use was approved in Argentina in a hepatitis B vaccine. It is often in combination with different delivery systems to enhance its solubility or improve its delivery without affecting immune stimulatory capacities [60]. In addition, several studies performed in experimental animals indicate that RC-529 is an efficient mucosal adjuvant against pathogens that lack effective vaccine therapy. For example, it elicited bactericidal antibodies after intranasal immunization with the *Streptococcus pneumoniae* protein PppA [61] and the meningogoccal protein P2086, [60] and it promoted high antibody titers in macaques at the nasal and genital mucosa against an HIV peptide immunogen [62]. Similarly, it reduced nasal colonization of nontypable *Haemophilus influenzae* and *Moraxella catarrhalis* in mice that were immunized via nasal with recombinant proteins, [63] and it conferred significant protection against lethal influenza challenge [64].

E6020 is an hexa-acylated acyclic backbone [65, 66] with higher biological activities than alum [65, 66] or MPL [65] and with a reduced toxicity [67]. Its simple structure allows the production of high-purity material than other synthetic TLR4 agonists [65]. Various works showed its high capacity to shift immune responses towards a Th1 profile when combined with conventional vaccines [65, 67, 68]. Generation of this immune profile is especially relevant in cancer vaccines. Indeed, E6020 in combination with a monoclonal antibody (trastuzumab) enhanced significantly protection against tumor growth in animal models [69].

GLA is a hexa-acyl synthetic lipid A derivative composed of a disaccharide backbone with a single phosphate group. Results show that it has even more powerful adjuvant abilities than MPL [70, 71], and it exhibited a good safety profile in animals and in Phase I trials [72]. MPL has a strong but not overwhelming ability to promote Th1 responses. Interestingly, GLA exhibits a strong ability to shift antigen-specific immune responses towards Th1 type [73, 74]; hence, it is being proposed as a better alternative to MPL to confer adequate protection against certain pathogens. In fact, significant protection in animals of experimentation was reported against *Toxoplasma gondii*, [75] *Mycobacterium tuberculosis*, [76, 77] or influenza virus [72].

ONO-4007 is a tri-acylated acyclic sulphonated backbone. This molecule induced tumor and metastases regression in animal models [78, 79]. This property was due to its strong ability to stimulate secretion of tumor necrosis factor (TNF-$\alpha$) by macrophages [80, 81]. Studies in rodents showed remarkable but selective efficacy against TNF-$\alpha$-sensitive tumors, which improved in combination with other antitumor therapies [82]. Unfortunately, only a primed state induction of TNF-$\alpha$ was detected in human cells [83]. Phase I clinical studies revealed a limited capacity and the antitumor studies were not continued [79]. In contrast, the molecule exhibits anti-leishmanial [84] and anti-allergy activities [85].

## Conclusions

Although LPS was long known as an immune stimulatory substance with potential adjuvant use, the large variety of unacceptable toxic effects drastically restricted its clinical use. However, the finding that MPL was safe and retained the desirable adjuvant properties of LPS opened new possibilities to treat pathogen diseases. In contrast to previous adjuvants, LPS-based adjuvants offer new benefits from their ability to enhance Th1-type responses and stimulate cytotoxic T lymphocytes. This activity is essential to confer protection against many pathogens and to develop prophylaxis therapies against other diseases as cancer or allergy. Indeed, this is supported by the efficacy of MPL in available vaccines whereas standard adjuvants failed to provide protection. Additionally, the high adjuvant capacity of LPS-based adjuvants has considerable and obvious benefits in mucosa protection, faster activation of protection, reduction of booster doses, functional immunization in elderly, or preparation of polyvalent vaccine formulations. Certain drawbacks were attributed to MPL, e.g., elevate production cost and possible activation/enhancement of TLR4-related autoimmune diseases. Synthetic lipid A analogues with similar biological activity like MPL have demonstrated considerable reduction of the production cost. In regard to the activation TLR4 autoimmune diseases, accumulated data till date from immunization in humans provides further evidence of safety. In summary, LPS-based adjuvants improve the current vaccination therapies and open possibilities to solve their existing challenges.

## References

1. Rietschel, E.T., et al.: Bacterial endotoxins: chemical structure, biological activity and role in septicaemia. Scand. J. Infect. Dis. Suppl. **31**, 8–21 (1982)
2. Guo, L., et al.: Lipid A acylation and bacterial resistance against vertebrate antimicrobial peptides. Cell **95**, 189–198 (1998)
3. Gibbons, H.S., Lin, S., Cotter, R.J., Raetz, C.R.: Oxygen requirement for the biosynthesis of the S-2-hydroxymyristate moiety in *Salmonella typhimurium* lipid A. Function of LpxO, A new Fe2+/alpha-ketoglutarate-dependent dioxygenase homologue. J. Biol. Chem. **275**, 32940–32949 (2000)
4. Reynolds, C.M., et al.: An outer membrane enzyme encoded by *Salmonella typhimurium lpxR* that removes the 3'-acyloxyacyl moiety of lipid A. J. Biol. Chem. **281**, 21974–21987 (2006)
5. Trent, M.S., Pabich, W., Raetz, C.R., Miller, S.I.: A PhoP/PhoQ-induced Lipase (PagL) that catalyzes 3-O-deacylation of lipid A precursors in membranes of *Salmonella typhimurium*. J. Biol. Chem. **276**, 9083–9092 (2001)
6. Kawai, T., Akira, S.: Toll-like receptors and their crosstalk with other innate receptors in infection and immunity. Immunity **34**, 637–650 (2011)
7. Palsson-McDermott, E.M., O'Neill, L.A.: Signal transduction by the lipopolysaccharide receptor, Toll-like receptor-4. Immunology **113**, 153–162 (2004)
8. Trinchieri, G., Sher, A.: Cooperation of Toll-like receptor signals in innate immune defence. Nat. Rev. Immunol. **7**, 179–190 (2007)
9. Kawai, T., Akira, S.: TLR signaling. Cell Death Differ. **13**, 816–825 (2006)
10. Miggin, S.M., O'Neill, L.A.: New insights into the regulation of TLR signaling. J. Leukoc. Biol. **80**, 220–226 (2006)
11. Alexander, C., Rietschel, E.T.: Bacterial lipopolysaccharides and innate immunity. J. Endotoxin Res. **7**, 167–202 (2001)
12. Morrison, D.C., Kline, L.F.: Activation of the classical and properdin pathways of complement by bacterial lipopolysaccharides (LPS). J. Immunol. **118**, 362–368 (1977)
13. Cooper, N.R., Morrison, D.C.: Binding and activation of the first component of human complement by the lipid A region of lipopolysaccharides. J. Immunol. **120**, 1862–1868 (1978)
14. Conti, P., et al.: Activation of human natural killer cells by lipopolysaccharide and generation of interleukin-1 alpha, beta, tumour necrosis factor and interleukin-6. Effect of IL-1 receptor antagonist. Immunology **73**, 450–456 (1991)
15. Kobayashi, M., et al.: Identification and purification of natural killer cell stimulatory factor (NKSF), a cytokine with multiple biologic effects on human lymphocytes. J. Exp. Med. **170**, 827–845 (1989)
16. Cinel, I., Dellinger, R.P.: Advances in pathogenesis and management of sepsis. Curr. Opin. Infect. Dis. **20**, 345–352 (2007)
17. Annane, D., Bellissant, E., Cavaillon, J.M.: Septic shock. Lancet **365**(63–78) (2005)
18. Vincent, J.L., et al.: Sepsis in European intensive care units: results of the SOAP study. Crit. Care Med. **34**, 344–353 (2006)
19. Martin, G.S., Mannino, D.M., Eaton, S., Moss, M.: The epidemiology of sepsis in the United States from 1979 through 2000. N. Engl. J. Med. **348**, 1546–1554 (2003)
20. Kotani, S., Takada, H.: Structural requirements of lipid A for endotoxicity and other biological activities – an overview. Adv. Exp. Med. Biol. **256**, 13–43 (1990)

21. Nowotny, A.: Molecular aspects of endotoxic reactions. Bacteriol. Rev. **33**, 72–98 (1969)
22. Miller, S.I., Ernst, R.K., Bader, M.W.: LPS, TLR4 and infectious disease diversity. Nat. Rev. Microbiol. **3**, 36–46 (2005)
23. Golenbock, D.T., Hampton, R.Y., Qureshi, N., Takayama, K., Raetz, C.R.: Lipid A-like molecules that antagonize the effects of endotoxins on human monocytes. J. Biol. Chem. **266**, 19490–19498 (1991)
24. Kotani, S., et al.: Synthetic lipid A with endotoxic and related biological activities comparable to those of a natural lipid A from an *Escherichia coli* re-mutant. Infect. Immun. **49**, 225–237 (1985)
25. Saitoh, S., et al.: Lipid A antagonist, lipid IVa, is distinct from lipid A in interaction with Toll-like receptor 4 (TLR4)-MD-2 and ligand-induced TLR4 oligomerization. Int. Immunol. **16**, 961–969 (2004)
26. David, S., Vermeer-de Bondt, P.E., van der Maas, N.A.: Reactogenicity of infant whole cell pertussis combination vaccine compared with acellular pertussis vaccines with or without simultaneous pneumococcal vaccine in the Netherlands. Vaccine **26**, 5883–5887 (2008)
27. Gustafsson, L., Hallander, H.O., Olin, P., Reizenstein, E., Storsaeter, J.: A controlled trial of a two-component acellular, a five-component acellular, and a whole-cell pertussis vaccine. N. Engl. J. Med. **334**, 349–355 (1996)
28. Brewer, J.M., et al.: Aluminium hydroxide adjuvant initiates strong antigen-specific Th2 responses in the absence of IL-4- or IL-13-mediated signaling. J. Immunol. **163**, 6448–6454 (1999)
29. Dubensky Jr., T.W., Reed, S.G.: Adjuvants for cancer vaccines. Semin. Immunol. **22**, 155–161 (2010)
30. Soghoian, D.Z., Streeck, H.: Cytolytic CD4(+) T cells in viral immunity. Expert Rev. Vaccines **9**, 1453–1463 (2010)
31. Ziegler, A., et al.: EpCAM, a human tumor-associated antigen promotes Th2 development and tumor immune evasion. Blood **113**, 3494–3502 (2009)
32. Akkoc, T., Akdis, M., Akdis, C.A.: Update in the mechanisms of allergen-specific immunotherapy. Allergy Asthma Immunol. Res. **3**, 11–20 (2011)
33. Chen, K., Cerutti, A.: Vaccination strategies to promote mucosal antibody responses. Immunity **33**, 479–491 (2010)
34. Lawson, L.B., Norton, E.B., Clements, J.D.: Defending the mucosa: adjuvant and carrier formulations for mucosal immunity. Curr. Opin. Immunol. **23**, 414–420 (2011)
35. O'Hagan, D.T.: MF59 is a safe and potent vaccine adjuvant that enhances protection against influenza virus infection. Expert Rev. Vaccines **6**, 699–710 (2007)
36. Arenas, J.: The role of bacterial lipopolysaccharides as immune modulator in vaccine and drug development. Endocr. Metab. Immune Disord. Drug Targets **12**(3), 221–235 (2012)
37. Baldridge, J.R., Crane, R.T.: Monophosphoryl lipid A (MPL) formulations for the next generation of vaccines. Methods **19**, 103–107 (1999)
38. Evans, J.T., et al.: Enhancement of antigen-specific immunity via the TLR4 ligands MPL adjuvant and Ribi.529. Expert Rev. Vaccines **2**, 219–229 (2003)
39. Ulrich, J.T., Myers, K.R.: Monophosphoryl lipid A as an adjuvant. Past experiences and new directions. Pharm. Biotechnol. **6**, 495–524 (1995)
40. Vajdy, M., Singh, M.: The role of adjuvants in the development of mucosal vaccines. Expert Opin. Biol. Ther. **5**, 953–965 (2005)
41. Drachenberg, K.J., Wheeler, A.W., Stuebner, P., Horak, F.: A well-tolerated grass pollen-specific allergy vaccine containing a novel adjuvant, monophosphoryl lipid A, reduces allergic symptoms after only four preseasonal injections. Allergy **56**, 498–505 (2001)
42. Drachenberg, K.J., Heinzkill, M., Urban, E., Woroniecki, S.R.: Efficacy and tolerability of short-term specific immunotherapy with pollen allergoids adjuvanted by monophosphoryl lipid A (MPL) for children and adolescents. Allergol. Immunopathol. (Madr.) **31**, 270–277 (2003)
43. Kundi, M.: New hepatitis B vaccine formulated with an improved adjuvant system. Expert Rev. Vaccines **6**, 133–140 (2007)
44. Aide, P., et al.: Four year immunogenicity of the RTS, S/AS02(A) malaria vaccine in Mozambican children during a phase IIb trial. Vaccine **29**, 6059–6067 (2011)
45. Von, E.K., et al.: The candidate tuberculosis vaccine Mtb72F/AS02A: Tolerability and immunogenicity in humans. Hum. Vaccin. **5**, 475–482 (2009)
46. Polhemus, M.E., et al.: Evaluation of RTS, S/AS02A and RTS, S/AS01B in adults in a high malaria transmission area. PLoS One **4**, e6465 (2009)
47. Brichard, V.G., Lejeune, D.: Cancer immunotherapy targeting tumour-specific antigens: towards a new therapy for minimal residual disease. Expert Opin. Biol. Ther. **8**, 951–968 (2008)
48. Garcon, N., Chomez, P., Van, M.M.: GlaxoSmithKline Adjuvant Systems in vaccines: concepts, achievements and perspectives. Expert Rev. Vaccines **6**, 723–739 (2007)
49. Fries, L.F., et al.: Liposomal malaria vaccine in humans: a safe and potent adjuvant strategy. Proc. Natl. Acad. Sci. U. S. A. **89**, 358–362 (1992)
50. Arenas, J., et al.: Coincorporation of LpxL1 and PagL mutant lipopolysaccharides into liposomes with *Neisseria meningitidis* opacity protein: influence on endotoxic and adjuvant activity. Clin. Vaccine Immunol. **17**, 487–495 (2010)
51. Vernacchio, L., et al.: Effect of monophosphoryl lipid A (MPL) on T-helper cells when administered as an adjuvant with pneumococcal-CRM197 conjugate vaccine in healthy toddlers. Vaccine **20**, 3658–3667 (2002)

52. Olson, K., et al.: Liposomal gD ectodomain (gD1-306) vaccine protects against HSV2 genital or rectal infection of female and male mice. Vaccine **28**, 548–560 (2009)
53. Hall, M.A., et al.: Intranasal immunization with multivalent group A streptococcal vaccines protects mice against intranasal challenge infections. Infect. Immun. **72**, 2507–2512 (2004)
54. Tana, W.S., Isogai, E., Oguma, K.: Induction of intestinal IgA and IgG antibodies preventing adhesion of verotoxin-producing *Escherichia coli* to Caco-2 cells by oral immunization with liposomes. Lett. Appl. Microbiol. **36**, 135–139 (2003)
55. Baldridge, J.R., Yorgensen, Y., Ward, J.R., Ulrich, J.T.: Monophosphoryl lipid A enhances mucosal and systemic immunity to vaccine antigens following intranasal administration. Vaccine **18**, 2416–2425 (2000)
56. Kidon, M.I., Shechter, E., Toubi, E.: Vaccination against human papilloma virus and cervical cancer. Harefuah **150**, 33–36 (2011). 68
57. Labadie, J.: Postlicensure safety evaluation of human papilloma virus vaccines. Int. J. Risk Saf. Med. **23**, 103–112 (2011)
58. Morello, C.S., Levinson, M.S., Kraynyak, K.A., Spector, D.H.: Immunization with herpes simplex virus 2 (HSV-2) genes plus inactivated HSV-2 is highly protective against acute and recurrent HSV-2 disease. J. Virol. **85**, 3461–3472 (2011)
59. Llanos-Cuentas, A., et al.: A clinical trial to evaluate the safety and immunogenicity of the LEISH-F1 + MPL-SE vaccine when used in combination with sodium stibogluconate for the treatment of mucosal leishmaniasis. Vaccine **28**, 7427–7435 (2010)
60. Zhu, D., Barniak, V., Zhang, Y., Green, B., Zlotnick, G.: Intranasal immunization of mice with recombinant lipidated P2086 protein reduces nasal colonization of group B *Neisseria meningitidis*. Vaccine **24**, 5420–5425 (2006)
61. Green, B.A., et al.: PppA, a surface-exposed protein of *Streptococcus pneumoniae*, elicits cross-reactive antibodies that reduce colonization in a murine intranasal immunization and challenge model. Infect. Immun. **73**, 981–989 (2005)
62. Egan, M.A., et al.: A comparative evaluation of nasal and parenteral vaccine adjuvants to elicit systemic and mucosal HIV-1 peptide-specific humoral immune responses in cynomolgus macaques. Vaccine **22**, 3774–3788 (2004)
63. Mason, K.W., et al.: Reduction of nasal colonization of nontypeable *Haemophilus influenzae* following intranasal immunization with rLP4/rLP6/UspA2 proteins combined with aqueous formulation of RC529. Vaccine **22**, 3449–3456 (2004)
64. Baldridge, J.R., et al.: Immunostimulatory activity of aminoalkyl glucosaminide 4-phosphates (AGPs): induction of protective innate immune responses by RC-524 and RC-529. J. Endotoxin Res. **8**, 453–458 (2002)
65. Ishizaka, S.T., Hawkins, L.D.: E6020: a synthetic Toll-like receptor 4 agonist as a vaccine adjuvant. Expert Rev. Vaccines **6**, 773–784 (2007)
66. Morefield, G.L., Hawkins, L.D., Ishizaka, S.T., Kissner, T.L., Ulrich, R.G.: Synthetic Toll-like receptor 4 agonist enhances vaccine efficacy in an experimental model of toxic shock syndrome. Clin. Vaccine Immunol. **14**, 1499–1504 (2007)
67. Przetak, M., et al.: Novel synthetic LPS receptor agonists boost systemic and mucosal antibody responses in mice. Vaccine **21**, 961–970 (2003)
68. Baudner, B.C., et al.: MF59 emulsion is an effective delivery system for a synthetic TLR4 agonist (E6020). Pharm. Res. **26**, 1477–1485 (2009)
69. Wang, S., et al.: Effective antibody therapy induces host-protective antitumor immunity that is augmented by TLR4 agonist treatment. Cancer Immunol. Immunother. **61**, 49–61 (2012)
70. Baldwin, S.L., et al.: Intradermal immunization improves protective efficacy of a novel TB vaccine candidate. Vaccine **27**, 3063–3071 (2009)
71. Bertholet, S., et al.: Optimized subunit vaccine protects against experimental leishmaniasis. Vaccine **27**, 7036–7045 (2009)
72. Coler, R.N., et al.: A synthetic adjuvant to enhance and expand immune responses to influenza vaccines. PLoS One **5**, e13677 (2010)
73. Xiao, L., et al.: A TLR4 agonist synergizes with dendritic cell-directed lentiviral vectors for inducing antigen-specific immune responses. Vaccine **30**, 2570–2581 (2012)
74. Pantel, A., et al.: A new synthetic TLR4 agonist, GLA, allows dendritic cells targeted with antigen to elicit Th1 T-cell immunity in vivo. Eur. J. Immunol. **42**, 101–109 (2012)
75. Cong, H., et al.: *Toxoplasma gondii* HLA-B*0702-restricted GRA7(20-28) peptide with adjuvants and a universal helper T cell epitope elicits CD8(+) T cells producing interferon-gamma and reduces parasite burden in HLA-B*0702 mice. Hum. Immunol. **73**, 1–10 (2012)
76. Baldwin, S.L., et al.: The importance of adjuvant formulation in the development of a tuberculosis vaccine. J. Immunol. **188**, 2189–2197 (2012)
77. Windish, H.P., et al.: Protection of mice from *Mycobacterium tuberculosis* by ID87/GLA-SE, a novel tuberculosis subunit vaccine candidate. Vaccine **29**, 7842–7848 (2011)
78. Ueda, H., Yamazaki, M.: Induction of tumor necrosis factor in a murine tumor by systemic administration of a novel synthetic lipid A analogue, ONO-4007. J. Immunother. **20**, 65–69 (1997)
79. Matsumoto, N., Aze, Y., Akimoto, A., Fujita, T.: Restoration of immune responses in tumor-bearing mice by ONO-4007, an antitumor lipid A derivative. Immunopharmacology **36**, 69–78 (1997)
80. Kuramitsu, Y., et al.: A new synthetic lipid A analog, ONO-4007, stimulates the production of tumor necrosis factor-alpha in tumor tissues, resulting in the

rejection of transplanted rat hepatoma cells. Anticancer Drugs **8**, 500–508 (1997)
81. Matsumoto, N., Oida, H., Aze, Y., Akimoto, A., Fujita, T.: Intratumoral tumor necrosis factor induction and tumor growth suppression by ONO-4007, a low-toxicity lipid A analog. Anticancer Res. **18**, 4283–4289 (1998)
82. Inagawa, H., et al.: Mechanisms by which chemotherapeutic agents augment the antitumor effects of tumor necrosis factor: involvement of the pattern shift of cytokines from Th2 to Th1 in tumor lesions. Anticancer Res. **18**, 3957–3964 (1998)
83. Matsumoto, N., Aze, Y., Akimoto, A., Fujita, T.: ONO-4007, an antitumor lipid A analog, induces tumor necrosis factor-alpha production by human monocytes only under primed state: different effects of ONO-4007 and lipopolysaccharide on cytokine production. J. Pharmacol. Exp. Ther. **284**, 189–195 (1998)
84. Khan, M.A., et al.: Inhibition of intracellular proliferation of *Leishmania* parasites in vitro and suppression of skin lesion development in BALB/c mice by a novel lipid A analog (ONO-4007). Am. J. Trop. Med. Hyg. **67**, 184–190 (2002)
85. Iio, J., et al.: Lipid A analogue, ONO-4007, inhibits IgE response and antigen-induced eosinophilic recruitment into airways in BALB/c mice. Int. Arch. Allergy Immunol. **127**, 217–225 (2002)

# Bacterial Toxins Are Successful Immunotherapeutic Adjuvants and Immunotoxins

## 34

Irena Adkins

## Contents

34.1  Introduction ................................................. 538
34.2  The Structure and Mode of Entry of Bacterial Toxins into Host Cells .............. 539
34.3  Bacterial Toxins Used as Adjuvants ........... 543
34.4  Immunotoxins ............................................... 545
34.5  Bacterial Toxins Used as Antiviral Agents ..... 545
34.6  Bacterial Toxins Used as Antigen and Drug Delivery Agents ........................... 546
34.7  Bacterial Toxins Used in Treatment of Neurological Disorders and Spinal Cord Injuries .............................. 546
34.8  Bacterial Toxins Commercially Used for Vaccination ..................................... 546
Conclusion ............................................................. 547
References ............................................................. 547

I. Adkins, PhD
Institute of Microbiology of the ASCR, v.v.i.,
Sotio a.s., Prague, Czech Republic
e-mail: irena.adkins@seznam.cz

## Abstract

Bacterial protein toxins of pathogenic bacteria play an important role in infectious diseases to hijack mammalian cells and manipulate host immune responses. Several bacterial toxins and their non-toxic mutants have been intensively studied over the past decades in order to harness their abilities to enter host cells for stimulation of adaptive T cell responses, direct elimination of cancer cells or to boost immunity as adjuvants. Moreover, another set of bacterial toxins called effectors, which are translocated into the host cells by specialized secretion systems, have been used to facilitate the delivery of heterologous antigens into antigen presenting cells. Some of the most powerful bacterial neurotoxins have found their application in treatment of various neurological disorders as well as in cosmetic industry. Whereas few bacterial toxins are currently in various phases of clinical cancer trials, the immunotherapeutic potential of others needs to be established further. This chapter will introduce bacterial toxins used in immunotherapy and present the major achievements in this field. Similarly, the hurdles and limitations of the use of bacterial toxins in human immunotherapy will be discussed together with other diagnostics applications.

## 34.1 Introduction

Bacterial toxins were perfected through evolution in order to enhance competitive potential and defense abilities of bacteria. They represent the principal virulence factors of pathogenic bacteria to hijack mammalian cells. Many bacterial toxins are encoded by chromosome, plasmids, or phages and can be broadly divided in two groups referred to as endotoxins and exotoxins. Endotoxins are cell-associated structural components of bacteria cell envelope represented by peptidoglycan constituent muramyl dipeptide (MDP) and muropeptides or lipopolysaccharide (LPS) or lipooligosaccharide (LOS) which are located in the outer membrane of Gram-negative bacteria. Endotoxins generally act in the vicinity of a bacterial cell. Exotoxins are mostly protein toxins which belong to the most powerful human poisons known. They are actively transported from bacteria or released by cell lysis induced by host defense mechanisms or by the action of antibiotics. Some bacteria produce many toxins like Staphylococci or Streptococci, whereas other bacteria like *Corynebacterium diphtheriae* produce just one.

Some toxins, e.g., pore-forming hemolysins or phospholipases, exhibit broad cytotoxic activity towards various cell types, while others like clostridial neurotoxins act specifically only on neurons. Some bacterial toxins may serve as invasins; others damage the cellular integrity by means of enzymatic activity or pore formation causing the cell death.

Over the last decades, bacterial toxins have played an important role in unraveling signaling pathways in eukaryotic cells. The cytotoxic activity of toxins has been utilized to treat various neurological disorders as well as to generate immunotoxins to kill malignant or virus-infected cells. Moreover, various bacterial toxins have been differentially modified either by single point mutations, insertion of specific sequences, or by a deletion of parts of the toxin molecule to alter their binding abilities and/or abolish their enzymatic activity to be used for vaccination, antigen/drug delivery, or, most importantly, as adjuvants. The most studied bacterial toxins used for immunotherapy are listed according to the mode of action in Table 34.1. The specific immunotherapeutic use of bacterial toxins is listed in Table 34.2.

**Table 34.1** Characteristics of the most studied bacterial toxins used in immunotherapy

| Toxin/origin/abbreviation | Disease | Mode of action | Target |
|---|---|---|---|
| Activate immune responses | | | |
| Muramyl dipeptide (MDP)/Muropeptides peptidoglycan components of bacterial cell wall | – | TLR ligand | NOD receptors |
| Enterotoxins A,B,C/*S. aureus* (SEA, SEB, SEC) | Food poisoning[b] | Superantigen | TCR and MHC II |
| Inhibit protein synthesis | | | |
| Shiga toxin/*Shigella dysenteriae* (Stx) | HC | N-glycosidase | 28S rRNA |
| Shiga-like toxin/ *Escherichia coli* (Stx1) | HUS | N-glycosidase | 28S rRNA |
| Exotoxin A/*Pseudomonas aeruginosa* (PE) | Pneumonia[b] | ADP-ribosyltransferase | Elongation factor 2 |
| Diphtheria toxin/*Corynebacterium diphtheriae* (DT) | Diphtheria | ADP-ribosyltransferase | Elongation factor 2 |
| Protease | | | |
| Lethal toxin/*Bacillus anthracis* | Anthrax | Zinc metalloprotease | MAPKKs |
| Tetanus toxin/*Clostridium tetani* (TeNT) | Tetanus | Zinc metalloprotease | SNARE proteins |
| Neurotoxins A-G/*Clostridium botulinum* (BoNTs) | Botulism | Zinc metalloprotease | SNARE proteins |

**Table 34.1** (continued)

| Toxin/origin/abbreviation | Disease | Mode of action | Target |
|---|---|---|---|
| Activate second messenger pathway | | | |
| Adenylate cyclase toxin/*Bordetella pertussis* (CyaA) | Whooping cough | Adenylate cyclase | ATP |
| Pertussis toxin/*B. pertussis* (PT) | Whooping cough | Pore formation | G-proteins |
| Dermonecrotic toxin/*B. pertussis* (DNT) | Whooping cough, rhinitis | ADP-ribosyltransferase | Rho G-proteins |
| Edema toxin/*B. anthracis* | Anthrax | Deamidase | ATP |
| Cholera toxin/*Vibrio cholerae* (CT) | Cholera | Adenylate cyclase | G-proteins |
| Zonula occludens toxin/*V. cholerae* (ZOT) | Cholera | ADP-ribosyltransferase | Actin polymerization via protein kinase C signaling |
| C2 toxin/*C. botulinum* | Botulism[c] | Reversible disruption of intercellular tight junctions | G-actin |
| C3 toxin/*C. botulinum* | Botulism[c] | Mimics zonulin | Rho G-protein |
| Toxin A/*C. difficile* | Diarrhea/PC | ADP-ribosyltransferase | Rho G-proteins |
| Cytotoxic necrotizing factor 1/*E. coli* (CNF1) | UTIs[c] | ADP-ribosyltransferase | Cholesterol |
| Heat-labile toxin/*E. coli* (LT) | Diarrhea | UDP-Glucosyltransferase | Rho G-proteins |
| | | Pore formation | G-proteins |
| | | Deamidase | |
| | | ADP-ribosyltransferase | |
| Damage membranes | | | |
| Hemolysin[a]/*E. coli* (HlyA) | UTIs | Pore formation | Plasma membrane |
| Listeriolysin O/*Listeria monocytogenes* (LLO) | Foodborne systematic illnesses, meningitis | Pore formation | Cholesterol |
| Perfringolysin O/*C. perfringens* | Gas gangrene[b,c] | Pore formation | Cholesterol |
| Aerolysin/*Aeromonas hydrophila* | Diarrhea | Pore formation | GPI-AP |
| Pneumolysin/*Streptococcus pneumoniae* | Pneumonia[b] | Pore formation | Cholesterol |
| Intermedilysin/*Staphylococcus intermedius* | Abscesses | Pore formation | Cholesterol |
| Crystal proteins/*Bacillus thuringiensis* (Cry) | – | Pore formation | Plasma membrane |

Adopted according to Schmitt et al.[1]
Abbreviations: *TLR* Toll-like receptor, *NOD* nucleotide oligomerization domain, *TCR* T cell receptor, *MHC II* major histocompatibility class II, *HC* hemorrhagic colitis, *HUS* hemolytic uremic syndrome, *MAPKKs* mitogen-activated protein kinase kinases, *PC* antibiotic-associated pseudomembranous colitis, *UTIs* urinary tract infections, *GPI-AP* glycosylphosphatidylinositol-anchored proteins
[a]Toxin produced also by other genera of bacteria
[b]Other diseases are also associated with the organism
[c]There is no clear causal relationship between toxin and the disease [1]

## 34.2 The Structure and Mode of Entry of Bacterial Toxins into Host Cells

The structure of bacterial toxins and their specific modification for immunotherapy are described in Table 34.3. The majority of toxins are referred as AB toxins where the A moiety generally possesses an enzymatic activity and the monomeric or multimeric B moiety binds the toxin to cell surface receptors. The B moiety can play a role in the translocation of the A moiety into the cytosol [65]. After receptor binding via B moiety, most protein toxins enter host

**Table 34.2** Experimental immunotherapeutic use of bacterial toxins

| Toxin | Experimental immunotherapeutic use | | | | | | | |
|---|---|---|---|---|---|---|---|---|
| | Adjuvant | Antigen delivery[a] | Drug/DNA delivery | IT | Antiviral agent | Diagnostics/Imaging | Cancer therapy | Therapy of neurological disorders/injury |
| Muramyl dipeptide/Muropeptides | x | | x | | x | | x | |
| Enterotoxins A,B,C | x | | | | | | x | |
| Shiga toxin | x | x | x | x | x | x | x | |
| Shiga-like toxin | x | x | | x | x | x | x | |
| Exotoxin A | x | x | | x | x | | x | |
| Diphtheria toxin | x | x | | x | x | | x | |
| Lethal toxin | | x | | x | | x | x | |
| Tetanus toxin | | x | x | | | | x | x |
| Neurotoxins | | | x | | | | x | x |
| Adenylate cyclase toxin | x | x | | | | | x | |
| Pertussis toxin | x | x | | | x | | x | |
| Dermonecrotic toxin | x | | | | | | | |
| Edema toxin | x | x | | | | | | |
| Cholera toxin | x | | | | | | x | |
| Zonula occludens toxin | x | x | x | | | | x | |
| C2 toxin | | | x | | | | x | |
| C3 toxin | | | | | | | | x |
| Toxin A | | | | | | | x | |
| CNF1 | x | | | | | | | x |
| Heat-labile toxin | x | | | | | | x | |
| Hemolysin | | x | | | | | | |
| Listeriolysin O | x | x | x | | | | x | |
| Perfringolysin O | x | x | x | | | | | |
| Aerolysin | | | | x | | x | x | |
| Pneumolysin | x | x | | | | | | |
| Intermedilysin | | | x | | x | | | |
| Crystal proteins | x | x | | | | | x | |

Abbreviations: *IT* immunotoxin
[a]Antigen genetically fused or chemically coupled to the toxin molecule

**Table 34.3** Structure of bacterial toxins and its modification for immunotherapy

| Toxin | Structure/receptor/host cell entry | Modification for immunotherapy |
|---|---|---|
| Muramyl dipeptide/ muropeptides | MDP is a peptide consisting of N-acetylmuramic acid attached to a short amino acid chain of l-Ala-d-isoGln/ NOD2 receptor | Unmodified, l-MTP-PE various chemical modifications reviewed [2] |
| | Muropeptides are various tripeptides and disaccharide tri- and tetrapeptides, may contain DAP/NOD2, DAP ligand is NOD1 receptor | |
| Enterotoxins A, B, C | 27–28 kDa, Staphylococcal enterotoxin type A (SEA), type B (SEB), and type C (SEC) share structure and sequence similarities, compact two domain globular proteins, highly stable and protease resistant, mitogenic activity is postulated to be in the N-terminal domain, the emetic activity is located in the central and the C-terminal part of the molecule | C215Fab-SEA [3] SEC2(T20L/G22E) [4] |
| Shiga toxin Shiga-like toxin | Holotoxin 70 kDa, the enzymatically active A moiety consisting of A1 and A2 domains is non-covalently associated with nontoxic homopentameric B subunits/Gb3 receptor/RME and retrograde transportation via GA and ER from where the enzymatic A moiety, proteolytically activated by protease furin, is translocated into the cytosol of the host cell | StxB or Stx1B with inserted or chemically coupled Ag or unmodified reviewed [5] |
| Exotoxin A | 66 kDa, the enzymatically active domain (A), the translocation and the binding domain (B)/CD91 receptor/ RME but also through caveosomes. PE is cleaved by protease furin in endosomes and transported in a retrograde fashion via GA to ER, where the active A moiety is translocated into the host cell cytosol | PE38 and others reviewed [6–8] |
| Diphtheria toxin | 58 kDa, the catalytic domain C (A), the translocation domain T, and the C-terminal receptor-binding domain (B). Proteolytic cleavage by cell-associated proteases prior to binding to the complex of heparin-binding epidermal growth factor (EGF)-like precursor and CD9 receptors/ RME, the A moiety is translocated from acidic endosomes into the host cell cytosol | DT388, DAB389, and others reviewed [6, 7] |
| Tetanus toxin Neurotoxins | ~150 kDa, posttranslational proteolytic cleavage to form a dichain molecule, a light chain (L, 50 kDa, catalytic) and a heavy chain (H, 100 kDa, translocation $H_N$ and binding $H_C$/polysialogangliosides/RME and/or synaptic vesicle reuptake, the enzymatic L chain is translocated from acidic endosomes into the neuronal cytoplasm | Nontoxic C-fragment ($H_C$) for drug delivery [9], or unmodified [10, 11] |
| Adenylate cyclase toxin | 200 kDa, N-terminal adenylate cyclase (AC) domain (~400 aa), hydrophobic domain, and C-terminal RTX domain (~1,306 aa)/CD11b receptor/translocation across plasma membrane into the host cell cytosol, $Ca^{2+}$ influx, cation-selective pore formation | Gly-Ser insertion at position 188 of AC domain or unmodified with inserted Ag [12] |
| Pertussis toxin | 105 kDa, composes of enzymatically active A subunit (S1) and five B subunits (S2, S3, two subunits of S4, and S5) which create a ring-like form and are responsible for binding via glycolipids or glycoproteins containing sialic acid/ RME, retrograde transportation via GA and ER from where the enzymatic A moiety is translocated into the host cell cytosol | PT9K/129G [13] alone or with inserted Ag or unmodified reviewed [5] |

(continued)

**Table 34.3** (continued)

| Toxin | Structure/receptor/host cell entry | Modification for immunotherapy |
|---|---|---|
| Dermonecrotic toxin | 160.6 kDa, single-chain polypeptide consisting of N-terminal binding domain (B), and C-terminal catalytic domain (A)/unknown receptor/dynamin-dependent endocytosis, cleaved by protease furin, translocated from acidic endosomes into the host cell cytosol | Unmodified [14] |
| Lethal toxin Edema toxin | Cleavage of the protective antigen (PA, 83 kDa) (B moiety) by protease furin exposes a binding site for the lethal factor (LF, 90 kDa) or the edema factor (EF, 89 kDa) (A moiety) which binds competitively. PA heptamer with LF or EF binds to TEM8 and CMG2 receptors/RME, LF, and EF are subsequently translocated from acidic endosomes into the host cell cytosol | LFn (1-255 aa), EFn (1-260 aa) with inserted or coupled Ag; PA with altered cleavage site unmodified reviewed [5] |
| Cholera toxin Heat-labile toxin | ~85 kDa, both toxins form oligomers (AB5) composed of a single A subunit (27 kDa, CTA or LTA) that contains two domains (A1 and A2) linked together by a disulfide bridge and five identical B subunits (11.5 kDa each, CTB or LTB)/GM1 ganglioside receptors/RME, retrograde transportation via GA and ER from where the enzymatic A moiety is translocated into the host cell cytosol | Various single point mutations, e.g., LTR72, R192G reviewed [15] CTA1-DD [16] CTB-CpG [17] |
| Zonula occludens toxin | A single polypeptide of 45 kDa, which undergoes cleavage. C-terminal fragment of 12 kDa (aa 288–293) is excreted from the bacteria and exerts its biological effect by reversibly disassembling intercellular tight junctions mimicking zonulin/zonulin receptor | Unmodified [18, 19] AT1001 [20] AT1002 [21] |
| C2 toxin | The binary AB type toxin which is composed of the binding component C2II (80.5–100 kDa, B) and the enzyme component C2I (49.4 kDa, A)/asparagine-linked carbohydrate structure/RME, A moiety is translocated from acidic endosomes into the host cell cytosol | C2IN-streptavidin fusion for the drug delivery [22] |
| C3 toxin | 25 kDa, single polypeptide without identified translocation domain. The toxin reaches the host cell cytosol; however, the specific cell entry machinery is not known | Unmodified or $C_{154\text{-}182}$ fragment [23, 24] |
| Toxin A | 308 kDa, family of large clostridial toxins, single-chain polypeptide consisting of N-terminal catalytic domain (A), followed by a hydrophobic translocation domain and the C-terminal binding domain (B)/various types of carbohydrate structures/RME, A moiety is translocated from acidic endosomes into the host cell cytosol; cholesterol-dependent pore formation | Formalin-treated toxin [25] |
| Cytotoxic necrotizing factor 1 | 113 kDa, single-chain polypeptide consisting of N-terminal binding domain (B) and C-terminal catalytic domain (A)/laminin/RME, A moiety is translocated from acidic endosomes into the host cell cytosol | Unmodified [14] |
| Hemolysin | 107 kDa, a single polypeptide that belongs to the RTX family of toxins. HlyA is synthesized as a nontoxic prohemolysin (proHlyA), which is further activated to a mature toxin by the co-synthesized fatty acid acyltransferase HlyC/HlyA targets endothelial cells and cells of the immune system/HlyA causes cell lysis by generating cation-selective pores in plasma membrane | Ag linked to HlyAs (C-terminal 50–60 aa) as a secretion signal reviewed [5] |
| Listeriolysin O | 58 kDa, member of the cholesterol-dependent cytolysin family. It is selectively activated within the acidic phagosome of cells that have phagocytized *L. monocytogenes*. It lyses the phagosome and aids bacteria to escape into the cytosol | LLO fused to tumor Ag or unmodified reviewed [5] |

**Table 34.3** (continued)

| Toxin | Structure/receptor/host cell entry | Modification for immunotherapy |
|---|---|---|
| Perfringolysin O Intermedilysin Pneumolysin | 53–54 kDa single-chain polypeptides, thiol-activated cholesterol-dependent cytolysins that form large homo-oligomeric pores (~50 monomer subunits) in the cellular plasma membrane | Unmodified, domain 4 of Intermedilysin, detoxified Pneumolysin with Ag (C428G, W433F, ΔA146) reviewed [5] |
| Aerolysin | Inactive proaerolysin (52 kDa) binds GPI-AP on the cell surface, before or after the proteolytic removal of C-terminal fragment by furin and other proteases; it is converted to aerolysin (50 kDa). Aerolysin oligomerizes to form heptamers that insert into the plasma membrane to form channels that cause target cell death | PRX302 contains PSA-cleavable sequence [26] R336A do not bind GPI-AC [27] |
| Crystal proteins | ~130 kDa, globular molecules containing three structural domains connected by single linkers. The N-terminal domain I (membrane insertion and pore formation), domain II (receptor-binding domain), and the C-terminal domain III (stability of the toxic fragment and regulation of the pore-forming activity). Cleaved by gut proteases yielding 60–70 kDa proteins/a cadherin-like protein, GPI-anchored aminopeptidase N, a GPI-anchored alkaline phosphatase and a 270 kDa glycoconjugate receptors/pore formation in midgut epithelial cells | Cry1Aa8 or unmodified [5] |

Abbreviations: *aa* amino acids, *Ag* antigen, *CMG2* capillary morphogenesis protein 2, *DAP* diaminopimelic acid, *ER* endoplasmic reticulum, *GA* Golgi apparatus, *GPI-AP* glycosylphosphatidylinositol-anchored proteins, *RTX* repeat-in-toxin, *TEM8* tumor endothelium marker 8, *RME* receptor-mediated endocytosis

cells predominantly via receptor-mediated endocytosis.

Toxins are transported via acidic endosomes or in a retrograde fashion through Golgi apparatus and endoplasmic reticulum into the host cell cytosol. Here the translocated enzymatic A moiety interferes with components of the protein synthesis machinery, various cellular signaling pathways, actin polymerization, and intracellular trafficking of vesicles causing, in most cases, the death of intoxicated cells (Table 34.3, Parts 1 and 2). On contrary, *Bordetella pertussis* adenylate cyclase toxin (CyaA) after binding to its receptor CD11b translocates its enzymatic adenylate cyclase domain directly across the cellular membrane into the host cell cytosol where it generates nonphysiological levels of cAMP thereby paralyzing innate immune functions of myeloid phagocytes and inducing cell death. After the receptor binding, pore-forming toxins act directly to form pores in the plasma membrane without the need of endocytosis causing lysis of the target cells [5].

The ability of bacterial toxins to bind their specific cellular receptors has been exploited as a diagnostic tool for profiling of human tumors in vivo. Toxin conjugation with a number of detectable moieties including fluorochromes, radionuclides, fluorescent proteins, or even magnetic resonance image contrast agents can provide real-time, noninvasive imaging of specific cells or cell-associated enzymatic activity [28–30].

## 34.3 Bacterial Toxins Used as Adjuvants

Adjuvants generally enhance immunity to vaccines or experimental antigens. They act on cells of innate immune system rather than directly on lymphocytes with the exception of superantigens whose adjuvant activity has also been documented [31]. Mainly in antigen-presenting cells like macrophages and dendritic cells, adjuvants induce expression of costimulatory and cell adhesion molecules as well as proteins of antigen-processing machinery thereby increasing their antigen-presenting capacity to T cells. This process is called maturation in dendritic cells. Moreover, adjuvants induce the production of cytokines and chemokines, which stimulates migration, proliferation, and differentiation of T cells. They further induce phagocytic and antimicrobial activity and facilitate antibody-mediated cytotoxicity. Most bacterial toxins possess an adjuvant activity towards host immune cells; however, they act concomitantly as immunogens eliciting the immune responses to themselves. Endotoxins as well as exotoxins have been exploited as adjuvants, mainly mutated toxins with ablated or reduced enzymatic activity or modified chemical structure to reduce their cytotoxic effects (Tables 34.2 and 34.3).

The most studied bacterial endotoxins for their adjuvanticity are LPS (Chap. 33), muramyl dipeptide (MDP), and muropeptides. MDP was discovered to be the minimal structure (Table 34.3) retaining adjuvant activity in Freund's complete adjuvant (FCA), however, not as efficient lacking the antigens present in the FCA [32]. Muropeptides are other breakdown products of peptidoglycan of Gram-negative and Gram-positive bacteria which as well as MDP express strong synergy with other Toll-like receptor (TLR) ligands like LPS. Numerous muropeptides and derivatives have been synthesized chemically to explore their adjuvant activity and found a variety of clinical uses. Murabutide, a synthetic immunomodulator derived from MDP, has been shown to enhance resistance to bacterial and viral infections as well as displayed antitumor effects in mice [2]. Other MDP-derived drugs like PolyG (a 10-mer polyguanylic acid) or paclitaxel (Taxol®) conjugated to MDP have shown antitumor activity as well as immunoenhancement effects [33, 34]. Most importantly, mifamurtide, a liposomal muramyl tripeptide phosphatidylethanolamine (L-MTP-PE) activating macrophages and monocytes, has been recently approved in Europe for the treatment of nonmetastatic osteosarcoma with chemotherapy [35]. Interestingly, a novel anti-inflammatory MDP derivative has been recently prepared [2].

From bacterial exotoxins, *V. cholerae* cholera toxin (CT) and *Escherichia coli* heat-labile toxin (LT) represent the most potent oral-mucosal immunogens and adjuvants for a variety of coadministered antigens. Their inherent immunomodulatory properties depend on the ADP-ribosylating activity and structural properties of the A subunit together with the activity of the pentameric B subunit to bind to widely expressed GM1 gangliosides. Because of their toxicity, mainly the nontoxic B subunits have been extensively used as mucosal immunogens in humans, e.g., cholera vaccine. As the B subunits alone were poor mucosal adjuvant, CT and LT mutants were introduced where the enzymatically A subunit was mutated. These mutants behaved as strong mucosal adjuvants for all antigens tested when given orally, intranasally, or vaginally, inducing strong systemic and mucosal antigen-specific antibody response [15, 36]. Several of these mutants like LTK63 or LTR192G were tested as adjuvants in phase I clinical trials [37]. Some other modified adjuvant based on CT have been also described like CTA1-DD, where fully active CTA1 and an *S. aureus* protein A derivative named DD are genetically fused [16] or CTB coupled with CpG oligonucleotide [17].

On the other hand, CTB and LTB with conjugated antigens were shown to efficiently induce oral tolerance [38]. In experimental models CTB-antigen conjugate was shown to suppress the development of several autoimmune diseases, type I allergies, and allograft rejection [5, 36]. The therapeutic tolerizing properties of CTB have been extended to patients with Behcet's disease in a proof-of-concept clinical trial [39]. An interesting alternative to CT and LT might be Cry proteins of entomopathogenic *B. thuringiensis* which exhibits similar potent mucosal and systemic immunogenicity and adjuvanticity to coadministered antigens [5, 40]. Although Cry proteins seems to be innocuous to mammalian cells, manifest stable properties, and can be cheaply produced in a large scale, detailed evaluation of their effects on mammalian cells is lacking.

Another group of toxins, which were exploited as adjuvants, is represented by pyrogenic toxin superantigens like enterotoxins A, B, and C of *S. aureus* (SEA, SEB, and SEC). These toxins bind to distinct regions outside the peptide-binding cleft of the major histocompatibility class II (MHC II) molecules expressed on antigen-presenting cells and to specific Vβ elements on the T cell receptor which induces massive T cell proliferation and cytokine release. Modified enterotoxins have been exploited for cancer immunotherapy. Enterotoxins were genetically engineered to reduce MHC II binding and were than fused to Fab parts of tumor-directed monoclonal antibodies. These tumor-targeted superantigens were able to mediate lysis of various tumor cells in vitro and in an animal model [4, 31]. Moreover, immunotherapy with C215Fab-SEA in combination with docetaxel resulted in synergistic antitumor effects [3].

Many other bacterial toxins have been shown to enhance an antibody response to fused or coadministered antigens and/or to exert adjuvant activity towards the immune cells potentiating T cell adaptive responses [5] (Table 34.2). However, their therapeutic potential as adjuvants has to be carefully evaluated, as, e.g., *B. pertussis* pertussis toxin exacerbated an experimental autoimmune encephalomyelitis in mouse model of multiple sclerosis [41].

## 34.4 Immunotoxins

Besides protein toxins used as adjuvants, also immunotoxins play an important role in cancer immunotherapy [6]. Immunotoxins are molecules consisting of a protein toxin and a ligand which is usually an antibody or its derivative, growth factor, or cytokine. On the surface of the target cell, the ligand binds to a tumor-associated antigen which then delivers the toxin into the cell cytosol, thereby causing the cell death. The most studied immunotoxins for cancer immunotherapy are based on *P. aeruginosa* exotoxin A (PE) and *C. diphtheriae* diphtheria toxin (DT). Interestingly, novel immunotoxins based on modified aerolysin of *A. hydrophila* have been recently described [26, 27]. Besides bacteria-derived immunotoxins (Table 34.2), plant-derived toxins like ricin, saporin, and pokeweed antiviral protein were used for immunotoxin preparation [6].

*P. aeruginosa* exotoxin A (PE) and *C. diphtheriae* diphtheria toxin (DT)-based immunotoxins [42] were shown to be effective mainly in hematological malignancies than in solid tumors and displayed major effects on immune responses after the failure of standard chemotherapy suggesting that they can be a useful tool for the treatment of minimal residual disease. A large number of PE- and DT-based immunotoxins directed against various tumor-associated antigens were constructed and tested in preclinical and/or phase I/II clinical trials [6–8]. The success of immunotoxins is emphasized by the FDA approval of a recombinant DT-based immunotoxin DAB(389)IL-2 (denileukin diftitox; ONTAK) for cutaneous T cell lymphoma. However, further improvement of immunotoxins is necessary to limit their side effects like vascular leak syndrome and hepatotoxicity, to increase their molecular specificity and transport through physiological barriers, and to enhance their capacity to withstand inactivation by the immune system [7, 8].

The generation of protease-activated toxins and the toxin-based suicide gene therapy represents additional approaches to toxin-based cancer immunotherapy [6]. Protease-activated toxins are engineered to be activated by cleavage of disease-related proteases after delivery to the target cells. Mainly modified protective antigen of anthrax toxin combined with lethal factor, PE or DT, has been shown to target matrix metalloproteinases or the urokinase plasminogen activator system overexpressing tumor cells in preclinical studies [6, 43, 44]. Similarly, prostate-specific antigen-cleavable sequence was introduced into proaerolysin to mediate its toxicity specifically against prostate cancer cells [26]. The toxin-based suicide gene therapy involves a DNA construct, encoding for the toxic moiety, whose expression is under the control of cancer-specific promoter. Many therapeutic vectors carrying DT subunit A were successfully tested in mice for treatment of various cancer types; however, only DTA-H19 [45] is currently tested in phase I/II study [6].

## 34.5 Bacterial Toxins Used as Antiviral Agents

Few bacterial toxins have shown a potential to be used as antiviral agents mainly against HIV infection [42] (Table 34.2). *S. dysenteriae* Shiga toxin, *P. aeruginosa* PE, and *C. diphtheriae* DT-based immunotoxins [42, 46] were shown to act directly by killing the infected cells. Especially, PE-based Env-targeting immunotoxins as a complementation of HAART might be reconsidered for clinical studies to deplete persisting HIV-infected reservoirs [46]. Pertussis toxin as well as its receptor-binding pentameric B subunits alone inhibited HIV infection in vitro and mediated a partial antiviral effect against other viruses in experimental animals [5, 42]. Similarly, *E. coli* producing Shiga-like toxin were shown to mitigate bovine leukemia virus infection in experimentally infected sheep [47]. The inherent ability of intermedilysin of *S. intermedius* to increase host cell susceptibility to a complement-mediated lysis has been exploited for an enhancement of HIV opsonization [48].

## 34.6 Bacterial Toxins Used as Antigen and Drug Delivery Agents

Several toxins and their derivatives were shown to serve as vectors for antigen delivery for treatment of viral, bacterial, and malignant diseases. Antigenic epitopes or polyepitopes can be genetically engineered or chemically coupled to the toxin molecule. Similarly, protein toxins can served also as drug or DNA delivery agents (Table 34.2). The most studied toxin is *B. pertussis* CyaA, specifically its genetically detoxified variant CyaA-AC$^-$ where the ability to catalyze conversion of cytosolic ATP to cAMP is disrupted [49, 50]. Similarly to modified *B. anthracis* anthrax toxins, CyaA-AC$^-$ was shown to accommodate large multiepitopes and stimulate antigen-specific T cell responses at very low concentrations in vivo without the need of adjuvant [12, 51–53]. Most importantly, CyaA-AC$^-$ toxoids have been demonstrated to elicit protective and therapeutic immunity against HPV-16-induced tumors and melanoma in mice [54, 55]. CyaA-AC$^-$-based vaccines for immunotherapy of cervical tumors and metastatic melanoma have entered phase I/II clinical trials [5]. CyaA-AC$^-$ carrying *M. tuberculosis* antigens was further proven to be a useful diagnostic tool for the diagnosis of latent tuberculosis [56]. In addition, *E. coli* α-hemolysin, *L. monocytogenes* Listeriolysin O (LLO), or effector proteins of *Yersinia* and *Salmonella* secreted by the type III secretion systems were used to facilitate the delivery of fused heterologous proteins or peptides for antigen presentation [5, 57–60]. Phase I study has shown that attenuated *L. monocytogenes* vaccine utilizing the LLO-based antigen delivery system is safe and well tolerated with an efficacy signal observed in terminal stage cancer patients [61].

## 34.7 Bacterial Toxins Used in Treatment of Neurological Disorders and Spinal Cord Injuries

*C. botulinum* neurotoxins represent a therapeutic paradigm as BoNT/A and BoNT/B are commercially used in the treatment of various neurological and muscle tone disorders or in the cosmetic industry [10]. However, there are some limitation in the use of botulinum neurotoxin like formation of antibodies and obliteration of response to type A toxin, diffusion of the toxin to neighboring muscles, or the need for repeated injection of toxin in chronic disorders [10, 11]. The therapeutic use has also been proposed for *C. tetani* tetanus toxin (TeNT) and its derivative nontoxic C-fragment as a carrier molecule to peripheral nerves [9]. *C. botulinum* C3 toxin, termed BA-210 (Centhrin®), reached phase IIb of clinical studies for the treatment of spinal cord injuries [62]. Interestingly, enzymatically inactive C3$_{154-182}$ peptide has recently been shown to be efficacious in posttraumatic neuro-regeneration in vivo [23, 24]. Furthermore, cytotoxic necrotizing factor 1 (CNF1) of *E. coli* has been reported to improve learning and memory [63].

## 34.8 Bacterial Toxins Commercially Used for Vaccination

Scientific advances in molecular engineering enabled the development of vaccines based on mutated bacterial toxins; however, only few are nowadays used for treatment and/or prevention of human diseases. Detoxified *C. diphtheriae* diphtheria toxoid or *C. tetani* tetanus toxoid are used for the universal vaccination against diphtheria or tetanus, respectively. The nontoxic B subunit of *V. cholerae* cholera toxin (CTB) is used as a protective antigen together with killed vibrios in a widely licensed oral cholera vaccine. Similarly, *B. pertussis* pertussis toxoid (PT9K/129G) [13] has been approved for human use as a component of a vaccine against *B. pertussis* infection [64]. Although diphtheria toxoid is successfully used as a carrier in glycoconjugate vaccines against variety of bacterial pathogens, four clinical trials failed to prove the benefit of *P. aeruginosa* exotoxin A in such vaccines.

### Conclusion

*C. botulinum* neurotoxins, MDP-derivatives represented by mifamurtide used as adjuvant, and immunotoxins represented by denileukin diftitox used in treatment of cutaneous T cell lymphoma depict the most successful achievement in the field of bacterial toxin immunotherapy. Many other bacterial toxins being evaluated in preclinical and/or in early phases of clinical studies have shown promising results. However, further research is required to improve not only their efficiency and safety but also to deepen our knowledge on bacterial toxin modes of action to harness their great potential for our own benefit.

## References

1. Schmitt, C.K., Meysick, K.C., O'Brien, A.D.: Bacterial toxins: friends or foes? Emerg. Infect. Dis. **5**, 224–234 (1999)
2. Ogawa, C., Liu, Y.J., Kobayashi, K.S.: Muramyl dipeptide and its derivatives: peptide adjuvant in immunological disorders and cancer therapy. Curr. Bioact. Compd. **7**, 180–197 (2011). doi:10.2174/157340711796817913
3. Sundstedt, A., Celander, M., Ohman, M.W., Forsberg, G., Hedlund, G.: Immunotherapy with tumor-targeted superantigens (TTS) in combination with docetaxel results in synergistic anti-tumor effects. Int. Immunopharmacol. **9**, 1063–1070 (2009). doi:10.1016/j.intimp.2009.04.013
4. Xu, M., et al.: An engineered superantigen SEC2 exhibits promising antitumor activity and low toxicity. Cancer Immunol. Immunother. **60**, 705–713 (2011). doi:10.1007/s00262-011-0986-6
5. Adkins, I., Holubova, J., Kosova, M., Sadilkova, L.: Bacteria and their toxins tamed for immunotherapy. Curr. Pharm. Biotechnol. **13**(8), 1446–1473 (2012)
6. Shapira, A., Benhar, I.: Toxin-based therapeutic approaches. Toxins (Basel) **2**, 2519–2583 (2010). doi:10.3390/toxins2112519
7. Kreitman, R.J.: Recombinant immunotoxins containing truncated bacterial toxins for the treatment of hematologic malignancies. BioDrugs **23**, 1–13 (2009)
8. Wolf, P., Elsasser-Beile, U.: Pseudomonas exotoxin A: from virulence factor to anti-cancer agent. Int. J. Med. Microbiol. **299**, 161–176 (2009)
9. Toivonen, J.M., Olivan, S., Osta, R.: Tetanus toxin C-fragment: the courier and the cure? Toxins (Basel) **2**, 2622–2644 (2010). doi:10.3390/toxins2112622
10. Truong, D.D., Jost, W.H.: Botulinum toxin: clinical use. Parkinsonism Relat. Disord. **12**, 331–355 (2006). doi:10.1016/j.parkreldis.2006.06.002
11. Johnson, E.A.: Clostridial toxins as therapeutic agents: benefits of nature's most toxic proteins. Annu. Rev. Microbiol. **53**, 551–575 (1999). doi:10.1146/annurev.micro.53.1.551
12. Fayolle, C., et al.: Delivery of multiple epitopes by recombinant detoxified adenylate cyclase of Bordetella pertussis induces protective antiviral immunity. J. Virol. **75**, 7330–7338 (2001)
13. Pizza, M., et al.: Mutants of pertussis toxin suitable for vaccine development. Science **246**, 497–500 (1989)
14. Munro, P., et al.: The Rho GTPase activators CNF1 and DNT bacterial toxins have mucosal adjuvant properties. Vaccine **23**, 2551–2556 (2005). doi:10.1016/j.vaccine.2004.11.042
15. Pizza, M., et al.: Mucosal vaccines: non toxic derivatives of LT and CT as mucosal adjuvants. Vaccine **19**, 2534–2541 (2001)
16. Lycke, N.: Targeted vaccine adjuvants based on modified cholera toxin. Curr. Mol. Med. **5**, 591–597 (2005)
17. Adamsson, J., et al.: Novel immunostimulatory agent based on CpG oligodeoxynucleotide linked to the nontoxic B subunit of cholera toxin. J. Immunol. **176**, 4902–4913 (2006)
18. Fasano, A., Uzzau, S.: Modulation of intestinal tight junctions by Zonula occludens toxin permits enteral administration of insulin and other macromolecules in an animal model. J. Clin. Investig. **99**, 1158–1164 (1997). doi:10.1172/JCI119271
19. Marinaro, M., Di Tommaso, A., Uzzau, S., Fasano, A., De Magistris, M.T.: Zonula occludens toxin is a powerful mucosal adjuvant for intranasally delivered antigens. Infect. Immun. **67**, 1287–1291 (1999)
20. Paterson, B.M., Lammers, K.M., Arrieta, M.C., Fasano, A., Meddings, J.B.: The safety, tolerance, pharmacokinetic and pharmacodynamic effects of single doses of AT-1001 in coeliac disease subjects: a proof of concept study. Aliment. Pharmacol. Ther. **26**, 757–766 (2007). doi:10.1111/j.1365-2036.2007.03413.x
21. Song, K.H., Fasano, A., Eddington, N.D.: Effect of the six-mer synthetic peptide (AT1002) fragment of zonula occludens toxin on the intestinal absorption of cyclosporin A. Int. J. Pharm. **351**, 8–14 (2008). doi:10.1016/j.ijpharm.2007.09.011
22. Fahrer, J., Rieger, J., van Zandbergen, G., Barth, H.: The C2-streptavidin delivery system promotes the uptake of biotinylated molecules in macrophages and T-leukemia cells. Biol. Chem. **391**, 1315–1325 (2010). doi:10.1515/BC.2010.132

23. Boato, F., et al.: C3 peptide enhances recovery from spinal cord injury by improved regenerative growth of descending fiber tracts. J. Cell Sci. **123**, 1652–1662 (2010). doi:10.1242/jcs.066050
24. Holtje, M., Just, I., Ahnert-Hilger, G.: Clostridial C3 proteins: recent approaches to improve neuronal growth and regeneration. Ann. Anat. **193**, 314–320 (2011). doi:10.1016/j.aanat.2011.01.008
25. Ghose, C., et al.: Transcutaneous immunization with Clostridium difficile toxoid A induces systemic and mucosal immune responses and toxin A-neutralizing antibodies in mice. Infect. Immun. **75**, 2826–2832 (2007). doi:10.1128/IAI.00127-07
26. Williams, S.A., et al.: A prostate-specific antigen-activated channel-forming toxin as therapy for prostatic disease. J. Natl. Cancer Inst. **99**, 376–385 (2007). doi:10.1093/jnci/djk065
27. Osusky, M., Teschke, L., Wang, X., Buckley, J.T.: A chimera of interleukin 2 and a binding variant of aerolysin is selectively toxic to cells displaying the interleukin 2 receptor. J. Biol. Chem. **283**, 1572–1579 (2008). doi:10.1074/jbc.M706424200
28. Janssen, K.P., et al.: In vivo tumor targeting using a novel intestinal pathogen-based delivery approach. Cancer Res. **66**, 7230–7236 (2006)
29. Engedal, N., Skotland, T., Torgersen, M.L., Sandvig, K.: Shiga toxin and its use in targeted cancer therapy and imaging. Microb. Biotechnol. **4**, 32–46 (2011). doi:10.1111/j.1751-7915.2010.00180.x
30. Battiwalla, M., et al.: Multiparameter flow cytometry for the diagnosis and monitoring of small GPI-deficient cellular populations. Cytometry B Clin. Cytom. **78**, 348–356 (2010). doi:10.1002/cyto.b.20519
31. Totterman, T.H., et al.: Targeted superantigens for immunotherapy of haematopoietic tumours. Vox Sang. **74**(Suppl 2), 483–487 (1998)
32. Ellouz, F., Adam, A., Ciorbaru, R., Lederer, E.: Minimal structural requirements for adjuvant activity of bacterial peptidoglycan derivatives. Biochem. Biophys. Res. Commun. **59**, 1317–1325 (1974)
33. Killion, J.J., Fidler, I.J.: Therapy of cancer metastasis by tumoricidal activation of tissue macrophages using liposome-encapsulated immunomodulators. Pharmacol. Ther. **78**, 141–154 (1998)
34. Li, X., et al.: Chemical conjugation of muramyl dipeptide and paclitaxel to explore the combination of immunotherapy and chemotherapy for cancer. Glycoconj. J. **25**, 415–425 (2008). doi:10.1007/s10719-007-9095-3
35. Ando, K., Mori, K., Corradini, N., Redini, F., Heymann, D.: Mifamurtide for the treatment of nonmetastatic osteosarcoma. Expert Opin. Pharmacother. **12**, 285–292 (2011). doi:10.1517/14656566.2011.543129
36. Sanchez, J., Holmgren, J.: Cholera toxin – a foe & a friend. Indian J. Med. Res. **133**, 153–163 (2011)
37. Lapa, J.A., et al.: Randomized clinical trial assessing the safety and immunogenicity of oral microencapsulated enterotoxigenic Escherichia coli surface antigen 6 with or without heat-labile enterotoxin with mutation R192G. Clin. Vaccine Immunol. **15**, 1222–1228 (2008)
38. Sun, J.B., Holmgren, J., Czerkinsky, C.: Cholera toxin B subunit: an efficient transmucosal carrier-delivery system for induction of peripheral immunological tolerance. Proc. Natl. Acad. Sci. U. S. A. **91**, 10795–10799 (1994)
39. Stanford, M., et al.: Oral tolerization with peptide 336-351 linked to cholera toxin B subunit in preventing relapses of uveitis in Behcet's disease. Clin. Exp. Immunol. **137**, 201–208 (2004)
40. Vazquez, R.I., Moreno-Fierros, L., Neri-Bazan, L., De La Riva, G.A., Lopez-Revilla, R.: Bacillus thuringiensis Cry1Ac protoxin is a potent systemic and mucosal adjuvant. Scand. J. Immunol. **49**, 578–584 (1999)
41. Munoz, J.J., Bernard, C.C., Mackay, I.R.: Elicitation of experimental allergic encephalomyelitis (EAE) in mice with the aid of pertussigen. Cell. Immunol. **83**, 92–100 (1984)
42. Alfano, M., Rizzi, C., Corti, D., Adduce, L., Poli, G.: Bacterial toxins: potential weapons against HIV infection. Curr. Pharm. Des. **11**, 2909–2926 (2005)
43. Liu, S., Bugge, T.H., Leppla, S.H.: Targeting of tumor cells by cell surface urokinase plasminogen activator-dependent anthrax toxin. J. Biol. Chem. **276**, 17976–17984 (2001)
44. Liu, S., Netzel-Arnett, S., Birkedal-Hansen, H., Leppla, S.H.: Tumor cell-selective cytotoxicity of matrix metalloproteinase-activated anthrax toxin. Cancer Res. **60**, 6061–6067 (2000)
45. Ohana, P., et al.: Use of H19 regulatory sequences for targeted gene therapy in cancer. Int. J. Cancer **98**, 645–650 (2002)
46. Berger, E.A., Pastan, I.: Immunotoxin complementation of HAART to deplete persisting HIV-infected cell reservoirs. PLoS Pathog. **6**, e1000803 (2010). doi:10.1371/journal.ppat.1000803
47. Ferens, W.A., Hovde, C.J.: The non-toxic A subunit of Shiga toxin type 1 prevents replication of bovine immunodeficiency virus in infected cells. Virus Res. **125**, 29–41 (2007)
48. Nagamune, H., et al.: The human-specific action of intermedilysin, a homolog of streptolysin O, is dictated by domain 4 of the protein. Microbiol. Immunol. **48**, 677–692 (2004)
49. Sebo, P., et al.: Cell-invasive activity of epitope-tagged adenylate cyclase of Bordetella pertussis allows in vitro presentation of a foreign epitope to CD8+ cytotoxic T cells. Infect. Immun. **63**, 3851–3857 (1995)
50. Osicka, R., et al.: Delivery of CD8(+) T-cell epitopes into major histocompatibility complex class I antigen presentation pathway by Bordetella pertussis adenylate cyclase: delineation of cell invasive structures and permissive insertion sites. Infect. Immun. **68**, 247–256 (2000)
51. Ballard, J.D., Collier, R.J., Starnbach, M.N.: Anthrax toxin-mediated delivery of a cytotoxic T-cell epitope in vivo. Proc. Natl. Acad. Sci. U. S. A. **93**, 12531–12534 (1996)

52. Fayolle, C., Sebo, P., Ladant, D., Ullmann, A., Leclerc, C.: In vivo induction of CTL responses by recombinant adenylate cyclase of Bordetella pertussis carrying viral CD8+ T cell epitopes. J. Immunol. **156**, 4697–4706 (1996)
53. Loucka, J., Schlecht, G., Vodolanova, J., Leclerc, C., Sebo, P.: Delivery of a MalE CD4(+)-T-cell epitope into the major histocompatibility complex class II antigen presentation pathway by Bordetella pertussis adenylate cyclase. Infect. Immun. **70**, 1002–1005 (2002)
54. Preville, X., Ladant, D., Timmerman, B., Leclerc, C.: Eradication of established tumors by vaccination with recombinant Bordetella pertussis adenylate cyclase carrying the human papillomavirus 16 E7 oncoprotein. Cancer Res. **65**, 641–649 (2005)
55. Fayolle, C., Ladant, D., Karimova, G., Ullmann, A., Leclerc, C.: Therapy of murine tumors with recombinant Bordetella pertussis adenylate cyclase carrying a cytotoxic T cell epitope. J. Immunol. **162**, 4157–4162 (1999)
56. Vordermeier, H.M., et al.: Recognition of mycobacterial antigens delivered by genetically detoxified Bordetella pertussis adenylate cyclase by T cells from cattle with bovine tuberculosis. Infect. Immun. **72**, 6255–6261 (2004)
57. Blight, M.A., Holland, I.B.: Heterologous protein secretion and the versatile Escherichia coli haemolysin translocator. Trends Biotechnol. **12**, 450–455 (1994)
58. Lee, K.D., Oh, Y.K., Portnoy, D.A., Swanson, J.A.: Delivery of macromolecules into cytosol using liposomes containing hemolysin from Listeria monocytogenes. J. Biol. Chem. **271**, 7249–7252 (1996)
59. Russmann, H., et al.: Protection against murine listeriosis by oral vaccination with recombinant Salmonella expressing hybrid Yersinia type III proteins. J. Immunol. **167**, 357–365 (2001)
60. Russmann, H., et al.: Delivery of epitopes by the Salmonella type III secretion system for vaccine development. Science **281**, 565–568 (1998)
61. Maciag, P.C., Radulovic, S., Rothman, J.: The first clinical use of a live-attenuated Listeria monocytogenes vaccine: a Phase I safety study of Lm-LLO-E7 in patients with advanced carcinoma of the cervix. Vaccine **27**, 3975–3983 (2009)
62. Fehlings, M.G., et al.: A phase I/IIa clinical trial of a recombinant Rho protein antagonist in acute spinal cord injury. J. Neurotrauma **28**, 787–796 (2011). doi:10.1089/neu.2011.1765
63. Diana, G., et al.: Enhancement of learning and memory after activation of cerebral Rho GTPases. Proc. Natl. Acad. Sci. U. S. A. **104**, 636–641 (2007). doi:10.1073/pnas.0610059104
64. Podda, A., et al.: Metabolic, humoral, and cellular responses in adult volunteers immunized with the genetically inactivated pertussis toxin mutant PT-9K/129G. J. Exp. Med. **172**, 861–868 (1990)
65. Sandvig, K., van Deurs, B.: Delivery into cells: lessons learned from plant and bacterial toxins. Gene Ther. **12**, 865–872 (2005). doi:10.1038/sj.gt.3302525

# Plant Heat-Shock Protein-Based Self-Adjuvanted Immunogens

## 35

Selene Baschieri

## Contents

35.1 Introduction .................................................... 551
35.2 Plants as "Biofactories" ................................. 552
35.3 Plants for the Production of Recombinant Antigens ............................ 554
35.4 Plants as "Biofactories" of Self-Adjuvanted Antigens ...................... 555
35.5 HSPs and Vaccine Development .................. 555
35.6 Plant HSPs and Their Immune Properties ..... 556
Conclusions ............................................................. 558
References ............................................................... 559

S. Baschieri, PhD
Biotechnology Laboratory ENEA,
Casaccia Research Center, Rome, Italy
e-mail: selene.baschieri@enea.it

### Abstract

Subunit vaccines are based on isolated pure or semi-pure microorganism components (antigens). As compared to traditional formulations based on whole pathogens (killed or attenuated), these vaccines are safer even if unable *per se* to boost immune responses unless supplemented by adjuvants. Nowadays, thanks to the development of high-performance gene engineering and biochemical procedures, subunit-based vaccines emerging on the market are formulated with recombinant antigens produced in bacterial, yeast or animal cells. Plant-based expression systems are turning out to be very attractive "biofactories" of recombinant antigens as well, since they ensure low-cost, rapid and easy manufacturing scaling up and intrinsic biosafety of the final product.

Nevertheless, also plant-produced recombinant antigens are *per se* poorly immunogenic. A few attempts to set-up strategies to obtain self-adjuvanted immunogens from plants have been made. This chapter will be mainly focused on the possibility to exploit to this aim plant heat-shock proteins.

### 35.1 Introduction

Vaccines should be ideally able to mimic, as much as possible, pathogen infections resulting in the activation of long-lasting protective immune responses but without inducing adverse side effects [1]. Traditional vaccine formulations based on whole killed or attenuated pathogens, albeit extremely efficient, are occasionally associated to undesired reactions [2]. To overcome these problems, the attempt to define safer formulations based only on those pathogen components (antigens) identified as able to initiate protective immune responses has been undertaken. In the near past, these subunit-based vaccines were prepared by purifying key antigens directly from the pathogens. Nowadays antigens start to be produced in their recombinant version using bacterial, yeast or animal cells [3].

The use of plants to this aim represents an opportunity that is growing out of advances in stable and transient cell transformation methods to over-express foreign genes [4]. Nonetheless, similarly to all recombinant antigens, after purification also plant-derived proteins require the cold chain to be preserved and adjuvants and injections to be delivered. Some attempts have been made to obtain intrinsically immunogenic antigens from plants thus exploiting the benefits of these expression systems at their best. To this aim, the immunogens to be expressed have been rationally designed to achieve direct delivery to antigen-presenting cells (APCs), dendritic cells (DCs) in particular, and to get the activation of both innate and pathogen-specific adaptive immune responses [5]. After a brief introduction to the methods commonly used to turn the plant into a "biofactory" of antigens, the possibility of using heat-shock proteins (HSPs) from plants as vaccine components will be herein considered.

## 35.2 Plants as "Biofactories"

Gene transfer aimed to the expression of recombinant proteins in plants can be obtained through stable or transient transformation methods [6].

To get stable transformation (and expression), a DNA fragment encoding the protein of interest and a selectable marker needs to be inserted into the nuclear or chloroplast genome in cells of leaf explants in culture [7]. The transfer of the heterologous sequence into the nuclear genome is efficiently obtained using the plant pathogen *Agrobacterium tumefaciens* transformed with binary vectors (i.e. artificial vectors including elements derived from naturally occurring tumour-inducing plasmids originally found in *A. tumefaciens* and required to transfer heterologous DNA into plant cells) [8] (Fig. 35.1), while the transfer into the plastid genome needs the use of direct DNA delivery methods, such as particle bombardment or treatment with polyethylene glycol [9].

In both cases, after transformation, the explants are used to regenerate whole plants, relying on the totipotency properties of plant cells, by growth on a selective medium that allows the survival (and plants regeneration) only to cells expressing the selectable marker gene. In order to obtain homozygosity and stable inheritance of the gene, plants transformed into the nuclear genome (and the seeds they produce through self-fertilization) are grown under selective conditions for a couple of generations, while, in the case of plastid transformation, the selection process necessary to achieve homoplasmy (the condition in which the transgene is integrated in every plastid genome, i.e. up to 10,000 per leaf cell) takes a longer time [10]. Despite this and the fact that recombinant proteins produced in chloroplasts may have "defects" similar to those produced in bacteria in terms of post-translational modification (such as differences in the glycosylation pattern), plastid transformation has some interesting features. Indeed, the foreign gene is inserted into a specific site of the genome by homologous recombination, an opportunity that eliminates problems typical of random gene insertion into the nuclear genome ("position effects"). For this reason and because gene silencing does not occur in chloroplasts, very high expression levels can be obtained by this transformation method. Moreover transplastomic plants are "naturally contained" because plastids are not transmitted through the pollen [11].

As compared to stable transformation methods, transient transformation is a much more rapid and easy high-throughput procedure to produce large quantities of recombinant proteins in plants but, as suggested by the name, the heterologous sequence is only temporary expressed and cannot be transmitted to the plant progeny as

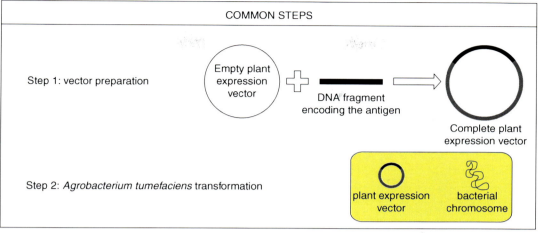
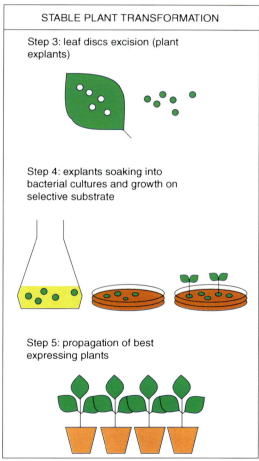
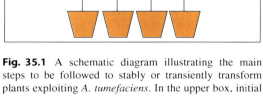
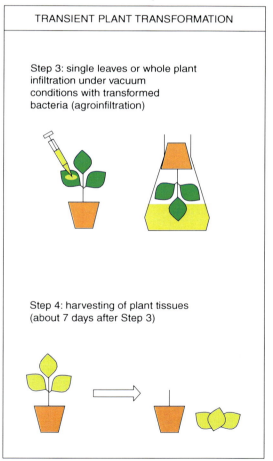

**Fig. 35.1** A schematic diagram illustrating the main steps to be followed to stably or transiently transform plants exploiting *A. tumefaciens*. In the upper box, initial genetic engineering and bacteria transformation steps common to both plant transformation procedures

transformation is performed in vivo on fully developed plants [12].

One of the methods that can be used to get transient transformation has the same initial step of stable nuclear transformation, i.e. the insertion of the sequence to be expressed into a binary vector that is then used to transform *A. tumefaciens* cells [13] (Fig. 35.1). In transient transformation however, a suspension of the transformed bacteria is then infiltrated onto the abaxial (i.e. underside) surface of leaves attached to the plant using a syringe deprived of the needle. This procedure, called agroinfiltration, allows the penetration of the suspension into the leaf airspaces through the stomata or through a tiny incision made on the leaf. Alternatively, the whole plant can be submerged upside down into a beaker containing the bacterial suspension and placed in a vacuum chamber. When the vacuum is applied, air is forced out of the mesophyll to be substituted by the bacterial suspension when vacuum is released. *Agrobacterium*-mediated transient expression is restricted to infiltrated plant tissues.

Instead of agroinfiltration, transient expression can be gained by inserting the heterologous sequence into the genome of a plant virus (or into an expression vector encoding a complete viral genome) and using the genetically modified plant pathogen to infect plants [14]. To this aim, few microliters of a solution containing recombinant plant virus particles or the expression vector encoding the recombinant viral genome are distributed on the adaxial (i.e. upper) surface of the leaf, and the expression of the heterologous protein proceeds throughout the plant together with viral infection.

Recently, several transient over-expression systems have been developed that conjugate the efficiency of *Agrobacterium*-mediated transformation with viral rate of expression [15]. These vectors encode for those few elements of the genome of a plant virus important in controlling gene expression and are vehiculated into plant cells by recombinant Agrobacteria. These modular vector systems besides ending up with very high expression levels in plant tissues are also intrinsically safe because plant viruses never assemble.

## 35.3 Plants for the Production of Recombinant Antigens

All the presented strategies have been extensively used to express recombinant antigens in plants [16]. Even if stable transformation approaches have the main disadvantage of being extremely time-consuming as compared to transient transformation, they are the methods of choice to produce recombinant antigens when the aim is to use the plant itself as a (edible) vaccine [17]. The enormous potentialities of this kind of application of plant "biofactories" have been demonstrated through several clinical trials in which a wide range of plant species expressing different antigens were orally delivered to humans [18]. Plant scientists are working hard to improve this technology, for example, by trying to target the insertion of the heterologous genes into specific sites of the nuclear genome through homologous recombination to obtain higher and reproducible expression levels of the encoded antigen in plant tissues and to guarantee that a dose of antigen sufficient to confer protection is ingested in a single serving [19]. The aim is to strengthen a technology that by reducing the production and preservation costs could represent a real breakthrough towards the elimination of still existing disparities between western and developing countries thus improving global health equity [20].

If the use of the plants as mere "biofactories" is intended, the adoption of transient transformation methods is probably more appropriate because these methods ensure higher expression levels in shorter times. A critical point concerns in this case the purification phase as the efficiency of this production step can be extremely variable [21]. The problem is less evident if plants are used to express self-assembling antigens for the production of virus-like particles (VLPs) [22]. This approach has been successfully implemented to produce influenza VLPs by transiently expressing hemagglutinin in *Nicotiana benthamiana* plants [23]. The use of this platform to produce influenza vaccines offers significant advantages in terms of time of production, safety and scale up. In particular plant "biofactories" can ensure the achievement of 1,500 doses from about 1 kg of agroinfiltrated leaves within 3 weeks from the release of a novel viral sequence while the egg-based supply of the global seasonal influenza vaccine requires 4–6 months [24].

## 35.4 Plants as "Biofactories" of Self-Adjuvanted Antigens

A major limitation of subunit-based vaccines is that purified antigens are *per se* unable to elicit protective wide-spectrum immune responses and to result in the stimulation of both innate and adaptive immune systems, activating both B- and T-lymphocytes [25]. For this reason these vaccines must be potentiated by immune-stimulatory molecules (adjuvants). Adjuvants currently accepted for human and veterinary use are very few, in that potent immune-stimulation can result in adverse reactions [26]. Studies are underway in an effort to identify novel delivery strategies or molecules working as real boosters of the immune response to antigens and devoid of unwanted side effects [27]. This effort is brought about also by plant biotechnologists that are trying to identify plant products/structures that could be used to vehicle recombinant antigens. Among the attempts made in this direction are the construction of chimeric plant viruses [28] or of chimeric plant organelles [29] displaying peptides or proteins and the production in plants of immune complexes made of antigens and antibodies [30]. Another approach has considered the possible use in vaccine formulations of plant HSPs.

## 35.5 HSPs and Vaccine Development

HSPs are a heterogeneous group of ubiquitous proteins classified into six major families, named in accordance to the molecular weight (with the exception of calreticulin, calmodulin and ubiquitin) [31]. Apart from small HSPs, they are among the most phylogenetically conserved proteins of living organisms [32]. Originally, HSP expression has been considered as strictly related to stress because it is strongly evident under stressful conditions (such as temperature shift), but it is now clear that HSPs are constitutively expressed housekeeping proteins. Indeed, they are essential in the control of cellular proteome homeostasis as involved in protein folding and unfolding, intra- and extracellular targeting, degradation and multimer assembly [33]. Each HSP family includes several highly homologous isoforms, associated to particular cell compartments and/or tissues and expressed in normal or stress conditions [34]. HSP90 and HSP70 are the most extensively studied HSP families [35].

HSP90 is the core component of a dynamic set of proteins (co-chaperones) whose number and roles have not yet been fully clarified. In eukaryotes it is involved in the maturation and activation of key signalling proteins such as hormone receptors, transcription factors and protein kinases. It works as a dimer and is structurally characterized by three domains [36]. The N-terminal domain is endowed of ATPase activity and fundamental for the biological function of the chaperone. The middle domain is involved in the interaction with client proteins. The C-terminal domain is essential for dimerization. Proteins associate to HSP90 dimers in the absence of ATP. Upon ATP binding, the N-terminal domains of the two HSP90 components rotate, come in contact with each other and produce the activation, stabilization or assembly of the client protein through a still undefined mechanism.

HSP70 controls the conformational states of proteins and is essential in protein biogenesis, transport and degradation [37]. The analysis of the structure of HSP70 isoforms from different organisms has revealed that the folding of this chaperone is similar due to the high degree of homology in the primary structure [38–40]. All the isoforms are characterized by two main domains. The N-terminal domain is endowed with ATPase activity (nucleotide-binding domain, NBD), binds ATP and hydrolyzes it to ADP. The C-terminal domain can be divided into two subdomains: (1) a pocket-shaped substrate-binding domain (SBD) formed by two β-sheets and involved in polypeptide binding [41] and (2) a "lid domain" formed by a pair of α-helices whose conformation influences substrate binding [42]. The communication between the two domains is bidirectional and HSP70 exists in two defined and stable molecular conformations. When the NBD is occupied by ATP the "lid" is open and the SBD is free, while when the NBD is occupied by ADP the lid is closed and the SBD is occupied by a polypeptide (unfolded, natively folded or aggregated) [43].

HSP90 and HSP70 besides being involved in controlling protein shaping are also endowed of immune-stimulating properties [44]. This feature has been revealed the first time by a study demonstrating that selected protein components isolated from murine sarcomas and later identified as HSP were able to confer protection to tumour challenge when injected systemically. The protection resulted by the activation of an adaptive immune response specific to the tumour [45]. Cancer vaccines based on tumour-derived HSPs are currently under evaluation in clinical trials [46, 47].

It has been reported that also HSPs extracted from virus-infected cells are able to activate CD8+ T-cell clones specific for viral peptides [48] and that HSPs purified from mammalian P815 cells expressing a heterologous protein are able to activate major histocompatibility complex (MHC) class I-restricted T-lymphocytes specific for the recombinant protein in H-$2^d$ mice (i.e. the same haplotype of P815 cells) as well as in mice of different haplotypes [49].

Once purified, HSPs can be also *in vitro* unloaded of naturally bound peptides, to be subsequently loaded with a peptide of choice. These complexes, delivered to mice without adjuvant, are able to elicit both antibody and cell-mediated immune responses specific to the carried peptide [50]. Peptide-loaded mycobacterial HSP70 are far more efficient in the generation of peptide-specific cytotoxic T-cell responses than peptide alone [51].

The main reasons why HSPs are endowed of immunological properties [52] are that (1) the peptides chaperoned by HSP are the fingerprinting of the proteins expressed within a cell at a given time, (2) the complexes formed by HSP and peptides are stable and can be purified, (3) HSP-peptide complexes can be up-taken by APCs through the interaction with specific receptors (such as CD91, expressed on DCs) [53, 54], and (4) inside APCs, the peptides are transferred from HSPs to MHC molecules to be transported to the cell membrane for presentation.

Beside their ability to deliver antigens to APCs (specifically DCs) via receptor binding and internalization, HSPs have been demonstrated to be endowed of adjuvant properties as they are active alone in inducing the release by DCs of pro-inflammatory cytokines [55]. The two functions (antigen delivery vs. DCs' stimulation) are structurally separated since a single mutation, able to prevent peptide binding to SBD, do not affect the ability to stimulate innate immunity [51].

Overall, these observations together with the evidence that HSPs derived from phylogenetically distant organisms are able to activate immune responses in mice [56] suggested that also plant HSPs could find an application in vaccinology as simple adjuvants or as carriers for antigen delivery.

## 35.6 Plant HSPs and Their Immune Properties

Plants are sessile organisms and environmental and biological stresses are a major challenge for their protein homeostasis. They respond to different abiotic stresses, such as low and high temperatures, drought, salinity or flooding, by producing HSPs. Also infection by plant viruses and agroinfiltration are strong HSP inducers [57] as these biological stresses result in the massive accumulation of misfolded proteins with severe consequences for cellular metabolism [58]. Moreover, HSPs have been demonstrated to actively participate in viral protein maturation and turnover during replication and/or movement [59].

Specific HSP90 and HSP70 proteins are found in different plant cell compartments (chloroplasts, mitochondria, endoplasmic reticulum). These plant chaperones are expected to be highly similar to those present in animal cells both structurally and functionally. Primarily, this hypothesis gains support from the evidence that amino acid sequence identities between HSPs from plants and from other eukaryotes are significantly high.

The percent identity among *Nicotiana tabacum* cytosolic HSP70 and the corresponding protein expressed in *Homo sapiens*, *Bos taurus* and *Mus musculus* exceed in average 75 %. The alignment of the sequences evidences that the motifs critical for ATP/ADP or peptide binding are perfectly conserved. The structural superimposability is definitely confirmed by the comparison of the solved structure of a mammalian HSP70 with the model of a plant HSP70 (designed adopting the homology modelling method) [57].

As it concerns functional aspects, it has been shown that similarly to what happens in mammalian cells, the downregulation of cytosolic HSP70 in *Arabidopsis thaliana* plants reduces thermotolerance [60]. Moreover geldanamycin, a drug that acts specifically on animal HSP90, is effective also on plant HSP90, while a complex formed by plant HSP90 and HSP70 is able to correctly assemble animal steroid receptors and oncogenic protein kinases [61].

On this basis, the possibility that the similarities between plant and mammalian HSPs could be extended also to immune properties has been explored.

One study was intended to determine if HSP90 from plants have the same adjuvant properties of their mammalian counterpart. In this study, the sequences encoding HSP90 in *A. thaliana* and *N. benthamiana* were cloned, independently expressed in *Escherichia coli* as fusion to a histidine tag and purified through nickel affinity chromatography. The purified proteins (that have around 68 % amino acid sequence identity with those in *H. sapiens* HSP90) were tested for their ability to stimulate *in vitro* the proliferation of murine splenocytes. The results of these experiments indicated that plant-derived HSP90 is able to bind to toll-like receptor 4 expressed on the plasma membrane of B-lymphocytes and is selectively mitogenic for these cells suggesting that this plant chaperone is endowed of immune-stimulatory properties [62].

Two more studies were intended to establish if plant HSP70 could be used to efficiently deliver plant-expressed proteins/antigens [57, 63]. To this aim, *N. benthamiana* plants were used to transiently express a reporter protein (the coat protein (CP) of the plant virus potato virus X (PVX)) or the highly conserved influenza A virus nucleoprotein (NP). This antigen was selected to take advantage of data available on the fine characterization of T-lymphocyte-mediated immune responses it triggers in genetically different strains of mice [64, 65]. NP was also intriguing for vaccine design as it is known to induce the activation of cross-protective immune responses against different viral variants, therefore potentially useful in the case of the outbreak of "new" influenza viruses against whom pre-existing humoral immunity is not (or poorly) effective [66].

Transient expression of the CP of PVX was induced by inoculating the plants with an expression vector encoding the complete viral genome while the transient expression of NP was obtained by infiltrating plant leaves with *A. tumefaciens* cells carrying a binary vector including an NP-encoding cassette. Seven days after viral infection or agroinfiltration (the time interval necessary to get in both cases maximal expression levels of the heterologous protein and also of endogenous plant HSP70), *N. benthamiana* leaves were sampled and used to purify plant HSP70. The chromatographic method used to this aim was essentially the same developed to purify mammalian HSP70 and was based on the affinity of this chaperone to ATP/ADP. To preserve pHSP70 binding to polypeptides, the purification was performed using an ADP-agarose resin as this approach prevents the release from the SBD of the chaperoned peptides.

The obtained recovery of highly pure plant HSP70 chaperoning polypeptides of approximately 23 μg plant HSP70 per gram of fresh leaves was reproducible. The purified plant HSP70-polypeptide complexes were tested for their ability to activate immune responses specific for the heterologous proteins expressed in plant tissues. The results of these experiments clearly evidenced that mice immunization without adjuvant was able to induce CP- and NP-specific antibody responses.

As predicted, the antibodies present in the sera were able to react also against a wild-type plant crude extract (depleted of plant HSP70). This was expected as plant HSP70 is likely to carry peptides derived not only from the recombinant proteins but also from endogenous plant cell proteins. Even so, the antibody responses against these plant targets were lower if compared to CP- or NP-specific responses probably because, at the sampling time, CP and NP were overrepresented in plant cells as compared to endogenous proteins. The antibody response induced in mice was directed also against plant HSP70. Also this result was expected. Nonetheless it was demonstrated that, despite the high similarity between mammalian and plant HSP70, anti-plant HSP70 antibodies are unable to cross-react with murine or human HSP70, thus are unable to work as possible trigger of autoimmune responses, an issue that cannot be excluded for HSP-based cancer vaccines prepared from autologous tumour cells that could induce the activation of undesired immune responses against self-antigens.

The ability of the purified plant HSP70-polypeptide complexes to prime T-lymphocyte-mediated immune responses specific to the NP antigen was also verified. In this case the immune response was evaluated by interferon-γ enzyme-linked immunosorbent spot (IFNγ ELISPOT) assay. This method allows to detect and enumerate antigen-specific CD8+ T-cells within *ex vivo* cultures of splenocytes, because these cells secrete IFNγ when stimulated with MHC class I-restricted synthetic peptides, whose sequences are extrapolated from the antigen. Briefly, C57BL/6 mice were immunized with pHSP70-polypeptide complexes purified from plant tissues expressing NP and their splenocytes stimulated *ex vivo* with five different NP-derived peptides (singly or in different combinations). All the peptides were known to specifically activate $H-2D^b$-restricted CD8+ T-cell responses. The sequence of one of the peptides was that of the NP immune-dominant epitope in C57BL/6 J mice [64] while the others were epitopes that are subdominant in this strain [65].

Overall the results clearly showed that plant HSP70-polypeptide complexes were able to prime T-cell-mediated immune responses to NP without the need of adjuvant co-delivery and that the response was polyclonal and directed against both dominant and subdominant epitopes. The higher response was obtained when stimulating the splenocytes with the complete pool of peptides (immune-dominant+subdominants). The polyclonality of the response was confirmed by experiments demonstrating that the same plant HSP70 preparation used to immunize C57BL/6 J mice was effective in inducing the activation of CD8+ T-cells also in Balb/c mice that have a different genetic background ($H-2K^d$), but in this case the T-cell response was directed against a different peptide. In both strains of mice, the efficacy of the pHSP70 preparation in inducing CD8+ T-cell activation was maintained even after endotoxin removal.

These data evidence a general ability of plant HSPs of stimulating immune responses and shed light also on a more specific use of HSP-polypeptide complexes derived from plant tissues expressing recombinant antigens. Plant HSP70-carrying peptides appear to induce both antibody production and CD8+ T-cell activation. The responses are in both cases polyclonal and, intriguingly, do not need adjuvant delivery. These data indicate that HSP70 derived from plants expressing recombinant antigens could be used as multiepitope vaccines to stimulate immune responses specific to the antigen in individuals of any given MHC type. The immunization strategy based on plant HSP70 complexes could be particularly attractive also for the development of vaccines against complex pathogens in which the protective antigens are either difficult to achieve or unknown. The fact that the expressed foreign polypeptide does not represent the final target of the extraction procedure is one of the major technological strong points of this approach as, in this way, all the benefits of plant-based expression systems are maintained by overcoming the often limiting steps of purification whose efficiency usually depends on intrinsic features of the recombinant protein.

### Conclusions

The use of plants in the field of vaccine development have been envisaged and could represent an opportunity to face some of still open issues such as efficacy, safety and, last but not least, ethical and social aspects of the research in vaccinology.

As it concerns specifically the possible use of plant HSPs, the results obtained till now are very promising and indicate that further exploration is desirable to refine the characterization of the immune properties of these natural plant products. An opportunity would be to test their efficacy in the veterinarian field by developing vaccines for one of the innumerable animal infections requiring appropriate and low-cost protective vaccinations.

The unconventional exploitation of plants in immunology is the demonstration that a multidisciplinary "cross-fertilizing" approach to current relevant issues related to vaccination can be the direct path to innovation.

## References

1. Germain, R.N.: Vaccines and the future of human immunology. Immunity **33**, 441–450 (2010)
2. Plotkin, S.A., Plotkin, S.L.: The development of vaccines: how the past let to the future. Nat. Rev. Immunol. **9**, 889–893 (2011)
3. Ulmer, J.B., Valley, U., Rappuoli, R.: Vaccine manufacturing: challenges and solutions. Nat. Biotechnol. **24**, 1377–1383 (2006)
4. Streatfield, S.J.: Approaches to achieve high-level heterologous protein production in plants. Plant Biotechnol. J. **5**, 2–15 (2007)
5. Lico, C., et al.: Plant-based vaccine delivery strategy. In: Baschieri, S. (ed.) Innovation in Vaccinology. From Design, Through to delivery and Testing. Springer, New York (2012)
6. Fischer, R., et al.: Plant-based production of biopharmaceuticals. Curr. Opin. Plant Biol. **7**, 152–158 (2004)
7. Newell, C.A.: Plant transformation technology. Mol. Biotechnol. **16**, 53–65 (2000)
8. Lee, L.-Y., Gelvin, S.B.: T-DNA binary vectors and systems. Plant Physiol. **146**, 325–332 (2008)
9. Bock, R., Khan, M.S.: Taming plastids for a green future. Trends Biotechnol. **22**, 311–318 (2004)
10. Meyers, B., et al.: Nuclear and plastid genetic engineering of plants: comparison of opportunities and challenges. Biotechnol. Adv. **28**, 747–756 (2010)
11. Bock, R.: Plastid biotechnology: prospects for herbicide and insect resistance, metabolic engineering and molecular farming. Curr. Opin. Biotechnol. **18**, 100–106 (2007)
12. Komarova, T.V., et al.: Transient expression systems for plant-derived biopharmaceuticals. Expert Rev. Vaccines **9**, 859–876 (2010)
13. Sheludko, Y.V.: Agrobacterium-mediated transient expression as an approach to production of recombinant proteins in plants. Recent Pat. Biotechnol. **2**, 198–208 (2008)
14. Pogue, G.P., et al.: Making an ally from an enemy: plant virology and the new agriculture. Annu. Rev. Phytopathol. **40**, 45–74 (2002)
15. Lico, C., Chen, Q., Santi, L.: Viral vectors for production of recombinant proteins in plants. J. Cell. Physiol. **216**, 366–377 (2008)
16. Rybicki, E.P.: Plant-made vaccines for humans and animals. Plant Biotechnol. J. **8**, 620–637 (2010)
17. Walmsley, A.M., Arntzen, C.J.: Plants for delivery of edible vaccines. Curr. Opin. Biotechnol. **11**, 126–129 (2000)
18. Tacket, C.O.: Plant-based oral vaccines: results of human trials. Curr. Top. Microbiol. Immunol. **332**, 103–117 (2009)
19. Butaye, K.M.J., et al.: Approaches to minimize variation of transgene expression in plants. Mol. Breed. **16**, 79–91 (2005)
20. Penney, C.A., et al.: Plant-made vaccines in support of the Millennium Development Goals. Plant Cell Rep. **30**, 789–798 (2011)
21. Wilken, L.R., Nikolov, Z.L.: Recovery and purification of plant-made recombinant proteins. Biotechnol. Adv. **30**, 419–433 (2012)
22. Santi, L., Huang, Z., Mason, H.: Virus-like particles production in green plants. Methods **40**, 66–76 (2006)
23. D'Aoust, M.A., et al.: Influenza virus-like particles produced by transient expression in *Nicotiana benthamiana* induce a protective immune response against a lethal viral challenge in mice. Plant Biotechnol. J. **6**, 930–940 (2008)
24. D'Aoust, M.A., et al.: The production of hemagglutinin-based virus-like particles in plants: a rapid, efficient and safe response to pandemic influenza. Plant Biotechnol. J. **8**, 607–619 (2010)
25. Plotkin, S.A.: Vaccines: the fourth century. Clin. Vaccine Immunol. **16**, 1709–1719 (2009)
26. Guy, B.: The perfect mix: recent progress in adjuvant research. Nat. Rev. Microbiol. **5**, 505–517 (2007)
27. Bachmann, M.F., Jennings, G.T.: Vaccine delivery: a matter of size, geometry, kinetics and molecular patterns. Nat. Rev. Immunol. **10**, 787–796 (2010)
28. Lico, C., et al.: Plant-produced potato virus X chimeric particles displaying an influenza virus-derived peptide activate specific CD8+ T cells in mice. Vaccine **27**, 5069–5076 (2009)
29. Capuano, F., et al.: LC-MS/MS methods for absolute quantification and identification of proteins associated with chimeric plant oil bodies. Anal. Chem. **83**, 9267–9272 (2011)
30. Chargelegue, D., et al.: Highly immunogenic and protective recombinant vaccine candidate expressed in transgenic plants. Infect. Immun. **73**, 5915–5922 (2005)
31. Feder, M.E., Hoffmann, G.E.: Heat-shock proteins, molecular chaperones, and the stress response: evolutionary and ecological physiology. Annu. Rev. Physiol. **61**, 243–282 (1999)
32. Lindquist, S., Craig, E.A.: The heat-shock proteins. Annu. Rev. Genet. **22**, 631–677 (1988)
33. Christis, C., Lubsen, N.H., Braakman, I.: Protein folding includes oligomerization – examples from the endoplasmic reticulum and cytosol. FEBS J. **275**, 4700–4727 (2008)

34. Rutherford, S.L.: Between genotype and phenotype: protein chaperones and evolvability. Nat. Rev. Genet. **4**, 263–274 (2003)
35. Wegele, H., Muller, L., Buchner, J.: Hsp70 and Hsp90 – a relay team for protein folding. Rev. Physiol. Biochem. Pharmacol. **151**, 1–44 (2004)
36. Pearl, L.H., Prodromou, C.: Structure and mechanism of the HSP90 molecular chaperone machinery. Annu. Rev. Biochem. **75**, 271–294 (2006)
37. Morano, K.A.: New tricks for an old dog. The evolving world of Hsp70. Ann. N. Y. Acad. Sci. **1113**, 1–14 (2007)
38. Zhu, X., et al.: Structural analysis of substrate binding by the molecular chaperone DnaK. Science **72**, 1606–1614 (1996)
39. Worrall, L.J., Walkinshaw, M.D.: Crystal structure of the C-terminal three-helix bundle subdomain of *C. elegans* HSP70. Biochem. Biophys. Res. Commun. **357**, 105–110 (2007)
40. Jiang, J., Lafer, E.M., Sousa, R.: Crystallization of a functionally intact HSC70 chaperone. Acta Crystallogr. Sect. F Struct. Biol. Cryst. Commun. **62**, 39–43 (2006)
41. Rudiger, S., Buchberger, A., Bukau, B.: Interaction of Hsp70 chaperones with substrates. Nat. Struct. Biol. **4**, 342–349 (1997)
42. Schlecht, R., et al.: Mechanics of Hsp70 chaperones enables differential interaction with client proteins. Nat. Struct. Mol. Biol. **18**, 345–351 (2011)
43. Nicolai, A., et al.: Human inducible Hsp70: structures, dynamics, and interdomain communication from all-atom molecular dynamics simulations. J. Chem. Theory Comput. **6**, 2501–2519 (2010)
44. Srivastava, P.K.: Roles of heat-shock proteins in innate and adaptive immunity. Nat. Rev. Immunol. **2**, 185–194 (2002)
45. Srivastava, P.K., Deleo, A.B., Old, L.J.: Tumor rejection antigens of chemically induced sarcomas of inbred mice. Proc. Natl. Acad. Sci. U. S. A. **83**, 3407–3411 (1986)
46. Murshid, A., Gong, J., Calderwood, S.K.: Heat-shock proteins in cancer vaccines: agents of antigen cross-presentation. Expert Rev. Vaccines **7**, 1019–1030 (2008)
47. Srivastava, R.M., Khar, A.: Dendritic cells and their receptors in antitumor immune response. Curr. Mol. Med. **6**, 708–724 (2009)
48. Nieland, T.J.F., et al.: Isolation of an immunodominant viral peptide that is endogenously bound to the stress protein GP96/GRP94. Proc. Natl. Acad. Sci. U. S. A. **93**, 6135–6139 (1996)
49. Arnold, D., et al.: Cross-priming of minor histocompatibility antigen-specific cytotoxic T cells upon immunization with the heat shock protein gp96. J. Exp. Med. **182**, 885–889 (1995)
50. Blachere, N.E., et al.: Heat shock protein-peptide complexes, reconstituted in vitro, elicit peptide-specific cytotoxic T lymphocyte response and tumor immunity. J. Exp. Med. **186**, 1315–1323 (1997)
51. MacAry, P.A., et al.: HSP70 peptide binding mutants separate antigen delivery from dendritic cell stimulation. Immunity **20**, 95–106 (2004)
52. Javid, B., MacAry, P.A., Lehner, P.J.: Structure and function: heat shock proteins and adaptive immunity. J. Immunol. **179**, 2035–2040 (2007)
53. Basu, S., et al.: CD91 is a common receptor for heat shock proteins gp96, hsp90, hsp70, and calreticulin. Immunity **14**, 303–313 (2001)
54. Basu, S., Matsutake, T.: Heat shock protein-antigen presenting cell interactions. Methods **32**, 38–41 (2004)
55. Asea, A., et al.: HSP70 stimulates cytokine production through a CD14-dependant pathway, demonstrating its dual role as a chaperone and cytokine. Nat. Med. **6**, 435–442 (2000)
56. Kumaraguru, U., et al.: Antigenic peptides complexed to phylogenetically diverse Hsp70s induce differential immune responses. Cell Stress Chaperones **8**, 134–143 (2003)
57. Buriani, G., et al.: Plant heat shock protein 70 as carrier for immunization against a plant-expressed reporter antigen. Transgenic Res. **20**, 331–344 (2011)
58. Sugio, A., et al.: The cytosolic protein response as a subcomponent of the wider heat shock response in *Arabidopsis*. Plant Cell **21**, 642–654 (2009)
59. Aparicio, F., et al.: Virus induction of heat shock protein 70 reflects a general response to protein accumulation in the plant cytosol. Plant Physiol. **138**, 529–536 (2005)
60. Lee, J.H., Schöffl, F.: An Hsp70 antisense gene affects the expression of HSP70/HSC70, the regulation of HSF, and the acquisition of thermotolerance in transgenic *Arabidopsis thaliana*. Mol. Gen. Genet. **252**, 11–19 (1996)
61. Kadota, Y., Shirasu, K.: The HSP90 complex of plants. Biochim. Biophys. Acta **1823**, 689–697 (2012)
62. Corigliano, M.G., et al.: Plant Hsp90 proteins interact with B-cells and stimulate their proliferation. PLoS One **6**, e21231 (2011)
63. Buriani, G., et al.: Heat-shock protein 70 from plant biofactories of recombinant antigens activate multiepitope-targeted immune responses. Plant Biotechnol. J. **10**, 363–371 (2012)
64. Townsend, A.R., et al.: The epitopes of influenza nucleoprotein recognized by cytotoxic T lymphocytes can be defined with short synthetic peptides. Cell **44**, 959–968 (1986)
65. Oukka, M., et al.: Protection against lethal viral infection by vaccination with non immunodominant peptides. J. Immunol. **157**, 3039–3045 (1996)
66. Doherty, P.C., Kelso, A.: Toward a broadly protective influenza vaccine. J. Clin. Invest. **118**, 3273–3275 (2008)

# Functionalised Nanoliposomes for Construction of Recombinant Vaccines: Lyme Disease as an Example

## 36

Jaroslav Turánek, Josef Mašek, Michal Křupka, and Milan Raška

## Contents

36.1   Introduction ............................................. 562
36.2   Lyme Diseases ........................................ 563
36.3   History ..................................................... 563
36.4   Aetiological Agents ................................ 563
36.5   Epidemiology, Clinical Disease, and Treatment ........................................ 563
36.6   Antigens and Immune Response ........... 564
36.7   Licensed Veterinary Lyme Disease Vaccines .................................................. 565
36.8   Ferritin ..................................................... 565
36.9   OspC Antigen .......................................... 565
36.10  Liposomes as Biocompatible and Versatile Carriers for Construction of Recombinant Vaccines ...... 566
36.11  Chemical, Physicochemical, and Structural Parameters Affecting Activity of Liposome-Based Vaccine ......... 566
36.12  Cationic Liposomes ................................ 568
36.13  Binding of Antigen to Liposomes .......... 569
36.14  Metallochelating Bond and Metallochelating Liposomes .......... 571
36.15  Adjuvants for Liposomal-Based Vaccines .................................................. 571
References ........................................................ 575

J. Turánek, Dr. Sc (✉) • J. Mašek, PharmDr
Department of Pharmacology and Immunotherapy,
Veterinary Research Institute,
Brno, Czech Republic
e-mail: turanek@vri.cz

M. Křupka, PhD • M. Raška, MD, PhD
Department of Immunology,
Faculty of Medicine and Dentistry,
Palacky University, Olomouc,
Czech Republic

### Abstract

Liposomes (phospholipid bilayer vesicles) represent an almost ideal carrier system for the preparation of synthetic vaccines due to their biodegradability and capacity to protect and transport molecules of different physicochemical properties (including size, hydrophilicity, hydrophobicity, and charge). Liposomal carriers can be applied by invasive (e.g. *i.m.*, *s.c.*, *i.d.*) as well as non-invasive (transdermal and mucosal) routes. In the last 15 years, liposome vaccine technology has matured and several vaccines containing liposome-based adjuvants have been approved for human and veterinary use or have reached late stages of clinical evaluation.

Given the intensifying interest in liposome-based vaccines, it is important to understand precisely how liposomes interact with the immune system and how they stimulate immunity. It has become clear that the physicochemical properties of liposomal vaccines – method of antigen attachment, lipid composition, bilayer fluidity, particle charge, and other properties – exert strong effects on the resulting immune response. In this chapter we will discuss some aspects of liposomal vaccines including the effect of novel and emerging immunomodulator incorporation. The application of metallochelating nanoliposomes for development of recombinant vaccine against Lyme disease will be presented as a suitable example.

## 36.1 Introduction

Vaccinology as a scientific field is undergoing dramatic development. Sophisticated techniques and rapidly growing in-depth knowledge of immunological mechanisms are at hand to exploit fully the potential for protecting from, as well as curing of diseases through vaccination. In spite of great successes like eradication of smallpox in the 1970s and in a lesser extent poliomyelitis (two important milestones in medical history), new challenges have arisen to be faced. Rapidly changing ecosystems and human behaviour, an ever-increasing density of human and farmed animal populations, a high degree of mobility resulting in rapid spreading of pathogens in infected people and animals, poverty and war conflicts in the third world, and many other factors contribute to the more frequent occurrence and rapid dissemination of new as well as some old infectious disease.

The three infections that most heavily afflict global health are AIDS, tuberculosis, and malaria. As an example of new viral pathogens, Ebola virus, SARS coronavirus, or new strains of influenza virus can be mentioned [1]. Rapid sequencing of the genome of pathogens leads to development of sensitive molecular diagnostic tools and augments identification and expression of recombinant antigenic targets for future vaccines [2]. Special field represents immunotherapy of cancer, where anticancer vaccines could be a powerful weapon for long-term effective treatment.

The progress in vaccine development is tightly connected not only with new findings in immunology but also in molecular biology and biotechnology. The new term "reverse vaccinology" was proposed by Rappuoli to describe a complex genome-based approach toward vaccine design [3]. In comparison with conventional approaches which require a laborious process of attenuation or inactivation of pathogens, or selection of individual components important for induction of immune response, reverse vaccinology offers the possibility of using genomic information derived from in silico analyses for direct design and production of protective antigen using recombinant technology.

This approach can significantly reduce the time necessary for the identification of antigens for development of candidate vaccine and enables systematic identification of all potential antigens even from pathogens which are difficult or currently impossible to culture. Of course, this approach is limited to identification of protein or glycoprotein antigens, omitting such important vaccine components such as polysaccharides and glycolipids. The principal question for reverse vaccinology consists in identification of protective antigen, which presents the main hurdle of this approach. Nevertheless, once the protective antigen is identified it enables scientists to systemically classify such antigens, and develop efficient preparations virtually against any pathogen that has had its genome sequence determined.

Subunit vaccines offer superior safety profiles and can be manufactured with minimal risk of contamination [4, 5]. When coupled with appropriate adjuvants, they can also focus the immune response on protective or highly conserved antigenic determinants that may not elicit a potent response during natural infection or after vaccination with an inactivated or attenuated pathogen [6, 7].

A common observation from the process of elicitation of the adaptive immune response is that the antigen by itself is not a stimulating agent. In other words, administration of absolutely pure recombinant protein antigens and synthetic peptide antigens generally does not induce specific immune response. Therefore there is a need for potent co-stimulation by co-administration of appropriate adjuvants, biocompatible carrier systems and application devices for vaccines consisting of highly purified antigens. These particulate systems are supposed to mediate efficient delivery to antigen-presenting cells and may induce inflammation through activation of innate immunity [8–10].

## 36.2 Lyme Diseases

Lyme disease is the most frequent zoonosis both in Europe and the United States. Disease may progress into a chronic form and cause damage of the nervous or cardiovascular system, joints, skin, or eyes. Infected patients can be affected by prolong work disability or even permanent invalidity. Prevention and treatment of disease thus becomes a long-term priority for medical research.

## 36.3 History

Some symptoms typical for Lyme disease have been known since the beginning of the twentieth century – acrodermatitis chronica atrophicans, erythema migrans, lymphocytoma, and meningopolyradiculoneuritis [11]. Nevertheless, the aetiological agent *Borrelia burgdorferi* was described quite recently on the basis of endemic juvenile arthritis accompanied with erythema migrans [12, 13].

Later it was shown that *B. burgdorferi* sensu lato is a complex of several sibling species. Currently, 12 species are distinguished and new variants are identified continuously, so the number of *Borrelia* species is probably not final. At least three species are known to be pathogenic for human – *B. burgdorferi* sensu stricto, *B. afzelii* and *B. garinii*. The number of annually reported Lyme disease cases continually increases in many geographical areas. This may be due to the actual spread of the disease or alternatively, to improved diagnostic methods [14].

## 36.4 Aetiological Agents

The genus *Borrelia* Swellengrebel 1907 is a member of the family Spirochaetaceae together with genera *Leptospira* and *Treponema*. In the USA, the only one causative agent – *B. burgdorferi* s.s. – was described. In Europe, two species – *B. afzelii* and *B. garinii* – are the most common.

*Borreliae* are typical for their spiral-shaped cells, 10–30 µm in length and 0.2–0.3 µm in diameter. This shape is caused by periplasmatic flagella responsible for the high motility in viscous environment like connective tissue. *Borrelia* lacks the rigid cell wall and the surface is composed mainly of lipoproteins with strictly controlled expression pattern, essential for adaptation for external conditions [15].

## 36.5 Epidemiology, Clinical Disease, and Treatment

The disease has a vector character. Most of the time, it is transmitted to humans by infected ticks of the genus *Ixodes*, but *Borrelia* has also been found in the midgut of the mosquitoes and other blood sucking insect. The role of haematophagous insect in disease transmission is still unclear [16, 17]. Reservoir competence has been described in the broad spectrum of wildlife animals, e.g. rodents (genera *Apodemus, Clethrionomys, Microtus, Rattus*, etc.), squirrels (*Sciurus sp.*), hares (*Lepus sp.*), and birds (genera *Turdus, Phasianus, Carduelis and Fringilla*, etc.). Human is a terminal host, incapable of further spreading of infection [18].

Lyme disease can have three stages – early localised infection, early disseminated infection, and late persistent infection, but not all stages develop in every infected individual. It appears that a significant portion of infections has asymptomatic course without clinical symptoms, but with elevated levels of anti-*Borrelia*-specific antibodies. The first stage is characterized by non-specific symptoms including fever, chills, headache, lethargy, and/or muscle and joints pain, about 70 % is accompanied by erythema migrans at the site of bacterial entry. If the infection is not eliminated by the host immune system or antibiotics treatment, it may disseminate and affect central nervous system (meningitis, radiculopathy, seventh cranial nerve palsy) or cardiovascular system (atrioventricular heart block, myopericarditis). The late persistent infection may develop months to years after the transmission and commonly affects the nervous system, skin, joints, or less frequently heart, eyes, or other organs [15].

Treatment of Lyme disease is based on the application of antibiotics. For adults administration of doxycycline is recommended. For children mainly beta-lactam antibiotics are prescribed. For the treatment of advanced forms, cephalosporins or penicillin G are recommended. Long-term usage of antibiotics in persistent forms of Lyme disease has only limited effect [19].

The first effort to design vaccine against Lyme disease followed shortly after the discovery of *Borrelia* as the aetiological agent causing this disease. First experiments demonstrated the immunogenicity of bacterial whole cell lysates. Later, specific proteins were identified to be recognised by the host immune system [20].

## 36.6 Antigens and Immune Response

Expression of surface antigens by *Borrelia* is highly variable. In the tick host, *Borrelia* expresses outer surface proteins A (OspA) and B (OspB). Genes coding both proteins are located within one operon and expression of these proteins is dependent on housekeeping sigma factor RpoD ($\sigma^{70}$). Presumed function of OspA and OspB is the adhesion to tick gut epithelium. When tick starts blood feeding on the vertebrate host, both proteins are downregulated. At the same time the expression of proteins dependent on alternative sigma factor RpoS ($\sigma^{38}$) is induced, e.g. (OspC) and OspF and decorin-binding proteins DbpA and DbpB. These proteins are required for *Borrelia* transmission to vertebrate and for initial stages of vertebrate infection. In later stages, OspC is downregulated and the expression of VlsE protein (Vmp-like sequence, expressed) is induced. It enables *Borrelia* escape from specific humoral response due to high VlsE variability.

Early humoral response to *Borrelia* is characterised by production of IgM antibodies specific to OspC, flagellins (p39 and p41), and BmpA (p39). Later IgG antibodies against p39, p41, p83/100, DbpA and VlsE arise. In some cases, antibodies against OspA and OspB can be detected in later stages of the disease [15].

*Subunit vaccines of first generation* were based on OspA antigen, which is highly expressed during cultivation of Lyme disease spirochetes in vitro. Preclinical trials with vaccines based on OspA antigen demonstrated protection in animal models. OspA antigen was produced either as a recombinant protein or lipoprotein, or it was expressed on the surface of *Escherichia coli*, *Salmonella typhimurium*, or *Mycobacterium bovis*.

Two OspA-based vaccine candidates were developed and tested. The ImuLyme vaccine (Pasteur Merieux-Connaught) based on purified recombinant OspA antigen expressed in *E. coli* and LYMErix (GlaxoSmithKline) containing purified recombinant OspA from *B. burgdorferi* s.s. [21]. In both vaccines protein was adsorbed on aluminium hydroxide. In clinical trials involving more than 10,000 people, LYMErix vaccine was found to confer protective immunity to *Borrelia* in 76 % of adults and 100 % of children with only mild or moderate and transient adverse effects [20]. Based on these results, LYMErix was approved by the Food and Drug Administration (FDA) on December 21, 1998.

Subsequently, hundreds of vaccinee reported the development of autoimmune side effects. Supported by some patient advocacy groups, a number of class-action lawsuits were filed against GlaxoSmithKline, alleging the vaccine had caused these health problems. These claims were investigated by the FDA and the US Centers for Disease Control (CDC), who found no connection between the vaccine and the autoimmune complaints [22, 23]. Despite the lack of evidence, sales plummeted and LYMErix was withdrawn from the US market by GlaxoSmithKline in February 2002 in the setting of negative media coverage and fears of vaccine side effects.

There are a number of reasons for the slow development of a Lyme disease vaccine, including reluctance by pharmaceutical companies to get burned in a similar way to LYMErix manufacturers. Others argue that it is simply more profitable for drug manufacturers to 'treat' the symptoms of Lyme palliatively, giving anti-inflammatory medications, pain relievers, and other medications where antibiotics fail to address a patient's ailments. The continued availability of a Lyme disease vaccine for dogs has made many people question the motives of drug companies and the psychological component of many symptoms often blamed on Lyme disease.

## 36.7 Licensed Veterinary Lyme Disease Vaccines

Unlike humane medicine, several licensed veterinary Lyme disease vaccines are available for companion pets. Three are formulations of *B. burgdorferi* bacterin (Merilym, Merial, Germany; Galaxy Lyme, Schering Plough, USA; LymeVax, Fort Dodge, USA). A European vaccine based on combination of bacterins from *B. garinii* and *B. afzelii* is produced for dogs (Biocan, Bioveta, Czech Republic) [24]. There are also available veterinary subunit vaccines based on recombinant OspC antigen (ProLyme, Intervet, USA; Recombitek Lyme, Merial, USA). All of the veterinary vaccines are applied in two-dose immunisation scheme with boosters recommended yearly. Safety data are limited and minor cross-reactivity between heterologous *Borrelia* species has been reported [25]. Thus, the poor cross-protective coverage is a problem of currently available vaccines.

*Second generation of OspA subunit vaccines* is based on recombinant OspA with genetically removed potentially cross-reactive T-cell epitopes, formerly suspected for the induction of autoimmunity [26]. Development of broadly cross-protective rOspA vaccine must address issues pertaining to sequence variation, because at least seven serotypes of OspA do exist. Multivalent-chimeric OspA proteins have been developed by molecular cloning incorporating the protective epitopes from several OspA serotypes [27]. Nevertheless, OspA antigen-based vaccines require repeated vaccination to keep high titre of specific antibodies in vaccinee serum to prevent transfer of bacteria from tick into human host.

## 36.8 Ferritin

New interesting antigen for vaccine development is tick ferritin 2, which is responsible for maintaining iron homeostasis. Iron is an essential but potentially toxic element; therefore, during feeding, ticks must deal with the challenge of an enormous iron supply in the blood meal. Ferritins, the iron storage proteins, play a pivotal role in this process. It was shown that vaccination with recombinant FER2 significantly reduces tick infestations in vaccinated rabbits infected with *Ixodes ricinus* and in cattle infected with *Rhipicephalus microplus* and *Rhipicephalus annulatus*. These results support the inclusion of FER2 as a promising candidate antigen for development of new anti-tick vaccines [28]. In spite of the fact that ferritin 2 is not the antigen derived from *Borrelia*, the vaccination with this antigen prevents long-term feeding of tick and transmission of *Borrelia* as well as other tick-transmitted pathogens into host. In this aspect ferritin 2 could be superior to OspA and is of interest to include it in development of combined multiantigen vaccine.

## 36.9 OspC Antigen

As mentioned above, several new antigens were discovered and studied as possible targets for new vaccines. We focus on OspC antigen, which received considerable attention in the effort to develop a broadly protective Lyme disease vaccine. OspC is an essential virulence factor that is critical for the establishment of early infection in mammals [29] and the sequence of the protein is not undergoing mutation during infection [30]. OspC is a 22 kDa immunodominant lipoprotein that is anchored in the spirochete outer membrane by an *N*-terminal tripalmitoyl-*S*-glyceryl-cysteine [31].

The factor complicating the exploitation of OspC as a vaccine candidate is the existence of OspC variation. Molecular phylogenetic analyses have revealed that OspC sequences form at least 38 distinct OspC types. This OspC variation arises primarily through genetic exchange and recombination and not by hypermutation with concomitant immune selection [32]. Owing to this sequence variation the protective range of single OspC-based vaccine is narrow. A potential approach to solve the high variability of the OspC is to prepare (a) polyvalent vaccine containing several OspC variants [33]; (b) generation of recombinant chimeric protein that consists of protective epitopes from those OspC types associated with disease [34]; and (c) combination of two or more different recombinant immunogens like OspC, DbpA, and fibronectin-binding protein (BBK32) [35].

From the biotechnological point of view the full-length OspC is difficult to express in a high yield and purity as a recombinant protein [36]. The removal of the lipidisation signal substantially increases both the yield (28 mg/l of the bacterial culture) and purity (93 %) of the recombinant OspC protein [37]. On the other side delipidised rOspC exerted very low immunogenicity [38, 39] and strong adjuvanticity of vaccine formulation is necessary to induce specific antibody response especially in IgG2 subclass (in experimental mice) important for effective complement activation and opsonisation.

In effort to adhere to all above requirements for OspC vaccine, we developed experimental vaccine based on functionalised metallochelating nanoliposomes and synthetic non-pyrogenic adjuvants derived from muramyl dipeptide. This formulation induced strong immune response in experimental mice, which was superior to aluminium hydroxide formulation. Owing to the versatility and safety as an adjuvant, liposomes thus represent promising platform for developing the clinically acceptable vaccine against Lyme disease. Next part of the chapter is focused on liposomal-based vaccines.

## 36.10 Liposomes as Biocompatible and Versatile Carriers for Construction of Recombinant Vaccines

Of the numerous particulate delivery systems that have been developed until now, phospholipid bilayer vesicles (liposomes) are among the most promising. Gregoriadis and Allison first reported the use of liposomes as immunological adjuvants in 1974 [40, 41]. Since that time, liposomes and related vesicular carriers have been established as robust systems for induction of humoral and cell-mediated immunity to a broad spectrum of infectious diseases and cancers [42].

Liposomes represent the oldest nanoparticle systems described for applications in biological studies as model membranes and in medicine. Over 44 years, liposomes have been shown to be suitable drug delivery systems for applications ranging from cosmetics and dermatology to medical applications such as therapy of infections, anticancer therapy, and veterinary vaccination [43].

Liposome-based vaccines have been around for approximately 30 years, and numerous liposome variants have been developed, some with evident immune-stimulatory properties and attractive safety profiles, resulting in registration of several products on the market and progressing of others to advanced stage of clinical testing. Epaxal® (Crucell, hepatitis A, formalin-inactivated hepatitis A virus adsorbed to virosomes) and Inflexal® V (Crucell, influenza, virosomes – reconstituted influenza viral membranes) are two examples of marketed liposome-based vaccines for human application. Inflexal® V is licensed in 43 countries with over 60 million doses applied [44, 45].

Furthermore, at least eight liposome-based adjuvant systems are currently approved for human use or undergoing clinical evaluation [46].

## 36.11 Chemical, Physicochemical, and Structural Parameters Affecting Activity of Liposome-Based Vaccine

Liposomes represent almost ideal carrier/delivery systems for the components of synthetic vaccines due to their biodegradability and their ability to retain/incorporate a variety of essential vaccine components simultaneously, even components possessing quite different physicochemical properties (different size, hydrophobicity, charge, etc.). Different synthetic vaccine components can be encapsulated within the aqueous cavities of liposomes (if hydrophilic) or associated with liposome bilayers (if at least partially hydrophobic in character). Furthermore, essential components can be attached to either internal or external outer leaflet membrane by electrostatic, covalent, or metallochelation interactions. Biochemically the

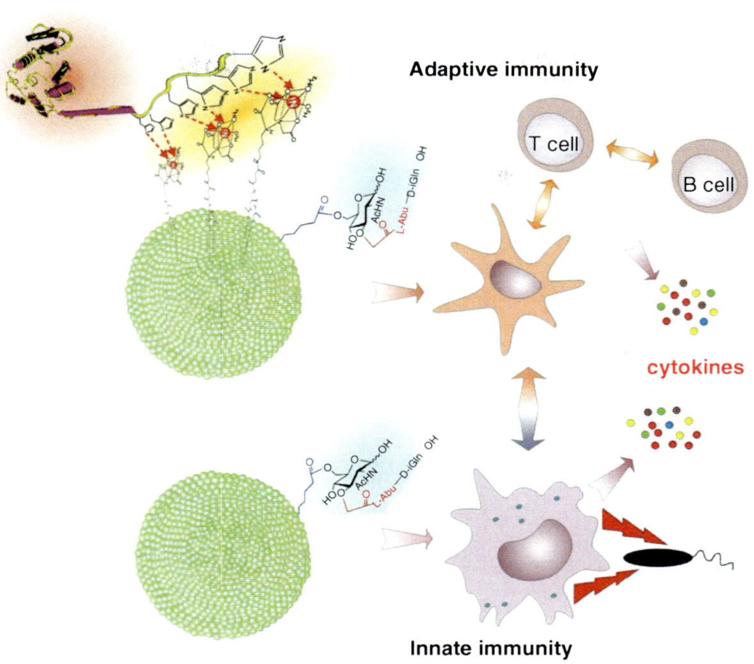

**Fig. 36.1** Interaction of liposome-based immunomodulators and proteoliposomal vaccines with immune cells. Functionalised nanoliposomes facilitate co-delivery of antigen to immune cells (dendritic cells, macrophages and B-cells). Highly oriented protein antigen on the liposomal surface is able to interact with membrane-bound immunoglobulins on B-cells followed by internalisation of whole complex liposomal antigen. Synthetic lipophilic derivative of muramyl dipeptide (norAbuMDP) provides the danger signal via intracellular receptors like NOD2 and NALP-3 to induce innate and adaptive immune responses. Moreover, glycopeptide part of the molecule exposed on the liposomal surface forms molecular pattern recognised by immune cells and enhancing internalisation of liposomes (Liposomes and immune cells are not in scale)

most diverse vaccine components are adjuvants required for efficacious activation of innate immunity cells including antigen-presenting cells (APC) (e.g. monophosphoryl lipid A, CpG oligonucleotides, muramyl dipeptide, and analogues). In addition, adjuvants can be combined with antigens such as soluble or membrane proteins to provoke strong specific immune response. Finally, the liposomes may present carrier for targeting of antigens and adjuvants to antigen-presenting cells [47]. The laboratory and industrial procedures for the liposome preparation have been established, and liposomes have been approved by the US Food and Drug Administration for biomedical applications. The potential for the participation of liposome-based recombinant vaccines in the human and veterinary vaccine market is very promising [48].

The beneficial effect of liposomal carriers consists in their ability to ensure (1) protection and stabilisation of the antigen, including reconstitution of its native conformation, (2) enhanced uptake by APC by passive or active targeting and (3) enhanced or controlled antigen processing after uptake.

Interaction of liposomal vaccine and liposomal immunomodulators with immune cells and cooperation between innate and adaptive branches of immune system is schematically represented in Fig. 36.1.

Liposome-based nanoparticle formulations used for vaccination can be broadly grouped into several classes:

*Conventional liposomes* composed of neutral lipids and cholesterol have been the most widely studied due to their greatest versatility – desired formulation parameters can be achieved through modification of the lipid composition or vesicle preparation method. These liposomes were used for steric entrapment of protein antigens in early stage of liposome-based vaccine research.

These classical liposomes are relatively ineffective at enhancing the immunogenicity of antigens because the phospholipids do not act as adjuvants. Adjuvant effect of these liposomes is based on multimeric presentation of antigen, and their adjuvant effect can be optimised by variation of lipid composition, co-incorporation of various adjuvant molecules (e.g. MLP-A, MDP, CpG oligonucleotides, trehalose dibehenate), and attachment of antigen.

*Transferosomes* are ultradeformable liposomes with enhanced skin penetration properties. They are composed of phosphatidylcholine and cholate (9:2 M ratio). Presence of cholate endowed these vesicles with high elasticity enables them to squeeze through pores in the stratum corneum. As a result of high elasticity of the bilayer, vesicles of the size of 200–300 nm can pass the highly compact stratum corneum and deliver their content into deep parts of the skin. Transferosomes are utilised as carrier system for topical application of vaccines [49]. Cationic transferosomes were also tested for noninvasive topical vaccination with pDNA vaccine against hepatitis B [50].

## 36.12 Cationic Liposomes

Surface adsorption on the cationic liposomes, via electrostatic interactions, is another useful method for attachment of antigens (primarily proteins) on liposomes. The recent work of Christensen, Perrie, and others with protein antigens adsorbed to cationic DDA/TDB (dimethyldioctadecylammonium/ trehalose-6,6-dibehenate) liposomes has demonstrated that formulations with surface-adsorbed antigens can be highly stable and elicit robust antibody and cell-mediated immune responses in mice and ferrets [42, 51–53]. Incorporation of synthetic immunostimulatory molecules like lipophilic TDB into cationic liposomes is able to potentiate significantly their adjuvant effect [52]. It was also observed that certain cationic lipids, originally synthesised as transfection reagents, induce robust IFN-γ and TNF-α response in serum and antitumour immunity in murine models of pulmonary metastasis and fibrosarcoma [54]. DOTAP is an example of such a cationic lipid used for many years as component of liposomal gene delivery system exerted an enantiospecific adjuvant effect in dendritic cells [55].

*Virosomes* are a special class proteoliposomes prepared from reconstituted influenza virus membranes supplemented with PC [56]. The physicochemical features of virosomes are constrained by their well-defined composition and method of preparation, but these vesicles benefit greatly from the inherent delivery properties (efficient cell binding, internalisation, and cytosolic release) and immunogenicity of the influenza virus.

*Archaeosomes* are liposomes prepared from special lipids isolated from archaebacteria, which are extremophiles, and their cell membranes contain lipids and proteins resistant to extreme temperatures, pH, and salt concentration. These lipids are less sensitive to oxidation and the membrane of archaeosomes is resistant to action of bile acid salts. Therefore they are suitable for enteral application. Immune responses induced by archaeosomes are comparable to those induced by complete Freund's adjuvants [57].

*Functionalised liposomes in general* represent various liposomal structures having their surface modified by various ligands (biotin, oligosaccharides, peptides), lipids with reactive head groups, pH or temperatures sensitive polymers, monoclonal antibodies or their fragments, lectins, etc. to tailor their functions with respect to selective interaction with targeted extra or intracellular structures (e.g. pathogenic microorganisms, tumour cells, immune cells, blood clothes). Targeting of antigen to antigen-presenting cells is of great importance for enhancement of efficacy of subunit recombinant vaccines. In this chapter we will introduce a new system for construction of liposomal-based vaccines. This system employs metallochelating lipids for binding of HIS-tagged recombinant antigens onto the surface of liposomes.

## 36.13 Binding of Antigen to Liposomes

One of the most critical parameters influencing the immunogenicity of liposomal vaccines is the method by which the antigen is physically or chemically associated with the formulation. The most common modes of association include covalent lipid conjugation (either pre- or post-vesicle formation), non-covalent surface attachment (via biotin, NTA-Ni-His, or antibody–epitope interactions), encapsulation, and surface adsorption (Fig. 36.2). Many of the early studies of liposomal peptide and protein antigenicity in mice compared encapsulated antigens to those conjugated to the surface of preformed liposomes. Early studies by Alving, Gregoriadis, Therien, and others confirmed that both methods are generally effective for inducing antibody and T-cell responses to associated protein antigens such as albumin and tetanus toxoid [58–60].

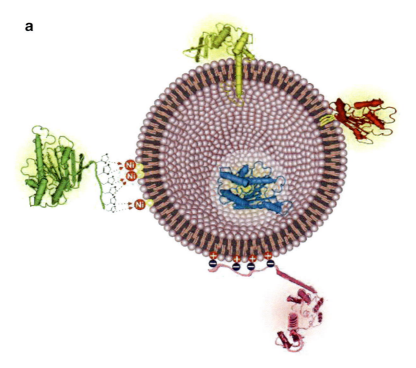

**Fig. 36.2** (a) Association of protein antigen with liposome. Protein antigen can be associated with liposome by various ways. Membrane protein can be reconstituted in their natural conformation in liposome membrane (*yellow*); lipophilised protein (e.g. palmitoylated protein) can be anchored in lipid membrane by lipidic residue (*red*). Functionalised liposomes can be prepared to facilitate covalent or non-covalent binding of protein onto the surface of liposomes. Selective metallochelating bond is applicable for His-tagged recombinant proteins (*green*) and this non-covalent bond can be transformed to a covalent one via carbodiimide conjugation. All above-mentioned examples represent highly oriented binding of protein onto liposomal surface. Covalent coupling of protein onto liposomal surface functionalised by, e.g. reactive maleiimide phospholipid headgroup, leads to random orientation of bound protein. Electrostatic interaction represents another way for association of proteins with liposomes. This association is non-specific and relatively labile in biological milieu (*pink*). Molecule of soluble protein can be also sterically entrapped in water compartment of liposome (*blue*). This is of importance especially for peptide antigens and protein antigens intended for direct delivery into the cytoplasm of immune cells to mimic virus-like pathways for antigen processing. (b) Binding of protein onto liposomal surface. A. Oriented non-covalent binding of HIS-tagged GFP (green fluorescent protein) onto liposomal surface. B. Transformation of metallochelating bond into covalent amide bond via carbodiimide chemistry preserving oriented binding of GFP. Ni+2 ions are removed by EDTA as a metallochelation agent. C. Example of random orientation of GFP bound onto liposomal surface by application of carbodiimide chemistry without pre-orientation via metallochelating bond

**Fig. 36.2** (continued)

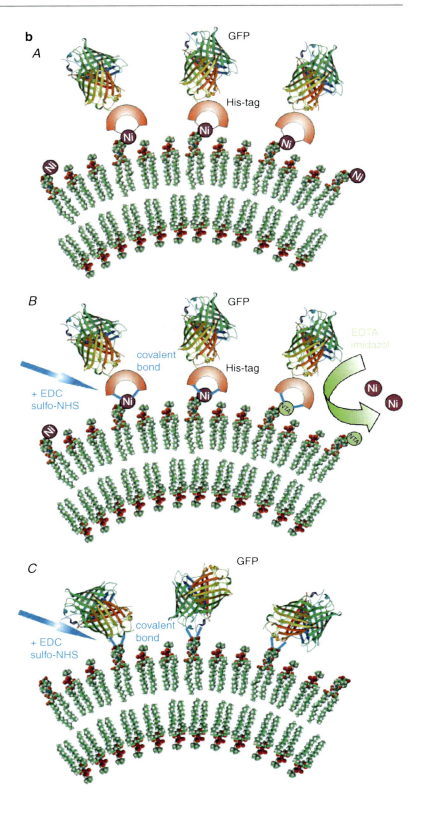

In some cases, covalent antigen conjugation results in superior antibody induction, which is not surprising because B-cell receptors can recognise intact antigen on the liposome surface [61]. As the size and complexity of the antigen decreases, the benefit of their surface conjugation for antibody induction becomes more pronounced, as found for the synthetic peptides for which surface conjugation provides immune responses superior to encapsulation [61, 62].

## 36.14 Metallochelating Bond and Metallochelating Liposomes

With respect to potential application of metallochelating liposomes for the construction of vaccines, the question of in vitro and especially in vivo stability is of great importance. This problem could be divided into two fields: first, the stability of the liposomes themselves and second, the effect of the components presented in biological fluids (e.g. proteins and ions) on the stability of the metallochelating bond. In the example of OspC proteoliposomes, the gel permeation chromatography data indicated a good in vitro stability of the OspC proteoliposomes during the chromatographic process, within which they experience a shear stress and dilution. Also after incubating OspC proteoliposomes in serum at 37 °C, we demonstrated the stability of the metallochelating bond linking the protein to the liposomal surface [37]. In fact, in vivo *fate* of liposomes after the intradermal application is different from the situation after intravenous injection. First, dilution of proteoliposomes after intradermal application is not so rapid and second, the ratio of tissue fluid proteins to proteoliposomes is more favourable to proteoliposomes owing to their relatively high concentration at the site of application. In the case of another route of application – intradermal – the flow rate of tissue fluid within intradermal extracellular matrix is even lower in comparison with muscle tissue or blood vessels. This fact is often overlooked. The stability of metallochelating bond probably depends also on the character of the particular protein. It was shown that Ni-NTA3–DTDA liposomes (containing three functional chelating lipid) with single-chain Fv fragments (anti CD11c) bound onto the liposomal surface were able to target dendritic cells in vitro as well as in vivo. The application of Ni-NTA3–DTDA probably endows the metallochelating bond with a higher in vivo stability [63], but this improved stability did not influence higher immunogenicity [64]. Generally, metal ions, physicochemical character of the metallochelating lipids, and their surface density belong to the factors that could be optimised to get a required in vivo stability and therefore strong immune response.

Binding of recombinant OspC (rOspC) onto metallochelating liposomes was confirmed by TEM, GPC, SDS PAGE and dynamic light scattering methods. Stability of the metallochelating bond in model biological fluids was studied by the incubation of rOspC liposomes with human serum. This study showed that after 1 h incubation at 37 °C, more than 60 % of rOspC was still associated with liposomes. Based on this data, therefore, the half-life of rOspC proteoliposomes in serum was estimated to be at least 1 h. Increase in stability of surface-exposed antigens on the liposomes can be achieved by chemical binding onto the outer liposomal surface that is appropriately functionalised. With respect to a potential application for the construction of vaccines, the question of in vitro and especially in vivo stability is of great importance.

There are only few references reporting the metallochelating bond implemented in the construction of supramolecular structures as vaccine carriers [65–69]. Various ways for association of protein antigen with liposome is depicted in Fig. 36.2a. Real structures revealed by various microscopy methods are presented in Fig. 36.3.

## 36.15 Adjuvants for Liposomal-Based Vaccines

The interaction between the innate and the adaptive immune responses is paramount in generating an antigen-specific immune response. The initiation of innate immune responses begins

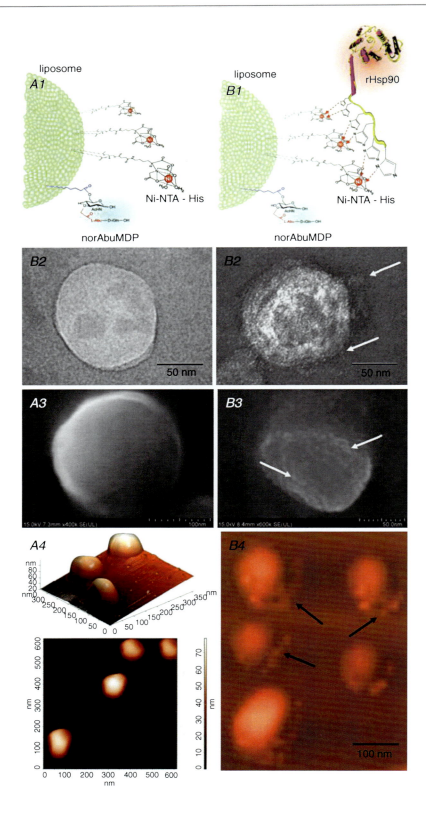

with the interaction of pathogen-associated molecular patterns (PAMPs) on the pathogen side with pattern-recognising receptors (PRR) such as Toll-like receptors (TLRs) on the host cells involved in the innate immunity (e.g. dendritic cells). A major functional criterion commonly used for the evaluation of various new adjuvants involves their ability to stimulate the innate immunity cells. This would include engaging other PRRs and the co-receptors and intracellular adaptor signalling proteins with which they are associated. PAMPs and their derivatives are utilised by adjuvant developers to harness the power of innate immunity to channel the immune response in a desired direction.

Based on the identification of several TLRs and PAMPs recognised by them, various PAMP agonists were tested as adjuvants. Examples of TLR–PAMP-specific interaction include bacterial or viral unmethylated immunostimulating CpG oligonucleotides interacting with TLR9 and liposaccharide and its component monophosphoryl lipid A (MPLA) interacting with TRL4. These two types of adjuvants are in advanced stage of testing in clinical trials and some already licensed vaccines contain MPLA in liposomal form [70, 71].

New lipophilic adjuvant (see Fig. 36.4) suitable for construction of liposome-based vaccines is the mycobacterial cord factor trehalose-6,6-dimycolate (TDM), and its synthetic analogue trehalose-6,6-dibehenate (TDB) are potent adjuvants for Th1/Th17 vaccination that activate Syk-Card9 signalling in APCs [72].

The last group of potent synthetic or semisynthetic adjuvants is derived from muramyl dipeptide (MDP).

*Muramyl dipeptide and other muropeptides* are derived from very specific group of PAMPs represented by peptidoglycans (PGN). Both Gram-positive and Gram-negative bacteria contain PGN which consists of numerous glycan chains that are cross-linked by oligopeptides. These glycan chains are composed of alternating N-acetylglucosamine (GlcNAc) and N-acetylmuramic acid (MurNAc) with the amino acids coupled to the muramic acid. Muropeptides are breakdown products of PGN that bear at least the MurNAc moiety and one amino acid [73]. One of the prominent muropeptides is muramyl dipeptide (MDP), which is known since the 1970s. We will deal with this compound in more detail.

Recently, the molecular bases for MDP recognition and subsequent stimulation of the host immune system have been uncovered. Myeloid immune cells (monocytes, granulocytes, neutrophils, and also DCs) possess two types of intracellular receptor for MDP/MDP analogues, namely, NOD2 and cryopyrin (inflammasome-NALP-3 complex) [74–76]. These two receptors recognise MDP/MDP analogues minimal recognition motifs for bacterial cell wall peptidoglycans [77]. Another recently reported sensor of MDP is cryopyrin (also known as CIAS1 and NALP3), which is a member of the NOD–LRR family [74]. Cryopyrin is a part of the inflammasome complex that is responsible for processing caspase-1 to its active form. Caspase-1 cleaves the precursors of interleukin IL-1β and IL-18, thereby activating these proinflammatory cytokines and promoting their secretion. IL-1β is known to be a strong endogenous pyrogen induced by MDP. We showed that norAbu-MDP and that norAbu-GMDP analogues

**Fig. 36.3** Structure of metallochelating liposomes and proteoliposomes revealed by TEM, SEM and AFM. Recombinant heat shock protein derived from *Candida albicans* was used as illustrative example. *A1.* Schematic presentation of a metallochelating liposome with metallochelating lipids and incorporated lipidized norAbuMDP molecules. *A2.* TEM photograph of a metallochelating liposome with metallochelating lipids and incorporated lipidized norAbuMDP molecules. *A3.* SEM photograph of a metallochelating liposome with metallochelating lipids and incorporated lipidized norAbuMDP molecules. *A4.* AFM photograph of a metallochelating liposome with metallochelating lipids and incorporated lipidized norAbuMDP molecules. *B1.* Schematic presentation of a metallochelating liposome (A1) with rHsp90 bound via metallochelating bond. *B2.* TEM photograph of a metallochelating liposome with rHsp90 bound via metallochelating bond. *B3.* SEM photograph of a metallochelating liposome with rHsp90 bound via metallochelating bond. *B4.* AFM photograph of a metallochelating liposome with rHsp90 bound via metallochelating bond

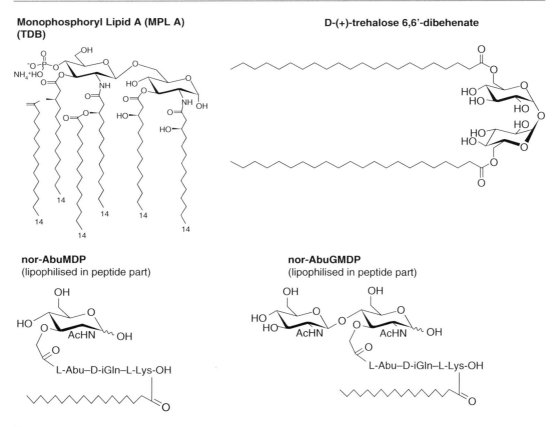

**Fig. 36.4** New lipophilic adjuvants for liposomal vaccines

were not pyrogenic even at a high concentration, much higher than the concentrations used for vaccination [15, 65, 66].

The expression of NOD2 in dendritic cells is of importance with respect to the application of MDP analogues as adjuvants. Nanoparticles like liposomes are able to provide a direct co-delivery of a danger signal (e.g., MDP) together with the recombinant antigen and therefore to induce an immune response instead of an immune tolerance. This is especially important for weak recombinant antigens or peptide antigens. Clearly, the recognition of MDP by DCs is crucial for the application of MDP analogues as adjuvants. Within the cell, MDP/MDP analogues trigger intracellular signalling cascades that culminate in the transcriptional activation of inflammatory mediators such as the nuclear transcription factor NF-κB. Liposomes probably play the role of efficient carriers for MDP and its analogues on the pathway from extracellular milieu into the cytosol, where they trigger intracellular signalling cascades that culminate in transcriptional activation of inflammatory mediators such as the nuclear transcription factor NF-κB pathway. Since the discovery and first synthesis of MDP, about one thousand various derivatives of MDP have been designed, synthesised, and tested to develop an appropriate drug for an immunotherapeutic application that would be free of the side effect exerted by MDP. The main side effects of MDP are pyrogenicity, rigor, headache, flue-like symptoms, hypertension, etc. Only several preparations reached the stage of clinical testing and only mifamurtide was approved for the treatment of osteosarcoma [77].

**Acknowledgements** This work was supported by grants: GAČR P304/10/1951 to J.T. and M.R. GAP503/12/G147 and the Ministry of Education, Youth and Sports of the Czech Republic (CZ.1.07/2.3.00/20.0164) to J.T. and M.R.

## References

1. Wack, A., Rappuoli, R.: Vaccinology at the beginning of the 21st century. Curr. Opin. Immunol. **17**, 411–418 (2005)
2. Stadler, K., et al.: SARS – beginning to understand a new virus. Nat. Rev. Microbiol. **1**, 209–218 (2003)
3. Rappuoli, R.: Reverse vaccinology. Curr. Opin. Microbiol. **3**, 445–450 (2000)
4. Zanetti, A.R., Van Damme, P., Shouval, D.: The global impact of vaccination against hepatitis B: a historical overview. Vaccine **26**, 6266–6273 (2008)
5. Munoz, N., et al.: Safety, immunogenicity, and efficacy of quadrivalent human papillomavirus (types 6, 11, 16, 18) recombinant vaccine in women aged 24–45 years: a randomised, double-blind trial. Lancet **373**, 1949–1957 (2009)
6. O'Hagan, D.T., Valiante, N.M.: Recent advances in the discovery and delivery of vaccine adjuvants. Nat. Rev. Drug Discov. **2**, 727–735 (2003)
7. Perrie, Y., Mohammed, A.R., Kirby, D.J., McNeil, S.E., Bramwell, V.W.: Vaccine adjuvant systems: enhancing the efficacy of sub-unit protein antigens. Int. J. Pharm. **364**, 272–280 (2008)
8. Storni, T., Kundig, T.M., Senti, G., Johansen, P.: Immunity in response to particulate antigen-delivery systems. Adv. Drug Deliv. Rev. **57**, 333–355 (2005)
9. Harris, J., Sharp, F.A., Lavelle, E.C.: The role of inflammasomes in the immunostimulatory effects of particulate vaccine adjuvants. Eur. J. Immunol. **40**, 634–638 (2010)
10. Marrack, P., McKee, A.S., Munks, M.W.: Towards an understanding of the adjuvant action of aluminium. Nat. Rev. Immunol. **9**, 287–293 (2009)
11. Stanek, G., Strle, F., Gray, J., Wormser, G. P.: History and characteristics of Lyme borreliosis. In: Gray, J.S., Kahl, O., Lane, R.S., Stanek, G. (eds.) Lyme Borreliosis – Biology, Epidemiology and Control, vol. 1, Ch. 1, pp. 1–28. CABI Publishing, Oxford (2002)
12. Steere, A.C., et al.: Erythema chronicum migrans and Lyme arthritis. The enlarging clinical spectrum. Ann. Intern. Med. **86**, 685–698 (1977)
13. Burgdorfer, W., et al.: Lyme disease-a tick-borne spirochetosis? Science **216**, 1317–1319 (1982)
14. Krupka, M., et al.: Biological aspects of Lyme disease spirochetes: Unique bacteria of the Borrelia burgdorferi species group. Biomed. Pap. Med. Fac. Univ. Palacky Olomouc Czech Repub. **151**, 175–186 (2007)
15. Krupka, M., Zachova, K., Weigl, E., Raska, M.: Prevention of lyme disease: promising research or sisyphean task? Arch. Immunol. Ther. Exp. (Warsz.) **59**, 261–275 (2011)
16. Zakovska, A., Nejedla, P., Holikova, A., Dendis, M.: Positive findings of Borrelia burgdorferi in Culex (Culex) pipiens pipiens larvae in the surrounding of Brno city determined by the PCR method. Ann. Agric. Environ. Med. **9**, 257–259 (2002)
17. Nejedla, P., Norek, A., Vostal, K., Zakovska, A.: What is the percentage of pathogenic borreliae in spirochaetal findings of mosquito larvae? Ann. Agric. Environ. Med. **16**, 273–276 (2009)
18. Gern, L., Humair, P.F.: Ecology of Borrelia burgdorferi sensu lato in Europe. In: Gray, J.S., Kahl, O., Lane, R.S., Stanek, G. (eds.) Lyme Borreliosis – Biology, Epidemiology and Control, vol. 1, Ch. 1, pp 149–174. CABI Publishing, Oxford (2002)
19. Klempner, M.S., et al.: Two controlled trials of antibiotic treatment in patients with persistent symptoms and a history of Lyme disease. N. Engl. J. Med. **345**, 85–92 (2001)
20. Poland, G.A., Jacobson, R.M.: The prevention of Lyme disease with vaccine. Vaccine **19**, 2303–2308 (2001)
21. Sigal, L.H., et al.: A vaccine consisting of recombinant Borrelia burgdorferi outer-surface protein A to prevent Lyme disease. Recombinant Outer-Surface Protein A Lyme Disease Vaccine Study Consortium. N. Engl. J. Med. **339**, 216–222 (1998)
22. Abbott, A.: Lyme disease: uphill struggle. Nature **439**, 524–525 (2006)
23. Nigrovic, L.E., Thompson, K.M.: The Lyme vaccine: a cautionary tale. Epidemiol. Infect. **135**, 1–8 (2007)
24. Tuhackova, J., et al.: Testing of the Biocan B inj. ad us. vet. vaccine and development of the new recombinant vaccine against canine borreliosis. Biomed. Pap. Med. Fac. Univ. Palacky Olomouc Czech Repub. **149**, 297–302 (2005)
25. Topfer, K.H., Straubinger, R.K.: Characterization of the humoral immune response in dogs after vaccination against the Lyme borreliosis agent A study with five commercial vaccines using two different vaccination schedules. Vaccine **25**, 314–326 (2007)
26. Trollmo, C., Meyer, A.L., Steere, A.C., Hafler, D.A., Huber, B.T.: Molecular mimicry in Lyme arthritis demonstrated at the single cell level: LFA-1 alpha L is a partial agonist for outer surface protein A-reactive T cells. J. Immunol. **166**, 5286–5291 (2001)
27. Gern, L., Hu, C.M., Voet, P., Hauser, P., Lobet, Y.: Immunization with a polyvalent OspA vaccine protects mice against Ixodes ricinus tick bites infected by Borrelia burgdorferi ss, Borrelia garinii and Borrelia afzelii. Vaccine **15**, 1551–1557 (1997)
28. Hajdusek, O., et al.: Characterization of ferritin 2 for the control of tick infestations. Vaccine **28**, 2993–2998 (2010)
29. Stewart, P.E., et al.: Delineating the requirement for the Borrelia burgdorferi virulence factor OspC in the mammalian host. Infect. Immun. **74**, 3547–3553 (2006)
30. Hodzic, E., Feng, S., Barthold, S.W.: Stability of Borrelia burgdorferi outer surface protein C under immune selection pressure. J. Infect. Dis. **181**, 750–753 (2000)

31. Brooks, C.S., Vuppala, S.R., Jett, A.M., Akins, D.R.: Identification of Borrelia burgdorferi outer surface proteins. Infect. Immun. **74**, 296–304 (2006)
32. Earnhart, C.G., Marconi, R.T.: OspC phylogenetic analyses support the feasibility of a broadly protective polyvalent chimeric Lyme disease vaccine. Clin. Vaccine Immunol. **14**, 628–634 (2007)
33. Earnhart, C.G., Marconi, R.T.: An octavalent lyme disease vaccine induces antibodies that recognize all incorporated OspC type-specific sequences. Hum. Vaccin. **3**, 281–289 (2007)
34. Buckles, E.L., Earnhart, C.G., Marconi, R.T.: Analysis of antibody response in humans to the type A OspC loop 5 domain and assessment of the potential utility of the loop 5 epitope in Lyme disease vaccine development. Clin. Vaccine Immunol. **13**, 1162–1165 (2006)
35. Brown, E.L., Kim, J.H., Reisenbichler, E.S., Hook, M.: Multicomponent Lyme vaccine: three is not a crowd. Vaccine **23**, 3687–3696 (2005)
36. Krupka, M., et al.: Isolation and purification of recombinant outer surface protein C (rOspC) of Borrelia burgdorferi sensu lato. Biomed. Pap. Med. Fac. Univ. Palacky Olomouc Czech Repub. **149**, 261–264 (2005)
37. Krupka, M., et al.: Enhancement of immune response towards non-lipidized Borrelia burgdorferi recombinant OspC antigen by binding onto the surface of metallochelating nanoliposomes with entrapped lipophilic derivatives of norAbuMDP. J. Control. Release **160**, 374–381 (2012)
38. Weis, J.J., Ma, Y., Erdile, L.F.: Biological activities of native and recombinant Borrelia burgdorferi outer surface protein A: dependence on lipid modification. Infect. Immun. **62**, 4632–4636 (1994)
39. Gilmore Jr., R.D., et al.: Inability of outer-surface protein C (OspC)-primed mice to elicit a protective anamnestic immune response to a tick-transmitted challenge of Borrelia burgdorferi. J. Med. Microbiol. **52**, 551–556 (2003)
40. Allison, A.G., Gregoriadis, G.: Liposomes as immunological adjuvants. Nature **252**, 252 (1974)
41. Gregoriadis, G., Allison, A.C.: Entrapment of proteins in liposomes prevents allergic reactions in pre-immunised mice. FEBS Lett. **45**, 71–74 (1974)
42. Henriksen-Lacey, M., Korsholm, K.S., Andersen, P., Perrie, Y., Christensen, D.: Liposomal vaccine delivery systems. Expert Opin. Drug Deliv. **8**, 505–519 (2011)
43. Gregoriadis, G.: Engineering liposomes for drug delivery: progress and problems. Trends Biotechnol. **13**, 527–537 (1995)
44. Bovier, P.A.: Epaxal: a virosomal vaccine to prevent hepatitis A infection. Expert Rev. Vaccines **7**, 1141–1150 (2008)
45. Herzog, C., et al.: Eleven years of Inflexal V-a virosomal adjuvanted influenza vaccine. Vaccine **27**, 4381–4387 (2009)
46. Watson, D.S., Endsley, A.N., Huang, L.: Design considerations for liposomal vaccines: influence of formulation parameters on antibody and cell-mediated immune responses to liposome associated antigens. Vaccine **30**, 2256–2272 (2012)
47. Altin, J.G., Parish, C.R.: Liposomal vaccines – targeting the delivery of antigen. Methods **40**, 39–52 (2006)
48. Adu-Bobie, J., Capecchi, B., Serruto, D., Rappuoli, R., Pizza, M.: Two years into reverse vaccinology. Vaccine **21**, 605–610 (2003)
49. Gupta, P.N., et al.: Non-invasive vaccine delivery in transfersomes, niosomes and liposomes: a comparative study. Int. J. Pharm. **293**, 73–82 (2005)
50. Mahor, S., et al.: Cationic transfersomes based topical genetic vaccine against hepatitis B. Int. J. Pharm. **340**, 13–19 (2007)
51. Henriksen-Lacey, M., et al.: Comparison of the depot effect and immunogenicity of liposomes based on dimethyldioctadecylammonium (DDA), 3beta-[N-(N′, N′-Dimethylaminoethane)carbomyl] cholesterol (DC-Chol), and 1,2-Dioleoyl-3-trimethylammonium propane (DOTAP): prolonged liposome retention mediates stronger Th1 responses. Mol. Pharm. **8**, 153–161 (2011)
52. Henriksen-Lacey, M., Devitt, A., Perrie, Y.: The vesicle size of DDA:TDB liposomal adjuvants plays a role in the cell-mediated immune response but has no significant effect on antibody production. J. Control. Release **154**, 131–137 (2011)
53. Henriksen-Lacey, M., et al.: Liposomal cationic charge and antigen adsorption are important properties for the efficient deposition of antigen at the injection site and ability of the vaccine to induce a CMI response. J. Control. Release **145**, 102–108 (2010)
54. Whitmore, M., Li, S., Huang, L.: LPD lipopolyplex initiates a potent cytokine response and inhibits tumor growth. Gene Ther. **6**, 1867–1875 (1999)
55. Vasievich, E.A., Chen, W., Huang, L.: Enantiospecific adjuvant activity of cationic lipid DOTAP in cancer vaccine. Cancer Immunol. Immunother. **60**, 629–638 (2011)
56. Daemen, T., et al.: Virosomes for antigen and DNA delivery. Adv. Drug Deliv. Rev. **57**, 451–463 (2005)
57. Conlan, J.W., Krishnan, L., Willick, G.E., Patel, G.B., Sprott, G.D.: Immunization of mice with lipopeptide antigens encapsulated in novel liposomes prepared from the polar lipids of various Archaeobacteria elicits rapid and prolonged specific protective immunity against infection with the facultative intracellular pathogen. Listeria monocytogenes. Vaccine **19**, 3509–3517 (2001)
58. Davis, D., Gregoriadis, G.: Liposomes as adjuvants with immunopurified tetanus toxoid: influence of liposomal characteristics. Immunology **61**, 229–234 (1987)
59. Shahum, E., Therien, H.M.: Liposomal adjuvanticity: effect of encapsulation and surface-linkage on antibody production and proliferative response. Int. J. Immunopharmacol. **17**, 9–20 (1995)
60. Shahum, E., Therien, H.M.: Immunopotentiation of the humoral response by liposomes: encapsulation versus covalent linkage. Immunology **65**, 315–317 (1988)
61. White, W.I., et al.: Antibody and cytotoxic T-lymphocyte responses to a single liposome-associated peptide antigen. Vaccine **13**, 1111–1122 (1995)
62. Frisch, B., Muller, S., Briand, J.P., Van Regenmortel, M.H., Schuber, F.: Parameters affecting the immunogenicity of a liposome-associated synthetic hexapeptide antigen. Eur. J. Immunol. **21**, 185–193 (1991)

63. van Broekhoven, C.L., Parish, C.R., Demangel, C., Britton, W.J., Altin, J.G.: Targeting dendritic cells with antigen-containing liposomes: a highly effective procedure for induction of antitumor immunity and for tumor immunotherapy. Cancer Res. **64**, 4357–4365 (2004)
64. Watson, D.S., Platt, V.M., Cao, L., Venditto, V.J., Szoka Jr., F.C.: Antibody response to polyhistidine-tagged peptide and protein antigens attached to liposomes via lipid-linked nitrilotriacetic acid in mice. Clin. Vaccine Immunol. **18**, 289–297 (2011)
65. Masek, J., et al.: Metallochelating liposomes with associated lipophilised norAbuMDP as biocompatible platform for construction of vaccines with recombinant His-tagged antigens: preparation, structural study and immune response towards rHsp90. J. Control. Release **151**, 193–201 (2011)
66. Masek, J., et al.: Immobilization of histidine-tagged proteins on monodisperse metallochelation liposomes: preparation and study of their structure. Anal. Biochem. **408**, 95–104 (2011)
67. Chikh, G.G., Li, W.M., Schutze-Redelmeier, M.P., Meunier, J.C., Bally, M.B.: Attaching histidine-tagged peptides and proteins to lipid-based carriers through use of metal-ion-chelating lipids. Biochim. Biophys. Acta **1567**, 204–212 (2002)
68. Malliaros, J., et al.: Association of antigens to ISCOMATRIX adjuvant using metal chelation leads to improved CTL responses. Vaccine **22**, 3968–3975 (2004)
69. Patel, J.D., O'Carra, R., Jones, J., Woodward, J.G., Mumper, R.J.: Preparation and characterization of nickel nanoparticles for binding to his-tag proteins and antigens. Pharm. Res. **24**, 343–352 (2007)
70. Alving, C.R., Rao, M., Steers, N.J., Matyas, G.R., Mayorov, A.V.: Liposomes containing lipid A: an effective, safe, generic adjuvant system for synthetic vaccines. Expert Rev. Vaccines **11**, 733–744 (2012)
71. Kim, D., Kwon, H.J., Lee, Y.: Activation of Toll-like receptor 9 and production of epitope specific antibody by liposome-encapsulated CpG-DNA. BMB Rep. **44**, 607–612 (2011)
72. Schoenen, H., et al.: Cutting edge: Mincle is essential for recognition and adjuvanticity of the mycobacterial cord factor and its synthetic analog trehalose-dibehenate. J. Immunol. **184**, 2756–2760 (2010)
73. Traub, S., von Aulock, S., Hartung, T., Hermann, C.: MDP and other muropeptides – direct and synergistic effects on the immune system. J. Endotoxin Res. **12**, 69–85 (2006)
74. Agostini, L., et al.: NALP3 forms an IL-1beta-processing inflammasome with increased activity in Muckle-Wells autoinflammatory disorder. Immunity **20**, 319–325 (2004)
75. Girardin, S.E., Philpott, D.J.: Mini-review: the role of peptidoglycan recognition in innate immunity. Eur. J. Immunol. **34**, 1777–1782 (2004)
76. McDonald, C., Inohara, N., Nunez, G.: Peptidoglycan signaling in innate immunity and inflammatory disease. J. Biol. Chem. **280**, 20177–20180 (2005)
77. Ando, K., Mori, K., Corradini, N., Redini, F., Heymann, D.: Mifamurtide for the treatment of nonmetastatic osteosarcoma. Expert Opin. Pharmacother. **12**, 285–292 (2011)

# Emerging Nanotechnology Approaches for Pulmonary Delivery of Vaccines

## 37

Amit K. Goyal, Goutam Rath, and Basant Malik

## Contents

| | | |
|---|---|---|
| 37.1 | Introduction | 580 |
| 37.2 | Mucosal Immune Response | 580 |
| 37.3 | Vaccine Delivery via Pulmonary Route | 582 |
| 37.4 | Lymphoid Tissue and Immune Response in the Respiratory Tract | 583 |
| 37.5 | Pulmonary Immunization | 584 |
| 37.6 | Challenges Against Pulmonary Immunization | 584 |
| 37.7 | Particle Deposition in Respiratory Tract | 585 |
| 37.7.1 | Impaction | 585 |
| 37.7.2 | Sedimentation | 585 |
| 37.7.3 | Interception | 586 |
| 37.7.4 | Diffusion | 586 |
| 37.8 | Implication of Nanotechnology for Pulmonary Vaccine Delivery | 586 |
| 37.8.1 | Nebulized Dispersions | 587 |
| 37.8.2 | Nanoscale Powders for Redispersion | 588 |
| 37.8.3 | Dry Powders for Inhalation | 588 |
| 37.8.4 | Nanodisperse Microspheres | 588 |
| 37.8.5 | Nanoaggregates | 589 |
| 37.9 | Selection of Suitable Polymers for Pulmonary Delivery | 589 |
| 37.10 | Lipid-Based Carrier Systems | 589 |
| 37.10.1 | Liposomes | 590 |
| 37.10.2 | Solid Lipid Nanoparticles | 591 |
| 37.10.3 | ISCOMs | 591 |
| 37.11 | Polymer-Based Nanoparticulate Delivery Systems | 591 |
| 37.12 | Chitosan | 592 |
| 37.13 | Alginates | 593 |
| 37.14 | Hyaluronic Acid | 593 |
| 37.15 | Carboxymethyl Cellulose | 594 |
| 37.16 | Cyclodextrin | 594 |
| 37.17 | Carbopol | 594 |
| 37.18 | Poly (ε-Caprolactone) (PCL) | 594 |
| 37.19 | Polylactic Acid (PLA) | 595 |
| 37.20 | Poly Lactic-co-Glycolic Acid (PLGA) | 595 |
| 37.21 | Metal-Based Nanoparticle Delivery | 595 |
| 37.21.1 | Iron | 595 |
| 37.21.2 | Gold | 596 |
| 37.21.3 | Zinc | 596 |
| 37.21.4 | Silica | 597 |
| Conclusion | | 597 |
| References | | 597 |

A.K. Goyal (✉) • G. Rath • B. Malik
Department of Pharmaceutics,
ISF College of Pharmacy, Ghal Kalan,
Moga, Punjab, India
e-mail: amitkumargoyal1979@yahoo.com

### Abstract

Pulmonary immunization has been recently explored as a suitable substitute for parenteral vaccination. Vaccine administered via pulmonary route can induce both systemic and local mucosal immune responses. The extensive population of dendritic cells (DC) in the respiratory epithelial lining and hub of macrophages (interstitium and the alveoli) plays important roles in the induction of strong immune response. There are several factors which have restricted the effectiveness of pulmonary immunization including poor deposition of the antigen at the alveolar region, low absorption from the epithelial barriers in the peripheral airways and the central lungs and the presence of a mucociliary escalator in the central and upper lung, which rapidly removes antigens or particles from the central respiratory tract. In past few years a number of highly effective novel nanocarriers have been developed for safe and effective vaccine delivery via pulmonary route. This chapter gives an overview of every aspects of pulmonary delivery of vaccines.

## 37.1 Introduction

Situation is changing day-by-day with the advancement in the field of antibiotics, vaccination programs, improved sanitation, and public awareness about the health. Now public is more health conscious and mortalities rate caused by infectious diseases throughout the world is reducing significantly. In developed countries, widespread immunization programs for a number of infectious diseases such as smallpox, diphtheria, pertussis, tetanus, poliomyelitis, measles, mumps, rubella, and H. influenza type B have significantly reduced the morbidity caused by these infectious diseases [1]. However, an alarming number of deadly infections are still severely prevalent in the world and some mutant strains are resistant even to new generation antibiotics. This is a major challenge in front of scientific society. These diseases are generally those against which we are not having any effective vaccine; moreover, they cannot be treated with antibiotics.

Most of such diseases follow the path through mucosal barrier itself to invade the body [1, 2]. A number of infectious microorganisms including Vibrio cholerae, enterotoxigenic and enteropathogenic *E. coli*, Salmonella, and various strains of Shigella, as well as viruses involved in sexually transmitted diseases, e.g., HIV, breach mucosal barrier to infect our body. The situation is more critical in developing countries or in the areas where the public is not much concerned about hygiene [3].

To tackle these mucosal breaches, strong mucosal barrier is essential to prevent the problem at the initial phases. This is the strong background behind a huge research on the development of successful mucosal vaccines. But unfortunately only a few mucosal vaccines are at commercial level; however, some of them are doing wonders, e.g., polio vaccine. In this chapter, we will discuss about the various sites for mucosal immunization in brief and significance of pulmonary immunization in detail. Furthermore, we will discuss about the role of nanotechnology for targeted antigen delivery via pulmonary route for optimum immune response.

## 37.2 Mucosal Immune Response

All the mucosal compartments are interconnected and this is called as common mucosal immune system (CIMS) (Fig. 37.1). Therefore, immunization at one mucosal site protects the other mucosal compartments as well. Every mucosal

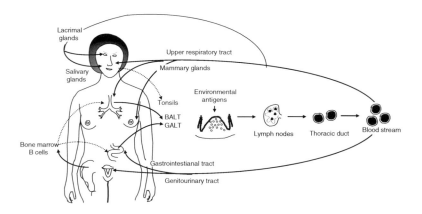

**Fig. 37.1** Sites of common mucosal immune systems

compartment shares some common features and can be subdivided in two general compartments commonly termed as inductive and effector sites which play special roles in immunity. Immune inductive sites are the primary sites for antigen sampling to activate immune cells. It includes organized mucosa-associated lymphoid tissue (MALT) and the regional mucosa-draining lymph nodes (LNs) (Fig. 37.1).

The effector sites are the venue where antibodies and cells of the immune system actually perform their real function in body's defense system. The effector sites include the lamina propria (LP) of various mucosa-associated lymphoid tissues, stroma of exocrine glands, surface epithelia, and the Peyer's patches (PPs) in the small intestine (Table 37.1) [4–6]. Moreover, MALT structures are enriched with T-cell zones, the B-cell follicles, and various kinds of antigen-presenting cells, including DCs and macrophages, continuously traffic these areas. All MALT structures are covered by a characteristic follicle-associated epithelium (FAE) containing microfold (M) cells which are actively involved in the sampling of exogenous antigens directly from the mucosal surfaces [7]. These specialized thin epithelial cells are also effectively involved in the transport of any particulate matter to the immune inductive sites. [8, 9] This is deeply explained in our previous articles that, why an

**Table 37.1** Various parts of MALT

| Anatomical system of the body | Inductive sites | Effector sites | Lymphoid structures |
|---|---|---|---|
| Respiratory tract | Nasopharynx-associated lymphoid tissue (NALT) | Nasopharyngeal mucosa | Waldeyer's pharyngeal ring |
| | | | Adenoids (pharyngeal tonsils) |
| | | | Palatine tonsils |
| | | | Lingual tonsils |
| | | | Tubal tonsils |
| | Bronchus-associated lymphoid tissue (BALT) | Bronchial mucosa | Peyer's patches |
| | | Lower respiratory tract | Isolated lymphoid follicles |
| | Larynx-associated lymphoid tissue (LALT) | Larynx | Laryngeal tonsils |
| | | | Lymphoid follicles |
| | | | Lymphoid follicles with germinal centers |
| Digestive system | Salivary duct-associated lymphoid tissue (DALT) | Salivary glands | Lymphoid follicles |
| | Gut-associated lymphoid tissue (GALT) | Gastrointestinal mucosa | Peyer's patches |
| | | | Lymphoglandular complexes |
| | | | Isolated lymphoid follicles |
| | | | Cryptopatches |
| | | | Appendix |
| Skin | Skin-associated lymphoid tissue (SALT) | Keratinocytes | Skin trophic T cells |
| | | Langerhans cells | Lymphatic endothelial cells |
| Ocular system | Conjunctiva-associated lymphoid tissue (CALT) | Conjunctiva | Lymphoepithelium |
| | | | Lymphoid follicles with B- and T-cell zones |
| | | | Adjacent blood vessels that have thickened endothelia |
| | | | Lymphoid vessels |
| | Lacrimal duct-associated lymphoid tissue (LDALT) | Ocular Tissues | Lymphoid follicles |
| Reproductive system | Vulvovaginal-associated lymphoid tissue (VALT) | Urogenital tract | Lymphoid follicles |
| Excretory system | Rectal lymphoepithelial tissue | Gastrointestinal mucosa | Lymphoid follicles |

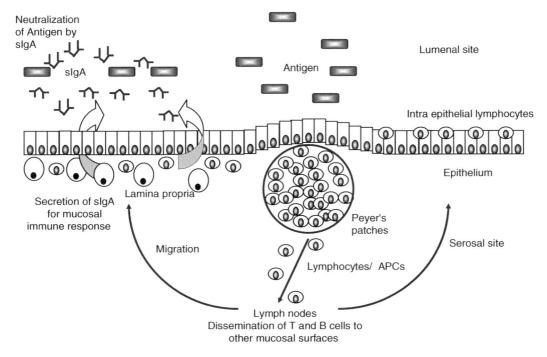

**Fig. 37.2** Induction of mucosal S–IgA response

advanced research is going on targeted antigen delivery to these specific cells viz. M cells and dendritic cells [2, 9].

## 37.3 Vaccine Delivery via Pulmonary Route

Today most of the commercial vaccines are administered by needles and syringes which are often accompanied with low patient compliance due to associated pain, chances of needle-based infections, requirement of trained personal, and significantly low mucosal immune response. To avoid these limitations, noninvasive immunization strategies have been extensively explored in the past few years with limited success rate. In the past few years, mucosal immunization has gained remarkable scientific attention. This may be due to a number of reasons such as awareness regarding the scientific advantages of the mucosal vaccines, advancement in the knowledge about the involvement of mucosa in pathogenesis of various diseases, the easy scalable production, and ease in vaccine delivery. Among the various routes, pulmonary immunization has received a special attention due to some special features of respiratory tract.

The respiratory epithelium is highly permeable and easily accessed by the inhaled particles. Respiratory epithelium contains a highly active immune system which is contentiously exposed to thousands of pathogens in our day-to-day life. It presents a strong innate immunity which includes mucociliary escalator, secretions with antimicrobial properties, chemokines, cytokines, and the mucus covering. Moreover, it also contains huge population of pulmonary macrophages and dendritic cells which are involved in both innate as well as adaptive immunity. Billions of macrophages are present in the lung periphery and interstitium, while DCs are located in the epithelial linings of conducting airways, submucosa below epithelial lining, within alveolar septum, and on the alveolar surface. These cells processed and presented the antigen to activate the immune system (Fig. 37.2) [10–12]. These special characteristics make the pulmonary route a potential site for vaccine administration.

## 37.4 Lymphoid Tissue and Immune Response in the Respiratory Tract

Depending upon the locations, the whole respiratory tract is divided into two parts, upper respiratory tract (URT) and lower respiratory tract (LRT). The immune system in the upper respiratory tract is architectured by the adenoids (nasopharyngeal tonsil) which are attached to the upper pharynx, the paired tubal tonsils at the opening of the Eustachian tube, the paired palatine tonsils at the oropharynx, and the lingual tonsil at the back of the tongue. Collectively, these structures constitute Waldeyer's ring which is also known as the nasal-associated lymphoid tissues (NALT).

Efferent lymphatics drain from NALT into the superficial cervical lymph nodes of the upper thorax, and these in turn are drained by the posterior cervical lymph nodes the involvement of afferent lymphatics. NALT also involves the aggregates of lymphoid follicles (B-cell areas), interfollicular areas (T-cell areas), macrophages, and dendritic cells (DCs). However, contrast to the NALT, the lymphoid tissue in the lower respiratory tract (LRT), also known as bronchus-associated lymphoid tissue (BALT), is not so organized. The architecture of BALT immune system is entirely different before and after stimulation by an external stimulus. Before stimulation, BALT is populated with undifferentiated lymphocytes and macrophages in a loose stromal network protected by a unique "lymphoepithelium" which is devoid of the goblet cells and cilia which are otherwise characteristic of the adjoining respiratory mucosal epithelium [13]. The lymphoepithelium contains specialized cells, similar to the "M" or microfold cells in Peyer's patches, which are trait in sampling antigen or particulate matter from the lumen of the airway (Fig. 37.2).

After stimulation, BALT changes its organization and completely converts to a dome shape, bulging slightly into the lumen of the airway beneath the epithelium to facilitate the encounter of the invading pathogen with immune cells [13]. Since most respiratory tract infections are initiated in the URT, it seems logical that the NALT should act as the first site of antigen recognition. There is also strong evidence that it is an important inductive site for immune responses which operate to clear infection from the URT. But the mucociliary escalator in the URT quickly removes particulate antigens from the mucosa to the exterior and eliminates the antigen without immune induction. This is the reason, why we are more concerned about the antigen targeting to deeper airways.

The particles with mucoadhesive property may be taken up by the microfold (M) cells present in the epithelium overlying the NALT. The basolateral surface of M cells is folded and invaginated to give pockets filled with lymphocytes and macrophages. After a long processing in the cytoplasm, the antigen is presented by antigen-presenting cells (APCs) [14]. The soluble antigens are taken up by the nasal epithelium and drain directly by means of the efferent lymphatics to the lymph nodes. The viral antigens and particulate carriers are taken up by macrophages and (DCs) in the area of the pockets, results in priming of IgA precursors which leave the NALT and enter the cervical lymph nodes for further amplification and differentiation into immunoglobulin producing cells. These cells then migrate to mucosal effector sites and to the respiratory mucosa.

The secretary IgA (sIgA) is actively transported to the mucosal surface by means polymeric Ig receptor (pIgR) [15–17]. It may be the reason why the infection at the mucosal surface results in systemic (serum IgG) and mucosal (IgA) antibody responses. Primed T and B cells also migrate to effector sites via lymph nodes and provide a long-term B- and T-cell memory which will differentiate into antibody-producing cells quickly after secondary infection. Various studies have reported that both natural viral infection and pulmonary vaccination result in the generation of antiviral cytotoxic T lymphocytes (CTL) [18]. It has been shown that NALT is a potent inductive site for virus-specific CTLs after intranasal infection [19]. The induction of a strong antiviral CTL response in addition to virus-specific IgA is a major goal for vaccination via pulmonary route. Mucosal responses in the LRT are induced in the

BALT by similar mechanisms. Therefore, we can target both URT and LRT, but targeting to LRT may skip the mucociliary escalator, and the particles can stay for a longer time period for a prolonged antigen delivery.

## 37.5 Pulmonary Immunization

Pulmonary delivery presents an attractive alternative to injection for the vaccination of children and adults due to some unique characteristics. Pulmonary immunization eliminates the use of syringe and needle which simplifies administration, reduces pain and suffering associated with injections, improves regimen compliance, reduces the risk of contaminated needle-based infections, and trims down a serious waste disposal problem [20]. As nearly all respiratory pathogens gain entry through mucosal membranes, establishing immunity at the port of entry might result in better first-line defenses [21, 22].

As explained above the respiratory tract has a large interrelated network of associated lymphoid tissues that readily share antigenic information and initiate quick immune response [1, 23]. Some unique features of the deep lungs, e.g., physiological pH and reduced mucociliary action, may circumvent many problems that exist with other noninvasive routes such as rapid clearance, poor absorption, exposure to digestive enzymes, and the antigenic tolerance [2, 24]. The extensive surface area which is almost equivalent to the surface of a single's tennis court for inhaled particles deposition is available. Therefore, an efficient system might induce good immune response at lower antigenic doses than conventional parenteral administration. It is not only a choice for only a few respiratory diseases and can be explored for greater diversity of antigens [12].

Aerosol-based vaccines have the additional benefit of being easily formulated as dry powders, using material science which has its own advantages [25, 26]. Development of dry powder-based vaccines can potentially reduce or may eliminate cold chain requirements, promote sterility, and increase the overall stability of antigens and thus reduce the overall cost of the product [27, 28]. For these reasons pulmonary vaccination has generated a great research interest, and furthermore, successful delivery of dry powder vaccines against influenza [29] and measles [30] generated new hopes.

## 37.6 Challenges Against Pulmonary Immunization

Several antigens have induced immunity after pulmonary administration and despite of the demonstrated efficacy, there are no commercial pulmonary vaccine [31, 32]. According to the literature the major challenge in the development of successful pulmonary vaccines is the development of appropriate tool to deliver the antigen to the deeper airways effectively. However, the thin epithelium barrier in the respiratory tract provides good permeability to the macromolecules. The conventional parenteral vaccines cannot be administered via respiratory route because the safety profile of these formulations has been established for other routes and approved for oral administration only. The use of excipients like permeability enhancers including surfactants and bile salts might improve penetration and uptake of the delivered antigen, but may lead to irreversible damage to the epithelium. Even more, it can cause short-term loss of integrity of the pulmonary epithelium that can initiate serious implications which are more difficult to treat [33]. Long-term toxicity studies are required to clearly identify the toxicity of novel molecules delivered to the lung which is not yet established. Although some recent studies using ISCOMATRIX™, a saponin-based adjuvant and chitosan, a nontoxic biodegradable polymer for pulmonary delivery, appears to be promising, but a lot of work is pending till date [34–36].

Toxicity associated with the use of saponins is still a matter of healthy debate. Muramyl dipeptide (MDP) and trehalose dibehenate (TDB) have also been screened in in vitro studies for potential use in pulmonary tuberculosis vaccines [37]. The optimal region in the respiratory tract for vaccine delivery remains to be defined. Recent studies in mice suggest that targeted antigen delivery to the lower airways is required to produce a strong immune response,

**Fig. 37.3** Particle deposition in respiratory tract and mechanisms involved

| Principle of sedimentation | Mass median Aerodynamic diameter | Part of respiratory tract |
|---|---|---|
| Impaction Interception | Mouth and nasal cavity MMAD 10–30 μm | |
| Interception | Upper respiratiry tract MMAD 2–10 μm | |
| Sedimentation Interception Diffusion | Bronchio-alveolar region MMAD <2 μm | |

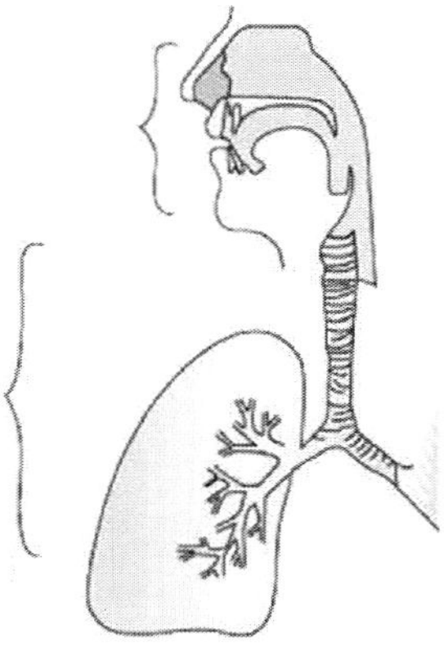

but results may vary in clinical phase [34, 38]. There are lots of dissimilarities in the structures of respiratory system of humans and rodents. Moreover, till date the exact mechanism of immune induction after antigen delivery via pulmonary route is not accurately clear.

## 37.7 Particle Deposition in Respiratory Tract

The deposition of inhaled particles in the different regions of the respiratory system is very complex, and may vary based upon number of factors. Some of the factors influencing respiratory deposition include:
- Breathing rate
- Mouth or nose breathing
- Lung volume
- Respiration volume
- Health of the individual
- Bifurcations in the airways result in a constantly changing hydrodynamic flow field.

Type of formulation may also impart significant effect on the biological activity of the delivered molecules by pulmonary route. Furthermore, the factors such as the particle size, particle density, morphology, and airflow are also effects in the efficiency of drug delivery by inhalation. There are several mechanisms such as impaction, sedimentation, interception, and diffusion (Fig. 37.3) that involved particle settling in the respiratory route.

### 37.7.1 Impaction

Impaction accounts for the majority of particle deposition on a mass basis. Impaction most commonly occurs in case of larger particles which are very close to airway walls and mainly near the first airway bifurcations. Impaction is highly dependent on aerodynamic diameter, and particle deposition by impaction is highest in the bronchial region [39]. Therefore, the larger particles (MMAD > 10 μm) are not suitable for pulmonary immunization, more specifically in case of diseases where antigen should be presented in deeper airways, e.g., tuberculosis. With the larger particles, antigen will be presented by the immune cells in the upper respiratory tract including immune cells in cervical lymph nodes, tonsils, pharynx, and larynx.

### 37.7.2 Sedimentation

Sedimentation is the settling out of particles in the smaller airways of the bronchioles and alveoli, where the airway dimensions are small and the inspired air flow is quite low. The mechanism of particle deposition via sedimentation depends upon the terminal settling velocity of the particles whose aerodynamic diameter is larger [40]. Hygroscopic particles may grow in size as they pass through the warm, humid air passages, thus gradually increasing the probability of deposition by sedimentation. Therefore, hygroscopic particles are not suitable for pulmonary antigen delivery even if they have initial very small particle size as it is prone to aggregation during inhalation.

### 37.7.3 Interception

Interception occurs when a particle contacts an airway surface. Interception plays its role in the particle deposition when the air streamline comes closer to reparatory wall and it is most likely to occur in smaller airways. The particles that are deposited by interception do not deviate from their air streamlines as in case of impaction. However, it is also reported that the role of interception in particle deposition is very less, but interception plays significantly for particles with irregular shape (elongated particles) like fibers, which easily contact airway surfaces due to their length [41]. Furthermore, fibers have small aerodynamic diameters relative to their size, so they can often reach the smallest airways.

### 37.7.4 Diffusion

Diffusion is the primary mechanism of deposition for particles in the nano-range (diameter <0.5 μm). This is mainly governed by geometric size rather than aerodynamic diameter. Diffusion is the movement of particles from a region of high concentration to a section of lower concentration by Brownian motion. This is the random motion of a particle due to the constant bombardment of air molecules. Diffusional deposition takes place when the particles just enters the nasopharynx, and most likely to occur in the smaller airways of the pulmonary (alveolar) region, where air flow is low. Nanocarriers can play important role in the delivery of vaccines via diffusion within the deeper airways [40].

Depending upon the above discussed factors and the principles behind the particle deposition in the respiratory tract, we can design the delivery tool according to the requirement. We can't even imagine any alteration in the patient anatomy and physiology, but can develop a formulation in accordance with the requirements.

## 37.8 Implication of Nanotechnology for Pulmonary Vaccine Delivery

Although nanoparticle-based formulations for other routes have gained success and reached up to the commercial scale, the use of inhaled nanoparticles is limited. Currently, no inhaled nanoparticle vaccine formulation has been approved for human use despite of some successful clinical studies and novel technologies. Studies demonstrated that deposition of nanoparticles within lungs is very inefficient and unpredictable [42]. During aerosolization, interparticulate interaction predominates over inertial separation leading to aggregation in unpredictable manner. Uniform filling of dry nanomaterial is itself a challenging task due to particle cohesion resulting in poor flow properties.

Moreover, these cohesive aggregates are difficult to redisperse with normal flow rates generated in passive dry powder inhalers. Additionally, up to 80 % of inhaled nanomaterial with particle size below 1 μm is exhaled due to the lack of inertial and gravitational forces needed for deposition [42]. Deposition of nanoparticles in the lungs is based upon the diffusive forces associated with Brownian motion which is a time-dependent process. In contrast to the inertia and sedimentation, diffusion is a very inefficient mean of lung deposition. Several studies have reported that decrease in particle size below 1 μm, the efficiency of lung deposition increases due to more rapid diffusion [42].

Studies using computer modeling shows that deposition in the lower airway begins to rise and more than 50 % of inhaled ultrafine(less than 100 nm) particles deposited in the alveolar region [43, 44]. Interestingly, the same results were achieved in clinical trials also, and 40–60 % of inhaled particles with a mean particle size of 40 nm were deposited in the lungs. These studies indicated that Brownian motion may be a possible mechanism of diffusion for very small-sized particles. However, some limitations are associated with nanoparticles like deposition variability due to deviation in breathing pattern, expensive production processes, and difficulties during handling of ultrafine material.

Furthermore, in vivo fate of the nanoparticles would also be variable. The biodegradable nanoparticles with entrapped antigens such as proteins, peptides, or DNA releases and represents the antigen in a controlled manner. Further, developed nanoparticles can be selectively targeted to antigen-presenting cells (APCs), e.g., dendritic cells (DCs) for desired immune response. These cells play an important role in body's defense system. The key steps in the induction of immune responses include capture, processing, and loading of exogenous antigen onto MHC class I and class II molecules by antigen-presenting cells (APCs) [45]. Studies have shown that soluble antigen rarely undergo cross-presentation, while the antigen in particulate form is considered critical for inducing robust $CD^{4+}$ and $CD^{8+}$ T-cell activation following vaccination [46]. Therefore, delivery of antigenic nanoparticulate carriers will result in the induction of both humoral and cellular immune responses, while soluble antigens result in the induction of dominant humoral response only.

Along with the effective results, it is also a bitter experience that the nanoparticles smaller than 1 μm do not efficiently deposit on the pulmonary epithelium and are extremely difficult to develop into a full-scale manufacturing process.

To overcome the problems associated with the nanoparticles, the pharmaceutical scientists have investigated a number of approaches including the use of both liquid and solid carriers, particle engineering to improve aerosolization (Fig. 37.4). As discrete nanoparticles are not well suitable for inhalation, much advancement has been done to tailor inhalable nanoparticles based formulation retaining the benefits of nanoparticles for effective delivery to lungs.

## 37.8.1 Nebulized Dispersions

Nebulization of an aqueous colloidal dispersion or suspension is an effective delivery strategy to enable the deposition of nanoparticles in the deep lungs. The drugs that cannot be delivered as an aqueous solution, nebulization of disperse nanoparticulates offers many advantages over

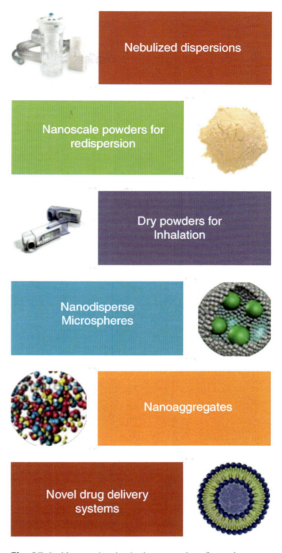

**Fig. 37.4** Nanotechnological approaches for pulmonary delivery

micronized drug dispersions. Nanoparticles can be distributed more uniformly throughout 1–5 μm droplets resulting in greater dose uniformity and more effective deep lung deposition [47, 48]. Moreover, due to their rheology almost identical to solutions, nebulized particles dispersions possess minimum negative influence on nebulizer function and aerosol droplet size compared to the microparticles [49]. The choice of stabilizing material for the delivery of an aqueous dispersion of therapeutic nanoparticles especially containing the protein drugs to the lungs is critical. Many synthetic surfactants and stabilizers that are normally used in oral and intravenous applications are not well suited for vaccine delivery, particularly in high concentrations. Along with, the disadvantages like associated toxicity, their tendency to foam during nebulization, and these surfactants can also alter the structural integrity of the proteins. Advancements in the dry powder-based formulations have limited the use of liquid formulations.

### 37.8.2 Nanoscale Powders for Redispersion

Most of the dispersions are likely to experience chemical and physical instability due to hydrolysis, particle settling, or aggregation or require the use of stabilizers that must be proven safe and nontoxic in the lungs before human use. On the other hand, nanoparticles for redispersion offer greater stability, but typically require a clinician for administration, and consequently would not be convenient for mass immunization.

The dispersion of nanoparticles into aqueous media has proven to be a suitable carrier in a range of preclinical and animal studies but failed at clinical phase. Most antigens are macromolecules, such as polysaccharides, proteins, and peptides, which are prone to chemical and physical degradation in liquid formulations [50]. Due to the stability problem associated with the protein in the liquid phase, some researchers used dry formulations which are intended for dispersion before administration. Nanoscale dry powder formulation eliminates the need of stabilizing surfactants and also reduces the contact time of the protein with the aqueous phase. A number of studies have reported the utility of lyoprotective agents such as lactose, mannitol, glucose, and sucrose to prolong the protein stability without significantly altering the particle size [51].

### 37.8.3 Dry Powders for Inhalation

Dry powder inhalers (DPIs) are available as simple, cheap, highly compact, and disposable devices in a single-use format, which are highly effective for vaccine delivery via pulmonary route. Delivery of macromolecules in the form of dry powder aerosols to the pulmonary system is the most successful alternate to parenteral immunization that provides improved stability and better immune response when compared with conventional liquid formulations [52]. Stability of proteins can be further increased by formulating them in a dry, solid state with additives such as mannitol, trehalose, sucrose, and inulin [53, 54]. The recent reports on dry powder measles vaccine formulations showed better stability even without refrigeration [53, 55]. The utility of dry powder aerosol vaccines has been demonstrated in several studies, e.g., dry powder influenza subunit vaccines prepared by spray-freeze-drying were shown to induce superior systemic and mucosal humoral and cell-mediated immune responses in mice after pulmonary delivery when compared with liquid vaccines administered via either the pulmonary or the intramuscular (IM) route [56]. Moreover, spray-dried nanoparticles of Bacille Calmette-Guerin (BCG) have shown better immune response [57]. It is also approved that dry particulate antigens are more efficientially taken up by the APCs, leading to a more powerful immune response [58]. The development of inhalable dry powder vaccines therefore appears to be a highly effective for pulmonary vaccination.

### 37.8.4 Nanodisperse Microspheres

For pulmonary delivery, microencapsulation may be used to form particles with the appropriate aerodynamic diameter for deep lung deposition. A common approach to produce powders suitable for inhalation is spray drying. This production method is often used to make respirable particles. Some studies have also explored microencapsulation to produce nanoparticles dispersed within carrier microparticles [59]. In a study investigating the feasibility of spray drying as a technique to incorporate nanoparticles into carrier microparticles, gelatin and polycyanoacrylate nanoparticles were dispersed in lactose matrix microparticles. After redispersion in aqueous media, nanoparticles remained unaggregated and the benefits of nano-sized particles would likely be present in vivo.

Additionally, impaction studies using cascade impactor showed that respirable lactose matrix particles could produce fine particle fractions ranging from 38 to 42 %. This approach is more beneficial for peptide and protein-based pharmaceuticals due to very low moisture content in the final formulation and the relatively low processing temperatures (40–45 °C). Nanoencapsulation offers a solution to in vivo degradation by the lung macrophage, esterases, proteases, and various epithelial metabolic pathways. Furthermore, dispersion of these nanoparticles in a solid microparticles matrix may provide a more suitable alternative to nebulization of an aqueous dispersion which imparts additional stability due to absence of water. In a study microspheres containing lipid/chitosan nanoparticles complexes were used for pulmonary administration of macromolecules. For spray drying, mannitol was chosen as microencapsulation excipient and insulin as the model protein. They reported that the developed systems can be successfully used for the delivery of therapeutic macromolecules by the pulmonary route [60].

### 37.8.5 Nanoaggregates

As mentioned previously, deposition of discrete aerosolized nanoparticles in the lower airways can be difficult due to the negligible effect of inertial and sedimentation forces needed for impaction. One of the very effective strategies to form low density aggregates of multiple nanoparticles. These nanoaggregates have low densities (<0.1 g/cm$^3$) and can have variable shapes such as hollow spheres [61, 62], spherical agglomerates [63], and nonspherical flocculates [64, 65], or aggregated plates [66]. A variety of techniques such as spray drying, salt flocculation, and rapid freezing processes have been invented for the production of nanoaggregates. Spray drying is oftenly used for the production of inhalable nanoaggregates as both hollow and solid spheres. Formation of hollow spheres composed of aggregated nanoparticles was initially proposed for pulmonary delivery by Tsapis et al. [61]. Kho et al., developed hollow spherical aggregates of biocompatible silica nanoparticles using spray drying technique to facilitate effective lung deposition. The large geometric size and the low density of the nanoaggregates imparted high aerosolization efficiency and an effective lung deposition. Based upon the promising results, they concluded that hollow spherical silica nanoaggregates are the potential candidates for inhaled drug delivery [67].

### 37.9 Selection of Suitable Polymers for Pulmonary Delivery

Different strategies have been developed and patented to facilitate and enhance the pulmonary drug delivery. Different lipid-based delivery systems, i.e., liposomes, immune stimulating complexes, solid lipid nanoparticles, as well as polymeric nanoparticles using diverse range of polymers, have been successfully developed. Several factors affect the performance of the developed formulations such as rate and mechanism of degradation, by-products, ease of antigen attachment or encapsulation, thermal stability, cost and availability, and safety profile. The ideal attributes of polymers are summarized in Table 37.2.

Each of these factors is an important criterion for the preparation of nanoparticles that are safe, capable of stimulating the immune system, and suitable for formulation into aerosols. A wide variety of polymers have been explored for use in drug delivery [68, 69].

### 37.10 Lipid-Based Carrier Systems

Despite a number of available more advanced drug delivery systems such as polymeric microparticles and nanoparticles and metallic nanoparticles, lipidic carriers are still considered to be promising candidates for pulmonary drug delivery due to numerous impressive characteristics (Fig. 37.5) [70]. Some earlier studies have demonstrated the safety and efficacy of lipidic carriers such as liposome, solid lipid nanoparticles, and ISCOMs for drug delivery to deeper lungs [71, 72].

**Table 37.2** Ideal characteristics for the selection of suitable polymer for pulmonary immunization

| S. No. | Properties | | Ideal requirements |
|---|---|---|---|
| 1. | Physical properties | Safety | The polymer should be nontoxic, biodegradable, and biocompatible in nature |
| | | Particle size | Particles with small aerodynamic diameter (optimally 2–5 µm) for alveolar localization |
| | | | Particles in nano-range can be suitable designed to facilitate the uptake by specific cells |
| | | Versatility | The carrier system and the polymer could be utilized for a wide range of antigens without altering their structural integrity |
| | | Moisture content | The final formulation should not be hygroscopic (moisture content (<5 %)) and should be stable for longer time period |
| 2. | Administration | | Should be patient compatible, should be easy to inhale, must not irritate respiratory mucosa |
| 3. | Immune response | | Must be capable of inducing both mucosal and systemic immune responses |
| | | | Should be devoid of tolerance |
| | | | Should eradicate the requirement of booster dose (controlled antigen delivery) |
| | | | Effective at low antigen dose |
| 4. | Manufacturing | | Should be cost-effective, able to sterilize, available in bulk, technique should be scalable, must eliminate cold chain requirement, and must match with regulatory requirements |

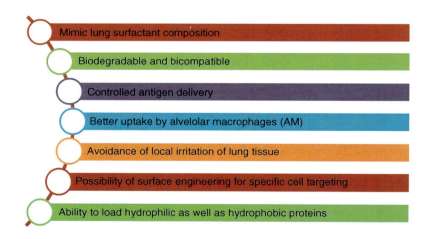

**Fig. 37.5** Advantages of lipidic carrier systems

## 37.10.1 Liposomes

Liposomes are vesicular carriers having one or more closed, concentric lipid bilayers alternating with aqueous compartments and the encapsulated drugs which provides an extended therapeutic response due to the depot action. It is well approved fact that the size and charge of liposomes have a critical influence on macrophage uptake. Chono et al. studied the influence of particle size on AM uptake. They prepared liposomes with varying particle size ranging from 0.1 to 2 µm using hydrogenated soya phosphatidylcholine (HSPC), cholesterol, and dicetyl phosphate (DCP). Delivery efficiency of the liposomes to AM increased with increase in particle size up to 1 µm, over which there was no significant change, and they provided a prolonged release of the drug for more than 24 h with the particles in nano-range [73].

Moreover, surface of liposomes can be anchored with a number of path navigating molecules for targeted drug delivery. It is reported that the mannose modified liposomes are better taken up by AM than non modified ones [74]. Similarly, our group also observed better AM disposition of mannosylated liposomes than non-modified liposomes [75]. These studies approved the potential of liposomes as a tool for targeted drug delivery to lung macrophages which actively participates in the lung immunity. Therefore, liposomes can be a better tool for pulmonary immunization.

### 37.10.2 Solid Lipid Nanoparticles

Solid lipid nanoparticles are submicron-sized lipidic nanocarriers. They are structurally different from other lipid-based vesicles due to the presence of a solid hydrophobic core coated with phospholipid monolayer. SLNs have also proved their potential for effective delivery of therapeutic agents via pulmonary route for systemic or localized action. Many studies such as human alveolar epithelial cell line (A549) and murine precision-cut lung slices (PCLS) cell lines have demonstrated a controlled release of a protein (lactose dehydrogenase, LDH) and proved the suitability of SLNs for drug delivery to lungs [76]. Liu et al. evaluated the potential of SLNs as a delivery tool for effective delivery of proteins via pulmonary route using fluorescent-labelled insulin. Studies revealed that SLNs were effectively and homogeneously distributed in the lung alveoli [77]. Moreover, surface modifications are also possible with SLNs to improve the stability, selective uptake, and bioavailability of encapsulated drugs, e.g., PEGylation of SLN can modify surface properties very easily and exhibited enhanced bioavailability to specific alveolar sites.

### 37.10.3 ISCOMs

Typically, ISCOMs are spherical, hollow, rigid, cage-like structures with particle size approximately 40 nm with a strong negative charge [78, 79]. ISCOMs are composed of phospholipid, cholesterol, saponin, and immunogen (antigen), usually a protein. These systems are actually designed for effective vaccine delivery. They mimic the virus particles in various aspects such as their size, orientation of surface proteins associated with a powerful immunostimulatory activity due to saponins [80]. These characteristics of ISCOMs furnish them with an ability to induce strong immune responses to a variety of antigens in a number of species. In contrast to many other carrier systems, ISCOMs promote a broad immune response by simultaneously inducting both high levels of antibody titer and strong T-cell response, including enhanced cytokine secretion and activation of cytotoxic T lymphocyte (CTL) responses [81].

The ability to stimulate CTL may be important in generating effective immune responses to virus-infected cells where strong T-cell response is desired. A further attraction of ISCOMs is their reduced toxicity and reactogenicity compared with other saponin-based formulations. Presently, ISCOM-based vaccine for veterinary use is available in the market against equine influenza. A critical question concerning the use of ISCOMs as a general vaccine adjuvant is whether the results in animal models such as the efficient induction of CTL responses can be reproduced in humans. The potential for ISCOMs to induce significant and sustained levels of CTL activity in humans may be of particular importance to viral vaccines [82]. Hopefully, the ongoing clinical trials of ISCOMs-based vaccines will provide superior results in forthcoming years.

### 37.11 Polymer-Based Nanoparticulate Delivery Systems

Polymeric nanoparticles are colloidal carriers ranging in size from 10 to 1,000 nm. The smaller size helps in targeting and maintaining the encapsulated particles. It represents an attractive means of delivering the proteins. The broad choice of polymeric materials is available which will affect the various physicochemical characteristics of the carrier constructs such as the drug release behavior, zeta potential, and hydrophobicity. The selected polymer must be safe, biocompatible, and biodegradable for use. The biodegradability is necessary for the release of the antigen and to

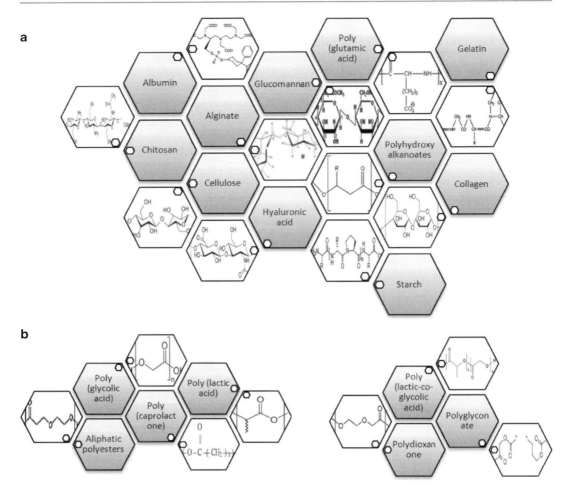

**Fig. 37.6** Commonly used polymers for pulmonary drug delivery (**a**) natural sources (**b**) synthetic sources (**a**) Polymers from natural sources (**b**) Polymers from synthetic sources

avoid the surgical step for the recovery of the depleted system. Various types of the polymers used for the formulation of the nanoparticles are chitosan, alginates, PLGA, and PLA etc. The natural polymers like chitosan, gelatin, albumin, and sodium alginate (Fig. 37.6) seem to be safer than the synthetic polymers. A number of polymers which are in continuous use for pulmonary delivery are briefly explained below.

## 37.12 Chitosan

Chitosan is a biocompatible biomaterial with a huge number of applications in the fields of biomedical engineering and drug/vaccine delivery. Chitosan is a copolymer of the alpha glucosamine and N-acetyl-D-glucosamine with mucoadhesive property due to its polycationic nature. Since chitosan easily forms microparticles and nanoparticles that can encapsulate a large amount of number of antigens, therefore it can be better utilized for controlled vaccine delivery to induce immunological response including both humoral as well as cellular immunity responses [9]. The cationic nature of chitosan imparts intrinsic immune stimulating activity and bioadhesive properties which improves the cellular uptake and permeation of the antigen. It also protects the antigen both in vitro and in vivo [83, 84].

The use of chitosan for the administration of proteins to the lungs has received great attention in past few years. The chitosan particles can be produced in a broad size range from 50 nm up to several microns depending upon diverse process variables [85]. In a study using whole inactivated influenza virus (WIV) coated with N, N, N-trimethyl chitosan (TMC) improved the delivery and immunogenicity of pulmonary vaccination against influenza [86]. The potential of chitosan nanoparticles for pulmonary immunization has been also approved by a study conducted by Benita et al. They found that pulmonary administration of the DNA plasmid loaded in chitosan nanoparticles resulted in increased levels of IFN-γ secretion compared to pulmonary delivery of plasmid in solution form or the more frequently used intramuscular immunization [36].

## 37.13 Alginates

Alginate is a nature gifted polymer derived from brown algae. It is a linear unbranched polysaccharide containing alternating residues of 1, 4′-linked β-D-mannuronic acid (M) and α-l-guluronic acid (G). The characteristics of the alginate-like molecular weight and chemical constitution depend upon the source from which the alginate is isolated and the proportion and sequence of M and G residues. Like chitosan alginates also possess strong mucoadhesive property which makes it a suitable choice for pulmonary vaccination. It quickly forms a gel in the presence of counterions such as $Ca^{++}$, which is generally used for cell immobilization, delivering drugs and antigens due to its nontoxic biodegradable nature [87]. Alginate nanoparticles have been prepared for efficient delivery of antitubercular drugs via pulmonary route. The relative bioavailability of the drugs in alginate was significantly higher compared with oral free drugs, and the results demonstrated that efficacy of 3 doses of alginates loaded nanoparticles nebulizers in 15 days was comparable to 45 daily doses of oral free drugs. These studies clearly indicated the potential of alginate nanoparticles for targeted drug delivery to the lungs [88].

## 37.14 Hyaluronic Acid

Hyaluronic acid (HA) is an excellent polymer for pulmonary immunization as it is endogenous to the pulmonary environment, it plays a function in various inflammatory mediators and agglutination of alveolar macrophages, and it inhibits phagocytosis. Moreover, it is a high molecular weight bioadhesive polymer which provides prolonged drug release and not actively taken up by mucociliary escalator [89]. Hyaluronic acid is a polyanionic non-sulfated polysaccharide that consists of N-acetyl-D-glucosamine and beta glucuronic acid. HA is an important component of the cell coat of many strains of bacteria. Hyaluronan is gaining great attention of the researchers due to its high biocompatibility and mucoadhesive nature. Lowest molecular weight (202 kD) hyaluronic acid exhibited better penetration enhancement properties compared with chitosan hydrochloride [90, 91]. HA particles have been prepared by spray drying techniques for targeted drug delivery to the lung and more specifically to alveolar macrophages.

Studies have suggested that HA particles were efficiently taken up by air-surface cultured alveolar macrophages (RAW 264.7) [92]. Spray-dried particles of hyaluronic acid have also been used for the lung delivery of macromolecules. Studies demonstrate the potential of HA-based dry powder drug delivery systems for controlled pulmonary delivery of insulin [93, 94]. Moreover, novel hybrid nanoparticles comprised of hyaluronic acid and iron oxide showed that iron complex hyaluronic acid particles can be efficiently utilized for delivery of peptides [95]. Low-density porous microparticles of hyaluronate were also prepared using positively charged lysozyme, negative-charged hyaluronate, and cyclodextrin derivative as a porogen, and the interaction of lysozyme and hyaluronate resulted in high protein encapsulation efficiency and also stabilized lysozyme against a denaturing organic solvent. Results suggested that porous microparticles may be applied in long-term pulmonary administration of protein or peptide drugs, especially in deeper lung epithelium [96].

## 37.15 Carboxymethyl Cellulose

Carboxymethyl cellulose (CMC) is anionic, biodegradable, and linear polymer cellulose ether. It has a number of fascinating characteristics including permeation-enhancing property, better transfection, and internalization of therapeutics within the mucosal cells and retarded release of loaded drugs/biomolecules. Therefore, CMC is widely used for drug delivery via mucosal route and also extensively used as a spray drying excipient in the preparation of inhalable formulations of proteins. Spray-dried particles of NaCMC with proteins and peptides may offer an alternative method for the preparation of stable formulation meant for pulmonary delivery [97]. Recently, studies indicated that spray drying of proteins with sodium carboxymethyl cellulose protects the structural integrity of proteins [97]. The studies indicated that the resultant spray-dried powders can be utilized for developing an aerosol formulation that exhibits high stability of encapsulated protein and high respirable fractions [98]. Moreover, some studies indicated that CMC also inhibits the mucociliary clearance; therefore, more and more antigen will be available for immune induction [99].

## 37.16 Cyclodextrin

Cyclodextrins are cyclic polymers of alpha-d-glucopyranose made up of sugar molecules bound together in a ring that are able to form host-guest complexes with hydrophobic molecules. By forming inclusion complexes, cyclodextrins have an ability to alter physical, chemical, and biological properties of guest molecules. Cyclodextrin possess high permeation-enhancing property which makes cyclodextrins ideal penetration enhancers. Furthermore, cyclodextrins can stabilize delicate drugs molecules by reducing exposure with external environment. Studies revealed that cyclodextrin-based porous particles can be explored to deliver insulin to deeper airways without altering its activity [100]. An inhalable dry powder of recombinant human growth hormone has been prepared using dimethyl-beta-cyclodextrin for systemic delivery of the protein, and the results showed improved aerosol performance of the spray-dried powders [101].

## 37.17 Carbopol

Carbopol are polymers of acrylic acid cross-linked with polyalkenyl ethers or divinyl glycol. Some studies strongly indicated that carbopols have excellent adjuvant property and is a part of some marketed animal vaccine [102, 103]. *Suvaxyn®  M. Hyo Vaccine* for pigs contains carbopol as an adjuvant. More interestingly, this vaccine is against a respiratory infection. It is designed for the vaccination of healthy pigs against clinical signs caused by Mycoplasma hyopneumoniae and it aids in the reduction of mycoplasma-induced respiratory disease and growth suppression. This is also documented that due to some important characteristics including the three-dimensional structure that allows for biological inertness, insolubility in water and their ability to swell in water can control the release of antigen and will prolong the immune response. A lot of research is going on for the development of successful vaccine not only for the pulmonary route but for the rest of mucosal routes also, viz. nasal, ophthalmic, vaginal, and rectal.

## 37.18 Poly (ε-Caprolactone) (PCL)

The synthetic biodegradable PCL, which is linear, hydrophobic, and partially crystalline polyester, can be utilized slowly by microorganisms with the enzyme production [104]. It has been reported that degradation of this aliphatic polyester in a living environment can result either from enzymatic attack or from chemical hydrolysis of ester bonds or both [105]. PCL is cost-effective, commonly available, synthetic polyester that has been employed in therapeutic delivery carrier [106]. The favorable characteristics which made it a suitable choice include prolonged stability in physiological fluids, negligible toxicity, and slow degradation in vivo compared to poly (lactic acid) and poly (lactic-co-glycolide acid) copolymers [107]. The hydrophobicity of PCL also makes it an ideal selection for mucosal vaccine delivery [108]. Spray drying of PCL has been explored in several studies [108, 109].

## 37.19 Polylactic Acid (PLA)

PLA is a biocompatible and biodegradable polymer which converts to monomeric units of lactic acid. Lactic acid is a natural intermediate/by product of anaerobic respiration which further results in the generation of glucose by the liver during the Cori cycle. Glucose is used as an energy source in the body. Therefore, use of the PLA nanoparticles is safe and devoid of any severe toxicity. PLA has broad biomedical applications, such as sutures, stents, dialysis media, and drug delivery devices. Anionic PEG-PLA nanoparticles produce low toxicity and proved highly effective drug carriers for therapeutics to lungs [110]. A spray-dried formulation of PLA has demonstrated satisfactory aerosol characteristics [111]. The solution upon inhalation was found to have 20 times higher drug concentration in macrophages than others. Due to better uptake by lung macrophages, PLA nanoparticles seem to be a suitable option for pulmonary immunization.

## 37.20 Poly Lactic-co-Glycolic Acid (PLGA)

PLGA has been used extensively in the past few years due to its excellent biocompatibility, biodegradability, and nontoxicity. Inhalation of the nanoparticles containing PLGA with mannitol has shown good aerosol performance. Moreover, the developed nanoparticles were easily recognized by the alveolar macrophages and exhibited increased uptake at alveolar region of lungs [112]. PLGA nanoparticles containing lactose and dipalmitoylphosphatidylcholine (DPPC) loaded with antibodies (human IgG, model antibody) resulted in prolonged release (>35 days) of protein [113]. The released antibody was stable and active without any structural alteration, which was confirmed by gel electrophoresis (SDS-PAGE), field-flow fractionation (FFF), and enzyme-linked immunosorbent assay (ELISA). Similarly, insulin-loaded large porous PLGA particles were prepared for pulmonary delivery, and in vivo data has shown that large porous PLGA particles reached up to alveoli and the released insulin was in its bioactive form [100]. Furthermore, surface-modified PLGA nanospheres showed slow elimination from the lungs than plane PLGA nanospheres. It may be attributed due to their adherence to the bronchial mucus and lung tissue and sustained drug release at the adherence site [114].

## 37.21 Metal-Based Nanoparticle Delivery

Metallic nanoparticles have received great attention in the past few years for pulmonary vaccine delivery. They have some unique characteristics (Fig. 37.7) which distinguish these carriers from other systems explored for pulmonary drug delivery. However, mixed reports are available in literature regarding the safety of the metallic nanoparticles can be explored in future for better results that ascertain the safety.

### 37.21.1 Iron

Iron nanoparticles are widely used in medical, diagnostic, and laboratory applications. They are highly reactive because of their large surface area and they oxidize in the presence of oxygen and water. It is still a prediction that iron may have high toxicity due to highly redox-active nature. However, in vitro cell line studies ensured the iron nanoparticles did not affect cell viability or the morphologic parameters of cell lines, while they potentiate the internalization of nanoparticles into cells through a macropinocytosis process [115]. Superparamagnetic iron oxide nanoparticles resulted in magnetic targeting of inhaled aerosols within the lungs. These particles are also used as contrast agents in magnetic resonance imaging (MRI), and these studies have shown better localization of superparamagnetic iron oxide nanoparticles into the lungs [116]. Based upon the cell membrane injury induced by nanoparticles studied using the lactate dehydrogenase assay, some studies also reported that

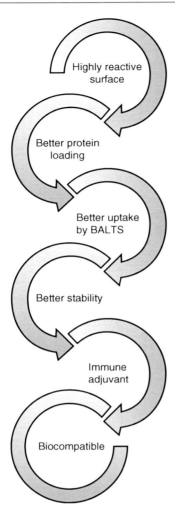

**Fig. 37.7** Advantages of metal nanoparticles

stability, high electron density, and affinity to biomolecules, making them drug carriers and a tool for diagnosis. Along with the associated benefits, these properties of nano-sized gold particles have also raised substantial questions about the safety of gold nanoparticles in the body. During the past decades, the mechanism involved in the interaction between gold nanoparticles and pulmonary structures has been investigated. Preliminary cytotoxic studies on the alveolar type II cell lines A549 and NCIH441 using gold nanoparticles reported that gold nanoparticles are internalized by the cells in a size-independent manner and can cause mild cytotoxicity [118]. In a study Sadauskas et al. studied the biodistribution of gold nanoparticles in mouse lung following intratracheal instillation and found that the inert gold nanoparticles were phagocytosed by lung macrophages and only a small fraction of the gold particles was translocated into systemic circulation depending on the particle size and greatest translocation was with the smallest particles (2 nm) included in the study. The study found that the instilled nanoparticles were internalized by the lung macrophages within 1 h after a single instillation [119]. Thus, it was demonstrated that the inert gold nanoparticles, administered intratracheally, are phagocytosed by lung macrophages. Therefore, gold nanoparticles can be suitable delivery tool for pulmonary immunization, but the complete evaluation of the safety related issues is necessary.

supermagnetic iron nanoparticles showed both concentration and time-dependent damage [117]. There is no doubt about the potential of iron nanoparticles for drug delivery to lungs, but their safety is still a matter of healthy debate and a huge research is pending in this area.

## 37.21.2 Gold

Gold nanoparticles are widely used in the areas of medicine due to some unique optical, electronic, and molecular recognition properties. Gold NPs have properties such as chemical

## 37.21.3 Zinc

Zinc nanoparticles also seem to be a suitable choice for pulmonary drug delivery. Studies have been conducted to develop in vitro screening assays to determine lung hazard potential of nanoscale (NZO) as well as the fine-sized zinc oxide (FZO). The effect of instillation vs. inhalation was also evaluated in rats. Lung inflammation, cytotoxicity, and histopathological endpoints were assessed at several time points postexposure using cultures of rat lung epithelial cells (L2), primary alveolar macrophages (AM), and alveolar macrophages-lung epithelial cells coculture

(AM-L2) based upon cytotoxicity biomarkers (LDH) and proinflammatory cytokines (MIP-2 and TNF-alpha). NZO or FZO particles produced potent but reversible inflammation which was resolved by 1-month postinstillation exposure. Moreover, as per the LDH results, L2 cells were found to be sensitive to the exposures of nano- or fine-sized ZnO. However, macrophages were resistant in dose-dependent manner [120]. The results of different studies involved in this experiment were not interrelated and variable results were achieved. Therefore, before concluding anything about the safety of zinc, nanoparticles extensive work would be required in this area to explore this tool for pulmonary drug delivery.

### 37.21.4 Silica

Exposure to silica as such has been demonstrated to cause pulmonary inflammation and damage to the lung tissue. Studies indicated that reactive oxygen and nitrogen species play a role in the acute phase pulmonary response from silica [121]. But in the past few years, mesoporous silica nanoparticles (MSNs) have gained increasing interest for use in vaccines. MSNs have become uniquely attractive because of some unique characteristics as follows such as nontoxic and biocompatible nature, adjustable pore size, large surface area, and a hexagonally ordered and well-defined internal structure and high thermal and chemical stability and controllable degradation rates [122, 123]. Due to the hollow nature, large surface area and low density mesoporous silica nanoparticles have ideal aerodynamic property. These characteristics made MSNs as ideal carrier system for pulmonary vaccine delivery. Moreover, in a study by our group, protocell of silica nanoparticles was prepared, and the spray-dried particles showed better aerodynamic behavior and can be utilized for pulmonary immunization [124].

#### Conclusion

Pulmonary immunization seems to be highly effective due to the presence of highly responsive immune response. However, delivery of the macromolecules/antigen to the deeper airways is a challenging task. A number of hurdles including presence of proteolytic enzymes, presence of mucociliary escalator, and particle deposition in upper airways make our path more difficult. Introduction of nanocarriers for pulmonary vaccine delivery imparted new hopes. A number of lipidic, polymeric, and metallic nanocarriers have been invented for the development of strong immune response after pulmonary immunization. These systems improved the immune response due to controlled antigen delivery to specific immune cells viz. dendritic cells and lung macrophages. However, a lot of research is required to justify the utility of nanocarriers more specifically the metallic nanoparticles at clinical stages.

**Acknowledgement** Author Dr. Amit K. Goyal is thankful to the Department of Biotechnology (DBT), New Delhi, India, for providing financial assistance to carry out research on the development of novel nanocarriers for pulmonary vaccine against tuberculosis.

## References

1. Ogra, P.L., Faden, H., Welliver, R.C.: Vaccination strategies for mucosal immune responses. Clin. Microbiol. Rev. **14**, 430–45 (2001). doi:10.1128/CMR.14.2.430-445.2001
2. Malik, B., Goyal, A.K., Mangal, S., Zakir, F., Vyas, S.P.: Implication of gut immunology in the design of oral vaccines. Curr. Mol. Med. **10**, 47–70 (2010)
3. Russell-Jones, G.J.: Oral vaccine delivery. J. Control. Release **65**, 49–54 (2000)
4. Cesta, M.F.: Normal structure, function, and histology of mucosa-associated lymphoid tissue. Toxicol. Pathol. **34**, 599–608 (2006). doi:10.1080/01926230600865531
5. Chadwick, S., Kriegel, C., Amiji, M.: Nanotechnology solutions for mucosal immunization. Adv. Drug Deliv. Rev. **62**, 394–407 (2010). doi:10.1016/j.addr.2009.11.012
6. McKenzie, B.S., Brady, J.L., Lew, A.M.: Mucosal immunity: overcoming the barrier for induction of proximal responses. Immunol. Res. **30**, 35–71 (2004). doi:10.1385/IR:30:1:035
7. Brandtzaeg, P., Pabst, R.: Let's go mucosal: communication on slippery ground. Trends Immunol. **25**, 570–7 (2004). doi:10.1016/j.it.2004.09.005
8. Moghaddami, M., Cummins, A., Mayrhofer, G.: Lymphocyte-filled villi: comparison with other lymphoid aggregations in the mucosa of the human small intestine. Gastroenterology **115**, 1414–25 (1998)

9. Malik, B., et al.: Microfold-cell targeted surface engineered polymeric nanoparticles for oral immunization. J. Drug Target. **20**, 76–84 (2012). doi:10.3109/1061186X.2011.611516
10. McCray Jr., P.B., Bentley, L.: Human airway epithelia express a beta-defensin. Am. J. Respir. Cell Mol. Biol. **16**, 343–9 (1997)
11. Larrick, J.W., et al.: Human CAP18: a novel antimicrobial lipopolysaccharide-binding protein. Infect. Immun. **63**, 1291–7 (1995)
12. Bivas-Benita, M., Ottenhoff, T.H., Junginger, H.E., Borchard, G.: Pulmonary DNA vaccination: concepts, possibilities and perspectives. J. Control. Release **107**, 1–29 (2005). doi:10.1016/j.jconrel.2005.05.028
13. Bienenstock, J.: Gut and bronchus associated lymphoid tissue: an overview. Adv. Exp. Med. Biol. **149**, 471–7 (1982)
14. Kuper, C.F., et al.: The role of nasopharyngeal lymphoid tissue. Immunol. Today **13**, 219–24 (1992)
15. Tamura, S., Kurata, T.: Defense mechanisms against influenza virus infection in the respiratory tract mucosa. Jpn. J. Infect. Dis. **57**, 236–47 (2004)
16. Zuercher, A.W.: Upper respiratory tract immunity. Viral Immunol. **16**, 279–89 (2003). doi:10.1089/088282403322396091
17. Perry, M., Whyte, A.: Immunology of the tonsils. Immunol. Today **19**, 414–21 (1998)
18. Cerwenka, A., Morgan, T.M., Dutton, R.W.: Naive, effector, and memory CD8 T cells in protection against pulmonary influenza virus infection: homing properties rather than initial frequencies are crucial. J. Immunol. **163**, 5535–43 (1999)
19. Zuercher, A.W., Coffin, S.E., Thurnheer, M.C., Fundova, P., Cebra, J.J.: Nasal-associated lymphoid tissue is a mucosal inductive site for virus-specific humoral and cellular immune responses. J. Immunol. **168**, 1796–803 (2002)
20. Giudice, E.L., Campbell, J.D.: Needle-free vaccine delivery. Adv. Drug Deliv. Rev. **58**, 68–89 (2006). doi:10.1016/j.addr.2005.12.003
21. Bouvet, J.P., Decroix, N., Pamonsinlapatham, P.: Stimulation of local antibody production: parenteral or mucosal vaccination? Trends Immunol. **23**, 209–13 (2002)
22. Stevceva, L., Abimiku, A.G., Franchini, G.: Targeting the mucosa: genetically engineered vaccines and mucosal immune responses. Genes Immun. **1**, 308–15 (2000). doi:10.1038/sj.gene.6363680
23. McNeela, E.A., Mills, K.H.: Manipulating the immune system: humoral versus cell-mediated immunity. Adv. Drug Deliv. Rev. **51**, 43–54 (2001)
24. Moyle, P.M., McGeary, R.P., Blanchfield, J.T., Toth, I.: Mucosal immunisation: adjuvants and delivery systems. Curr. Drug Deliv. **1**, 385–96 (2004)
25. Valente, A.X., Langer, R., Stone, H.A., Edwards, D.A.: Recent advances in the development of an inhaled insulin product. BioDrugs **17**, 9–17 (2003)
26. Bosquillon, C., Preat, V., Vanbever, R.: Pulmonary delivery of growth hormone using dry powders and visualization of its local fate in rats. J. Control. Release **96**, 233–44 (2004). doi:10.1016/j.jconrel.2004.01.027
27. Huang, J., et al.: A novel dry powder influenza vaccine and intranasal delivery technology: induction of systemic and mucosal immune responses in rats. Vaccine **23**, 794–801 (2004). doi:10.1016/j.vaccine.2004.06.049
28. LiCalsi, C., Christensen, T., Bennett, J.V., Phillips, E., Witham, C.: Dry powder inhalation as a potential delivery method for vaccines. Vaccine **17**, 1796–803 (1999)
29. Smith, D.J., Bot, S., Dellamary, L., Bot, A.: Evaluation of novel aerosol formulations designed for mucosal vaccination against influenza virus. Vaccine **21**, 2805–12 (2003)
30. LiCalsi, C., et al.: A powder formulation of measles vaccine for aerosol delivery. Vaccine **19**, 2629–36 (2001)
31. Amorij, J.P., et al.: Pulmonary delivery of an inulin-stabilized influenza subunit vaccine prepared by spray-freeze drying induces systemic, mucosal humoral as well as cell-mediated immune responses in BALB/c mice. Vaccine **25**, 8707–17 (2007). doi:10.1016/j.vaccine.2007.10.035
32. Wee, J.L., et al.: Pulmonary delivery of ISCOMATRIX influenza vaccine induces both systemic and mucosal immunity with antigen dose sparing. Mucosal Immunol. **1**, 489–96 (2008). doi:10.1038/mi.2008.59
33. Weers, J.G., Tarara, T.E., Clark, A.R.: Design of fine particles for pulmonary drug delivery. Expert Opin. Drug Deliv. **4**, 297–313 (2007). doi:10.1517/17425247.4.3.297
34. Sanders, M.T., Deliyannis, G., Pearse, M.J., McNamara, M.K., Brown, L.E.: Single dose intranasal immunization with ISCOMATRIX vaccines to elicit antibody-mediated clearance of influenza virus requires delivery to the lower respiratory tract. Vaccine **27**, 2475–82 (2009). doi:10.1016/j.vaccine.2009.02.054
35. Vujanic, A., et al.: Combined mucosal and systemic immunity following pulmonary delivery of ISCOMATRIX adjuvanted recombinant antigens. Vaccine **28**, 2593–7 (2010). doi:10.1016/j.vaccine.2010.01.018
36. Bivas-Benita, M., et al.: Pulmonary delivery of chitosan-DNA nanoparticles enhances the immunogenicity of a DNA vaccine encoding HLA-A*0201-restricted T-cell epitopes of Mycobacterium tuberculosis. Vaccine **22**, 1609–15 (2004). doi:10.1016/j.vaccine.2003.09.044
37. Wang, C., et al.: Screening for potential adjuvants administered by the pulmonary route for tuberculosis vaccines. AAPS J. **11**, 139–47 (2009). doi:10.1208/s12248-009-9089-0
38. Minne, A., et al.: The delivery site of a monovalent influenza vaccine within the respiratory tract impacts on the immune response. Immunology **122**, 316–25 (2007). doi:10.1111/j.1365-2567.2007.02641.x
39. Heyder, J.: Deposition of inhaled particles in the human respiratory tract and consequences for regional targeting in respiratory drug delivery. Proc.

40. Rogueda, P.G., Traini, D.: The nanoscale in pulmonary delivery. Part 1: deposition, fate, toxicology and effects. Expert Opin. Drug Deliv. **4**, 595–606 (2007). doi:10.1517/17425247.4.6.595
41. Hinds, W.C.: Aerosol Technology: Properties, Behavior, and Measurement of Airborne Particles, 2nd edn. Wiley, New York (1999)
42. Watts, A.B., Williams, R.O., 3rd: Chapter 15: Nanoparticles for Pulmonary Delivery. In: Smyth, H.D., Hickey, A.J. (eds.) Controlled Pulmonary Drug Delivery, Springer, New York (2011) p 335
43. Byron, P.R.: Prediction of drug residence times in regions of the human respiratory tract following aerosol inhalation. J. Pharm. Sci. **75**, 433–8 (1986)
44. Yang, W., Peters, J.I., Williams III, R.O.: Inhaled nanoparticles – a current review. Int. J. Pharm. **356**, 239–47 (2008). doi:10.1016/j.ijpharm.2008.02.011
45. Kurts, C., Robinson, B.W., Knolle, P.A.: Cross-priming in health and disease. Nat. Rev. Immunol. **10**, 403–14 (2010). doi:10.1038/nri2780
46. Shen, H., et al.: Enhanced and prolonged cross-presentation following endosomal escape of exogenous antigens encapsulated in biodegradable nanoparticles. Immunology **117**, 78–88 (2006). doi:10.1111/j.1365-2567.2005.02268.x
47. Ostrander, K.D., Bosch, H.W., Bondanza, D.M.: An in-vitro assessment of a NanoCrystal beclomethasone dipropionate colloidal dispersion via ultrasonic nebulization. Eur. J. Pharm. Biopharm. **48**, 207–15 (1999)
48. Wiedmann, T.S., DeCastro, L., Wood, R.W.: Nebulization of NanoCrystals: production of a respirable solid-in-liquid-in-air colloidal dispersion. Pharm. Res. **14**, 112–6 (1997)
49. Dailey, L.A., et al.: Nebulization of biodegradable nanoparticles: impact of nebulizer technology and nanoparticle characteristics on aerosol features. J. Control. Release **86**, 131–44 (2003)
50. Mahler, H.C., Muller, R., Friess, W., Delille, A., Matheus, S.: Induction and analysis of aggregates in a liquid IgG1-antibody formulation. Eur. J. Pharm. Biopharm. **59**, 407–17 (2005). doi:10.1016/j.ejpb.2004.12.004
51. Packhaeuser, C.B., et al.: Stabilization of aerosolizable nano-carriers by freeze-drying. Pharm. Res. **26**, 129–38 (2009). doi:10.1007/s11095-008-9714-0
52. Schule, S., Schulz-Fademrecht, T., Garidel, P., Bechtold-Peters, K., Frieb, W.: Stabilization of IgG1 in spray-dried powders for inhalation. Eur. J. Pharm. Biopharm. **69**, 793–807 (2008). doi:10.1016/j.ejpb.2008.02.010
53. Geeraedts, F., et al.: Preservation of the immunogenicity of dry-powder influenza H5N1 whole inactivated virus vaccine at elevated storage temperatures. AAPS J. **12**, 215–22 (2010). doi:10.1208/s12248-010-9179-z
54. Maury, M., Murphy, K., Kumar, S., Mauerer, A., Lee, G.: Spray-drying of proteins: effects of sorbitol and trehalose on aggregation and FT-IR amide I spectrum of an immunoglobulin G. Eur. J. Pharm. Biopharm. **59**, 251–61 (2005). doi:10.1016/j.ejpb.2004.07.010
55. Ohtake, S., et al.: Heat-stable measles vaccine produced by spray drying. Vaccine **28**, 1275–84 (2010). doi:10.1016/j.vaccine.2009.11.024
56. Saluja, V., et al.: A comparison between spray drying and spray freeze drying to produce an influenza subunit vaccine powder for inhalation. J. Control. Release **144**, 127–33 (2010). doi:10.1016/j.jconrel.2010.02.025
57. Garcia-Contreras, L., et al.: Immunization by a bacterial aerosol. Proc. Natl. Acad. Sci. U. S. A. **105**, 4656–60 (2008). doi:10.1073/pnas.0800043105
58. Thomas, C., Gupta, V., Ahsan, F.: Particle size influences the immune response produced by hepatitis B vaccine formulated in inhalable particles. Pharm. Res. **27**, 905–19 (2010). doi:10.1007/s11095-010-0094-x
59. Sham, J.O., Zhang, Y., Finlay, W.H., Roa, W.H., Lobenberg, R.: Formulation and characterization of spray-dried powders containing nanoparticles for aerosol delivery to the lung. Int. J. Pharm. **269**, 457–67 (2004)
60. Grenha, A., Remunan-Lopez, C., Carvalho, E.L., Seijo, B.: Microspheres containing lipid/chitosan nanoparticles complexes for pulmonary delivery of therapeutic proteins. Eur. J. Pharm. Biopharm. **69**, 83–93 (2008). doi:10.1016/j.ejpb.2007.10.017
61. Tsapis, N., Bennett, D., Jackson, B., Weitz, D.A., Edwards, D.A.: Trojan particles: large porous carriers of nanoparticles for drug delivery. Proc. Natl. Acad. Sci. U. S. A. **99**, 12001–5 (2002). doi:10.1073/pnas.182233999
62. Hadinoto, K., Phanapavudhikul, P., Kewu, Z., Tan, R.B.: Dry powder aerosol delivery of large hollow nanoparticulate aggregates as prospective carriers of nanoparticulate drugs: effects of phospholipids. Int. J. Pharm. **333**, 187–98 (2007). doi:10.1016/j.ijpharm.2006.10.009
63. Hu, T., Chiou, H., Chan, H.K., Chen, J.F., Yun, J.: Preparation of inhalable salbutamol sulphate using reactive high gravity controlled precipitation. J. Pharm. Sci. **97**, 944–9 (2008)
64. McConville, J.T., et al.: Targeted high lung concentrations of itraconazole using nebulized dispersions in a murine model. Pharm. Res. **23**, 901–11 (2006). doi:10.1007/s11095-006-9904-6
65. Plumley, C., et al.: Nifedipine nanoparticle agglomeration as a dry powder aerosol formulation strategy. Int. J. Pharm. **369**, 136–43 (2009). doi:10.1016/j.ijpharm.2008.10.016
66. Richardson, P.C., Boss, A.H.: Technosphere insulin technology. J. Diabetes Sci. Technol. **9**(Suppl 1), S65–72 (2007). doi:10.1089/dia.2007.0212
67. Katherine, K., Kunn, H.: Aqueous re-dispersibility characterization of spray-dried hollow spherical silica nano-aggregates. Powder Technol. **198**, 354–63 (2010)
68. Nair, L.S., Laurencin, C.T.: Polymers as biomaterials for tissue engineering and controlled drug delivery. Adv. Biochem. Eng. Biotechnol. **102**, 47–90 (2006)
69. Langer, R.: New methods of drug delivery. Science **249**, 1527–33 (1990)

70. Gaspar, M.M., Bakowsky, U., Ehrhardt, C.: Inhaled liposomes – current strategies and future challenges. J. Biomed. Nanotechnol. **4**, 1–13 (2008)
71. Henderson, A., Propst, K., Kedl, R., Dow, S.: ucosal immunization with liposome-nucleic acid adjuvants generates effective humoral and cellular immunity. Vaccine **29**, 5304–12 (2011). doi:10.1016/j.vaccine.2011.05.009
72. Videira, M.A., et al.: Lymphatic uptake of pulmonary delivered radiolabelled solid lipid nanoparticles. J. Drug Target. **10**, 607–13 (2002). doi:10.1080/1061186021000054933
73. Chono, S., Tanino, T., Seki, T., Morimoto, K.: Influence of particle size on drug delivery to rat alveolar macrophages following pulmonary administration of ciprofloxacin incorporated into liposomes. J. Drug Target. **14**, 557–66 (2006). doi:10.1080/10611860600834375
74. Chono, S., Tanino, T., Seki, T., Morimoto, K.: Uptake characteristics of liposomes by rat alveolar macrophages: influence of particle size and surface mannose modification. J. Pharm. Pharmacol. **59**, 75–80 (2007). doi:10.1211/jpp.59.1.0010
75. Vyas, S.P., Quraishi, S., Gupta, S., Jaganathan, K.S.: Aerosolized liposome-based delivery of amphotericin B to alveolar macrophages. Int. J. Pharm. **296**, 12–25 (2005). doi:10.1016/j.ijpharm.2005.02.003
76. Nassimi, M., et al.: Low cytotoxicity of solid lipid nanoparticles in in vitro and ex vivo lung models. Inhal. Toxicol. **21**(Suppl 1), 104–9 (2009). doi:10.1080/08958370903005769
77. Liu, J., et al.: Solid lipid nanoparticles for pulmonary delivery of insulin. Int. J. Pharm. **356**, 333–44 (2008). doi:10.1016/j.ijpharm.2008.01.008
78. Rimmelzwaan, G.F., Osterhaus, A.D.: A novel generation of viral vaccines based on the ISCOM matrix. Pharm. Biotechnol. **6**, 543–58 (1995)
79. Morein, B.: Iscom – an immunostimulating complex. Arzneimittelforschung **37**, 1418 (1987)
80. Morein, B., Sundquist, B., Hoglund, S., Dalsgaard, K., Osterhaus, A.: Iscom, a novel structure for antigenic presentation of membrane proteins from enveloped viruses. Nature **308**, 457–60 (1984)
81. Barr, I.G., Mitchell, G.F.: ISCOMs (immunostimulating complexes): the first decade. Immunol. Cell Biol. **74**, 8–25 (1996). doi:10.1038/icb.1996.2
82. Sjolander, A., Cox, J.C., Barr, I.G.: ISCOMs: an adjuvant with multiple functions. J. Leukoc. Biol. **64**, 713–23 (1998)
83. Khatri, K., Goyal, A.K., Gupta, P.N., Mishra, N., Vyas, S.P.: Plasmid DNA loaded chitosan nanoparticles for nasal mucosal immunization against hepatitis B. Int. J. Pharm. **354**, 235–41 (2008). doi:10.1016/j.ijpharm.2007.11.027
84. Marcinkiewicz, J., Polewska, A., Knapczyk, J.: Immunoadjuvant properties of chitosan. Arch. Immunol. Ther. Exp. (Warsz.) **39**(127–132) (1991)
85. Wright, I.K., Higginbotham, A., Baker, S.M., Donnelly, T.D.: Generation of nanoparticles of controlled size using ultrasonic piezoelectric oscillators in solution. ACS Appl. Mater. Interfaces **2**, 2360–4 (2010). doi:10.1021/am100375w
86. Hagenaars, N., et al.: Physicochemical and immunological characterization of N, N, N-trimethyl chitosan-coated whole inactivated influenza virus vaccine for intranasal administration. Pharm. Res. **26**, 1353–64 (2009). doi:10.1007/s11095-009-9845-y
87. Jain, A., Gupta, Y., Jain, S.K.: Perspectives of biodegradable natural polysaccharides for site-specific drug delivery to the colon. J. Pharm. Pharm. Sci. **10**, 86–128 (2007)
88. Ahmad, Z., Khuller, G.K.: Alginate-based sustained release drug delivery systems for tuberculosis. Expert Opin. Drug Deliv. **5**, 1323–34 (2008). doi:10.1517/17425240802600662
89. Rouse, J.J., Whateley, T.L., Thomas, M., Eccleston, G.M.: Controlled drug delivery to the lung: Influence of hyaluronic acid solution conformation on its adsorption to hydrophobic drug particles. Int. J. Pharm. **330**, 175–182 (2007). doi:10.1016/j.ijpharm.2006.11.066
90. Fraser, J.R., Laurent, T.C., Laurent, U.B.: Hyaluronan: its nature, distribution, functions and turnover. J. Intern. Med. **242**, 27–33 (1997)
91. Liao, Y.H., Jones, S.A., Forbes, B., Martin, G.P., Brown, M.B.: Hyaluronan: pharmaceutical characterization and drug delivery. Drug Deliv. **12**, 327–42 (2005). doi:10.1080/10717540590952555
92. Hwang, S.M., Kim, D.D., Chung, S.J., Shim, C.K.: Delivery of ofloxacin to the lung and alveolar macrophages via hyaluronan microspheres for the treatment of tuberculosis. J. Control. Release **129**, 100–6 (2008). doi:10.1016/j.jconrel.2008.04.009
93. Surendrakumar, K., Martyn, G.P., Hodgers, E.C., Jansen, M., Blair, J.A.: Sustained release of insulin from sodium hyaluronate based dry powder formulations after pulmonary delivery to beagle dogs. J. Control. Release **91**, 385–94 (2003)
94. Morimoto, K., Metsugi, K., Katsumata, H., Iwanaga, K., Kakemi, M.: Effects of low-viscosity sodium hyaluronate preparation on the pulmonary absorption of rh-insulin in rats. Drug Dev. Ind. Pharm. **27**, 365–71 (2001). doi:10.1081/DDC-100103737
95. Kumar, A., et al.: Development of hyaluronic acid-Fe2O3 hybrid magnetic nanoparticles for targeted delivery of peptides. Nanomedicine **3**, 132–7 (2007). doi:10.1016/j.nano.2007.03.001
96. Lee, E.S., Kwon, M.J.: Protein release behavior from porous microparticle with lysozyme/hyaluronate ionic complex. Colloids Surf. B Biointerfaces **55**, 125–30 (2007). doi:10.1016/j.colsurfb.2006.11.024
97. Li, H.Y., Song, X., Seville, P.C.: The use of sodium carboxymethylcellulose in the preparation of spray-dried proteins for pulmonary drug delivery. Eur. J. Pharm. Sci. **40**, 56–61 (2010). doi:10.1016/j.ejps.2010.02.007
98. Li, H.Y., Seville, P.C.: Novel pMDI formulations for pulmonary delivery of proteins. Int. J. Pharm. **385**, 73–8 (2010). doi:10.1016/j.ijpharm.2009.10.032
99. Dailey, L.A., et al.: Surfactant-free, biodegradable nanoparticles for aerosol therapy based on the branched polyesters. DEAPA-PVAL-g-PLGA. Pharm. Res. **20**, 2011–20 (2003)

100. Ungaro, F., et al.: Insulin-loaded PLGA/cyclodextrin large porous particles with improved aerosolization properties: in vivo deposition and hypoglycaemic activity after delivery to rat lungs. J. Control. Release **135**, 25–34 (2009). doi:10.1016/j.jconrel.2008.12.011
101. Jalalipour, M., Najafabadi, A.R., Gilani, K., Esmaily, H., Tajerzadeh, H.: Effect of dimethyl-beta-cyclodextrin concentrations on the pulmonary delivery of recombinant human growth hormone dry powder in rats. J. Pharm. Sci. **97**, 5176–85 (2008). doi:10.1002/jps.21353
102. Krashias, G., et al.: Potent adaptive immune responses induced against HIV-1 gp140 and influenza virus HA by a polyanionic carbomer. Vaccine **28**, 2482–9 (2010). doi:10.1016/j.vaccine.2010.01.046
103. Robert, E.: Chapter 1: An Overview of Adjuvant Use. In: O'Hagan, D.T. (ed.). Vaccine Adjuvants: Preparation Methods and Research Protocols. Humana Press, Totowa (2000) p 5
104. Singh, R.P., Pandey, J.K., Rutot, D., Degee, P., Dubois, P.: Biodegradation of poly(epsilon-caprolactone)/starch blends and composites in composting and culture environments: the effect of compatibilization on the inherent biodegradability of the host polymer. Carbohydr. Res. **338**, 1759–69 (2003)
105. Khatiwala, V., Shekhar, N., Aggarwal, S., Mandal, U.: Biodegradation of Poly($\varepsilon$-caprolactone) (PCL) Film by *Alcaligenes faecalis*. J Polym Environ **16**, 61–7 (2008). doi:10.1007/s10924-008-0104-9
106. Balmayor, E.R., Tuzlakoglu, K., Azevedo, H.S., Reis, R.L.: Preparation and characterization of starch-poly-epsilon-caprolactone microparticles incorporating bioactive agents for drug delivery and tissue engineering applications. Acta Biomater. **5**, 1035–45 (2009). doi:10.1016/j.actbio.2008.11.006
107. Sinha, V.R., Trehan, A.: Formulation, characterization, and evaluation of ketorolac tromethamine-loaded biodegradable microspheres. Drug Deliv. **12**, 133–9 (2005)
108. Baras, B., Benoit, M.A., Gillard, J.: Influence of various technological parameters on the preparation of spray-dried poly(epsilon-caprolactone) microparticles containing a model antigen. J. Microencapsul. **17**, 485–98 (2000). doi:10.1080/026520400405732
109. Zalfen, A.M., et al.: Controlled release of drugs from multi-component biomaterials. Acta Biomater. **4**, 1788–96 (2008). doi:10.1016/j.actbio.2008.05.021
110. Harush-Frenkel, O., et al.: A safety and tolerability study of differently-charged nanoparticles for local pulmonary drug delivery. Toxicol. Appl. Pharmacol. **246**, 83–90 (2010). doi:10.1016/j.taap.2010.04.011
111. Muttil, P., et al.: Inhalable microparticles containing large payload of anti-tuberculosis drugs. Eur. J. Pharm. Sci. **32**, 140–50 (2007). doi:10.1016/j.ejps.2007.06.006
112. Ohashi, K., Kabasawa, T., Ozeki, T.: One-step preparation of rifampicin/poly(lactic-co-glycolic acid) nanoparticle-containing mannitol microspheres using a four-fluid nozzle spray drier for inhalation therapy of tuberculosis. J. Control. Release **135**, 19–24 (2009). doi:10.1016/j.jconrel.2008.11.027
113. Kaye, R.S., Purewal, T.S., Alpar, H.O.: Simultaneously manufactured nano-in-micro (SIMANIM) particles for dry-powder modified-release delivery of antibodies. J. Pharm. Sci. **98**, 4055–68 (2009). doi:10.1002/jps.21673
114. Yamamoto, H., Kuno, Y., Sugimoto, S., Takeuchi, H., Kawashima, Y.: Surface-modified PLGA nanosphere with chitosan improved pulmonary delivery of calcitonin by mucoadhesion and opening of the intercellular tight junctions. J. Control. Release **102**, 373–81 (2005). doi:10.1016/j.jconrel.2004.10.010
115. Canete, M., et al.: The endocytic penetration mechanism of iron oxide magnetic nanoparticles with positively charged cover: a morphological approach. Int. J. Mol. Med. **26**, 533–9 (2010)
116. Hasenpusch, G., et al.: Magnetized aerosols comprising superparamagnetic iron oxide nanoparticles improve targeted drug and gene delivery to the lung. Pharm. Res. **29**, 1308–18 (2012). doi:10.1007/s11095-012-0682-z
117. Naqvi, S., et al.: Concentration-dependent toxicity of iron oxide nanoparticles mediated by increased oxidative stress. Int. J. Nanomed. **5**, 983–9 (2010). doi:10.2147/IJN.S13244
118. Uboldi, C., et al.: Gold nanoparticles induce cytotoxicity in the alveolar type-II cell lines A549 and NCIH441. Part. Fibre Toxicol. **6**, 18 (2009). doi:10.1186/1743-8977-6-18
119. Sadauskas, E., et al.: Biodistribution of gold nanoparticles in mouse lung following intratracheal instillation. Chem. Cent. J. **3**, 16 (2009). doi:10.1186/1752-153X-3-16
120. Sayes, C.M., Reed, K.L., Warheit, D.B.: Assessing toxicity of fine and nanoparticles: comparing in vitro measurements to in vivo pulmonary toxicity profiles. Toxicol. Sci. **97**, 163–80 (2007). doi:10.1093/toxsci/kfm018
121. DiMatteo, M., Antonini, J.M., Van Dyke, K., Reasor, M.J.: Characteristics of the acute-phase pulmonary response to silica in rats. J. Toxicol. Environ. Health A **47**, 93–108 (1996). doi:10.1080/009841096161951
122. Balas, F., Manzano, M., Horcajada, P., Vallet-Regi, M.: Confinement and controlled release of bisphosphonates on ordered mesoporous silica-based materials. J. Am. Chem. Soc. **128**, 8116–7 (2006). doi:10.1021/ja062286z
123. Lai, C.Y., et al.: A mesoporous silica nanosphere-based carrier system with chemically removable CdS nanoparticle caps for stimuli-responsive controlled release of neurotransmitters and drug molecules. J. Am. Chem. Soc. **125**, 4451–9 (2003). doi:10.1021/ja028650l
124. Kaur, G., Rath, G., Heer, H., Goyal, A.K.: Optimization of protocell of silica nanoparticles using 3(2) factorial designs. AAPS Pharm. Sci. Tech. **13**, 167–73 (2012). doi:10.1208/s12249-011-9741-8

# Antigen Delivery Systems as Oral Adjuvants

## 38

Carlos Gamazo and Juan M. Irache

## Contents

| | | |
|---|---|---|
| 38.1 | Introduction | 603 |
| 38.2 | The Mucosal Immune System | 604 |
| 38.3 | **Oral Adjuvants** | 608 |
| 38.3.1 | Exotoxins | 608 |
| 38.4 | **Microbial Colonisation Factors** | 610 |
| 38.4.1 | Microbial-Associated Molecular Patterns | 610 |
| 38.5 | **Nonmicrobial Ligands** | 611 |
| 38.6 | **Nanoparticulate Systems for Oral Delivery** | 612 |
| 38.6.1 | Nanoparticles | 612 |
| 38.6.2 | Bioadhesion | 612 |
| 38.6.3 | Direct Interactions on the Immune Cells | 613 |
| | Conclusion | 616 |
| | References | 616 |

C. Gamazo, PhD (✉)
Department of Microbiology, University of Navarra, Pamplona, Spain
e-mail: cgamazo@unav.es

J.M. Irache, PhD
Department of Pharmacy and Pharmaceutical Technology, University of Navarra, Pamplona, Spain

## Abstract

Animals, including humans, have evolved a sophisticated mucosal immune system. By understanding mucosal immune activation, we can rationally design adjuvants to elicit a long-lasting protective immune response. In this chapter we will discuss about what is known about mucosal immunity and oral adjuvants. In particular, we will specially focus on polymeric nanoparticles as mucosal adjuvants that have been developed to stimulate lifelong memory for vaccination purposes.

## 38.1 Introduction

Typically, vaccines containing live microorganisms are effective and low cost, but they carry the risk of reversion to virulence and the induction of disease in immunocompromised individuals; in contrast, vaccines based on subunit pathogens are safer but require the use of adjuvants to induce protection. Under this disjunctive position, the industry, especially veterinary and pharmaceutical companies, prefers to bet for easiness and profits more than for safety. However, this is the big motivation for many researchers, to find the right antigen complex adjuvant binomial with convincing costs and even more, susceptible to be used by the oral route, a terrific challenge [1].

In the eyes of most vaccinologists, the key is in the adjuvant. This chapter deals with adjuvants ready to be used orally. If we take a look to the current market situation, very few adjuvants fulfil this requirement in spite that the history of adjuvants is almost centennial. In 1925–1926, Ramon, Glenny and col. published their observations where demonstrated that the immunisation with toxoids could be improved by adsorption to "inert" substances, such as charcoal, collodion particles, dextran or aluminium potassium sulphate [2, 3].

These substances were called adjuvants (lat., adjuvare: "to help"). Following Glenny's discovery, aluminium salts were used in vaccine preparations with tetanus and diphtheria toxoids. In general, adjuvants have been used empirically since then, and, despite the many adjuvants being developed, insoluble aluminium salts are still the most used worldwide. Besides, if we add to the challenge the oral route, the goal is rather difficult to get. The induction of mucosal immunity through vaccination is a hard task. To illustrate the complexity of mucosal vaccine development, we can just take a look on the very limited number of oral vaccines has been approved for human use: attenuated polio vaccine; attenuated cholera vaccine; attenuated typhoid vaccine; attenuated *Mycobacterium tuberculosis* (BCG); attenuated rotavirus vaccine; and, the only nonliving oral vaccine, a *Vibrio cholerae* bacterin containing the cholera toxin B subunit. It appears clear that living-attenuated vaccines which replicate in the host are the light to follow; in fact, the current tendency is to prepare adjuvants that help nonliving vaccine to mimic the live attenuated ones, but without the intrinsic dangerous nature of the live ones. Potent mucosal adjuvants including special delivery systems are required for successful oral vaccination to face the peculiar nature of the mucosal immune system [4].

## 38.2 The Mucosal Immune System

The host is open to the outer world with an extensive internal coating, the so-called mucosae. Consequently, the immune defence is centralised on it, and so, about 60 % of total body lymphocytes are in organised follicles along mucosae called broadly the mucosa-associated lymphoid tissue (MALT) [5]. Following its physical situation is designated as gut-associated lymphoid tissue (GALT), nasopharynx-associated lymphoid tissue (NALT), bronchus-associated lymphoid tissue (BALT) or conjunctiva-associated lymphoid tissue (CALT), among other major systems. Regardless of anatomical location, MALT sites are functionally connected being recognised as a "common mucosal immune system". In consequence, after the administration of a vaccine through the oral route, we can expect to find specific memory cells and effector responses, such as specific IgA at the intestinal, vaginal and other mucosal tracts.

Concerning oral vaccination, a big challenge for the immune system is to reach the right equilibrium between tolerance and inflammatory response at mucosal level. This is the result of multifactorial processes, dependent in first instance of the peculiar anatomy of GALT. The intestine is the largest lymphoid organ in mammals, contains more immune cells and the largest concentration of antibodies than any other organ including the spleen and liver [6]. Antigens may follow different routes to cross the epithelial cell barrier. In any case, dendritic cells (DCs) act as sentinels, uptaking the antigens and then migrating to the subepithelial dome (SED) in Peyer's patches (PP) or to the mesenteric lymph nodes (MLN) to activate naive T cells (Fig. 38.1). In fact, intestinal DCs have been recognised as the critical cells involved in the decision of tolerance vs. inflammation [7]. There are a large variety of subsets of DCs that differ in their degree of "natural" or inducible predisposition to express and release determined repertoire of cytokines and, subsequently, to mount a specific immune response. We can find DCs at lamina propria (LP) and PP in different degrees of maturation. Some are "resident", whereas others are "immigrant". In general terms, resident mucosal DCs are microenvironmentally conditioned to favour tolerogenic responses, whereas newly recruited inflammatory DCs may avoid it and drive inflammatory responses (Fig. 38.2). This is a main item to consider in mucosal vaccination.

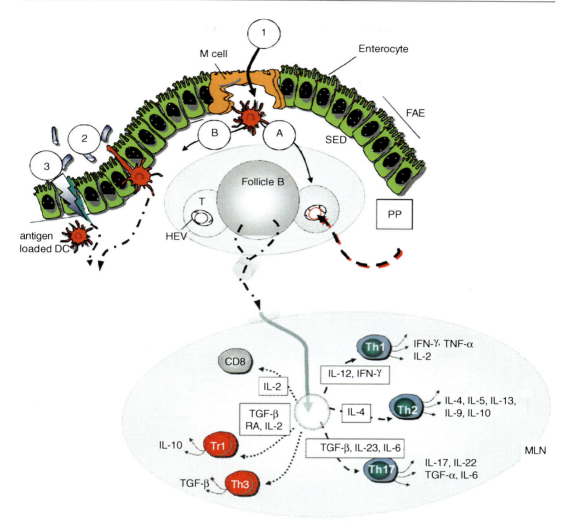

**Fig. 38.1** Antigen sampling and cellular traffic at GALT

Basically, immature DCs are present in peripheral tissues and are mainly phagocytic cells; mature DCs are found in PP and MLN, and are specialised in antigen presentation, as they are characterised by the high surface expression of co-stimulatory molecules that are required for T-cell activation. Mature DCs derive from immature cells after a maturation process that is initiated by inflammatory stimuli and that leads to a massive migration of these DCs to the draining lymph nodes. In this case, an inflammatory response is induced. However, when immature or semimature DCs arrive from peripheral tissues to draining lymph nodes, tolerance induction is expected (Figs. 38.2 and 38.3).

Apart of the maturation effect on tolerogenic or inflammatory response, different subsets of DCs have been described in mouse PPs with specialised functions [8, 9]. For instance, the phenotype $CD11b^+CD8\alpha^-CCR6^+$ found primarily in the follicle-associated epithelium in the SED, and therefore are the first to uptake luminal antigens transported by M cells, preferentially induces a Th2 differentiation and IgA plasmablasts [10]. Mucosal $CXCR1^+CD11b^+$ DCs are the most important initiators of adaptive immune

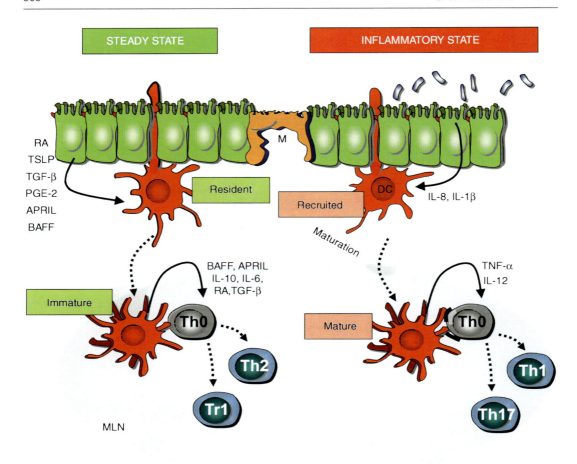

**Fig. 38.2** Gut-associated lymphoid tissues

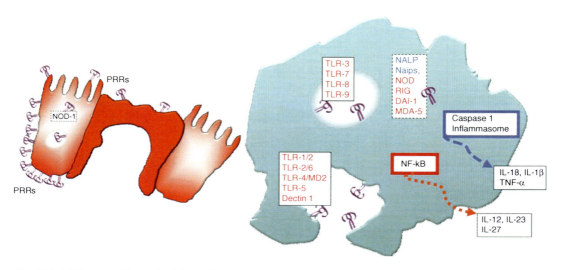

**Fig. 38.3** PRR recognition in dendritic cells and enterocytes

responses and vaccine-induced protective immunity. However, CD103+ DCs in the lamina propria induce regulatory T cells, which are important for the induction and maintenance of immunological tolerance [11]. Antigen sampling in the mucosal surfaces are performed both by professional APCs like CXCR1+CD11b+ DCs and through a specific microfold (M).

It may be inferred that a different repertoire of conditioned or not conditioned DC subsets will process the antigens depending on the route used to cross the intestinal epithelium, eliciting a different immune response [12]. The adjuvant used plays a major role in the trailed route as we will discuss in the following section.

The inductive sites of GALT comprise the appendix, isolated lymphoid follicles, Peyer's patches (PP) (in the small intestine) and the lymphoglandular complexes (large intestine). Isolated lymphoid follicles and lymphoglandular complexes are similar to PP (depicted in the figure) although smaller. Overall, they are histologically organised like the well-known lymph nodes, containing a network of APCs, B-cell and T-cell zones and connexions with the host system through efferent lymphatic vessels. The successful initiation of the specific immune responses depends on this strategic organisation being evolved to regulate the activation of T and B cells.

The follicle-associated epithelium (FAE) is a monolayer containing enterocytes and the microfold cells (M cells). The M cells comprise 10 % of FAE in mice and 1 % in human, evolving from normal enterocytes for the capture and transport of unmodified antigens. Thus, M cells contain low microvilli, few lysosomes, lack membrane-associated hydrolytic enzymes and conform like an open pocket where B and T cells, and APCs are located in order to facilitate the capture of the transcytosed antigens. Underneath is positioned a net of resident DC and also the subepithelial dome (SED) containing B-cell follicles separated by T-cell areas, known as interfollicular regions, also enriched in APCs. The T area also includes high endothelial venules (HEV) for the circulation of the immune cells. B-cell follicles are five times more extensive than T-cell areas, which is in agreement to the importance of the IgA response at mucosal sites.

The lamina propria, as a highly vascular layer of loose connective tissue underlying the mucosal epithelium, is considered the effector site within the GALT.

In the figure are represented the three known routes by which inert particles may gain access into the lamina propria from the lumen: (1) through M cells, (2) by DCs expanding dendrites to the lumen and (3) through disrupted or injured epithelium.

A main route is through the M cells in the FAE. Receptors (PRR and pIgR) are important for antigen recognition and transport. After transfer to DCs, antigen may be either presented directly to T cells in the Peyer's patch (A), or, alternatively, antigen or antigen-loaded DCs gain access to mesenteric lymph node (MLN) with the subsequent T-cell recognition (B). In both cases, DCs enter the interfollicular T-cell rich zone via the afferent lymphatic zones where DC finds lymphocytes that arrived via HEV. DC prime specific T cells that then either leave the area to become effector cells or memory T cells or they can migrate to the marginal zone and provide help to naïve and memory B cells. Then, B cells migrate to the germinal centres where a process of somatic hypermutation and switch class is activated. Finally B cells differentiate into memory B cells or plasma cells and leave the germinal centre through HEV. In turn, the activated T cells acquire expression of integrins and chemokine receptors in order to leave the MLN and after entering the bloodstream exit into the lamina propria of villi and to the different MALT effector sites. A similar process of antigen or antigen-loaded DC transport to MLNs might occur if antigen enters through the effector site (villi).

Currently, we assume (to simplify) that CD4+ T cells are subdivided into T-helper cells (Th1, Th2, Th17) and regulatory T cells (including, Tr1 and Th3). It has been shown that the presence of signature cytokines during T-cell activation directs the resulting response. Thus, Th1 cells are inflammatory cells that release IFN-γ as a key signature and are mainly involved in inflammation and protective immunity against intracellular pathogens. Th2 cells are primarily involved in B-cell help as they release B-cell growth factors like IL-4. Th17 cells may exacerbate inflammation and play a critical role in host defence against a variety of bacteria and fungi. On the other hand, there are several subsets of T regulatory cells that suppress the function of effector T cells and are then prominent in the maintenance of mucosal immunity and to control inflammatory diseases.

## 38.3 Oral Adjuvants

The default responses in the gut after mucosal immunisation are to drive a tolerogenic or a non-inflammatory response (Fig. 38.2); however, through the use of appropriate adjuvants, it is possible to abrogate it, eliciting the often desired balance Th1/Th2/Th17 responses. In few words, under mucosal tolerogenic pressure, oral adjuvants activate DCs in the proper manner. However, before that encounter takes places, oral adjuvants have a dramatic tour de force, facing with acidic conditions, hydrophobic environment, a huge diversity of hydrolytic enzymes, a network of mucoproteins and a dense microbiota covering their targets, as well as peristaltic forces that push forward (see Box 38.1). It is not surprising then the low number of oral adjuvants are currently in the market.

Basically, intestinal epithelial mucosa may be found in two main states: steady state-conditioned and inflammatory condition. Both situations are dependent mainly of the microenvironment dictated by the enterocytes and the type of subset and activation state of the DCs. In a steady state, nonactivated mucosa, the uptake of luminal antigens is mediated by immature DCs located largely in the lamina propria (see Fig. 38.1). DCs regulate the activation and differentiation of these T-cell subsets through the pattern of cytokines secreted. For example, IL-12 (p70) promotes Th1 cells; IL-10 favours Th2 or Treg responses and stimuli that induce DC to release TGF-β; IL-6 and IL-23 promote Th17 differentiation.

In addition, several important cytokines are secreted by enterocytes (see Fig. 38.3) that "push" intestinal DCs to induce Treg differentiation. Enterocytes may inhibit the generation of inflammatory responses through the release of these so-called "conditioning factors" that drive the differentiation to tolerogenic immature DCs. Thus, retinoic acid (RA), the thymic stromal lymphopoietin (TSLP), PGE-2 and TGF-β, promote tolerogenic DCs and favour the release of a pattern of cytokines [including, BAFF (B-cell activating factor) and APRIL (cytokine proliferation-inducing ligand)], driving a Treg or Th2 response and IgA-producing cells. In addition, the own regulatory cytokines released either from the enterocytes or from these tolerogenic DCs may promote the switch from IgA to IgB cells in a T-independent "archaic" manner. In the presence of challenging stimuli or disruption of the epithelial barrier, epithelial cells and newly recruited DCs release IL-1 and IL-8 to recruit DCs and macrophages in order to contain the challenge. The characteristic recruitment of immune cells to damaged areas is in part dependent of the chemotactic IL-8 released by epithelial cells. Antigens are then transported by DCs to the MLN where they induce an inflammatory immune response.

When teaching microbiology, we often use analogue terms to describe the harsh conditions that need to face the enteropathogens during its travel after being ingested. Evolution has modulated pathogen tools to successfully operate, winning in the face of the fences that are encountering centimetre after centimetre. So, it is no surprising, even it is obvious, that most successful oral adjuvants are those mimicking enteropathogens. Exotoxins, lectins and other structural components from microorganisms are being studied as oral adjuvants:

### 38.3.1 Exotoxins

Exotoxins are proteins, many of them with catalytic activity, secreted by some bacteria to gain access through the host. The enterotoxins, exotoxins that affect cells lining the gastrointestinal tract, secreted by *Vibrio cholerae* (cholera toxin, CT) and *Escherichia coli* (heat-labile toxin, LT) are the best studied mucosal adjuvants [13]. CT and LT have two subunits forming a complex with the formula A-5B: subunit A, with enzymatic ADP-ribosyltransferase active and subunit B, the one that binds to the GM1 ganglioside receptor on mammalian cells, including APCs [14]. They can be administered via mucosal or systemic routes with effects on antibody and cellular immune responses, including cytotoxic CD8 T-cell responses. Specifically, CT exposure of APCs has an augmenting effect on IL-1, IL-6 and IL-10 production and downregulation of IL-12 and TNF-α, indicating both pro- and anti-inflammatory functions [15, 16]. Besides, CT-adjuvanted antigens generated long-term-specific memory B cells after oral immunisation [17–19] since it has a potent ability to promote large numbers of germinal centre and CD86 upregulation. This is important considering that the formation of GCs after immunisation is correlated with the differentiation to B memory cells and maturation to high-affinity and switch-class antibodies [20].

However, a main drawback of exotoxins is, evidently, toxicity. CT and LT induce diarrhoea and neurological disorders since GM1 ganglioside receptor for subunit B is found on all nucleated cells, including epithelial cells and nerve cells [21].

An essential feature for the host is the discrimination between potential pathogens from the non-pathogenic, symbiotic microbiota. This property resides in a series of pattern recognition receptors (PRR), proteins able to recognise conserved molecular markers that are shared by pathogens but absent in the host. These signatures of danger may reside in the pathogen as pathogen-associated molecular pattern (PAMP) or may be a product of its action as damage-associated molecular patterns (DAMP). PAMPs and DAMPs are detected directly by the PRRs expressed by dendritic cells (DCs) and enterocytes which then release either tolerogenic or pro-inflammatory cytokines (see Fig. 38.2). Thus, microorganisms that express virulence factors that disrupt the host homeostasis trigger the rapid mobilisation of multiple PRR systems leading to inflammatory responses. In contrast, normal microbiota leads to tolerogenic responses [22]. The outcome of antigen presentation by the DCs is, therefore, a function of the level of maturation: less mature DCs will more likely result in tolerance, whereas fully mature DCs will prime strong T-cell immunity [23].

Toll-like receptors (TLRs) are transmembrane PRRs, whereas Nods, Naips, Nalps, RIG-I, MDA-5 and DAI-1 are in the cytosol. Ligation of TLR initiates a signalling cascade that results in the activation of the transcription factor NF-κB and subsequent upregulation of co-stimulatory molecules as well as inflammatory cytokines and chemokines and hormonal factors. On the other hand, members of Nalp and Naip PRRs control activation of the inflammasome, a multiprotein complex responsible for the processing and secretion of IL-1b.

Currently, there have been described 13 TLRs in mammals, with different capacities involved in the elicited immune responses [24]. Thus, the specific recognition of the corresponding PAMPs will direct differential innate and, further, adaptive immune responses. TLR3 (recognises double-stranded RNA from virus), TLR4 (bacterial lipopolysaccharide), TLR5 (bacterial flagellin), TLR7 (single-stranded RNA from virus) and TLR9 (bacterial CpG-containing DNA) are related with a preferential Th1 response. In contrast, TLR2 (recognises bacterial peptidoglycan and lipopeptides) elicits a Th2 response, although in combination with TLR6 elicit a regulatory response (Th3/Treg) [25].

Enterocytes transcribe mRNA for all types of TLRs [26]. Actually, the expression of TLRs in the intestinal epithelium is greater than that of other major organs, like the liver. However, to avoid a permanent inflammation, several mechanisms operate to maintain tolerance towards normal microbiota. Thus, the number and location of TLR in the enterocytes have a great impact. For example, TLR-4 is expressed internally, only in the Golgi apparatus, meaning that LPS will activate enterocytes only if it can penetrate in these cells. In addition, most TLRs are expressed basolaterally and only in limited number on the apical region. Cytoplasmic PRRs, such as NOD-like receptors (NLR), also play an important role in the intestinal epithelium. M cells express also several PRRs, such as TLR-4, PAF-R (platelet-activating factor receptor) and the α5β1 integrin [27]. Both TLR and NLR are expressed differentially along the intestine. This is again a critical-evolving consequence to cover the expected pathogen signals along the epithelial intestinal surfaces and has a direct effect in vaccinology since each intestinal section may respond differently to the same antigenic stimulus [7].

Both, binding and enzymatic subunits, contribute to immunomodulation, although the enzymatic activity appears to be key one [28]. CTA1-DD is a nontoxic mutant that lacks the B subunit, and the enzymatic A subunit is linked to the protein A of *Staphylococcus aureus* (antibody-binding domains) [29], maintaining adjuvant activity after nasal administration [30, 31] but being inactive orally. Other CT and LT mutants in the A1 subunit have been proposed (LTK63, LT196/211, LTR72 or the CT112 [32] retains adjuvant function by oral route, although LTK63 and LTR196/211 are still able to bind GM1 ganglioside and therefore may be toxic) [33].

*Vaccine administration*: simple to increase drug efficacy and ensure patient compliance.
*Resistance to low gastric pH and hydrolytic digestive enzymes.*
*Protection against maternal neutralising antibodies* [35].
*Increase uptake by enterocytes*, i.e. decorating with ligands for specific receptors, since it has been observed that the poor transport of antigens across the intestinal epithelium to reach the GALT renders immune unresponsiveness [36, 37].
*Induce long-term protective immunity*: by helping cross-presentation of antigens by APC, via both MHC class I and II pathways [38], to lead the development of Th1, Th2, Th17, $T_{FH}$ or $T_{reg}$ cells [39, 40].

---

**Box 38.1 Tips for a Good Oral Vaccination**
To Avoid
*Tolerance*: Dose and dose regimes are major dilemma during oral vaccination. The oral administration of a single high dose or repeated low doses of antigens induces tolerance characterised by regulatory Tr1 and Th3 subsets. On the other hand, after a big challenge, such as the one produced by virulent pathogens that make a breach in the epithelium, or by using potent adjuvants, a high inflammatory response may be induced (Th1/Th2/Th17) (Fig. 38.1) [7].
*Micronutrient deficiency states in the host*: such as retinoic acid (vitamin A) or zinc, particularly found in developing countries, can reduce the immune response by affecting discrete subpopulations of intestinal DCs and T cells [34].

To Succeed
*Vaccine final product*: safe, nontoxic and contaminant free.
*Vaccine preparation process*: standardised and reproducible, relatively inexpensive and easy to scale up.

## 38.4 Microbial Colonisation Factors

Another strategy is based on mimicking microorganism colonisation processes. Enteropathogens use diverse structures for attachment and invasion through M cells and enterocytes, resulting in their transcytosis across the intestinal epithelium [41]. This biomimetic approach have led to the use of reovirus capsid protein [42]; fimbriae from different bacteria [43–45] or outer membrane proteins from *Yersinia* [46]. The main limitations of this approach are that the efficacy depends on the host expression of the specific receptors, and that microbial adhesins are immunogenics, and so, pre-existing antibodies may neutralise them.

### 38.4.1 Microbial-Associated Molecular Patterns

A similar mimetic approach leads the use of other structures known as microbial-associated molecular patterns (MAMPs) that include PAMPs (pathogen-associated molecular patterns). These markers that may be microbial structures or metabolites are sensed as danger molecules to the host, eliciting the release of cytokines and chemokines that initiate an inflammatory defensive response. The special receptors for MAMPs on the cells, immune and epithelial cells, are known as PRRs, including TLR and NLR (non-TLR receptors). Therefore, modifying vaccines with adjuvants that carry PRR agonists may be a way out to face the natural mucosal tolerogenic tendency [47].

A main PAMP is the lipopolysaccharide (LPS) from Gram-negative bacteria [48]. The heterodimer TLR4/MD-2 recognises LPS, mediating bacterial translocation through enterocytes and M cells [49] making it suitable as oral adjuvant. However, LPS, known as endotoxin, may have potent biological activity with deleterious side effects. Natural or synthetic derivatives of LPS with reduced toxicity are being searched [48]. The LPS from *Brucella ovis* has been used due to its low toxicity but maintained capacity to induce Th1 responses to the carried antigens [50]. Monophosphoryl lipid A (MPL) from *Salmonella enterica* serovar. Minnesota is included in the FDA-approved AS04 adjuvant used in human papillomavirus vaccine, Cervarix™, combined with alum [51]. The AS02 adjuvant also contains MPL but combined with QS-21 in an oil-in-water emulsion. AS01, which contains MPL and saponin in liposomes, has demonstrated to enhance systemic and mucosal immunity in an HIV vaccine study with non-human primates [52].

The flagellin is a main danger molecule as it is exclusive of bacteria. Flagellin is the monomeric protein that conforms the bacterial flagellum, and so, it is being extensively investigated as a PAMP, since flagellin binds TLR5 [53] expressed by enterocytes is suitable to be used in oral adjuvants [54, 55]. Flagellin induces the maturation of intestinal DC, activates CD4$^+$ T cells in vivo and promotes the development of mixed effector Th cell responses [56]. As a mucosal adjuvant, flagellin is almost as potent as CT or LT but much safer [53].

Microorganisms are enriched in unmethylated CpG oligodeoxynucleotides (CpG) and so are recognised as danger signatures by the host. CpG domains interact with TLR9, inducing strong Th1 responses. They have shown to be effective in animal vaccine studies when delivered mucosally [57]. Human clinical trials demonstrate CpG domains and have a good safety profile, which provides a scientific ground to expand the CpG application towards mucosal vaccines [58].

Other PRR agonists efficacious as oral adjuvants include α-Galactosylceramide, a CD1d ligand and NKT cell activator [59] and fungal or bacterial structures that bind non-TLR PRRs on APCs, such as dectin-1 [60], mannose [61–63] or the bacterial second messenger 3′,5′-Cyclic diguanylic acid recognised as a danger signal to the host as a PAMP and effective mucosal adjuvant [64].

## 38.5 Nonmicrobial Ligands

*Lectins* are proteins which bind to specific carbohydrate. Plant lectins are the most used since they are more resistant to intestinal degradation and, therefore, are also susceptible to be used as ligands for targeting to M cells and enterocytes, covered by glycoproteins and glycolipids. Among several binomial plants lectin carbohydrate, the most investigated lectin is *Ulex europaeus* agglutinin 1 (UEA-1) that binds to α-L-fucose on M cells, goblet and Paneth cells [65, 66]. However, toxicity and anti-nutritional properties are major drawbacks that limit its use for oral vaccination [67].

*Antibodies* directed against cell surface antigens, for example, monoclonal antibodies to M cells [68].

Several *cytokines* have been exploited for their immunostimulatory properties. These include the recombinant IL-1 family cytokines (IL-1α/β, IL-18 and IL-33) or the thymic stromal lymphopoietin (TSLP) [69, 70]. However, cost/effect ratio is not convincing.

The above strategies present advantages and inconvenience that needs to be solved. Antigen delivery systems are ready to overcome all the barriers that still challenge oral vaccination as we discuss next.

## 38.6 Nanoparticulate Systems for Oral Delivery

Pharmaceutical technocrats are developing numerous particle systems for antigen delivery to the host. As main characteristic properties it can be highlighted by the following: protect antigens from digestive enzymes; facilitated uptake by enterocytes specially if decorated with ligands specific for epithelial receptors [11]; facilitate the antigen uptake by APCs or by increasing the influx of professional APCs into the injection site and further, the antigenic cross-presentation by APC via both MHC class I and II pathways [38]. Among the different types of particulate delivery systems are liposomes, archaeoliposomes, produced with lipids from prokaryotic *Archaea* [71]. Iscomatrix comprising Quil A, cholesterol and phospholipids, but it is not for hydrophilic antigens; VLP prepared from the capsid proteins of mucosal infecting virus, liposomes, immunestimulating complexes (ISCOMs), yeast ghosts and polymeric micro- or nanoparticles, cochleates and polymeric particulate delivery systems [72–78].

### 38.6.1 Nanoparticles

Oral vaccination using antigens loaded or encapsulated in particles and specifically nanoparticles appears to have a sound scientific rationale based on the protection of an antigen from exposure to extreme pH conditions, bile and pancreatic secretions [79, 80] and provides a depot effect [81]. At the same time, advantage is taken of the inherent inclination of particulate to be naturally captured by mucosal APCs as part of its duty as a sentinel in triggering of mucosal immunity against pathogens [82–85]. In fact, it has been clearly demonstrated that nanoparticles can interact with different components of the mucosa.

### 38.6.2 Bioadhesion

In order to increase the residence of nanoparticles in close contact with the epithelial surface of the mucosa and, then, improve their capability to reach specific immunocompetent cells and the GALT, different strategies have been developed. One of these strategies may be to potentiate the bioadhesive properties of these carriers by modifying or decorating their surface with ligands capable to (1) cross the mucus layer protecting the mucosal surface and/or (2) target specific receptors on the immunocompetent cells located in the mucosa.

One of the most popular techniques for modification of the nanoparticle's surface is PEGylation. The adequate coating of nanoparticles with poly(ethylene glycol) (PEG) permits to decrease the interaction of nanoparticles with the mucins and cross through the mucus layer on the road to the mucosal epithelium [86, 87]. In addition it was demonstrated that pegylation of poly(anhydride) nanoparticles yielded carriers with a high ability to minimise adhesive interactions within the stomach mucosa and to concentrate them in the small intestine mucosa of animals [88].

Another possibility can be the use of biomimetic approaches such us the decoration of nanoparticle compounds or molecules involved in either the colonisation process of microorganisms or the activation of the host immune system. Microorganisms can invade and colonise the host tissue by using a number of different specific adherence factors including outer membrane proteins [89], flagella [90], fimbriae and pili [91], lectins [92] and glycoproteins [55]. Most of these adhesive factors are also considered as immunomodulators, and they are included in the generic denomination of PAMPs (pathogen-associated molecular patterns).

In the recent past, our group of research demonstrated that the association of either flagellin from *Salmonella* Enteritidis flagella or mannosamine to Gantrez AN nanoparticles could enhance the bioadhesive capabilities of the resulted decorated nanoparticles. These nanoparticles demonstrated a high ability to colonise the gut of animals with a characterised ileum tropism and high affinity to Peyer's patches [62, 93] (Fig. 38.4).

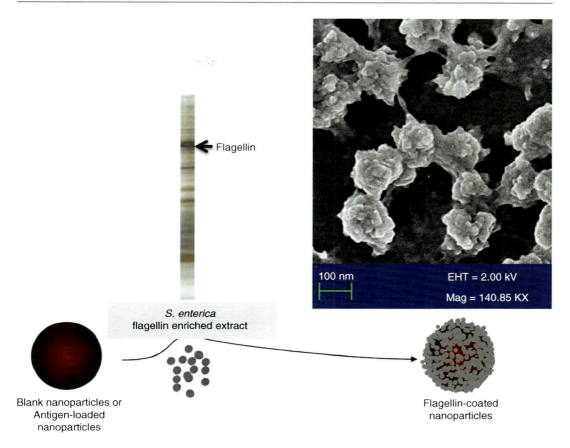

**Fig. 38.4** Scheme on flagellin-containing nanoparticles

## 38.6.3 Direct Interactions on the Immune Cells

In addition, the use of nanoparticles offers another advantage such as the possibility to load oral adjuvants, such as those described above (i.e. CT, lectins, PRR agonists) with the antigen in order to increase and induce the more appropriate immune response [94, 95]. CTA1-DD, the already mentioned nontoxic mutant that renders a cholera toxin that lacks the B subunit, is not stable for oral administration, but protected when incorporated in particulate delivery systems [96–98].

"*Salmonella*-like" nanoparticles have been obtained by the association with the flagellin of *Salmonella* Enteritidis to PVM/MA nanoparticles [99]. These carriers displayed an important tropism for the ileum, and their distribution within the gut correlated well with the described colonisation profile for the bacteria, including a broad concentration in Peyer's patches [62]. Using ovalbumin as model antigen, "*Salmonella*-like" nanoparticles induced a strong and balanced secretion of both IgG2a- (Th1) and IgG1 (Th2)-specific antibodies. In addition, these nanoparticles were able to induce a much more strong mucosal IgA response than control nanoparticles [99–103].

Table 38.1 summarises some other examples related with the development of nanoparticles as adjuvants for mucosal vaccination. Several factors contribute to the efficacy of particulates as oral adjuvants. The adequate combination between the polymer and the antigens as well as the preparative method employed for antigen loading have a great influence on the physicochemical properties of the resulting particulates,

**Table 38.1** Examples of oral immunisations using antigen-loaded biodegradable[a] nanoparticles

| Antigen | Polymer | Observations | Refs |
|---|---|---|---|
| BSA | Hydrophobic carbon nanoparticles | Serum IgG titre close to that observed for BSA emulsified in FCA and i.m. administered | [104] |
| *Eimeria* recombinant profilin | Montanide IMS nanoparticles | Decreased faecal parasite excretion. Reduced intestinal lesions in infected animals (chickens) | [105] |
| F4 fimbriae *Escherichia coli* | Gantrez AN nanoparticles | Increased levels of F4-specific antibody-secreting cells in the spleen | [106] |
| Ovalbumin | PLGA nanoparticles containing MPLA | Strong IgG immune response. Strong IgA titres | [107] |
| DNA encoding for the major capsid protein of LCDV | PLGA nanoparticles | Reduced fish infection rate | [108] |
| HBsAg | *Lotus tetragonolobus* lectin-coated PLGA nanoparticles | Strong mucosal and systemic immune responses | [109] |
| HBsAg | Pegylated PLA nanoparticles | Effective levels of cellular Th-1 response and mucosal humoral immunity (IgA) | [110] |
| Influenza A antigen | Bilosomes (lipid vesicles) | Higher antibody production when administered orally than by the i.m. route | [111] |
| DNA-encoding house dust mite allergen Der p2 | Chitosan nanoparticles | Induction of Th-1 responses with high serum levels of IFN-g | [112] |

*BSA* bovine serum albumin, *FCA* Freund's complete adjuvant, *MPLA* monophosphoryl lipid A, *LCDV* Lymphocystis disease virus, *HBsAg* Hepatitis B surface antigen, *PLGA* poly(D,L-lactide-co-glycolide), *PLA* poly(D,L-lactic acid)
[a]Under physiological conditions, nanoparticles derived from PLGA, chitosan, lipids or Gantrez AN suffer from biodegradation or bioerosion. Concerning hydrophobic carbon nanoparticles, which are derived from silica, there are is no information about their degradability, and we think that they are not degraded under physiological conditions. In any case, it would be necessary also to take also into account that the proposed route of administration for all of these nanoparticles is the oral/mucosal route. In the oral, physiological process such as peristaltism, enterocytes/mucosa and mucus turnover are very important and plays an important role in the elimination of the gut contents. Thus, the degradation of nanoparticles will be less important than the adequate release of the loaded antigen

determining their size and surface characteristics [113]. Although the correlation between particle size and their adjuvant activity has been controversial [111, 114–121], particles in the nanoscale size rather than microscale size appear to be more adapted for cellular uptake and immune stimulation in the GI tract [122–124].

On the other hand, the polymer used not only determines the antigen loading in the resulting nanoparticles, but also the stability of the resulting particles in the gastrointestinal tract as well their interaction with components of the mucosa. It is important to note that the stability may be an important factor influencing the release of the loaded antigen.

Bacterial flagellin in addition to being a rational target of the specific immune system can directly activate innate immune cells that, as a consequence, secrete inflammatory cytokines. This immunostimulatory capacity is mediated by surface receptor TLR-5 expressed by monocyte/macrophages and dendritic cells, but also by the enterocytes in the gut. This innate ability makes flagellin as an effective adjuvant. Furthermore, flagellin is an adhesive factor used by mucosal bacterial pathogens for invasion of host surfaces and hence colonises the gut. Considering both properties, flagellin may be used as a great synergistic mucosal adjuvant. The figure shows a schematic representation of the preparative process of flagellin-coated poly(anhydride) nanoparticles. Poly(anhydride) nanoparticles were prepared by a desolvation method followed by a drying step by lyophilisation. Briefly, flagellin from *Salmonella enteritidis* was incubated with the polymer (copolymer of methyl vinyl ether and maleic anhydride, Gantrez AN) in acetone. Then, nanoparticles were obtained by the addition of a mixture of ethanol and water. The organic solvents were eliminated under reduced pressure and the resulting nanoparticles cross-linked by incubation with 100 μg 1,3-Diaminopropane for 5 min. Then, the suspensions were purified by centrifugation, and, finally, the nanoparticles were lyophilised using sucrose (5 %) as cryoprotector. The SEM microphotograph shows the morphology and shape of the resulting nanoparticles where flagellin proteins are on their surface for direct presentation to the enterocytes and after to the immunocompetent cells.

Among others, poly(D,L-lactide-co-glycolide) (PLGA), chitosan, lipids, starch, phosphazene, poly(epsilon-caprolactone), poly(anhydride) and cationic cross-linked polysaccharides have also been proposed as antigen delivery nanoparticles for oral vaccination.

In this way, a great number of antigens have been successfully encapsulated in PLGA particles with a full maintenance of structural, antigenic integrity and immunogenicity after oral administration [125, 126], even in clinical trials [127].

Chitosan may also have an immunomodulatory effect as it has been shown to stimulate production of cytokines from immune cells in vitro [128] and enhance a naturally Th2/Th3-biassed microenvironment at the mucosal level in absence of antigen [129].

Poly(anhydride) nanoparticulate systems made by the copolymers of sebacic acid, 1,6-bis-(*p*-carboxyphenoxy)hexane, and 1,8-bis-(*p*-carboxyphenoxy)-3,6-dioxaocatane have demonstrated biocompatibility both in vivo and in vitro at concentrations expected for human use [63, 130–133].

Some amphiphilic polyanhydrides have also been reported to exhibit adjuvant characteristics [134]. In particular, poly(anhydride) nanoparticulate systems made by the copolymer of methyl vinyl ether and maleic anhydride (PVM/MA) have demonstrated their efficacy as adjuvants to induce Th1 immune responses [50, 135–138]. Furthermore, PVM/MA nanoparticles may be exploited as oral adjuvants since they can enhance the delivery of the loaded antigen to the gut lymphoid cells due to their special bioadhesive properties [79].

Interactions between particles and DCs depend on particle characteristics such as size and shape, charge and hydrophobicity, but the mechanisms responsible for DC maturation may be mostly related to TLR recognition in DCs. Many reports show how the stimulation with TLR agonists induces the surface expression of co-stimulatory molecules and hence the phenotypic modulation to the typical feature of a mature DC. It has been shown that poly(anhydride) nanoparticles of PMVA induced innate immune responses mediated by a TLR-2 and TLR-4 dependent manner [128, 138]. These nanoparticles induced maturation of DC with a significant upregulation of CD40, CD80 and CD86 and a biassed Th1-present response in animal models. This is an important finding since it has been recently shown that the use of multiple TLR agonists carried by nanoparticles influences the induction of long-term memory cells, being the ultimate goal for any vaccine the stimulation of long-lasting protective immunological memory [139, 140].

Little information is available on adjuvant effects and memory development in general, and even less is known about mucosal adjuvant effects and memory responses. Long-term memory paradigm is offered by key successful examples including smallpox, measles, mumps, polio, rubella and the paradigmatic yellow fever [141, 142]. Cell-mediated immunity may be maintained in immunised people with live attenuated vaccines for decades. This last vaccine (YF-17D) contains agonists for several Toll-like receptors. The impact on the activation of innate immunity could explain why this vaccine effectively stimulated long-term memory. In fact, a new approach for successful vaccine adjuvant development is based on the use of TLR agonist [143].

Other polymers differ in their potential capacity to be modified with chemicals and ligands (see above). For instance, Poly(lactic-co-glycolic acid) and polystyrene microparticles were found to activate caspase-1 through NLRP3 in vitro as efficiently as alum [144]. Other experimental adjuvants have also been shown to mediate an NLRP-dependent IL-1 release, including Quil A, a saponin extracted from the bark of the *Quillaja saponaria* tree, and chitosan [145].

## Conclusion

Vaccination coverage, especially in developing countries, is a main challenge for our society [146]. Several key issues related with vaccine development should be considered as follows: safety, low-cost production, mass vaccination and convenient mucosal administration, among others. Mucosal administration of subunit vaccines offers security and convenience, but they require potent adjuvants to overcome the epithelial barriers. Nanoparticles upsurge as a group of delivery systems with interesting abilities as adjuvants. However, a better knowledge of human mucosal immune responsiveness is required to establish the usefulness of these adjuvants regardless of the optimistic presages on nanotechnology-based antigen delivery (see Box 38.1). Currently, there are no pure subunit vaccines formulated and licensed for mucosal administration. This is mainly due to the regrettable fact that few adjuvants have been approved for human use [147, 148]. Major drawbacks on toxins as mucosal adjuvants have induced to a major research on the development of new mucosal adjuvants [149, 150]. Therefore, this is our major challenge, the development of adjuvants that enhance the potency of future subunit vaccines administered by different mucosal routes.

Subunit vaccine's main challenge is to induce lifelong memory are live attenuated does, such as vaccinia virus vaccine, measles, mumps, polio, rubella or yellow fever. Therefore, future adjuvant design may pay attention to the attenuated live vaccine's mechanisms of action. The default response in the gut after mucosal immunisation is to drive a tolerogenic or a noninflammatory response. This is an evolving consequence of food ingestion. The continuous sampling of the intestinal contents by immature DCs, which do not have many co-stimulatory molecules, will render an anti-inflammatory response. However, pathogens and attenuated vaccines activate multiple pathways and present "danger molecules" (PAMPs) that include toll-like receptors, C-type lectin receptors, RIG-1-like receptors and NOD-like receptors [151] that bind mucosal epithelium and cause epithelial cells to release cytokines and chemokines that attract immune cells for induction of an inflammatory response. Therefore, modifying the vaccines with immunological adjuvants that carry PAMPs or other PRR agonists may be a way out to face the natural mucosal tolerogenic tendency. Taken together, the identification of ways to modulate Th1, Th2, Th17, $T_{FH}$ or $T_{reg}$ will most certainly be key to the successful development of novel vaccine adjuvants.

Thus, nanoparticles have revealed as good oral adjuvants, being able to induce simultaneously a peripheral and a mucosal immunity. Through the use of nanoparticulated delivery systems, it is possible to abrogate tolerance and elicit the desired balance Th1/Th2/Th17-type responses in both mucosal and systemic sites, opening new dimensions in the adjuvant field for mucosal vaccination.

## References

1. Nepom, G.T.: Mucosal matters. Foreword. Nat. Rev. Immunol. **8**, 409 (2008)
2. Ramon, G.: Sur l'augmentation anormale de l'antitoxine chez les chevaux producteurs de serum antidiphterique. Bull. Soc. Centr. Med. Vet. **101**, 227–234 (1925)
3. Glenny, A., Pope, C., Waddington, H., Wallace, V.: The antigenic value of toxoid precipitated by potassium-alum. J. Path. Bacteriol. **29**, 38–45 (1926)
4. Holmgren, J., Czerkinsky, C.: Mucosal immunity and vaccines. Nat. Med. **11**, S45–S53 (2005)
5. Czerkinsky, C., Holmgren, J.: Mucosal delivery routes for optimal immunization: targeting immunity to the right tissues. Curr. Top. Microbiol. Immunol. **354**, 1–18 (2012)
6. Mayer, L.: Mucosal immunity and gastrointestinal antigen processing. J. Pediatr. Gastroenterol. Nutr. **30**, S4–S12 (2000)

7. Ng, S.C., Kamm, M.A., Stagg, A.J., Knight, S.C.: Intestinal dendritic cells: their role in bacterial recognition, lymphocyte homing, and intestinal inflammation. Inflamm. Bowel Dis. **16**, 1787–1807 (2010)
8. Persson, K.E., Jaensson, E., Agace, W.W.: The diverse ontogeny and function of murine small intestinal dendritic cell/macrophage subsets. Immunobiology **215**, 692–697 (2010)
9. Rescigno, M., Di Sabatino, A.: Dendritic cells in intestinal homeostasis and disease. J. Clin. Invest. **119**, 2441–2450 (2009)
10. Sato, A., Hashiguchi, M., Toda, E., Iwasaki, A., Hachimura, S., Kaminogawa, S.: CD11b+Peyer's patch dendritic cells secrete IL-6 and induce IgA secretion from naive B cells. J. Immunol. **171**, 3684–3690 (2003)
11. Devriendt, B., Devriendt, B., De Geest, B.G., Goddeeris, B.M., Cox, E.: Crossing the barrier: targeting epithelial receptors for enhanced oral vaccine delivery. J. Control. Release **160**, 431–439 (2012)
12. Dudziak, D., Kamphorst, A.O., Heidkamp, G.F., Buchholz, V.R., Trumpfheller, C., Yamazaki, S., Cheong, C., Liu, K., Lee, H.W., Park, C.G., Steinman, R.M., Nussenzweig, M.C.: Differential antigen processing by dendritic cell subsets in vivo. Science **315**, 107–111 (2007)
13. Freytag, L.C., Clements, J.D., Eliasson, D.G., Lycke, N.: Use of genetically or chemically detoxified mutants of cholera and *Escherichia coli* heat-labile enterotoxins as mucosal adjuvants. In: Levine, M.M. (ed.) New Generation Vaccines, 4th edn, pp. 273–283. Informa Healthcare USA, New York (2010)
14. Fan, E., O'Neal, C.J., Mitchell, D.D., Robien, M.A., Zhang, Z., Pickens, J.C., Tan, X.J., Korotkov, K., Roach, C., Krumm, B., Verlinde, C.L., Merritt, E.A., Hol, W.G.: Structural biology and structure-based inhibitor design of cholera toxin and heat-labile enterotoxin. Int. J. Med. Microbiol. **294**, 217–223 (2004)
15. Lavelle, E.C., Jarnicki, A., McNeela, E., Armstrong, M.E., Higgins, S.C., Leavy, O., Mills, K.H.: Effects of cholera toxin on innate and adaptive immunity and its application as an immunomodulatory agent. J. Leukoc. Biol. **75**, 756–763 (2004)
16. Lavelle, E.C., McNeela, E., Armstrong, M.E., Leavy, O., Higgins, S.C., Mills, K.H.: Cholera toxin promotes the induction of regulatory T cells specific for bystander antigens by modulating dendritic cell activation. J. Immunol. **171**, 2384–2392 (2003)
17. Lycke, N.: Targeted vaccine adjuvants based on modified cholera toxin. Curr. Mol. Med. **5**, 591–597 (2005)
18. Vajdy, M., Lycke, N.Y.: Cholera toxin adjuvant promotes long-term immunological memory in the gut mucosa to unrelated immunogens after oral immunization. Immunology **75**, 488–492 (1992)
19. Soenawan, E., Srivastava, I., Gupta, S., Kan, E., Janani, R., Kazzaz, J., Singh, M., Shreedhar, V., Vajdy, M.: Maintenance of long-term immunological memory by low avidity IgM-secreting cells in bone marrow after mucosal immunizations with cholera toxin adjuvant. Vaccine **22**, 1553–1563 (2004)
20. Schwickert, T.A., Alabyev, B., Manser, T., Nussenzweig, M.C.: Germinal center reutilization by newly activated B cells. J. Exp. Med. **206**, 2907–2914 (2009)
21. Fujihashi, K., Koga, T., van Ginkel, F.W., Hagiwara, Y., McGhee, J.R.: A dilemma for mucosal vaccination: efficacy versus toxicity using enterotoxin-based adjuvants. Vaccine **20**, 2431–2438 (2002)
22. Miron, N., Cristea, V.: Enterocytes: active cells in tolerance to food and microbial antigens in the gut. Clin. Exp. Immunol. **167**, 405–412 (2011)
23. Blander, M.J., Sander, L.E.: Beyond pattern recognition: five immune checkpoints for scaling the microbial threat. Nat. Rev. Immunol. **12**, 215–225 (2012)
24. Takeda, K., Kaisho, T., Akira, S.: Toll-like receptors. Annu. Rev. Immunol. **21**, 335–376 (2003)
25. Yamamoto, M., Sato, S., Mori, K., Hoshino, K., Takeuchi, O., Takeda, K., Akira, S.: Cutting edge: role of Toll-like receptor 1 in mediating immune response to microbial lipoproteins. J. Immunol. **169**, 10–14 (2002)
26. Gewirtz, T.: Intestinal epithelial toll-like receptors: to protect. And serve? Curr. Pharm. Des. **9**, 1–5 (2003)
27. Tyrer, P., Foxwell, A.R., Cripps, A.W., Apicella, M.A., Kyd, J.M.: Microbial pattern recognition receptors mediate M-Cell uptake of a gram-negative bacterium. Infect. Immun. **74**(1), 625–631 (2006)
28. Kawamura, Y.I., Kawashima, R., Shirai, Y., Kato, R., Hamabata, T., Yamamoto, M., Furukawa, K., Fujihashi, K., McGhee, J.R., Hayashi, H., Dohi, T.: Cholera toxin activates dendritic cells through dependence on GM1-ganglioside which is mediated by NF-kappaB translocation. Eur. J. Immunol. **33**, 3205–3212 (2003)
29. Agren, L., Löwenadler, B., Lycke, N.: A novel concept in mucosal adjuvanticity: the CTA1-DD adjuvant is a B cell-targeted fusion protein that incorporates the enzymatically active cholera toxin A1 subunit. Immunol. Cell Biol. **76**, 280–287 (1998)
30. van Ginkel, F.W., Jackson, R.J., Yuki, Y., McGhee, J.R.: Cutting edge: the mucosal adjuvant cholera toxin redirects vaccine proteins into olfactory tissues. J. Immunol. **165**, 4778–4782 (2000)
31. Eliasson, D.G., El Bakkouri, K., Schön, K., Ramne, A., Festjens, E., Löwenadler, B., Fiers, W., Saelens, X., Lycke, N.: CTA1-M2e-DD: a novel mucosal adjuvant targeted influenza vaccine. Vaccine **26**, 1243–1252 (2008)
32. Giuliani, M.M., Del Giudice, G., Giannelli, V., Dougan, G., Douce, G., Rappuoli, R., Pizza, M.: Mucosal adjuvanticity and immunogenicity of LTR72, a novel mutant of Escherichia coli heat-labile enterotoxin with partial knockout of ADP-ribosyltransferase activity. J. Exp. Med. **187**, 1123–1132 (1998)
33. Summerton, N.A., Welch, R.W., Bondoc, L., Yang, H.H., Pleune, B., Ramachandran, N., Harris, A.M.,

Bland, D., Jackson, W.J., Park, S., Clements, J.D., Nabors, G.S.: Toward the development of a stable, freeze-dried formulation of Helicobacter pylori killed whole cell vaccine adjuvanted with a novel mutant of Escherichia coli heat-labile toxin. Vaccine **28**, 1404–1411 (2010)
34. Spencer, S.P., Belkaid, Y.: Dietary and commensal derived nutrients: shaping mucosal and systemic immunity. Curr. Opin. Immunol. **24**, 379–384 (2012)
35. Holmgren, J., Svennerholm, A.M.: Vaccines against mucosal infections. Curr. Opin. Immunol. **24**, 343–353 (2012)
36. Dunne, A., Marshall, N.A., Mills, K.H.: TLR based therapeutics. Curr. Opin. Pharmacol. **11**, 404–411 (2011)
37. Higgins, S.C., Mills, K.H.: TLR, NLR agonists, and other immune modulators as infectious disease vaccine adjuvants. Curr. Infect. Dis. Rep. **12**, 4–12 (2010)
38. Men, Y., Audran, R., Thomasin, C., Eberl, G., Demotz, S., Merkle, H.P., Gander, B., Corradin, G.: MHC class I- and class II-restricted processing and presentation of microencapsulated antigens. Vaccine **17**, 1047–1056 (1999)
39. Lefrancois, L.: Development, trafficking, and function of memory T-cell subsets. Immunol. Rev. **211**, 93–103 (2006)
40. Sallusto, F., Mackay, C.R., Lanzavecchia, A.: The role of chemokine receptors in primary, effector, and memory immune responses. Annu. Rev. Immunol. **18**, 593–620 (2000)
41. Khader, S.A., Gaffen, S.L., Kolls, J.K.: Th17 cells at the crossroads of innate and adaptive immunity against infectious diseases at the mucosa. Mucosal Immunol. **2**, 403–411 (2009)
42. Rubas, W., Banerjea, A.C., Gallati, H., Speiser, P.P., Joklik, W.K.: Incorporation of the reovirus M cell attachment protein into small unilamellar vesicles: incorporation efficiency and binding capability to L929 cells in vitro. J. Microencapsul. **7**, 385–395 (1990)
43. Hase, K., Kawano, K., Nochi, T., Pontes, G.S., Fukuda, S., Ebisawa, M., Kadokura, K., Tobe, T., Fujimura, Y., Kawano, S., Yabashi, A., Waguri, S., Nakato, G., Kimura, S., Murakami, T., Iimura, M., Hamura, K., Fukuoka, S., Lowe, A.W., Itoh, K., Kiyono, H., Ohno, H.: Uptake through glycoprotein 2 of FimH+ bacteria by M cells initiates mucosal immune response. Nature **462**, 226–230 (2009)
44. Verdonck, F., De Hauwere, V., Bouckaert, J., Goddeeris, B.M., Cox, E.: Fimbriae of enterotoxigenic *Escherichia coli* function as a mucosal carrier for a coupled heterologous antigen. J. Control. Release **104**, 243–258 (2005)
45. Melkebeek, V., Rasschaert, K., Bellot, P., Tilleman, K., Favoreel, H., Deforce, D., De Geest, B.G., Goddeeris, B.M., Cox, E.: Targeting aminopeptidase N, a newly identified receptor for F4ac fimbriae, enhances the intestinal mucosal immune response. Mucosal Immunol. **5**(6), 635–645 (2012)
46. Suzuki, T., Yoshikawa, Y., Ashida, H., Iwai, H., Toyotome, T., Matsui, H., Sasakawa, C.: High vaccine efficacy against shigellosis of recombinant non-invasive *Shigella* mutant that expresses *Yersinia* invasion. J. Immunol. **177**, 4709–4717 (2006)
47. Chadwick, S., Kriegel, C., Amiji, M.: Delivery strategies to enhance mucosal vaccination. Expert Opin. Biol. Ther. **9**, 427–440 (2009)
48. Fox, C.B., Friede, M., Reed, S.G., Ireton, G.C.: Synthetic and natural TLR4 agonists as safe and effective vaccine adjuvants. Subcel. Biochem. **53**, 303–321 (2010)
49. Neal, M.D., Leaphart, C., Levy, R., Prince, J., Billiar, T.R., Watkins, S., Li, J., Cetin, S., Ford, H., Schreiber, A., Hackam, D.J.: Enterocyte TLR4 mediates phagocytosis and transcytosis of bacteria across the intestinal barrier. J. Immunol. **176**, 3070–3079 (2006)
50. Gómez, S., Gamazo, C., Roman, B.S., Ferrer, M., Sanz, M.L., Irache, J.M.: Gantrez® AN nanoparticles as an adjuvant for oral immunotherapy with allergens. Vaccine **25**, 5263–5271 (2007)
51. Schwarz, T.: Clinical update of the AS04-adjuvanted human papillomavirus-16/18 cervical cancer vaccine, cervarix®. Adv. Ther. **26**, 983–998 (2009)
52. Cranage, M.P., Fraser, C.A., Cope, A., McKay, P.F., Seaman, M.S., Cole, T., Mahmoud, A.N., Hall, J., Giles, E., Voss, G., Page, M., Almond, N., Shattock, R.J.: Antibody responses after intravaginal immunisation with trimeric HIV-1CN54 clade C gp140 in Carbopol gel are augmented by systemic priming or boosting with an adjuvanted formulation. Vaccine **4**, 1421–1430 (2011)
53. Mizel, S.B., Bates, J.T.: Flagellin as an adjuvant: cellular mechanisms and potential. J. Immunol. **185**, 5677–5682 (2010)
54. Salman, H.H., Irache, J.M., Gamazo, C.: Immunoadjuvant capacity of flagellin and mannosamine-coated poly(anhydride) nanoparticles in oral vaccination. Vaccine **27**, 4784–4790 (2009)
55. Salman, H.H., Gamazo, C., Campanero, M.A., Irache, J.M.: Salmonella-like bioadhesive nanoparticles. J. Control. Release **106**, 1–13 (2005)
56. Uematsu, S., Fujimoto, K., Jang, M.H., Yang, B.G., Jung, Y.J., Nishiyama, M., Sato, S., Tsujimura, T., Yamamoto, M., Yokota, Y., Kiyono, H., Miyasaka, M., Ishii, K.J., Akira, S.: Regulation of humoral and cellular gut immunity by lamina propria dendritic cells expressing Toll-like receptor 5. Nat. Immunol. **9**, 769–776 (2008)
57. Harandi, A.M.J., Holmgren, J.: CpG DNA as a potent inducer of mucosal immunity: implications for immunoprophylaxis and immunotherapy of mucosal infections. Curr. Opin. Invest. Drugs **5**, 141–145 (2004)
58. Lawson, L.B., Norton, E.B., Clements, J.D.: Defending the mucosa: adjuvant and carrier formulations for mucosal immunity. Curr. Opin. Immunol. **23**, 414–420 (2011)
59. Courtney, A.N., Nehete, P.N., Nehete, B.P., Thapa, P., Zhou, D., Sastry, K.J.: Alpha-galactosylceramide

is an effective mucosal adjuvant for repeated intranasal or oral delivery of HIV peptide antigens. Vaccine 27, 3335–3341 (2009)
60. Agrawal, S., Gupta, S., Agrawal, A.: Human dendritic cells activated via dectin-1 are efficient at priming Th17, cytotoxic CD8T and B cell responses. PLoS One 5(10), e13418 (2010)
61. Da Costa Martins, R., Gamazo, C., Sánchez-Martínez, M., Barberán, M., Peñuelas, I., Irache, J.M.: Conjunctival vaccination against *Brucella ovis* in mice with mannosylated nanoparticles. J. Control. Release 162, 553–560 (2012)
62. Salman, H.H., Gamazo, C., Campanero, M.A., Irache, J.M.: Bioadhesive mannosylated nanoparticles for oral drug delivery. J. Nanosci. Nanotechnol. 6, 3203–3209 (2006)
63. Carrillo-Conde, B., Song, E.H., Chavez-Santoscoy, A., Phanse, Y., Ramer-Tait, A.E., Pohl, N.L., Wannemuehler, M.J., Bellaire, B.H., Narasimhan, B.: Mannose-functionalized "pathogen-like" polyanhydride nanoparticles target C-type lectin receptors on dendritic cells. Mol. Pharm. 8, 1877–1886 (2011)
64. Chen, W., Kuolee, R., Yan, H.: The potential of 3′,5′-cyclic diguanylic acid (c-di-GMP) as an effective vaccine adjuvant. Vaccine 28, 3080–3085 (2010)
65. Clark, M.A., Hirst, B.H., Jepson, M.A.: Lectin-mediated mucosal delivery of drugs and microparticles. Adv. Drug Deliv. Rev. 43, 207–223 (2000)
66. Jepson, M.A., Clark, M.A., Hirst, B.H.: M cell targeting by lectins: a strategy for mucosal vaccination and drug delivery. Adv. Drug Deliv. Rev. 56, 511–525 (2004)
67. Rajapaksa, T.E., Lo, D.D.: Microencapsulation of vaccine antigens and adjuvants for mucosal targeting. Curr. Immunol. Rev. 6, 29–37 (2010)
68. Nochi, T., Yuki, Y., Matsumura, A., Mejima, M., Terahara, K., Kim, D.Y., Fukuyama, S., Iwatsuki-Horimoto, K., Kawaoka, Y., Kohda, T., Kozaki, S., Igarashi, O., Kiyono, H.: A novel M cell-specific carbohydrate-targeted mucosal vaccine effectively induces antigen-specific immune responses. J. Exp. Med. 204, 2789–2796 (2007)
69. Kayamuro, H., Yoshioka, Y., Abe, Y., Arita, S., Katayama, K., Nomura, T., Yoshikawa, T., Kubota-Koketsu, R., Ikuta, K., Okamoto, S., Mori, Y., Kunisawa, J., Kiyono, H., Itoh, N., Nagano, K., Kamada, H., Tsutsumi, Y., Tsunoda, S.: Interleukin-1 family cytokines as mucosal vaccine adjuvants for induction of protective immunity against influenza virus. J. Virol. 84, 12703–12712 (2010)
70. Tovey, M.G., Lallemand, C.: Adjuvant activity of cytokines. Methods Mol. Biol. 626, 287–309 (2010)
71. Gmajner, D., Ota, A., Sentjurc, M., Ulrih, N.P.: Stability of diether C25,25liposomes from the hyperthermophilic archaeon *Aeropyrum pernix* K1. Chem. Phys. Lipids 164, 236–245 (2011)
72. Nordly, P., Madsen, H.B., Nielsen, H.M., Foged, C.: Status and future prospects of lipid-based particulate delivery systems as vaccine adjuvants and their combination with immunostimulators. Expert Opin. Drug Deliv. 6, 657–672 (2009)
73. Kersten, G.F., Crommelin, D.J.: Liposomes and ISCOMs. Vaccine 21, 915–920 (2003)
74. Rao, R., Squillante, E.I.I.I., Kim, K.H.: Lipid-based cochleates: a promising formulation platform for oral and parenteral delivery of therapeutic agents. Crit. Rev. Ther. Drug Carrier Syst. 24, 41–61 (2007)
75. McDermott, M.R., Heritage, P.L., Bartzoka, V., Brook, M.A.: Polymer-grafted starch microparticles for oral and nasal immunization. Immunol. Cell Biol. 76, 256–262 (1998)
76. Eldridge, J.H., Gilley, R.M., Staas, J.K., Moldoveanu, Z., Meulbroek, J.A., Tice, T.R.: Biodegradable microspheres: vaccine delivery system for oral immunization. Curr. Top. Microbiol. Immunol. 146, 59–66 (1989)
77. Payne, L.G., Jenkins, S.A., Andrianov, A., Roberts, B.E.: Water-soluble phosphazene polymers for parenteral and mucosal vaccine delivery. Pharm. Biotechnol. 6, 473–493 (1995)
78. Payne, L.G., Jenkins, S.A., Woods, A.L., Grund, E.M., Geribo, W.E., Loebelenz, J.R., Andrianov, A.K., Roberts, B.E.: Poly[di(carboxylatophenoxy) phosphazene] (PCPP) is a potent immunoadjuvant for an influenza vaccine. Vaccine 16, 92–98 (1998)
79. Arbós, P., Campanero, M.A., Arangoa, M.A., Renedo, M.J., Irache, J.M.: Influence of the surface characteristics of PVM/MA nanoparticles on their bioadhesive properties. J. Control. Release 89, 19–30 (2003)
80. Olbrich, C., Müller, R.H., Tabatt, K., Kayser, O., Schulze, C., Schade, R.: Stable biocompatible adjuvants – a new type of adjuvant based on solid lipid nanoparticles: a study on cytotoxicity, compatibility and efficacy in chicken. Altern. Lab. Anim. 30, 443–458 (2002)
81. Storni, T., Kündig, T.M., Senti, G., Johansen, P.: Immunity in response to particulate antigen-delivery systems. Adv. Drug Deliv. Rev. 57, 333–355 (2005)
82. Krahenbuhl, J.P., Neutra, M.R.: Epithelial M cells: differentiation and function. Annu. Rev. Cell Dev. Biol. 16, 301–332 (2000)
83. Espuelas, S., Irache, J.M., Gamazo, C.: Synthetic particulate antigen delivery systems for vaccination. Inmunologia 24, 207–223 (2005)
84. O'Hagan, D.T., Jeffery, H., Maloy, K.J., Mowat, A.M., Rahman, D., Challacombe, S.J.: Biodegradable microparticles as oral vaccines. Adv. Exp. Med. Biol. 371B, 1463–1467 (1995)
85. Vila, A., Sánchez, A., Tobío, M., Calvo, P., Alonso, M.J.: Design of biodegradable particles A. for protein delivery. J. Control. Release 78, 15–24 (2002)
86. Yoncheva, K., Gómez, S., Campanero, M.A., Gamazo, C., Irache, J.M.: Bioadhesive properties of pegylated nanoparticles. Expert Opin. Drug Deliv. 2(2), 205–218 (2005)
87. Yoncheva, K., Lizarraga, E., Irache, J.M.: Pegylated nanoparticles based on poly(methyl vinyl ether-co-maleic anhydride): preparation and evaluation of

their bioadhesive properties. Eur. J. Pharm. Sci. **24**, 411–419 (2005)
88. Fadl, A.A., Venkitanarayanan, K.S., Khan, M.I.: Identification of Salmonella enteritidis outer membrane proteins expressed during attachment to human intestinal epithelial cells. J. Appl. Microbiol. **92**, 180–186 (2002)
89. Allen-Vercoe, E., Woodward, M.J.: The role of flagella, but not fimbriae, in the adherence of *Salmonella enterica* serotype Enteritidis to chick gut explant. J. Med. Microbiol. **48**, 771–780 (1999)
90. Humphries, A.D., Townsend, S.M., Kingsley, R.A., Nicholson, T.L., Tsolis, R.M., Bäumler, A.J.: Role of fimbriae as antigens and intestinal colonization factors of Salmonella serovars. FEMS Microbiol. Lett. **201**, 121–125 (2001)
91. Kaltner, H., Stierstorfer, B.: Animal lectins as cell adhesion molecules. Acta Anat. (Basel) **161**, 162–179 (1998)
92. Lloyd, D.H., Viac, J., Werling, D., Rème, C.A., Gatto, H.: Role of sugars in surface microbe-host interactions and immune reaction modulation. Vet. Dermatol. **18**, 197–204 (2007)
93. Conway, M.A., Madrigal-Estebas, L., McClean, S., Brayden, D.J., Mills, K.H.: Protection against *Bordetella pertussis* infection following parenteral or oral immunization with antigens entrapped in biodegradable particles: effect of formulation and route of immunization on induction of Th1 and Th2 cells. Vaccine **19**, 1940–1950 (2001)
94. Carcaboso, A.M., Hernández, R.M., Igartua, M., Gascón, A.R., Rosas, J.E., Patarroyo, M.E., Pedraz, J.L.: Immune response after oral administration of the encapsulated malaria synthetic peptide SPf66. Int. J. Pharm. **260**, 273–282 (2003)
95. Sundling, C., Schön, K., Mörner, A., Forsell, M.N., Wyatt, R.T., Thorstensson, R., Karlsson Hedestam, G.B., Lycke, N.Y.: CTA1-DD adjuvant promotes strong immunity against human immunodeficiency virus type 1 envelope glycoproteins following mucosal immunization. J. Gen. Virol. **89**, 2954–2964 (2008)
96. Helgeby, A., Robson, N.C., Donachie, A.M., Beackock-Sharp, H., Lövgren, K., Schön, K., Mowat, A., Lycke, N.Y.: The combined CTA1-DD/ISCOM adjuvant vector promotes priming of mucosal and systemic immunity to incorporated antigens by specific targeting of B cells. J. Immunol. **176**, 3697–3706 (2006)
97. Smith, R.E., Donachie, A.M., Grdic, D., Lycke, N., Mowat, A.M.: Immune-stimulating complexes induce an IL-12-dependent cascade of innate immune responses. J. Immunol. **162**, 5536–5546 (1999)
98. Dalle, F., Jouault, T., Trinel, P.A., Esnault, J., Mallet, J.M., d'Athis, P., Poulain, D., Bonnin, A.: Beta-1,2- and alpha-1,2-linked oligomannosides mediate adherence of Candida albicans blastospores to human enterocytes in vitro. Infect. Immun. **71**, 7061–7068 (2003)
99. Irache, J.M., Salman, H.H., Gamazo, C., Espuelas, S.: Mannose-targeted systems for the delivery of therapeutics. Expert Opin. Drug Deliv. **5**, 703–724 (2008)
100. Jack, D.L., Turner, M.W.: Anti-microbial activities of mannose-binding lectin. Biochem. Soc. Trans. **31**, 753–775 (2003)
101. Uemura, K., Hiromatsu, K., Xiong, X., Sugita, M., Buhlmann, J.E., Dodge, I.L., Lee, S.Y., Roura-Mir, C., Watts, G.F., Roy, C.J., Behar, S.M., Clemens, D.L., Porcelli, S.A., Brenner, M.B.: Conservation of CD1 intracellular trafficking patterns between mammalian species. J. Immunol. **169**, 6945–6958 (2002)
102. Wagner, S., Lynch, N.J., Walter, W., Schwaeble, W.J., Loos, M.: Differential expression of the murine mannose-binding lectins A and C in lymphoid and nonlymphoid organs and tissues. J. Immunol. **170**, 1462–1465 (2003)
103. Tamayo, I., Irache, J.M., Mansilla, C., Ochoa-Repáraz, J., Lasarte, J.J., Gamazo, C.: Poly(anhydride) nanoparticles as adjuvants for mucosal vaccination. Front. Biosci. (Schol. Ed.) **2**, 876–890 (2010)
104. Wang, T., Zou, M., Jiang, H., Ji, Z., Gao, P., Cheng, G.: Synthesis of a novel kind of carbon nanoparticle with large mesopores and macropores and its application as an oral vaccine adjuvant. Eur. J. Pharm. Sci. **44**(5), 653–659 (2011)
105. Jang, S.I., Lillehoj, H.S., Lee, S.H., Lee, K.W., Lillehoj, E.P., Bertrand, F., Dupuis, L., Deville, S.: TMIMS 1313 N VG PR nanoparticle adjuvant enhances antigen-specific immune responses to profilin following mucosal vaccination against Eimeria acervulina. Vet. Parasitol. **182**(2–4), 163–170 (2011)
106. Vandamme, K., Melkebeek, V., Cox, E., Adriaensens, P., Van Vlierberghe, S., Dubruel, P., Vervaet, C., Remon, J.P.: Influence of polymer hydrolysis on adjuvant effect of Gantrez® AN nanoparticles: implications for oral vaccination. Eur. J. Pharm. Biopharm. **79**(2), 392–398 (2011)
107. Sarti, F., Perera, G., Hintzen, F., Kotti, K., Karageorgiou, V., Kammona, O., Kiparissides, C., Bernkop-Schnürch, A.: In vivo evidence of oral vaccination with PLGA nanoparticles containing the immunostimulant monophosphoryl lipid A. Biomaterials **32**(16), 4052–4057 (2011)
108. Tian, J., Yu, J.: Poly(lactic-co-glycolic acid) nanoparticles as candidate DNA vaccine carrier for oral immunization of Japanese flounder (Paralichthys olivaceus) against lymphocystis disease virus. Fish Shellfish Immunol. **30**(1), 109–117 (2011)
109. Mishra, N., Tiwari, S., Vaidya, B., Agrawal, G.P., Vyas, S.P.: Lectin anchored PLGA nanoparticles for oral mucosal immunization against hepatitis B. J. Drug Target. **19**(1), 67–78 (2011)
110. Jain, A.K., Goyal, A.K., Mishra, N., Vaidya, B., Mangal, S.: Vyas SP PEG-PLA-PEG block copolymeric nanoparticles for oral immunization against hepatitis B. Int. J. Pharm. **387**(1–2), 253–262 (2010)
111. Mann, J.F., Shakir, E., Carter, K.C., Mullen, A.B., Alexander, J., Ferro, V.A.: Lipid vesicle size of an

oral influenza vaccine delivery vehicle influences the Th1/Th2 bias in the immune response and protection against infection. Vaccine **27**, 3643–3649 (2009)

112. Li, G.P., Liu, Z.G., Liao, B., Zhong, N.S.: Induction of Th1-type immune response by chitosan nanoparticles containing plasmid DNA encoding house dust mite allergen Der p 2 for oral vaccination in mice. Cell. Mol. Immunol. **6**(1), 45–50 (2009)

113. Oyewumi, M.O., Kumar, A., Cui, Z.: Nano-microparticles as immune adjuvants: correlating particle sizes and the resultant immune responses. Expert Rev. Vaccines **9**, 1095–1107 (2010)

114. Gutierro, I., Hernández, R.M., Igartua, M., Gascón, A.R., Pedraz, J.L.: Size dependent immune response after subcutaneous, oral and intranasal administration of BSA loaded nanospheres. Vaccine **21**, 67–77 (2002)

115. Kanchan, V., Panda, A.K.: Interactions of antigen-loaded polylactide particles with macrophages and their correlation with the immune response. Biomaterials **28**, 5344–5357 (2007)

116. Jung, T., Kamm, W., Breitenbach, A., Hungerer, K.D., Hundt, E., Kissel, T.: Tetanus toxoid loaded nanoparticles from sulfobutylated poly(vinyl alcohol)-graft-poly(lactide-co-glycolide): evaluation of antibody response after oral and nasal application in mice. Pharm. Res. **18**, 352–360 (2001)

117. Wendorf, J., Chesko, J., Kazzaz, J., Ugozzoli, M., Vajdy, M., O'Hagan, D., Singh, M.: A comparison of anionic nanoparticles and microparticles as vaccine delivery systems. Hum. Vaccin. **4**, 44–49 (2008)

118. Fifis, T., Gamvrellis, A., Crimeen-Irwin, B., Pietersz, G.A., Li, J., Mottram, P.L., McKenzie, I.F., Plebanski, M.: Size-dependent immunogenicity: therapeutic and protective properties of nano-vaccines against tumors. J. Immunol. **173**, 3148–3154 (2004)

119. Kalkanidis, M., Pietersz, G.A., Xiang, S.D., Mottram, P.L., Crimeen-Irwin, B., Ardipradja, K., Plebanski, M.: Methods for nano-particle based vaccine formulation and evaluation of their immunogenicity. Methods **40**, 20–29 (2006)

120. Mottram, P.L., Leong, D., Crimeen-Irwin, B., Gloster, S., Xiang, S.D., Meanger, J., Ghildyal, R., Vardaxis, N., Plebanski, M.: Type 1 and 2 immunity following vaccination is influenced by nanoparticle size: formulation of a model vaccine for respiratory syncytial virus. Mol. Pharm. **4**, 73–84 (2007)

121. Caputo, A., Castaldello, A., Brocca-Cofano, E., Voltan, R., Bortolazzi, F., Altavilla, G., Sparnacci, K., Laus, M., Tondelli, L., Gavioli, R., Ensoli, B.: Induction of humoral and enhanced cellular immune responses by novel core-shell nanosphere- and microsphere-based vaccine formulations following systemic and mucosal administration. Vaccine **27**, 3605–3615 (2009)

122. Estevan, M., Irache, J.M., Grilló, M.J., Blasco, J.M., Gamazo, C.: Encapsulation of antigenic extracts of Salmonella enterica serovar. Abortus-ovis into polymeric systems and efficacy as vaccines in mice. Vet. Microbiol. **118**, 124–132 (2006)

123. Shakweh, M., Ponchel, G., Fattal, E.: Particle uptake by Peyer's patches: a pathway for drug and vaccine delivery. Expert Opin. Drug Deliv. **1**, 141–163 (2004)

124. Peppoloni, S., Ruggiero, P., Contorni, M., Morandi, M., Pizza, M., Rappuoli, R., Podda, A., Del Giudice, G.: Mutants of the Escherichia coli heat-labile enterotoxin as safe and strong adjuvants for intranasal delivery of vaccines. Expert Rev. Vaccines **2**, 285–293 (2003)

125. Maloy, K.J., Donachie, A.M., O'Hagan, D.T., Mowat, A.M.: Induction of mucosal and systemic immune responses by immunization with ovalbumin entrapped in poly(lactide-co-glycolide) microparticles. Immunology **81**, 661–667 (1994)

126. Kim, S.Y., Doh, H.J., Jang, M.H., Ha, Y.J., Chung, S.I., Park, H.J.: Oral immunization with Helicobacter pylori-loaded poly(D, L-lactide-co-glycolide) nanoparticles. Helicobacter **4**, 33–39 (1999)

127. Katz, D.E., DeLorimier, A.J., Wolf, M.K., Hall, E.R., Cassels, F.J., van Hamont, J.E., Newcomer, R.L., Davachi, M.A., Taylor, D.N., McQueen, C.E.: Oral immunization of adult volunteers with micro-encapsulated enterotoxigenic Escherichia coli (ETEC) CS6 antigen. Vaccine **21**, 341–346 (2003)

128. Otterlei, M., Vårum, K.M., Ryan, L., Espevik, T.: Characterization of binding and TNF-alpha-inducing ability of chitosans on monocytes: the involvement of CD14. Vaccine **12**, 825–832 (1994)

129. Porporatto, C., Bianco, I.D., Correa, S.G.: Local and systemic activity of the polysaccharide chitosan at lymphoid tissues after oral administration. J. Leukoc. Biol. **78**, 62–69 (2005)

130. Kumar, N., Langer, R.S., Domb, A.J.: Polyanhydrides: an overview. Adv. Drug Deliv. Rev. **54**, 889–910 (2002)

131. Torres, M.P., Wilson-Welder, J.H., Lopac, S.K., Phanse, Y., Carrillo-Conde, B., Ramer-Tait, A.E., Bellaire, B.H., Wannemuehler, M.J., Narasimhan, B.: Polyanhydride microparticles enhance dendritic cell antigen presentation and activation. Acta Biomater. **7**, 2857–2864 (2011)

132. Petersen, L.K., Ramer-Tait, A.E., Broderick, S.R., Kong, C.S., Ulery, B.D., Rajan, K., Wannemuehler, M.J., Narasimhan, B.: Activation of innate immune responses in a pathogen-mimicking manner by amphiphilic polyanhydride nanoparticle adjuvants. Biomaterials **32**, 6815–6822 (2011)

133. Ulery, B.D., Phanse, Y., Sinha, A., Wannemuehler, M.J., Narasimhan, B., Bellaire, B.H.: Polymer chemistry influences monocytic uptake of polyanhydride nanospheres. Pharm. Res. **26**, 683–690 (2009)

134. Ulery, B.D., Kumar, D., Ramer-Tait, A.E., Metzger, D.W., Wannemuehler, M.J., Narasimhan, B.: Design of a protective single-dose intranasal nanoparticle-based vaccine platform for respiratory infectious diseases. PLoS One **6**(3), e17642 (2011). doi:10.1371/journal.pone.0017642

135. Ochoa, J., Irache, J.M., Tamayo, I., Walz, A., DelVecchio, V.G., Gamazo, C.: Protective immunity

of biodegradable nanoparticle-based vaccine against an experimental challenge with Salmonella enteritidis in mice. Vaccine **25**, 4410–4419 (2007)
136. Gómez, S., Gamazo, C., San Roman, B., Vauthier, C., Ferrer, M., Irachel, J.M.: Development of a novel vaccine delivery system based on Gantrez nanoparticles. J. Nanosci. Nanotechnol. **6**, 3283–3289 (2006)
137. Tamayo, I., Irache, J.M., Mansilla, C., Ochoa-Repáraz, J., Lasarte, J.J., Gamazo, C.: Poly(anhydride) nanoparticles act as active Th1 adjuvants through toll-like receptor exploitation. Clin. Vaccine Immunol. **17**, 1356–1362 (2010)
138. Camacho, A.I., Da Costa Martins, Tamayo, I., de Souza, J., Lasarte, J.J., Mansilla, C., Esparza, I., Irache, J.M., Gamazo, C.: Poly(methyl vinyl ether-co-maleic anhydride) nanoparticles as innate immune system activators. Vaccine **22**, 7130–7135 (2011)
139. Lycke, N., Bemark, M.: Mucosal adjuvants and long-term memory development with special focus on CTA1-DD and other ADP-ribosylating toxins. Mucosal Immunol. **3**, 556–566 (2010)
140. Kasturi, S.P., Skountzou, I., Albrecht, R.A., Koutsonanos, D., Hua, T., Nakaya, H.I., Ravindran, R., Stewart, S., Alam, M., Kwissa, M., Villinger, F., Murthy, N., Steel, J., Jacob, J., Hogan, R.J., García-Sastre, A., Compans, R., Pulendran, B.: Programming the magnitude and persistence of antibody responses with innate immunity. Nature **470**, 543–547 (2011)
141. Amanna, I.J., Carlson, N.E., Slifka, M.K.: Duration of humoral immunity to common viral and vaccine antigens. N. Engl. J. Med. **357**, 1903–1915 (2007)
142. Hammarlund, E., Lewis, M.W., Hansen, S.G., Strelow, L.I., Nelson, J.A., Sexton, G.J., Hanifin, J.M., Slifka, M.K.: Duration of antiviral immunity after smallpox vaccination. Nat. Med. **9**, 1131–1137 (2003)
143. Querec, T.D., Akondy, R.S., Lee, E.K., Cao, W., Nakaya, H.I., Teuwen, D., Pirani, A., Gernert, K., Deng, J., Marzolf, B., Kennedy, K., Wu, H., Bennouna, S., Oluoch, H., Miller, J., Vencio, R.Z., Mulligan, M., Aderem, A., Ahmed, R., Pulendran, B.: Systems biology approach predicts immunogenicity of the yellow fever vaccine in humans. Nat. Immunol. **10**, 116–125 (2009)
144. Sharp, F.A., Ruane, D., Claass, B., Creagh, E., Harris, J., Malyala, P., Singh, M., O'Hagan, D.T., Pétrilli, V., Tschopp, J., O'Neill, L.A., Lavelle, E.C.: Uptake of particulate vaccine adjuvants by dendritic cells activates the NALP3 inflammasome. Proc. Natl. Acad. Sci. U. S. A. **106**, 870–875 (2009)
145. Li, H., Willingham, S.B., Ting, J.P., Re, F.: Cutting edge: inflammasome activation by alum and alum's adjuvant effect are mediated by NLRP3. J. Immunol. **181**, 17–21 (2008)
146. Correia-Pinto, J.F., Csaba, N., Alonso, M.J.: Vaccine delivery carriers: insights and future perspectives. Int. J. Pharm. **440**(1), 27–38 (2013). doi:10.1016/j.bbr.2011.03.031
147. Lambrecht, B.N., Kool, M., Willart, M.A., Hammad, H.: Mechanism of action of clinically approved adjuvants. Curr. Opin. Immunol. **21**, 23–29 (2009)
148. Tagliabue, A., Rappuoli, R.: Vaccine adjuvants: the dream becomes real. Hum. Vaccin. **4**, 347–349 (2008)
149. Chen, W., Patel, G.B., Yan, H., Zhang, J.: Recent advances in the development of novel mucosal adjuvants and antigen delivery systems. Hum. Vaccin. **6**, 706–714 (2010)
150. Holmgren, J., Czerkinsky, C., Eriksson, K., Mharandi, A.: Mucosal immunisation and adjuvants: a brief overview of recent advances and challenges. Vaccine **21**, S89–S95 (2003)
151. Pulendran, B.: Learning immunology from the yellow fever vaccine: innate immunity to systems vaccinology. Nat. Rev. Immunol. **9**, 741–747 (2009)

# Chitosan-Based Adjuvants

## 39

Guro Gafvelin and Hans Grönlund

## Contents

| | | |
|---|---|---|
| 39.1 | **Disease/Application Area** | 624 |
| 39.2 | **Vaccine Delivery** | 624 |
| 39.2.1 | Formulation/Chemistry | 624 |
| 39.2.2 | Microtechnology/Nanotechnology | 625 |
| 39.2.3 | Working Mechanism | 626 |
| 39.2.4 | Preclinical Development, Safety, and Efficacy | 626 |
| 39.2.5 | Clinical Development | 627 |
| 39.2.6 | Strength and Weakness | 628 |
| **References** | | 629 |

G. Gafvelin, PhD
Department of Clinical Neuroscience,
Therapeutic Immune Design Unit,
Center for Molecular Medicine,
Karolinska Institutet,
Karolinska University Hospital,
Stockholm, Sweden

Viscogel AB, Gunnar Asplunds Allé, Solna, Sweden

H. Grönlund, PhD (✉)
Therapeutic Immune Design Unit,
Department of Clinical Neuroscience,
Center for Molecular Medicine, Karolinska Institutet,
Karolinska University Hospital,
Stockholm, Sweden
e-mail: hans.gronlund@ki.se

## Abstract

Safe and efficient prophylactic and therapeutic vaccines are lacking for a number of infectious diseases and immunologically related disorders. Novel adjuvants, able to enhance and modify immune responses, thus have the potential to improve both prophylactic and therapeutic vaccines. In the search for new types of adjuvants the carbohydrate polymer chitosan has gained increasing interest. Chitosan derives from the natural product chitin and chemically consists of N-acetyl-glucosamine and glucosamine. Attractive characteristics of chitosan as an adjuvant candidate include its natural origin, full biodegradability, non-toxicity and low cost of goods. Chitosan has the ability to stimulate innate immunity but the mechanism of action is not fully understood. The adjuvant effect may also be mediated by improved antigen presentation to immune cells and by enhancing the antigen uptake. Because of its cationic character chitosan adheres to mucosal surfaces. Various chemical derivatives and particulate variants of chitosan have been explored for mucosal administration of vaccines. Available data from clinical trials support chitosan as being a useful adjuvant- and delivery system for intranasal vaccines. In order to develop chitosan based adjuvants for different vaccine applications and administration routes it is crucial to provide high-quality medical grade chitosan. In this chapter we present published data on chitosan as an adjuvant and specifically focus on a recently developed chitosan based adjuvant, ViscoGel.

## 39.1 Disease/Application Area

The undisputable success of vaccines and vaccine programs in the battle against a number of infectious diseases has highlighted novel vaccine strategies as a solution to a wide range of immunologically related conditions. However, several challenges remain, both for the development of therapeutic vaccines to, e.g., cancer, autoimmune diseases, and allergy, as well as for infectious diseases. Examples of diseases in which efficient vaccines are urgently needed include malaria, tuberculosis, and HIV infection/AIDS, collectively responsible for millions of deaths every year [1, 2]. A group of patients with a particular need of efficient vaccines is the worldwide growing elderly population. Individuals over the age of 65 often get insufficient protection from current vaccination to, e.g., seasonal influenza and pneumococcal infections [2–4]. Thus there is a need for vaccines able to elicit efficient and appropriate immune responses.

One way to improve suboptimal vaccines is to employ more efficient adjuvants, i.e., enhancers of the immune response. Adjuvants also provide means to direct the response into a specific direction, e.g., a humoral or a cellular immune response. The route of administration is known to affect the immune response stimulated and, in addition, adjuvants can be designed to act as vehicles or delivery systems for vaccine administration.

The most widely used, and until recently the only adjuvant approved for human use, are aluminum salts (alum). Alum mainly enhances the humoral arm of the immune system, and particularly in mice it strongly promotes Th2-type immune responses [5]. To meet the extensive demand for vaccines able to elicit cellular responses, several new types of adjuvants have been developed or are currently investigated [6, 7]. There is also a need for novel adjuvants in vaccines designed for non-parenteral administration routes.

In the search for new types of adjuvants, the carbohydrate polymer chitosan has gained increasing interest, due to its demonstrated immunostimulatory effect [8, 9]. Chitosan derives from the natural product chitin. Chitin, being one of the most abundant polymers in nature, is found as a structural element in crustacean shells, exoskeletons of insects, and cell walls of microorganisms. Chemically, chitosan consists of $N$-acetylglucosamine and glucosamine. Because of its cationic character, it efficiently adheres to mucosal surfaces [10]. Additional attractive characteristics of chitosan as an adjuvant candidate include its natural origin, full biodegradability, nontoxicity, and low cost of goods. Here we will describe chitosan as an adjuvant and specifically focus on a recently developed chitosan-based adjuvant, ViscoGel.

## 39.2 Vaccine Delivery

### 39.2.1 Formulation/Chemistry

Chitosan is commercially manufactured from the exoskeletons of crustaceans, recovered as a side product from the seafood industry. The manufacturing process involves strong alkaline treatment of chitin, a homopolymer consisting of 1-4-β-linked-$N$-acetylglucosamine units, to remove acetyl groups and form water-soluble chitosan. Chitosan is thus composed of $N$-acetylglucosamine and glucosamine (Fig. 39.1) at various ratios. The percentage of glucosamines in the chitosan is termed degree of deacetylation (DD). Standard manufacturing processes typically lead to a DD of 80–95 %, but the DD may vary considerably. The quality of chitosans applied in published studies is often inadequately defined, making it difficult to link physical and chemical composition of chitosans to stimulation of certain immune responses. Existing data mostly refer to chitosan in the 80–95 % DD range, since commercially available chitosans are generally highly deacetylated.

Due to its high content of glucosamines, chitosan is cationic [11] and muco-adherent [10, 12]. Moreover, chitosan possesses permeation-enhancing properties [10, 13]. These characteristics are of importance for formulation of mucosal vaccines. However, one limitation for the medical application of chitosan is that it is only soluble in its protonated form, at pH below 6–6.5.

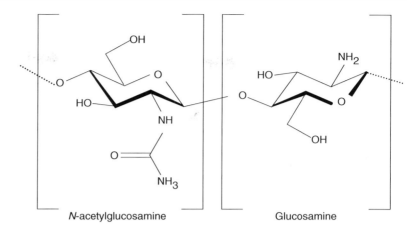

**Fig. 39.1** Chitosan is a carbohydrate polymer composed of N-acetylglucosamine and glucosamine units. The polymer length, as well as the distribution and overall composition of N-acetylglucosamine and glucosamine units, varies between chitosans. The percentage of glucosamine units in the polymer denotes the degree of deacetylation (DD)

In order to keep chitosans in solutions at physiological conditions (neutral pH, salts, etc.), various chemical modifications have been attempted [10, 14]. The most common soluble chitosan derivatives for mucosal applications are N-trimethylated and carboxymethylated chitosans [15, 16]. Chitosan and its derivatives have been produced in a number of different physical forms to match the various demands on vaccine delivery systems, e.g., as gels, powders, and nanoparticles [16, 17].

A chitosan that is soluble at physiological conditions has been developed and is manufactured under the name Viscosan. The water solubility is promoted by a random distribution of acetyl groups and a DD close to 50 %. In addition, Viscosan is rapidly degraded in a biological environment, and it has a high purity. Viscosan can be processed into a viscoelastic hydrogel, ViscoGel, which possesses useful properties for adjuvant and vaccine delivery applications. The gel is formed by cross-linking of chitosan chains to achieve a physically stable hydrogel consisting of 99 % water and 1 % Viscosan. ViscoGel is further processed into particles of defined sizes. Finally, ViscoGel can be mixed or covalently linked with different antigens or immunomodulators to generate vaccine-adjuvant formulations. Formulations with covalently linked antigens may be very advantageous, since it has been shown that antigens physically linked to the adjuvant generate superior immune responses compared to non-associated antigen/adjuvants [18, 19].

### 39.2.2 Microtechnology/Nanotechnology

Particulate systems offer several benefits to vaccine technology. The particulate form enables exposure of a large surface area and efficient display of antigen. Microparticles exhibit similar sizes as microorganisms, which the immune system has evolved to combat. Thus, they are efficiently phagocytosed and activate antigen-presenting cells (APC) [20]. ViscoGel provides a microparticulate system and can be processed into gel particles in the size range from a few micrometers up to 200 μm. Notably, ViscoGel does not precipitate but remains in the form of clear transparent gel particles when subjected to physiological pH.

Nanoparticulate chitosan systems have especially been studied for mucosal vaccine applications. For example, chitosan nanospheres (approx 350 nm) have been applied as vaccine carriers in preclinical testing of nasal delivery of a tetanus toxoid [21, 22]. Superior systemic and local immune responses were induced by the chitosan-nanoparticle-formulated tetanus toxoid compared to fluid vaccine. Chitosan nanoparticles (350 nm) loaded with fluorescently labelled albumin (OVA) have been shown to be internalized by nasal epithelial cells and transported to the submucosal layer after intranasal administration to rats [23]. In a mouse model for sublingual allergen-specific immunotherapy, it was on the other hand shown that treatment with the model allergen OVA formulated with chitosan microparticles (1–3 μm) was superior to a nanoparticle (300–800 nm)-formulated vaccine. The uptake of OVA by sublingual APCs and the activation of OVA-specific T-cell responses were increased for the microparticle-formulated OVA, and, furthermore, the airway hyperresponsiveness and inflammation in the lung was reduced after treatment with the microparticulate vaccine [24].

### 39.2.3 Working Mechanism

Immune activation by chitosan has been demonstrated after intranasal [8, 21], oral [25], subcutaneous [9, 26], and intramuscular antigen administration [9, 27]. The oral route is expected to induce tolerogenic immune responses. Chitosan given perorally without any protein antigen stimulated a Th2/regulatory T-cell-promoting milieu in the intestinal mucosa, characterized by secretion of IL-10 and expression of IL-4 and TGF-β [25]. When given by the subcutaneous or intramuscular route, chitosan stimulates both strong humoral and cellular responses to coadministered antigens. A mixed Th1/Th2 type of response as well as CTL activation has been demonstrated in mice [9, 27]. It has been proposed that chitin, the precursor to chitosan, activates innate immune responses through two main mechanisms: activation of pathogen pattern recognition receptors (PRRs) and activation of chitinases [28]. The PRRs suggested to be activated by chitin include the mannose receptor, TLR-2, and dectin-1 [29–32].

Chitosan is defined as partially deacetylated chitin and may share some of the immune-activating mechanisms demonstrated for chitin. There are, however, few reports on how chitosan acts as an adjuvant. In an early study chitosan was shown to activate macrophages [33]. Chitosan has later been demonstrated to provide an antigen depot effect, which has been suggested to enhance adjuvant activity [9]. It has also been reported that chitosan induces inflammasome activation in an NLRP3-dependent fashion, a mechanism that may be shared among particulate adjuvants, including alum [34]. Interestingly, in an *in vitro* study performed on murine bone marrow-derived dendritic cells (DC), it was shown that chitosan, in contrast to alum, did not inhibit stimulated production of the Th1-polarizing cytokine IL-12 from activated DCs, providing a possible explanation to why chitosan stimulates Th1 responses more readily than alum [35]. These and other studies suggest that chitosan activates innate immune signaling, but so far no candidate PRR specifically interacting with chitosan has been identified.

ViscoGel contains low-deacetylated chitosan with a random distribution of acetyl groups [26]. Its specific characteristics suggest that even though some mechanisms of immune activation will be common to other chitosans, ViscoGel will activate the immune system in a unique way. The ViscoGel-stimulated immune response will depend on the nature of the ViscoGel preparation, e.g., its particle size, DD, and means to mix or link an antigen to the ViscoGel. These factors should be evaluated for each vaccine formulation in order to obtain an optimal response towards any given disease. Importantly, the versatility of the ViscoGel system provides a possibility to design vaccines with tailored properties for specific applications.

### 39.2.4 Preclinical Development, Safety, and Efficacy

A number of chitosan-based candidate adjuvants have been preclinically evaluated with various vaccines and administration routes, although mostly for mucosal vaccine applications [16]. Here we describe the preclinical development of ViscoGel from proof-of-concept in mice to a product approved for use in a clinical trial in man. Initially a study was performed in mice to show proof-of-concept for ViscoGel as an adjuvant. A glycoconjugate vaccine to *Haemophilus influenzae* type b (Act-HIB) was used as model antigen [26]. The model vaccine was chosen because it is approved for human use, and the outcome of vaccination can be measured as antibody titers in standardized assays [36]. The vaccine was formulated as a mixture of Act-HIB with 200 μm ViscoGel particles. Groups of mice were immunized with the adjuvant-formulated vaccine and compared to non-adjuvanted Act-HIB [26]. Significantly enhanced IgG1 and IgG2a titers were found in sera from mice immunized with ViscoGel-formulated Act-HIB. In fact, the antigen dose could be reduced tenfold when ViscoGel was used as an adjuvant. Furthermore, the Act-HIB-specific cellular response was significantly stronger and of a mixed Th1/Th2/Th17 type, based on the cytokine profile. Similar effects were seen when the adjuvant was administered either subcutaneously or intramuscularly. The successful demonstration of adjuvant properties in mice led to preparations for clinical proof-of-concept. Medical grade ViscoGel mixed with Act-HIB was produced in a validated process. A number of quality control analyses were then performed to record physical and chemical characteristics and to evaluate purity and stability.

The safety profile is of vital importance for an adjuvant to be applied in human vaccines. The adjuvant should enhance antigen-specific responses without promoting strong nonspecific inflammation or immune responses to the adjuvant. The toxicity of ViscoGel, as well as for the formulated vaccine consisting of Act-HIB mixed with ViscoGel, was evaluated in mice and rabbits. Local and systemic toxic effects were carefully investigated after three intramuscular administrations of ViscoGel. No systemic toxic effects were recorded. The local reactions observed were mild and associated with the immune-stimulating effect, which was confirmed by measurement of antibodies raised to Act-HIB by the vaccination. The possibility to develop an immune response to chitosan, and possibly an allergic, IgE-mediated response, has been addressed earlier [27]. In mice given intramuscular injections with chitosan alone or in combination either with an influenza vaccine or an added adjuvant (Freunds incomplete adjuvant), no significant anti-chitosan antibody responses were detected. Moreover, no increase of IgE to chitosan or to influenza antigen was recorded in sera from rats injected with chitosan or chitosan vaccine [27]. The question if seafood allergic individuals can react to chitosan has been raised [37]. This concern applies to the possible presence of trace amounts of allergenic proteins in chitosan preparations. Well-characterized chitosans of high purity, e.g., ViscoGel, are not associated with this risk.

### 39.2.5 Clinical Development

Even though several chitosan-based adjuvants have been developed and tested preclinically [15, 16], few studies have evaluated chitosan as an adjuvant in man (Table 39.1). So far, the clinical trials where the immunostimulatory effect has been studied involve mucosal vaccinations. A couple of studies have been designed to compare the adjuvant effect of chitosan to vaccination without adjuvant or with another adjuvant. Intranasal administration of a detoxified diphtheria toxin vaccine ($CRM_{197}$) formulated with chitosan glutamate generated significantly higher levels of neutralizing antitoxin serum antibodies compared to the non-adjuvanted intranasal vaccine and to intramuscular alum-formulated $CRM_{197}$ [38]. Only the chitosan-formulated vaccine induced an antitoxin IgA response [38], and the nasal $CRM_{197}$/chitosan vaccine also generated a stronger Th2 type of cell response to diphtheria toxin than the vaccine given alone or by the intramuscular route with alum [39]. In another study, intranasal vaccination with a trivalent influenza vaccine formulated with chitosan glutamate was compared to vaccination by the conventional intramuscular route [40]. The intranasal immunization resulted in satisfactory protective responses that were not significantly different from those generated by the intramuscular vaccine. A third study investigated a *Neisseria meningitidis* glycoconjugate vaccine administered intranasally with chitosan in comparison to the same vaccine formulated with alum and administered intramuscular [41]. The intranasal chitosan-formulated vaccine was well tolerated and generated protective antibody responses comparable to the parenteral vaccine. In addition, trials employing chitosan as adjuvant and vaccine vehicle, but without comparison to other administration routes or non-adjuvanted vaccines, have been conducted with a Norwalk viruslike particle

**Table 39.1** Clinical trials where chitosan has been applied as adjuvant

| Antigen | Vaccination route | Subjects | Study description | Reference |
|---|---|---|---|---|
| Inactivated mutant diphtheria toxoid $CRM_{197}$ | Intranasal (intramuscular control group) | Healthy adults | Two i.n. admin of $CRM_{197}$ with or without chitosan ($n=10$/group), one i.m. admin of $CRM_{197}$ + alum ($n=5$) | Mills et al. [38] McNeela et al. [39] |
| Trivalent subunit influenza vaccine | Intranasal (intramuscular control group) | Healthy adults | Two i.n. admin with chitosan, two dose groups ($n=23$), vaccine i.m ($n=22$) | Read et al. [40] |
| $CRM_{197}$-based meningococcal C oligosaccharide conjugate | Intranasal (intramuscular control group) | Healthy adults | Two i.n. admin with chitosan (divided into subgroups, total $n=30$), one i.m. admin of conj. vaccine + alum ($n=6$) | Huo et al. [41] |
| Norwalk viruslike particle vaccine, coformulated with MPL and chitosan | Intranasal | Healthy adults | Randomized, placebo controlled in two steps: 1. Three vaccine doses ($n=5+5+10$, $2+2+4$ adjuvant controls), two i.n. admin 2. Two vaccine doses ($n=20+20$, 10 adjuvant control, 11 placebo), two i.n. admin | El-Kamary et al. [42] Atmar et al. [43] |
| HIV virus coat protein | Intranasal, intramuscular, and intravaginal admin | Healthy adult women | Three vaccine admin, i.n. vaccine formulated with chitosan ($n=36$) | MUCOVAC2 [44] |
| Laser immunotherapy of metastatic breast cancer | Local intratumoral | Adult late-stage breast cancer patients | Chitosan admin post laser immunotherapy ($n=10$), three treatments | Li et al. [45] |

vaccine formulated with the adjuvant monophosphoryl lipid A (MLP) and chitosan for intranasal delivery. This vaccine was evaluated in a safety and dose-finding study [42], followed by a randomized, double-blind, placebo-controlled, multicenter study [43]. It was concluded that the nasal vaccine was well tolerated, highly immunogenic, and provided increased protection to experimentally induced Norwalk virus gastroenteritis. Chitosan has also been applied for intranasal vaccination in a Phase I trial (MUCOVAC2), evaluating an HIV viral coat vaccine administered via the intranasal, intramuscular, and vaginal routes [44]. Finally, chitosan has been employed as an immunoactivator in cancer immunotherapy. In a safety-efficacy study on late-stage metastatic breast cancer patients, glycated chitosan was applied in combination with laser immunotherapy [45]. The clinical outcome of this study was promising, exhibiting systemic antitumor responses with milder side effects than conventional treatments.

In conclusion, available clinical data support chitosan as being a useful adjuvant and delivery system, in particular for intranasal vaccines. Clinical trials using chitosan-based adjuvants, including ViscoGel, for additional vaccines and administration routes are under way.

### 39.2.6 Strength and Weakness

Chitosan possesses many benefits as adjuvant and vaccine vehicle, but some limitations remain to be solved. Table 39.2 lists advantages and restrains associated with the application

**Table 39.2** Chitosan as adjuvant: strengths and weaknesses

| Strength | Weakness |
|---|---|
| Natural product; nontoxic, biodegradable | Natural product; formulation and characterization methods are not standardized |
| Controlled manufacturing process enables high purity without traces of allergenic components in high-quality chitosans | Chitosans of inferior qualities that contain impurities generate non-reliable results |
| May be manufactured in various forms, e.g., gel, powder, nano/microparticles, chemical derivatives | Most non-modified chitosans are insoluble at physiological pH |
| Cationic and mucoadhesive; application as vehicle/adjuvant for mucosal vaccination, formulation for various administration routes possible | Few studies available on other administration routes than intranasal |
| Low cost of goods, "unlimited" access to raw material | Few medical grade, high-quality chitosans available |
| Different chemical (e.g., DD) and physical (e.g., particle size) characteristics determine immune response; versatility allows specific design of vaccines | Chemical and physical characteristics that determine immune responses are ill defined |
| Stimulates humoral and cellular immune responses | Mechanisms of action are poorly understood |

of chitosan in vaccine development. Access to high-quality chitosan is crucial for the elucidation of immunostimulatory mechanisms and for the progress of clinical applications. Obstacles linked to problems to provide highly purified, well-characterized, medical grade chitosan can be solved, as shown by the ViscoGel system. Thus the development of new chitosan forms and derivatives, produced by controlled manufacturing processes, opens for designed vaccines for various disease targets and administration routes.

## References

1. Nossal, G.J.: Vaccines of the future. Vaccine **29 Suppl 4**, D111–D115 (2011). doi:10.1016/j.vaccine.2011.06.089. S0264-410X(11)00983-2 [pii]
2. Leroux-Roels, G.: Unmet needs in modern vaccinology: adjuvants to improve the immune response. Vaccine **28 Suppl 3**, C25–C36 (2010). doi:10.1016/j.vaccine.2010.07.021. S0264-410X(10)01004-2 [pii]
3. McElhaney, J.E.: Prevention of infectious diseases in older adults through immunization: the challenge of the senescent immune response. Expert Rev. Vaccines **8**, 593–606 (2009). doi:10.1586/erv.09.12
4. Schubert, C.: New vaccine tailored to the weakened elderly immune system. Nat. Med. **16**, 137 (2010). doi:10.1038/nm0210-137a. nm0210-137a [pii]
5. Brewer, J.M., et al.: Aluminium hydroxide adjuvant initiates strong antigen-specific Th2 responses in the absence of IL-4- or IL-13-mediated signaling. J. Immunol. **163**, 6448–6454 (1999). ji_v163n12p6448 [pii]
6. Mbow, M.L., De Gregorio, E., Valiante, N.M., Rappuoli, R.: New adjuvants for human vaccines. Curr. Opin. Immunol. **22**, 411–416 (2010). doi:10.1016/j.coi.2010.04.004. S0952-7915(10)00068-3 [pii]
7. Schijns, V.E., Lavelle, E.C.: Trends in vaccine adjuvants. Expert Rev. Vaccines **10**, 539–550 (2011). doi:10.1586/erv.11.21
8. Jabbal-Gill, I., Fisher, A.N., Rappuoli, R., Davis, S.S., Illum, L.: Stimulation of mucosal and systemic antibody responses against Bordetella pertussis filamentous haemagglutinin and recombinant pertussis toxin after nasal administration with chitosan in mice. Vaccine **16**, 2039–2046 (1998). S0264410X98000772 [pii]
9. Zaharoff, D.A., Rogers, C.J., Hance, K.W., Schlom, J., Greiner, J.W.: Chitosan solution enhances both humoral and cell-mediated immune responses to subcutaneous vaccination. Vaccine **25**, 2085–2094 (2007). doi:10.1016/j.vaccine.2006.11.034. S0264-410X(06)01225-4 [pii]
10. Bonferoni, M.C., Sandri, G., Rossi, S., Ferrari, F., Caramella, C.: Chitosan and its salts for mucosal and transmucosal delivery. Expert Opin. Drug Deliv. **6**, 923–939 (2009). doi:10.1517/17425240903114142
11. Sorlier, P., Denuziere, A., Viton, C., Domard, A.: Relation between the degree of acetylation and the electrostatic properties of chitin and chitosan. Biomacromolecules **2**, 765–772 (2001)
12. Kumar, M.N., Muzzarelli, R.A., Muzzarelli, C., Sashiwa, H., Domb, A.J.: Chitosan chemistry and pharmaceutical perspectives. Chem. Rev. **104**, 6017–6084 (2004). doi:10.1021/cr030441b
13. Illum, L., Farraj, N.F., Davis, S.S.: Chitosan as a novel nasal delivery system for peptide drugs. Pharm. Res. **11**, 1186–1189 (1994)
14. Alves, N.M., Mano, J.F.: Chitosan derivatives obtained by chemical modifications for biomedical and environmental applications. Int. J. Biol. Macromol. **43**, 401–414 (2008). doi:10.1016/j.ijbiomac.2008.09.007. S0141-8130(08)00208-0 [pii]
15. Arca, H.C., Gunbeyaz, M., Senel, S.: Chitosan-based systems for the delivery of vaccine antigens. Expert Rev. Vaccines **8**, 937 (2009). doi:10.1586/erv.09.47
16. Jabbal-Gill, I., Watts, P., Smith, A.: Chitosan-based delivery systems for mucosal vaccines. Expert Opin.

Drug Deliv. **9**, 1051–1067 (2012). doi:10.1517/17425247.2012.697455

17. Luppi, B., Bigucci, F., Cerchiara, T., Zecchi, V.: Chitosan-based hydrogels for nasal drug delivery: from inserts to nanoparticles. Expert Opin. Drug Deliv. **7**, 811–828 (2010). doi:10.1517/17425247.2010.495981

18. Gronlund, H., et al.: Carbohydrate-based particles: a new adjuvant for allergen-specific immunotherapy. Immunology **107**, 523–529 (2002). 1535 [pii]

19. Kamath, A.T., et al.: Synchronization of dendritic cell activation and antigen exposure is required for the induction of Th1/Th17 responses. J. Immunol. **188**, 4828–4837 (2012). doi:10.4049/jimmunol.1103183. jimmunol.1103183 [pii]

20. Kovacsovics-Bankowski, M., Clark, K., Benacerraf, B., Rock, K.L.: Efficient major histocompatibility complex class I presentation of exogenous antigen upon phagocytosis by macrophages. Proc. Natl. Acad. Sci. U.S.A. **90**, 4942–4946 (1993)

21. Vila, A., Sanchez, A., Tobio, M., Calvo, P., Alonso, M.J.: Design of biodegradable particles for protein delivery. J. Control. Release **78**, 15–24 (2002). S0168365901004862 [pii]

22. Vila, A., et al.: Low molecular weight chitosan nanoparticles as new carriers for nasal vaccine delivery in mice. Eur. J. Pharm. Biopharm. **57**, 123–131 (2004). S0939641103001620 [pii]

23. Amidi, M., et al.: Preparation and characterization of protein-loaded N-trimethyl chitosan nanoparticles as nasal delivery system. J. Control. Release **111**, 107–116 (2006). doi:10.1016/j.jconrel.2005.11.014. S0168-3659(05)00652-8 [pii]

24. Saint-Lu, N., et al.: Targeting the allergen to oral dendritic cells with mucoadhesive chitosan particles enhances tolerance induction. Allergy **64**, 1003–1013 (2009). doi:10.1111/j.1398-9995.2009.01945.x. ALL1945 [pii]

25. Porporatto, C., Bianco, I.D., Correa, S.G.: Local and systemic activity of the polysaccharide chitosan at lymphoid tissues after oral administration. J. Leukoc. Biol. **78**, 62–69 (2005). doi:10.1189/jlb.0904541. jlb.0904541 [pii]

26. Neimert-Andersson, T., et al.: Improved immune responses in mice using the novel chitosan adjuvant ViscoGel, with a haemophilus influenzae type b glycoconjugate vaccine. Vaccine **29**, 8965–8973 (2011). doi:10.1016/j.vaccine.2011.09.041. S0264-410X(11)01447-2 [pii]

27. Ghendon, Y., et al.: Evaluation of properties of chitosan as an adjuvant for inactivated influenza vaccines administered parenterally. J. Med. Virol. **81**, 494–506 (2009). doi:10.1002/jmv.21415

28. Lee, C.G., Da Silva, C.A., Lee, J.Y., Hartl, D., Elias, J.A.: Chitin regulation of immune responses: an old molecule with new roles. Curr. Opin. Immunol. **20**, 684–689 (2008). doi:10.1016/j.coi.2008.10.002. S0952-7915(08)00185-4 [pii]

29. Shibata, Y., Metzger, W.J., Myrvik, Q.N.: Chitin particle-induced cell-mediated immunity is inhibited by soluble mannan: mannose receptor-mediated phagocytosis initiates IL-12 production. J. Immunol. **159**, 2462–2467 (1997)

30. Da Silva, C.A., Hartl, D., Liu, W., Lee, C.G., Elias, J.A.: TLR-2 and IL-17A in chitin-induced macrophage activation and acute inflammation. J. Immunol. **181**, 4279–4286 (2008). 181/6/4279 [pii]

31. Da Silva, C.A., Pochard, P., Lee, C.G., Elias, J.A.: Chitin particles are multifaceted immune adjuvants. Am. J. Respir. Crit. Care Med. **182**, 1482–1491 (2010). doi:10.1164/rccm.200912-1877OC. 200912-1877OC [pii]

32. Mora-Montes, H.M., et al.: Recognition and blocking of innate immunity cells by Candida albicans chitin. Infect. Immun. **79**, 1961–1970 (2011). doi:10.1128/IAI.01282-10. IAI.01282-10 [pii]

33. Peluso, G., et al.: Chitosan-mediated stimulation of macrophage function. Biomaterials **15**, 1215–1220 (1994)

34. Li, H., Willingham, S.B., Ting, J.P., Re, F.: Cutting edge: inflammasome activation by alum and alum's adjuvant effect are mediated by NLRP3. J. Immunol. **181**, 17–21 (2008). 181/1/17 [pii]

35. Mori, A., et al.: The vaccine adjuvant alum inhibits IL-12 by promoting PI3 kinase signaling while chitosan does not inhibit IL-12 and enhances Th1 and Th17 responses. Eur. J. Immunol. (2012). doi:10.1002/eji.201242372

36. Hallander, H.O., Lepp, T., Ljungman, M., Netterlid, E., Andersson, M.: Do we need a booster of Hib vaccine after primary vaccination? A study on anti-Hib seroprevalence in Sweden 5 and 15 years after the introduction of universal Hib vaccination related to notifications of invasive disease. APMIS **118**, 878–887 (2010). doi:10.1111/j.1600-0463.2010.02674.x

37. Muzzarelli, R.A.: Chitins and chitosans as immunoadjuvants and non-allergenic drug carriers. Mar. Drugs **8**, 292–312 (2010). doi:10.3390/md8020292

38. Mills, K.H., et al.: Protective levels of diphtheria-neutralizing antibody induced in healthy volunteers by unilateral priming-boosting intranasal immunization associated with restricted ipsilateral mucosal secretory immunoglobulin a. Infect. Immun. **71**, 726–732 (2003)

39. McNeela, E.A., et al.: Intranasal immunization with genetically detoxified diphtheria toxin induces T cell responses in humans: enhancement of Th2 responses and toxin-neutralizing antibodies by formulation with chitosan. Vaccine **22**, 909–914 (2004). doi:10.1016/j.vaccine.2003.09.012. S0264410X03006728 [pii]

40. Read, R.C., et al.: Effective nasal influenza vaccine delivery using chitosan. Vaccine **23**, 4367–4374 (2005). doi:10.1016/j.vaccine.2005.04.021. S0264-410X(05)00463-9 [pii]

41. Huo, Z., et al.: Induction of protective serum meningococcal bactericidal and diphtheria-neutralizing antibodies and mucosal immunoglobulin A in volunteers by nasal insufflations of the Neisseria meningitidis serogroup C polysaccharide-CRM197 conjugate vaccine mixed with chitosan. Infect. Immun. **73**, 8256–8265 (2005). doi:10.1128/IAI.73.12.8256-8265.2005. 73/12/8256 [pii]

42. El-Kamary, S.S., et al.: Adjuvanted intranasal Norwalk virus-like particle vaccine elicits antibodies

and antibody-secreting cells that express homing receptors for mucosal and peripheral lymphoid tissues. J. Infect. Dis. **202**, 1649–1658 (2010). doi:10.1086/657087

43. Atmar, R.L., et al.: Norovirus vaccine against experimental human Norwalk Virus illness. N. Engl. J. Med. **365**, 2178–2187 (2011). doi:10.1056/NEJMoa1101245

44. Mucovac2. http://public.ukcrn.org.uk/Search/StudyDetail.aspx?StudyID=11679 (2012)

45. Li, X., et al.: Preliminary safety and efficacy results of laser immunotherapy for the treatment of metastatic breast cancer patients. Photochem. Photobiol. Sci. **10**, 817–821 (2011). doi:10.1039/c0pp00306a

# Mechanism of Adjuvanticity of Aluminum-Containing Formulas

## 40

Mirjam Kool and Bart N. Lambrecht

## Contents

| | | |
|---|---|---|
| 40.1 | **Introduction** | 633 |
| 40.2 | **Vaccine Delivery** | 634 |
| 40.2.1 | Formulation of Aluminum-Containing Adjuvants | 634 |
| 40.2.2 | Working Mechanism | 635 |
| 40.3 | **Concluding Remark** | 639 |
| **References** | | 639 |

### Abstract

For well over 80 years, alum is the most widely used adjuvant. The use of alternative adjuvants has been explored, however, aluminum adjuvants will continue to be used for many years. This is due to their good track record of safety, low cost, and adjuvanticity with a variety of antigens. Surprisingly, its mechanism of action remains largely unknown.

In this book chapter we will describe the different alum formulations and our current understanding of its working mechanism, although alum's final mode of action is not definite yet.

## 40.1 Introduction

Vaccinations have been given for well over a 100 years at the moment. The first reported vaccination was done by Edward Jenner in 1796 [1, 2]. He inoculated a young boy with cowpox virus and thereby rendered him resistant to a subsequent challenge with smallpox virus, an experiment that today would most certainly not be approved by regulatory agencies. Protection by vaccination can be achieved by giving inactivated microbes of virus particles, live attenuated virus, or subunit vaccine. However, subunit vaccination does not induce a strong immune response, which can be achieved by the administration of an adjuvant (Latin verb *adjuvare* means to help/aid). In immunology, an adjuvant is an agent that may stimulate the immune system and increase the response to a vaccine, without having any specific antigenic effect in it.

M. Kool, PhD (✉)
Department Pulmonary Medicine, Erasmus MC, Rotterdam, The Netherlands

Laboratory of Immunoregulation and Mucosal Immunology, Ghent University, Ghent, Belgium
e-mail: m.kool@erasmusmc.nl

B.N. Lambrecht, MD, PhD
Department Pulmonary Medicine, Erasmus MC, Rotterdam, The Netherlands

Department for Molecular Biomedical Research, VIB, Ghent, Belgium

Laboratory of Immunoregulation and Mucosal Immunology, Ghent University, Ghent, Belgium

**Table 40.1** Overview of different aluminum-containing adjuvant formulations

| Commercial name | Chem. formulation | Chemical name | Used as adjuvant in |
|---|---|---|---|
| Alum | AlK$(SO_4)_2 \cdot 12H_2O$ | Aluminum potassium sulfate | Not used anymore |
| Imject-alum | $Al(OH)_3 + Mg(OH)_2$ | Aluminum hydroxide and magnesium hydroxide | Experimental immunology |
| Alhydrogel | $Al(OH)_3$ | Aluminum hydroxide | Human and veterinary vaccines |
| Adju-Phos | $Al(PO_4)_3$ | Aluminum phosphate | Human and veterinary vaccines |

In 1926, Alexander Glenny and colleagues reported that toxoid precipitated with aluminum potassium sulfate, referred to as alum, induced a stronger antibody production when injected into guinea pigs than soluble toxoid alone [3, 4]. Today, alum is the most common adjuvant used in approved prophylactic vaccines because of its excellent safety profile and ability to enhance protective humoral immune responses. However, the long history behind the use of alum as an adjuvant contrasts with our poor understanding of its mechanism of action, which has only recently begun to unravel.

Currently, alum is still the only licensed vaccine adjuvant in the USA. In Europe, however, since the 1990s of the last century, MF59, AS03 (both oil-in-water emulsions), and MPL (monophospholipid A; an LPS analog) formulated in alum are also approved for use [5–7]. Besides being used in human vaccinations, aluminum adjuvants are used in both experimental immunology [8, 9] and to produce murine monoclonal antibodies as well as polyclonal antisera. In veterinary medicine aluminum adjuvants have been used in a large number of vaccine formulations against viral and bacterial diseases [10, 11].

Because vaccines are administered to healthy individuals including infants and children and there are potential safety concerns with novel adjuvants, extensive preclinical safety studies, including local reactogenicity and systemic toxicity testing, are required. While a significant number of adjuvants are noticeably more potent than alum, they have largely had a higher toxicity level, which has been the main reason to exclude them as adjuvants for human vaccine formulations.

## 40.2 Vaccine Delivery

### 40.2.1 Formulation of Aluminum-Containing Adjuvants

Numerous inaccuracies are found in literature when referred to the term "alum." The most used term "alum" only applies to aluminum potassium sulfate [12–15] (Table 40.1). Since problems have occurred during manufacturing, alum is not used anymore in vaccines. Instead, several other insoluble aluminum salts are used, such as aluminum hydroxide and aluminum phosphate as they can be prepared in a more standardized manner and capture antigen by direct adsorption. Two different adsorption forces prevail adsorptive interaction between the antigen and the adjuvant [8, 13, 16, 17]. First is electrostatic interactions, which occur most strongly between negatively charged proteins and aluminum hydroxide and between positively charged proteins and aluminum phosphate. Anions present at time of preparation may coprecipitate and change the characteristics of "pure" aluminum hydroxide. One very critical example of an anion having such influence is phosphate. Exposure of aluminum hydroxide adjuvant and phosphorylated antigens to phosphate ions in the formulation should be minimized to produce maximum adsorption of the antigen [7, 18, 19]. The second adsorption force, anionic ligand exchange, is a covalent interaction that occurs when an antigen contains a phosphate group that can displace a hydroxyl group on the adjuvant surface. This forms an inner-sphere surface complex with aluminum that is the inorganic equivalent of a covalent bond. Ligand exchange is the strongest adsorption force and can occur even when an electrostatic repulsive force is present [7, 20, 21]. Importantly, in mice the strength of the adsorption force is inversely related to antibody titer and T cell activation [1, 22–24]. Antigens that adsorb to aluminum-containing adjuvants by electrostatic attraction are more likely to elute upon intramuscular or subcutaneous administration than antigens that adsorb by ligand exchange [3]. Antigen released from the adjuvant, antigen trapped in void spaces within the adjuvant, and antigen adsorbed to aluminum-containing adjuvants are all present for proper uptake of antigen by dendritic cells (DCs) [5, 7]. Of note, absorbed antigen to either aluminum hydroxide or aluminum phosphate adjuvant can be completely eluted within 4 h or 15 min, respectively [8].

### 40.2.2 Working Mechanism

Three potential mechanisms are frequently cited to explain how adjuvants increase humoral immunity, although scarce experimental evidence is publicly available: firstly, the formation of a depot by which the antigen is slowly released to enhance the antibody production [10]; secondly, the induction of inflammation, thus recruiting and activating antigen-presenting cells that capture the antigen [12, 14, 15]; and thirdly, the conversion of soluble antigen into a particulate form so that it is phagocytosed by antigen-presenting cells such as macrophages, DCs, and B cells. The main theory has been that depot formation and the associated slow release of antigen [8, 16] are responsible for alum enhancement of antigen presentation and subsequent T and B cell responses. However, recently several papers show that the depot formation is dispensable for the adjuvanticity of alum [18, 19]. Although aluminum precipitates from the peritoneal cavity could transfer the adjuvant effect for as long as 1 month after injection when transplanted from one mouse to another, this is not the predominant working mechanism of alum. It was observed that around alum nodules, an accumulation of neutrophils, CD11c$^+$ DCs, and antigen-specific T cells was present, suggesting that depot formation could be involved in the maintenance of a memory pool (MK and BNL 2008, [20, 21]).

#### 40.2.2.1 Cellular Pathways Activated by Aluminum-Containing Adjuvants

Our immune system is composed of two major subdivisions, the innate or nonspecific immune system and the adaptive or specific immune system. The innate immune system is our first line of defense against invading organisms, while the adaptive immune system acts as a second line of defense and also provides protection against reexposure to the same pathogen. The fast-acting innate immune responses provide a necessary first-line defense. In contrast, the adaptive immunity uses selection and clonal expansion of immune cells recognizing antigens from the pathogen, thereby providing specificity and long-lasting immunological memory. Although these two arms of the immune system have distinct functions, there is interplay between these systems (i.e., components of the innate immune system influence the adaptive immune system and vice versa). DCs form a bridge between innate and adaptive immune response [22–24] and DCs play a significant role in adjuvant-mediated increases in vaccine-antigen immunogenicity.

The efficacy of different adjuvants can be explained by differences in signaling in DCs to undergo complex maturation events that are required for initiation of T cell differentiation and activation [22]. Injection of alum induces inflammation at the injection site, being it subcutaneous, intramuscular, or in an experimental setting, intraperitoneally (Fig. 40.1). Within several hours after injection, there is a significant increase in the number of conventional and plasmacytoid DCs, neutrophils, NK cells, eosinophils, and inflammatory monocytes, and a disappearance of mast cells and macrophages [25–28]. The recruitment of these innate cells was accompanied by the secretion of several necessary chemokines, such as monocyte chemotactic MCP1 (CCL2), MIP-1α (CCL2), MIP-1β (CCL4), the neutrophil chemotaxin KC/IL-8 (CXCL8), the eosinophil chemotaxin1 (CCL11), IL-5, and RANTES (CCL5) [25, 27, 29].

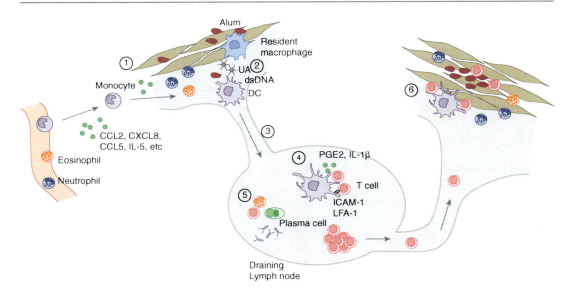

**Fig. 40.1** Overview of cellular immune response induced by alum adjuvant. (*1*) After an alum-formulated antigen is injected intramuscularly, there is an immediate response of muscle cells to release chemokines and cytokines, like CCL2, CCL5, IL-5, and CXCL8 that attract cells of the innate immune system, like monocytes, eosinophils, and neutrophils to the site of injection. (*2*) In addition, there is the local release of uric acid (UA) and dsDNA (at least in some models) by local tissue damage or active production by macrophages. (*3*) Recruited monocytes differentiate into dendritic cells (DCs), take up antigen, and process it on MHCI and MHCII molecules on their way to the draining nodes. (*4*) In the nodes, mature DCs select antigen-specific T cells that differentiate into Th effector cells. Particularly in the mouse, this response is Th2 biased due to PGE2 and IL1β production and the expression of ICAM-1 and LFA-1. (*5*) In the spleen, and possibly in the draining nodes, there is also recruitment of Gr1[+], IL-4[+] eosinophils that stimulate B cell responses. Alum mainly induces long-lived B cell responses. (*6*) Effector T cells are recruited back to the site of injection and found around the depot of alum and antigen, where also DCs, neutrophils, and eosinophils reside. Possibly, depot formation could be involved in stimulating long-lasting memory

It has been less clear if and how aluminum-containing adjuvants can induce DC mobilization and maturation. Most likely the mode of action of alum *in vivo* is upstream of the direct activation of DCs, as confirmed by *in vitro* data [30]. At least *in vitro*, alum did not enhance costimulatory molecule expression and DC maturation, leading to the proposal that they might induce activation of DCs only *in vitro*, upon release of endogenous danger signals by structural cells or inflammatory cells [31–33]. It has been demonstrated that both uric acid and dsDNA are released upon alum injection acting as danger signals [25, 34]. In view of the crucial role of DCs in activation of adaptive immunity, several studies focused on the effects of alum on DCs and their monocytic precursors *in vivo* following intraperitoneal and intramuscular injection of antigen in alum [25, 35]. Monocytes consequently took up antigen and migrated to the draining lymph nodes where they became monocytes-derived DCs that expressed high levels of MHC class II and CD86 and induced vigorous proliferation of antigen-specific T cells.

Importantly, all the adjuvant effects of alum on T cell responses and humoral immunity were abolished in mice conditionally depleted of CD11c[high] DCs, yet adoptive transfer of sorted Ly6C[high] monocytes restored these effects. In humans, it also appears that alum-formulated vaccines mainly act at the levels of the monocytes, inducing phenotypic and functional maturation [36, 37]. Together, these experiments demonstrate that monocyte-derived DCs are necessary and sufficient to mediate the adjuvant effects of alum. Furthermore, alum also enhances the magnitude and duration of expression of peptide-to-MHCII complexes on the DC surface, with an accompanying increase in MHCII expression [38]. This together with increased ICAM-1 expression will lead to an enhance contact time with T cells, resulting in a strong CD4+ T cell response [39].

Exposure of DCs to alum with antigen does not result in the phagocytosis of the particulate antigen; DCs acquire only soluble antigen via endocytosis [39]. This pathway of antigen uptake does not lead to cross-presentation (e.g. presentation of exogenous antigens in MHC I) [40]. It is a common knowledge that aluminum-containing adjuvants predominantly induce humoral immunity while sparing cellular immunity. This is further supported by the fact that alum induces B cell priming via a splenic Gr-1+ myeloid IL-4-producing cell type, most likely representing eosinophils [41–43]. Classical cell-mediated immunity measured by DTH responses and induction of CD8+ CTL responses to a range of polypeptide and protein antigens is poorly induced by alum, because of a lack of cross-priming [13, 44, 45]. Nonetheless, proliferative response of CD4+ T cells and Th2 cytokine production have been found to be enhanced in a number of murine and human studies, suggesting that alum boosts humoral immunity by providing Th2 help to follicular B cells [12, 46–48].

While animal studies have been extremely important to unravel the fundamental mechanisms that govern the immune response, we are still facing the challenge of understanding the human immune system in its complexity and genetic heterogeneity and unraveling the sophisticated escape mechanism used by human pathogens. This knowledge will help to design new vaccine strategies.

### 40.2.2.2 Molecular Pathways Activated in DCs by Aluminum-Containing Adjuvants

How exactly aluminum-containing adjuvants are recognized by antigen-presenting cells has been – and continues to be – a puzzle (Fig. 40.2). Several high-profile papers published over the last years have shown that Toll-like receptors (TLR) and TLR signaling through the MyD88 or TRIF adaptor pathway, classical activators of innate immunity, and the DC network *in vivo* were not always necessary for alum to act as an adjuvant for humoral immunity [49–52]. However, the innate immune response depended on MyD88 as a crucial adaptor molecule [25]. It is imperative in this context to annotate that both TLR signaling and interleukin-1 receptor (IL-1R) signaling depend on MyD88 for efficient intracellular signaling.

Additionally, it has become clear that an alternative pathogen detection system exists, which relies on a family of intracellular receptors, called NOD-like receptors (NLRs) [53, 54]. NLRs have a variety of functions in regulating inflammatory and apoptotic responses. Whereas TLRs sense extracellular nonself-motifs of infectious organisms, NLRs sense stimuli of microbial origin as well as endogenous signals (DAMPs) rather than microbial patterns [55]. They can recognize stress, abnormal self, or danger signals such as DNA, RNA, and uric acid [31]. NLRP3 (NALP3), a member of the NLR family, along with ASC and caspase-1 forms a molecular platform called the inflammasome [56]. NLRP3 can be activated by several agonists, like endogenous ATP, uric acid, silica, and alum adjuvant, leading to the processing and release of IL-1β and IL-18 [57].

The release of these proinflammatory cytokines requires two signals. The first signal is derived from TLR agonist such as lipopolysaccharide (LPS) and inflammatory cytokines, activating the NF-κB pathway [58], thereby initiating the transcription and accumulation of the precursor proteins. And secondly, cleavage and secretion of the cytokines is mediated by the inflammasome. IL-1β, in turn, triggers another cascade of molecular events that result in inflammation [59]. IL-1β is a potent inflammatory cytokine, which is implicated in acute and chronic inflammatory disorders.

Pioneering was the paper of Fabio Re's group showing that human DCs and macrophages *in vitro* produce IL-1β and IL-18 by a combination stimulus of alum and a TLR agonist. This cytokine production was dependent on caspase-1 and NLRP3 [33]. It was later shown that also mouse DCs behaved similar upon alum stimulation (Fig. 40.2) [27, 60–63]. In contrast to the *in vitro* data showing that alum activates the NLRP3

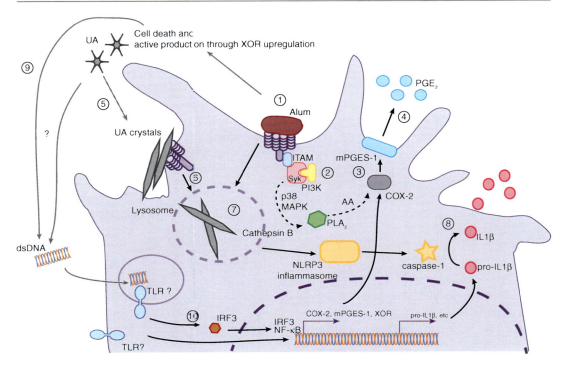

**Fig. 40.2** The molecular pathways induced by alum in DCs and macrophages. (*1*) Alum binds to the cell membrane and thereby induces lipid raft formation. (*2*) In the lipid raft, ITAM-containing receptors cluster and activate the Syk-PI3Kδ pathway. (*3*) This pathway in turn activates cytosolic phospholipase A2 (*cPLA2*), probably via p38 MAP kinase, resulting in the release of arachidonic acid from membrane lipids. COX-2 and membrane-associated PGE synthase-1 (*mPGES-1*) converts arachidonic acid to prostaglandin E2 (*PGE$_2$*). (*4*) PGE$_2$ is released from the cell and instructs the Th2 response. (*5*) Furthermore, alum induces the release of uric acid via the induction XOR. (*6*) Upon phagocytosis of alum or uric acid crystals lysosomal damage is induced. (*7*) This activates the NLRP3 inflammasome resulting from the release of enzymes like cathepsin B into the cytoplasm. (*8*) Activated caspase-1 mediates the proteolytic cleavage of pro-IL1β in biologically active IL1β. (*9*) Besides the release of uric acid, alum also induces the release of dsDNA, probably through cell death. (*10*) dsDNA activates the immune system and more specifically monocytes through *Irf3* which is critical for the migration of inflammatory monocytes

inflammasome, there is considerable controversy if the inflammasome-IL-1β pathway is necessary for the humoral immune response induced by alum. *In vitro*, there appears to be a discrepancy between the innate and adaptive immune response in the necessity of the NLRP3 inflammasome. Antigen uptake, recruitment of immune cells to the injection site, migration of DCs to the draining lymph nodes, and CD4[+] T cells proliferation are decreased, but still present, in NLRP3 deficient mice [60]. The results on the requirement of NLRP3 for alum adjuvanticity and humoral response are conflicting. It was shown to be either abrogated in the absence of NLRP3 [61, 62], no need for NLRP3 [27, 63], whereas our studies [60] only showed a need for the NLRP3 inflammasome for IgE production. The difference in results obtained by the different groups is unexplained at the moment; however, different types of alum and immunization protocols are used which could contribute to the different phenotypes seen. To add to the confusion, the need for NLRP3 is challenged as the adaptor molecules MyD88 and TRIF of the TLR4 and IL-1R pathway are not required for the alum-induced humoral response [49, 51].

Recently it has been shown that alum can activate DCs through lipid raft formation due to its interaction with membrane lipids, such as sphingomyelin and cholesterol [39]. Through lipid sorting, immune-receptor signaling motif (ITAM)-containing receptors are aggregated which leads to ITAM phosphorylation. Phosphorylated ITAM motifs recruit Syk and subsequently activate PI3 kinase. This pathway will finally lead to phagocytosis and cytokine production [64–66]. In contrast to uric acid crystals, which also induce the lipid raft formation and activation of Syk [67], alum is not phagocytosed by DCs. However, DCs that have been primed with alum interact stronger with CD4$^+$ T cells, which are antigen independent, but were dependent on ICAM-1 and LFA-1 expression [39, 65]. Also, alum-induced prostaglandin E2 (PGE$_2$) release by macrophages was crucial for the Th2 polarization [68]. PGE$_2$, a well-characterized proinflammatory lipid mediator, is an arachidonic acid metabolite that suppresses Th1 responses by elevating intracellular cAMP concentrations in DCs, macrophages, and Th1 cells. The PGE$_2$ production was independent of inflammasome activation but dependent on the Syk and P38 MAP kinase pathway. Likewise, Syk and PI3 kinase signaling played an important role in the activation and Th2 polarization in DCs [69]. Finally, alum is a potent inhibitor of LPS-induced IL-12 production by DC, which results from alum-induced PI3 kinase signaling [70].

### 40.2.2.3 The Involvement of Endogenous Danger Signals

Besides direct effects, alum adjuvant exerts some level of cytotoxicity [15]. In humans an increase in necrotic cells, a potential source of endogenous danger signals, or damage-associated molecular patterns (DAMPs) can be found at the injection site [71]. Dying cells release numerous molecules that act as DAMPs and which can alert the innate immune system through the activation of various PRR signaling pathways. Aluminum-containing adjuvants induce the endogenous release of uric acid (UA) and host dsDNA that will activate the immune system indirectly [25, 34]. UA is a danger signal produced during the catabolism of purines and is the end product in ureotelic mammals. Additionally, UA can be released from injured cells, after degradation of RNA and DNA and permutation of the liberated purines into UA.

Host dsDNA is exposed to the immune system through the release from necrotic cells. Both DAMPs alert the immune system for cell death. After administration of alum in mice, both UA and dsDNA levels are rapidly increased at the site of injection [25, 34]. Both UA and dsDNA are able to act as an adjuvant and boost Th2-associated humoral responses [25, 34, 69]. UA crystals induce Th2 immunity by activating the Syk/PI3 kinase pathway [69], whereas dsDNA signals via Tbk1 and Irf3 [34]. Research on how these two pathways interact is needed to fully understand the mechanism of alum adjuvant in future.

## 40.3 Concluding Remark

As this chapter shows, there are complex interactions between multiple large systems; immunology is all about complexity and control. However, vaccine development is all about coordinated leverage of complexity. Modern vaccine development endeavors to render complexity into simplicity, and practical realization of immunological leverage for optimal vaccine development can often be influenced by vaccine adjuvants [72]. Therefore, our understanding on the working mechanism of the oldest and most often used adjuvant alum is crucial for the development of new adjuvants.

## References

1. Hansen, B., Sokolovska, A., HogenEsch, H., Hem, S.L.: Relationship between the strength of antigen adsorption to an aluminum-containing adjuvant and the immune response. Vaccine **25**, 6618–6624 (2007)
2. Jenner, E.: The Three Original Publications on Vaccination Against Smallpox. Harvard Classics. P.F. Collier & Son, New York (1909)
3. Jiang, D., Morefield, G.L., HogenEsch, H., Hem, S.L.: Relationship of adsorption mechanism of antigens by aluminum-containing adjuvants to in vitro elution in interstitial fluid. Vaccine **24**, 1665–1669 (2006)
4. Glenny, A.: Insoluble precipitates in diphtheria and tetanus immunization. Br. Med. J. **2**, 244–245 (1930)
5. Romero Méndez, I.Z., Shi, Y., HogenEsch, H., Hem, S.L.: Potentiation of the immune response to

non-adsorbed antigens by aluminum-containing adjuvants. Vaccine 25, 825–833 (2007)
6. Tritto, E., Mosca, F., De Gregorio, E.: Mechanism of action of licensed vaccine adjuvants. Vaccine 27, 3331–3334 (2009)
7. Iyer, S., HogenEsch, H., Hem, S.L.: Effect of the degree of phosphate substitution in aluminum hydroxide adjuvant on the adsorption of phosphorylated proteins. Pharm. Dev. Technol. 8, 81–86 (2003)
8. Shi, Y., HogenEsch, H., Hem, S.L.: Change in the degree of adsorption of proteins by aluminum-containing adjuvants following exposure to interstitial fluid: freshly prepared and aged model vaccines. Vaccine 20, 80–85 (2001)
9. Kool, M., et al.: Alum adjuvant boosts adaptive immunity by inducing uric acid and activating inflammatory dendritic cells. J. Exp. Med. 205, 869–882 (2008)
10. Glenny, A., Pope, C., Waddington, H., Wallace, U.: Immunological notes. XVII. The antigenic value of toxoid precipitated by potassium alum. J. Path. and Bact 29, 31–40 (1926)
11. Heegaard, P.M.H., et al.: Adjuvants and delivery systems in veterinary vaccinology: current state and future developments. Arch. Virol. 156, 183–202 (2011)
12. Mannhalter, J.W., Neychev, H.O., Zlabinger, G.J., Ahmad, R., Eibl, M.M.: Modulation of the human immune response by the non-toxic and non-pyrogenic adjuvant aluminium hydroxide: effect on antigen uptake and antigen presentation. Clin. Exp. Immunol. 61, 143–151 (1985)
13. Hem, S.L., HogenEsch, H.: Relationship between physical and chemical properties of aluminum-containing adjuvants and immunopotentiation. Expert Rev. Vaccines 6, 685–698 (2007)
14. Goto, N., Akama, K.: Histopathological studies of reactions in mice injected with aluminum-adsorbed tetanus toxoid. Microbiol. Immunol. 26, 1121–1132 (1982)
15. Goto, N., et al.: Local tissue irritating effects and adjuvant activities of calcium phosphate and aluminium hydroxide with different physical properties. Vaccine 15, 1364–1371 (1997)
16. Gupta, R.K., Chang, A.C., Griffin, P., Rivera, R., Siber, G.R.: In vivo distribution of radioactivity in mice after injection of biodegradable polymer microspheres containing 14C-labeled tetanus toxoid. Vaccine 14, 1412–1416 (1996)
17. Hem, S.L., HogenEsch, H., Middaugh, C.R., Volkin, D.B.: Preformulation studies – the next advance in aluminum adjuvant-containing vaccines. Vaccine 28, 4868–4870 (2010)
18. Hutchison, S., et al.: Antigen depot is not required for alum adjuvanticity. FASEB J. 26, 1272–1279 (2011)
19. Noe, S.M., Green, M.A., HogenEsch, H., Hem, S.L.: Mechanism of immunopotentiation by aluminum-containing adjuvants elucidated by the relationship between antigen retention at the inoculation site and the immune response. Vaccine 28, 3588–3594 (2010)
20. Munks, M.W., et al.: Aluminum adjuvants elicit fibrin-dependent extracellular traps in vivo. Blood 116, 5191–5199 (2010)
21. Lambrecht, B.N., Kool, M., Willart, M.A.M., Hammad, H.: Mechanism of action of clinically approved adjuvants. Curr. Opin. Immunol. 21, 23–29 (2009)
22. Steinman, R.M., Pope, M.: Exploiting dendritic cells to improve vaccine efficacy. J. Clin. Invest. 109, 1519–1526 (2002)
23. Pashine, A., Valiante, N.M., Ulmer, J.B.: Targeting the innate immune response with improved vaccine adjuvants. Nat. Med. 11, S63–S68 (2005)
24. Bendelac, A., Medzhitov, R.: Adjuvants of immunity: harnessing innate immunity to promote adaptive immunity. J. Exp. Med. 195, F19–F23 (2002)
25. Kool M., et al.: Cutting edge: Alum adjuvant stimulates inflammatory dendritic cells through activation of the NALP3 inflammasome. J. Immunol. 181, 3755–3759 (2008)
26. Seubert, A., et al.: Adjuvanticity of the oil-in-water emulsion MF59 is independent of Nlrp3 inflammasome but requires the adaptor protein MyD88. Proc. Natl. Acad. Sci. 108, 11169–11174 (2011)
27. McKee, A.S., et al.: Alum induces innate immune responses through macrophage and mast cell sensors, but these sensors are not required for alum to act as an adjuvant for specific immunity. J. Immunol. 183, 4403–4414 (2009)
28. Calabro, S., et al.: Vaccine adjuvants alum and MF59 induce rapid recruitment of neutrophils and monocytes that participate in antigen transport to draining lymph nodes. Vaccine 29, 1812–1823 (2011)
29. Didierlaurent, A.M., et al.: AS04, an aluminum salt- and TLR4 agonist-based adjuvant system, induces a transient localized innate immune response leading to enhanced adaptive immunity. J. Immunol. 183, 6186–6197 (2009)
30. Mosca, F., et al.: Molecular and cellular signatures of human vaccine adjuvants. Proc. Natl. Acad. Sci. 105, 10501–10506 (2008)
31. Shi, Y., Evans, J.E., Rock, K.L.: Molecular identification of a danger signal that alerts the immune system to dying cells. Nature 425, 516–521 (2003)
32. Sun, H., Pollock, K.G.J., Brewer, J.M.: Analysis of the role of vaccine adjuvants in modulating dendritic cell activation and antigen presentation in vitro. Vaccine 21, 849–855 (2003)
33. Li, H., Nookala, S., Re, F.: Aluminum hydroxide adjuvants activate caspase-1 and induce IL-1beta and IL-18 release. J. Immunol. 178, 5271–5276 (2007)
34. Marichal, T., et al.: DNA released from dying host cells mediates aluminum adjuvant activity. Nat. Med. 17, 996–1002 (2011)
35. Langlet, C., et al.: CD64 expression distinguishes monocyte-derived and conventional dendritic cells and reveals their distinct role during intramuscular immunization. J. Immunol. 188, 1751–1760 (2012)
36. Ulanova, M., Tarkowski, A., Hahn-Zoric, M., Hanson, L.A.: The common vaccine adjuvant aluminum

36. hydroxide up-regulates accessory properties of human monocytes via an interleukin-4-dependent mechanism. Infect. Immun. **69**, 1151–1159 (2001)
37. Seubert, A., Monaci, E., Pizza, M., O'Hagan, D.T., Wack, A.: The adjuvants aluminum hydroxide and MF59 induce monocyte and granulocyte chemoattractants and enhance monocyte differentiation toward dendritic cells. J. Immunol. **180**, 5402–5412 (2008)
38. Ghimire, T.R., Benson, R.A., Garside, P., Brewer, J.M.: Alum increases antigen uptake, reduces antigen degradation and sustains antigen presentation by DCs in vitro. Immunol. Lett. **147**, 55–62 (2012)
39. Flach, T.L., et al.: Alum interaction with dendritic cell membrane lipids is essential for its adjuvanticity. Nat. Med. **17**, 479–487 (2011)
40. Burgdorf, S., Kautz, A., Böhnert, V., Knolle, P.A., Kurts, C.: Distinct pathways of antigen uptake and intracellular routing in CD4 and CD8 T cell activation. Science **316**, 612–616 (2007)
41. Jordan, M.B., Mills, D.M., Kappler, J., Marrack, P., Cambier, J.C.: Promotion of B cell immune responses via an alum-induced myeloid cell population. Science **304**, 1808–1810 (2004)
42. Wang, H.-B., Weller, P.F.: Pivotal advance: eosinophils mediate early alum adjuvant-elicited B cell priming and IgM production. J. Leukoc. Biol. **83**, 817–821 (2008)
43. McKee, A.S., et al.: Gr1 + IL-4-producing innate cells are induced in response to Th2 stimuli and suppress Th1-dependent antibody responses. Int. Immunol. **20**, 659–669 (2008)
44. Wijburg, O.L., et al.: The role of macrophages in the induction and regulation of immunity elicited by exogenous antigens. Eur. J. Immunol. **28**, 479–487 (1998)
45. Bomford, R.: The comparative selectivity of adjuvants for humoral and cell-mediated immunity. II. Effect on delayed-type hypersensitivity in the mouse and guinea pig, and cell-mediated immunity to tumour antigens in the mouse of Freund's incomplete and complete adjuvants, alhydrogel, Corynebacterium parvum, Bordetella pertussis, muramyl dipeptide and saponin. Clin. Exp. Immunol. **39**, 435–441 (1980)
46. Brewer, J.M., et al.: Aluminium hydroxide adjuvant initiates strong antigen-specific Th2 responses in the absence of IL-4- or IL-13-mediated signaling. J. Immunol. **163**, 6448–6454 (1999)
47. Grun, J.L., Maurer, P.H.: Different T helper cell subsets elicited in mice utilizing two different adjuvant vehicles: the role of endogenous interleukin 1 in proliferative responses. Cell. Immunol. **121**, 134–145 (1989)
48. Serre, K., et al.: IL-4 directs both CD4 and CD8 T cells to produce Th2 cytokines in vitro, but only CD4 T cells produce these cytokines in response to alum-precipitated protein in vivo. Mol. Immunol. **47**, 1914–1922 (2010)
49. Gavin, A.L., et al.: Adjuvant-enhanced antibody responses in the absence of toll-like receptor signaling. Science **314**, 1936–1938 (2006)
50. Schnare, M., et al.: Toll-like receptors control activation of adaptive immune responses. Nat. Immunol. **2**, 947–950 (2001)
51. Nemazee, D., Gavin, A., Hoebe, K., Beutler, B.: Immunology: toll-like receptors and antibody responses. Nature **441**, (2006)
52. Palm, N.W., Medzhitov, R.: Immunostimulatory activity of haptenated proteins. Proc. Natl. Acad. Sci. **106**, 4782–4787 (2009)
53. Martinon, F., Tschopp, J.: Inflammatory caspases and inflammasomes: master switches of inflammation. Cell Death Differ. **14**, 10–22 (2007)
54. Ting, J.P.Y., Willingham, S.B., Bergstralh, D.T.: NLRs at the intersection of cell death and immunity. Nat. Rev. Immunol. **8**, 372–379 (2008)
55. Mariathasan, S., Monack, D.M.: Inflammasome adaptors and sensors: intracellular regulators of infection and inflammation. Nat. Rev. Immunol. **7**, 31–40 (2007)
56. Martinon, F., Gaide, O., Pétrilli, V., Mayor, A., Tschopp, J.: NALP inflammasomes: a central role in innate immunity. Semin. Immunopathol. **29**, 213–229 (2007)
57. Fritz, J.H., Ferrero, R.L., Philpott, D.J., Girardin, S.E.: Nod-like proteins in immunity, inflammation and disease. Nat. Immunol. **7**, 1250–1257 (2006)
58. Kawai, T., Akira, S.: Signaling to NF-κB by Toll-like receptors. Trends Mol. Med. **13**, 460–469 (2007)
59. Arend, W.P., Palmer, G., Gabay, C.: IL-1, IL-18, and IL-33 families of cytokines. Immunol. Rev. **223**, 20–38 (2008)
60. Kool, M., et al.: Cutting edge: alum adjuvant stimulates inflammatory dendritic cells through activation of the NALP3 inflammasome. J. Immunol. **181**, 3755–3759 (2008)
61. Eisenbarth, S.C., Colegio, O.R., O'Connor, W., Sutterwala, F.S., Flavell, R.A.: Crucial role for the Nalp3 inflammasome in the immunostimulatory properties of aluminium adjuvants. Nature **453**, 1122–1126 (2008)
62. Li, H., Willingham, S.B., Ting, J.P.Y., Re, F.: Cutting edge: inflammasome activation by alum and alum's adjuvant effect are mediated by NLRP3. J. Immunol. **181**, 17–21 (2008)
63. Franchi, L., Núñez, G.: The Nlrp3 inflammasome is critical for aluminium hydroxide-mediated IL-1β secretion but dispensable for adjuvant activity. Eur. J. Immunol. **38**, 2085–2089 (2008)
64. Nakashima, K., et al.: A novel Syk kinase-selective inhibitor blocks antigen presentation of immune complexes in dendritic cells. Eur. J. Pharmacol. **505**, 223–228 (2004)
65. Greenberg, S., Chang, P., Wang, D.C., Xavier, R., Seed, B.: Clustered syk tyrosine kinase domains trigger phagocytosis. Proc. Natl. Acad. Sci. U.S.A. **93**, 1103–1107 (1996)
66. Turner, M., Schweighoffer, E., Colucci, F., Di Santo, J.P., Tybulewicz, V.L.: Tyrosine kinase SYK: essential functions for immunoreceptor signalling. Immunol. Today **21**, 148–154 (2000)

67. Ng, G., et al.: Receptor-independent, direct membrane binding leads to cell-surface lipid sorting and Syk kinase activation in dendritic cells. Immunity **29**, 807–818 (2008)
68. Kuroda, E., et al. Silica crystals and aluminum salts regulate the production of prostaglandin in macrophages via NALP3 inflammasome-independent mechanisms. Immunity. **34**, 1–13 (2011)
69. Kool, M., et al.: An unexpected role for uric acid as an inducer of T helper 2 cell immunity to inhaled antigens and inflammatory mediator of allergic asthma. Immunity **34**, 527–540 (2011)
70. Mori, A., et al.: The vaccine adjuvant alum inhibits IL-12 by promoting PI3 kinase signaling while chitosan does not inhibit IL-12 and enhances Th1 and Th17 responses. Eur. J. Immunol. **42**, 2709–2719 (2012)
71. Goto, N., Akama, K.: Local histopathological reactions to aluminum-adsorbed tetanus toxoid. Naturwissenschaften **71**, 427–428 (1984)
72. Alving, C.R., Peachman, K.K., Rao, M., Reed, S.G.: Adjuvants for human vaccines. Curr. Opin. Immunol. **24**, 310–315 (2012)

# From Polymers to Nanomedicines: New Materials for Future Vaccines

# 41

Philipp Heller, David Huesmann, Martin Scherer, and Matthias Barz*

## Contents

| | | |
|---|---|---|
| 41.1 | **Introduction** | 644 |
| 41.2 | **Favorable Properties of Nanoparticles** | 646 |
| 41.2.1 | Stabilization of Nanoparticles | 646 |
| 41.3 | **Types of Nanoparticulate Carriers and Fabrication Methods** | 647 |
| 41.4 | **Polymeric Materials** | 654 |
| 41.4.1 | Poly(meth)acrylates and Acrylamides | 654 |
| 41.4.2 | Polypeptides | 655 |
| 41.4.3 | Polyglycerols | 657 |
| 41.4.4 | Polyoxazolines | 658 |
| 41.5 | **Characterization of Nanoparticles** | 659 |
| 41.6 | **Nanoparticulate Vaccine Formulations** | 660 |
| 41.6.1 | Advantages of Particle-Based Vaccines | 660 |
| 41.6.2 | Approaches | 661 |
| 41.7 | **Summary and Conclusions** | 663 |
| References | | 663 |

## Abstract

Nanomedicine is the medical application of nanotechnology and therefore covers various kinds of nanoparticles. In this chapter, we would like to provide a brief introduction and overview of nanoparticles for the modulation of the immune system. In general, these nano-sized objects can be inorganic colloids, organic colloids (synthesized by emulsion polymerization or mini-/nanoemulsion techniques), polymeric aggregates (micelles or polymersomes), core cross-linked aggregates (nanohydrogels, crosslinked micelles, or polyplexes), multi-functional polymer coils, dendritic polymers or perfect dendrimers. A special focus is set on polymeric materials, because the chemical composition of the particle corona will shape particle properties by providing steric stabilization, avoiding protein adsorption and particle aggregation *in vivo*. Besides synthesis of new materials, particle characterization is equally important and might be the key to a more detailed understanding of the behavior of nano-sized systems. In addition, we would like to highlight approaches towards nanoparticle-based immunotherapies.

P. Heller, Dipl. Chemist • D. Huesmann,
Dipl. Chemist • M. Scherer, Dipl. Chemist
M. Barz, PhD (✉)
Institute of Organic Chemistry,
Johannes Gutenberg-University
Mainz, Mainz, Germany
e-mail: barz@uni-mainz.de

*The authors Philipp Heller, David Huesmann, Martin Scherer, Matthias Barz have contributed equally.

## 41.1 Introduction

Nanoparticles – as the name suggests – are objects in a size range typically from 1 to a few hundred nanometers [1]. They can be made from a variety of materials and, because of their large surface to volume ratio, often possess different properties from the bulk material they are made of. Nanoparticles can be synthesized by bottom up or top down approaches. In general, these nano-sized systems can be inorganic colloids, organic colloids (synthesized by emulsion polymerization or mini-/nanoemulsion techniques), polymeric aggregates (micelles or polymersomes), core cross-linked aggregates (nanohydrogels, cross-linked micelles, or polyplexes), multifunctional polymer coils, dendritic polymers, or perfect dendrimers.

For example, a paramagnetic iron oxide core can be used for magnetic resonance imaging (MRI) or a biodegradable core to release a bioactive by diffusion while degrading. Due to their size, nanoparticle can be used to combine different functionalities among one object leading to multifunctionality or multivalency. Additionally, those systems are able to encapsulate bioactive compounds. Shielding the cargo requires core-shell structures; while the core encapsulates the cargo, the shell needs to provide solubility and suppresses protein adsorption or aggregation.

When nano-sized objects are applied to diagnose or treatment of diseases, they become so-called nanomedicines, an umbrella term encompassing nanopharmaceuticals, nanoimaging agents, and theranostics [2].

Already at the beginning of this article, we would like to mention a point of great importance, which is – much to our surprise – quite often overlooked. The interaction of a nanoparticle with its environment takes place at the interface of particle and surrounding media – the particle surface. Thus, the molecules stabilizing the particle surface towards its surroundings determine the interaction with biological matter and therefore the in vivo fate of the particle. In many cases the characterization of nanoparticles is difficult as described later. In respect to this fact, we feel that it is necessary to carefully consider, understand, and design nanoparticles – in particular when they are utilized for complex in vivo applications.

---

**Terminology in Nanoparticle Synthesis**

*Polymers*: Chemical compounds consisting of basic structural repeating units called monomers. A polymer may be formed from one or multiple kinds of monomers during a process called polymerization. The material properties are determined not only by the chemical composition of the monomer(s) but also by the number of monomer molecules connected to each other to form the macromolecule (degree of polymerization) and the structural alignment.

*Dispersity*: Apart from a few examples (proteins, DNA, or perfect dendrimers), polymers often consist of a mixture of macromolecules differing in molecular weight/degree of polymerization due to a variance in the number of monomers the individual polymer molecule consists of. The dispersity $Đ$ (formally known as polydispersity index, PDI) is defined as $Đ = \frac{M_w}{M_n}$ where $M_w$ is the mass-average molar mass $M_w = \frac{\sum M_i^2 \cdot N_i}{\sum M_i \cdot N_i}$ and $M_n$ is the number-average molar mass $M_n = \frac{\sum M_i \cdot N_i}{\sum N_i}$.

A low dispersity (near 1) means that a large fraction of molecules have similar degrees of polymerization.

*Block Copolymers*: Block copolymers are an important type of copolymers (polymers consisting of more than one monomeric species). This term is applied when a sequence of a single type of monomer (homopolymer) is followed by a sequence formed from another species. Block copolymers may consist of two (diblock copolymer), three (triblock copolymer), or more blocks (multiblock copolymer). The chemical nature of the different blocks in a covalent macromolecule can lead to unique properties (such as tenside-like amphiphilicity in the case of hydrophilic/hydrophobic block copolymers) or can enable the block-specific functionalization (e.g., with dyes, receptor molecules).

*Colloids*: Colloids are particles ranging from 1 to 1,000 nm which are microscopically dispersed in a continuous medium. Colloidal solutions are an interstage between true solutions and suspensions and exhibit no colligative properties. Colloidal systems can be made of organic as well as inorganic particles. To prevent colloids from coagulation – which is thermodynamically favored – particles must be kept separate from each other by electrostatic or steric repulsion.

*Amphiphiles*: Amphiphiles are molecules having both hydrophilic (polar) and hydrophobic (nonpolar) properties. When in contact with water, amphiphiles will self-assemble into different superstructures in order to minimize the interphase energy. In this way, various aggregates such as micelles, polymersomes (polymer-based liposome-like structures), or double-layer membranes can be formed. The precise structure depends on the ratio of hydrophilic versus hydrophobic domains as well as environmental conditions during preparation and storage.

*Dendrimers/Dendrons*: Dendrimers are polymeric molecules which are repetitively branched and exhibit a highly symmetrical and well-defined structure. They are synthesized in repeated cycles where new generations of monomers are added to a core exhibiting two or more reactive sites. The reactive groups (again, more than one per unit) of the newly added outer sphere then act as a basis for the addition of the next generation. Alternatively, in a convergent approach, dendrons (branched molecules with a single reactive group, the so-called focal point) are synthesized which are then coupled to the multifunctional core in a final step. In both cases, monodisperse dendrimers can be synthesized, providing a large number of functional peripheral groups.

*Micelles*: Micelles are an important type of colloidal aggregate formed from amphiphilic molecules such as low molecular weight surfactants or amphiphilic polymers. Above the *critical micelle concentration* (CMC), the surfactant molecules will spontaneously arrange in an order where the polar head groups are directed towards the water phase (or other polar solvent) and the tails form the hydrophobic inner core which can be employed to encapsulate hydrophobic cargo.

*Surfactants*: Surfactants (*surface active agents*) are molecules lowering the surface tension of a liquid or the interfacial tension between two immiscible phases. They are amphiphilic compounds which will adsorb at the interface between hydrophilic and hydrophobic phase with the polar head group arranged in the hydrophilic environment and the nonpolar part extending into the hydrophobic phase, respectively. Surfactants are widely applied in interphase science and technology, serving as detergents or dispersants, wetting agents, and as stabilizers for emulsification.

Apart from both natural and synthetic small-molecule surfactants, polymer-based surfactants have attracted huge interest. Due to the variability of block materials and the possibility for post-polymerization modification, such amphiphiles can be specifically tailored for the intended application. Their high molecular weights result in stronger adsorption and superior stabilization properties since – in contrast to small-molecule amphiphiles – steric stabilization becomes an important aspect.

*Emulsions*: Emulsions are disperse systems which consist of droplets of one liquid

> phase (dispersed phase) which are dispersed in a second, immiscible liquid phase. Typically, a stable emulsion requires the addition of surfactants and a co-stabilizer which is dissolved in the dispersed phase. Via different methods, droplet sizes in the order of around 30 to a few hundred nanometers can be achieved. This so-called *miniemulsion* fabrication has become an interesting tool in nanoparticle synthesis. Such particles can either be obtained by *miniemulsion polymerization* or by (polymer) coating and stabilization of preformed droplets (*miniemulsion technique*).

## 41.2 Favorable Properties of Nanoparticles

During the last decades, nanoparticles have shown great potential for the application in drug delivery and vaccination [3], and they are expected to provide solutions to various current issues in this field. The first one is the fundamental problem of solubility: according to Torchilin [4], about half of potentially valuable drug candidates identified by high throughput screening technologies including those with the highest activities demonstrate poor solubility in water and – for this reason – never enter further development. By encapsulation or binding to a nanoparticulate carrier, this problem can be overcome.

Additionally, encapsulation of sensitive agents such as DNA, RNA, or peptides protects them against premature enzymatic or proteolytic degradation.

In combination with a "stealth" surface of the carrier, which prevents opsonization through steric repulsion [5], a prolonged circulation can be achieved. Moreover, through chemical modification, defined recognition and site-specific release are possible.

One main objective in using nanocarriers is to increase the specificity towards certain cells or tissues. This can be achieved either by passive targeting as in the accumulation in well-vascularized tumors driven by the EPR (enhanced permeability and retention) effect [6] or actively by binding of the particles to specific receptors presented by certain cells. Such targeted localization is enabled by functionalizing the particle surface with ligand molecules binding to those receptors, as has been carried out using antibodies [7], sugar moieties [8], transferrin [9], or folates [10], for instance.

In combination, encapsulation and targeting can improve the bioavailability of the respective therapeutic agent. Furthermore, the toxic side effects associated with many drugs are minimized if the active component is shielded from the environment and only released at the designated site of action. As an example, by employing doxorubicin as HPMA copolymer conjugates PK1/PK2, the maximum tolerated dose (MTD) of the drug was increased threefold [11]. Controlled release can be triggered by factors such as pH [12], redox potential [13], enzyme activity [14], temperature [15], or magnetic [16], electric, and ultrasonic signals [17, 18].

As pointed out by Little, another promising aspect of nanoparticles in vaccine development is the multiple presentation of subunit antigens [19]. Synthetic particles can be designed to exhibit a repetitive orientation of the antigen on the surface, giving the potential of generating B cell receptor cross-linking and enhanced activation [20].

### 41.2.1 Stabilization of Nanoparticles

In employing nanoparticles as carriers in drug delivery, several aspects need to be taken into consideration: not only must the carrier itself remain in a colloidal form, but nonspecific aggregation in the body has to be avoided or at least controlled [21].

The two main mechanisms concerning the stabilization of dispersed or emulsified systems are electrostatic and steric stabilization. The electrostatic interaction is described by the *DLVO theory* named after Derjaguin, Landau, Verwey, and Overbeek. Steric stabilization results from polymers adsorbed or covalently linked to the particle surface, which is mostly due to entropic effects together with a minor enthalpic contribution via solvation energies [21].

For in vivo applications, electrostatic stabilization appears difficult to achieve, since serum proteins include polycationic as well as –anionic macromolecules, which can adsorp onto oppositely charged colloids. The human body has developed a series of protective mechanisms against foreign material far beyond the scope of this article. One of them is the removal of foreign particles larger than the renal threshold from the bloodstream. However, this includes the removal of nano-sized drug carriers, a process initiated by opsonization by the mononuclear phagocyte system (MPS). Macrophages of the MPS can remove particles within seconds of intravenous administration which impedes their application as drug carriers due to premature clearance [22, 23].

Even though the exact mechanism of this complicated process is not fully understood yet, it is widely accepted that opsonization starts with absorption of a series of proteins including immunoglobulins and components of the complement system and other serum proteins. The (incomplete) list of relevant proteins identified so far contains C3, C4, C5, laminin, fibronectin, C-reactive protein, type I collagen, and many more [24, 25]. Those proteins approach the nanocarrier by random Brownian motion and, once close enough to the surface, may bind to the particle via different forces; van der Waals or ionic forces, just to name two of them [23]. The step of opsonin binding is followed by the attachment of phagocytes to the nanoparticle. Phagocytes cannot recognize the carrier themselves but will dock to the particle via surface-bound opsonins. Two alternative mechanisms are the stimulation of phagocytosis by nonspecific adherence to the opsonins and complement activation, which results in binding of the foreign particle by mononuclear phagocytes [23].

In a final step the particle is taken up by phagocytes, typically by phagocytosis. Once the particle is endocytosed, it is degraded by various enzymes and chemical substances or – in case of a nondegradable material – stored and enriched in the body.

As of today, there is no universal solution to completely prevent the opsonization of particles. However, a general trend is observed that hydrophilic but neutral ligands hamper the adsorption of proteins and therefore slow down the clearance from the bloodstream [26].

Thus, steric stabilization appears to be the method of choice for in vivo applications. Steric stabilization can be achieved by the use of commercial polymeric surfactants (e.g., Tween 20/40/60/80, Pluronics, Cremophor) or other amphiphilic polymers. Block copolymers seem to form more stable layers at the interface of particle and aqueous solution as reported by Scheibe et al. [27] and Kelsch et al. [28] and can suppress aggregation in blood serum. In addition, the polymer can be modified with functional groups, which can later on be addressed for further functionalization.

## 41.3 Types of Nanoparticulate Carriers and Fabrication Methods

As stated above, the location of the bioactive compound does have an influence on its action. For example, in vaccination the immune response can be modulated by the location of the antigen. Depending on the specific aim and the nature of the agent, it can be advantageous to encapsulate the antigen to prevent it from premature degradation and to present it within the endosomal pathway (intracellular receptor). In other cases the presentation in a repetitive orientation on the outside of the carrier (extracellular receptor) optimized immune response may be preferred. Consequently, a variety of different colloidal carriers can be employed. Among the most frequently used systems are liposomes, solid lipid nanoparticles (SLNs), immunostimulating complexes (ISCOMs), micelles, hydrogels, polymer colloids prepared by miniemulsions, and polymer nanoparticles such as those fabricated by particle replication in non-wetting templates (PRINT) (see Fig. 41.1).

Polymeric micelles [29], core cross-linked micelles [30], nanohydrogels [31], and polyplexes [32, 33] are mostly based on the assembly of block copolymers into super structures, which is mainly driven by four forces (hydrophobic, electrostatic, metal complexation and hydrogen bonds) [34]. Often, hydrophobic (micelles and core cross-linked micelles) and electrostatic (polyplexes) interactions are involved, while metal complexation and the formation of H bonds are only applicable in specific cases. For *hydrophobic interactions*, a different polarity of the blocks leads to a phase separation, resulting in the formation of polymeric micelles where the hydrophobic part aggregates to avoid contact to water. In core-shell structures the hydrophobic core can then be used for the transport of a hydrophobic payload. In this case the release of the encapsulated compound is based on passive diffusion or disintegration of the micelle. *Electrostatic interactions* on the other hand rely on the interaction of oppositely charged molecules to form superstructures. A typical example for the formation of so-called polyplexes is the interaction of a positively charged polymer (e.g., polylysine) and negatively charged nucleic acids (DNA or RNA). The driving force is mainly a gain in entropy, which is due to the release of counter ions during complex formation.

*Polymeric micelles* form above their critical micelle concentration (CMC), but are always in equilibrium with free unimer. The CMC can be considered as the maximum concentration of free unimer in solution underlining the dynamics of these systems. Upon injection into the blood, the micellar solution is rapidly diluted, potentially leading to disassembly of micelles [35]. Although the CMC of polymeric micelles is much lower

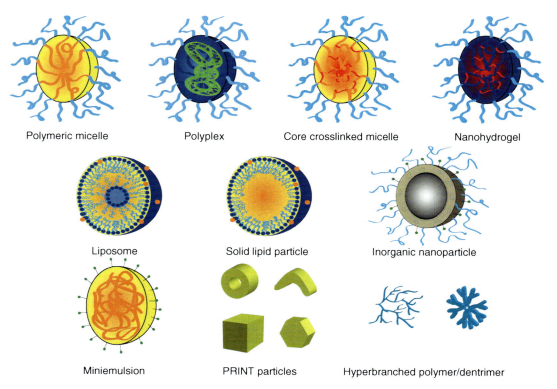

**Fig. 41.1** Nanoparticulate carriers

**Table 41.1** Controlled release strategies

| Trigger | Linkage (examples) | Site of release |
|---|---|---|
| pH [12] | Cleavage of covalent bonds in acetales, hydrazones, esters | Acidic environment (e.g., endosomal/lysosomal compartment) |
| Reduction potential [13] | Cleavage of disulfides or benzoquinones | Reductive environment (cytosol) |
| Enzyme activity [14] | Cleavage of peptide bonds | Specific cells/tissues/diseases |
| Temperature [15] | Phase transition or change in solubility | Hyperthermic tissue/where hyperthermia is applied |
| Magnetic signal [16] | Pore formation through movement of magnets | Where signal is applied |
| Electric signal [17] | Pore formation through electrically stimulated polymer swelling or piezoelectric elements | Where signal is applied |
| Ultrasonic signal [18] | Polymer degradation | Where signal is applied |

than that of surfactant micelles [36], this might still pose a problem.

Therefore, micelles that are further stabilized are of interest. The easiest way to further stabilize micelles is to cross-link the core chemically [37]. These *core cross-linked micelles* find themselves in the dilemma of both stably encapsulating the cargo but also needing to be capable of releasing it, once the vesicle has reached its destination. To achieve this, the cross-linking has to be reversible, and a stimulus at the destination has to break the stabilizing bonds. This can be realized with various stimuli-responsive cross-linkers. The stimuli used most often are pH or redox potential, but also temperature, enzymes, light, and others have been realized (see Table 41.1) [38]. pH-labile bonds are cleaved during endocytosis in the endosome, while redox-labile bonds are cleaved in the reducing environment of the cytoplasm, or during endocytosis in antigen presenting cells. Thus leading to a release of the cargo after uptake by the cell.

*Liposomes* are spherical structures composed of one or more phospholipid bilayers surrounding an aqueous core, mimicking the basic structure of cell membranes. Depending on the number of bilayers, their size ranges from around 20 nm to several μm [39]. Their structural composition gives liposomes great versatility with respect to the way of loading pharmaceutically active agents and the agent's properties. Antigens or immunopotentiators can be encapsulated within the core, embedded in the lipid bilayer, or linked to the outer surface [40]. Liposomes are commonly produced by thin-film hydration, solvent injection, or reverse-phase evaporation. In order to produce smaller vesicles and narrow the size distribution, sonication and extrusion techniques can be applied [41]. Liposomes are often modified with polymers to shield the liposome and reduce protein adsorption (e.g., PEGylation) [42] or to integrate pH-responsive elements [43] and molecules for the specific targeting of cells of the immune system like mannose derivatives [44]. Liposomes can be designed to carry not only antigens but also adjuvants to stimulate the innate immune response [45, 46]. Liposomal formulations currently in clinical use include virosomal influenza and hepatitis A vaccines [47, 48]. Despite the aforementioned benefits, employment of liposomes is still limited by several issues. These include long-term physical instability and the low entrapment rate of molecules but also the difficult large-scale production resulting in high costs [49]. Some of these issues can be tackled by *polymersomes* [50, 51], vesicular carriers built up analogue to liposomes but from block copolymer amphiphiles instead of lipids. Due to their molecular weight, which exceeds the molecular weight of lipids by up to two orders of magnitude, polymersomes have superior material properties and storage capacities [50]. In contrast to lipids most polymers, however do not have GRAS (generally regarded as safe) status by the FDA.

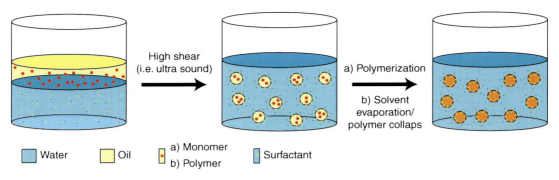

**Fig. 41.2** Miniemulsion polymerization (*a*) and miniemulsion technique (*b*)

Another type of lipid-based nanoparticles is *immunostimulating complexes* (*ISCOMs*). They have a spherical open lattice structure which, additional to the antigen, is formed from phospholipids, cholesterol, and Quil A saponin, a strong adjuvant. Such ISCOMs can carry a variety of antigens and have been reported to be about ten times more immunogenic than similar colloidal particles because of the saponin [52]. It also has to be mentioned that there are concerns regarding the toxicity of Quil A due to its hemolytical activity [53].

*Solid lipid nanoparticles* (SLNs) are also lipid based, but exhibit a structure different from liposomes. They consist of a core formed from lipids, which are in a solid state under physiological conditions, surrounded by an emulsifier shell. The approach here is to combine the advantages of conventional lipid-based carriers (good biocompatibility, ease of production/scaling up) with those of polymeric nanoparticles (solid matrix for physical and chemical stabilization) [54, 55]. SLNs are mainly produced via high pressure homogenization and miniemulsion methods. For processing the solid lipids, elevated temperatures are required, but the use of organic solvents can be avoided. A major issue of SLNs is polymorphic transitions due to alteration of the lipid packing (e.g., crystallization), affecting the particle shape which has an influence on the interaction with the physiological environment [56].

*Emulsion techniques* are versatile methods for particle formation, giving rise to different classes of colloid carriers in vaccination and drug delivery. This includes nano- and microemulsion particles directly administered [57–59] as well as polymer or inorganic nanoparticles produced by emulsion polymerization or processing of previously formed emulsion droplets [3, 53, 60, 61].

*(Mini)emulsions* (see Fig. 41.2) are composed of droplets of one liquid (dispersed phase) in another, immiscible liquid (continuous phase). Depending on their composition and way of fabrication, miniemulsions occupy a size range of about 30–500 nm [62].

By the use of surfactants, resulting in electrostatic or steric repulsion, the droplets are stabilized against coalescence. The destabilization of dispersed droplets by net mass flux of molecules due to the different Laplace pressure of differently sized particles can be prevented by a second additive mixed into the dispersed phase. Such additives are of a highly hydrophobic nature, keeping them from diffusing from one droplet to another and therefore establishing an osmotic pressure counteracting the Laplace pressure [63]. The effectiveness of this process is governed by the hydrophobicity of the additive. Many different types of molecules have been employed as hydrophobes, including alkanes [64], dyes [65], and hydrophobic initiators for polymerization [66]. Given the aforementioned mechanisms and their nanometer size providing kinetic stability against sedimentation, miniemulsions are considered as thermodynamically stable [46, 58], and time scales for the stability of such emulsions are reported to be in the order of months [62].

Nano- or miniemulsification – as well as dispersion – of one species in another is a process

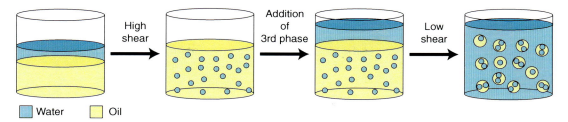

**Fig. 41.3** Preparation of a double emulsion

comprising two major aspects: the disruption of the to-be-emulsified phase resulting in droplets of high specific surface areas and the stabilization of these interfaces by a surfactant. In the emulsification process, the oil phase consisting of the molecules to be emulsified and the hydrophobe is dispersed in the water phase by means of an adequate surfactant and then homogenized. The required high shear forces are provided by ultrasonication, rotor-stator dispersers, microfluidizers, or high pressure homogenizers [62]. After homogenization, a second charge of surfactant is added to poststabilize the system effectively [67].

Apart from the facts already mentioned, an important advantage of emulsion techniques is the relatively easy scale-up. While this is a critical point with any kind of highly ordered, self-assembled structures, the bulk production of emulsion formulations affords homogenous and highly stable results in short periods of time.

Among the most common emulsification methods for antigen encapsulation are solvent evaporation and coacervation. Both methods include the formation of a double emulsion (water phase 1/oil phase/water phase 2, see Fig. 41.3) and share the first step, where an aqueous solution of the respective agent is emulsified in an organic polymer solution. In solvent evaporation, this first W/O emulsion is then further dispersed in an aqueous solution containing an appropriate emulsifier – poly(vinyl alcohol) is frequently employed for this purpose [68]. Finally, the organic solvent is evaporated under reduced pressure, leading to hardening of the polymer and formation of the desired particle.

Coacervation is a phase separation process in which a nonsolvent for the polymer (and antigen) is added to the previously formed W/O emulsion. This results in a phase separation to form a coacervate phase encapsulating the antigen.

Another possibility for the formation of nanocarriers is nanoprecipitation, also known as solvent displacement method. Here, stabilized nanoparticles are formed by the interfacial deposition of a pre-formed polymer onto emulsion droplets. An inverse emulsion is formed in a mixed solvent/nonsolvent medium containing a surfactant and the polymer for shell formation. Adding nonsolvent leads to the precipitation of the polymer and deposition on the large surface droplets. After precipitation, the particles can be transferred into water.

However, the formation of nano-sized carriers requires more harsh conditions compared to their micrometer-sized analogues [69]. Consequently, sensitive molecules such as DNA might be structurally damaged and utmost care needs to be taken when engineering the conditions for forming such particles. Such measures may include the complexation of the agent with cationic amphiphilic molecules (hydrophobic ion pairing, HIP) [70], the judicious choice of solvents and the adsorption of antigens on the surface of pre-formed particles [71].

A class of polymeric materials widely used to form drug carriers are polyesters such as poly (D,L-lactide) (PLA) and poly (D,L-lactic-co-glycolic acid) (PLGA). They are biodegradable through hydrolysis, exhibit excellent biocompatibility and sustained release properties [72]. To stabilize such polymers during mini- or nano-emulsion processes, PEG-b-PLA copolymers appear advantageous since the PLA block mixes well with the polyester core and collapses during solvent evaporation. While the PEG block forms a hydrophilic, protein-resistant corona. Whenever a corona with higher functionality is required, multifunctional PHPMA-b-PLA copolymers as developed by Barz and co-workers can be applied [73, 74].

However, natural or semisynthetic polymers are also widely employed. Such polymers are usually polysaccharide or protein based, and to review them all is beyond the scope of this chapter. Therefore, only dextrans and chitosan shall be mentioned here as two important examples.

Dextrans are polymers mainly consisting of α-1,6-glycosides and a small percentage of 1,3-linkages leading to branching [75]. They are well soluble in water and relatively stable due to the glycosidic bonds, being hydrolyzed only under extremely acidic or alkaline conditions. The hydroxyl groups allow for derivatization, for example, a partial hydrophobization to obtain amphiphilic molecules [76]. Chitosan is not water soluble at physiological pH and is therefore chemically modified at the amine or hydroxyl group to obtain soluble derivatives. Chitosans have shown great potential for mucosal delivery due to their mucoadhesive properties [77, 78].

A good overview on materials used for biopolymeric nanoparticles is given by Kundu et al. [79]. Current research on colloidal nanoparticles includes the development of vaccines [80] against various infectious diseases like hepatitis B [81] and HIV [82] as well as cancers [83].

An interesting approach to forming quasi-monodisperse and shape-specific polymeric nanoparticles is a method of preparation called PRINT (particle replication in non-wetting templates), developed by the group of DeSimone [84] (see Fig. 41.1). In this "top down" approach, particles are formed in a mold made from perfluoropolyethers. The molecules the particle is made from – such as poly(ethylene glycol) trisacrylate or p-hydroxystyrene – are filled into the mold as a solution or melt and then cured by UV radiation or chemical means. This is enabled by the low surface energy of perfluoropolyethers which allows for filling the cavities trough capillary action, but prevents the deposition of material between the cavities [85]. Loading of the particles with therapeutic agents can be achieved either by inclusion into the particle matrix or by post-fabrication functionalization. As an example, siRNA for gene silencing was recently incorporated into PRINT particles [86].

Apart from particles made from organic materials, colloidal particles from inorganic matter have also been explored. Here, research has mainly focused on particles consisting of gold and magnetite ($Fe_3O_4$). A broad variety of routes for particle synthesis has been described in the literature [87]. Metal nanoparticles are typically prepared by reduction of metal salts using suitable reducing agents while metal oxide nanoparticles are usually prepared through base hydrolysis of molecular precursors [61]. Stabilization of the colloids is achieved by adequate ligands, e.g., chelating polymers providing steric stabilization, which are either added directly during the initial synthesis, or post-synthesis by modification of the ligand sphere by exchange or polymer coating [88, 89].

Gold nanoparticles have so far essentially been studied for immunotherapy of tumors. Absorbing near-infrared (NIR) light, gold nanoparticles can locally increase the temperature when targeted to tumors via tumor-specific antibodies and thus have a perspective use in photothermal cancer therapy [90]. Recently, gold nanoparticle-epitope-Fc conjugates were also successfully employed as antigen carriers, exhibiting better activities and immunological responses compared to liposomes [91].

Magnetite nanoparticles too have been investigated for hyperthermia therapy with a focus on heat shock protein expression for antitumor T-lymphocyte-mediated immunity [92]. Furthermore, they have been investigated as MRI contrast agents [93]. In all the abovementioned cases, it has to be kept in mind that a stabilizing and solubilizing agent is required to make inorganic nanoparticles applicable to in vivo applications. For the same reasons stated for organic colloids, polymers are preferable over tensides or other low molecular weight components to fulfill this task.

In summary, there are many kinds of nanoparticles, and not all of them are useful as carriers in the body. To be considered for transporting cargo in humans, the particles have to fulfill some basic requirements. Obviously they need to be biocom-

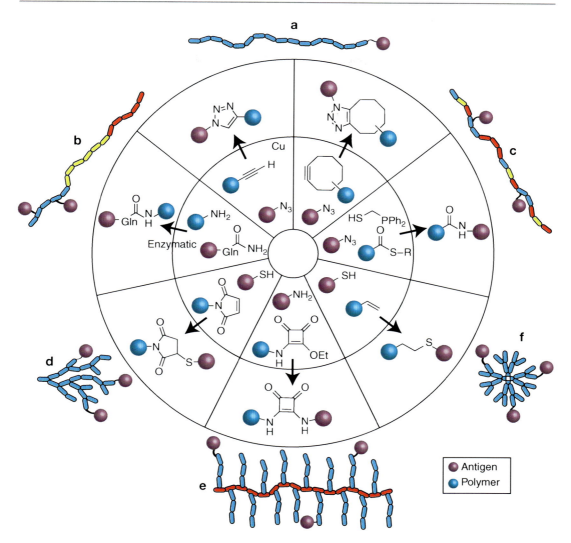

**Fig. 41.4** Polymer architectures and possible antigen conjugation. (**a**) homopolymer; (**b**) block copolymer; (**c**) random copolymer; (**d**) hyperbranched polymer; (**e**) polymer brush; (**f**) dendrimer. Antigens can be attached using (clockwise from *top left*) copper-catalyzed azid-alkin-cycloaddition [101], copper-free azid-alkin-cycloaddition [101], traceless Staudinger ligation [102], thiol-ene chemistry [101], squaric acid [103], maleimide [101], enzymatic conjugation [104], and other chemical methods [105]

patible or, better yet, biodegradable to avoid acute toxicity or storage diseases as reported for PEG [94–96]. They need to be water soluble, and they need to show little or no interaction with proteins in order to ensure smooth circulation in the blood. Materials that do not interact with proteins need to be hydrophilic, electrically neutral, and have hydrogen bond acceptors but no hydrogen bond donors [26, 97, 98]. This concerns the outer material of the particle that is in contact with the body and therefore determines how the body reacts to it (or not), for example, through uptake by the MPS. While in many cases the particles are shielded by PEG, there are some interesting alternatives available, which we would like to introduce in the following sections.

Since many nanoparticles are made up of polymers or use them to form a shell around the particle, a brief description of polymer structures available appears reasonable (see Fig. 41.4).

Macromolecules, consisting of multiple repeating units, can have different levels of complexity. The simplest structure is a linear homopolymer that has only one type of repeating unit. Block copolymers and random copolymers consist of two or more repeating units. But while the repeating units are ordered in distinct blocks in the first case, they are randomly distributed along the chain in the second. When multifunctional monomers are used, branched structures and brushes are also available. A special case is the perfect dendrimer, a branched structure where every generation of branches is synthesized in a different step. In contrast to classical polymers, which have a distribution in the number of repeating units per molecule, (molecular weight distribution) dendrimers are defined like proteins. The high density of functional groups on their surface allows them to engage in multivalent interactions [99]. Glycopeptide dendrimers, for example, have been employed in immunotherapy [100]. Synthesis of perfect dendrimers, however, is demanding and appears difficult to perform on a larger scale.

But besides variations in polymer architecture, the material of which the polymers consist of can be varied as well. Within the next paragraph we would like to introduce polymers, which are of interest as materials building a hydrophilic corona around nanoparticles and thus making them applicable to the development of systems for in vivo applications, e.g., drug delivery, imaging, or vaccination.

## 41.4 Polymeric Materials

### 41.4.1 Poly(meth)acrylates and Acrylamides

The development of controlled radical polymerization (CRP) techniques has had a tremendous impact on synthetic polymer chemistry, since it has enabled the controlled polymerization of (meth)acrylates and acrylamides (Table 41.2). The CRP techniques were developed to reduce termination and can be divided into three subgroups: stable free radical polymerization (e.g., NMP), degenerative transfer polymerization (e.g., RAFT, MADIX), and transition metal-mediated controlled radical polymerization (e.g., ATRP). Among these, atom transfer radical polymerization (ATRP) and RAFT (reversible addition-fragmentation chain transfer) polymerization are arguably the most commonly used and most versatile processes. There have been various reviews describing mechanism as well as recent developments of either RAFT [106, 107] or ATRP [108, 109].

The CRP techniques can be used in the synthesis of complex polymer architectures, e.g., (multi)block copolymers, branched polymers, or hybrid systems [110–113].

A group of (meth)acrylates being particularly interesting are systems with oligoethylene glycol side chains (OEGMA) – e.g., diethylene glycol methacrylate (DEGMA) or polyethylene glycol methacrylate (PEGMA). The interest in these

**Table 41.2** Comparison of free radical polymerization (FRP) to CRP

|               | FRP | CRP |
|---------------|-----|-----|
| Advantages    | Compatible with a broad variety of monomers | Low dispersity/narrow molecular weight distribution |
|               | Insensitive to impurities | Good control over polymer architecture (chain topology, end group functionality) |
|               | Employed in industry for decades | |
| Disadvantages | Lack of control over chain topology (no block architecture, no defined end groups!) | Inferior reaction rate due to limited number of active species |
|               | Little control over dispersity/broad molecular weight distribution | Higher sensitivity to impurities |

systems has grown recently. Polymers based on these monomers possess beneficial properties, such as a high solubility in water, non-immunogenicity and low toxicity, a lower critical solution temperature (LCST), and enhanced blood circulation times [114–116]. The LCST can be nicely tuned by copolymerization of both monomers. It was reported by Lutz and Hoth that the LCST can be adjusted from 26 to 90 °C by changing the ratio of OEGMA to DEGMA units in the copolymer [117].

Cytotoxicity was investigated in various cell lines, such as Caco-2, HT29-MTX-E12, or HepG2, ensuring nontoxic behavior up to a concentration of 5 mg mL$^{-1}$ and 72-h exposure [118, 119].

Another interesting polymer is poly(2-(meth) acryloyloxyethyl phosphorylcholine) (PMPC). The monomer structure is highly bio-inspired since the side chain contains the head group of natural PC-phospholipids, ensuring high biocompatibility [120, 121].

PMPC-based polymersomes have been synthesized and applied to study diffusion across oral epithelium [122]. Additionally, PMPC-based block copolymers have been used for transfection by Battaglia et al. [123]. PMPC polymersomes showed pronounced cellular uptake as well as nontoxic behavior up to 3 mg mL$^{-1}$. Furthermore, PMPC polymers were used for protein conjugation by Lewis and co-workers [124]. A reduced tissue migration compared to PEG-protein conjugates of the same hydrodynamic volume was observed. Thus, an improved depot effect in the tissue as well as subsequent longer elimination half-life may lead to improved pharmacokinetics. These findings underline the potential of PMPC-based polymeric systems.

Finally, also well-established polymers such as 2-(hydroxypropyl) methacrylamide (HPMA) can be prepared in a controlled manner by the CRP methods. HPMA was the first polymer entering clinical trials as polymer drug conjugate in the 1990s [125], and HPMA-based copolymers can be synthesized directly or using post-polymerization modifications [126, 127].

Those three types of monomers may be the material future nanomedicines are made of. However, it has to be kept in mind that neither the polymer backbone of poly(meth)acrylates nor acrylamides can be degraded in vivo. Thus, polymers need to be excreted, limiting the molecular weight of the single polymer as well as the potential application.

## 41.4.2 Polypeptides

Polypeptides are one of the major building blocks in nature and a very versatile class of materials. They naturally have multiple functionalities, which allow them to respond to stimuli like change in pH or oxidation potential.

There are various methods of synthesizing polypeptides. The use of biotechnological methods that allow production of peptides in cells or the solid phase peptide synthesis leads to sequence-defined polypeptides, but are limited in amino acids, expensive, and can only be performed on a relatively small scale and therefore will not be discussed here. Where large quantities of homo or block copolypeptides are required, the polymerization of α-amino acid-$N$-carboxyanhydrides (NCAs) is a suitable production method (see Fig. 41.5). The polymerization of NCAs is known since the beginning of the last century [128], and the work since then has been summarized in excellent books and reviews [129, 130]. But only in recent years, different ways of controlling this type of polymerization have been developed [131].

Since synthetic polypeptides, just like natural proteins, are made up of amino acids, they can be nontoxic, biocompatible, and degradable in the body while they remain stable in aqueous solution. Because of their similarity to natural proteins, immunogenicity is a concern that has to be addressed. Although immunogenicity has not been observed for homo and block copolypeptides [132, 133], new polypeptides have to be studied in this respect especially when various amino acids are incorporated.

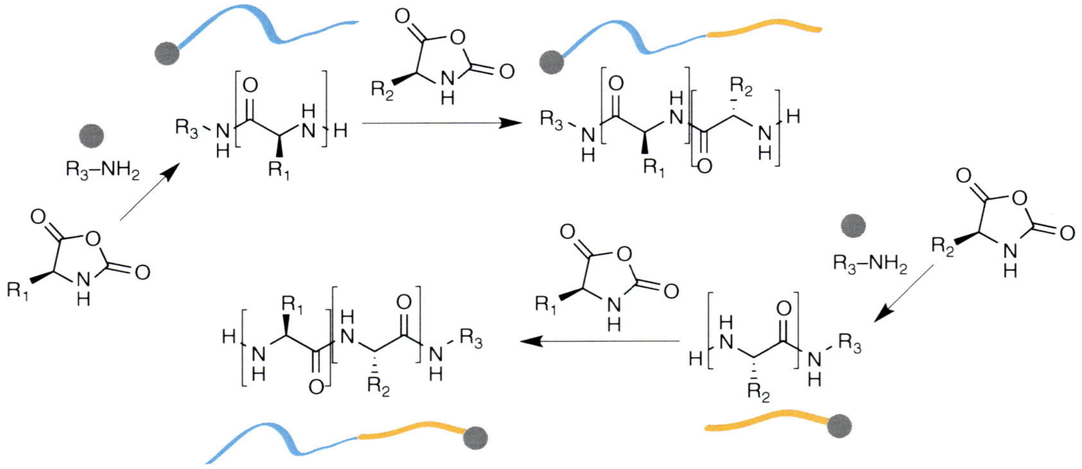

**Fig. 41.5** Synthesis of block copolypeptides

The multitude of different side chains enable the design of peptidic superstructures like polyion complexes [33, 134], polymer micelles [135, 136], polymer vesicles [137, 138], nanofibers or nanotubes [139], and hydrogels [140].

All these different structures give rise to a wide range of biomedical applications. In the field of DNA and RNA transport, peptidic polyplexes have evolved from simple polylysine to highly complex structures that contain a removable hydrophilic shell [141] and respond to stimuli like change in pH [142], oxidation potential [143], or ATP concentration [144].

Polypeptides are also used to encapsulate hydrophobic drugs. Probably, the most prominent example is PEG-block-poly(aspartate) where the Asp is modified with 4-phenyl-1-butanol to make it more hydrophobic [145]. Loaded with paclitaxel, this micellar formulation is currently in a phase III clinical trial for breast cancer under the name NK105.

Polypeptides have also found their way into the field of diagnostics. For example, PEGylated polylysine with an NIR dye was used as an in vivo imaging probe to visualize tumor-associated proteases in nude mice [146].

Polypeptides can also work as bioactive agents themselves. The immunomodulator drug Copaxone is a random copolymer of alanine, lysine, glutamic acid, and tyrosine and is approved for the treatment of multiple sclerosis [147]. The exact mode of action of Copaxone is still under discussion, but it is believed that due to the similarity of Copaxone and the myelin basic protein (MBP), the drug can divert the autoimmune response away from the myelin sheath of neurons.

Also the covalent attachment of drugs to polypeptides is very promising. In Opaxio (formerly known as Xyotax or CT-2103), paclitaxel is conjugated to polyglutamic acid (PGA) via an ester bond, and the conjugate is currently in phase III clinical trials for ovarian and non-small cell lung cancer [133]. Here paclitaxel is only released when the polymer is degraded by tumor-associated proteases leading to a selective release of the chemotherapeutic agent [148].

Lysine-block-leucine copolypeptides were used by Deming for the preparation of the first stable double emulsions below 100 nm [149]. These double emulsions consist of water droplets in oil droplets in a water phase and have the capacity to carry hydrophobic as well as hydrophilic drugs, which makes them attractive for delivery applications.

Polypeptides in clinical trials, as carriers or as active agents, are summarized in Table 41.3 [150–152].

Although most of the polypeptides in the clinics are for the treatment of cancer, they underline the potential of polypeptides as carriers. Their

**Table 41.3** Polypeptides in clinical trials

| Name | Polypeptide | Drug | Indication | Phase |
|---|---|---|---|---|
| NK105 | PEG-b-Asp(buPh) | Paclitaxel | Stomach cancer | III |
|  |  |  | Breast cancer | II |
| NK012 | PEG-b-Glu | SN-38 | Small cell lung cancer | II |
|  |  |  | Breast cancer | II |
| NC-6004 | PEG-b-Glu | Cisplatin | Pancreatic cancer | II |
|  |  |  | Solid cancer | I |
| NC-4016 | PEG-b-Glu | DACH-platin | Solid cancer | I |
| Opaxio | p(Glu) | Paclitaxel | Ovarian cancer | III |
|  |  |  | Non-small cell Lung cancer | III |
| CT-2106 | p(Glu) | Camptothecin | Ovarian cancer | II |
|  |  |  | Solid cancer | I |
| Copaxone | Ala-Lys-Glu-Tyr copolymer | Peptide itself | Multiple sclerosis | FDA/EMA/PMDA-approved |
| Vivagel | Modified p(Lys) dendrimer | Peptide itself | Bacterial vaginosis | III |

multifunctionality, biocompatibility, and biodegradability make them ideal materials for the transport of bioactive substances in the body.

### 41.4.3 Polyglycerols

An important class of biocompatible branched polymers are polyglycerols (PGs), which can be synthesized as dendrons, dendrimers, or with hyperbranched architectures [153–155].

When compared to linear polymers of similar molecular weight, dendrimers and hyperbranched polymers have interesting properties like globular structure and low intrinsic viscosities [156]. The chemical structure of PG is similar to that of PEG with which PGs share their low affinity to proteins [157]. But while PEG only offers two sites for modification, PG has many free OH-groups which might be used to attach bioactive molecules or targeting moieties.

While dendrons and dendrimers are synthesized in multistep processes [158, 159], well defined hyperbranched polyglycerols can be prepared via anionic polymerization of glycidol [160, 161]. The polymers can be modified in the core [162] or in the shell by chemically addressing the alcohol groups in the corona [163].

For linear and hyperbranched PG (molecular weight 6.4 kg mol$^{-1}$) complement activation, red blood cell aggregation and coagulation were not observed to a significant extend. Cytotoxicity of the PGs was comparable to the controls PEG and hetastarch (80 % cell viability at a concentration of 10 mg mL$^{-1}$; L-929 fibroblasts). Hyperbranched polyglycerols were found to be well tolerated by mice when injected in doses of up to 62.5 mg kg$^{-1}$ [164].

The biocompatibility of larger hyperbranched PGs (160 and 540 kg mol$^{-1}$) was investigated in vitro [165] and in vivo [166]. The in vitro tests were similar to the ones performed with low molecular weight PG and came to a similar conclusion, i.e., that PGs are biocompatible. These polymers were tested in vivo in mice for their pharmacokinetic behavior. The plasma half-lives were 32 h and 57 h, respectively. No toxicity was observed with a dose of 1 g kg$^{-1}$. But due to the high molecular weight of the PGs, urinary excretion was very low, and significant accumulation by the mononuclear phagocyte system was observed. PGs were detectable in the liver and spleen for at least 30 days. This again highlights the need for biodegradable materials.

Due to their good biocompatibility, there are plenty of biomedical applications of PGs [38, 155, 167]. Bioactive compounds can be transported by PG either through covalent attachment or encapsulation.

For covalent attachment, there has been great interest to add functionality to the alcohol-terminated PG. The introduction of amine groups

[163], for example, allows the attachment of acids via an amide, while thiols on the PG surface have been used to attach drugs via a biodegradable peptide linker [168]. Doxorubicin has been attached to PG via an acid cleavable linker. The conjugate was tested in vitro and in an ovarian carcinoma A2780 xenograft model, showing a distinct increase in tolerability and antitumor efficacy with tumor remissions for up to 30 days [169]. Also, alkyne groups have been introduced for site-selective *click chemistry*, which has been employed to synthesize sugar-terminated PGs with high binding affinities to selectins [170]. Equal results have been found for negatively charged PG sulfates [171, 172]. Such systems use their multivalency to achieve high binding affinities.

The non-degradability of PG was partially addressed by synthesizing PG in an inverse miniemulsion process via an acid-catalyzed ring-opening polyaddition of disulfide containing polyols and polyepoxides [173]. These nanoparticles can be cleaved in a reductive environment yielding smaller fragments that can be excreted from the body.

The application of PGs also includes hydrogel scaffolds for tissue engineering, which will not be discussed here.

Since dendrimers are quite small (1–5 nm) and hyperbranched architectures are not much bigger (2–20 nm), assembly into superstructures is desirable to get into a size range of 20–100 nm.

Microgels were prepared using the inverse miniemulsion technique [174]. The particles were between 23 and 80 nm in diameter and seemed to accumulate in the perinuclear region of human lung cancer cells. An MTT essay showed no influence of the microgel on human hematopoietic U-937 cell metabolic activity below 0.5 mg mL$^{-1}$.

Crosslinked, hyperbranched PG, so-called megamers, were prepared from PG modified with azides and alkynes, using the miniemulsion technique [175]. Hydrophilic as well as hydrophobic megamers were prepared with diameters of 25–90 nm.

With their low protein adsorption and their good biocompatibility, polyglycerols are a promising class of materials, which have already proven their benefits in a wide range of biomedical applications. In the future they promise to lead to more advanced, highly functionalizable transport systems of bioactive molecules.

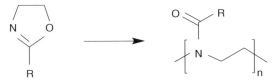

**Fig. 41.6** Synthesis of polyoxazolines

### 41.4.4 Polyoxazolines

Poly(2-oxazoline)s (POx) can also be expected to be one of the upcoming polymer platforms for nanomedicines [176, 177]. POx meet the specific requirements needed for the development of next-generation nanomedicines such as biocompatibility, high modulation of solubility, variation of size, architecture, as well as chemical functionality. POx are synthesized by cationic ring-opening polymerization of 2-oxazolines (see Fig. 41.6). This methodology offers precise control over molecular weight and chain-end functionality ($Đ \leq 1.2$) [178] and therefore allows chemists to adjust the polymers to meet the necessary requirements.

POx have an LCST, a temperature below which the polymers are fully soluble in aqueous solution. Above this temperature they become insoluble, leading to polymer precipitation. The LCST of POx depends on the polymer composition and structure and can be adjusted by copolymerization of the monomers nPrOx with iPrOx from about 26 to 38.7 °C and with the even more hydrophilic EtOx from 35 to 75.1 °C. Those properties may enable the specific accumulation by precipitation of polymers in parts of the body with higher temperature induced externally (hyperthermia). Furthermore, POx fulfill several important aspects of polymers used for in vivo applications. Considering in vitro cytotoxicity, different POx-based polymers were found to be nonhazardous in general [179–182]. In addition to that, PEtOx exhibited no hemolytic effects at 20 g L$^{-1}$ [183]. Results for hybrid systems with a

POx part are variable. The triblock copolymers poly(L-lactide)-*b*-PEtOx-*b*-poly(L-lactide) and PEtOx-*b*-PCL-*b*-PEtOx were found to be noncytotoxic at concentrations of up to 50 g L$^{-1}$ and 10 g L$^{-1}$, respectively [184, 185]. Contrary to that, PEtOx-*b*-poly(ε-caprolactone) copolymers showed explicit cytotoxicity at concentration of below 1 g L$^{-1}$, although in another experiment of the same study, no hemolytic effects of PEtOx-b-PCL were observed at 10 g L$^{-1}$ [186].

Previous studies on in vivo toxicity of poly(2-oxazoline)s focused mainly on PEtOx and PMeOx. Viegas et al. intravenously injected repeated doses of 10 kg mol$^{-1}$ PEtOx (2 g kg$^{-1}$) and 50 kg mol$^{-1}$ (50 mg kg$^{-1}$) into rats. No adverse effects were observed, and histology of kidney, liver, and spleen showed no abnormalities [183].

Similar to other polymer systems, the susceptibility of poly(2-oxazoline)s to interaction with proteins depends on their hydrophilicity or hydrophobicity. Overall, hydrophilic poly(2-oxazoline)s like PMeOx and PEtOx show low affinity to protein adsorption [187–190]. Consequently, they exhibit stealth properties and may be used to mediate them, for example, as coating moiety for liposomes [191, 192]. The biodistribution profile of PMeOx and PEtOx is consistent with their stealth properties, i.e., low organ uptake and rapid or slow elimination through the kidneys depending on the molecular weight [193, 194].

With regard to potential immunogenicity, different block copolymers of PMeOx, PEtOx, PBuOx, and PPheOx were screened for their capability to activate the complement system. Compared to negative controls, levels of complement proteins were slightly elevated [181, 195]. Although a lot of research has been conducted, further clinical data is necessary to underline the potential of POx, making it a substitute for PEG in general.

## 41.5 Characterization of Nanoparticles

Besides developing different kinds of nanoparticles based on novel materials, another focus needs to be set on the characterization of nanomedicines, especially when those systems are applied in targeted delivery or vaccination.

The synthesized particles can be characterized in solution by (cryo-) transmission electron microscopy (TEM), scanning electron microscopy (SEM), or atomic force microscopy (AFM). Furthermore, light scattering techniques as well as fluorescence correlation spectroscopy can be used to determine hydrodynamic diameters of particles. All these tools, however, are somehow limited to artificial conditions and thus often cannot provide information regarding protein adsorption or protein-induced aggregation.

Especially in vaccination, when a nanoparticle should deliver an antigen and adjuvant to an antigen-presenting cell, a protein corona may cause an immune reaction against the undesirably absorbed serum proteins causing autoimmune diseases. To tackle those important questions, novel methodologies are required and some of them will be briefly described.

As stated above, the absorption of proteins might not be hindered by charge-stabilized nanoparticles, since proteins can be regarded as polyelectrolytes containing anionic or cationic groups. In salt-containing solution, charges are compensated by counterions. Those ions are set free whenever a multivalent binding partner is present since the systems entropy is increased by the freely diffusing ions. Though, enthalpy contributions to the system's energy also play a role, for macromolecules, usually entropic contributions are the major factor. Thus, steric stabilization of nanoparticles appears reasonable, but in all cases protein adsorption needs to be investigated carefully.

Often, particles are characterized in aqueous solution at low ionic strength and in absence of proteins by measuring and interpreting electrokinetic phenomena. In many cases zeta (ζ) potential measurements are carried out using commercial systems. The ζ-potential itself cannot be measured but has to be derived from electrokinetic signals using sophisticated models considering surface properties such as nature of the surface, roughness, charge, the electrolyte concentration, and the nature of the electrolyte and of the solvent used in the measurements. Since those parameters differ easily from one particle to the other, an IUPAC technical report highly recommends to use this methodology carefully and not to over-interpret the results obtained [196]. However, the ζ-potential can be a helpful tool in quality control of the particles synthesized.

More suitable methodologies may be dynamic light scattering in serum to determine changes in hydrodynamic radius as reported by Rausch and co-workers [28, 197]. This method allows studying the aggregation behavior of nanoparticles in serum. Even small traces of aggregated particles can be identified. The drawback of this methodology is that the adsorption of a single or few proteins cannot be detected since the changes in hydrodynamic radius are below the accuracy of measurements.

In a more artificial setup, isothermal titration calorimetry (ITC) can be used to measure the heat flow and therefore quantify an average adsorption energy over all particles [198–200]. In combination with quantitative label-free liquid chromatography and mass spectrometry (MS), adsorbed proteins can be identified and quantified [201].

In addition, in vivo imaging of the biodistribution of nanoparticles is necessary to monitor local accumulation at the target side or to ensure excretion from the organism. This requires nanoparticles to be labeled with near-infrared dyes [202] or radioactive isotopes [203] for single photon emission computed tomography (SPECT) or positron emission tomography (PET). During the last decade, various methodologies have been developed to label nanoparticles. Labeling methods for polymers using prosthetic groups for $^{18}$F as well as $^{72/74}$Ar have been established [204–206].

In summary, novel methodologies have been developed and may shine light on the behavior of nanoparticles in vivo and therefore facilitate the development of next-generation nanomedicines.

## 41.6 Nanoparticulate Vaccine Formulations

### 41.6.1 Advantages of Particle-Based Vaccines

The next years will see an increasing use of synthetic subunit antigen vaccines because they provide many advantages over traditional vaccines derived from live attenuated, killed, or inactivated pathogens [207]. However, it is a well-known fact that synthetic peptides often lack the immunogenicity to induce a strong immune response. Therefore, efficient vaccination requires the use of adjuvants [208]. In the field of tumor immunotherapy, adjuvants are especially important to reverse tumor immunosuppression mediated, for example, by regulatory T cells or myeloid-derived suppressor cells.

Adjuvants, among other things, promote the maturation of antigen-presenting cells by interaction with receptors of the innate immune system, e.g., dendritic cells. Only mature dendritic cells are able to generate co-stimulatory signals, which together with the presented antigen lead to the desired effector T cell responses. Thus, it is important that antigen and adjuvant act on dendritic cells concomitantly. Most of today's human vaccine formulations use aluminum-based compounds as adjuvants. However, as recent years have seen intense research for more potent and specific alternatives [209], new means of co-delivery are required.

Another problem of subunit antigen vaccines is their susceptibility to enzymatic degradation. To increase their lifetime and biological activity, protective carrier molecules are needed. Nanoparticles have become of great interest as an instrument to protect antigens and adjuvants against blood serum proteins and to ensure their safe and simultaneous delivery to dendritic cells.

Alongside their capability for protection [210] and co-delivery of antigen and adjuvant, nanoparticles show several other advantages:
- They can deliver an antigen together with combinations of adjuvants overcoming immune suppression by regulatory T cells [211–213].
- They trap their cargo, thus preventing systemic distribution and toxicity [214].
- They can be used as a platform to present multiple copies of an antigen on their surface which especially improves B cell activation [215].
- They can promote the cytosolic accumulation of antigens in dendritic cells. This is important

for cross-presentation, the key mechanism for CD8+ T cell activation.
- Modification of nanoparticles with inert substances like poly(ethylene glycol) that mediate stealth properties [216] can minimize protein absorption. This increases their blood circulation time [217].
- Surface functionalization of nanoparticles with receptor ligands or antibodies enables specific tissue or cellular targeting [217]. Thereby, antigen and adjuvant are retained at the site of action so that unspecific distribution is reduced, lower doses are needed, and adverse reactions to vaccination are minimized.
- Nanoparticles provide a sustained release of adjuvants and thus a prolonged exposure to the immune system which is essential for a proper activation of dendritic cells [218].

## 41.6.2 Approaches

Big potential for nanoparticulate vaccine formulations most likely lies in cancer immunotherapy. In addition to this, applications for vaccination against various infectious diseases are investigated. These include, for example, nanoparticle-based vaccines against influenza [219], West Nile encephalitis [220], and malaria [221].

The components of a typical nanoparticulate vaccine formulation combine an antigen (T cell and/or B cell epitopes), adjuvant, and targeting moiety on one particle [214]. A wide variety of antigens such as peptides/proteins [222], lipoproteins [223], glycopeptides [224], tumor-associated carbohydrate antigens (TACA) [225], and carbohydrate mimetic peptides [226] are tested in combination with nanoparticulate carriers.

Two basic concepts can be distinguished for targeting dendritic cells with antigens: ex vivo pulsing and subsequent administration of matured dendritic cells or direct in vivo targeting of immature dendritic cells. Nanoparticles appear valuable for in vivo strategies [227].

The group of adjuvants mainly studied in the context of nanoparticulate vaccines are several Toll-like receptor ligands. LPS, the major component of the outer membrane of Gram-negative bacteria, is a pathogen-derived TLR 4 ligand. While being far too toxic for systemic application as an immune potentiator for classic vaccines, LPS shows promising results in vitro and in mouse models, when administered within PLGA nanoparticles [228]. 7-acyl lipid A, a component of LPS, is also investigated in mice as an additive for PLGA-based vaccines [229]. In addition to pathogen-derived TLR ligands, synthetic molecules mimicking pathogen-associated molecular patterns (PAMPs) are evaluated. Representatives of this group are CpG oligodeoxynucleotides (CpG ODN), binding to TLR 9, and polyribosinic:polyribocytidic acid (Poly I:C), a TLR 3 ligand [211]. Alongside the general activation of dendritic cells, the application of certain TLR ligands like CpG oligodeoxynucleotides and LPS together with nanoparticulate carriers might be of special interest as they, contrary to alum, elicit a Th1-biased immune response [230].

Demento et al. developed PLGA nanoparticles loaded with a recombinant envelope protein antigen from the West Nile virus and surface functionalized with CpG ODN. In several experiments they compared this system with unmodified nanoparticles and an aluminum hydroxide vaccine formulation. In mice, CpG ODN modified nanoparticles produced a Th1 immune response biased antibody profile in contrast to a Th2 profile elicited by aluminum hydroxide. Moreover, immunization of mice with modified nanoparticles induced a higher number of circulating effector T cells and an enhanced antigen-specific IL-2 and IFN-γ production compared to control groups. In a mouse model of West Nile encephalitis, modified nanoparticles showed a superior protection in comparison to aluminum hydroxide [220].

Studies suggest that some nanoparticles, probably by binding to Toll-like receptors, can act as adjuvants themselves [231]. This has been shown for high molecular weight γ-PGA particles [232], PLGA nano- and microparticles [233], poly(anhydride) nanoparticles [234], liposomes [235], polystyrene nano- and microparticles [236], and acid-degradable cationic polyacrylamide nanoparticles [237]. Though, particle size and concentration seem to be critical for activation of dendritic cells as in other experiments, PLGA particles failed to induce dendritic cell maturation [238].

In one experiment, mucosal immunization of mice against highly pathogenic influenza A H5N1 virus was carried out. The vaccine consisted of recombinant influenza hemagglutinin (rHA) antigen and γ-PGA/chitosan nanoparticles. This system was compared to two other formulations: inactivated virus with γ-PGA/chitosan nanoparticles and cholera toxin with recombinant influenza hemagglutinin antigen. The γ-PGA/chitosan nanoparticles in combination with rHA antigen led to potent virus-specific humoral and cellular immune responses. Compared to cholera toxin the adjuvant properties of the γ-PGA/chitosan nanoparticles were equally good. Immunized mice were protected against challenge with a lethal dose of highly pathogenic influenza A H5N1 virus [239].

A different concept for nanoparticle-based vaccines investigated in multiple studies [240, 241] involves the delivery of nucleic acids (mRNA or pDNA) encoding a defined antigen into dendritic cells. In contrast to peptide antigens, mRNA and pDNA have to remain intact and functional all along the endocytic route until they have reached the cytosol or the nucleus, respectively. After being endocytosed, nanoparticles functionalized with buffering groups or cell penetrating peptides can protect nucleic acids against hydrolytic enzymes and foster endosomal escape [242]. Once nucleic acids have reached their destination inside the cell, expression of the encoded peptide antigen can take place.

One advantage of pDNA is that it can encode for various peptide antigens. Thus, one nanoparticulate system that is able to transport pDNA can be used for a broad range of vaccinations. Further, pDNA can function as an immunostimulant because it contains CpG sequences recognized by TLR 9 [243].

Sudowe et al. investigated the potential of in vivo pDNA vaccination to generate an immune response against β-Gal as model antigen. Plasmids encoding β-Gal under the control of DC-specific, keratinocyte-specific, or unspecific promoters were adsorbed on gold nanoparticles and applied to the skin of mice via gene gun. As a result, Th1-biased immune responses were observed. Moreover, with regard to CD8$^+$ T cell priming, it was shown that the presentation of endogenous antigen produced by transfected skin dendritic cells is equally effective as the cross-presentation of exogenous antigen [244].

As mentioned above, the surface of nanocarriers can be functionalized with a multitude of different ligands, including proteins, peptides, antibodies, antibody fragments, aptamers, and small molecules [217]. By targeting defined cell surface structures with receptor ligands or antibodies attached to nanoparticles, an increased and/or more specific cellular uptake might be achieved. Considering vaccine formulations in particular, the configuration of the carrier system with ligands for pattern recognition receptors, e.g., Toll-like receptors or C-type lectin receptors, might augment its immunogenicity and contribute to dendritic cell activation [241]. Different targeting concepts have been realized with various particulate carriers and have shown stronger immune responses compared to nontargeted carriers [245]. Mannose is a ligand for a C-type lectin receptor expressed on dendritic cells. Multiple research projects have examined its performance for targeting dendritic cells and macrophages [44, 235, 246–251]. In experiments comparing mannose-decorated PLGA particles with plain PLGA particles, it was shown that mannose only causes a slight increase in cellular uptake. Presumably, because depending on the experimental conditions, nonspecific uptake usually predominates specific uptake [252]. However, in another study it was demonstrated that mannan, a natural polymannose derived from the cell wall of *Saccharomyces cerevisiae*, when conjugated to PLGA particles leads to increased antigen-specific CD4$^+$ and CD8$^+$ T cell responses [253].

Other methods for addressing dendritic cells reported in literature are the deployment of Fc receptor ligands or anti-DEC-205 antibodies. Raghuwanshi et al. evaluated a two-component vaccine formulation, one element being biotinylated PLGA nanoparticles loaded with ovalbumin as model antigen and the other a bifunctional fusion protein consistent of truncated core-streptavidin and an anti-DEC-205 single-chain antibody. They observed a twofold increase in receptor-mediated uptake of functionalized nanoparticles compared to nontargeted nanoparticles. OVA-specific IgG responses of mice immunized with anti-DEC-205-modified nanoparticles in connection with αCD40 as adjuvant were considerably higher than those of control groups [254].

To trigger an antibody response, an antigen has to be recognized and processed not only by dendritic cells but also by B cells. Consequently, alongside T cell epitopes, the inclusion of B cell epitopes into a vaccine formulation is a central approach. As mentioned before, nanoparticles can be used to present multiple copies of epitopes on their surface [215]. This enables the induction of T cell-independent type I antibody responses as has been shown for virus-like particles [255]. T cell-independent type II antibody responses can be realized through co-delivery of adjuvants suited for B cell activation [256].

Especially in tumor immunotherapy, the delivery of B cell epitopes derived from glycopeptides, TACA, or carbohydrate mimetic peptides is of great interest. In several studies Kunz et al. investigated the B cell activating potential of different synthetic tumor-associated MUC1 glycopeptides which contain the Thomsen-Friedenreich antigen, its precursor or its sialylated derivatives. They chemically linked the glycopeptides to bovine serum albumin (BSA) or tetanus toxoid as carrier molecules. These vaccine systems were applied to Balb/c mice. In the case of the BSA formulations, complete Freud's adjuvant was co-administered. For all vaccines strong specific antibody responses against the MUC1 antigens were observed [257–259].

Brinãs et al. used gold nanoparticles to co-deliver MUC4 glycopeptides containing the Thomsen-Friedenreich antigen together with a peptide derived from the complement protein C3d as B cell activating adjuvant. Mice immunized with these nanoparticles showed statistically significant immune responses [256].

## 41.7 Summary and Conclusions

Nanoparticles have already underlined their potential in medical application. Their particulate properties may be especially attractive in the stimulation of the immune system. Nanoparticle-based vaccines can be expected to have a major impact on future vaccines, because they bear the potential of directed delivery of sensitive antigens or combinations of antigen or antigen/adjuvant in one moiety. Thus, immune modulation may be enhanced. However, the synthesis of nanoparticles is still demanding, but new tools increase the control of particle properties and may allow the synthesis of optimized immune modulating nanomedicines for effective vaccination approaches.

## References

1. Duffus, J.H., Nordberg, M., Templeton, D.M.: Glossary of terms used in toxicology, 2nd edition (IUPAC Recommendations 2007). Pure Appl. Chem. **79**, 1153–1344 (2007)
2. Duncan, R., Gaspar, R.: Nanomedicine(s) under the microscope. Mol. Pharm. **8**, 2101–2141 (2011)
3. Johnston, A.P.R., Such, G.K., Ng, S.L., Caruso, F.: Challenges facing colloidal delivery systems: from synthesis to the clinic. Curr. Opin. Colloid In.. **16**, 171–181 (2011)
4. Torchilin, V.P.: Micellar nanocarriers: pharmaceutical perspectives. Pharm. Res. **24**, 1–16 (2007)
5. Lasic, D.D., Martin, F.J.: Stealth Liposomes. CRC Press, Boca Raton (1995)
6. Hobbs, S.K., et al.: Regulation of transport pathways in tumor vessels: role of tumor type and microenvironment. Proc. Natl. Acad. Sci. U. S. A. **95**, 4607–4612 (1998)
7. Torchilin, V.P.: Targeted polymeric micelles for delivery of poorly soluble drugs. Cell. Mol. Life Sci. **61**, 2549–2559 (2004)
8. Nagasaki, Y., Yasugi, K., Yamamoto, Y., Harada, A., Kataoka, K.: Sugar-installed block copolymer micelles: their preparation and specific interaction with lectin molecules. Biomacromolecules **2**, 1067–1070 (2001)
9. Ogris, M., Brunner, S., Schuller, S., Kircheis, R., Wagner, E.: PEGylated DNA/transferrin-PEI complexes: reduced interaction with blood components, extended circulation in blood and potential for systemic gene delivery. Gene Ther. **6**, 595–605 (1999)
10. Leamon, C.P., Weigl, D., Hendren, R.W.: Folate copolymer-mediated transfection of cultured cells. Bioconjug. Chem. **10**, 947–957 (1999)

11. Hopewell, J.W., Duncan, R., Wilding, D., Chakrabarti, K.: Preclinical evaluation of the cardiotoxicity of PK2: a novel HPMA copolymer–doxorubicin–galactosamine conjugate antitumour agent. Hum. Exp. Toxicol. **20**, 461–470 (2001)
12. Ahmed, F., et al.: Shrinkage of a rapidly growing tumor by drug-loaded polymersomes: pH-triggered release through copolymer degradation. Mol. Pharm. **3**, 340–350 (2006)
13. Saito, G., Swanson, J.A., Lee, K.-D.: Drug delivery strategy utilizing conjugation via reversible disulfide linkages: role and site of cellular reducing activities. Adv. Drug Deliv. Rev. **55**, 199–215 (2003)
14. Thornton, P.D., Mart, R.J., Webb, S.J., Ulijn, R.V.: Enzyme-responsive hydrogel particles for the controlled release of proteins: designing peptide actuators to match payload. Soft Matter **4**, 821–827 (2008)
15. Ishida, O., Maruyama, K., Yanagie, H., Iwatsuru, M., Eriguchi, M.: Targeting chemotherapy to solid tumors with long circulating thermosensitive liposomes and local hyperthermia. Jpn. J. Cancer Res. **91**, 118–126 (2000)
16. Edelman, E.R., Kost, J., Bobeck, H., Langer, R.: Regulation of drug release from polymer matrices by oscillating magnetic fields. J. Biomed. Mater. Res. **19**, 67–83 (1985)
17. Langer, R.: New methods of drug delivery. Science **249**, 1527–1533 (1990)
18. Uhrich, K.E., Cannizzaro, S.M., Langer, R.S., Shakesheff, K.M.: Polymeric systems for controlled drug release. Chem. Rev. **99**, 3181–3198 (1999)
19. Little, S.R.: Reorienting our view of particle-based adjuvants for subunit vaccines. Proc. Natl. Acad. Sci. **109**, 999–1000 (2012)
20. Moon, J.J., et al.: Enhancing humoral responses to a malaria antigen with nanoparticle vaccines that expand Tfh cells and promote germinal center induction. Proc. Natl. Acad. Sci. U. S. A. **109**, 1080–1085 (2012)
21. Prokop, A., Davidson, J.M.: Nanovehicular intracellular delivery systems. J. Pharm. Sci. **97**, 3518–3590 (2008)
22. Gref, R., et al.: Biodegradable long-circulating polymeric nanospheres. Science **263**, 1600–1603 (1994)
23. Owens, D.E., Peppas, N.A.: Opsonization, biodistribution, and pharmacokinetics of polymeric nanoparticles. Int. J. Pharm. **307**, 93–102 (2006)
24. Frank, M., Fries, L.: The role of complement in inflammation and phagocytosis. Immunol. Today **12**, 322–326 (1991)
25. Johnson, R.J.: The complement system. In: Ratner, B.D., Hoffman, A.S., Schoen, F.J., Lemons, J.E. (eds.) Biomaterials Science: An Introduction to Materials in Medicine, pp. 318–328. Elsevier/Academic, Amsterdam (2004)
26. Ostuni, E., Chapman, R.G., Holmlin, R.E., Takayama, S., Whitesides, G.M.: A survey of structure–property relationships of surfaces that resist the adsorption of protein. Langmuir **17**, 5605–5620 (2001)
27. Scheibe, P., Barz, M., Hemmelmann, M., Zentel, R.: Langmuir-Blodgett films of biocompatible poly(HPMA)-block-poly(lauryl methacrylate) and poly(HPMA)-random-poly(lauryl methacrylate): influence of polymer structure on membrane formation and stability. Langmuir **26**, 5661–5669 (2010)
28. Kelsch, A., et al.: HPMA copolymers as surfactants in the preparation of biocompatible nanoparticles for biomedical application. Biomacromolecules **13**, 4179–4187 (2012)
29. Riess, G.: Micellization of block copolymers. Prog. Polym. Sci. **28**, 1107–1170 (2003)
30. O'Reilly, R.K., Hawker, C.J., Wooley, K.L.: Cross-linked block copolymer micelles: functional nanostructures of great potential and versatility. Chem. Soc. Rev. **35**, 1068–1083 (2006)
31. Kabanov, A.V., Vinogradov, S.V.: Nanogels as pharmaceutical carriers: finite networks of infinite capabilities. Angew. Chem. Int. Ed. Engl. **48**, 5418–5429 (2009)
32. Christie, R.J., Nishiyama, N., Kataoka, K.: Delivering the code: polyplex carriers for deoxyribonucleic acid and ribonucleic acid interference therapies. Endocrinology **151**, 466–473 (2010)
33. Miyata, K., Nishiyama, N., Kataoka, K.: Rational design of smart supramolecular assemblies for gene delivery: chemical challenges in the creation of artificial viruses. Chem. Soc. Rev. **41**, 2562–2574 (2012)
34. Kataoka, K., Harada, A., Nagasaki, Y.: Block copolymer micelles for drug delivery: design, characterization and biological significance. Adv. Drug Deliv. Rev. **47**, 113–131 (2001)
35. Torchilin, V.P.: Structure and design of polymeric surfactant-based drug delivery systems. J. Control. Release **73**, 137–172 (2001)
36. Gaucher, G., et al.: Block copolymer micelles: preparation, characterization and application in drug delivery. J. Control. Release **109**, 169–188 (2005)
37. Nuhn, L., et al.: Cationic nanohydrogel particles as potential siRNA carriers for cellular delivery. ACS Nano **6**, 2198–2214 (2012)
38. Fleige, E., Quadir, M.A., Haag, R.: Stimuli-responsive polymeric nanocarriers for the controlled transport of active compounds: concepts and applications. Adv. Drug Deliv. Rev. **64**, 866–884 (2012)
39. Jesorka, A., Orwar, O.: Liposomes: technologies and analytical applications. Annu. Rev. Anal. Chem. **1**, 801–832 (2008)
40. Torchilin, V.P.: Recent advances with liposomes as pharmaceutical carriers. Nat. Rev. Drug Discov. **4**, 145–160 (2005)
41. Szoka, F.C.: Comparative properties and methods of preparation of lipid vesicles (liposomes). Ann. Rev. Biophys. Bioeng. **9**, 467–508 (1980)
42. Lasic, D.D.: Sterically stabilized vesicles. Angew. Chem. Int. Ed. Engl. **33**, 1685–1698 (1994)
43. Allen, T.M., Cullis, P.R.: Drug delivery systems: entering the mainstream. Science **303**, 1818–1822 (2004)
44. White, K.L., Rades, T., Furneaux, R.H., Tyler, P.C., Hook, S.: Mannosylated liposomes as antigen deliv-

ery vehicles for targeting to dendritic cells. J. Pharm. Pharmacol. **58**, 729–737 (2006)
45. Drummond, D.C., Meyer, O., Hong, K., Kirpotin, D.B., Papahadjopoulos, D.: Optimizing liposomes for delivery of chemotherapeutic agents to solid tumors. Pharmacol. Rev. **51**, 691–744 (1999)
46. Krishnamachari, Y., Geary, S.M., Lemke, C.D., Salem, A.K.: Nanoparticle delivery systems in cancer vaccines. Pharm. Res. **28**, 215–236 (2011)
47. Gluck, R.: Immunopotentiating reconstituted influenza virosomes (IRIVs) and other adjuvants for improved presentation of small antigens. Vaccine **10**, 915–919 (1992)
48. Moser, C., et al.: Influenza virosomes as a combined vaccine carrier and adjuvant system for prophylactic and therapeutic immunizations. Expert Rev. Vaccines **6**, 711–721 (2007)
49. Kayser, O., Olbrich, C., Croft, S.L., Kiderlein, A.F.: Formulation and biopharmaceutical issues in the development of drug delivery systems for antiparasitic drugs. Parasitol. Res. **90**, S63–S70 (2003)
50. Discher, D.E., Ahmed, F.: Polymersomes. Annu. Rev. Biomed. Eng. **8**, 323–341 (2006)
51. Levine, D.H., et al.: Polymersomes: a new multifunctional tool for cancer diagnosis and therapy. Methods **46**, 25–32 (2008)
52. Osterhaus, A., Rimmelzwaan, G.F.: Induction of virus-specific immunity by ISCOMs. Dev. Biol. Stand. **92**, 49–58 (1998)
53. Saupe, A., McBurney, W., Rades, T., Hook, S.: Immunostimulatory colloidal delivery systems for cancer vaccines. Expert Opin. Drug Deliv. **3**, 345–354 (2006)
54. Westesen, K., Siekmann, B.: Biodegradable colloidal drug carrier systems based on solid lipids. In: Benita, S. (ed.) Microencapsulation, pp. 213–258. Marcel Dekker, New York (1996)
55. Bunjes, H.: Lipid nanoparticles for the delivery of poorly water-soluble drugs. J. Pharm. Pharmacol. **62**, 1637–1645 (2010)
56. Petersen, S., Steiniger, F., Fischer, D., Fahr, A., Bunjes, H.: The physical state of lipid nanoparticles affects their in vitro cell viability. Eur. J. Pharm. Biopharm. **79**, 150–161 (2011)
57. Shi, R., et al.: Enhanced immune response to gastric cancer specific antigen peptide by coencapsulation with CpG oligodeoxynucleotides in nanoemulsion. Cancer Biol. Ther. **4**, 218–242 (2005)
58. Gupta, S., Moulik, S.P.: Biocompatible microemulsions and their prospective uses in drug delivery. J. Pharm. Sci. **97**, 22–45 (2008)
59. Fanun, M.: Microemulsions as delivery systems. Curr. Opin. Colloid In. **17**, 306–313 (2012)
60. Sailaja, A.K., Amareshwar, P., Chakravarty, P.: Chitosan nanoparticles as a drug delivery system. Res. J. Pharm. Biol. Chem. Sci. **1**, 474–484 (2010)
61. Lohse, S.E., Murphy, C.J.: Applications of colloidal inorganic nanoparticles: from medicine to energy. J. Am. Chem. Soc. **134**, 15607–15620 (2012)
62. Landfester, K.: Synthesis of colloidal particles in miniemulsions. Annu. Rev. Mater. Res. **36**, 231–279 (2006)
63. Klinger, D., Landfester, K.: Stimuli-responsive microgels for the loading and release of functional compounds: Fundamental concepts and applications. Polymer **53**, 5209–5231 (2012)
64. Ugelstad, J., Mork, P.C., Kaggerud, K.H., Ellingsen, T., Berge, A.: Swelling of oligomer-polymer particles: new method of preparation of emulsions and polymer dispersions. Adv. Colloid Interface Sci. **13**, 101–140 (1980)
65. Chern, C.S., Chen, T.J., Liou, Y.C.: Miniemulsion polymerization of styrene in the presence of a water-insoluble blue dye. Polymer **37**, 3767–3777 (1998)
66. Reimers, J.L., Schork, F.J.: Lauroyl peroxide as a cosurfactant in miniemulsion polymerization. Ind. Eng. Chem. Res. **36**, 1085–1087 (1997)
67. Landfester, K.: Recent developments in miniemulsions – formation and stability mechanisms. Macromol. Symp. **150**, 171–178 (2000)
68. Tamber, H., Johansen, P., Merkle, H.P., Gander, B.: Formulation aspects of biodegradable polymeric microspheres for antigen delivery. Adv. Drug Deliv. Rev. **57**, 357–376 (2005)
69. Mok, H., Park, T.G.: Direct plasmid DNA encapsulation within PLGA nanospheres by single oil-in-water emulsion method. Eur. J. Pharm. Biopharm. **68**, 105–111 (2008)
70. Meyer, J.D., Manning, M.C.: Hydrophobic ion pairing: altering the solubility properties of biomolecules. Pharm. Res. **15**, 188–193 (1998)
71. Kazzaz, J., Neidleman, J., Singh, M., Ott, G., O'Hagan, D.T.: Novel anionic microparticles are a potent adjuvant for the induction of cytotoxic T lymphocytes against recombinant p55 gag from HIV-1. J. Control. Release **67**, 347–356 (2000)
72. Schwendeman, S.P.: Recent advances in the stabilization of proteins encapsulated in injectable PLGA delivery systems. Crit. Rev. Ther. Drug Carr. Syst. **19**, 73–98 (2002)
73. Barz, M., et al.: Synthesis, characterization and preliminary biological evaluation of P(HPMA)-b-P(LLA) copolymers: a new type of functional biocompatible block copolymer. Macromol. Rapid Comm. **31**, 1492–1500 (2010)
74. Barz, M., et al.: P(HPMA)-block-P(LA) copolymers in paclitaxel formulations: polylactide stereochemistry controls micellization, cellular uptake kinetics, intracellular localization and drug efficiency. J. Control. Release **163**, 63–74 (2012)
75. Aspinall, G.O.: The Polysaccharides 35. Academic, New York (1982)
76. Leonard, M., et al.: Preparation of polysaccharide-covered polymeric nanoparticles by several processes involving amphiphilic polysaccharides. ACS Symp. Ser. **996**, 322–340 (2008)
77. Artursson, P., Lindmark, T., Davis, S., Illum, L.: Effect of chitosan on the permeability of monolayers of intestinal epithelial-cells (Caco-2). Pharm. Res. **11**, 1358–1361 (1994)

78. Domard, A., Gey, C., Rinaudo, M., Terrassin, C., et al.: C-13 and H-1-NMR spectroscopy of chitosan and Ntrimethyl chloride derivates. Int. J. Biol. Macromol. **9**, 233–237 (1987)
79. Sundar, S., Kundu, J., Kundu, S.C.: Biopolymeric nanoparticles. Sci. Technol. Adv. Mater. (11) (2010)
80. Schultze, V., et al.: Safety of MF59(TM) adjuvant. Vaccine **26**, 3209–3222 (2008)
81. Makidon, P.E., et al.: Pre-clinical evaluation of a novel nanoemulsion-based hepatitis B mucosal vaccine. PLoS One **3**, e2954 (2008)
82. Bielinska, A.U., et al.: Nasal immunization with a recombinant HIV gp120 and nanoemulsion adjuvant produces Th1 polarized responses and neutralizing antibodies to primary HIV type 1 isolates. AIDS Res. Hum. Retrov. **24**, 271–281 (2008)
83. Ge, W., et al.: The antitumor immune responses induced by nanoemulsion encapsulated MAGE1-HSP70/SEA complex protein vaccine following different administration routes. Oncol. Rep. **22**, 915–920 (2009)
84. Rolland, J.P., et al.: Direct fabrication and harvesting of monodisperse. Shape-specific nanobiomaterials. J. Am. Chem. Soc. **127**, 10096–10100 (2005)
85. Gratton, S.E., et al.: Nanofabricated particles for engineered drug therapies: a preliminary biodistribution study of PRINT nanoparticles. J. Control. Release **121**, 10–18 (2007)
86. Dunn, S.S., et al.: Reductively responsive siRNA-conjugated hydrogel nanoparticles for gene silencing. J. Am. Chem. Soc. **134**, 7423–7430 (2012)
87. Laurent, S., et al.: Magnetic iron oxide nanoparticles: synthesis, stabilization, vectorization, physicochemical characterizations, and biological applications. Chem. Rev. **108**, 2064–2110 (2008)
88. Dahl, J.A., Maddux, B.L.S., Hutchison, J.E.: Toward greener nanosynthesis. Chem. Rev. **107**, 2228–2269 (2007)
89. Carageorgheopol, A., Chechik, V.: Mechanistic aspects of ligand exchange in Au nanoparticles. Phys. Chem. Chem. Phys. **10**, 5029–5041 (2008)
90. Bernardi, R.J., Lowery, A.R., Thompson, P.A., Blaney, S.M., West, J.L.: Immunonanoshells for targeted photothermal ablation in medulloblastoma and glioma: an in vitro evaluation using human cell lines. J. Neurooncol **86**, 165–172 (2008)
91. Cruz, L.J., et al.: Targeting nanosystems to human DCs via Fc receptor as an effective strategy to deliver antigen for immunotherapy. Mol. Pharm. **8**, 104–116 (2011)
92. Ito, A., Honda, H., Kobayashi, T.: Cancer immunotherapy based on intracellular hyperthermia using magnetite nanoparticles: a novel concept of "heat-controlled necrosis" with heat shock protein expression. Cancer Immunol. Immunother. **55**, 320–328 (2006)
93. Masoudi, A., Madaah Hosseini, H.R., Shokrgozar, M.A., Ahmadi, R., Oghabian, M.A.: The effect of poly(ethylene glycol) coating on colloidal stability of superparamagnetic iron oxide nanoparticles as potential MRI contrast agent. Int. J. Pharm. **433**, 129–141 (2012)
94. Webster, R. et al.: PEG and PEG conjugates toxicity: towards an understanding of the toxicity of PEG and its relevance to PEGylated biologicals. In: PEGylated Protein Drugs: Basic Science and Clinical Applications. Birkhäuser Verlag, Basel (2009) pp. 127–146
95. Bendele, A., Seely, J., Richey, C., Sennello, G., Shopp, G.: Short communication: renal tubular vacuolation in animals treated with polyethylene-glycol-conjugated proteins. Toxicol. Sci. **42**, 152–157 (1998)
96. Young, M.A., Malavalli, A., Winslow, N., Vandegriff, K.D., Winslow, R.M.: Toxicity and hemodynamic effects after single dose administration of MalPEG-hemoglobin (MP4) in rhesus monkeys. Transl. Res. **149**, 333–342 (2007)
97. Chapman, R.G., et al.: Surveying for surfaces that resist the adsorption of proteins. J. Am. Chem. Soc. **122**, 8303–8304 (2000)
98. Zhou, M., et al.: High throughput discovery of new fouling-resistant surfaces. J. Mater. Chem. **21**, 693 (2011)
99. Fasting, C., et al.: Multivalency as a chemical organization and action principle. Angew. Chem. Int. Ed. Engl. **51**, 10472–10498 (2012)
100. Niederhafner, P., Reinis, M., Sebestík, J., Jezek, J.: Glycopeptide dendrimers, part III: a review. Use of glycopeptide dendrimers in immunotherapy and diagnosis of cancer and viral diseases. J. Pept. Sci. **14**, 556–587 (2008)
101. Günay, K.A., Theato, P., Klok, H.A.: Standing on the shoulders of Hermann Staudinger: post-polymerization modification from past to present. J. Polym. Sci. A1 **51**, 1–28 (2013)
102. Grandjean, C., Boutonnier, A., Guerreiro, C., Fournier, J.-M., Mulard, L.A.: On the preparation of carbohydrate-protein conjugates using the traceless Staudinger ligation. J. Org. Chem. **70**, 7123–7132 (2005)
103. Xu, P., et al.: Simple, direct conjugation of bacterial O-SP-core antigens to proteins: development of cholera conjugate vaccines. Bioconjugate Chem. **22**, 2179–2185 (2011)
104. Scaramuzza, S., et al.: A new site-specific monoPE-Gylated filgrastim derivative prepared by enzymatic conjugation: production and physicochemical characterization. J. Control. Release **164**, 355–363 (2012)
105. Jung, B., Theato, P.: Chemical strategies for the synthesis of protein – polymer conjugates. Bio-synth. Polym. Conjugates **253**, 37–70 (2013)
106. Moad, G., Rizzardo, E., Thang, S.H.: Living radical polymerization by the RAFT process. Aust. J. Chem. **58**, 379 (2005)
107. Moad, G., Rizzardo, E., Thang, S.H.: Radical addition–fragmentation chemistry in polymer synthesis. Polymer **49**, 1079–1131 (2008)
108. Braunecker, W.A., Matyjaszewski, K.: Controlled/living radical polymerization: features, developments, and perspectives. Prog. Polym. Sci. **32**, 93–146 (2007)

109. Matyjaszewski, K., Xia, J.: Atom transfer radical polymerization. Chem. Rev. **101**, 2921–2990 (2001)
110. York, A.W., Kirkland, S.E., McCormick, C.L.: Advances in the synthesis of amphiphilic block copolymers via RAFT polymerization: stimuli-responsive drug and gene delivery. Adv. Drug Deliv. Rev. **60**, 1018–1036 (2008)
111. Gao, H., Matyjaszewski, K.: Synthesis of functional polymers with controlled architecture by CRP of monomers in the presence of cross-linkers: from stars to gels. Prog. Polym. Sci. **34**, 317–350 (2009)
112. Marsden, H.R., Kros, A.: Polymer-peptide block copolymers – an overview and assessment of synthesis methods. Macromol. Biosci. **9**, 939–951 (2009)
113. Tizzotti, M., Charlot, A., Fleury, E., Stenzel, M., Bernard, J.: Modification of polysaccharides through controlled/living radical polymerization grafting-towards the generation of high performance hybrids. Macromol. Rapid Comm. **31**, 1751–1772 (2010)
114. Lutz, J.F.: Polymerization of oligo(ethylene glycol) (meth)acrylates: toward new generations of smart biocompatible materials. J. Polym. Sci. A1 **46**, 3459–3470 (2008)
115. Lutz, J.-F., Akdemir, O., Hoth, A.: Point by point comparison of two thermosensitive polymers exhibiting a similar LCST: is the age of poly(NIPAM) over? J. Am. Chem. Soc. **128**, 13046–13047 (2006)
116. Tao, L., Mantovani, G., Lecolley, F., Haddleton, D.M.: Alpha-aldehyde terminally functional methacrylic polymers from living radical polymerization: application in protein conjugation "pegylation". J. Am. Chem. Soc. **126**, 13220–13221 (2004)
117. Lutz, J.-F., Hoth, A.: Preparation of Ideal PEG analogues with a tunable thermosensitivity by controlled radical copolymerization of 2-(2-Methoxyethoxy) ethyl methacrylate and oligo(ethylene glycol) methacrylate. Macromolecules **39**, 893–896 (2006)
118. Ryan, S.M., et al.: Conjugation of salmon calcitonin to a combed-shaped end functionalized poly(poly(ethylene glycol) methyl ether methacrylate) yields a bioactive stable conjugate. J. Control. Release **135**, 51–59 (2009)
119. Lutz, J.-F., Andrieu, J., Üzgün, S., Rudolph, C., Agarwal, S.: Biocompatible, thermoresponsive, and biodegradable: simple preparation of "all-in-one" biorelevant polymers. Macromolecules **40**, 8540–8543 (2007)
120. Ishihara, K., Ziats, N.P., Tierney, B.P., Nakabayashi, N., Anderson, J.M.: Protein adsorption from human plasma is reduced on phospholipid polymers. J. Biomed. Mater. Res. A **25**, 1397–1407 (1991)
121. Salvage, J.P., et al.: Novel biocompatible phosphorylcholine-based self-assembled nanoparticles for drug delivery. J. Control. Release **104**, 259–270 (2005)
122. Murdoch, C., et al.: Internalization and biodistribution of polymersomes into oral squamous cell carcinoma cells in vitro and in vivo. Nanomedicine **5**, 1025–1036 (2010)
123. Lomas, H., et al.: Non-cytotoxic polymer vesicles for rapid and efficient intracellular delivery. Faraday Discuss. **139**, 143–159 (2008)
124. Lewis, A., Tang, Y., Brocchini, S., Choi, J.-W., Godwin, A.: Poly(2-methacryloyloxyethyl phosphorylcholine) for protein conjugation. Bioconjugate Chem. **19**, 2144–2155 (2008)
125. Kopecek, J., Kopecková, P.: HPMA copolymers: origins, early developments, present, and future. Adv. Drug Deliv. Rev. **62**, 122–149 (2010)
126. Barz, M., et al.: From defined reactive diblock copolymers to functional HPMA-based self-assembled nanoaggregates. Biomacromolecules **9**, 3114–3118 (2008)
127. Barz, M., Canal, F., Koynov, K., Zentel, R., Vicent, M.J.: Synthesis and in vitro evaluation of defined HPMA folate conjugates: influence of aggregation on folate receptor (FR) mediated cellular uptake. Biomacromolecules **11**, 2274–2282 (2010)
128. Leuchs, H.: Über die Glycin-carbonsäure. Ber. Dtsch. Chem. Ges. **39**, 857–861 (1906)
129. Kricheldorf, H.R.: α-Amino acid-N-Carboxy-Anhydrides and Related Heterocycles: Syntheses, Properties, Peptide Synthesis, Polymerization. Springer, Berlin/Heidelberg/New York (1987)
130. Kricheldorf, H.R.: Polypeptides and 100 years of chemistry of alpha-amino acid N-carboxyanhydrides. Angew. Chem. Int. Ed. Engl. **45**, 5752–5784 (2006)
131. Hadjichristidis, N., Iatrou, H., Pitsikalis, M., Sakellariou, G.: Synthesis of well-defined polypeptide-based materials via the ring-opening polymerization of alpha-amino acid N-carboxyanhydrides. Chem. Rev. **109**, 5528–5578 (2009)
132. Bogdanov, A.A., et al.: A new macromolecule as a contrast agent for MR angiography: preparation, properties, and animal studies. Radiology **187**, 701–706 (1993)
133. Singer, J.W., et al.: Paclitaxel poliglumex (XYOTAX; CT-2103): an intracellularly targeted taxane. Anticancer Drug. **16**, 243–254 (2005)
134. Harada, A., Kataoka, K.: Formation of polyion complex micelles in an aqueous milieu from a pair of oppositely-charged block copolymers with poly(ethylene glycol) segments. Macromolecules **28**, 5294–5299 (1995)
135. Carlsen, A., Lecommandoux, S.: Self-assembly of polypeptide-based block copolymer amphiphiles. Curr. Opin. Colloid In. **14**, 329–339 (2009)
136. Deng, J., et al.: Self-assembled cationic micelles based on PEG-PLL-PLLeu hybrid polypeptides as highly effective gene vectors. Biomacromolecules **13**, 3795–3804 (2012)
137. Bellomo, E.G., Wyrsta, M.D., Pakstis, L., Pochan, D.J., Deming, T.J.: Stimuli-responsive polypeptide vesicles by conformation-specific assembly. Nat. Mater. **3**, 244–248 (2004)
138. Holowka, E.P., Sun, V.Z., Kamei, D.T., Deming, T.J.: Polyarginine segments in block copolypeptides drive both vesicular assembly and intracellular delivery. Nat. Mater. **6**, 52–57 (2007)

139. Kanzaki, T., Horikawa, Y., Makino, A., Sugiyama, J., Kimura, S.: Nanotube and three-way nanotube formation with nonionic amphiphilic block peptides. Macromol. Biosci. **8**, 1026–1033 (2008)
140. Nowak, A.P., et al.: Rapidly recovering hydrogel scaffolds from self-assembling diblock copolypeptide amphiphiles. Nature **417**, 424–428 (2002)
141. Takae, S., et al.: PEG-detachable polyplex micelles based on disulfide-linked block catiomers as bioresponsive nonviral gene vectors. J. Am. Chem. Soc. **130**, 6001–6009 (2008)
142. Uchida, H., et al.: Odd-even effect of repeating aminoethylene units in the side chain of N-substituted polyaspartamides on gene transfection profiles. J. Am. Chem. Soc. **133**, 15524–15532 (2011)
143. Sanjoh, M., et al.: Dual environment-responsive polyplex carriers for enhanced intracellular delivery of plasmid DNA. Biomacromolecules **13**, 3641–3649 (2012)
144. Naito, M., et al.: A phenylboronate-functionalized polyion complex micelle for ATP-triggered release of siRNA. Angew. Chem. Int. Ed. Engl. **124**, 10909–10913 (2012)
145. Hamaguchi, T., et al.: NK105, a paclitaxel-incorporating micellar nanoparticle formulation, can extend in vivo antitumour activity and reduce the neurotoxicity of paclitaxel. Brit. J. Cancer **92**, 1240–1246 (2005)
146. Weissleder, R., Tung, C.H., Mahmood, U., Bogdanov, A.: In vivo imaging of tumors with protease-activated near-infrared fluorescent probes. Nat. Biotechnol. **17**, 375–378 (1999)
147. Arnon, R.: The development of Cop 1 (Copaxone), an innovative drug for the treatment of multiple sclerosis: personal reflections. Immunol. Lett. **50**, 1–15 (1996)
148. Shaffer, S.A., et al.: In vitro and in vivo metabolism of paclitaxel poliglumex: identification of metabolites and active proteases. Cancer Chemother. Pharmacol. **59**, 537–548 (2007)
149. Hanson, J.A., et al.: Nanoscale double emulsions stabilized by single-component block copolypeptides. Nature **455**, 85–88 (2008)
150. Matsumura, Y.: Poly (amino acid) micelle nanocarriers in preclinical and clinical studies. Adv. Drug Deliv. Rev. **60**, 899–914 (2008)
151. http://www.clinicaltrials.gov.
152. Li, C., Wallace, S.: Polymer-drug conjugates: recent development in clinical oncology. Adv. Drug Deliv. Rev. **60**, 886–898 (2008)
153. Wilms, D., Stiriba, S.-E., Frey, H.: Hyperbranched polyglycerols: from the controlled synthesis of biocompatible polyether polyols to multipurpose applications. Acc. Chem. Res. **43**, 129–141 (2010)
154. Quadir, M.A., Haag, R.: Biofunctional nanosystems based on dendritic polymers. J. Control. Release **161**, 484–495 (2012)
155. Khandare, J., Calderón, M., Dagia, N.M., Haag, R.: Multifunctional dendritic polymers in nanomedicine: opportunities and challenges. Chem. Soc. Rev. **41**, 2824–2848 (2012)
156. Mourey, T.H., et al.: Unique behavior of dendritic macromolecules: intrinsic viscosity of polyether dendrimers. Macromolecules **25**, 2401–2406 (1992)
157. Wyszogrodzka, M., Haag, R.: Study of single protein adsorption onto monoamino oligoglycerol derivatives: a structure-activity relationship. Langmuir **25**, 5703–5712 (2009)
158. Wyszogrodzka, M., et al.: New approaches towards monoamino polyglycerol dendrons and dendritic triblock amphiphiles. Eur. J. Org. Chem. **2008**, 53–63 (2008)
159. Haag, R., Sunder, A., Stumbé, J.-F.: An approach to glycerol dendrimers and pseudo-dendritic polyglycerols. J. Am. Chem. Soc. **122**, 2954–2955 (2000)
160. Sunder, A., Krämer, M., Hanselmann, R., Mülhaupt, R., Frey, H.: Molecular nanocapsules based on amphiphilic hyperbranched polyglycerols. Angew. Chem. Int. Ed. Engl. **38**, 3552–3555 (1999)
161. Wilms, D., et al.: Hyperbranched polyglycerols with elevated molecular weights: a facile two-step synthesis protocol based on polyglycerol macroinitiators. Macromolecules **42**, 3230–3236 (2009)
162. Barriau, E., et al.: Systematic investigation of functional core variation within hyperbranched polyglycerols. J. Polym. Sci. A1 **46**, 2049–2061 (2008)
163. Roller, S., Zhou, H., Haag, R.: High-loading polyglycerol supported reagents for Mitsunobu- and acylation-reactions and other useful polyglycerol derivatives. Mol. Divers. **9**, 305–316 (2005)
164. Kainthan, R.K., Janzen, J., Levin, E., Devine, D.V., Brooks, D.E.: Biocompatibility testing of branched and linear polyglycidol. Biomacromolecules **7**, 703–709 (2006)
165. Kainthan, R.K., Hester, S.R., Levin, E., Devine, D.V., Brooks, D.E.: In vitro biological evaluation of high molecular weight hyperbranched polyglycerols. Biomaterials **28**, 4581–4590 (2007)
166. Kainthan, R.K., Brooks, D.E.: In vivo biological evaluation of high molecular weight hyperbranched polyglycerols. Biomaterials **28**, 4779–4787 (2007)
167. Calderón, M., Quadir, M.A., Sharma, S.K., Haag, R.: Dendritic polyglycerols for biomedical applications. Adv. Mater. **22**, 190–218 (2010)
168. Calderón, M., Graeser, R., Kratz, F., Haag, R.: Development of enzymatically cleavable prodrugs derived from dendritic polyglycerol. Bioorg. Med. Chem. Lett. **19**, 3725–3728 (2009)
169. Calderón, M., et al.: Development of efficient acid cleavable multifunctional prodrugs derived from dendritic polyglycerol with a poly(ethylene glycol) shell. J. Control. Release **151**, 295–301 (2011)
170. Papp, I., Dernedde, J., Enders, S., Haag, R.: Modular synthesis of multivalent glycoarchitectures and their unique selectin binding behavior. Chem. Commun. **4**, 5851–5853 (2008)
171. Türk, H., Haag, R., Alban, S.: Dendritic polyglycerol sulfates as new heparin analogues and potent inhibitors of the complement system. Bioconjug. Chem. **15**, 162–167 (2003)

172. Dernedde, J., et al.: Dendritic polyglycerol sulfates as multivalent inhibitors of inflammation. Proc. Natl. Acad. Sci. U. S. A. **44**, 19679–19684 (2010)
173. Steinhilber, D., et al.: Synthesis, reductive cleavage, and cellular interaction studies of biodegradable polyglycerol nanogels. Adv. Funct. Mater. **20**, 4133–4138 (2010)
174. Sisson, A.L., et al.: Biocompatible functionalized polyglycerol microgels with cell penetrating properties. Angew. Chem. Int. Ed. Engl. **48**, 7540–7545 (2009)
175. Sisson, A.L., Papp, I., Landfester, K., Haag, R.: Functional nanoparticles from dendritic precursors: hierarchical assembly in miniemulsion. Macromolecules **42**, 556–559 (2009)
176. Luxenhofer, R., et al.: Poly(2-oxazoline)s as polymer therapeutics. Macromol. Rapid Comm. **33**, 1613–1631 (2012)
177. Viegas, T.X., et al.: Polyoxazoline: chemistry, properties, and applications in drug delivery. Bioconjugate Chem. **22**, 976–986 (2011)
178. Knop, K., Hoogenboom, R., Fischer, D., Schubert, U.S.: Poly(ethylene glycol) in drug delivery: pros and cons as well as potential alternatives. Angew. Chem. Int. Ed. Engl. **49**, 6288–6308 (2010)
179. Kempe, K., et al.: Multifunctional poly(2-oxazoline) nanoparticles for biological applications. Macromol. Rapid Comm. **31**, 1869–1873 (2010)
180. Luxenhofer, R., et al.: Structure-property relationship in cytotoxicity and cell uptake of poly(2-oxazoline) amphiphiles. J. Control. Release **153**, 73–82 (2011)
181. Donev, R., Koseva, N., Petrov, P., Kowalczuk, A., Thome, J.: Characterisation of different nanoparticles with a potential use for drug delivery in neuropsychiatric disorders. World J. Biol. Psychiatry **12**, 44–51 (2011)
182. Tong, J., et al.: Neuronal uptake and intracellular superoxide scavenging of a fullerene (C60)-poly(2-oxazoline)s nanoformulation. Biomaterials **32**, 3654–3665 (2011)
183. Viegas, T.X., et al.: Polyoxazoline: chemistry, properties, and applications in drug delivery. Bioconjug. Chem. **22**, 976–986 (2011)
184. Wang, X., et al.: Synthesis, characterization and biocompatibility of poly(2-ethyl-2-oxazoline)-poly(D, L-lactide)-poly(2-ethyl-2-oxazoline) hydrogels. Acta Biomater. **7**, 4149–4159 (2011)
185. Wang, C.-H., et al.: Extended release of bevacizumab by thermosensitive biodegradable and biocompatible hydrogel. Biomacromolecules **13**, 40–48 (2012)
186. Cheon Lee, S., Kim, C., Chan Kwon, I., Chung, H., Young Jeong, S.: Polymeric micelles of poly(2-ethyl-2-oxazoline)-block-poly(epsilon-caprolactone) copolymer as a carrier for paclitaxel. J. Control. Release **89**, 437–446 (2003)
187. Konradi, R., Pidhatika, B., Mühlebach, A., Textor, M.: Poly-2-methyl-2-oxazoline: a peptide-like polymer for protein-repellent surfaces. Langmuir **24**, 613–616 (2008)

188. Pidhatika, B., et al.: The role of the interplay between polymer architecture and bacterial surface properties on the microbial adhesion to polyoxazoline-based ultrathin films. Biomaterials **31**, 9462–9472 (2010)
189. Zhang, N., et al.: Tailored poly(2-oxazoline) polymer brushes to control protein adsorption and cell adhesion. Macromol. Biosci. **12**, 926–936 (2012)
190. Wang, H., Li, L., Tong, Q., Yan, M.: Evaluation of photochemically immobilized poly(2-ethyl-2-oxazoline) thin films as protein-resistant surfaces. ACS Appl. Mater. Interfaces **3**, 3463–3471 (2011)
191. Woodle, M.C., Engbers, C.M., Zalipsky, S.: New amphipathic polymer-lipid conjugates forming long-circulating reticuloendothelial system-evading liposomes. Bioconjug. Chem. **5**, 493–496 (1994)
192. Zalipsky, S., Hansen, C.B., Oaks, J.M., Allen, T.M.: Evaluation of blood clearance rates and biodistribution of poly(2-oxazoline)-grafted liposomes. J. Pharm. Sci. **85**, 133–137 (1996)
193. Gaertner, F.C., Luxenhofer, R., Blechert, B., Jordan, R., Essler, M.: Synthesis, biodistribution and excretion of radiolabeled poly(2-alkyl-2-oxazoline)s. J. Control. Release **119**, 291–300 (2007)
194. Goddard, P., Hutchinson, L.: Soluble polymeric carriers for drug delivery. Part 2. Preparation and in vivo behaviour of N-acylethylenimine copolymers. J. Control. Release **10**, 5–16 (1989)
195. Luxenhofer, R., et al.: Doubly amphiphilic poly(2-oxazoline)s as high-capacity delivery systems for hydrophobic drugs. Biomaterials **31**, 4972–4979 (2010)
196. Delgado, A.V., González-Caballero, F., Hunter, R.J., Koopal, L.K., Lyklema, J.: Measurement and interpretation of electrokinetic phenomena. J. Colloid Interface Sci. **309**, 194–224 (2007)
197. Rausch, K., Reuter, A., Fischer, K., Schmidt, M.: Evaluation of nanoparticle aggregation in human blood serum. Biomacromolecules **11**, 2836–2839 (2010)
198. Olsen, S.N.: Applications of isothermal titration calorimetry to measure enzyme kinetics and activity in complex solutions. Thermochim. Acta **448**, 12–18 (2006)
199. Gourishankar, A., Shukla, S., Ganesh, K.N., Sastry, M.: Isothermal titration calorimetry studies on the binding of DNA bases and PNA base monomers to gold nanoparticles. J. Am. Chem. Soc. **126**, 13186–13187 (2004)
200. Cedervall, T., et al.: Understanding the nanoparticle-protein corona using methods to quantify exchange rates and affinities of proteins for nanoparticles. Proc. Natl. Acad. Sci. U. S. A. **104**, 2050–2055 (2007)
201. Tenzer, S., et al.: Nanoparticle size is a critical physicochemical determinant of the human blood plasma corona: a comprehensive quantitative proteomic analysis. ACS Nano **5**, 7155–7167 (2011)
202. Ghoroghchian, P.P., Therien, M.J., Hammer, D.A.: In vivo fluorescence imaging: a personal perspective. Wiley Interdiscip. Rev. Nanomed. Nanobiotechnol. **1**, 156–167 (2009)

203. Herzog, H., Rösch, F.: PET- und SPECT-Technik: Chemie und Physik der Bildgebung. Pharm. Unserer Zeit **34**, 468–473 (2005)
204. Herth, M.M., et al.: Radioactive labeling of defined HPMA-based polymeric structures using [18F] FETos for in vivo imaging by positron emission tomography. Biomacromolecules **10**(4), 1697–1703 (2009)
205. Devaraj, N.K., Keliher, E.J., Thurber, G.M., Nahrendorf, M., Weissleder, R.: 18F labeled nanoparticles for in vivo PET-CT imaging. Bioconjugate Chem. **20**, 397–401 (2009)
206. Herth, M.M., Barz, M., Jahn, M., Zentel, R., Rösch, F.: 72/74As-labeling of HPMA based polymers for long-term in vivo PET imaging. Bioorg. Med. Chem. Lett. **20**, 5454–5458 (2010)
207. Fujita, Y., Taguchi, H.: Current status of multiple antigen-presenting peptide vaccine systems: application of organic and inorganic nanoparticles. Chem. Cent. J. **5**, 1–8 (2011)
208. Vogel, F.R.: Immunologic adjuvants for modern vaccine formulations. Ann. N. Y. Acad. Sci. **754**, 153–160 (1995)
209. Petrovsky, N., Aguilar, J.: Vaccine adjuvants: current state and future trends. Immunol. Cell Biol. **82**, 488–496 (2004)
210. Panyam, J., Labhasetwar, V.: Biodegradable nanoparticles for drug and gene delivery to cells and tissue. Adv. Drug Deliv. Rev. **55**, 329–347 (2003)
211. Lee, Y.-R., Lee, Y.-H., et al.: Biodegradable nanoparticles containing TLR3 or TLR9 agonists together with antigen enhance MHC-restricted presentation of the antigen. Arch. Pharm. Res. **33**, 1859–1866 (2010)
212. Stone, G.W., et al.: Nanoparticle-delivered multimeric soluble CD40L DNA combined with Toll-like receptor agonists as a treatment for melanoma. PLoS One **4**, e7334 (2009)
213. Kasturi, S.P., et al.: Programming the magnitude and persistence of antibody responses with innate immunity. Nature **470**, 543–547 (2011)
214. Malyala, P., O'Hagan, D.T., Singh, M.: Enhancing the therapeutic efficacy of CpG oligonucleotides using biodegradable microparticles. Adv. Drug Deliv. Rev. **61**, 218–225 (2009)
215. O'Hagan, D.T., Singh, M., Ulmer, J.B.: Microparticle-based technologies for vaccines. Methods **40**, 10–19 (2006)
216. Sherman, M. R. et al.: Conjugation of high-molecular weight poly(ethylene glycol) to cytokines: granulocyte-macrophage colony-stimulating factors as model substrates. ACS Symposium Series, Vol. 680, pp. 155–169. (1997)
217. Alexis, F., Pridgen, E., Molnar, L.K., Farokhzad, O.C.: Factors affecting the clearance and biodistribution of polymeric nanoparticles. Mol. Pharm. **5**, 505–515 (2008)
218. Yang, Y., Huang, C.-T., Huang, X., Pardoll, D.M.: Persistent Toll-like receptor signals are required for reversal of regulatory T cell-mediated CD8 tolerance. Nat. Immunol. **5**, 508–515 (2004)
219. Galloway, A.L., et al.: Development of a nanoparticle-based influenza vaccine using the PRINT® technology. Nanomedicine **9**(4), 523–531 (2013)
220. Demento, S.L., et al.: TLR9-targeted biodegradable nanoparticles as immunization vectors protect against West Nile encephalitis. J. Immunol. **185**, 2989–2997 (2010)
221. Tyagi, R.K., Garg, N.K., Sahu, T.: Vaccination strategies against malaria: novel carrier(s) more than a tour de force. J. Control. Release **162**, 242–254 (2012)
222. Brandt, E.R., et al.: New multi-determinant strategy for a group A streptococcal vaccine designed for the Australian Aboriginal population. Nat. Med. **6**, 455–459 (2000)
223. Shi, L., et al.: Pharmaceutical and immunological evaluation of a single-shot hepatitis B vaccine formulated with PLGA microspheres. J. Pharm. Sci. **91**, 1019–1035 (2002)
224. Pejawar-Gaddy, S., et al.: Generation of a tumor vaccine candidate based on conjugation of a MUC1 peptide to polyionic papillomavirus virus-like particles. Cancer Immunol. Immunother. **59**, 1685–1696 (2010)
225. Sundgren, A., Barchi, J.: Varied presentation of the Thomsen–Friedenreich disaccharide tumor-associated carbohydrate antigen on gold nanoparticles. Carbohydr. Res. **343**, 1594–1604 (2008)
226. Monzavi-Karbassi, B., Pashov, A., Jousheghany, F., Artaud, C., Kieber-Emmons, T.: Evaluating strategies to enhance the anti-tumor immune response to a carbohydrate mimetic peptide vaccine. Int. J. Mol. Med. **17**, 1045–1052 (2006)
227. Reddy, S., Swartz, M., Hubbell, J.: Targeting dendritic cells with biomaterials: developing the next generation of vaccines. Trends Immunol. **27**, 573–579 (2006)
228. Demento, S.L., et al.: Inflammasome-activating nanoparticles as modular systems for optimizing vaccine efficacy. Vaccine **27**, 3013–3021 (2009)
229. Hamdy, S., et al.: Co-delivery of cancer-associated antigen and Toll-like receptor 4 ligand in PLGA nanoparticles induces potent CD8+ T cell-mediated anti-tumor immunity. Vaccine **26**, 5046–5057 (2008)
230. Barton, G.M., Medzhitov, R.: Control of adaptive immune responses by Toll-like receptors. Curr. Opin. Immunol. **14**, 380–383 (2002)
231. Akagi, T., Baba, M., Akashi, M.: Biodegradable nanoparticles as vaccine adjuvants and delivery systems: regulation of immune responses by nanoparticle-based vaccine. Adv. Polym. Sci. **247**, 31–64 (2011)
232. Lee, T.Y., et al.: Oral administration of poly-gamma-glutamate induces TLR4- and dendritic cell-dependent antitumor effect. Cancer Immunol Imm. **58**, 1781–1794 (2009)

233. Yoshida, M., Babensee, J.E.: Poly(lactic-co-glycolic acid) enhances maturation of human monocyte-derived dendritic cells. J. Biomed. Mater. Res. A **71**, 45–54 (2004)
234. Tamayo, I., et al.: Poly(anhydride) nanoparticles act as active Th1 adjuvants through Toll-like receptor exploitation. Clin. Vaccine Immunol. **17**, 1356–1362 (2010)
235. Copland, M.J., et al.: Liposomal delivery of antigen to human dendritic cells. Vaccine **21**, 883–890 (2003)
236. Matsusaki, M., et al.: Nanosphere induced gene expression in human dendritic cells. Nano Lett. **5**, 2168–2173 (2005)
237. Kwon, Y.J., Standley, S.M., Goh, S.L., Fréchet, J.M.J.: Enhanced antigen presentation and immunostimulation of dendritic cells using acid-degradable cationic nanoparticles. J. Control. Release **105**, 199–212 (2005)
238. Sun, H., Pollock, K.G.J., Brewer, J.M.: Analysis of the role of vaccine adjuvants in modulating dendritic cell activation and antigen presentation in vitro. Vaccine **21**, 849–855 (2003)
239. Moon, H.-J., et al.: Mucosal immunization with recombinant influenza hemagglutinin protein and poly gamma-glutamate/chitosan nanoparticles induces protection against highly pathogenic influenza A virus. Vet. Microbiol. **160**, 277–289 (2012)
240. Geall, A., Verma, A.: Nonviral delivery of self-amplifying RNA vaccines. Proc. Natl. Acad. Sci. U. S. A. **109**, 14604–14609 (2012)
241. Van den Berg, J.H., et al.: Shielding the cationic charge of nanoparticle-formulated dermal DNA vaccines is essential for antigen expression and immunogenicity. J. Control. Release **141**, 234–240 (2010)
242. Varkouhi, A.K., Scholte, M., Storm, G., Haisma, H.J.: Endosomal escape pathways for delivery of biologicals. J. Control. Release **151**, 220–228 (2011)
243. Krieg, A.M.: CpG motifs in bacterial DNA and their immune effects. Annu. Rev. Immunol. **20**, 709–760 (2002)
244. Sudowe, S., et al.: Uptake and presentation of exogenous antigen and presentation of endogenously produced antigen by skin dendritic cells represent equivalent pathways for the priming of cellular immune responses following biolistic DNA immunization. Immunology **128**, e193–e205 (2009)
245. Joshi, M.D., Unger, W.J., Storm, G., Van Kooyk, Y., Mastrobattista, E.: Targeting tumor antigens to dendritic cells using particulate carriers. J. Control. Release **161**, 25–37 (2012)
246. Arigita, C., et al.: Liposomal meningococcal B vaccination: role of dendritic cell targeting in the development of a protective immune response. Infect. Immun. **71**, 5210–5218 (2003)
247. Espuelas, S., Haller, P., Schuber, F., Frisch, B.: Synthesis of an amphiphilic tetraantennary mannosyl conjugate and incorporation into liposome carriers. Bioorg. Med. Chem. Lett. **13**, 2557–2560 (2003)
248. Espuelas, S., Thumann, C., Heurtault, B., Schuber, F., Frisch, B.: Influence of ligand valency on the targeting of immature human dendritic cells by mannosylated liposomes. Bioconjug. Chem. **19**, 2385–2393 (2008)
249. Sheng, K.-C., et al.: Delivery of antigen using a novel mannosylated dendrimer potentiates immunogenicity in vitro and in vivo. Eur. J. Immunol. **38**, 424–436 (2008)
250. Chenevier, P., et al.: Grafting of synthetic mannose receptor-ligands onto onion vectors for human dendritic cells targeting. Chem. Commun. **20**, 2446–2447 (2002)
251. Saraogi, G.K., et al.: Mannosylated gelatin nanoparticles bearing isoniazid for effective management of tuberculosis. J. Drug Target. **19**, 219–227 (2011)
252. Brandhonneur, N., et al.: Specific and non-specific phagocytosis of ligand-grafted PLGA microspheres by macrophages. Eur. J. Pharm. Sci. **36**, 474–485 (2009)
253. Hamdy, S., Haddadi, A., Shayeganpour, A., Samuel, J., Lavasanifar, A.: Activation of antigen-specific T cell-responses by mannan-decorated PLGA nanoparticles. Pharm. Res. **28**, 2288–2301 (2011)
254. Raghuwanshi, D., Mishra, V., Suresh, M.R., Kaur, K.: A simple approach for enhanced immune response using engineered dendritic cell targeted nanoparticles. Vaccine **30**, 7292–7299 (2012)
255. Fehr, T., Skrastina, D., Pumpens, P., Zinkernagel, R.M.: T cell-independent type I antibody response against B cell epitopes expressed repetitively on recombinant virus particles. Proc. Natl. Acad. Sci. U. S. A. **95**, 9477–9481 (1998)
256. Brinãs, R.P., et al.: Design and synthesis of multifunctional gold nanoparticles bearing tumor-associated glycopeptide antigens as potential cancer vaccines. Bioconjug. Chem. **23**, 1513–1523 (2012)
257. Hoffmann-Röder, A., et al.: Synthetic antitumor vaccines from tetanus toxoid conjugates of MUC1 glycopeptides with the Thomsen-Friedenreich antigen and a fluorine-substituted analogue. Angew. Chem. Int. Ed. Engl. **49**, 8498–8503 (2010)
258. Gaidzik, N., et al.: Synthetic antitumor vaccines containing MUC1 glycopeptides with two immunodominant domains-induction of a strong immune response against breast tumor tissues. Angew. Chem. Int. Ed. Engl. **50**, 9977–9981 (2011)
259. Cai, H., et al.: Variation of the glycosylation pattern in MUC1 glycopeptide BSA vaccines and its influence on the immune response. Angew. Chem. Int. Ed. Engl. **51**, 1719–1723 (2012)

# Part VII

# In Silico and Delivery Systems

## Overview of Part VII

**In silico**
- Genomics, gene prediction and gene annotation
- Transcriptome and proteome analyses
- Reverse vaccinology

**Microneedles**
- Immune function of skin
- Vaccination *via* the skin using microneedles
- Dissolving/soluble microneedles

**Dry Powder**
- Nasal associated lymphoid tissue
- Nasal vaccine technology
- Spray dried and spray freeze dried powders

**Bacterial vector**
- Lactococcus lactis and Lactobacillus spp.
- Lactobacillus plantarum a potent adjuvant
- Safe and effective plague vaccine

**Electroporation**
- Electric field and transmembrane potential
- DNA electrotransfer principles
- Pulse generators and applications

**i.m. - why?**
- The inflammatory response into the muscle
- The adaptive immune response into the muscle
- Muscle cells as non-professional APC

In addition to the appropriate adjuvant, the delivery technology is another key element for vaccine development. The current intense research activity in this area will be presented in the following chapters. Pain-free and needle-less safe devices are the upcoming alternatives to conventional multiple injections Moreover, new vaccines are designed to elicit a CTL or a mucosal response. Unlike most of the conventional vaccines, such modern vaccines require the recruitment of cellular effector mechanisms and therefore necessitate new routes of administration, combined with new adjuvants (see Part VI). New delivery technologies should also meet the need for thermostable vaccines with the great chance to turn the enormous cost-intensive cold chain down.

*Vaccine design.* The future of vaccinology is moving into obtaining a global picture of the various factors involved in protective immunity with the development of systems biology and bioinformatics tools capable of integrating various types of data. The prediction of epitopes remains a crucial step in the screening of pathogens protein-coding sequences before experimental confirmation of immunogenicity.

*Microneedles.* Intradermal vaccination using MN is one of the most attractive approaches for delivering an antigen to the dermal layer of the skin without using hypodermic injections, which are associated with transmission of infection and inappropriate disposal in the developing world.

*Nasal delivery.* The nasal mucosa is complex and includes important elements of the immune system. Therefore, it is an ideal route of delivery for a noninvasive vaccine delivery. Spray-dried, freeze-dried, and spray freeze-dried powders have demonstrated equivalent immunogenicity to conventional liquid formulations.

*Nanotechnology and delivery.* Nano-vectors bear the advantage of being similar to a pathogen in terms of size; thus, they are efficiently recognized by antigen-presenting cells of skilled immune system. ISCOMS are characterized by a cage-like structure that incorporates the antigen. They consist of lipids and Quil A, the active component of the saponin derived from the plant *Quillaja saponaria* that has adjuvant activity. PADRE-PAMAM dendrimer nano-vaccine delivery may fulfill some of the major challenges facing robust protective vaccines: it provides high transfection efficiency, proper targeting to the APCs, and an adjuvant effect

*Bacterial vectored vaccines.* This induction of mucosal immunity through vaccination is a rather difficult task. Lactic acid bacteria have Generally Recognized As Safe (GRAS) status and have been developed in the past decade as potent adjuvants for mucosal delivery of vaccine antigens. Both *Lactococcus lactis* and *Lactobacillus spp.* have been used.

*Electroporation.* Application of an external electric field to a single cell, to cell suspensions, or to biological tissue generates a change in the cell transmembrane potential, resulting in changes to the membrane structure that render the membrane permeable to otherwise non-permeate molecules, a phenomenon termed electroporation. Electroporation technology is based on pulse generators that use different applicator electrodes, e.g., matrix of needles or plates to deliver suitable electric pulses to the target tissues.

*The Classical i.m. immunization.* Muscle cells are able to actively participate in the induction of immunity and to behave as nonprofessional APC.

Muscle cells express receptors for cytokines and PAMPs that enable them to respond to an inflammatory ***milieu***, by secreting cytokines and chemokines and expressing adhesion molecules. Some membrane proteins necessary to the APC function, like class I and class II MHC molecules and costimulatory molecules, have been observed in several experimental systems.

# Considerations for Vaccine Design in the Postgenomic Era

## 42

Christine Maritz-Olivier and Sabine Richards

## Contents

| | | |
|---|---|---|
| 42.1 | **Understanding the Pathogen: Biology of Invasion, Infection and Survival** | 678 |
| 42.1.1 | The Blueprint of a Pathogen: Genomics, Gene Prediction and Gene Annotation | 678 |
| 42.1.2 | Transcriptome and Proteome Analyses | 679 |
| 42.2 | **Understanding the Host Immune System** | 680 |
| 42.3 | **Vaccinomics: Towards Understanding the Host-Pathogen Interactions and Type of Immunity Induced During Infection and Vaccination** | 682 |
| 42.3.1 | The Challenges of Cross-Reactivity and Autoimmunity | 684 |
| 42.4 | **Reverse Vaccinology** | 685 |
| 42.4.1 | Prediction of Subcellular Localization | 685 |
| 42.4.2 | Prediction of Immunogenicity and Epitopes | 687 |
| 42.4.3 | Reverse Vaccinology Platforms | 688 |
| 42.5 | **Validation of Vaccine Candidates** | 689 |
| 42.5.1 | Immunological Biomarkers | 689 |
| 42.5.2 | Protein Expression, Adjuvants and Vaccine Trials | 689 |
| **References** | | 692 |

C. Maritz-Olivier, PhD (✉)
Faculty of Natural and Agricultural Sciences, Department of Genetics, University of Pretoria, Lynnwood Road, Pretoria, South Africa
e-mail: christine.maritz@up.ac.za

S. Richards, MSc
Faculty of Natural and Agricultural Sciences, Department of Genetics, University of Pretoria, Lynnwood Road, Pretoria, South Africa

## Abstract

The foundations of vaccination were laid in the eighteenth century by Edward Jenner and in the nineteenth century by Louis Pasteur. During the 1930s and 1940s, live attenuated and inactivated vaccines dominated the field. This was followed by the purification of antigens from pathogens grown in culture using biochemical methods, bringing the era of subunit vaccines to the fore. With the explosion in next-generation sequencing technologies and the availability first genomes, the field of reverse vaccinology alongside the associated "omics"-revolutions became of age. This allowed for the identification of promising antigens, not previously exploited for their protective abilities. By combining the latter technologies with immunogenetics and immunogenomics, insight into the immune response during infection/vaccination is providing a global picture of the various factors involved in protective immunity. In this post-genomic era, vaccine development is moving away from a trial-and-error approach to a knowledge-based vaccine development approach.

The foundations of vaccination were laid in the eighteenth century by Edward Jenner and in the nineteenth century by Louis Pasteur. During the 1930s and 1940s, live attenuated and inactivated vaccines dominated the field with the two most prominent examples being propagation of influenza viruses in chick embryos and polio and other viruses in cultures of human and monkey tissues [1]. This was followed by the purification of antigens from pathogens grown in culture using biochemical methods, bringing the era of subunit vaccines to the fore. With the availability of the first bacterial genome, Rappuoli invented the field of reverse vaccinology in 2000, which has since exploded alongside the development of next-generation sequencing technologies and the associated "omics"-revolution.

The first vaccine obtained through reverse vaccinology, the serogroup B meningococcus vaccine, is already in final trials and could substantiate reverse vaccinology in the very near future [2, 3]. In 2007 Poland and colleagues initiated vaccinomics, a field that aims to combine immunogenetics and immunogenomics, providing insight into the immune response during infection or during vaccination. The future of vaccinology is moving into obtaining a global picture of the various factors involved in protective immunity with the development of systems biology and bioinformatics tools capable of integrating various types of data.

Despite all of this progress, critical fundamentals remain to be understood in order to move away from a trial-and-error approach to a knowledge-based vaccine development approach. This includes a thorough understanding of pathogen biology, the host immune system, the network of immune reactions activated during host-pathogen interactions or during vaccination, the principles of reverse vaccinology and the software used during this approach, identification of immunological biomarkers and proper vaccine formulation and trials. This forms the focus of this chapter.

## 42.1 Understanding the Pathogen: Biology of Invasion, Infection and Survival

Pathogens for which vaccines are required range from viruses, prokaryotic bacteria, eukaryotic intracellular endoparasites (such as Plasmodium causing malaria) and extracellular parasites (such as helminths) to ectoparasites (such as ticks and mites). Each of these has its own unique biology and effect on the host immune system which must be understood if a successful vaccine is to be developed.

### 42.1.1 The Blueprint of a Pathogen: Genomics, Gene Prediction and Gene Annotation

The ability to obtain genomic DNA sequence information within a short time frame and at an affordable price is without doubt shaping modern biology. In the field of vaccinology, genomes hold the key to understanding pathogen genes, the host immune genes and the interplay between genes involved in host-pathogen interactions during infection and vaccination. Next-generation sequencing has been used with great success to gain insight into detection of both pathogen and host genetic variation, evidence of selection pressures, immune escape mechanisms, vaccine safety, the diversity of T- and B-cell repertoires, immune regulation and assessment of the quality of vaccine stocks [4].

It is nowadays becoming routine to start analyses with more than one genome of closely related organisms (pan-genomic analyses). This allows for the identification of immunogenic and conserved antigens suitable for use in a combinatorial vaccine that offers protection against genetically diverse strains. The best example is that of the Bexsero vaccine containing four immunogenic and conserved antigens, protecting against 77 % of more than 800 genetically diverse disease-causing MsnB strains [2]. Comparative studies on 8 g positive-pathogenic *Streptococcus agalactiae* genomes revealed both a core genome and a dispensable genome. Interestingly, the proteins found to be protective did not all originate from the core genome. In contrast, only one core genome antigen and three dispensable genome antigens were found to be protective in combination against a large panel of strains [5]. Subtractive genomics between pathogenic and nonpathogenic strains have been used with great success for the identification of the ECOK I-3385 antigen from pathogenic *Escherichia coli* [5].

In this postgenomic era, genome sequence assembly and gene prediction remain a bottleneck, especially for complex eukaryotic pathogens lacking gene scaffolds. Subsequent gene and open reading frame (ORF) prediction is one of the most fundamental and important steps in genome analysis having major implications for any downstream analysis. One method for the identification of ORFs is BLAST [6]. Otherwise, ab initio gene prediction can be performed utilizing mathematical models if no other evidence of a gene is present [7]. A number of programs optimized for prokaryotic ORF predictions are available, such as Glimmer [8], GeneMark.hmm [9] and Prodigal [10]. Pipelines providing increased accuracy of gene prediction by evaluating gene models of prokaryotes and detecting anomalies are also coming of age [11].

In eukaryotes, this is not a trivial process, as recent observation from the ENCODE project revealed that genes have overlapping transcripts as well as transcription from both DNA strands [12]. Furthermore, mathematical models are trained on organism-specific genomic traits often using classic model organisms such as *Drosophila melanogaster* [7]. The accuracy of these predictions is then dependent on how distantly the organism of interest is related to these eukaryotes [7]. Some gene predictors can also be trained using pipelines such as MAKER [13]. Other programs are capable of incorporating external evidence (such as EST and RNA-Seq data) into an ab initio approach for the improvement of prediction. Such software includes AUGUSTUS [14] and GNOMON (http://0-www.ncbi.nlm.nih.gov.innopac.up.ac.za/genome/guide/gnomon.shtml). Other strategies, such as SAGE libraries tag-to-gene mapping (mapping tags to a genome via BLAST and then mapping tags to their genes using a data-driven probability distribution), were required to obtain insight into the Leishmania genome structure and transcription [15].

Subsequent functional annotation of a gene product is also not a trivial process. Homology searches using BLAST [16] and BLAST-based algorithms have proved successful, while many online databases provide tools and information for annotation, e.g. PIR [17], Pfam [18], SWISS-PROT [19], UniProt [20] and COG [21]. Unfortunately most of these approaches rely on the availability of accurately characterized protein information. In addition, functional similarity cannot always be inferred from homology, and additional information is required to increase confidence in functional predictions [18]. The use of Gene Ontology (GO) [22] is used very widely and successfully, but can be limited by a dependence on un-curated, experimentally unverified and computationally derived annotations [23]. One alternative that is increasingly being incorporated is the use of protein-protein interactions to predict and identify functional relationships. A main limitation is the absence of protein interaction data [24], though the geometric relationship in the alignment of predicted secondary structural elements could be used to model and evaluate putative structural interactions for functional predictions [25]. Combinations of data, such as protein sequence, gene expression and protein interaction data, could be amalgamated to infer and confirm functional annotation predictions [26].

### 42.1.2 Transcriptome and Proteome Analyses

A transcriptome is a highly dynamic system, representing a defined set of genes that are expressed at a specific time under a specific set of conditions by a cell. Transcriptome analyses provide researchers with clues regarding rare transcripts, alternate splicing, copy number variation, sequence variation, regulatory mechanisms and biochemical pathways involved in cellular function. RNA-Seq is the choice of technique in the postgenomic era for obtaining large transcriptome datasets, as it has been estimated to detect 25 % more transcripts than conventional DNA microarrays [4]. To date, RNA-Seq and conventional transcriptome profiling have been used with great success to study between-subject differences in immune responses to a pathogen or a candidate vaccine [4], the role of miRNAs in regulating mRNA expression [4], epigenetic responses to viral vaccines [4], understanding of molecular processes and survival mechanisms (e.g. the heartworm *Dirofilaria immitis* for identification of vaccine and drug targets [27]), identification of additional surface-expressed vaccine candidates (e.g. *Neisseria meningitidis* [28]), identification of expressed surface and secreted proteins in complex pathogens (e.g. the apicomplexan parasite *Eimeria tenella* [29] and the ectoparasitic cattle tick, *R. microplus* [30]) which can now be exploited as vaccine candidates.

A proteome represents the complete protein complement expressed by a cell, allowing a more direct measure of cellular response than nucleic acid-based analyses and insight into post-transcriptional modifications. Typically proteomes were evaluated by two-dimensional electrophoresis, but this presented with the obstacle that few hydrophobic proteins were recovered for downstream analyses. Nowadays high-throughput mass spectrometry is accelerating membrane protein analysis [31] and revealing the presence of plasma membrane proteins that are unknown in databases despite extensive genome sequencing. Proteome analysis is vital to overcoming the limitations of in silico localization prediction software. As surface-exposed proteins are mostly targeted during vaccination, surfaceome analyses have received quite a bit of attention. Novel approaches include analysis of membrane proteins using "shaving", whereby cells are treated with proteases and the resulting peptides analysed by LC-MS-MS [5] and advanced mass spectrometry based identification of viral peptides from high- or low-responder-associated HLA molecules as candidate antigens [32].

To date, proteomics provided insight into the neglected malaria parasite *Plasmodium vivax* which identified 153 proteins that show no homology to previously identified products as well as 29 new proteins [33], insight into the pathogen biology, genes and metabolic pathways of *Trypanosoma cruzi* [34] as well as providing a platform for the development of epitope-driven vaccines and structural proteomics in *Mycobacterium tuberculosis* [35]. The availability of host proteome data and its integration with other "omics"-data (predominantly transcriptome data) are vital to the success of future vaccines (see Sect. 16.3 of this chapter). A glossary relevant to the multidisciplinary fields of vaccine development in the postgenomic era is presented in Table 42.1.

## 42.2 Understanding the Host Immune System

Pathogens are recognized by a wide variety of immune cells, depending on how and where the immune cells are encountered first. Predominantly, they are recognized by cells of the innate immune system that do not have antigen-specific receptors, but rather contain pathogen recognition receptors (PRRs) such as C-type lectins, Toll-like receptors (TLRs) or NOD (nuclear-binding oligomerization domain) proteins. These are present in innate immune cells such as dendritic cells (DCs), macrophages, mast cells and neutrophils as well as endothelial cells and fibroblasts (collectively referred to as antigen-presenting cells, APCs). Binding results in the activation of signalling pathways and the subsequent synthesis and release of pro-inflammatory cytokines and chemokines. Activated antigen-presenting cells (mainly DCs but also macrophages and B cells) can then process a pathogen-derived antigen for presentation to adaptive immune cells on MHC class II.

This is mediated by either macropinocytosis or endocytosis of exogenous protein, which is delivered to acidic endosomes where it is cleaved into peptides that combine with $\alpha$ and $\beta$ subunits of the MHC II complex and are expressed on the cell surface. Activated and mature APCs migrate to lymph nodes where they mediate antigen-specific activation and co-stimulatory signals to naïve T (via interaction between MHC and T-cell receptor) and B cells, bridging the gap between innate and adaptive immunity. T cells differentiate into CD4$^+$ T helper cells (T$_H$) or CD8$^+$ cytotoxic T lymphocytes (CTL). CD4$^+$ T$_H$ cells can be directed into a cellular Th1, a humoral Th2 or a tolerance Treg response, depending on contact with the antigen-presenting cells (APCs) and induction by specific cytokines [43]. In most cases, peptides associated with MHC II will

**Table 42.1** Glossary relevant to the multidisciplinary fields of vaccine development in the postgenomic era

| Term | Published definition/*popular explanation* | References |
| --- | --- | --- |
| Immunogenetics | Genetic analysis of the immune system | [36] |
|  | *The study of genes involved in the immune system* |  |
| Immunogenomics | Genomics technology combined with immunology | [37], [38] |
|  | *The study of sets of genes and their expression profile involved in immune responses* |  |
| Systems biology | Systems biology aims to understand quantitatively how properties of biological systems can be understood as functions of the characteristics of and interactions between their macromolecular components | [39] |
|  | *The study of the complete system of organisms including all interactions between biological elements and their surrounding environment* |  |
| Immunoproteomics | Defines the subset of proteins that induce an immune response (immunoproteome) | [40] |
|  | *The study of all proteins involved in the immune response, their identification and function* |  |
| Vaccinomics | The term refers to immunogenetics and immunogenomics as applied to vaccine immune responses and the mechanisms underlying heterogeneity in both the pathogen and the host response to vaccination | [41] |
|  | *An integrative approach incorporating immunogenetics and immunogenomics for the study of immune responses to vaccines or pathogens* |  |
| Reverse vaccinology | Genomic-based approaches to vaccine development | [42] |
|  | *The methodology of vaccine development based on genomic information of organisms* |  |
| Immunotherapy | The complex network of interactions between immune cells during manipulation of the host immune system | [43] |
|  | *Treatment of disease by inducing, enhancing or suppressing an immune response* |  |

induce CD4+ T-cell and antibody responses [1]. A subset of effector T cells will differentiate into memory cells capable of inducing a response against the pathogen during subsequent infection. It is this stimulation of crosstalk between the innate and adaptive immune systems that represents a major obstacle when using subunit vaccines. Most simple subunit vaccines only induce CD4+ responses (which activates B cells and antibody production) which are not sufficient to protect against chronic infections and intracellular pathogens, which require effective activation of cellular adaptive CD8+ T cells (via MHC I or MHC II) responses as well [1].

To provide insight into the vast amount of genes expressed within the human immune cells, the Immunological Genome Project [44] as well as IRIS [45] (immune response in silico) database was developed. These include profiles of virtually all human and murine genes expressed from key immune cell types and provide researcher with tools to combine immunology and computational biology.

This data provides insight into the differentiation of immune cells, their roles in inflammation and immunity, cellular localization of proteins and signalling pathways of the immune system [45]. Large-scale studies into T-cell receptor diversity were made possible with next-generation sequencing and have since provided insight into the role of rearrangement of segments (predominantly in the VDJ region) in the TCR gene and how this brings about the vast repertoire of antigen-specific T cells. This also expanded into B-cell studies with the identification of 14 new allelic variants in the human heavy chain immunoglobulin variable regions [4] as well as improved understanding of vaccines targeting dendritic cell subsets [46]. ChIP-Seq technology has also been used with success to provide insight into the transcription factors involved in the development of B- and T-cell responses [4].

Findings on the role of polymorphisms in the genome of immune cells insight into the heterogeneity of responses against vaccines are fast expanding and paving the way forward to personalized vaccines [47, 48]. Evidence for genetic host polymorphisms in determining responses against vaccines in humans was provided by the classical studies of Jacobson and colleagues as well as Tan and colleagues in which monozygotic and dizygotic twin pairs were evaluated in response to the MMR vaccine, hepatitis B, oral polio, tetanus toxoid and diphtheria vaccines [47]. Subsequently, a number of polymorphisms (predominantly SNPs) in the human leukocyte antigen (HLA) genes and their contribution to vaccine immunity towards MMR and rubella have been published (reviewed in [47]). Such linkages remain to be elucidated for other vaccines, but are currently limited by the vast amount of data required for large-scale genome-wide linkage studies. Polymorphisms in non-HLA genes such as SNPs in cytokines and cytokine receptors, Toll-like receptors, signalling molecules, vitamin A and D receptors, antiviral effectors and genes associated with innate and adaptive immunity have also been published for rubella and a number of adjuvants and how these affect vaccine efficacy [32, 47].

## 42.3 Vaccinomics: Towards Understanding the Host-Pathogen Interactions and Type of Immunity Induced During Infection and Vaccination

Upon infection, pathogens activate a great number of immune triggers in both the innate and adaptive immune responses mediating protective immunity and the progression to long-lived protection against the pathogen [49]. This has been termed "the immune response network theory", which aims to integrate environmental, host and pathogen or vaccination factors [47]. Our ability to understand the gene players involved in infection and vaccination and how they correlate to protective immunity will provide essential insights for improved vaccine development (Fig. 42.1). The latter is clearly stated by Bernstein and colleagues as "Future vaccines will focus not only on the identification of a single correlate (or multiple, independent corre-

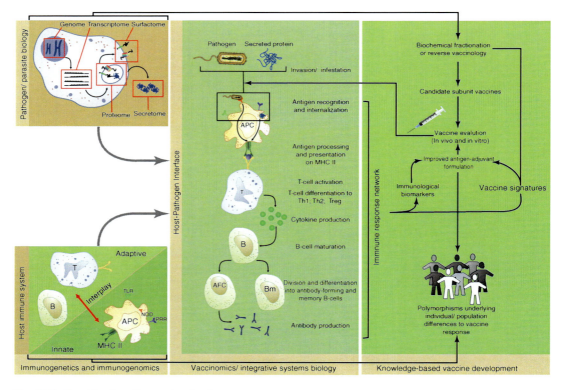

**Fig. 42.1** Considerations for vaccine development

lates), but on the identification of multifactorial signatures associated with immune protection (i.e. pathogen, host and interacting signatures)" [49], paving the way forward to the ultimate goal of universal vaccine signatures.

Vaccinomics was first described by Poland in 2007 as the application of immunogenetics and immunogenomics, which is the study of genetic and epigenetic determinants and pathways relevant to vaccine-induced responses. In 2011 it was expanded to include the mechanisms underlying heterogeneity in both the pathogen and the host response to vaccination, which aims at providing insight into the observed variability in response to vaccination [50].

Integrative approaches have been used successfully to provide insight into future vaccines. For microbial biology this has been excellently reviewed by Zhang et al. [51]. Other examples include simultaneous analysis of host and pathogen transcriptomes in high and low antibody responders to rubella, smallpox and influenza vaccines [32] as well as integrating genotypes/haplotypes and phenotypes for measles, mumps, rubella, influenza and smallpox vaccines [32]. The combination of genome-wide expression data with interactome (yeast two hybrid) data in influenza infection resulted in the identification of novel viral genes regulating interferon production and viral replication [50]. It has also been used in understanding T-cell responses to yellow fever vaccination (identifying signatures of neutralizing antibody responses and two additional signatures of $CD8^+$ T cells) as well as identifying two distinct IFN responses by neutrophils in tuberculosis patients [50].

Global transcriptional profiling of patients receiving the MAGE-A3 peptide vaccine for non-small-cell lung cancer revealed signatures correlating with responsiveness to treatment [43]. Immunogenetics studies focussing on the host OAS gene family and West Nile virus (WNV) revealed that polymorphisms in OASL were linked to susceptibility in WNV infection and that therapeutic agents/adjuvants capable of inducting OAS1 activity could enhance WNV vaccine efficacy [50]. Integrated transcriptome studies in zebrafish vaccinated with *Edwardsiella tarda* live attenuated vaccines revealed pathways involved in antigen processing and acute phase responses which provided insight into the mechanisms underlying zebrafish immune responses and directions for improved vaccines. Interestingly, they found that MHC I pathways were upregulated and MHC II pathways were downregulated during vaccination [52].

In another study, gene expression profiling of dendritic cells in the presence of *Staphylococcus aureus* revealed a unique role for the inflammatory process and T helper cell polarization as well as 204 differentially expressed genes between susceptible and resistant animals, providing explanations for the difference in susceptibility towards *S. aureus* infection [53]. Genome-wide digital gene expression (DGE) was successfully used to study the host response to the zoonotic pathogen, *Brucella melitensis*.

An in-depth understanding of the parasite/pathogen biology as well as the host immune system is required to fully comprehend the processes underlying infection and development of protective immunity. This is achieved by the combination of genomic, transcriptome and proteome data which in turn drives reverse vaccinology approaches and subunit vaccine development. Integrative systems biology approaches provide insight into the complex networks underlying infection and the development of protective immunity. By linking polymorphisms to heterogeneities observed in responses to vaccines and survival/virulence of pathogens, insight is gained for the development of improved vaccines. Abbreviations corresponds to T-cell receptor (TLR), nuclear-binding oligomerization domain-containing proteins (NOD), pathogen recognition receptor (PRR), major histocompatibility complex II (MHC), antigen-presenting cell (APC), T helper type 1, 2 or tolerance response (Th1, Th2 or Treg, respectively), B cell (B), antibody-forming cell (AFC) and memory B cells (Bm)

This study revealed, amongst others, a strong role for macrophages during infection and the induction of anti-inflammatory and anti-apoptotic factors in the survival of strains with differing virulence, directing research into development of new attenuated vaccines with enhanced efficacy [54].

To accomplish insight into the complex traits underlying immunity and vaccine reactivity, an integrated systems biology approach that combines different "omics"-datasets with powerful analytical tools, providing scoring and correlation between datasets, is required [50]. This remains a serious bottleneck in a systems biology approach to vaccine design, but the field is expanding rapidly. Several good reviews on in silico tools to combine "omics"-data have been published [51, 55–58]. To date, software packages and databases for integration of "omics"-data include RefDIC [58] (the reference genomics database of immune cells which combines transcriptome, proteome and immunogenetic data), IIDB [59] (the innate immune database), WIBL [60] (workbench for integrative biological learning), a guided clustering package for use in R [61] (integration of microarray, genome-wide chromatin immunoprecipitation and cell perturbation assays) and EchoBasE [62] (for *E. coli*). Several public available databases such as integrOmics [63], bioPIXIE [57] (currently optimized for yeast biological processes), VESPA [64] (integration of genomic, transcriptome and proteome data from prokaryotes), openBIS [65] (integration of next-generation sequencing, metabolomics and proteomics), GPS-Prot [66] (integration of HIV-human interaction networks, in the process of being extended to other host-pathogen systems), Booly [67] and PARE [68] are also available.

Finally, structural biology has started to provide insight into how the VRC01 antibody neutralizes some 90 % of HIV-1 strains, indicating a vast diversity in neutralizing antibodies directed against autologous HIV-envelope sequences across many vaccine recipients and infected individuals [4]. Structural vaccinology incorporates structural biology with vaccinology and can be used for the improvement of vaccine design. As neutralizing antibodies need to recognize specific antigen architectures, the identification of these regions using high-resolution structural analysis is important. By keeping these neutralizing architectures intact, while changing other regions to make an antigen more stable, cost-effective and eliminate variable regions, vaccines can be improved.

Furthermore, the knowledge about the structure of certain antigenic regions will allow the improvement of our understanding of immunogenicity and immunodominance thus helping with future vaccine design [69]. The structural vaccinology approach was followed in several studies, for example, constructing a chimeric protein as vaccine against group B *Streptococcus* [70]. Some proteins of these microbes have evolved to contain antigenic regions with non-similar properties in different isolates, however, with same functional properties thus helping evade the host immune system. Therefore, Nuccitelli et al. constructed a chimeric protein containing the domains of six antigenic variants resulting in protection against all six tested isolates [70].

### 42.3.1 The Challenges of Cross-Reactivity and Autoimmunity

Cross-protection against related pathogenic strains remains an attribute of a good vaccine. Although vaccinating against conserved proteins seems promising, antigens conserved across genomes are less likely to result in an effective immune response as they are generally exposed to the host immune system and thus are under strong selection pressure [71]. If the antigen shares similarity with a host protein, a poor immune response is most likely the consequence due to epitope mimicry, and autoimmunity can be triggered, targeting both the antigen and host protein(s) [72–74]. Not only proteins can elicit the latter but also polysaccharides. This became evident from vaccines in meningococcal bacteria which could not be used for *Neisseria meningitidis* because of the induction of polysaccharide-based autoimmunity [2].

Similarity searches between the identified putative antigens and the host proteins should be performed with tools such as BLAST. This allows the comparison of predicted ORFs with known genes and proteins in databases. MHC ligands can be very short with only nine residues which might allow a short region to result in an autoimmune response or result in low immunogenicity [75]. A subtractive approach can also be followed using similarity searches in order to identify proteins present in pathogenic strains, but not in non-pathogenic strains [76]. Furthermore, an epitope conservancy tool is available, allowing the analysis of how conserved or variable a specific protein is [77].

## 42.4 Reverse Vaccinology

Reverse vaccinology is a process whereby genome and transcriptome sequences are analysed by bioinformatics tools to identify potentially protective secreted and surface-displayed antigenic proteins [2]. These proteins are then expressed, used to immunize a suitable host to validate their immunogenicity and protective abilities before proceeding to production and formulation (Fig. 42.2). A main concern associated with this approach is that vaccinologists have not been able to improve the initial bioinformatics of candidate selection steps allowing improved prediction of antigenicity and even more so, their ability to induce protective immunity [2]. Although reverse vaccinology has been applied to a number of human and animal pathogens (see list in [71]), it must be noted that it may not be applicable to all vaccine development projects.

### 42.4.1 Prediction of Subcellular Localization

Most protective antigens require that the antigenic determinants are displayed on the surface of the pathogen as immune factors may be unable

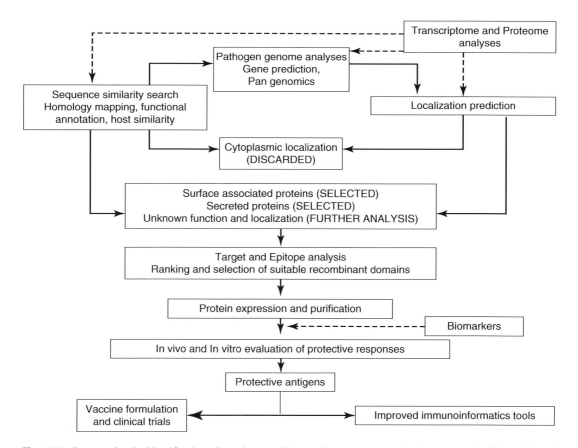

**Fig. 42.2** Strategy for the identification of vaccine candidates using a reverse vaccinology approach (Adapted from [78, 79])

to cytoplasmic or inner-membrane proteins [42, 80, 81]. To date, vaccines have targeted an array of surface-displayed adhesins, fimbrial proteins, GPI-anchored antigens, toxins, invasins and porins [82]. Knowledge on the localization of an antigen additionally provides researchers with clues to the biological function of a protein and its involvement in biological pathways [83].

A vast amount of protein subcellular localization predictors are freely available, and choosing a program suitable for your organism of choice can be challenging. In many cases, a combinatorial approach is required to identify secreted proteins as was the case in the parasitic nematode, *Strongyloides ratti* [84]. The software used for subcellular localization prediction can be divided into two groups. The first uses amino acid sequence for the prediction, while the second is annotation based [85]. PSORT, the first computer program based on protein sorting signals, was published in 1990 [86]. Several improvements have followed, such as the incorporation of machine-learning techniques, in order to increase the accuracy of localization prediction in various organisms. The latest version of PSORT, PSORTb 3.0, was published in 2010 and represents one of the most accurate subcellular localization prediction programs for bacteria and can be used for all prokaryotes [87].

A study conducted in 2006 compared several mammalian subcellular localization predictors and determined that no method was sensitive enough to predict the protein localization in both tested datasets accurately. Furthermore, proteins targeted to the secretory pathway were found to be the most difficult to predict [88]. Newer predictors for eukaryotic protein localization include iLoc-Euk, which is the most accurate software taking into account that some proteins can be present at more than one subcellular localization and even move between these [83]. Other software, such as SherLoc2, includes sequence-based and annotation-based features, phylogenetic profiles as well as Gene Ontology terms obtained from the protein sequence [85]. The Cell-PLoc package contains six predictors, Euk-mPLoc, Hum-mPLoc, Plant-PLoc, Gpos-PLoc, Gneg-PLoc and Virus-PLoc. These predict the subcellular localization of proteins in eukaryotes, human, plants, gram-positive bacteria, gram-negative bacteria and viruses, respectively, combining different web servers with various approaches [89].

The process starts with the analysis of the transcriptome and the proteome of a parasite/pathogen from which predicted surface-associated and secreted proteins are identified following localization prediction and sequence similarity searches. Homology mapping of neutralizing antibodies, functional annotation and host similarity searches for the prevention of autoimmunity follows. Identified targets may be ranked and used to predict epitopes, suitable recombinant domains and analyse B- and T-cell epitope variability. The chosen proteins then have to be expressed and purified followed by evaluation of immune responses using biomarker information if available. Protective antigens can then be used for vaccine formulation and clinical trials and the obtained data can be used to improve available immunoinformatics tools.

The subcellular localization of viral proteins in infected cells is of interest as this is directly linked to the health of the host and antiviral drug design [90]. Virus-mPLoc is one of the newer predictors and uses Gene Ontology, functional domain and evolutionary information. One of the main improvements to previous viral protein localizers is the ability of this program to detect proteins which can be found at multiple localizations [91]. Another web-based software with allegedly improved features is iLoc-Virus [90].

Most proteins are synthesized in the cytosol following sorting to different organelles based on N-terminal sorting signals [92]. For the use in a reverse vaccinology pipeline (see Sect. 42.4), signal peptides are of interest as these are responsible for the transport of proteins through the secretory pathway in both pro- and eukaryotes, thus, able to result in a localization on or outside the cell surface [93, 94]. A study conducted in 2009 identified SignalP 3.0 as the most accurate predictor compared to 11 others, followed by Rapid Prediction of Signal Peptides (RPSP) [95]. Viruses require the host cell machinery for replication and protein synthesis and thus often contain eukaryotic-targeting signals and functional domains in order to exploit the host localization mechanisms [96]. Therefore, appropriate eukaryotic subcellular localization predictors can be used. Both signal peptides and transmembrane regions have hydrophobic regions and can thus be misinterpreted by software when they occur in the N-terminal region of a protein [97]. To overcome this obstacle SignalP version 4.0 was trained on datasets differentiating between transmembrane region and signal peptide [97]. SignalP versions 3.0 and 4.0 can be used for eukaryotes, gram-positive and gram-negative bacteria. Other software, such as the Phobius web server [98, 99] or SPOCTOPUS [100], are available also taking the difficulty of correct prediction between signal peptide and transmembrane region into account.

Furthermore, it is important to consider the number of transmembrane helices in a protein. Firstly, if only part of a protein is expressed due to experimental constraints, an extracellular domain should be chosen due to its accessibility to the immune system. Secondly, if complete proteins are expressed, it has to be taken into consideration that proteins containing more than one transmembrane domain have been found to be more difficult to express [80]. Therefore, if one can choose an antigen from a number of available proteins, preference should be given to single-spanning compared to multi-spanning proteins.

### 42.4.2 Prediction of Immunogenicity and Epitopes

The prediction of epitopes (protein sequences that can be recognized by the host immune system and elicit an immune response) remains a crucial step in the screening of pathogens protein-coding sequences before experimental confirmation of immunogenicity. Several computational methods are available to date (reviewed in [101–104]) based on either sequence and/or structure.

Sequence-based approaches assume that a similar amino acid sequence leads to a similar protein structure and function. The ability to detect B- and T-cell epitopes also depends on a protein's sequence pattern and physicochemical properties such as flexibility, accessibility, hydrophilicity and charge [105]. In a comparative study conducted by Yu and colleagues, the use of sequence and protein motifs was found to be most accurate for the prediction of MHC binding. However, with increasing data volume, software utilizing machine-learning techniques becomes more precise [106]. Although structure-based predictions are possible with methods like homology modelling, these approaches are relatively new, costly and require the knowledge of the three-dimensional structure of proteins. Therefore, the latter approach is not yet commonly used [107]. The most accurate prediction would include several techniques as the strengths of each method can then be combined [105].

The prediction of T-cell epitopes can be indirectly predicted by the detection of MHC binders. Due to the different conformations of MHC I and MHC II proteins, they will bind different molecules [108, 109]. MHC II, in contrast to MHC I, can bind variable peptide lengths resulting in MHC II-binding predictions to have a much lower accuracy than for MHC I [105]. These predictions are not always accurate as some proteins, even though bound to MHC, do not elicit an immune response [108]. For both MHC classes ligands can be identified either by using predictions for the binding of the peptide to the MHC complex or by predicting the peptide's processing pathway before binding [108]. One of the newest software, POPISK, focuses on the prediction of the MHC I-peptide complex sequence which is recognized by the T-cell receptors. This recognition should then result in T-cell activation and further immune responses [110]. One of the most accurate predictors for cytotoxic T lymphocyte epitopes is NetCTLpan [111]. However, it is known that many different factors play a role in whether a protein results in an immune response or not. Binding affinity to MHC I, for example, is necessary but not sufficient to result in a T-cell response [110]. To therefore be able to fully exploit epitope prediction methods, more knowledge about the complex interplay of all factors involved in the immune system is required.

The prediction of B-cell epitopes is a more difficult task than for T cells. One reason for this is that MHC proteins are highly specific in the molecules they bind, while a vast amount of short, linear peptides have the potential to be able to bind to antibodies [108]. Due to the fact that two types of B-cell epitopes can be distinguished – continuous and discontinuous – different prediction methods must be applied. Methods identifying continuous B-cell epitopes are similar to T-cell epitope predictors based on protein properties. In contrast, discontinuous B-cell epitope prediction requires the knowledge of the 3D structure of the antigen-antibody complex [107].

This type of prediction is thus very difficult and current software for this purpose shows relatively poor performance. Bioinformatics tools for the prediction of discontinuous B-cell epitopes started with the obtainment of residue solvent accessibility (CEP server) [112]. Some of the newer software use a large amount of physiochemical as well as structural-geometrical properties included in a Bayesian analysis [113], consensus scoring from different functions (propensity, conservation, energy, contact, surface planarity and secondary structure composition) [114] and a logistic regression algorithm using B factor and accessible surface area as structural features in addition to taking the spatial environment for each residue into consideration [115]. Another approach for the identification of discontinuous B-cell epitopes uses a new concept of spatial characteristics of antigen residues (a distance-based feature) and three-dimensional structures [116].

The use of deep-panning combining the strengths of phage display with deep sequencing may aid in our understanding of the humoral response to disease, by mapping polyclonal antisera specificities [117]. For the prediction of protective bacterial, viral and tumour antigens, an alignment-independent method was published in the form of the VaxiJen server based on an autocross covariance method [118]. A method including support vector machine classification showed improved results thus having potential for the improvement of reverse vaccinology approaches [81]. Protein microarrays can also be used to identify proteins expressed during host infection and for the prediction of protein antigenicity [119, 120].

### 42.4.3 Reverse Vaccinology Platforms

In an attempt to simplify the reverse vaccinology approach, several platforms have been created. This enables a user to repeat the process as often as required with altered stringency conditions in order to obtain less vaccine candidates for further evaluation. NERVE (New Enhanced Reverse Vaccinology Environment) is a server used for automated reverse vaccinology analyses [121]. It does have the drawback of being computationally complicated [2]. Several filters are used which allow the prediction of a protein's localization (PSORTb) [122] (required to be non-cytoplasmic) and the identification of adhesins (SPAAN) [123]. Furthermore, the number of transmembrane helices (required to be <2) is identified according to HMMTOP [124]. Similarity searches are done in order to prevent significant similarity to human proteins, identify homology to known antigens and aid functional annotation.

Vaxign is a web-based platform that searches for antigens from over 70 pathogenic genomes, identifying protein conservation between genomes and excluding sequences present in nonpathogenic organisms. It includes a number of different programs for the prediction of a protein's subcellular localization, number of transmembrane helices, adhesion probabilities, functional analysis and immune epitopes (MHC classes I and II). Furthermore, the server allows comparison of the input sequences to the host in order to prevent any cross-reactions and consequent harming of the host [73]. Currently, this server only allows the processing of high-throughput data of microbes [73]. Systems allowing insight into eukaryotic organisms are not available to date, but VaxiJen has been able to show impressive prediction accuracy of up to 70–89 % for bacterial, viral and tumour antigens and 78–97 % accuracy for endoparasitic and fungal antigens [125, 126].

## 42.5 Validation of Vaccine Candidates

During a reverse vaccinology approach, hundreds of antigens are identified which could induce either humoral and/or cellular immunity. Relevant markers are required for in vitro and in vivo assays in order to assess the vaccine efficacy and to optimize future vaccine formulation as well as immunological biomarkers capable of evaluating immunogenicity.

In vitro approaches to date mainly consist of ELISAs and variations thereof, determining the amount of antibodies present in serum. A platform that sets the pace is MATS (meningococcal antigen typing system) which is a sandwich ELISA system capable of measuring the amount of antigen expressed by a strain, cross-reactivity with the antigen in a vaccine and correlating the data with large strain panels from many countries [5]. ELISPOT assays can quantitatively measure chemokine and cytokine profiles. Antisera raised can also be analysed using flow cytometry or in bactericidal assays [5]. A number of conventional techniques are still being used, such as agglutination and precipitation tests to detect antibodies, complement fixation tests, neutralization tests for antitoxins and fluorescent antibody tests to visualize antigen-AB complexes. These do not however provide insight into the complex cellular environment that is providing protective immunity. For in vivo evaluation, one approach is to pool antigens prior to immunizing mice to evaluate their immunogenicity [2]. Ex vivo immune system models that can provide an alternative to costly clinical trials, the testing of immunogenicity and vaccine responses are in dire need and require extensive research [32].

### 42.5.1 Immunological Biomarkers

Biomarkers are measurable products from organisms which can be used as indicators of a biological state, pathological processes or pharmacological responses to therapeutic interventions. They can be used for several applications including the prediction and evaluation of drugs for the improvement of current and the development of new drugs [127]. Although many vaccines may protect via multiple immune response mechanisms [128], immunological biomarkers have the potential to correlate a specific immune response with vaccine efficacy [129].

The use of biomarkers in cancer patients has shown that multiple markers can be useful to profile disease progression and correct treatment options [130]. Immunological biomarkers indicative of disease status and treatment options are however still limited [127]. The cytokine interferon gamma (IFN-γ), for example, is known to be involved in a protective immune response against various pathogens, including *Mycobacterium tuberculosis* [131] and *Ehrlichia ruminantium* [132]. This biomarker level can be tested before and after vaccination in order to determine if the current vaccine antigen is promising. Knowledge about the interaction of the host immune system and the pathogen is vital for the development of immune biomarkers and effective vaccines [128].

### 42.5.2 Protein Expression, Adjuvants and Vaccine Trials

To allow for the evaluation of the protective abilities of a subunit vaccine, recombinant proteins need to be recombinantly expressed and purified. This brings along a number of obstacles in itself, of which the most concerning is the presence of host cell endotoxins for which a number of approaches have been developed [71]. A noteworthy approach for initial analysis may be the use of an in vitro translation systems (these will be too expensive for large-scale vaccine production). In a study by Cardoso and colleagues, they expressed almost an entire proteome from Plasmodium species using in vitro transcription and translation systems. These proteins were then analysed for their ability to induce T-cell immunity using IFN-γ ELISpot and cytometric bead array cytokine assays, paving the way to high-throughput approaches to identify T-cell targets of complex pathogens [133].

Most inactivated vaccines or subunit vaccines cannot effectively activate the innate immune system, and this necessitates the addition of an adjuvant to improve their immunogenicity [43, 134]. Three types of adjuvants can be differentiated. Firstly, depot adjuvants are used to prolong the lifespan of the injected antigens by preventing them from being degraded. Examples of this type are Freund's incomplete adjuvant and Montanide ISA 50V2 (mineral oil-based adjuvant). Secondly, particulate adjuvants, such as saponin, help with the delivery of the antigens to antigen-presenting cells, thus resulting in an enhanced immune response.

Finally, immunostvimulatory adjuvants increase cytokine production thus promoting T helper cell responses by co-stimulating the host immune system in addition to the antigen. Such compounds contain, for example, complex microbial products which can easily be recognized by dendritic cells and macrophages. An example for this type of adjuvant is Muramyl dipeptide. Very potent adjuvants can also be produced when combining a particulate or depot adjuvant with an immunostimulatory one. An example for this mixture is Freund's complete adjuvant [135]. Two insightful reviews on the different types of adjuvants and their use in current vaccines have been published [136, 137].

The addition of an adjuvant to a vaccine formulation potentially has the advantages of resulting in sufficient immune responses to achieve protection, prolonging the duration of the immune response therefore resulting in lower number of immunizations, the requirement of lower antigen doses, preventing antigen competition when multiple compounds are injected, increasing the stability of the antigen as well as the breadth of the response and overcoming limited immune responses in some instances [137]. The most commonly used adjuvants are alums (insoluble aluminium salts). Its mode of action is currently seen as involving increased uptake of antigens by antigen-presenting cells and destabilizing them resulting in easier processing and presentation. Furthermore, alum can activate the NLRP3 inflammasome thus allowing the secretion of pro-inflammatory cytokines. However, the complete mechanism and induction of a resulting Th2 response is not fully understood [137]. The web-based Vaxjo is a database and analysis system allowing the curation, storage and analysis of vaccine adjuvants and their uses in vaccine development [136].

Vaccine trials must be planned beforehand to provide feedback and allow the development of further hypotheses. This is already the principle in drug-development trials but is missing to great extent in vaccine trials due to the lack of the availability of biomarkers. The availability of knockout mice containing human variants of the murine gene opens an exciting new approach to studying vaccine responses and linking genome-wide association studies into functional studies [50].

In conclusion, vaccine development has come a long way from trial-and-error testing of live attenuated pathogens to the development of the first subunit vaccines. In the postgenomic era, the availability of a vast array of "omics"-data coupled with the ability to integrate this data, is fast enabling researchers to word towards a multidisciplinary, exciting, yet challenging future for vaccine design. A glossary of software and databases mentioned in this chapter is given in Table 42.2.

**Table 42.2** Glossary of software and databases mentioned in this chapter

| Name | Description | Reference |
|---|---|---|
| BLAST | Similarity searches for various applications | [16] |
| ORF identification | | |
|   Glimmer | Bacteria, archaea, viruses | [8] |
|   GeneMark.hmm | Bacteria | [9] |
|   Prodigal | Prokaryotes | [10] |
|   AUGUSTUS | Eukaryotes | [14] |
|   GNOMON | Eukaryotes | http://0-www.ncbi.nlm.nih.gov.innopac.up.ac.za/genome/guide/gnomon.shtml |

**Table 42.2** (continued)

| Name | Description | Reference |
|---|---|---|
| MAKER | Eukaryotes, annotation pipeline | [13] |
| **Functional annotation** | | |
| PIR | Protein information resource | [17] |
| Pfam | Protein families database | [18] |
| SWISS-PROT | Manually annotated and reviewed section of UniProt knowledge base | [19] |
| UniProt | Knowledge base | [20] |
| COG | Database | [21] |
| GO | Database | [22] |
| **Subcellular localization** | | |
| PSORT | Bacteria, eukaryotes | [86] |
| PSORTb v3.0 | Prokaryotes | [87] |
| iLoc-Euk | Eukaryotes | [83] |
| SherLoc2 | Eukaryotes | [85] |
| Cell-PLoc | Eukaryotes, prokaryotes, viruses | [89] |
| Virus-mPloc | Viruses | [91] |
| iLoc-Virus | Viruses | [90] |
| **Signal peptide** | | |
| SignalP 4.0 | Signal peptide prediction | [97] |
| Phobius | Signal peptide prediction | [99] |
| SPOCTOPUS | Signal peptide prediction | [100] |
| **Epitopes** | | |
| POPISK | MHC I predictor | [110] |
| NetCTLpan | CTL predictor | [111] |
| CEP | Prediction of conformational epitopes | [112] |
| VaxiJen | Viral, tumour, bacterial protective antigens | [118] |
| Epitope conservancy tool | Database of conserved epitopes | [77] |
| **Reverse vaccinology platforms** | | |
| NERVE | Pipeline | [121] |
| Vaxign | Pipeline | [73] |
| **Software packages and databases for integration of "omics"-data** | | |
| RefDIC | Reference genomics database of immune cells | [58] |
| IIDB | Innate immune database | [59] |
| WIBL | Workbench for integrative biological learning | [60] |
| Guided clustering | Combines experimental and clinical data | [61] |
| EchoBasE | Database for *E. coli* | [62] |
| integrOmics | | [63] |
| bioPIXIE | Network predictions for *S. cerevisiae* | [57] |
| VESPA | Genomic annotation of prokaryotes | [64] |
| openBIS | | [65] |
| GPS-Prot | Platform for integrating host-pathogen interaction data | [66] |
| Booly | Data integration platform | [67] |
| PARE | Protein abundance and mRNA expression | [68] |
| **Other** | | |
| IRIS | Immune response in silico | [45] |
| Vaxjo | Curation, storage and analysis of vaccine adjuvants | [136] |

# References

1. Powell, M.F., Newman, M.J.: Vaccine Design – The Subunit and Adjuvant Approach, vol. 6, 1st edn. Plenum Press, New York (1995)
2. Jones, D.: Reverse vaccinology on the cusp. Nat. Rev Drug Discov. **11**, 175–176 (2012). doi:10.1038/nrd3679
3. Adu-Bobie, J., Arico, B., Giuliani, M.M.: Serruto, D., Chapter 9: The first vaccine obtained through reverse vaccinology: the serogroup B Meningococcus Vaccine. In: Rappuoli, R., Serruto, D., Rappuoli, R. (eds.). Vaccine Design – Innovative and Novel Strategies, vol. 1, pp. 225–241. Caister Academic. Press, Norfolk (2011)
4. Luciani, F., Bull, R.A., Lloyd, A.R.: Next generation deep sequencing and vaccine design: today and tomorrow. Trends Biotechnol. **30**, 443–452 (2012)
5. Seib, K.L., Zhao, X., Rappuoli, R.: Developing vaccines in the era of genomics: a decade of reverse vaccinology. Clin. Microbiol. Infect. **18**, 1–8 (2012)
6. Cheng, H., Chan, W.S., Wang, D., Liu, S., Zhou, Y.: Small open reading frames: current prediction techniques and future prospect. Curr. Protein Pept Sci. **12**, 503–507 (2011)
7. Yandell, M., Ence, D.: A beginner's guide to eukaryotic genome annotation. Nat. Rev. **13**, 329–342 (2012)
8. Delcher, A.L., Bratke, K.A., Powers, E.C., Salzberg, S.L.: Identifying bacterial genes and endosymbiont DNA with Glimmer. Bioinformatics **23**, 673–679 (2007). doi:10.1093/bioinformatics/btm009
9. Lukashin, A.V., Borodovsky, M.: GeneMark.hmm: new solutions for gene finding. Nucleic Acids Res. **26**, 1107–1115 (1998). doi:10.1093/nar/26.4.1107
10. Hyatt, D., et al.: Prodigal: prokaryotic gene recognition and translation initiation site identification. BMC Bioinformatics **11**, 119 (2010)
11. Pinheiro, C.S., et al.: Computational vaccinology: an important strategy to discover new potential *S. mansoni* vaccine candidates. J. Biomed. Biotechnol. **2011**, 503068 (2011). doi:10.1155/2011/503068
12. Ecker, J.R., et al.: Genomics: ENCODE explained. Nature **489**, 52–55 (2012). doi:10.1038/489052a
13. Cantarel, B.L., et al.: MAKER: an easy-to-use annotation pipeline designed for emerging model organism genomes. Genome Res. **18**, 188–196 (2008). doi:10.1101/gr.6743907
14. Keller, O., Kollmar, M., Stanke, M., Waack, S.: A novel hybrid gene prediction method employing protein multiple sequence alignments. Bioinformatics (2011). doi:10.1093/bioinformatics/btr010
15. Smandi, S., et al.: Methodology optimizing SAGE library tag-to-gene mapping: application to Leishmania. BMC Res. Notes **5**, 74 (2012). doi:10.1186/1756-0500-5-74
16. Altschul, S.F., Gish, W., Miller, W., Myers, E.W., Lipman, D.J.: Basic local alignment search tool. J. Mol. Biol. **215**, 403–410 (1990)
17. Barker, W.C., et al.: The protein information resource (PIR). Nucleic Acids Res. **28**, 41–44 (2000)
18. Punta, M., et al.: The Pfam protein families database. Nucleic Acids Res. **40**, D290–D301 (2012). doi:10.1093/nar/gkr1065
19. Boeckmann, B., et al.: The SWISS-PROT protein knowledgebase and its supplement TrEMBL in. Nucleic Acids Res. **31**, 365–370 (2003). doi:10.1093/nar/gkg095 (2003)
20. Magrane, M., Consortium, U.: UniProt Knowledgebase: a hub of integrated protein data. Database (2011). doi:10.1093/database/bar009 (2011)
21. Tatusov, R.L., Galperin, M.Y., Natale, D.A., Koonin, E.V.: The COG database: a tool for genome-scale analysis of protein functions and evolution. Nucleic Acids Res. **28**, 33–36 (2000). doi:10.1093/nar/28.1.33
22. Gene Ontology Consortium: The Gene Ontology (GO) database and informatics resource. Nucleic Acids Res. **32**, D258–D261 (2004). doi:10.1093/nar/gkh036
23. Yon Rhee, S., Wood, V., Dolinski, K., Draghici, S.: Use and misuse of the gene ontology annotations. Nat. Rev. Genet. **9**, 509–515 (2008)
24. Gomez, A., et al.: Gene ontology function prediction in mollicutes using protein-protein association networks. BMC Syst. Biol. **5**, 49 (2011). doi:10.1186/1752-0509-5-49
25. Zhang, Q.C., et al.: Structure-based prediction of protein-protein interactions on a genome-wide scale. Nature **490**, 556–560 (2012)
26. Wass, M.N., Barton, G., Sternberg, M.J.E.: CombFunc: predicting protein function using heterogeneous data sources. Nucleic Acids Res. **40**, W466–W470 (2012). doi:10.1093/nar/gks489
27. Fu, Y., et al.: Novel insights into the transcriptome of *Dirofilaria immitis*. PLoS One **7**, e41639 (2012). doi:10.1371/journal.pone.0041639
28. Hedman, A.K., Li, M.S., Langford, P.R., Kroll, J.S.: Transcriptional profiling of serogroup B *Neisseria meningitidis* growing in human blood: an approach to vaccine antigen discovery. PLoS One **7**, e39718 (2012). doi:10.1371/journal.pone.0039718
29. Amiruddin, N., et al.: Characterisation of full-length cDNA sequences provides insights into the *Eimeria tenella* transcriptome. BMC Genomics **13**, 21 (2012). doi:10.1186/1471-2164-13-21
30. Maritz-Olivier, C., van Zyl, W., Stutzer, C.: A systematic, functional genomics and reverse vaccinology approach to the identification of vaccine candidates in the cattle tick. *Rhipicephalus microplus*. Ticks Tick Borne Dis. **3**, 179–189 (2012)
31. Savas, J.N., Stein, B.D., Wu, C.C., Yates, J.R.: Mass spectrometry accelerates membrane protein analysis. Trends Biochem. Sci. **36**, 388–396 (2011). doi:10.1016/j.tibs.2011.04.005
32. Haralambieva, I.H., Poland, G.A.: Vaccinomics, predictive vaccinology and the future of vaccine development. Future Microbiol. **5**, 1757–1760 (2010). doi:10.2217/fmb.10.146
33. Acharya, P., et al.: Clinical proteomics of the neglected human malarial parasite *Plasmodium vivax*. PLoS One **6**, e26623 (2011). doi:10.1371/journal.pone.0026623
34. Minning, T.A., Weatherly, D.B., Atwood, J., Orlando, R., Tarleton, R.L.: The steady-state transcriptome of the four major life-cycle stages of *Trypanosoma cruzi*.

BMC Genomics **10**, 370 (2009). doi:10.1186/1471-2164-10-370
35. Jagusztyn-Krynicka, E.K., Roszczenko, P., Grabowska, A.: Impact of proteomics on anti-Mycobacterium *tuberculosis* (MTB) vaccine development. Pol. J. Microbiol. **58**, 281–287 (2009)
36. Lawrence, E.: Henderson's Dictionary of Biology. Pearson Education Limited, Harlow (2005)
37. Lillehoj, H.S., Kim, C.H., Keeler, C.L., Zhang, S.: Immunogenomic approaches to study host immunity to enteric pathogens. Poult. Sci. **86**, 1491–1500 (2007)
38. Ohara, O.: From transcriptome analysis to immunogenomics: current status and future direction. FEBS Lett. **583**, 1662–1667 (2009)
39. Snoep, J.L., Bruggeman, F., Olivier, B.G., Westerhoff, H.V.: Towards building the silicon cell: a modular approach. Biosystems **83**, 207–216 (2006)
40. Tjalsma, H., Schaeps, R.M.J., Swinkels, D.W.: Immunoproteomics: from biomarker discovery to diagnostic applications. Proteomics Clin. Appl. **2**, 167–180 (2008)
41. Poland, G.A., Ovsyannikova, I.G., Jacobson, R.M., Smith, D.I.: Heterogeneity in vaccine immune response: the role of immunogenetics and the emerging field of vaccinomics. Clin. Pharmacol. Ther. **82**, 653–664 (2007). doi:10.1038/sj.clpt.6100415
42. Rappuoli, R.: Reverse vaccinology. Curr. Opin. Microbiol. **3**, 445–450 (2000)
43. Buonaguro, L., Wang, E., Tornesello, M.L., Buonaguro, F.M., Marincola, F.M.: Systems biology applied to vaccine and immunotherapy development. BMC Syst. Biol. **5**, 146–157 (2011)
44. Heng, T.S., Painter, M.W.: The Immunological Genome Project: networks of gene expression in immune cells. Nat. Immunol. **9**, 1091–1094 (2008). doi:10.1038/ni1008-1091
45. Abbas, A.R., et al.: Immune response in silico (IRIS): immune-specific genes identified from a compendium of microarray expression data. Genes Immun. **6**, 319–331 (2005). doi:10.1038/sj.gene.6364173
46. Banchereau, J., et al.: Harnessing human dendritic cell subsets to design novel vaccines. Ann. N Y Acad. Sci. **1174**, 24–32 (2009). doi:10.1111/j.1749-6632.2009.04999.x
47. Ovsyannikova, I.G., Poland, G.A.: Vaccinomics: current findings, challenges and novel approaches for vaccine development. AAPS J. **13**, 438–444 (2011). doi:10.1208/s12248-011-9281-x
48. Poland, G.A., Kennedy, R.B., Ovsyannikova, I.G.: Vaccinomics and personalized vaccinology: is science leading us toward a new path of directed vaccine development and discovery? PLoS Pathog. **7**, e1002344 (2011). doi:10.1371/journal.ppat.1002344
49. Bernstein, A., Pulendran, B., Rappuoli, R.: Systems vaccinomics: the road ahead for vaccinology. OMICS **15**, 529–531 (2011). doi:10.1089/omi.2011.0022
50. Kennedy, R.B., Poland, G.A.: The top five "game changers" in vaccinology: toward rational and directed vaccine development. OMICS **15**, 533–537 (2011). doi:10.1089/omi.2011.0012
51. Zhang, W., Li, F., Nie, L.: Integrating multiple 'omics' analysis for microbial biology: application and methodologies. Microbiology **156**, 287–301 (2010). doi:10.1099/mic.0.034793-0
52. Yang, D., et al.: RNA-seq liver transcriptome analysis reveals an activated MHC-I pathway and an inhibited MHC-II pathway at the early stage of vaccine immunization in zebrafish. BMC Genomics **13**, 319 (2012). doi:10.1186/1471-2164-13-319
53. Toufeer, M., et al.: Gene expression profiling of dendritic cells reveals important mechanisms associated with predisposition to *Staphylococcus* infections. PLoS One **6**, e22147 (2011). doi:10.1371/journal.pone.0022147
54. Wang, F., et al.: Deep-sequencing analysis of the mouse transcriptome response to infection with *Brucella melitensis* strains of differing virulence. PLoS One **6**, e28485 (2011). doi:10.1371/journal.pone.0028485
55. Kaleta, C., de Figueiredo, L.F., Heiland, I., Klamt, S., Schuster, S.: Special issue: integration of OMICs datasets into metabolic pathway analysis. Biosystems **105**, 107–108 (2011). doi:10.1016/j.biosystems.2011.05.008
56. Joyce, A.R., Palsson, B.O.: The model organism as a system: integrating 'omics' data sets. Nat. Rev. Mol. Cell Biol. **7**, 198–210 (2006). doi:10.1038/nrm1857
57. Myers, C.L., Chiriac, C., Troyanskaya, O.G.: Discovering biological networks from diverse functional genomic data. Methods Mol. Biol. **563**, 157–175 (2009). doi:10.1007/978-1-60761-175-2_9
58. Hijikata, A., et al.: Construction of an open-access database that integrates cross-reference information from the transcriptome and proteome of immune cells. Bioinformatics **23**, 2934–2941 (2007). doi:10.1093/bioinformatics/btm430
59. Korb, M., et al.: The Innate Immune Database (IIDB). BMC Immunol. **9**, 7 (2008). doi:10.1186/1471-2172-9-7
60. Lesk, V., Taubert, J., Rawlings, C., Dunbar, S., Muggleton, S.: WIBL: Workbench for Integrative Biological Learning. JIB **8**, 156 (2011). doi:10.2390/biecoll-jib-2011-156
61. Maneck, M., Schrader, A., Kube, D., Spang, R.: Genomic data integration using guided clustering. Bioinformatics **27**, 2231–2238 (2011). doi:10.1093/bioinformatics/btr363
62. Misra, R.V., Horler, R.S.P., Reindl, W., Goryanin, I.I., Thomas, H.G.: *Echo*BASE: an integrated postgenomic database for *Escherichia coli*. Nucleic Acids Res. (2005). doi:10.1093/nar/gki028
63. Le Cao, K.A., Gonzalez, I., Dejean, S.: IntegrOmics: an R package to unravel relationships between two omics datasets. Bioinformatics **25**, 2855–2856 (2009). doi:10.1093/bioinformatics/btp515
64. Peterson, E.S., et al.: VESPA: software to facilitate genomic annotation of prokaryotic organisms through integration of proteomic and transcriptomic data. BMC Genomics **13**, 131 (2012). doi:10.1186/1471-2164-13-131
65. Bauch, A., et al.: OpenBIS: a flexible framework for managing and analyzing complex data in biology

research. BMC Bioinformatics **12**, 468 (2011). doi:10.1186/1471-2105-12-468
66. Fahey, M.E., et al.: GPS-Prot: a web-based visualization platform for integrating host-pathogen interaction data. BMC Bioinformatics **12**, 298 (2011). doi:10.1186/1471-2105-12-298
67. Do, L.H., Esteves, F.F., Karten, H.J., Bier, E.: Booly: a new data integration platform. BMC Bioinformatics **11**, 513 (2010). doi:10.1186/1471-2105-11-513
68. Yu, E.Z., Burba, A.E.C., Gerstein, M.: PARE: a tool for comparing protein abundance and mRNA expression data. BMC Bioinformatics **8**, 309 (2007). doi:10.1186/1471-2105-8-309
69. Dormitzer, P.R., Ulmer, J.B., Rappuoli, R.: Structure-based antigen design: a strategy for next generation vaccines. Trends Biotechnol. **26**, 659–667 (2008)
70. Nuccitelli, A., et al.: Structure-based approach to rationally design a chimeric protein for an effective vaccine against Group B *Streptococcus* infections. Proc. Natl. Acad. Sci. **108**, 10278–10283 (2011). doi:10.1073/pnas.1106590108
71. Bagnoli, F., et al.: Designing the next generation of vaccines for global public health. OMICS **15**, 545–566 (2011)
72. Fujinami, R.S., Oldstone, M.B., Wroblewska, Z., Frankel, M.E., Koprowski, H.: Molecular mimicry in virus infection: crossreaction of measles virus phosphoprotein or of herpes simplex virus protein with human intermediate filaments. Proc. Natl. Acad. Sci. **80**, 2346–2350 (1983)
73. He, Y., Xiang, Z., Mobley, H.: Vaxign: the first web-based vaccine design program for reverse vaccinology and applications for vaccine development. J. Biomed. Biotechnol. (2010). doi:10.1155/2010/297505 (2010)
74. Oldstone, M.B.A.: Molecular mimicry and immune-mediated diseases. FASEB J. **12**, 1255–1265 (1998)
75. Schatz, M.M., et al.: Characterizing the N-terminal processing motif of MHC class I ligands. J. Immunol. **180**, 3210–3217 (2008)
76. Moriel, D.G., et al.: Identification of protective and broadly conserved vaccine antigens from the genome of extraintestinal pathogenic *Escherichia coli*. Proc. Natl. Acad. Sci. **107**, 9072–9077 (2010). doi:10.1073/pnas.0915077107
77. Bui, H.H., Li, W., Fusseder, N., Sette, A.: Development of an epitope conservancy analysis tool to facilitate the design of epitope-based diagnostics and vaccines. BMC Bioinformatics **8**, 361 (2007). doi:10.1186/1471-2105-5-361
78. Bagnoli F., Norauis, N., Ferlenghi, I., Scarselli, M., Danati, C., Savina, S., Barocchi, M.A., Rappuoli, R. Chapter 2: deigning Vaccine in the Era of Genomics. In: Rappuoli, R., Serruto, D., Rappuoli, R. (eds.). Vaccine Design – Innovative and Novel Strategies, vol. 1, pp. 21–53. Caister Academic. Press, Norfolk (2011)
79. Sollner, J., et al.: Concept and application of a computational vaccinology workflow. Immunome Res. **6**(Suppl 2), S7 (2010). doi:10.1186/1745-7580-6-s2-s7
80. Pizza, M., et al.: Identification of vaccine candidates against serogroup *B.* m*eningococcus* by whole-genome sequencing. Science **287**, 1816–1820 (2000). doi:10.1126/science.287.5459.1816
81. Bowman, B., et al.: Improving reverse vaccinology with a machine learning approach. Vaccine **29**, 8156–8164 (2011)
82. Vivona, S., et al.: Computer-aided biotechnology: from immuno-informatics to reverse vaccinology. Trends Biotechnol. **26**, 190–200 (2008). doi:10.1016/j.tibtech.2007.12.006
83. Chou, K.C., Wu, Z.C., Xiao, X.: iLoc-Euk: a multi-label classifier for predicting the subcellular localization of singleplex and multiplex eukaryotic proteins. PLoS One **6**, e18258 (2011). doi:10.1371/journal.pone.0018258
84. Garg, G., Ranganathan, S.: *In silico* secretome analysis approach for next generation sequencing transcriptomic data. BMC Genomics **12**, 514–524 (2011)
85. Briesemeister, S., et al.: SherLoc2: a high-accuracy hybrid method for predicting subcellular localization of proteins. J. Proteome Res. **8**, 5363–5366 (2009). doi:10.1021/pr900665y
86. Nakai, K., Horton, P.: PSORT: a program for detecting sorting signals in proteins and predicting their subcellular localization. Trends Biochem. Sci. **24**, 34–35 (1999)
87. Yu, N.Y., et al.: PSORTb 3.0: improved protein subcellular localization prediction with refined localization subcategories and predictive capabilities for all prokaryotes. Bioinformatics **26**, 1608–1615 (2010). doi:10.1093/bioinformatics/btq249
88. Sprenger, J., Fink, J., Teasdale, R.: Evaluation and comparison of mammalian subcellular localization prediction methods. BMC Bioinformatics **7**, S3 (2006). doi:10.1186/1471-2105-7-S5-S3
89. Chou, K.-C., Shen, H.-B.: Cell-PLoc: a package of web servers for predicting subcellular localization of proteins in various organisms. Nat. Protoc. **3**, 153–162 (2008)
90. Xiao, X., Wu, Z.-C., Chou, K.-C.: iLoc-Virus: a multi-label learning classifier for identifying the subcellular localization of virus proteins with both single and multiple sites. J. Theor. Biol. **284**, 42–51 (2011)
91. Shen, H.-B., Chou, K.-C.: Virus-mPLoc: a fusion classifier for viral protein subcellular location prediction by incorporating multiple sites. J. Biomol. Struct. Dyn. **28**, 175–186 (2010)
92. Bannai, H., Tamada, Y., Maruyama, O., Nakai, K., Miyano, S.: Extensive feature detection of N-terminal protein sorting signals. Bioinformatics **18**, 298–305 (2002). doi:10.1093/bioinformatics/18.2.298
93. Lodish, H., Berk, A., Zipursky, S.L.: Molecular Cell Biology, 4th edn. W. H. Freeman, New York (2000)
94. Nielsen, H., Engelbrecht, J., Brunak, S., von Heijne, G.: Identification of prokaryotic and eukaryotic signal peptides and prediction of their cleavage sites. Protein Eng. **10**, 1 (1997). doi:10.1093/protein/10.1.1

95. Choo, K.H., Tan, T.W., Ranganathan, S.: A comprehensive assessment of N-terminal signal peptides prediction methods. BMC Bioinformatics 10, S3 (2009). doi:10.1186/1471-2105-10-S15-S2
96. Scott, M.S., Oomen, R., Thomas, D.Y., Hallett, M.T.: Predicting the subcellular localization of viral proteins within a mammalian host cell. J. Virol. 3, 24 (2006). doi:10.1186/1743-422X-3-24
97. Petersen, T.N., Brunak, S., Von Heijne, G., Nielsen, H.: SignalP 4.0: discriminating signal peptides from transmembrane regions. Nat. Methods 8, 784–786 (2011). doi:10.1038/nmeth.1701
98. Käll, L., Krogh, A., Sonnhammer, E.L.L.: A combined transmembrane topology and signal peptide prediction method. J. Mol. Biol. 338, 1027–1036 (2004)
99. Käll, L., Krogh, A., Sonnhammer, E.: Advantages of combined transmembrane topology and signal peptide prediction - the Phobius web server. Nucleic Acids Res. 35, W429–W432 (2007). doi:10.1093/nar/gkm256
100. Viklund, H.K., Bernsel, A., Skwark, M., Elofsson, A.: SPOCTOPUS: a combined predictor of signal peptides and membrane protein topology. Bioinformatics 24, 2928–2929 (2008). doi:10.1093/bioinformatics/btn550
101. Chen, P., Rayner, S., Hu, K.H.: Advances of bioinformatics tools applied in virus epitopes prediction. Virol. Sin. 26, 1–7 (2011). doi:10.1007/s12250-011-3159-4
102. Flower, D.R. Chapter 5: Vaccines: data driven prediction of binders, epitopes and immunogenicity. In: Flower, D.R. (ed.) Bioinformatics for Vaccinology, pages 167–216. Wiley-Blackwell, Chichester (2008)
103. Iurescia, S., Fioretti, D., Fazio, V.M., Rinaldi, M.: Epitope-driven DNA vaccine design employing immunoinformatics against B-cell lymphoma: a biotech's challenge. Biotechnol. Adv. 30, 372–383 (2012)
104. Sirskyj, D., Diaz-Mitoma, F., Golshani, A., Kumar, A., Azizi, A.: Innovative bioinformatic approaches for developing peptide-based vaccines against hypervariable viruses. Immunol. Cell Biol. 89, 81–89 (2011). doi:10.1038/icb.2010.65
105. Yang, X., Yu, X.: An introduction to epitope prediction methods and software. Rev. Med. Virol. 19, 77–96 (2009)
106. Yu, K., Petrovsky, N., Schönbach, C., Koh, J., Brusic, V.: Methods for prediction of peptide binding to MHC molecules: a comparative study. Mol. Med. 8, 137–148 (2002)
107. Davydov, Y.I., Tonevitsky, A.G.: Prediction of linear B-cell epitopes. Mol. Biol. 43, 150–158 (2009)
108. Zhang, Q., et al.: Immune epitope database analysis resource (IEDB-AR). Nucleic Acids Res. 36, W513–W518 (2008). doi:10.1093/nar/gkn254
109. Van Bergen, J., et al.: Get into the groove! Targeting antigens to MHC class II. Immunol. Rev. 172, 87–96 (1999)
110. Tung, C.W., Ziehm, M., Kamper, A., Kohlbacher, O., Ho, S.Y.: POPISK: T-cell reactivity prediction using support vector machines and string kernels. BMC Bioinformatics 12, 446 (2011). doi:10.1186/1471-2105-12-446
111. Stranzl, T., Larsen, M., Lundegaard, C., Nielsen, M.: NetCTLpan: pan-specific MHC class I pathway epitope predictions. Immunogenetics 62, 357–368 (2010). doi:10.1007/s00251-010-0441-4
112. Kulkarni-Kale, U., Bhosle, S., Kolaskar, A.S.: CEP: a conformational epitope prediction server. Nucleic Acids Res. 33, W168–W171 (2005). doi:10.1093/nar/gki460
113. Rubinstein, N.D., Mayrose, I., Pupko, T.: A machine-learning approach for predicting B-cell epitopes. Mol. Immunol. 46, 840–847 (2009)
114. Liang, S., Zheng, D., Zhang, C., Zacharias, M.: Prediction of antigenic epitopes on protein surfaces by consensus scoring. BMC Bioinformatics 10, 302 (2009). doi:10.1186/1471-2105-10-302
115. Liu, R., Hu, J.: Prediction of discontinuous B-cell epitopes using logistic regression and structural information. J. Proteomics Bioinformatics 4, 010–015 (2011)
116. Zhang, W., et al.: Prediction of conformational B-cell epitopes from 3D structures by random forests with a distance-based feature. BMC Bioinformatics 12, 341 (2011). doi:10.1186/1471-2105-12-341
117. Ryvkin, A., et al.: Deep panning: steps towards probing the IgOme. PLoS One 7, e41469 (2012). doi:10.1371/journal.pone.0041469
118. Doytchinova, I.A., Flower, D.R.: Identifying candidate subunit vaccines using an alignment-independent method based on principal amino acid properties. Vaccine 25, 856–866 (2007)
119. Magnan, C.N., et al.: High-throughput prediction of protein antigenicity using protein microarray data. Bioinformatics 26, 2936–2943 (2010). doi:10.1093/bioinformatics/btq551
120. Grandi, G.: Genomics and proteomics in reverse vaccines. Methods Biochem. Anal. 49, 379–393 (2006)
121. Vivona, S., Bernante, F., Filippini, F.: NERVE: new enhanced reverse vaccinology environment. BMC Biotechnol. 6, 35 (2006). doi:10.1186/1472-6750-6-35
122. Gardy, J.L., et al.: PSORT-B: improving protein subcellular localization prediction for gram-negative bacteria. Nucleic Acids Res. 31, 3613–3617 (2003). doi:10.1093/nar/gkg602
123. Sachdeva, G., Kumar, K., Jain, P., Ramachandran, S.: SPAAN: a software program for prediction of adhesins and adhesin-like proteins using neural networks. Bioinformatics 21, 483–491 (2005). doi:10.1093/bioinformatics/bti028
124. Tusnády, G.E., Simon, I.: Principles governing amino acid composition of integral membrane proteins: application to topology prediction. J. Mol. Biol. 283, 489–506 (1998)
125. Doytchinova, I.A., Flower, D.R.: VaxiJen: a server for prediction of protective antigens, tumour antigens and subunit vaccines. BMC Bioinformatics 8, 4 (2007). doi:10.1186/1471-2105-8-4
126. Flower, D.R., Macdonald, I.K., Ramakrishnan, K., Davies, M.N., Doytchinova, I.A.: Computer aided selec-

tion of candidate vaccine antigens. Immunome Res. **6**(Suppl 2), S1 (2010). doi:10.1186/1745-7580-6-s2-s1
127. Plotkin, S.A.: Correlates of protection induced by vaccination. Clin. Vaccine Immunol. **17**, 1055–1065 (2010)
128. Thakur, A., Pedersen, L.E., Jungersen, G.: Immune markers and correlates of protection for vaccine induced immune responses. Vaccine **30**, 4907–4920 (2012). doi:10.1016/j.vaccine.2012.05.049
129. Whelan, M., Ball, G., Beattie, C., Dalgleish, A.: Biomarkers for development of cancer vaccines. Future Med. **3**, 79–88 (2006)
130. Mou, Z., He, Y., Wu, Y.: Immunoproteomics to identify tumor-associated antigens eliciting humoral response. Cancer Lett. **278**, 123–129 (2009). doi:10.1016/j.canlet.2008.09.009
131. Walzl, G., Ronacher, K., Hanekom, W., Scriba, T.J., Zumla, A.: Immunological biomarkers of tuberculosis. Nat. Rev. Immunol. **11**, 343–354 (2011). doi:10.1038/nri2960
132. Liebenberg, J., et al.: Identification of *Ehrlichia ruminantium* proteins that activate cellular immune responses using a reverse vaccinology strategy. Vet. Immunol. Immunopathol. **145**, 340–349 (2012). doi:10.1016/j.vetimm.2011.12.003
133. Cardoso, F.C., Roddick, J.S., Groves, P., Doolan, D.L.: Evaluation of approaches to identify the targets of cellular immunity on a proteome-wide scale. PLoS One **6**, e27666 (2011). doi:10.1371/journal.pone.0027666
134. Coffman, R.L., Sher, A., Seder, R.A.: Vaccine adjuvants: putting innate immunity to work. Immunity **33**, 492–503 (2010)
135. Tizard, I.A.: Veterinary Immunology – An Introduction, 8th edn. Philadelphia, Saunders (2008)
136. Sayers, S., Ulysse, G., Xiang, Z., He, Y.: Vaxjo: a web-based vaccine adjuvant database and its application for analysis of vaccine adjuvants and their uses in vaccine development. J. Biomed. Biotechnol. **2012**, 13 (2012). doi:10.1155/2012/831486
137. Skibinski, D.A.G., O´Hagan, D.T., Chapter 6: Adjuvants. In: Rappuoli, R., Serruto, D., Rappuoli, R. (eds.). Vaccine Design – Innovative and Novel Strategies, vol. 1, pp.139–169. Caister Academic Press, Norfolk (2011)

# Vaccine Delivery Using Microneedles

Ryan F. Donnelly, Sharifa Al-Zahrani, Marija Zaric, Cian M. McCrudden, Cristopher J. Scott, and Adrien Kissenpfenning

## Contents

| | | |
|---|---|---|
| 43.1 | Vaccination | 697 |
| 43.2 | Skin Structure and Function | 698 |
| 43.3 | Immune Function of Skin | 700 |
| 43.4 | Vaccination via the Skin Using Microneedles | 700 |
| 43.4.1 | Poke and Patch | 701 |
| 43.4.2 | Coat and Poke | 703 |
| 43.4.3 | Poke and Flow | 705 |
| 43.4.4 | Dissolving/Soluble Microneedles | 705 |
| 43.4.5 | Intradermal Gene Delivery | 706 |
| 43.5 | Microneedle-Mediated Delivery of Nanoparticles as a Vehicle for Improved Antigen Targeting to Skin DCs | 708 |
| Conclusions | | 709 |
| References | | 711 |

R.F. Donnelly, PhD (✉) • S. Al-Zahrani, BSc
C.M. McCrudden, PhD • C.J. Scott, PhD
School of Pharmacy, Queen's University Belfast,
Belfast, UK
e-mail: r.donnelly@qub.ac.uk

M. Zaric, MB • A. Kissenpfenning, PhD
Centre for Infection and Immunology,
Queen's University Belfast, Belfast, UK

### Abstract

Breaching the skin's *stratum corneum* barrier raises the possibility of administration of vaccines, gene vectors, antibodies, photosensitisers and even nanoparticles, all of which have at least their initial effect on populations of skin cells. Intradermal vaccine delivery, in particular, holds enormous potential for improved therapeutic outcomes for patients, particularly those in the developing world. Various microneedle-based vaccine delivery strategies have been employed, and here we discuss each one in turn. We also describe the importance of cutaneous immunobiology on the effect produced by microneedle-mediated intradermal vaccination.

## 43.1 Vaccination

The importance of vaccination in limiting morbidity from infectious disease cannot be overestimated and, according to the World Health Organisation, saves the lives of over 2.5 million children per year. The aim of vaccination is to build individuals' immunity against a specific disease. Vaccines traditionally correspond to one of four types: those containing a dead microorganism, those with live-attenuated microorganisms, those with protein subunits and those with inactivated toxic components (toxoid). A number of innovative vaccines are in development such as recombinant vector and DNA vaccines. These agents resemble a disease-causing microorganism and stimulate the body's immune system to recognise the agent as foreign, destroy it and 'remember' it, so that the immune system can more easily challenge these microorganisms upon subsequent encounters.

The critical component in guaranteeing successful vaccination is appropriate vaccine administration. While intramuscular (M) and subcutaneous (SC) routes are used for the majority of vaccines, this method is not without its problems. These routes require highly trained personnel for administration and are associated with pain and distress which might lead to reduced patient compliance. Additionally, in developing countries, hypodermic injections are associated with a high risk of cross-contamination between patients due to the possibility of needlestick injuries or reuse of contaminated needles. When mass vaccination is necessary, issues of production and/or supply may also arise [1, 2].

The majority of vaccines are delivered either into subcutaneous fat or muscle. Delivery vaccines into the dermis are rare [3], and even topical or transcutaneous application to the surface of the skin [4, 5], also termed epicutaneous application [5], are rarer still. The above routes of application are each effective only because of the ability of dendritic cells (DCs) to uptake, process and present the antigen to T lymphocytes in the draining lymphoid organs. Whereas subcutaneous fat and muscle tissue contain relatively few DCs, the dermis and the epidermis are densely populated by different subsets of DCs. Consequently, antigen delivery by hypodermic injection will bypass the skin's immune cells leading to less efficient vaccination. For this reason, the skin represents an ideal site for vaccine delivery, as vaccination at this site will evoke strong immune responses at much lower doses of antigen than intramuscular vaccines [6]. The potential of skin immunisation was observed in a clinical trial where epidermal influenza vaccination induced influenza-specific CD8 T cell response, while classical intramuscular route did not [7]. Dose-sparing approaches are critical to ensuring a sufficient supply of certain vaccines, especially in pandemic diseases [8].

## 43.2 Skin Structure and Function

As the largest and one of the most complex organs in the human body, the skin (Fig. 43.1) is responsible for a varied range of functions [9, 10]. The barrier properties of the skin afford protection against physical, microbial or chemical invasions. The skin is made up of three layers: the epidermis, dermis and subcutaneous tissue. The epidermis consists of the viable epidermis and the *stratum corneum*. The viable epidermis consists of four histologically distinct layers: the *stratum germinativum*, *stratum spinosum*, *stratum granulosum* and *stratum lucidum*. The epidermis is not of uniform thickness, varying from 60 µm on the eyelids to 800 µm on the palms [11]. The layers of the epidermis are a vascular and receive nutrients by diffusion of substances from the underlying dermal capillaries.

The dermis (or corium) resides atop the subcutaneous fat layer and is approximately 3–5 mm thick [12] and consists of a mucopolysaccharide matrix within which exists a network of elastin and collagen fibres, providing both elasticity and structure to the skin [13, 14]. The dermis is maintained physiologically by a network of nerve endings, lymphatics and blood vessels [15]. The cutaneous blood supply delivers nutrients and oxygen to the skin and allows waste products to be removed. Beneath the dermis lies the subcutaneous fat layer, subcutis, subdermis or hypodermis [13]. The hypodermis, consisting mainly of adipose tissue, acts as an insulator (due to the high content of adipose tissue), as well as supporting the dermis and epidermis physically and nutritionally. The hypodermis also carries the main blood vessels and nerves to the skin and may contain sensory organs [16]. Resting beneath the vascular dermis, the role of the hypodermis in drug delivery is not considered to be major [17].

The *stratum corneum*, or horny layer, is the outermost layer of the epidermis and thus the skin. It has now well accepted that this layer constitutes the principal barrier for penetration of most drugs. The horny layer represents the final stage of epidermal cell differentiation. The thickness of this layer is typically 10 µm, but a number of factors, including the degree of hydration and

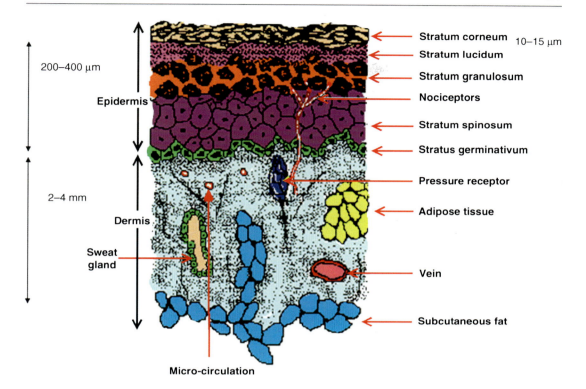

**Fig. 43.1** Diagrammatic representation of the major features of skin anatomy

skin location, influence this. For example, the stratum corneum on the palms and soles can be, on average, 400–600 μm thick while hydration can result in a four-fold increase in thickness.

The stratum corneum consists of 10–25 rows of dead keratinocytes, now called corneocytes, embedded in the secreted lipids from lamellar bodies. The corneocytes are flattened, elongated, dead cells, lacking nuclei and other organelles. The cells are joined together by desmosomes, maintaining the cohesiveness of this layer. The heterogeneous structure of the *stratum corneum* is composed of approximately 75–80 % protein, 5–15 % lipid and 5–10 % other substances on a dry weight basis.

The majority of protein present in the stratum corneum is keratin and is located within the corneocytes. The keratins are a family of α-helical polypeptides. Individual molecules aggregate to form filaments (7–10 nm diameter and many microns in length) that are stabilised by insoluble disulphide bridges. These filaments are thought to be responsible for the hexagonal shape of the corneocyte and provide mechanical strength for the stratum corneum [12]. Corneocytes possess a protein-rich envelope around the periphery of the cell, formed from precursors, such as involucrin, loricrin and cornifin. Transglutaminases catalyse the formation of γ-glutamyl cross-links between the envelope proteins that render the envelope resistant and highly insoluble. The protein envelope links the corneocyte to the surrounding lipid-enriched matrix.

The main lipids located in the *stratum corneum* are ceramides, fatty acids, cholesterol, cholesterol sulphate and sterol/wax esters. These lipids are arranged in multiple bilayers called lamellae (Fig. 43.2). Phospholipids are largely absent, a unique feature for a mammalian membrane. The ceramides are the largest group of lipids in the stratum corneum, accounting for approximately half of the total lipid mass, and are crucial to the lipid organisation of the *stratum corneum*.

The bricks and mortar model of the *stratum corneum* are a common representation of this

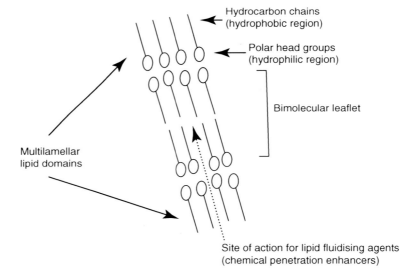

**Fig. 43.2** Arrangement of lipids in the *stratum corneum*

layer. The bricks correspond to parallel plates of dead keratinised corneocytes, and the mortar represents the continuous interstitial lipid matrix. It is important to note that the corneocytes are not actually brick shaped but rather are polygonal, elongated and flat (0.2–1.5 μm thick and 34.0–46.0 μm in diameter). The 'mortar' is not a homogenous matrix. Rather, lipids are arranged in the lamellar phase (alternating layers of water and lipid bilayers), with some of the lipid bilayers in the gel or crystalline state. The extracellular matrix is further complicated by the presence of intrinsic and extrinsic proteins, such as enzymes. The barrier properties of the stratum corneum have been assigned to the multiple lipid bilayers residing in the intercellular space. These bilayers prevent desiccation of the underlying tissues by inhibiting water loss and limit the penetration of substances from the external environment.

## 43.3 Immune Function of Skin

Vaccination development remains an important field in both research and pharma, whereby in addition to extending the spectrum of antigens for novel vaccines, developing improved administration strategies to ameliorate vaccine efficacy remains a challenge. The concept of delivery of vaccines through or into the skin has been gathering momentum in the past decade, largely due to the increasing recognition that a tight semi-contiguous network of immunoregulatory cells that reside in the different skin layers is an ideal target for vaccine administration. Dendritic cells (DCs), macrophages and neutrophilic granulocytes are the principal phagocytes in the skin (Fig. 43.3), while numerous cells of the adaptive immune system, such are CD8+ T cells and the full spectrum of CD4+ T cells, can be found in normal skin [3–7, 18–44]. A detailed description of dendritic cells is provided in Chap. 2 of this book.

## 43.4 Vaccination via the Skin Using Microneedles

Intradermal (ID) vaccination using MN is one of the most attractive approaches for delivering an antigen to the dermal layer of the skin without using hypodermic injections, which are associated with transmission of infection and inappropriate disposal in the developing world. MN arrays (Fig. 43.4) consist of a multiplicity of microprojections ranging from 25 to 2,000 μm in height, attached to a base support [45]. These microprojections can create aqueous transport pathways at the micron scale, painlessly breaking

**Fig. 43.3** Immune structure of skin

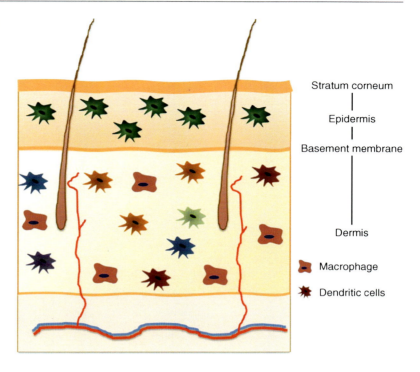

**Fig. 43.4** Laser engineered, micromoulded, microneedle array prepared from aqueous blends of the structure-forming copolymer poly(methyl vinyl ether-co-maleic acid)

through the *stratum corneum* barrier when the microneedles are applied to the skin surface [46]. The micropores created by MNs readily permit transport of a wide range of micromolecules and macromolecules, such as immunotherapeutic agents, including vaccines and proteins [47]. Importantly, MN insertion does not cause bleeding.

MNs were first described by Henry et al. in 1998 and have since been the subject of continuous research [48]. MNs are fabricated from various materials such as metals, glass, silicon and FDA-approved polymers [49]. Donnelly et al. detail a range of methods for the development and fabrication of various MNs [50]. There are four main modes of action of MNs, which are poke and patch, coat and poke, poke and release and poke and flow [51] (Fig. 43.5).

### 43.4.1 Poke and Patch

The poke and patch approach is based on using solid MNs to puncture the skin followed by applying antigen to the treated area in order to

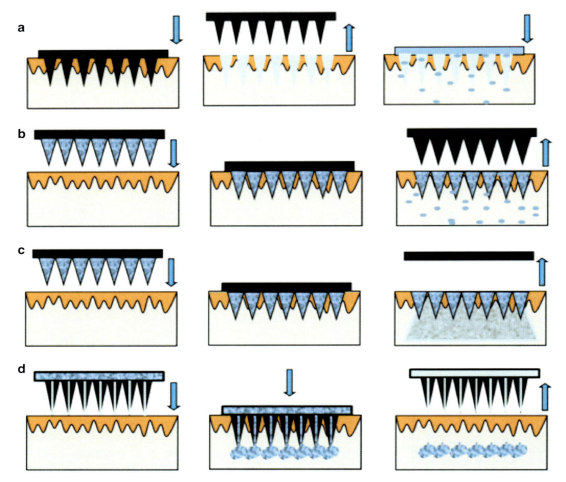

**Fig. 43.5** Schematic representation of methods of MN application to the skin to achieve enhanced transdermal drug delivery. (**a**) (Poke and patch) shows solid MN that are applied and removed to create transient micropores, followed by application of a traditional transdermal patch. (**b**) (Coat and poke) shows solid MN coated with drug for instant delivery. MN is removed after the coating material dissolves. (**c**) (Dissolving microneedles) show soluble polymeric/carbohydrate MN-containing drug that dissolves in skin interstitial fluid over time, thereby delivering the drug. (**d**) (Poke and flow) shows hollow MN for delivery of fluids containing drug

diffuse antigen into the skin. Substantial work using this approach has been carried out by the Bouwstra Group, where microneedles are inserted into the skin in order to increase its permeability, after which the vaccine is applied. The Bouwstra Group, in one report, studied mouse immune responses after transcutaneous immunisation (TCI) using two model antigens: diphtheria toxoid (DT) and influenza subunit vaccine [52]. Stainless steel MN arrays (16 × 300 μm MNs) were used to perforate the mouse skin followed by application of a DT formulation with or without cholera toxin (CT). The application of DT to MN-treated skin resulted in significantly higher serum IgG and toxin-neutralising antibody titres than in unperforated skin. The presence of CT increased the immune response to similar levels as observed after subcutaneous injection of AlPO4-adsorbed DT (DT-alum). Unlike in the DT case, MN array pretreatment showed no effect on the immune response to the influenza vaccine alone. The addition of CT strongly improved the immune response, independent of MN treatment. The conclusion drawn by the authors of the study was that that TCI of DT in the presence of CT after MN treatment results in similar protection to injection of DT-alum.

A study investigated the effect of co-administration of various adjuvants with DT in the modulation of the immune response in TCI mice after the application of MN arrays [53]. Mice were treated with DT co-administrated with lipopolysaccharide (LPS), Quil-A, CpG oligodeoxynucleotide (CpG) or CT as adjuvants. The MN array pretreatment group displayed high serum IgG levels, and these were remarkably improved by co-administration of adjuvants. The group treated with DT co-administered with CT showed similar IgG levels to those treated subcutaneously with DT-alum. N-trimethyl chitosan also proved beneficial in boosting the immune response to DT following MN pretreatment when in solution with the antigen, although no improvement in immune response was seen from DT-loaded nanoparticles of N-trimethyl chitosan [54].

The impact of transdermal vaccination on the development of melanoma was reported by Bhowmik et al. (2011), who delivered a novel microparticulate vaccine to the skin following puncture using of MN-based Dermaroller® [55]. Mice were grouped into four groups; group one was treated using Dermaroller® microneedles, followed by application of microparticles containing encapsulated antigen obtained from S-91 melanoma cancer cells. Group two was injected subcutaneously with the same dose of microparticles containing encapsulated vaccine. Group three was given blank microparticles administered in the same way as for the transdermal group. The last group was injected with saline subcutaneously. Eight weeks following vaccination, the mice were challenged with live melanoma cells. In the group which were treated using Dermaroller® MNs and SC injection of vaccine, there was no measurable tumour growth 35 days after tumour injection. A significant increase in IgG antibody levels was seen for both transdermal and subcutaneous vaccinated groups when compared with control groups. However, slightly increased IgG antibody levels were seen in the transdermally vaccinated group when compared to the SC group. This can be attributed to the presence of Langerhans cells (LCs) in the epidermis layers which are activated when exposed to antigen. The authors concluded that the developed formulation for melanoma cancer which can be administered using MNs technology gives rise to new approaches to the prevention of melanoma cancer.

### 43.4.2 Coat and Poke

The coat and poke approach involves the coating of solid MNs with an antigen of choice, which can be delivered in a one-step process. This could be a very attractive approach for mediating ID vaccine delivery since the smaller amount of antigen which coats the microneedles should be sufficient to induce a strong immune response. Since the antigen coating on the MNs is in a solid form [56], long-term stability should be improved, ensuring optimal shelf life [57].

The Prausnitz Group at the Georgia Institute of Technology has carried out immunisation studies using stainless steel monument-shaped arrays of 5 MNs dip-coated in vaccine. The MNs used were approximately 700 μm in length, and their manufacture was achieved by laser-cutting stainless steel sheets. Plasmid encoding hepatitis C virus and seasonal influenza: H1N1, H3N2, inactivated virus, influenza virus-like particles and recently BCG were engaged successfully in MN-mediated ID immunisation. After optimisation of the coating formulation that led to the inclusion of trehalose as an antigen stabiliser [58], MNs were coated and inserted into the skin of mice. The results showed that the coated MNs stimulated a robust immune response, providing complete protection against the lethal influenza virus challenge similar to conventional intramuscular injection. The study concluded that effective vaccination was achieved due to the inclusion of the stabilising agent in the coating formulation, as this acted to protect antigen activity. In light of this, in subsequent studies, trehalose was included in the coating formulation.

In a study by Koutsonanos et al., a single MN immunisation with inactivated H3N2 influenza virus induced significantly higher hemagglutination inhibition (HI) titres in comparison with that observed by IM injection [2]. Solid metal MNs coated with inactivated influenza virus were found to be at least as effective as the conventional IM route in inducing similar levels of functional antibodies at low or high antigen concentrations, in clearing the virus from the lungs of infected mice, in conferring protection and in inducing short lived as well as memory B immune responses. While serum IgG responses in IM injection were seen to be dependent upon dose, this was not the case with MN delivery, which produced similar responses at low or high antigen loadings. This finding suggests that the skin has a higher capacity to produce an immunologic response. The same system was used to evaluate the potential of BCG-coated MN vaccine patches [59]. The viability of BCG was maintained by adding 15 % trehalose to the coating solution, which improved the stability of the live BCG vaccine. BCG-coated 10 microneedle patches ($5 \times 10^4$ CFUs of BCG) were applied to the skin of guinea pigs. Another group of guinea pigs were injected intradermally with BCG ($5 \times 10^4$ CFUs) using a 26 gauge needle and 1 ml tuberculin syringe. The results of this study indicated that BCG vaccine-coated MNs can induce a strong antigen-specific cellular immune response in both the lung and spleen of guinea pigs, comparable to that induced by using a 26 gauge needle. It was found that MN BCG vaccination elicited similar frequencies of TNF-α secreting or both IFN-γ and TNF-α cytokine secreting bifunctional CD4+ T cells to that induced by hypodermic injection. A strong IgG response was produced by both vaccination methods.

The capacity to produce a protective immune response of modified recombinant trimeric soluble influenza virus hemagglutinin (sHA GCN4pII)-coated MNs was assessed by the group [60]. Comparison was made between results from the modified and unmodified protein (sHA). Mice vaccinated with MN-coated sHA trimeric induced fully protective immune response against influenza virus challenge. Both sHA- and sHA GCN4pII-coated MNs induced improved clearance of replicating virus compared to the SC route. The MNs coated with sHA GCN4pII induced a stronger Th1 response in mice suggested by the ratio of IFN-γ+CD4+ T cell to IL-4+. The study concluded that MNs coated with stabilised sHA trimers were as effective in inducing a protective immune response and afforded equal level of protection as conventional subcutaneous administration.

Professor Mark Kendall's research has pioneered the development of Nanopatch™ technology. Nanopatch™ devices are fabricated from silicon using a process of deep reactive-ion etching. The projections are solid silicon, sputter coated with a thin (~100 nm) layer of gold, and contain 3,364 densely packed projections. These arrays are smaller, by two orders of magnitude, than standard needles, and are also much smaller than typical microneedle arrays. The Nanopatches™ have been used to target ID vaccination against West Nile virus and chikungunya virus in mice. The efficiency of Nanopatch™ immunisation was demonstrated using an inactivated whole chikungunya virus vaccine and a DNA-delivered attenuated West Nile virus vaccine [61]. Nanopatch™ technology was also used as an alternative delivery system to the IM prophylactic vaccine against cervical cancer Gardasil® and succeeded in delivering up to 300 ng of vaccine to the ears of mice. Moreover, in terms of virus-neutralising response, results were found equal to those of a control group of IM-vaccinated mice in mouse serum samples from mice vaccinated using Nanopatch™ technology [62]. Similarly impressive results have also been reported when Nanopatch™ coating was used with the influenza vaccine, Fluvax® [63].

Research at Zosano Pharma (formerly ALZA Corporation) assessed the performance of the Macroflux®, another device containing an array of microprojections, for intracutaneous delivery of a model antigen, ovalbumin. The findings revealed that at all antigen doses, administration of OVA-coated Macroflux® resulted in immune response comparable to that observed after intradermal administration of the same doses of antigen. In addition, it was observed that application of 1 and 5 µg of antigen via Macroflux® resulted in 10- and 50-fold increases in anti-ovalbumin levels in comparison to delivery of equivalent doses via intramuscular or subcutaneous routes [64]. Follow-up mechanistic studies revealed that the immunologic response was unaffected by MN height (225–600 µm) and density (140 and 657/cm$^2$), but was dependent on the dose of antigen delivered [65].

### 43.4.3 Poke and Flow

The poke and flow approach is based on diffusion of vaccine through conduits of solid MNs. The antigen can be delivered either by passive diffusion, pressure or electrical-driven flow [45, 66]. However, this approach is vulnerable to some of the same problems as conventional vaccination techniques including the requirement of a cold chain and possible need for trained personnel [62]. The narrow bore of MNs and the dense elastic nature of the skin may also limit fluid flow. Partial retraction of the MN array before the introduction of fluid was found by Wang et al. to partially compensate for this limitation; however [67], Frost suggested that the co-administration of the enzyme hyaluronidase, which degrades hyaluronic acid in the extracellular matrix of the skin, might reduce skin resistance [68]. Martanto et al. provided many explanations of the factors affecting the flow rate through hollow MNs. Hollow MNs are the subject of increasing interest because of their potential for use for TCI or ID vaccination [69].

MicronJet® is a novel device developed by NanoPass specifically for intradermal delivery, consisting of an array of four MNs, each 450 µm in length. The needles are of silicon crystal bonded to a plastic adapter which can be mounted on any standard syringe. Van Damme et al. investigated the safety and efficacy of this device for dose-sparing intradermal influenza vaccination in healthy adults [70]. α-RIX® (GSK Biologicals) influenza vaccines were delivered using a hollow MN device (MicronJet®) and their safety and immunogenicity assessed. In the trial, which was carried out with a group of 180 healthy adult subjects, low-dose influenza vaccines delivered by MicronJet® elicited immune responses similar to those elicited by 15 µg HA per strain injected IM in human volunteers. The prevalence of local reaction presented a limitation however, although such reactions were transient in nature. Similar developments have been made at BD Technologies. A 34G stainless steel MN with an inner diameter of 76 µm, an outer diameter of 178 µm and an exposed length of 1 mm was used to deliver three distinct influenza vaccines. Results suggested that this mode of delivery was capable of producing the full immunological response at a dose at least ten-fold lower than with IM administration and up to 100-fold more potent, depending on the nature of the antigen [71]. The same researchers went on to uncover dose-sparing benefits of MN-delivered anthrax vaccinations over intramuscular administration in a subsequent study [72].

### 43.4.4 Dissolving/Soluble Microneedles

Dissolving MNs may present an innovative approach for vaccine delivery. Such MNs are based upon water-soluble polymers or carbohydrate material which encapsulates the drug within the needle matrix. On piercing the skin, the MNs dissolve completely to deliver their contents. Dissolvable MNs show promise in vaccine delivery breakthrough for many reasons. Since the MNs will dissolve after insertion into the patient's skin, the possibility of cross infection is eliminated. Also, the approach eliminates the necessity of any special disposal mechanism, as no sharp biohazardous waste is produced. The solid state nature of the contained/encapsulated vaccine should also reduce the need for cold chain storage and transport. In addition, MN patches offer the possibility of self-administration of vaccine during pandemics as well as rendering mass immunisation programs in developing countries. As these self-disabling MNs lack many of the disadvantages of conventional vaccination techniques and also some of those associated with the

MN strategies mentioned to this point, poke and patch MNs are receiving increasing attention for their value in vaccination applications.

Dissolvable MN patches were first taken up for the administration of vaccines by the Prausnitz group. Sullivan et al. manufactured MN patches and investigated possibilities for influenza vaccination using a simple patch-based system. Patches based on 650 μm poly(vinylpyrrolidone) MNs containing 3 μg lyophilised inactivated influenza virus vaccine generated robust antibody and cellular immune responses which gave complete protection against lethal influenza challenge [73]. In fact, in comparison with IM vaccination, MN delivery displayed better results following vaccination in terms of lung virus clearance and cellular recall responses.

The Kendall group described the micromoulding of dissolving MN arrays from master templates of one of their Nanopatch™ designs [74]. Replica MNs were formed from carboxymethylcellulose by means of multiple castings into poly(dimethylsiloxane) moulds. Dual-layer MNs containing both the model antigen ovalbumin and the adjuvant Quil-A elicited post-immunisation schedule antibody levels in mice which were similar to an IM ovalbumin/Quil-A immunisation group at day 28 and superior to the IM group at day 102, despite using a lower antigen dose in the MNs. Research using an influenza vaccine gave similar findings.

## 43.4.5 Intradermal Gene Delivery

Gene therapy may be defined as the insertion, alteration or removal of genes from an individual's cells and biological tissues to treat disease. Replacement of a mutated gene via the insertion of functional genes into an unspecified genomic location is the most frequently used gene therapy. Other approaches exist however which may involve either directly correcting the mutation or modifying normal genes. While the technology has yet to be developed to its full potential, it has been used with some success. The most common form of genetic engineering involves insertion of a functional gene into host cells. This is accomplished by isolating and copying the gene of interest, generating a construct containing all the genetic elements for correct expression and then inserting this construct into the host organism. Localised delivery and expression of gene therapeutics within the skin may provide novel treatment options for a number of pathological conditions, including skin disorders of genetic origin as well as nonsurgical management of malignancy. If genetic manipulation could be used to influence epidermal cells to produce and secrete antigenic molecules, then the potent immunostimulatory properties of the skin could be harnessed to provide efficient protection from the disease concerned.

The Birchall Group at Cardiff University in Wales have pioneered epidermal gene delivery using various microneedle arrays to penetrate the stratum corneum, which would form a barrier to intradermal delivery of genetic constructs. The group initially set out to establish whether silicon-based microneedles (150 μm in height, base width 45–50 μm), microfabricated via an isotropic etching/BOSCH reaction process, were capable of creating microchannels in the skin large enough to permit the passage of lipid/polycation/plasmid DNA (LPD) nonviral gene therapy vectors [75]. Scanning electron microscopy was used to visualise the microconduits created in heat-separated human epidermal sheets after application of the microneedles.

Following confirmation of particle size and particle surface charge by photon correlation spectroscopy and microelectrophoresis, respectively, the diffusion of fluorescent polystyrene nanospheres and LPD complexes through heat-separated human epidermal sheets was established in vitro using a Franz-type diffusion cell. In vitro cell culture with quantification by flow cytometry was used to determine gene expression in human keratinocytes (HaCaT cells). Membrane treatment with microneedles was found to notably enhance the diffusion through epidermal sheets of 100 nm diameter fluorescent polystyrene nanospheres, used as a readily quantifiable predictive model for LPD complexes. The delivery of LPD complexes either into or through the membrane microchannels was also demonstrated.

In both cases, considerable interaction between the particles and the epidermal sheet was observed. LPD complexes were shown through in vitro cell culture to mediate efficient reporter gene expression in human keratinocytes in culture when formulated at the appropriate surface charge.

The group's next study made use of platinum-coated silicon microneedles. These were used to produce microconduits, approximately 50 μm in diameter and extending through the stratum corneum and viable epidermis [76]. Following optimisation of skin explant culturing techniques and confirmation of tissue viability, it was demonstrated that gene expression could be mediated through use of the microneedles to transmit the beta-galactosidase reporter gene. Preliminary studies confirmed localised delivery, cellular internalisation and subsequent gene expression of pDNA following microneedle disruption of skin.

Following this result, it was decided to establish whether microfabricated silicon microneedle arrays could effectively achieve localised delivery of charged macromolecules and plasmid DNA (pDNA) [77]. Microconduits of 10–20 μm in diameter were found in human epidermal membrane following treatment with the microneedles. The delivery of the marker biomolecule beta-galactosidase and of a 'nonviral gene vector mimicking' charged fluorescent nanoparticle to the viable epidermis of microneedle-treated tissue was demonstrated using light and fluorescent microscopy.

Track-etched permeation profiles generated using 'Franz-type' diffusion cell methodology and a model synthetic membrane showed that >50 % of a colloidal particle suspension permeated through membrane pores in approximately 2 h. These findings were taken by the group to indicate cutaneous delivery of lipid/polycation/pDNA (LPD) gene vectors, and other related vectors, to the viable epidermis, could be achieved via microneedle treatment. Preliminary gene expression studies confirmed that naked pDNA can be expressed in excised human skin following microneedle disruption of the SC barrier. The presence of a limited number of microchannels, positive for gene expression, points to the value of further investigation aimed at optimisation of the morphology of the microneedle device, its method of application and the pDNA formulation.

Aqueous solutions loaded with gene delivery vehicles would not remain in situ on the surface of the skin and therefore would be of limited use in clinical practice. For this reason, the group then examined the possibility of using sustained release pDNA hydrogel formulations with their microneedle delivery system to improve delivery of plasmid DNA (pDNA) in skin [78]. Microneedles were again fabricated by wet etching silicon in potassium hydroxide. Hydrogels based on Carbopol® polymers and thermosensitive PLGA-PEG-PLGA triblock copolymers were prepared. Freshly excised human skin was used to characterise microneedle penetration (microscopy and skin water loss), gel residence in microchannels, pDNA diffusion and reporter gene (beta-galactosidase) expression. Upon application of the microneedles, channels of approximately 150–200 μm depth increased transepidermal water loss in skin. pDNA hydrogels were shown to harbour and gradually release pDNA. Following microneedle-assisted delivery of pDNA hydrogels to human skin, expression of the pCMV beta reporter gene was displayed in the viable epidermis proximal to the microchannels. It was concluded that targeted pDNA hydrogels can potentially provide sustained gene expression in the viable epidermis.

Microneedle-mediated intradermal gene delivery has also been recently explored by the Prausnitz Group [79]. In this study, specific cytotoxic T lymphocytes (CTLs) were effectively primed through vaccination with a plasmid encoding hepatitis C virus nonstructural 3/4A protein using coated microneedles. Importantly, the minimally invasive microneedles were as efficient in priming CTLs as more complicated or invasive delivery techniques, such as gene guns and hypodermic needles. The Kendall Group has also investigated gene delivery using coated Nanopatches™ [80], as described above.

Although Prausnitz had proposed the use of microneedles in combination with electroporation in 2005 [81], it was in 2007 that this was first achieved, when it was reported by Hooper et al., the efficient delivery of an experimental smallpox DNA vaccine consisting of four vaccinia virus genes (4pox) by means of a novel method involving skin electroporation using plasmid DNA-coated microneedle arrays [82]. Electroporation is a process where cells are transiently permeabilised using high-intensity electric field pulses. The Easy Vax® delivery system employed consisted of 80 electrically conducting microneedles each around 1 mm in height coated with dried vaccine DNA.

The pulse protocol consisted of six pulses of 100 V, 100 mS pulse duration and 125 mS pulse interval. Four arrays, each coated with 30 µg of a separate plasmid, were used to deliver the 4pox DNA vaccine to mice. A separate site was used for each array was administered to a separate site (inner and outer right and left thigh). Mice vaccinated with the negative control plasmids were administered using one array to an inner thigh. The smallpox DNA vaccine stimulated robust antibody responses against the four immunogens of interest, including neutralising antibody titres which were greater than were produced by the conventional live virus vaccine administered by scarification. Furthermore, complete protection was found in the vaccinated mice against a lethal intranasal challenge with vaccinia virus strain IHD-J.

This study represented the first demonstration of a protective immune response being elicited by microneedle-mediated skin electroporation and suggests that there is scope for further exploration of this area. Indeed, Daugimont et al [83]. followed up on this by investigating the potential of hollow conductive microneedles for needle-free intradermal injection and electric pulse application in order to generate an electric field in the superficial layers of the skin sufficient for electroporation. Microneedle arrays along with a vibratory inserter were employed to disrupt the stratum corneum, thus piercing the skin. Effective injection of proteins into the skin was achieved, resulting in an immune response directed to the model antigen ovalbumin. However, the dual function of microneedle electrode seemed to pose certain drawbacks for DNA electrotransfer. This could be due to the distribution of the electric field in the skin as shown by numerical calculations and/or the low dose of DNA injected. The authors concluded that these parameters require further study in order to optimise minimally invasive DNA electrotransfer in the skin.

## 43.5 Microneedle-Mediated Delivery of Nanoparticles as a Vehicle for Improved Antigen Targeting to Skin DCs

In the past few years, particle-based vaccines have been proposed for successful immunisation. They have been used to protect antigen stability in vivo and to deliver it in a controlled and sustained manner to the site of action [84].

Drug-loaded nanoparticles (NPs) are colloidal systems, typically 10–1,000 nm in diameter, with a therapeutic payload entrapped, adsorbed or chemically coupled to an orbicular matrix [85, 86]. Nanoparticles are widely used for controlled delivery of small molecule drugs, oligonucleotides and protein antigens to a variety cell types, including dendritic cells [87]. Among the different parameters that need to be considered in design of particle-based vaccines, the particle size and their physicochemical properties are particularly important for skin vaccination. It has been demonstrated that polymeric nanoparticles <500 nm in diameter have high rate of intracellular uptake by variety of APC [88].

Several groups have demonstrated that nanoparticles have adjuvant effects comparable to those of CFA or ALUM and, as synthetic adjuvants, can activate DCs to induce T cell immune responses against encapsulated antigens [89–91]. An important advance was the demonstration that nanoparticles as the adjuvants promote activation of the NLRP3 inflammasome [92].

Nanoparticles have been extensively studied for oral and parenteral administration owing to their sustained drug release [93, 94]. This property of nanoparticles could also be utilised for topical antigen administration to target skin DCs with antigen over a prolonged period. Researchers had attempted to use nanoparticles for topical drug delivery, and they found that the drug permeation was enhanced by gradual drug release from the nanoparticles on the skin surface, but did not optimise way to deliver nanoparticles inside the skin [95–97]. This suggested that as a drug delivery vehicle, the nanoparticle could sustain drug release, but, if it was applied as a drug reservoir to treat the skin disease, it must be delivered into the skin layers instead of remaining on the skin surface.

Some other researchers tried to verify the penetration of nanoparticles across the skin, but found that only small numbers of NPs were able to permeate into the skin passively through the hair follicles while most NPs were restricted by stratum corneum and unable to penetrate the skin [98, 99]. In the penetration experiments in vitro using the full-thickness skin, it was found that NPs could diffuse into the dermis as well as into the epidermis [100]. To investigate if the microconduits on the epidermis produced by microneedles could be the channels for NPs to penetrate the skin, the researches in vitro have been designed and proved that nanoparticles could pass through the human epidermal membrane and get into skin layers [101, 102]. Moreover, Bal and colleagues showed that in intradermal antigen delivery in vivo to skin pretreated by metal MNs antigen was more efficiently taken up by skin DCs when it was encapsulated into polymeric NPs, comparing when delivered in a soluble form [103]. All of the studies above indicated that microneedles may be an effective vehicle for the intradermal delivery of antigen encapsulated nanoparticles in vivo.

## Conclusions

Dendritic cells, as key regulators of immune responses, play a critical role in the design of modern vaccines [24, 99]. The skin harbours a network of these cells and for that reason is recognised as an attractive target for immunisation. However, one important element that has not been dealt with successfully yet is the functional heterogeneity of DCs subsets. Researchers are now trying to improve intracutaneous vaccination by harnessing specific properties of particular DCs subsets, as they became known. In the future, it might be possible to deliver antigen alongside a specific adjuvant to a particular DC subpopulation, while avoiding others with opposite effects. Targeting of antigen to the specific, functionally defined subsets of skin DCs is a promising strategy to further develop not only protective but also therapeutic vaccines.

Given the ever-increasing evidence available within the academic and patent literature that MN of a wide variety of designs are capable of achieving successful intradermal and transdermal delivery of vaccines, it is envisaged that the already concerted industrial effort into development of MN devices will now intensify. Furthermore, novel applications of MN technology are likely to come to the forefront. The ability of MN arrays to extract bodily fluids for determination of efficacy of vaccination is particularly interesting. As technological advances continue, MN arrays may well become the pharmaceutical dosage forms and monitoring devices of the near future. However, there are a number of barriers that will firstly need to be addressed in order for microneedle technology to progress.

The ultimate commercial success of microneedle-based delivery and monitoring devices will depend upon not only on the ability of the devices to perform their intended function but also on their overall acceptability by both healthcare professionals (e.g. doctors, nurses and pharmacists) and patients. Accordingly, efforts to ascertain the views of these end users will be essential moving forward. The seminal study by the Birchall Group [104] in this regard was highly informative. The majority of healthcare professionals and members of the public recruited into this focus-group-centred study were able to appreciate the potential advantage of using microneedles, including reduced pain, tissue damage, risk of transmitting infections and needle stick injuries, feasibility for self-administration and use in children, needlephobes and/or diabetics. However, some concerns regarding effectiveness mean to confirm successful drug delivery (such as a visual dose indicator), delayed onset of action, cost of the delivery system, possible accidental use, misuse or abuse were also raised. Healthcare professionals were also concerned about inter-individual variation in skin thickness, problems associated with injecting small volumes and risk of infection. Several other possible issues (accidental or errors based) and interesting doubts regarding microneedle use were discussed in this study. Overall, the group reported that 100 % of the public participants and 74 % of the healthcare professional participants were optimistic about the future of microneedle technology [104]. Such studies, when appropriately planned to capture the necessary demographics, will undoubtedly aid industry in taking necessary action to address concerns and develop informative labelling and patient counselling strategies to ensure safe and effective use of microneedle-based devices. Marketing strategies will, obviously, also be vitally important in achieving maximum market shares relative to existing and widely used conventional delivery systems.

In order to gain acceptance from healthcare professionals, patients and, importantly, regulatory authorities (e.g. the US FDA and the MHRA in the UK), it appears a strong possibility that an applicator aid and a 'dosing indicator' be included within the overall microneedle 'package', with the microneedle array itself being disposable and the applicator/dosing indicator reusable. While a wide variety of applicator designs have been disclosed within the patent literature, only a few, relatively crude, designs based upon high impact/velocity insertion, or rotary devices, have been described. Application force has a significant role to play in microneedle insertion depth. Clearly, patients cannot 'calibrate' their hands and, so, will apply microneedles with different forces. Unless a large-scale study can be done showing consistent rates and extents of microneedle-mediated drug delivery when the microneedles have been inserted by hand, then, for consistent dosing across the population, applicator devices will need to be supplied. Moreover, patients will need a level of assurance that the microneedle device has actually been inserted properly into their skin. This would be especially true in cases of global pandemics or bioterrorism incidents where self-administration of microneedle-based vaccines becomes a necessity. Accordingly, a suitable means of confirming that skin puncture has taken place may need to be included within an applicator device or the microneedle product itself.

From a regulatory point of view, currently little is known about the safety aspects that would be involved with long-term usage of microneedle devices. In particular, studies will need to be conducted to assess the effect that repeated microporation has upon recovery of skin barrier function. However, given the minimally invasive nature of the micropores created within the skin following microneedle application, especially in comparison to the use of a hypodermic needle and the fact that statistically it is highly unlikely that microneedles would be inserted at exactly the same sites more than once in a patient's lifetime, it is envisaged that microneedle technology will be shown to have a favourable safety profile. Indeed, skin barrier function is known to completely recover within a few hours of microneedle removal, regardless of how long the microneedles were in place. Local irritation or erythema (reddening) of the skin may be an issue for some patients. Since the skin is a potent immunostimulatory organ, it would be interesting to know whether repeated microneedle use would ever cause an immune reaction to the drug, excipients of microneedle materials and whether such an effect would be so significant as to cause problems for patients.

Infection is an issue that has long been discussed in relation to use of microneedle-based systems, since they, by necessity, puncture the skin's protective *stratum corneum* barrier. However, as we have shown [105], microbial penetration through microneedle-induced holes is minimal. Indeed, there have never been any reports of microneedles causing skin or systemic infections. This may be because of the above-mentioned immune component of the skin or the skin's inherent nonimmune, enzyme-based, defences. Alternatively, since the micropores are aqueous in nature, microorganisms may be more inclined to remain on the more hydrophobic *stratum corneum*.

Whether skin cleansing before microneedle application is necessary remains to be seen and is a vital question. Ideally, this would not have to be done, so as to avoid unnecessarily inconveniencing patients and making the use of the product in the domiciliary setting appear more akin to a self-administered injection than application of a conventional transdermal patch. Regulators will ultimately make the key decisions based on the weight of available evidence. Depending upon the application (e.g. drug/vaccine/active cosmeceutical ingredient delivery or minimally invasive monitoring), microneedle-based devices may be classed as drug delivery stems, consumer products or medical devices. From a delivery perspective, it will be important if microneedles are considered as injections rather than topical/transdermal/intradermal delivery systems, since this will determine whether the final product will need to be sterilised, prepared under aseptic conditions or simply host a low bioburden. Any contained microorganisms may need to be identified and quantified, as may the pyrogen content. Should sterilisation be required, then the method chosen will be crucial, since the most commonly employed approaches (moist heat, gamma or microwave radiation, ethylene oxide) may adversely affect the microneedles themselves and/or any contained active ingredient (e.g. biomolecules).

## References

1. Hegde, N.R., Kaveri, S.V., Bayry, J.: Recent advances in the administration of vaccines for infectious diseases: microneedles as painless delivery devices for mass vaccination. Drug Discov Today **16**, 1061–1068 (2011)
2. Koutsonanos, D., del Pilar Martin, M., Zarnitsyn, V., Sullivan, S., Compans, R., Skountzou, I., et al.: Transdermal influenza immunization with vaccine-coated microneedle arrays. PLoS One **4**, e4773 (2009)
3. Nicolas, J., Guy, B.: Intradermal, epidermal and transcutaneous vaccination: from immunology to clinical practice. Expert Rev Vaccines **7**, 1201–1214 (2008)
4. Warger, T., Schild, H., Rechtsteiner, G.: Initiation of adaptive immune responses by transcutaneous immunization. Immunol Lett **109**, 13–20 (2007)
5. Stoitzner, P., Sparber, F., Tripp, C.H.: Langerhans cells as targets for immunotherapy against skin cancer. Immunol Cell Biol **88**, 431–437 (2010)
6. Kenney, R.T., Yu, J., Guebre-Xabier, M., Frech, S.A., Lambert, A., Heller, B.A., et al.: Induction of protective immunity against lethal anthrax challenge with a patch. J Infect Dis **190**, 774–782 (2004)
7. Combadiere, B., Vogt, A., Mahe, B., Costagliola, D., Hadam, S.: Preferential amplification of CD8 effector-T cells after transcutaneous application of an inactivated influenza vaccine: a randomized phase I trial. PLoS One **5**, e10818 (2010)
8. Lambert, P.H., Laurent, P.E.: Intradermal vaccine delivery: will new delivery systems transform vaccine administration? Vaccine **26**, 3197–3208 (2008)
9. Wysocki, A.B.: Skin anatomy, physiology, and pathophysiology. Nurs Clin North Am **34**, 777–797 (1999)
10. Chuong, C.M., Nickoloff, B.J., Elias, P.M., Goldsmith, L.A., Macher, E., Maderson, P.A., et al.: What is the 'true' function of skin? Exp Dermatol **11**, 159–187 (2002)
11. Williams, A.C., Barry, B., Barry, B.W.: Skin absorption enhancers. Crit Rev Ther Drug Carrier Syst **9**, 305–353 (1992)
12. Wiechers, J.W.: The barrier function of the skin in relation to percutaneous absorption of drugs. Pharmaceutisch weekblad Scientific **11**, 185–198 (1989)
13. Tobin, D.: Biochemistry of human skin–our brain on the outside. Chem Soc Rev **35**, 52–67 (2006)
14. Asbill, C.S., El Kattan, A.F., Michniak, B.: Enhancement of transdermal drug delivery: chemical and physical approaches. Crit Rev Ther Drug Carrier Syst **17**, 621–658 (2000)
15. Menon, G.: New insights into skin structure: scratching the surface. Adv Drug Deliv Rev **54**, S3–S17 (2002)
16. Siddiqui, O.: Physicochemical, physiological, and mathematical considerations in optimizing percutaneous absorption of drugs. Crit Rev Ther Drug Carrier Syst **6**, 1–38 (1989)
17. Scheuplein, R.J.: Permeability of the skin: a review of major concepts. Curr Probl Dermatol **7**, 172–186 (1978)
18. Steinman, R.M., Hawiger, D., Nussenzweig, M.C.: Tolerogenic dendritic cells. Annu Rev Immunol **21**, 685–711 (2003)

19. Banchereau, J., Briere, F., Caux, C., Davoust, J., Lebecque, S., Liu, Y.J., et al.: Immunobiology of dendritic cells. Annu Rev Immunol **18**, 767–811 (2000)
20. Kleijwegt, F.S., Jansen, D.T., Teeler, J., Joosten, A.M., Laban, S., Nikolic, T., Roep, B.O.: Tolerogenic dendritic cells impede priming of naïve CD8(+) T cells and deplete memory CD8(+) T cells. Eur J Immunol **43**(1), 85–92 (2012). doi:10.1002/eji.201242879
21. Ginhoux, F., Ng, L.G., Merad, M.: Understanding the murine cutaneous dendritic cell network to improve intradermal vaccination strategies. Curr Top Microbiol Immunol **351**, 1–24 (2012)
22. Teunissen, M.B., Haniffa, M., Collin, M.P.: Insight into the immunobiology of human skin and functional specialization of skin dendritic cell subsets to innovate intradermal vaccination design. Curr Top Microbiol Immunol **351**, 25–76 (2012)
23. Steinman, R.M., Hemmi, H.: Dendritic cells: translating innate to adaptive immunity. Curr Top Microbiol Immunol **311**, 17–58 (2006)
24. Merad, M., Ginhoux, F., Collin, M.: Origin, homeostasis and function of Langerhans cells and other langerin-expressing dendritic cells. Nat Rev Immunol **8**, 935–947 (2008)
25. Romani, N., Clausen, B.E., Stoitzner, P.: Langerhans cells and more: langerin-expressing dendritic cell subsets in the skin. Immunol Rev **234**, 120–141 (2010)
26. Takahara, K., Omatsu, Y., Yashima, Y., Maeda, Y., Tanaka, S., Iyoda, T., et al.: Identification and expression of mouse Langerin (CD207) in dendritic cells. Int Immunol **14**, 433–444 (2002)
27. Valladeau, J., Saeland, S.: Cutaneous dendritic cells. Semin Immunol **17**, 273–283 (2005)
28. Steinman, R.M., Nussenzweig, M.C.: Avoiding horror autotoxicus: the importance of dendritic cells in peripheral T cell tolerance. Proc. Natl. Acad. Sci. U. S. A. **99**, 351–358 (2002)
29. Larregina, A.T., Falo Jr., L.D.: Changing paradigms in cutaneous immunology: adapting with dendritic cells. J Invest Dermatol **124**, 1–12 (2005)
30. Bursch, L.S., Wang, L., Igyarto, B., Kissenpfennig, A., Malissen, B., Kaplan, D.H., et al.: Identification of a novel population of Langerin+dendritic cells. J Exp Med **204**, 3147–3156 (2007)
31. Ginhoux, F., Collin, M.P., Bogunovic, M., Abel, M., Leboeuf, M., Helft, J., et al.: Blood-derived dermal langerin+dendritic cells survey the skin in the steady state. J Exp Med **204**, 3133–3146 (2007)
32. Poulin, L.F., Henri, S., de Bovis, B., Devilard, E., Kissenpfennig, A., Malissen, B.: The dermis contains langerin+dendritic cells that develop and function independently of epidermal Langerhans cells. J Exp Med **204**, 3119–3131 (2007)
33. Romani, N., Koide, S., Crowley, M., Witmer-Pack, M., Livingstone, A.M., Fathman, C.G., et al.: Presentation of exogenous protein antigens by dendritic cells to T cell clones. Intact protein is presented best by immature, epidermal Langerhans cells. J Exp Med **169**, 1169–1178 (1989)
34. Stoitzner, P., Tripp, C.H., Eberhart, A., Price, K.M., Jung, J.Y., Bursch, L., et al.: Langerhans cells cross-present antigen derived from skin. Proc. Natl. Acad. Sci. U. S. A. **103**, 7783–7788 (2006)
35. Stoitzner, P., Green, L.K., Jung, J.Y., Price, K.M., Tripp, C.H., Malissen, B., et al.: Tumor immunotherapy by epicutaneous immunization requires langerhans cells. J Immunol **180**, 1991–1998 (2008)
36. Cunningham, A.L., Carbone, F., Geijtenbeek, T.B.: Langerhans cells and viral immunity. Eur J Immunol **38**, 2377–2385 (2008)
37. Kautz-Neu, K., Meyer, R.G., Clausen, B.E., von Stebut, E.: Leishmaniasis, contact hypersensitivity and graft-versus-host disease: understanding the role of dendritic cell subsets in balancing skin immunity and tolerance. Exp Dermatol **19**, 760–771 (2010)
38. Bennett, C.L., van Rijn, E., Jung, S., Inaba, K., Steinman, R.M., Kapsenberg, M.L., et al.: Inducible ablation of mouse Langerhans cells diminishes but fails to abrogate contact hypersensitivity. J Cell Biol **169**, 569–576 (2005)
39. Stoecklinger, A., Eticha, T.D., Mesdaghi, M., Kissenpfennig, A., Malissen, B., Thalhamer, J., et al.: Langerin+dermal dendritic cells are critical for CD8+ T cell activation and IgH gamma-1 class switching in response to gene gun vaccines. J Immunol **186**, 1377–1383 (2011)
40. Angel, C.E., Lala, A., Chen, C.J., Edgar, S.G., Ostrovsky, L.L., Dunbar, P.R.: CD14+ antigen-presenting cells in human dermis are less mature than their CD1a+counterparts. Int Immunol **19**, 1271–1279 (2007)
41. Angel, C.E., Chen, C.J., Horlacher, O.C., Winkler, S., John, T., Browning, J., et al.: Distinctive localization of antigen-presenting cells in human lymph nodes. Blood **113**, 1257–1267 (2009)
42. Klechevsky, E., Morita, R., Liu, M., Cao, Y., Coquery, S., Thompson-Snipes, L., et al.: Functional specializations of human epidermal Langerhans cells and CD14+ dermal dendritic cells. Immunity **29**, 497–510 (2009)
43. van der Aar, A.M., de Groot, R., Sanchez-Hernandez, M., Taanman, E.W., van Lier, R.A., Teunissen, M.B., et al.: Cutting edge: virus selectively primes human langerhans cells for CD70 expression promoting CD8+ T cell responses. J Immunol **187**, 3488–3492 (2011)
44. Garland, M., Migalska, K., Mahmood, T.M.T., Singh, T.R.R., Woolfson, A.D., Donnelly, R.: Microneedle arrays as medical devices for enhanced transdermal drug delivery. Expert Rev Med Devices **8**, 459–482 (2011)
45. Donnelly, R., Majithiya, R., Singh, T., Morrow, D., Garland, M., Demir, Y.: Design, optimization and characterisation of polymeric microneedle arrays prepared by a novel laser-based micromoulding technique. Pharm Res **28**, 41–57 (2011)
46. Chen, X., Fernando, G.J.P., Crichton, M., Flaim, C., Yukiko, S., Corbett, H.J., et al.: Improving the reach

of vaccines to low-resource regions, with a needle-free vaccine delivery device and long-term thermostabilization. J Controlled Release. **152**, 349–355 (2011)
47. Henry, S., McAllister, D.V., Allen, M.G., Prausnitz, M.R.: Microfabricated microneedles: a novel approach to transdermal drug delivery. J Pharm Sci **87**, 922–925 (1998)
48. Prausnitz, M., Mikszta, J., Cormier, M., Andrianov, A.: Microneedle-based vaccines. Curr Top Microbiol Immunol **333**, 369–393 (2009)
49. Donnelly, R., Raj Singh, T.R., Woolfson, A.D.: Microneedle-based drug delivery systems: microfabrication, drug delivery, and safety. Drug Deliv **17**, 187–207 (2010)
50. Zhou, C., Liu, Y., Wang, H., Zhang, P., Zhang, J.: Transdermal delivery of insulin using microneedle rollers in vivo. Int J Pharm **392**, 127–133 (2010)
51. Ding, Z., Verbaan, F.J., Bivas-Benita, M., Bungener, L., Huckriede, A., Kersten, G., et al.: Microneedle arrays for the transcutaneous immunization of diphtheria and influenza in BALB/c mice. J Controlled Release. **136**, 71–78 (2009)
52. Ding, Z., Van Riet, E., Romeijn, S., Kersten, G.F.A., Jiskoot, W., Bouwstra, J.A.: Immune modulation by adjuvants combined with diphtheria toxoid administered topically in BALB/c mice after microneedle array pretreatment. Pharm Res **26**, 1635–1643 (2009)
53. Bal, S.M., Slütter, B., van Riet, E., Kruithof, A.C., Ding, Z., Kersten, G.F.A., et al.: Efficient induction of immune responses through intradermal vaccination with N-trimethyl chitosan containing antigen formulations. J Controlled Release. **142**, 374–383 (2010)
54. Bhowmik, T., D'Souza, B., Shashidharamurthy, R., Oettinger, C., Selvaraj, P., D'Souza, M.: A novel microparticulate vaccine for melanoma cancer using transdermal delivery. J Microencapsul **28**, 294–300 (2011)
55. Cleary, G.: Microneedles for drug delivery. Pharm Res **28**, 1–6 (2011)
56. Shah, U.U., Roberts, M., Orlu Gul, M., Tuleu, C., Beresford, M.W.: Needle-free and microneedle drug delivery in children: a case for disease-modifying antirheumatic drugs (DMARDs). Int J Pharm **416**, 1–11 (2011)
57. Kim, Y., Quan, F., Compans, R.W., Kang, S., Prausnitz, M.R.: Formulation and coating of microneedles with inactivated influenza virus to improve vaccine stability and immunogenicity. J Controlled Release. **142**, 187–195 (2010)
58. Hiraishi, Y., Nandakumar, S., Choi, S., Lee, J., Kim, Y., Prausnitz, M.R., et al.: Bacillus Calmette-Guérin vaccination using a microneedle patch. Vaccine **29**, 2626–2636 (2011)
59. Weldon, W., Martin, M., Zarnitsyn, V., Wang, B., Koutsonanos, D., Skountzou, I.: Microneedle vaccination with stabilized recombinant influenza virus hemagglutinin induces improved protective immunity. Clinical and vaccine immunol. **18**, 647–654 (2011)
60. Prow, T.: Nanopatch-targeted skin vaccination against West Nile Virus and Chikungunya virus in mice. Small **6**, 1776–1784 (2010)
61. Corbett, H., Chen, X., Frazer, I.: Skin vaccination against cervical cancer associated human papillomavirus with a novel micro-projection array in a mouse model. PLoS One **5**, e13460 (2010)
62. Fernando, G.J.P., Chen, X., Prow, T., Crichton, M., Fairmaid, E.: Potent immunity to low doses of influenza vaccine by probabilistic guided micro-targeted skin delivery in a mouse model. PLoS One **5**, e10266 (2010)
63. Matriano, J., Cormier, M., Johnson, J., Young, W., Buttery, M., Cormier, M., et al.: Macroflux microprojection array patch technology: a new and efficient approach for intracutaneous immunization. Pharm Res **19**, 63–70 (2002)
64. Widera, G., Johnson, J., Kim, L., Libiran, L., Nyam, K., Daddona, P.E., et al.: Effect of delivery parameters on immunization to ovalbumin following intracutaneous administration by a coated microneedle array patch system. Vaccine **24**, 1653–1664 (2006)
65. Escobar-Chvez, J., Bonilla-Martinez, D., Villegas-González, M.A., Molina Trinidad, E., Casas Alancaster, N., et al.: Microneedles: a valuable physical enhancer to increase transdermal drug delivery. J Clin Pharmacol **51**, 964–977 (2011)
66. Amorij, J., Frijlink, H., Wilschut, J., Huckriede, A.: Needle-free influenza vaccination. Lancet Infect Dis **10**, 699–711 (2010)
67. Wang, P., Cornwell, M., Hill, J., Prausnitz, M.: Precise microinjection into skin using hollow microneedles. J Invest Dermatol **126**, 1080–1087 (2006)
68. Frost, G.I.: Recombinant human hyaluronidase (rHuPH20): an enabling platform for subcutaneous drug and fluid administration. Expert Opin Drug Deliv **4**, 427–440 (2007)
69. Bal, S., Ding, Z., van Riet, E., Jiskoot, W., Bouwstra, J.: Advances in transcutaneous vaccine delivery: Do all ways lead to Rome? J Controlled Release. **148**, 266–282 (2010)
70. Van Damme, P., Oosterhuis-Kafeja, F., Van der Wielen, M., Almagor, Y., Sharon, O., Levin, Y.: Safety and efficacy of a novel microneedle device for dose sparing intradermal influenza vaccination in healthy adults. Vaccine **27**, 454–459 (2009)
71. Alarcon, J., Hartley, A., Harvey, N., Mikszta, J.: Preclinical evaluation of microneedle technology for intradermal delivery of influenza vaccines. Clin Vaccine Immunol **14**, 375–381 (2007)
72. Mikszta, J., Dekker, J., Harvey, N., Dean, C., Brittingham, J., Huang, J., et al.: Microneedle-based intradermal delivery of the anthrax recombinant protective antigen vaccine. Infect Immun **74**, 6806–6810 (2006)
73. Sullivan, S., Koutsonanos, D., Del Pilar Martin, M., Lee, J., Zarnitsyn, V., Compans, R.W., et al.:

Dissolving polymer microneedle patches for influenza vaccination. Nat Med **16**, 915–920 (2010)
74. Raphael, A., Prow, T., Crichton, M., Chen, X., Fernando, G.J.P., Prow, T.: Targeted, needle-free vaccinations in skin using multilayered, densely-packed dissolving microprojection arrays. Small **6**, 1785–1793 (2010)
75. Chabri, F., Bouris, K., Jones, T., Barrow, D., Hann, A., Allender, C., et al.: Microfabricated silicon microneedles for nonviral cutaneous gene delivery. Br J Dermatol **150**, 869–877 (2004)
76. Prow, T.W., Chen, X., Prow, N.A., Fernando, G.J., Tan, C.S., Raphael, A.P., et al.: Nanopatch-targeted skin vaccination against West Nile Virus and Chikungunya virus in mice. Small **16**, 1776–1784 (2010)
77. Birchall, J., Coulman, S., Pearton, M., Allender, C., Brain, K., Coulman, S., et al.: Cutaneous DNA delivery and gene expression in ex vivo human skin explants via wet-etch micro-fabricated microneedles. J Drug Target **13**, 415–421 (2005)
78. Coulman, S.A., Barrow, D., Anstey, A., Gateley, C., Morrissey, A., Wilke, N., et al.: Minimally invasive cutaneous delivery of macromolecules and plasmid DNA via microneedles. Curr Drug Deliv **3**, 65–75 (2006)
79. Pearton, M., Allender, C., Brain, K., Anstey, A., Gateley, C., Wilke, N., et al.: Gene delivery to the epidermal cells of human skin explants using microfabricated microneedles and hydrogel formulations. Pharm Res **25**, 407–416 (2008)
80. Gill, H.S., Soderholm, J., Prausnitz, M.R., Sallberg, M., Sderholm, J., Sllberg, M.: Cutaneous vaccination using microneedles coated with hepatitis C DNA vaccine. Gene Ther **17**, 811–814 (2010)
81. Choi, S. O., Park, J. H., Gill, H. S., Choi, Y., Allen, M.G., M. R.: Prausnitz. Microneedles electrode array for electroporation of skin for gene therapy. Controlled Release Society 32nd Annual Meeting & Exposition Transactions. 318 (2005)
82. Hooper, J., Golden, J., Ferro, A., King, A.: Smallpox DNA vaccine delivered by novel skin electroporation device protects mice against intranasal poxvirus challenge. Vaccine **25**, 1814–1823 (2007)
83. Daugimont, L., Baron, N., Vandermeulen, G., Pavselj, N., Miklavcic, D., Jullien, M., et al.: Hollow microneedle arrays for intradermal drug delivery and DNA electroporation. J Membr Biol **236**, 117–125 (2010)
84. Levine, M.M., Sztein, M.B.: Vaccine development strategies for improving immunization: the role of modern immunology. Nat Immunol **5**, 460–464 (2004)
85. Soppimath, K.S., Aminabhavi, T.M., Kulkarni, A.R., Rudzinski, W.E.: Biodegradable polymeric nanoparticles as drug delivery devices. J Control Release **70**, 1–20 (2001)
86. Delie, F., Blanco-Prieto, M.J.: Polymeric particulates to improve oral bioavailability of peptide drugs. Molecules **10**, 65–80 (2005)
87. McCarron, P.A., Donnelly, R.F., Marouf, W.: Celecoxib-loaded poly(D, L-lactide-co-glycolide) nanoparticles prepared using a novel and controllable combination of diffusion and emulsification steps as part of the salting-out procedure. J Microencapsul **23**, 480–498 (2006)
88. Eniola, A.O., Hammer, D.A.: Artificial polymeric cells for targeted drug delivery. J Control Release **87**, 15–22 (2003)
89. Jaganathan, K.S., Vyas, S.P.: Strong systemic and mucosal immune responses to surface-modified PLGA microspheres containing recombinant hepatitis B antigen administered intranasally. Vaccine **24**, 201–4211 (2006)
90. Gutierro, I., Hernandez, R.M., Igartua, M., Gascon, A.R., Pedraz, J.L.: Size dependent immune response after subcutaneous, oral and intranasal administration of BSA loaded nanospheres. Vaccine **21**, 67–77 (2002)
91. Lu, D., Garcia-Contreras, L., Xu, D., Kurtz, S.L., Liu, J., Braunstein, M., et al.: Poly (lactide-co-glycolide) microspheres in respirable sizes enhance an in vitro T cell response to recombinant Mycobacterium tuberculosis antigen 85B. Pharm Res **24**, 1834–1843 (2007)
92. Sharp, F.A., Ruane, D., Claass, B., Creagh, E., Harris, J., Malyala, P., et al.: Uptake of particulate vaccine adjuvants by dendritic cells activates the NALP3 inflammasome. Proc. Natl. Acad. Sci. U. S. A. **106**, 870–875 (2009)
93. de Jalon, E.G., Blanco-Prieto, M.J., Ygartua, P., Santoyo, S.: PLGA microparticles: possible vehicles for topical drug delivery. Int J Pharm **226**, 181–184 (2001)
94. Jenning, V., Gysler, A., Schafer-Korting, M., Gohla, S.H.: Vitamin A loaded solid lipid nanoparticles for topical use: occlusive properties and drug targeting to the upper skin. Eur J Pharm Biopharm **49**, 211–218 (2000)
95. Alvarez-Roman, R., Naik, A., Kalia, Y.N., Guy, R.H., Fessi, H.: Enhancement of topical delivery from biodegradable nanoparticles. Pharm Res **21**, 1818–1825 (2004)
96. Alvarez-Roman, R., Naik, A., Kalia, Y.N., Guy, R.H., Fessi, H.: Skin penetration and distribution of polymeric nanoparticles. J Control Release **99**, 53–62 (2004)
97. Luengo, J., Weiss, B., Schneider, M., Ehlers, A., Stracke, F., Konig, K., et al.: Influence of nanoencapsulation on human skin transport of flufenamic acid. Skin Pharmacol Physiol **19**, 190–197 (2006)
98. Lademann, J., Richter, H., Teichmann, A., Otberg, N., Blume-Peytavi, U., Luengo, J., et al.: Nanoparticles–an efficient carrier for drug delivery into the hair follicles. Eur J Pharm Biopharm **66**, 159–164 (2007)
99. Toll, R., Jacobi, U., Richter, H., Lademann, J., Schaefer, H., Blume-Peytavi, U.: Penetration profile of microspheres in follicular targeting of terminal hair follicles. J Invest Dermatol **123**, 168–176 (2004)

100. Coulman, S.A., Anstey, A., Gateley, C., Morrissey, A., McLoughlin, P., Allender, C., et al.: Microneedle mediated delivery of nanoparticles into human skin. Int J Pharm Jan. **366**, 190–200 (2009)
101. McAllister, D.V., Wang, P.M., Davis, S.P., Park, J.H., Canatella, P.J., Allen, M.G., et al.: Microfabricated needles for transdermal delivery of macromolecules and nanoparticles: fabrication methods and transport studies. Proc. Natl. Acad. Sci. U. S. A. **100**, 13755–13760 (2003)
102. Bal, S.M., Slutter, B., Jiskoot, W., Bouwstra, J.A.: Small is beautiful: N-trimethyl chitosan-ovalbumin conjugates for microneedle-based transcutaneous immunisation. Vaccine **29**, 4025–4032 (2011)
103. Ueno, H., Schmitt, N., Klechevsky, E., Pedroza-Gonzalez, A., Matsui, T., Zurawski, G., et al.: Harnessing human dendritic cell subsets for medicine. Immunol Rev **234**, 199–212 (2010)
104. Birchall, J.C., Clemo, R., Anstey, A., John, D.N.: Microneedles in clinical practice–an exploratory study into the opinions of healthcare professionals and the public. Pharm Res **28**, 95–106 (2011)
105. Donnelly, R.F., Singh, T.R., Tunney, M.M., Morrow, D.I., McCarron, P.A., O'Mahony, C., Woolfson, A.D.: Microneedle arrays allow lower microbial penetration than hypodermic needles in vitro. Pharm Res **26**, 2513–2522 (2009)

# Nasal Dry Powder Vaccine Delivery Technology

## 44

Anthony J. Hickey, Herman Staats, Chad J. Roy, Kenneth G. Powell, Vince Sullivan, Ginger Rothrock, and Christie M. Sayes

## Contents

| | | |
|---|---|---|
| 44.1 | Introduction | 717 |
| 44.2 | Target Diseases and Antigens | 718 |
| 44.3 | **Nasal Vaccine Technology** | 719 |
| 44.3.1 | Formulation | 719 |
| 44.3.2 | Devices | 720 |
| 44.3.3 | Adjuvants | 720 |
| 44.3.4 | Mechanism | 722 |
| 44.3.5 | Animal Models | 722 |
| 44.4 | Strengths/Weaknesses | 724 |
| Conclusion | | 724 |
| References | | 725 |

A.J. Hickey, PhD (✉) • G. Rothrock, PhD • C.M. Sayes, PhD
Center for Aerosols & Nanomaterials Engineering, RTI International, Research Triangle Park, NC, USA
e-mail: ahickey@rti.org

H. Staats, PhD
Departments of Pathology and Immunology, Duke University Medical Center, Durham, NC, USA

C.J. Roy, PhD
Departments of Microbiology and Immunology, Tulane University School of Medicine, New Orleans, LA, USA

K.G. Powell • V. Sullivan
BD Technologies, Research Triangle Park, NC, USA

### Abstract

Nasal delivery of vaccines occurred over a millennium ago in China, where ground scabs from small pox lesions, presumably containing live virus, were sniffed. This practice was the basis for early vaccination with live virus in Europe in the eighteenth century. In the past decade, a number of reports have focused on new antigens, adjuvants, and delivery systems, but few approaches have entered development as a clinical candidate. This chapter outlines the critical steps needed to create a comprehensive integrated strategy adopting an antigen, with candidate physical and biochemical adjuvants, in a delivery system. There is a need to define unique formulation and device performance properties, evaluate dose sparing achieved through novel construction and adjuvancy, and develop rapid screening methods to identify toxic formulations.

## 44.1 Introduction

The nasal mucosa is complex and includes important elements of the immune system. Therefore, it is an ideal route of delivery for a noninvasive vaccine delivery. The anatomy of the nasal cavity is shown in Fig. 44.1. The ciliated surfaces of the nasal turbinates deliver mucus to the floor of the nasal cavity and from there to the throat, site of the Waldeyer's ring, consisting

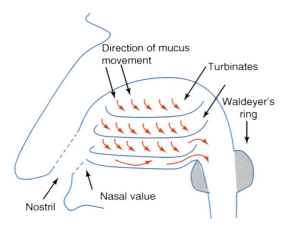

**Fig. 44.1** Schematic of the nasal cavity

notably of adenoids and tonsils [1]. Mucociliary clearance disposes of ambient particulates that deposit on the nasal epithelium. Normal mucociliary clearance rates are approximately 45–90 min from the proximal to distal points in the nasopharynx [2]. The intent of any nasal vaccine delivery system is to achieve a residence time at the nasal mucosa sufficient to elicit the desired immunological and hopefully protective response.

Nasal vaccine delivery proposes to target the nasal-associated lymphoid tissue (NALT) which is one of the most accessible of the mucosal-associated lymphoid tissues (MALT). The principle on which this approach depends is to present the selected antigen to the NALT and Waldeyer's ring and ultimately to the immunologically important M cells and dendritic cells. These cells are responsible for producing an immune response through antigen presentation to circulating lymphocytes.

The two most common (and effective) nasal delivery devices are either multiple- or single-dose pumps, examples of which deliver calcitonin, to treat osteoporosis, and corticosteroids, to treat allergic rhinitis, respectively [3]. However, these FDA-approved products require solution or suspension formulations which are not the most desirable preparations for vaccines. Recently, the use of powder aerosols for vaccination has been proposed with attendant advantages of increased residence time and greater opportunity for antigen presentation [3, 4].

## 44.2 Target Diseases and Antigens

The mortality associated with pulmonary anthrax, along with recent concerns regarding the possibility of the causative microorganism being used as a weapon, necessitates the development of a vaccine that is easily administered and requires minimal storage and distribution controls that can be rapidly deployed. The causative organism of anthrax is *Bacillus anthracis* (*B. anthracis*). This organism is associated with infections of animals and results in skin and pulmonary disease in humans following contact with infected animals or animal products [5]. Pulmonary anthrax is a particularly insidious and, if untreated, fatal form of the disease. A unique component of the lifecycle of *B. anthracis* is the formation of spores as a method of withstanding stressful environmental conditions. Anthrax spores, in common with those of other organisms, are exquisitely suited to dispersion in air. Spores then deposit throughout the respiratory tract. The ability of these spores to retain their viability through extremes of environmental stress for decades or centuries poses a public health risk. The virulence, ease of delivery, and long-term stability of *B. anthracis* spores rendered them a source of great concern throughout the middle of the twentieth century at the height of state-sponsored biological weapon development and, more recently, regarding bioterrorism.

After the terrorist action of September 11, 2001, the demand for biodefense measures increased following the incident in which letters contained "weaponized" anthrax powders [6, 7]. The immediate demand was for drugs (e.g., ciprofloxacin) to treat infected individuals. The need for vaccines that could be stockpiled and quickly distributed with the minimum storage considerations (special packaging or temperature) in quantities sufficient to contain an outbreak, particularly for health-care providers and military personnel, also became a research and development focus [8]. In this context, vaccines for protection and drugs for treatment have been developed for this specific application over the past decade.

In general, the practice of presenting live organisms as a vaccine system, albeit in an attenuated state, is an elevated health risk. The previously mentioned example for smallpox treatment in

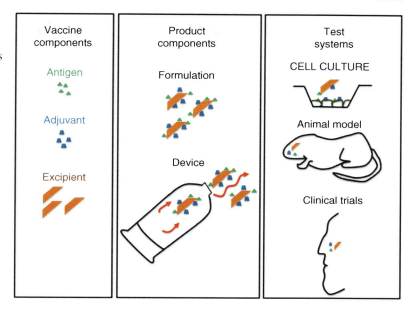

**Fig. 44.2** Schematic of the key elements of the vaccine product and the methods of evaluation as presented in this chapter

China and Europe was undoubtedly a risky procedure because the prospects of contracting disease were significant. Similarly, for anthrax, the use of a virulent organism would be a questionable strategy. The current vaccine, BioThrax (Emergent Solutions), uses culture filtrate, which presents all of the important antigens (lethal factor [LF], protective antigen [PA], and edema factor [EF]) [8]. The emphasis for novel immunization strategies for protection against anthrax has been the use of recombinant protective antigen, which has demonstrated protection in animal models of disease [9]. Indeed, it has just been announced that an rPA vaccine, SparVaxTM (PharmAthene), is entering further Phase II trials. However, it has been suggested that adjuvancy is required to ensure a sufficiently robust response. This may be achieved through coadministration of an adjuvant molecule of which there are many but most notably monophosphoryl lipid A (MPL) – an FDA-approved drug – referred to as Cervarix and safely used in human studies with a nasal norovirus vaccine.

Alternatively, a physical approach may be taken using small targeting particles (nanoparticles) that would deliver the antigen effectively to the NALT. Nanoparticles have a particular propensity for this purpose and have been shown for many years to be effective as carriers and adjuvants [10–12]. It is important to note that despite the mucosal route of administration, an IgA response is not thought to be significant in protection from anthrax [9].

However, centuries ago, Voltaire observed "...*the Chinese go about it* (inoculation) *in a different fashion they make no incision but give smallpox through the nose, like snuff. This way is pleasanter* (than cutaneous delivery), *but it comes to the same thing*...." [13]

Currently, we might add not only is it "pleasanter" (needle-free) and that it "amounts to the same thing" (a protective immune response), but with current technology, it is stable on storage, easy to use, and inexpensive. The delivery of particles in a stable dispersible form lends itself to the use of dry particles. The ability to spray dry particle suspensions has been demonstrated for a number of vaccine applications [14]. Specific to this discussion, Fig. 44.2 outlines the components needed for the development of a safe and effective technologically advanced nasal vaccine system.

## 44.3 Nasal Vaccine Technology

### 44.3.1 Formulation

In order to develop vaccine formulations that would have potential for worldwide distribution, dry powders must be able to be stored long term without refrigeration and be administered

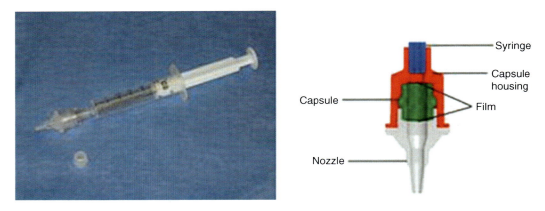

**Fig. 44.3** The BD Solovent® device for nasal vaccine administration [16]. This device depends on "burstable membranes" that rupture under syringe air pressure during use. This results in a gentle pop and the expulsion of the dose

without the use of needles. Spray-dried, freeze-dried, and spray-freeze-dried powders have demonstrated equivalent immunogenicity to conventional liquid formulations with the potential advantage that dry powders can circumvent the problem of the cold chain which is often associated with liquid vaccine formulations. These techniques can be paired with conventional and novel nanoparticle fabrication methodologies such as precipitation, homogenization, and top-down manufacturing to generate dry powders of loose nanoparticle aggregates having aerodynamic diameters suitable for mucosal delivery as has previously been demonstrated for inhaled formulations.

### 44.3.2 Devices

There are only a few nasal powder delivery devices available commercially or in clinical development [15]. One such device is shown in Fig. 44.3. While formulation and powder morphology play key roles in targeting deposition to the nasal tissues, it is essential that the powder be deaggregated and efficiently delivered from the device.

Key device attributes fall into two categories: those affecting clinical efficacy and those human factors reflective of the human-device interaction. First, the emitted dose fraction from the device needs to be as large as possible. Secondly, the emitted dose must be in a form that reaches the targeted tissue. Particles smaller than 10 µm can be carried deeper into the lung tissue which may not be desirable, while particles larger than 50 µm may have difficulty reaching immunoresponsive tissues. The challenge is in obtaining the proper deaggregation of the particles without excessive energy input. Moving air is generally used to deaggregate and propel the particles into the airway. However, excess airflow can force the particles past the targeted region so that they "*pass through*" without settling and eliciting an immune response. From the human factor perspective, the device should be intuitive and easy to handle. Devices are currently targeted for administration by clinicians, but possible applications include biodefense vaccines where delivery by untrained personnel may be desired. Finally, the device should be unit dose; there is no need for a multidose format and it would be unnecessarily risky to use the same device on multiple people. Furthermore, the device should not be "*reloadable*" for unintended use.

### 44.3.3 Adjuvants

Aluminum salts (alum) have been used as adjuvants with great success for almost a century and have been particularly effective at promoting protective humoral immunity. However, alum is not optimally effective for diseases where cell-mediated immunity is required for protection. Furthermore, classical biological adjuvants, such as Freund's adjuvant, have shown improved efficacy for protection in animal models, yet the toxicity profiles prevent widespread adoption for human use. In particular, there is demand for safe and nontoxic adjuvants able to stimulate cellular immunity without undue toxicity and adjuvants suitable for use with mucosally delivered recombinant and subunit vaccines. Novel particle and biological adjuvants offer a means to overcome the deficiencies of classical adjuvant technologies.

A human study was recently performed using a nasally delivered dry powder vaccine containing a Norwalk viruslike particle (VLP) immunogen (i.e., monophosphoryl lipid A [MPL]) as the adjuvant and chitosan as a mucoadhesive [17]. The Norwalk VLP was well tolerated and immunogenic in humans, suggesting that a dry powder vaccine formulation for use in humans is feasible. Polymeric nanoparticles have also demonstrated dual functionality as antigen delivery systems with added adjuvant effect [18]. We postulate that dry powder polymeric nanoparticle aerosols containing recombinant protective antigen (rPA) delivered from a commercial device and targeted to the nasal mucosa will protect non-human primates from challenge with airborne anthrax.

Adjuvants, in the context of vaccines, may be functionally defined as any substance able to augment the immunogenicity of the vaccine antigen. Adjuvants are beneficial since they are able to enhance the percentage of vaccines that achieve protective immunity at the completion of the vaccination regimen [19]; they are able to increase the magnitude of immune responses induced [19], reduce the antigen dose required to induce protective immunity [20], and/or reduce the number of immunizations needed to induce protective immunity [21]. In general terms, adjuvants either enhance antigen delivery to or enhance the activation state of dendritic cells, the primary antigen-presenting cell that initiates the development of adaptive immunity after vaccination [22]. For the pusposes of this review, we will briefly discuss (1) vaccine adjuvants that enhance the immunogenicity of the vaccine antigen by increasing delivery of the antigen to the immune system and (2) adjuvants that specifically activate the innate immune system. More information on the current status of vaccine adjuvants [23] and adjuvants for mucosally administered vaccines [24] is discussed elsewhere.

Adjuvants can increase the immunogenicity of nasally delivered vaccines by enhancing antigen delivery to the host. A recent study by Kiyono et al [25]. described the use of cationic nanogels as adjuvants for nasally delivered protein antigens. Although the cationic nanogels did not enhance the activation state of nasal dendritic cells, the cationic nanogels significantly enhanced the immunogenicity of the nasal vaccine due to improved antigen retention in the nasal cavity that was associated with better antigen delivery to nasal dendritic cells [25]. A nanoemulsion adjuvant system is also being developed as a nasal vaccine adjuvant [26–28] and has been described as "......*does not contain a proinflammatory component but penetrates mucosal surfaces to load antigens into dendritic cells.*" [28]

Dry powders produced by spray-freeze-drying (SFD) liquid vaccine formulations also provide a vaccine adjuvant strategy to enhance the immunogenicity of nasally delivered vaccines [29] when compared to nasal immunization with liquid vaccine formulations. Since other studies have demonstrated that increased nasal retention of vaccine formulations correlates with increased immunogenicity of the vaccine [25, 30, 31], it is possible that dry powder vaccine formulations augment the immunogenicity of nasally delivered vaccines by increasing antigen retention in the nasal cavity to ultimately improve antigen delivery to the nasal dendritic cells for better induction of antigen-specific adaptive immunity.

Adjuvants may also activate the host innate immune system to enhance the induction of antigen-specific adaptive immunity after nasal immunization [23, 24]. Adjuvants that activate the innate immune system are typically thought to activate dendritic cells [32] to increase their ability to induce antigen-specific adaptive immune responses. A wide range of mucosal adjuvants able to activate the innate immune system include proinflammatory cytokines [33, 34] and Toll-like receptor ligands (TLR ligands; monophosphoryl lipid A, CpG oligodeoxynucleotides, imiquimod) [33–35]. Although cholera toxin and related toxins are known to provide effective mucosal adjuvant activity, adverse events associated with the use of toxin-based adjuvants [36, 37] will likely limit their use in human vaccines. Cytokines and TLR ligand adjuvants have been used as nasal vaccine adjuvants in human clinical studies with acceptable safety profiles, suggesting that adjuvanted nasally administered vaccines may be feasible with additional optimization and safety evaluations [34, 38].

### 44.3.4 Mechanism

Many of the markers of toxicity are also pro-inflammatory mediators. This type of induced response is desirable to achieve a protective immunity from vaccines. Individual parameters may be associated with specific phenomena (e.g., caspase and oxygenase or extracellular cytoplasm leakage caused by peroxy radical formation on cellular membranes). However, these acellular markers may not be indicative of the entire tissue system – it is important to consider the combined response in terms of magnitude and duration of each cellular response as a signature of transition from reversible "transient" to irreversible "toxic" response. For example, the oxidative stress as indicated by the redox (i.e., reduction or oxidation) state of glutathione has been used to indicate this transition from normal transient phenomenon to toxic cell response [39–41].

Conclusions from the literature have led to the premise that there are two primary physicochemical properties of small particles, including nanoparticles, which influence their stability, mobility, and toxicity. These two primary physicochemical properties are as follows: (1) surface modification or conjugation to other molecules and (2) physical contact with cells. The former influences particle aggregation state, determines individual particle suspendability, and affects susceptibility to degradation. The latter affects cellular uptake mechanism substantially. Particles that are less than 20 nm can transport across the cytoplasmic membrane and accumulate in a variety of subcellular components. Particles (or aggregates of particles) between 20 and 200 nm have the potential to be taken up into cells via active transport mechanism of clathrin or caveolin pit formation. Particles (or aggregates of particles) larger than 200 nm are readily endocytosed via macropinocytosis. Ultimately, most particles sequester in membrane-bound vesicles where the acidification process begins. The acidification process usually breaks the particle down into their ions or molecules, releasing them in a concentrated dose to specific tissues most vulnerable to particle deposition. While not all mechanisms are known within this general field of study, most agree that cellular uptake plays a large role in a cell's and a tissue's response to particles.

Studies have shown that exposure to particles in both in vitro and in vivo systems may cause production of inflammatory biomarkers [40, 42–45]. After engineered particles are internalized by cell, such as phagocytes, the inflammatory cascade may be triggered. Inflammation is the complex biological response of cells and tissues to harmful pathogens and other toxicants. It is both a proactive mechanism to remove these harmful pathogens and to initiate production of repair enzymes. Unchecked inflammation can lead to a host of diseases, such as asthma, atherosclerosis, and rheumatoid arthritis; therefore, it is normally tightly regulated by the body. Figure 44.4 graphically depicts the four distinct steps in a particle-induced inflammatory cascade.

### 44.3.5 Animal Models

The in vivo testing and evaluation of a new vaccine formulated for aerosol delivery requires careful consideration of factors relating to the appropriateness of the species and delivery modality as it relates to the immunogen and the corresponding infectious disease agent. The rationale underpinning evaluation in a particular species is determined at least in part upon the predicted serological and/or mucosal immune

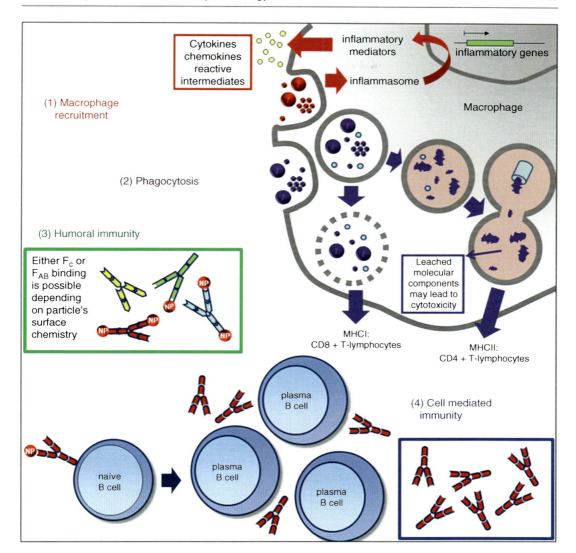

**Fig. 44.4** General mechanisms of cellular toxicity after inhalation of particles. Presumed steps in triggering either cell-mediated or humoral immunities include the following: (*1*) inhaled particles induce macrophage recruitment to mediate inflammation with cytokines and nitric oxide, causing local tissue damage, (*2*) individual particles are phagocytosed and acidified in macrophage endocytic pathway, causing cytotoxic release molecular components, (*3*) specific to nanometer-sized particles, binding to nonspecific subsets of naïve humoral repertoire is possible, and (*4*) particles are recognized by a specific B cell paratope, which clonally expands plasma cells, secreting particle-specific antibodies, clearing the particles

response in that particular species. There is a narrow range of species available for use in these types of studies because the availability of well-characterized and documented infectious disease models is generally lacking. In addition, matching the selected modality of vaccination in the animal model with the ultimate clinical administration route can further limit choices. Many delivery platforms do not cater to a seamless adaptation in a preclinical setting; this is especially the case where an inhalation device requires any sort of respiratory compliance maneuver to assure dosing. Thus, one is faced with the logistical challenge of providing a surrogate delivery platform for vaccination delivery for the animal experiments.

Specific to vaccination and the immune response, selection of the species that will produce the desired immune response is many times predicated upon past studies using similar immunogens with familiar targets. Selection of the rabbit in the establishment of a clinical correlate of immunity against anthrax, for example, uses a previous characterized animal model with robust serological antibody production from a culture-filtrate vaccine product (anthrax vaccine adsorbed, AVA) [46]. Inhalation anthrax in the rabbit has been well documented and provided a well-characterized model of infection that can be used to test efficacy of a particular vaccine formulation [47]. In this particular example, the model and corresponding immune response was used to establish an in vitro correlate of immunity ($\alpha$-PA IgG) predictive of clinical efficacy of the vaccine formulation. The establishing of a correlate of immunity would have not been possible if the basis of the model (inhalation anthrax in the rabbit) and the corresponding marker of immunity ($\alpha$-PA IgG) had not been first characterized.

Choice of species in the use of animal models also requires critical assessment of the interspecies-specific biological responses prior to selection. The available animal models for investigation of superantigen-associated shock (Sag), for example, are heavily dependent upon the richness of MHC-II receptors in the animal species. Direct binding of Sag toxins to circulating T cells is mediated through the MHC-II receptor and therefore represents a critical pathway in the subsequent inflammatory response. Rodents and rodent-like species are generally nonresponsive to challenge with purified staphylococcal and streptococcal enterotoxemic shock due to the lack of circulating MHC-II receptors. Although these receptors are found in great abundance in phylogenetically higher life forms such as the primate, larger animals are generally used as disease models much more sparingly and only at selected intervals compared to rodent species. Rodent models are still used as a disease model of Sag-induced toxicity through the use of co-stimulatory mediators such as LPS or through the use of transgenics that express for MHC-II [48, 49]. The particular example of animal models of Sag-induced toxicity punctuates the requirement for critical assessment of predictive response prior to design of a study that may incorporate challenge for determination of efficacy of a vaccine product.

## 44.4 Strengths/Weaknesses

A vaccine that has the greatest chance of tackling emerging infectious diseases and agents that are a threat to public health must have a method for rapid distribution and ease of delivery for health-care workers and military personnel as a risk mitigation strategy in the event of a threat. The critical performance parameters would, potentially, include particles that are less than 100 nm in diameter (a.k.a. nanoparticles) that contain an antigen that is released over a 1-week time period, a spray-freeze-dried aggregate of approximately 20 μm in aerodynamic diameter, and the spray-freeze-dried aggregates that should deaggregate into primary particles in less than 15 min. This short time period is due to the fact that mucociliary clearance rates of 45–90 min require rapid deaggregation to allow individual particles to access the mucosa. Ideally, the product should be room temperature stable for a few years.

### Conclusion

The intent of this review is to give a foundational overview for those considering nasal delivery of vaccines. There should be sufficient detail to introduce those interested in this approach to the basic principles that can be pursued in more detail through reference to the literature citations. The delivery of vaccines has been a protective strategy against disease for millennia. A variety of routes of administration have been employed among them the nasal mucosa. In modern times technologies and knowledge of the immune system have evolved, and we can now apply, in an optimal manner, engineering and immunology principles to take advantage of the potential of this important route of administration.

We have discussed important considerations in designing a nasal vaccine technology including the basic anatomy, physiology, cell biology, and immunology of the nasal mucosa. We have presented the important components of a nasal vaccine in the foregoing text including formulation (antigen, adjuvant, and additives), metering system, and device. Finally, we have concluded with in vitro cell culture and animal models that are the basis on which they would be selected to evaluate nasal vaccine technologies.

The future of nasal vaccine delivery rests on the increasing knowledge of the disposition of vaccine from the nasal mucosa and its implications for both efficacy and safety. The significant increase in literature on this topic over the last decade demonstrates the burgeoning interest in this route of administration. Hopefully a critical threshold will eventually be reached where industrial and regulatory confidence in efficacy and safety makes this a commonly employed route rather than simply an alternative to other routes of administration.

## References

1. Hickey, A.J., Garmise, R.J.: Dry powder nasal vaccines as an alternative to needle-based delivery. Crit. Rev. Ther. Drug. Carrier. Syst. **26**, 1–27 (2009)
2. Garmise, R.J., Staats, H.F., Hickey, A.J.: Novel dry powder preparations of whole inactivated influenza virus for nasal vaccination. AAPS PharmSciTech **8**, E81 (2007)
3. Hickey, A.J., Swift, D.: Aerosol measurement, principles, techniques and applications, 3rd edn, pp. 805–820. Wiley, Oxford (2011)
4. Garmise, R.J., et al.: Formulation of a dry powder influenza vaccine for nasal delivery. AAPS PharmSciTech **7**, E19 (2006)
5. Alcomo, I.E.: Fundamentals of microbiology, pp. 263–266. Jones and Bartlett Publishers, Sudbury, Mass (2001)
6. Krasner, R.I.: Biological weapons, the microbial challenge, pp. 335–360. ASM Press, Washington, D.C (2002)
7. Federal Bureau of Investigation, Famous Cases & Criminals: Amerithrax or Anthrax Investigation. www.fbi.gov/about-us/history/famous-cases/anthrax-amerithrax (2011)
8. AHFS-DI, 'Anthrax Vaccine Adsorbed', in Bioterrorism Resource Manual, American Society of Health-System Pharmacists, Bethesda, MD, 360–377 (2002)
9. Mantis, N. J., Morici, L. A., Roy, C. J: Mucosal vaccines for biodefense: critical factors in manufacture and delivery, pp. 181–195. Springer New York (2012)
10. Csaba, N., Garcia-Fuentes, M., Alonso, M.J.: Nanoparticles for nasal vaccination. Adv. Drug Deliv. Rev. **61**, 140–157 (2008)
11. Koping-Hoggard, M., Sanchez, A., Alonso, M.J.: Nanoparticles as carriers for nasal vaccine delivery. Expert Rev. Vaccines **4**, 185–196 (2005)
12. Espueles, S., Gamazo, C., Blanco-Prieto, M.J., Irache, J.M.: Nanoparticles as adjuvant-vectors for vaccination, pp. 317–325. Informa Healthcare, New York (2007)
13. Voltaire, F.M.A.: Philosophical letters. Dover, Mineola (2003)
14. El-Kamary, S.S., et al.: Adjuvanted intranasal Norwalk virus-like particle vaccine elicits antibodies and antibody-secreting cells that express homing receptors for mucosal and peripheral lymphoid tissues. J. Infect. Dis. **202**, 1649–1658 (2010)
15. Administration, U. S. F. a. D.: Efficacy Testing and Surrogate Markers of Immunity Workshop Vol. http://www.fda.gov/downloads/BiologicsBloodVaccines/NewsEvents/WorkshopsMeetingsConferences/TranscriptsMinutes/UCM054606.pdf. Center for Biologics Research and Evaluation (2002)
16. Sewall, H.: The role of epithelium in experimental immunization. Science **62**, 293–299 (1925)
17. Wang, S.H., Kirwan, S.M., Abraham, S.N., Staats, H.F., Hickey, A.J.: Stable dry powder formulation for nasal delivery of anthrax vaccine. J. Pharm. Sci. **101**, 31–47 (2011). doi:10.1002/jps.22742
18. Jain, S., O'Hagan, D.T., Singh, M.: The long-term potential of biodegradable poly(lactide-co-glycolide) microparticles as the next-generation vaccine adjuvant. Expert Rev. Vaccines **10**, 1731–1742 (2011)
19. Jacques, P., et al.: The immunogenicity and reactogenicity profile of a candidate hepatitis B vaccine in an adult vaccine non-responder population. Vaccine **20**, 3644–3649 (2002)
20. Keitel, W., et al.: Dose ranging of adjuvant and antigen in a cell culture H5N1 influenza vaccine: safety and immunogenicity of a phase 1/2 clinical trial. Vaccine **28**, 840–848 (2010). doi:10.1016/j.vaccine.2009.10.019
21. Nevens, F., et al.: Immunogenicity and safety of an experimental adjuvanted hepatitis B candidate vaccine in liver transplant patients. Liver Transpl. **12**, 1489–1495 (2006)
22. Soloff, A.C., Barratt-Boyes, S.M.: Enemy at the gates: dendritic cells and immunity to mucosal pathogens. Cell Res. **20**, 872–885 (2010). doi:10.1038/cr.2010.94
23. Levitz, S.M., Golenbock, D.T.: Beyond empiricism: informing vaccine development through innate immunity research. Cell **148**, 1284–1292 (2012). doi:10.1016/j.cell.2012.02.012
24. Lycke, N.: Recent progress in mucosal vaccine development: potential and limitations. Nat. Rev. Immunol. **12**, 592–605 (2012). doi:10.1038/nri3251

25. Nochi, T., et al.: Nanogel antigenic protein-delivery system for adjuvant-free intranasal vaccines. Nat. Mater. **9**, 572–578 (2010). doi:10.1038/nmat2784
26. Stanberry, L.R., et al.: Safety and immunogenicity of a novel nanoemulsion mucosal adjuvant W805EC combined with approved seasonal influenza antigens. Vaccine **30**, 307–316 (2012). doi:10.1016/j.vaccine.2011.10.094
27. Makidon, P.E., et al.: Nanoemulsion mucosal adjuvant uniquely activates cytokine production by nasal ciliated epithelium and induces dendritic cell trafficking. Eur. J. Immunol. **42**, 2073–2086 (2012). doi:10.1002/eji.201142346
28. Bielinska, A.U., et al.: Mucosal immunization with a novel nanoemulsion-based recombinant anthrax protective antigen vaccine protects against Bacillus anthracis spore challenge. Infect. Immun. **75**, 4020–4029 (2007). doi:10.1128/iai.00070-07
29. Amorij, J.P., et al.: Pulmonary delivery of an inulin-stabilized influenza subunit vaccine prepared by spray-freeze drying induces systemic, mucosal humoral as well as cell-mediated immune responses in BALB/c mice. Vaccine **25**, 8707–8717 (2007)
30. Gwinn, W.M., et al.: Effective induction of protective systemic immunity with nasally administered vaccines adjuvanted with IL-1. Vaccine **28**, 6901–6914 (2010). doi:10.1016/j.vaccine.2010.08.006
31. Jaganathan, K.S., Vyas, S.P.: Strong systemic and mucosal immune responses to surface-modified PLGA microspheres containing recombinant hepatitis B antigen administered intranasally. Vaccine **24**, 4201–4211 (2006)
32. Thompson, A.L., et al.: Maximal adjuvant activity of nasally delivered IL-1alpha requires adjuvant-responsive CD11c(+) cells and does not correlate with adjuvant-induced in vivo cytokine production. J. Immunol. **188**, 2834–2846 (2012). doi:10.4049/jimmunol.1100254
33. Thompson, A.L., Staats, H.F.: Cytokines: the future of intranasal vaccine adjuvants. Clin. Dev. Immunol. **2011**, 289597 (2011). doi:10.1155/2011/289597
34. Couch, R.B., et al.: Contrasting effects of type I interferon as a mucosal adjuvant for influenza vaccine in mice and humans. Vaccine **27**, 5344–5348 (2009). doi:10.1016/j.vaccine.2009.06.084. S0264-410X(09)00962-1 [pii]
35. Duthie, M.S., Windish, H.P., Fox, C.B., Reed, S.G.: Use of defined TLR ligands as adjuvants within human vaccines. Immunol. Rev. **239**, 178–196 (2011). doi:10.1111/j.1600-065X.2010.00978.x
36. Lewis, D.J., et al.: Transient facial nerve paralysis (Bell's palsy) following intranasal delivery of a genetically detoxified mutant of Escherichia coli heat labile toxin. PLoS One **4**, e6999 (2009). doi:10.1371/journal.pone.0006999
37. Mutsch, M., et al.: Use of the inactivated intranasal influenza vaccine and the risk of Bell's palsy in Switzerland.[see comment]. N. Engl. J. Med. **350**, 896–903 (2004)
38. Atmar, R.L., et al.: Norovirus vaccine against experimental human Norwalk Virus illness. N. Engl. J. Med. **365**, 2178–2187 (2011). doi:10.1056/NEJMoa1101245
39. Hayes, J.D., Pulford, D.J.: The glutathione S-transferase supergene family: regulation of GST and the contribution of the isoenzymes to cancer chemoprotection and drug resistance. Crit. Rev. Biochem. Mol. Biol. **30**, 445–600 (1995). doi:10.3109/10409239509083491
40. Sayes, C.M., Reed, K.L., Warheit, D.B.: Assessing toxicity of fine and nanoparticles: comparing in vitro measurements to in vivo pulmonary toxicity profiles. Toxicol. Sci. **97**, 163–180 (2007). doi:10.1093/toxsci/kfm018
41. Mittler, R.: Oxidative stress, antioxidants and stress tolerance. Trends Plant Sci. **7**, 405–410 (2002). doi:10.1016/s1360-1385(02)02312-9. Pii s1360-1385(02)02312-9
42. Warheit, D., et al.: Comparative pulmonary toxicity assessment of single-wall carbon nanotubes in rats. Toxicol. Sci. **77**, 117–125 (2004)
43. Gurr, J.R., Wang, A.S.S., Chen, C.H., Jan, K.Y.: Ultrafine titanium dioxide particles in the absence of photoactivation can induce oxidative damage to human bronchial epithelial cells. Toxicology **213**, 66–73 (2005). doi:10.1016/j.tox.2005.05.007
44. Sayes, C.M., et al.: Correlating nanoscale titania structure with toxicity: a cytotoxicity and inflammatory response study with human dermal fibroblasts and human lung epithelial cells. Toxicol. Sci. **92**, 174–185 (2006). doi:10.1093/toxsci/kfj197
45. Zhu, S.Q., Oberdorster, E., Haasch, M.L.: Toxicity of an engineered nanoparticle (fullerene, C-60) in two aquatic species, Daphnia and fathead minnow. Mar. Environ. Res. **62**, S5–S9 (2006). doi:10.1016/j.marenvres.2006.04.059
46. Pitt, M.L., et al.: In vitro correlate of immunity in a rabbit model of inhalational anthrax. Vaccine **19**, 4768–4773 (2001)
47. Zaucha, G.M., Pitt, L.M., Estep, J., Ivins, B.E., Friedlander, A.M.: The pathology of experimental anthrax in rabbits exposed by inhalation and subcutaneous inoculation. Arch. Pathol. Lab. Med. **122**, 982–992 (1998)
48. Roy, C.J., et al.: Human leukocyte antigen-DQ8 transgenic mice: a model to examine the toxicity of aerosolized staphylococcal enterotoxin B. Infect. Immun. **73**, 2452–2460 (2005)
49. LeClaire, R.D., et al.: Potentiation of inhaled staphylococcal enterotoxin B-induced toxicity by lipopolysaccharide in mice. Toxicol. Pathol. **24**, 619–626 (1996)

# Nanotechnology in Vaccine Delivery

Martin J. D'Souza, Suprita A. Tawde,
Archana Akalkotkar, Lipika Chablani,
Marissa D'Souza, and Maurizio Chiriva-Internati

## Contents

| | | |
|---|---|---|
| 45.1 | Introduction | 727 |
| 45.2 | Need for Particulate Vaccines | 729 |
| 45.3 | Polymeric Nanoparticles | 730 |
| 45.4 | Liposomes | 731 |
| 45.5 | Immunostimulatory Complexes (ISCOMs) | 733 |
| 45.6 | Virus-like Particles | 734 |
| 45.7 | Polymeric Micelles | 735 |
| 45.8 | Dendrimers | 735 |
| 45.9 | Carbon Nanotubes | 735 |
| 45.10 | Challenges and Future Directions | 737 |
| 45.10.1 | Advantages and Disadvantages of Nanoparticles as Vaccines | 737 |
| References | | 738 |

M.J. D'Souza, PhD (✉) • A. Akalkotkar
M. D'Souza
Vaccine Nanotechnology Laboratory,
Department of Pharmaceutical Sciences,
College of Pharmacy and Health Sciences,
Mercer University, Atlanta, GA, USA
e-mail: dsouza_mj@mercer.edu

S.A. Tawde, PhD
Akorn Inc, Vernon Hills, IL 60031, USA

L. Chablani, PhD
Department of Pharmaceutical Sciences,
St. John Fisher College, Wegmans School of Pharmacy,
Rochester, NY 14618, USA

M. Chiriva-Internati, PhD
Texas Tech University Health Science Center,
School of Medicine, Lubbock, TX 79430, USA

## Abstract

Immunotherapy has been a trusted therapy for centuries to eliminate infectious diseases. However, the successful immunotherapy depends on several factors such as nature of pathogen, vaccine delivery system, route of administration, and immune system of the host. With the advances in nanotechnology, immunotherapy is now targeting different challenging disorders including cancer as well as infectious diseases. Along with the evolution of several adjuvants to enhance immune response to vaccines, nanotechnology plays an important role by acting as self-adjuvant in form of particles.

## 45.1 Introduction

Advances in nanotechnology have led to innumerable ways for prevention or treatment of various diseases. Its impact on immunotherapy potentiates the vaccine delivery and efficacy.

Immunotherapy is a specialized way of eliminating diseases, where it prepares the immune system to combat the attack of foreign antigens (in case of infectious diseases) or self-antigens (in case of cancer). It has been proved very well for centuries that immunotherapy has been a cost-effective tool to prevent the disease.

With the evolution of different challenging diseases, there is an urgent need of vaccine development against them to save lives of millions all throughout the world. Moreover, in case of existing vaccines, there is still a need to address issues with respect to safety, effectiveness, ease of administration, time of preparation, and, most importantly, the cost.

Recent developments in immunology and molecular biology explore new vaccine materials and aim at triggering memory response to vaccines, which protect host's immune system against the disease attack for a longer period of time. Vaccine efficacy depends on its ability to induce memory T-cells and B-cells through Th1 and Th2 immune pathways, respectively. Conventional vaccine materials include whole foreign organism vaccine (live/attenuated/killed/lysate), cellular fragments of pathogens such as bacterial polysaccharides, and bacterial toxins [1]. On the other hand, development of recombinant technology and RT-PCR allows to obtain specific antigen expression or synthetic peptide on larger scale and to use as vaccines. DNA vaccines are recently developed type of immunotherapy which has shown encouraging results in some clinical trials [2–4].

There are two major approaches for vaccination: prophylactic or therapeutic. Prophylactic vaccines find their applications in the prevention of viral, bacterial, or parasitic infectious diseases such as influenza, HIV, tuberculosis, malaria, pneumonia, polio, and smallpox, which are caused by foreign antigens. However, in case of cancer which is caused by self-altered cells, vaccine formulation is a challenging task as it requires immune response against self-cell antigens without causing autoimmune response. There are very few prophylactic cancer vaccines available on market such as Gardasil® (Merck) and Cervarix® (GSK) vaccine for human papillomavirus infection causing cervical cancer. Prophylactic cancer vaccines can prevent the tumor development based on the use of overexpressed or mutated proteins, mutated oncogenic growth factor receptors, heat-shock proteins, or other tumor-associated antigens [5]. In case of therapeutic approach, vaccines are given in order to trigger immune response against existing residual tumor cells mostly in combination with surgery or chemotherapy and thus aiming at preventing or prolonging the relapse [6]. Currently, there is only one therapeutic cancer vaccine, Provenge® (Dendreon), approved recently by FDA for treatment of prostate cancer. On the other hand, studies are being carried out for melanoma and colorectal cancer [7]. Various other clinical trials have been reported utilizing DNA/dendritic cell (DC)/viral vector-based vaccines depicting the continuous growth in the field of cancer immunotherapy [8, 9].

Vaccine efficacy depends mainly on the immunogenicity of antigen. It can further be enhanced by the use of vaccine adjuvants which activate immune cells. Various adjuvants are being explored for their effectiveness to trigger humoral, cellular, and/or mucosal immunity against several antigens. Humoral immune response was found to be elicited mostly with the use of protein adjuvants. Cytotoxic T-cell responses were found to be triggered by ISCOMs, Montanide™, Montanide ISA720, ISA 51, and viral vectors. MF59 and MPL® (monophosphoryl lipid) were shown to enhance Th1 responses. Viruslike particles, nondegradable nanoparticles, and liposomes produced cellular immunity. Douglas et al. incorporated Montanide ISA720 as an adjuvant to obtain both T-cell and B-cell response equal or higher than the response obtained with viral or protein adjuvants alone against Plasmodium falciparum MSP1. In case of commercially available cancer vaccines, monophosphoryl lipid A (MPL) is being used in Cervarix® as a TLR-4-targeted adjuvant, while Gardasil® contains alum as an adjuvant. Compound AS04 (a combination of alum and monophosphoryl lipid A) has also been used in human vaccines against hepatitis B virus.

Adjuvants which are approved for human use include alum, compound AS04 (a combination of alum and monophosphoryl lipid A), AS03, and MF59. Among these, alum is used in many vaccines such as HAV, HBV, HPV, diphtheria, tetanus, Haemophilus influenzae type B, and pneumococcal conjugate vaccines. However, alum is a poor adjuvant for triggering Th1 response. A list of adjuvants tested in animal models includes bacterial toxins such as cholera toxin, heat-labile enterotoxin of E. coli, nontoxic B subunit of cholera toxin, Toll-like receptor

(TLR) 9 agonist, and cytosine phosphoguanosine (CpG) dinucleotides [10]. Montanide, PLG, flagellin, QS21, AS01, AS02, RC529, ISCOM, IC31, CpG, MF59 with MTP-PE, immunostimulatory sequences (ISS), and 1018 ISS are some of the adjuvants which are in clinical trials against various disorders such as malaria, cancer, flu, hepatitis B, hepatitis C, HIV, and TB [11]. Heffernan et al. found that the co-formulation of chitosan and IL-12 induced Th1, IgG2a, and IgG2b antibody immune response to a model protein vaccine. Denisov et al. evaluated various adjuvants (larifan, polyoxidonium, natrium thiosulphate, TNF-β, and Ribi adjuvant system) for their ability to enhance immune responses to the live brucellosis vaccine, Brucella abortus strain 82-PS (penicillin sensitive) in guinea pigs, and they found that the highest protection was offered by combining TNF-β or polyoxidonium with S82-PS. The recent findings by Chen et al. inferred that a compound 3′ 5′-cyclic diguanylic acid (c-di-GMP), which is a bacterial intracellular signaling molecule, can act as a vaccine adjuvant and has shown immunostimulatory properties. In a study by Skountzou et al., bacterial flagellins from *Escherichia coli* and *Salmonella*, coadministered intranasally with inactivated A/PR/8/34 (PR8) virus, were found to be enhancing the efficacy of influenza vaccines in mice. Thus, they can be termed as good candidates as mucosal vaccine adjuvants to improve protection against influenza epidemics as well as other infectious diseases. On the other hand, cancer vaccine efficacy has also been enhanced by the use of cytokines as adjuvants such as interleukins, IL-2, IL-12, and GM-CSF [12–15].

Another way of enhancing vaccine efficacy is with the use of nanotechnology. Nanotechnology has been explored for its different applications in delivering small molecules, proteins, and peptides. Recently, vaccine delivery has been achieved through various pharmaceutical approaches to establish enhanced efficacy and ease of delivery and to address the issues related to stability. Vaccine material has been formulated into nanocarriers such as liposomes, polymeric nanoparticles, ISCOMs, dendrimers, micelles, VLPs, and carbon nanotubes.

## 45.2 Need for Particulate Vaccines

Currently, there are no particulate vaccines available in the market, but extensive research is going on in this field that would eventually bring particulate vaccine approach from bench to clinical interphase. Nanovaccine against notorious diseases is an attractive option as it can elicit both humoral and cellular immunity [16]. Nanotechnology has also proven to offer mucosal immunity which can be targeted for infectious diseases caused by mucosal entry of pathogen [17]. These nanovectors bear the advantage of being similar to a pathogen in terms of size; thus, they are efficiently recognized by antigen-presenting cells (APCs) of skilled immune system [18]. Further, they will be drained into the nearby lymph nodes where they can activate the immune cells of the body. These immune cells are drained towards the epithelial gatekeeper cells receiving various chemokine signals [19].

In contrast to natural infections, vaccines alone are incapable of producing a high antibody response [20]. The approach of using nanoparticles as vaccines which can incorporate multiple antigens in a single entity will lead to an enhanced humoral response as well as provide cellular immunity [16, 21–24]. Uddin, Lai, and Yeboah et al. have successfully formulated and tested oral vaccines using the particulate vaccine delivery system for typhoid, melanoma, and tuberculosis, respectively [25–27]. In all these studies, significantly higher mucosal and serum antibody titers (IgA and IgG) were obtained for orally administered particulate vaccine than those observed for the oral solution vaccine. The duration of antigen presentation also plays an important role to enhance the immune response [28]. The release of antigen must be in a pulsatile fashion to decrease the number of booster doses required. The persistence of antigens can be obtained only if the particles are remaining intact and are protected from degradation in the harsh acidic gastric conditions.

Also, it is possible to modify the outer surface of the nanoparticles to increase its uptake by the APCs. It can be conjugated with either an immunostimulatory or targeting moiety; else the

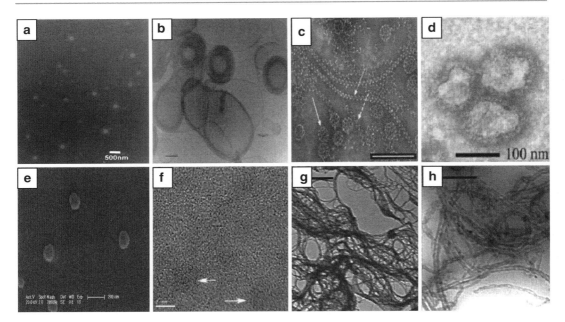

**Fig. 45.1** Different nanocarriers for vaccine delivery: (**a**) TEM image of PEG-PLGA nanoparticle (scale bar corresponds to 500 nm) (Reproduced from Bharali et al. [34] with permission from Elsevier). (**b**) Cryo-EM image of cationic liposomes entrapping DNA (scale bar corresponds to 200 nm) (Reproduced from Perrie et al. [35] with permission from Elsevier). (**c**) TEM image of ISCOMs of different types such as typical cage-like (*solid arrow*), helices (*dashed arrow*), and double helices (*dotted arrows*) (scale bar corresponds to 200 nm) (Reproduced from Sun et al. [36] with permission from Elsevier). (**d**) EM image of influenza H1N1 viruslike particles (scale corresponds to 100 nm) (Reproduced from Quan et al. [37], open-access article). (**e**) SEM image of PEG-PEI-PBLG copolymeric micelles (scale corresponds to 200 nm) (Reproduced from Tian et al. [38] with permission from Elsevier). (**f**) TEM image of PAMAM dendrimer (Reproduced from Jackson et al. [39] with permission from ACS). (G, H) TEM images of single-walled (**g**) and multi-walled (**h**) carbon nanotubes (scale bar corresponds to 1 μm and 250 nm, respectively) (Reproduced from Klumpp et al. [40] with permission from Elsevier)

inherent property of the surface itself can be modified. Surface charge also plays a vital role in uptake of the particles and affects the levels of immune response. It is shown that cationic particles are promising for uptake into macrophages and dendritic cells (DCs) [29].

Another useful property of nanoparticles is incorporation of various immunopotentiators to enhance the immune response to a further extent. This also includes targeting ligands, which can help to minimize the adverse effects of the vaccines. Some of the examples of these targeting ligands include aleuria aurantia lectin (AAL), ulex europaeus agglutinin 1 lectin (UEA-I), and wheat germ agglutinin (WGA) which act as targeting ligand to the M-cells present in Peyer's patches. This will eventually help to increase the uptake of particles through small intestine and bypass oral tolerance [30, 31]. Various co-stimulatory molecules like interleukins or cytokines can be included in the formulation to increase the immune response. It has been shown that DCs have receptors for both IL-2 and IL-12; hence, they have the capacity to present exogenous antigens and activate both MHC class I (cross-presentation) and MHC class II pathways by vaccination [32, 33].

This review aims to discuss the role of these nanocarriers as potential vaccine delivery vehicles as shown in Fig. 45.1. A brief description of each one is as follows:

## 45.3 Polymeric Nanoparticles

Polymeric nanoparticles (as shown in Fig. 45.1a) as vaccine delivery vehicles have been explored widely as they can act as adjuvants themselves.

Polymeric nanoparticles can offer the protection to proteins and peptides against gastric degradation upon oral administration and therefore vaccines are definitely one of the major applications for such particles [41]. The particles of size less than 1 μm offer adaptive immunity by facilitating their targeted uptake and extended presentation by APCs [42]. Nevertheless, the immune response also depends on rate of dissolution, surface morphology, charge, and size [43].

Oral administration is the most preferred route of administration as it is more patient compliant. Intestinal uptake of these particles is the key factor for determining the efficiency of oral vaccines. The usage of nanoparticles versus the use of microparticles as vaccine carriers for oral delivery is always debatable. There are conflicting reports as to which size can be considered as the optimum size range for eliciting a stronger and lasting immune response [44].

In this study by Desai et al., it was shown that particles of 100 nm showed increased uptake across the intestine in a rat model when compared to particles of 500, 1, and 10 μm size. These particles were prepared of polylactic/polyglycolic acid copolymer (50:50). Conventional nanoparticles are susceptible for entrapment in mucus due to steric as well as adhesive interactions. These interactions can be overwhelmed by tailoring the size of nanoparticles, which allows the particles to diffuse through mucus [45]. Here, Primard et al. reported that nanoparticles of size greater than 300 nm are less effective to move across the mucus lining of the intestine, when given orally. Therefore, particles in size range 200–250 nm were found to be taken up in M-cells of Peyer's patch of small intestine.

In contrast, a study conducted by Gutierro et al. showed that 1,000 nm particles of bovine serum albumin as a model protein incorporated in PLGA elicited higher IgG response when compared to 200 and 500 nm particles, and the immune response induced by 200 and 500 nm particles was comparable to each other by oral and subcutaneous route of administration [46].

However, in a contradictory study by Wendorf et al., poly(lactide-co-glycolide) nanoparticles of size 110 and 800–900 nm were compared for their efficacy and were found to be offering comparable immune response [47]. In a study by van den Berg J. et al., cationic nanoparticles containing DNA vaccines were evaluated via dermal route. It was seen that these cationic nanoparticles blocked vaccination-induced antigen expression in mice and ex vivo human skin due to immobilization of the nanoparticles in extracelvlular matrix caused by electrostatic interactions. Therefore, shielding the surface charge of the nanoparticles by PEGylation improved in vivo antigen expression [48]. Polylactic acid is one of the widely used biodegradable polymers in vaccine delivery. However, the use is restricted due to hydrophobic nature and generation of acidic microenvironment upon its degradation, rendering it unfavorable to the encapsulated antigen. In a study by Jain et al., PEG-derivatized block copolymers of PLA were used for development of nanoparticles encapsulating HBsAg for mucosal vaccination against hepatitis B. These polymers were found to produce better sIgA mucosal immune response [49], while in case of cancer, T-cell immune response can also be altered with the use of nanoparticles [50]. Several other polymers have been tried to formulate vaccine nanoparticles as listed in Table 45.1. Interesting uptake study performed by Primard et al. showed the poly(lactic acid) nanoparticles traversed from intestinal mucosa to Peyer's patch and then interacted with underlying B-cells and dendritic cells upon oral administration [45]. Due to all these advantages of polymeric nanoparticles, they remain a potential vaccine delivery system.

## 45.4 Liposomes

Although there are various nanocarriers available for vaccine delivery, liposomes play a prominent role as drug and potential vaccine delivery vehicles. Liposomes were introduced by Bangham et al. in 1960s [51], almost a decade later Allison et al. elicited their role as an immunological adjuvant [52], and since then, many studies have been done to exploit this approach. These are nanostructures (as shown in Fig. 45.1b) composed of phospholipids having a capacity to encapsulate both hydrophilic and hydrophobic

**Table 45.1** Summary of current polymeric nanoparticles under research for various vaccines tested using animal models

| Polymer used | Size (nm) | Charge (mV) | Vaccine preparation | Immune response | Route | Method of preparation | Animal model | Reference |
|---|---|---|---|---|---|---|---|---|
| Methoxypolyethylene glycol-poly(lactide-co-glycolide) | 150–200 | NA | Recombinant hepatitis B surface antigen (HBsAg) | Anti-HBs antibodies | IP | Double emulsion/solvent evaporation | BALB/cNCr mice | [34] |
| An inner hard core of poly(methyl-methacrylate) (PMMA) and hydrophilic tentacular shell and poly(ethylene-glycol) chains | 960±38 | 32.2±0.6 | HIV-1 Tat | Humoral, cellular responses, Th1-type T cell responses and CTLs | IM | Emulsion polymerization | BALB/c (H2kd) mice | [53] |
| Poly(γ-glutamic acid) | 200 | NA | gp120 of HIV-1 | Efficiently taken up by dendritic cells, CD8+ T cell responses, not effective for protection After the challenge inoculation with SHIV | IN, SC | Solvent evaporation | Rhesus macaques | [54] |
| Inner core of poly(methyl-methacrylate) (PMMA) and hydrophilic outer shell of Eudragit L100-55 | NA | NA | HIV-1 Tat | Long-lasting cellular and humoral responses | IM, SC, IN | Emulsion polymerization | BALB/c mice | [55] |
| Polypropylene sulfide | 50 | NA | Ovalbumin as a model antigen | Cytotoxic T lymphocytic responses in lung and spleen tissues, as well as humoral response in mucosal airways | IN | NA | C57BL/6 mice | [56] |
| Polypropylene sulfide | NA | NA | Ovalbumin as a model antigen | Cytotoxic and helper T cell responses | ID | Emulsion polymerization | C57Bl/6 mice | [57] |
| N-trimethyl chitosan and poly(lactic-co-glycolic acid) | 500 | 24.5±0.90 | Ovalbumin as a model antigen | Humoral immune response, mucosal response in case of intranasal administration | IM, IN | Emulsification/solvent extraction | Balb/c mice. | [58] |
| Polysaccharide chitosan | 160–200 | 6–10 | Recombinant hepatitis B surface antigen (rHBsAg) | Anti-HBsAg IgG levels | IM | Mild ionic gelation technique | BALB/cmice | [59] |

*IM* intramuscular, *SC* subcutaneous, *IP* intraperitoneal, *ID* intradermal, *IN* intranasal, *NA* not available drugs as well as vaccine antigens of various origins. They not only act as carriers to protect these bioactive moieties but also possess immunogenic properties, thus acting as a potential adjuvant [35, 60–62]. Conventional liposomes have been unsuccessful as vaccine particles due to their rapid clearance from the body because of their uptake by reticuloendothelial system [63], although, with the advent of stealth/PEGylated liposomes, increased half-life of these circulating nanocarriers has been achieved [64]. Doxil®, a PEGylated liposome of doxorubicin, is a marketed product utilizing this application and is used for the treatment of cancers. Other liposomal marketed formulations include Ambisome® (Gilead), Myocet® (Elan), and Depocyt® (SkyePharma).

To enhance the immunogenicity of these carriers, various other approaches have been employed. Mohammed et al. describe the use of cationic liposomes leading to improved stability and sustained immunological effects against Mycobacterium tuberculosis [65]. Further the use of adjuvants incorporated in the liposome has been explored to provide immune-stimulant effect; recently, the efficacy of monophosphoryl lipid A integrated dimethyldioctadecylammonium (DDA) and trehalose 6,6′-dibehenate (TDB) liposomes has been shown to induce cellular immunity along with the humoral response [66, 67]. Altin et al. further review the use of liposomes and plasma membrane vesicles (PMV) as a carrier for targeted delivery of antigens [63]. There are various other forms of liposomes which have been found to be promising as antigen carriers such as virosomes, archaesomes, and proteosomes [68–70].

Apart from imparting immunomodulatory properties, the physical properties of these nanocarriers are also important to act as a potent vaccine delivery vehicle. Xiang et al. discuss the role of size in development of particulate vaccines and describe various particle size range and their respective uptake mechanism; this can be useful as smaller liposomes mimic the uptake mechanisms of viruses whereas larger liposomes can follow a pathway as used by the bacteria [43]. As discussed previously, surface charge of these species also dictates their efficacy as a particulate vaccine; for example, cationic liposomes have shown better efficacy than others [71]. Such modifications in physical properties, use of immunoadjuvants, and stealth properties of these carriers potentiate their use as a particulate vaccine [72]. Considering the success of liposomal products in the market, it is promising to have a liposomal vaccine soon.

## 45.5 Immunostimulatory Complexes (ISCOMs)

ISCOMs (immunostimulatory complexes), as shown in Fig. 45.1c, are particulate vaccine nanocarriers of 40 nm size which are made up of cholesterol, phospholipid, and saponin along with antigen/s. However, ISCOMATRIX™ is now available without antigen and having the same composition as ISCOMs. This matrix provides incorporation of antigen which can be used as ISCOMATRIX™ vaccine with similar immunostimulatory activity as seen with ISCOMs. The immunostimulatory property is imparted to these complexes due to Quil A which is a purified less toxic extract from Quillaja saponin. These complexes have been reported to produce immune responses against variety of antigens such as viral, bacterial, parasitic, or tumor antigens [73, 74].

Some researchers have tried to enhance the immunostimulatory properties of these complexes by varying or replacing some of the components such as phospholipids or Quil A [36]. Several ISCOM™ and ISCOMATRIX™ vaccines have shown to induce humoral and cellular response in animal models (as shown in Table 45.2). These systems can access both the MHC I and MHC II pathways and act as a potent immunomodulator of both the innate and adaptive immune systems. Intranasal delivery of influenza ISCOMATRIX™ vaccine in humans has shown to induce systemic and mucosal responses, and therefore the ISCOMATRIX™ adjuvant can be used as a mucosal adjuvant [75]. Antigen-specific CTL, T-helper cells, and antibodies can be induced by ISCOM and ISCOMATRIX™

**Table 45.2** Summary of various Immunostimulatory Complexes (ISCOMs) based vaccines under research using animal models

| Vaccine delivery system | Vaccine preparation | Immune response | Route of administration | Animal model | Reference |
|---|---|---|---|---|---|
| ISCOM | Influenza viruses, H3N2 | Humoral and cellular immunity | NA | *Cynomolgus Macaques* | [76] |
| ISCOM | Used as a adjuvant for human norovirus GII.4 HS66 strain vaccine | Th2 biased responses with significantly elevated IgM, IgA and IgG antibody-secreting cells | Oral/IN | Gnotobiotic pigs | [77] |
| ISCOM | Avian influenza A viruses of the H5N1 subtype | Strong antibody responses | IM | Roosters | [78] |
| ISCOM | A/PR8/34 Influenza virus | Strong mucosal as well as systemic antibody and cytotoxic T-lymphocyte responses | IN | BALB/c mice | [79] |
| ISCOM | Virosomal influenza A H5N1 | Th1 CD4+ cells and strong antibody responses | IM | BALB/c mice | [80] |
| ISCOMATRIX™ | MEM influenza antigen | Mucosal and serum antibody response | IN | BALB/c mice | [81] |
| ISCOMATRIX™ and ISCOM™ | *H. pylori* | Reduction in *H. pylori* colonization | IN/SC | mice | [82] |
| ISCOM | Recombinant NcSRS2, of the intracellular protozoan parasite *Neospora caninum* | *N. caninum* specific antibodies and cellular response | SC | BALB/c mice | [83] |

*IM* intramuscular, *SC* subcutaneous, *IN* intranasal, *NA* not available

vaccines for cancer and infectious diseases [36]. Another modification of ISCOM with regard to charge resulted in cationic ISCOM derivatives (PLUSCOMs), which offered high anionic antigen loading and therefore enhanced T-cell response in comparison to classic anionic ISCOMs against a model protein antigen (ovalbumin) [84]. Moreover, these complexes can reduce the dose of antigen required to induce immune response [85]. Table 45.2 lists different ISCOMs and ISCOMATRIX™ which have been studied in vivo against various infections.

## 45.6 Virus-like Particles

Along with a range of nanocarriers available for the vaccine delivery, viruslike particles (VLPs), as shown in Fig. 45.1d, are one of the most potent ones [86]. As the name indicates, these are particles resembling size range of a virus from 22 to 150 nm and contain self-assembled envelopes/proteins of various viruses. As they lack the genetic material, they are regarded noninfectious. Noad et al. detail that for more than 30 different infectious viruses, VLPs have been produced, eliciting the need of this approach [87]. Due to various advantages of this delivery system, currently there are VLP-based vaccines commercially available against two diseases—HBV and HPV [88, 89]. Also, various clinical trials are in progress utilizing this particulate delivery system. Recently, Buonaguro et al. discussed the role of VLPs as particulate vaccines, their contribution to current vaccines and clinical trials, and also the immune response elicited by these particles [90]. Also, a detailed review by Grgacic et al. describes the role of VLPs as vaccine particles to elicit immune response [91]. Structurally, VLPs can be defined as enveloped or non-enveloped depending upon the presence or absence of their lipid envelope, surrounding the capsid protein.

VLPs of human papillomaviruses (HPV) are a good example of single-capsid non-enveloped VLPs consisting of L1 as major capsid protein. These VLPs can be produced in yeast (Gardasil) as well as in insect cells infected with baculovirus (Cervarix). Schiller et al. review the clinical trials performed using these HPV-VLPs and describe the efficacy of these systems against HPV [92]. The review also emphasizes that there are limited safety issues related to the vaccine as seen during the clinical trials.

On the other hand, various enveloped VLPs are available against influenza A, hepatitis B, hepatitis C, and several retroviruses. Recently, Kang et al. showed the possibility of influenza A vaccination through transdermal route using VLP-coated microneedle, thus enhancing the compliance towards these nanocarriers [93]. Considering the wide applications of these VLPs and their success as a commercial particulate vaccine, they continue to remain potential nanocarriers for future vaccines [37, 94–99].

## 45.7 Polymeric Micelles

Polymeric micelles (as shown in Fig. 45.1e) are a well-organized nano-sized assembly of synthetic polymers. These fall in the category of association colloids that are formed spontaneously when the amphiphilic molecules or hydrophilic regions are maintained at an appropriate concentration and temperature [38, 100]. They are not held together by covalent bonds and hence can be dissociated easily. This property of micelles can be exploited as per their applications [17]. They have shown high stability in vitro as well as in vivo [101]. Physical and chemical properties of polymeric micelles can be manipulated by selection of suitable hydrophilic and hydrophobic polymers [102]. In a study by Morein et al., a 30S protein subunit micellar vaccine induced a detectable antibody titer as well as protective immunity in a challenge study against pneumonia caused by the PI-3 virus [103]. Prabakaran et al. performed similar studies where they used soya phosphatidylcholine micelles against H5N1 infection [104]. Higher levels of serum IgG, mucosal IgA, and HI titers were observed when compared to the free antigen. Hence, micelles can serve as a promising carrier for vaccine antigens.

## 45.8 Dendrimers

Dendrimers (as shown in Fig. 45.1f) are highly branched, monodispersed polymeric nanoparticles. Dendrimers are composed of three different components: an initiator core, branches, and terminal functional groups. The initiator core is the main component of dendrimers and the branches extend in the outer directions. The terminal groups can be modified based on charge/hydrophilic/lipophilic properties [105]. They are similar to polymeric micelles but are linked covalently unlike micelles and thus have more stronger bonds and do not tend to dissociate easily [106]. The external surface can be easily modified and alterations of the internal cavity make dendrimers a promising carrier for various biomedical and industrial applications [107]. Recent work by Baker et al. involves coupling of various functional molecules including sensing units, MRI contrast agents, triggering devices, and targeting molecules to the surface of a generation 5 dendritic polymer (MW 25,000 Da, diameter 5 nm) [108]. A specific class of dendrimers called as multiple antigenic peptide (MAP) systems has been used widely for the vaccine purposes. MAP-based malaria vaccine has been tested in phase I clinical trials [109–111]. Having the potential to enter the clinical trials, these delivery systems are expected to be available on market shortly. Table 45.3 lists some of the dendrimeric systems currently under research.

## 45.9 Carbon Nanotubes

Recently, inorganic nanomaterials such as nanocrystals, nanowires, and nanotubes have been receiving an increasing amount of attention for vaccine delivery. Carbon nanotubes (as shown in

**Table 45.3** Summary of various dendrimer based vaccines under research in animal models

| Polymer used | Size (nm) | Charge (mV) | Vaccine preparation | Immune response | Route of administration | Method of preparation | Animal model | Reference |
|---|---|---|---|---|---|---|---|---|
| Polyetherimine (PETIM) | NA | NA | PETIM-pDNA complexes against rabies | virus neutralizing antibody responses | IM | NA | Swiss albino mice | [112] |
| Poly(propylene-imine) | NA | 21.3 ± 0.33 | Plasmid DNA encoding pRc/CMV-HBs[S] | Total IgG and its subclasses – IgG1, IgG2a, IgG2b | IM | Complexation | Female Balb/c mice | [113] |
| Polyacrylate core with a minimal B-cell epitope J14 | 20 | −16 | J14 B-cell epitope | IgG subclass Ab response – IgG1, IgG2b, and IgG3 | SC | Dialysis followed by "click" reaction | Murine model | [114] |

*IM* intramuscular, *SC* subcutaneous, *NA* not available

Fig. 45.1g–h) are explored as a vehicle for vaccines because of their capacity to link to an antigen while maintaining their conformation and thus inducing antigen-specific antibody response. They can also be modified in a non-immunogenic material [115]. Functionalized carbon nanotubes can be used as nanovectors for the delivery of antigen/s by forming covalent bonds or supramolecular assemblies based on non-covalent interactions [40]. Though carbon nanotubes remain an area of interest for current researchers, still extensive work is required before they can enter the clinical trials.

## 45.10 Challenges and Future Directions

With a wide range of nanocarriers available for vaccine delivery, nanotechnology not only gets the well-deserved limelight but also attracts attention of regulatory bodies and bears certain challenges that need to be considered before marketing these nanocarriers.

### 45.10.1 Advantages and Disadvantages of Nanoparticles as Vaccines

Nanovaccines have its own pros and cons as a delivery system. They are made up of biodegradable polymers and hence are considered safe for administration. Nanoparticulate vaccine can be administered easily by different routes such as parenteral, oral, transdermal, nasal, and even pulmonary route. Thus, being noninvasive, delivery systems other than parenteral allow pain-free delivery of vaccines over conventional vaccines [116]. They can trigger the immune system efficiently as described earlier. Moreover, release of antigen at a controlled rate and time in a desirable fashion can be achieved by nanoparticles [117].

The cost of production and storage of these vaccines is a basic concern. But the reproducibility of nanovaccines is a greater question [70]. On the other hand, nanoparticles of size larger than 300 nm are reported to be less efficient to traverse across the mucosal lining of intestine and hence result in lower particle uptake through Peyer's patches in the intestine and lesser immune response for vaccine particles [45]. Thus, size and charge of the nanoparticles play a critical role in determining the efficacy of vaccine formulation. Therefore, the reproducibility of vaccines during manufacturing should be ensured, which needs critical evaluation of the particles. Another issue is to address the sterilization performed by nonthermal methods needs to be taken care of [118]. Also, small nanoparticles are cleared rapidly from the body, whereas the larger aggregates might get accumulated in the organs and cause toxicity issues.

The "nano" size which makes these carriers so promising is also the reason behind the concerns of these delivery systems. Researchers propose that the smaller the carrier, the better it functions and remains protected by the body's RES system; also ways have been devised to impart stealth properties to these carriers to avoid their uptake by such phagocytic cells. Although all these properties make the nanocarrier a potential delivery system, it also makes it harder to be cleared from the body, thus adding to "nanotoxicity." Comparatively extensive studies have been done to determine the toxicity profile of nano-sized molecules than nanocarriers. Little is known about the toxic effects of such nanocarriers which have been used for vaccine delivery. Even though the use of these carriers remains questionable, various researches are being done to answer these concerns and regulatory authorities remain to be a part of these hassles.

**Declaration of Interest** As indicated in the affiliations, Suprita A. Tawde, Archana Akalkotkar, and Lipika Chablani were graduate students and Marissa D'Souza was a summer research student working under Prof. Dr. Martin J. D'Souza in the Vaccine Nanotechnology Laboratory, Department of Pharmaceutical Sciences, College of Pharmacy and Health Sciences, Mercer University, Atlanta, Georgia.

# References

1. Heffernan, M.J., Zaharoff, D.A., Fallon, J.K., Schlom, J., Greiner, J.W.: In vivo efficacy of a chitosan/IL-12 adjuvant system for protein-based vaccines. Biomaterials 32, 926–932 (2011)
2. Guimaraes-Walker, A., et al.: Lessons from IAVI-006, a phase I clinical trial to evaluate the safety and immunogenicity of the pTHr.HIVA DNA and MVA. HIVA vaccines in a prime-boost strategy to induce HIV-1 specific T-cell responses in healthy volunteers. Vaccine 26, 6671–6677 (2008)
3. Beckett, C.G., et al.: Evaluation of a prototype dengue-1 DNA vaccine in a Phase 1 clinical trial. Vaccine 29(5), 960–968 (2011)
4. Martin, J.E., et al.: A SARS DNA vaccine induces neutralizing antibody and cellular immune responses in healthy adults in a Phase I clinical trial. Vaccine 26, 6338–6343 (2008)
5. Finn, O.J., Forni, G.: Prophylactic cancer vaccines. Curr. Opin. Immunol. 14, 172–177 (2002)
6. Arlen, P.M., Madan, R.A., Hodge, J.W., Schlom, J., Gulley, J.L.: Combining vaccines with conventional therapies for cancer. Update. Cancer. Ther. 2, 33–39 (2007)
7. Tartaglia, J., et al.: Therapeutic vaccines against melanoma and colorectal cancer. Vaccine 19, 2571–2575 (2001)
8. Anderson, R.J., Schneider, J.: Plasmid DNA and viral vector-based vaccines for the treatment of cancer. Vaccine 25(Suppl 2), B24–B34 (2007)
9. Sheng, W.Y., Huang, L.: Cancer immunotherapy and nanomedicine. Pharm. Res. 28(2), 200–214 (2011)
10. Harandi, A.M., Medaglini, D., Shattock, R.J.: Vaccine adjuvants: a priority for vaccine research. Vaccine 28, 2363–2366 (2010)
11. Mbow, M.L., De Gregorio, E., Valiante, N.M., Rappuoli, R.: New adjuvants for human vaccines. Curr. Opin. Immunol. 22, 411–416 (2010)
12. Nguyen, C.L., et al.: Mechanisms of enhanced antigen-specific T cell response following vaccination with a novel peptide-based cancer vaccine and systemic interleukin-2 (IL-2). Vaccine 21, 2318–2328 (2003)
13. Toubaji, A., et al.: The combination of GM-CSF and IL-2 as local adjuvant shows synergy in enhancing peptide vaccines and provides long term tumor protection. Vaccine 25, 5882–5891 (2007)
14. Yin, W., et al.: A novel therapeutic vaccine of GM-CSF/TNFalpha surface-modified RM-1 cells against the orthotopic prostatic cancer. Vaccine 28, 4937–4944 (2010)
15. Germann, T., Rude, E., Schmitt, E.: The influence of IL12 on the development of Th1 and Th2 cells and its adjuvant effect for humoral immune responses. Res. Immunol. 146, 481–486 (1995)
16. Fifis, T., et al.: Size-dependent immunogenicity: therapeutic and protective properties of nanovaccines against tumors. J. Immunol. 173, 3148–3154 (2004)
17. Chadwick, S., Kriegel, C., Amiji, M.: Nanotechnology solutions for mucosal immunization. Adv. Drug Deliv. Rev. 62, 394–407 (2010)
18. Randolph, G.J., Inaba, K., Robbiani, D.F., Steinman, R.M., Muller, W.A.: Differentiation of phagocytic monocytes into lymph node dendritic cells in vivo. Immunity 11, 753–761 (1999)
19. Malik, B., Goyal, A.K., Mangal, S., Zakir, F., Vyas, S.P.: Implication of gut immunology in the design of oral vaccines. Curr. Mol. Med. 10, 47–70 (2010)
20. Ada, G.: Vaccines and vaccination. N. Engl. J. Med. 345, 1042–1053 (2001)
21. Fehr, T., Skrastina, D., Pumpens, P., Zinkernagel, R.M.: T cell-independent type I antibody response against B cell epitopes expressed repetitively on recombinant virus particles. Proc. Natl. Acad. Sci. U.S.A. 95, 9477–9481 (1998)
22. Bachmann, M.F., et al.: The influence of antigen organization on B cell responsiveness. Science 262, 1448–1451 (1993)
23. Bachmann, M.F., Zinkernagel, R.M.: Neutralizing antiviral B cell responses. Annu. Rev. Immunol. 15, 235–270 (1997)
24. O'Hagan, D.T., Singh, M., Ulmer, J.B.: Microparticle-based technologies for vaccines. Methods 40, 10–19 (2006)
25. Uddin, A.N., Bejugam, N.K., Gayakwad, S.G., Akther, P., D'Souza, M.J.: Oral delivery of gastro-resistant microencapsulated typhoid vaccine. J. Drug Target. 17, 553–560 (2009)
26. Yeboah, K.G., D'Souza, M.J.: Evaluation of albumin microspheres as oral delivery system for Mycobacterium tuberculosis vaccines. J. Microencapsul. 26, 166–179 (2009)
27. Lai, Y.H., D'Souza, M.J.: Formulation and evaluation of an oral melanoma vaccine. J. Microencapsul. 24, 235–252 (2007)
28. Storni, T., Ruedl, C., Renner, W.A., Bachmann, M.F.: Innate immunity together with duration of antigen persistence regulate effector T cell induction. J. Immunol. 171, 795–801 (2003)
29. Thiele, L., Merkle, H.P., Walter, E.: Phagocytosis and phagosomal fate of surface-modified microparticles in dendritic cells and macrophages. Pharm. Res. 20, 221–228 (2003)
30. Akande, J., et al.: Targeted delivery of antigens to the gut-associated lymphoid tissues: 2. Ex vivo evaluation of lectin-labelled albumin microspheres for targeted delivery of antigens to the M-cells of the Peyer's patches. J. Microencapsul. 27, 325–336 (2010)
31. Lai, Y.H., D'Souza, M.J.: Microparticle transport in the human intestinal M cell model. J. Drug Target. 16, 36–42 (2008)
32. Pulendran, B., Banchereau, J., Maraskovsky, E., Maliszewski, C.: Modulating the immune response with dendritic cells and their growth factors. Trends Immunol. 22, 41–47 (2001)
33. Banchereau, J., Steinman, R.M.: Dendritic cells and the control of immunity. Nature 392, 245–252 (1998)

34. Bharali, D.J., Pradhan, V., Elkin, G., Qi, W., Hutson, A., Mousa, S.A., Thanavala, Y.: Novel nanoparticles for the delivery of recombinant hepatitis B vaccine. Nanomedicine **4**, 311–317 (2008)
35. Perrie, Y., Mohammed, A.R., Kirby, D.J., McNeil, S.E., Bramwell, V.W.: Vaccine adjuvant systems: enhancing the efficacy of sub-unit protein antigens. Int. J. Pharm. **364**, 272–280 (2008)
36. Sun, H.X., Xie, Y., Ye, Y.P.: ISCOMs and ISCOMATRIX. Vaccine **27**, 4388–4401 (2009)
37. Quan, F.S., Vunnava, A., Compans, R.W., Kang, S.M.: Virus-like particle vaccine protects against 2009 H1N1 pandemic influenza virus in mice. PLoS One **5**, e9161 (2010)
38. Tian, H.Y., et al.: Biodegradable cationic PEG-PEI-PBLG hyperbranched block copolymer: synthesis and micelle characterization. Biomaterials **26**, 4209–4217 (2005)
39. Jackson, C.L., et al.: Visualization of dendrimer molecules by transmission electron microscopy (TEM): staining methods and cryo-TEM of vitrified solutions. Macromolecules **31**, 6259–6265 (1998)
40. Klumpp, C., Kostarelos, K., Prato, M., Bianco, A.: Functionalized carbon nanotubes as emerging nanovectors for the delivery of therapeutics. Biochim. Biophys. Acta **1758**, 404–412 (2006)
41. des Rieux, A., Fievez, V., Garinot, M., Schneider, Y.J., Preat, V.: Nanoparticles as potential oral delivery systems of proteins and vaccines: a mechanistic approach. J. Control. Release **116**, 1–27 (2006)
42. Rice-Ficht, A.C., Arenas-Gamboa, A.M., Kahl-McDonagh, M.M., Ficht, T.A.: Polymeric particles in vaccine delivery. Curr. Opin. Microbiol. **13**, 106–112 (2010)
43. Xiang, S.D., et al.: Pathogen recognition and development of particulate vaccines: does size matter? Methods **40**, 1–9 (2006)
44. Desai, M.P., Labhasetwar, V., Amidon, G.L., Levy, R.J.: Gastrointestinal uptake of biodegradable microparticles: effect of particle size. Pharm. Res. **13**, 1838–1845 (1996)
45. Primard, C., et al.: Traffic of poly(lactic acid) nanoparticulate vaccine vehicle from intestinal mucus to sub-epithelial immune competent cells. Biomaterials **31**, 6060–6068 (2010)
46. Gutierro, I., Hernandez, R.M., Igartua, M., Gascon, A.R., Pedraz, J.L.: Size dependent immune response after subcutaneous, oral and intranasal administration of BSA loaded nanospheres. Vaccine **21**, 67–77 (2002). S0264410X02004358 [pii]
47. Wendorf, J., et al.: A comparison of anionic nanoparticles and microparticles as vaccine delivery systems. Hum. Vaccin. **4**, 44–49 (2008)
48. van den Berg, J.H., et al.: Shielding the cationic charge of nanoparticle-formulated dermal DNA vaccines is essential for antigen expression and immunogenicity. J. Control. Release **141**, 234–240 (2010)
49. Jain, A.K., et al.: Synthesis, characterization and evaluation of novel triblock copolymer based nanoparticles for vaccine delivery against hepatitis B. J. Control. Release **136**, 161–169 (2009)
50. Demento, S., Steenblock, E.R., Fahmy, T.M.: Biomimetic approaches to modulating the T cell immune response with nano- and micro- particles. Conf. Proc. IEEE Eng. Med. Biol. Soc. **2009**, 1161–1166 (2009)
51. Bangham, A.D., Standish, M.M., Miller, N.: Cation permeability of phospholipid model membranes: effect of narcotics. Nature **208**, 1295–1297 (1965)
52. Allison, A.G., Gregoriadis, G.: Liposomes as immunological adjuvants. Nature **252**, 252 (1974)
53. Castaldello, A., Brocca-Cofano, E., Voltan, R., Triulzi, C., Altavilla, G., Laus, M., Sparnacci, K., Ballestri, M., Tondelli, L., Fortini, C., Gavioli, R., Ensoli, B., Caputo, A.: DNA prime and protein boost immunization with innovative polymeric cationic core-shell nanoparticles elicits broad immune responses and strongly enhance cellular responses of HIV-1 tat DNA vaccination. Vaccine **24**, 5655–5669 (2006)
54. Himeno, A., Akagi, T., Uto, T., Wang, X., Baba, M., Ibuki, K., Matsuyama, M., Horiike, M., Igarashi, T., Miura, T., Akashi, M.: Evaluation of the immune response and protective effects of rhesus macaques vaccinated with biodegradable nanoparticles carrying gp120 of human immunodeficiency virus. Vaccine **28**, 5377–5385 (2010)
55. Caputo, A., Castaldello, A., Brocca-Cofano, E., Voltan, R., Bortolazzi, F., Altavilla, G., Sparnacci, K., Laus, M., Tondelli, L., Gavioli, R., Ensoli, B.: Induction of humoral and enhanced cellular immune responses by novel core-shell nanosphere- and microsphere-based vaccine formulations following systemic and mucosal administration. Vaccine **27**, 3605–3615 (2009)
56. Stano, A., van der Vlies, A.J., Martino, M.M., Swartz, M.A., Hubbell, J.A., Simeoni, E.: PPS nanoparticles as versatile delivery system to induce systemic and broad mucosal immunity after intranasal administration. Vaccine **29**(4), 804–812 (2011)
57. Hirosue, S., Kourtis, I.C., van der Vlies, A.J., Hubbell, J.A., Swartz, M.A.: Antigen delivery to dendritic cells by poly(propylene sulfide) nanoparticles with disulfide conjugated peptides: cross-presentation and T cell activation. Vaccine **28**, 7897–7906 (2010)
58. Slutter, B., Bal, S., Keijzer, C., Mallants, R., Hagenaars, N., Que, I., Kaijzel, E., van Eden, W., Augustijns, P., Lowik, C., Bouwstra, J., Broere, F., Jiskoot, W.: Nasal vaccination with N-trimethyl chitosan and PLGA based nanoparticles: nanoparticle characteristics determine quality and strength of the antibody response in mice against the encapsulated antigen. Vaccine **28**, 6282–6291 (2010)
59. Prego, C., Paolicelli, P., Diaz, B., Vicente, S., Sanchez, A., Gonzalez-Fernandez, A., Alonso, M.J.: Chitosan-based nanoparticles for improving immunization against hepatitis B infection. Vaccine **28**, 2607–2614 (2010)

60. Gregoriadis, G.: Liposomes as immunoadjuvants and vaccine carriers: antigen entrapment. Immunomethods **4**, 210–216 (1994)
61. Wang, D., et al.: Liposomal oral DNA vaccine (mycobacterium DNA) elicits immune response. Vaccine **28**, 3134–3142 (2010)
62. Karkada, M., Weir, G.M., Quinton, T., Fuentes-Ortega, A., Mansour, M.: A liposome-based platform, VacciMax, and its modified water-free platform DepoVax enhance efficacy of in vivo nucleic acid delivery. Vaccine **28**, 6176–6182 (2010)
63. Altin, J.G., Parish, C.R.: Liposomal vaccines–targeting the delivery of antigen. Methods **40**, 39–52 (2006)
64. Immordino, M.L., Dosio, F., Cattel, L.: Stealth liposomes: review of the basic science, rationale, and clinical applications, existing and potential. Int. J. Nanomedicine **1**, 297–315 (2006)
65. Mohammed, A.R., Bramwell, V.W., Kirby, D.J., McNeil, S.E., Perrie, Y.: Increased potential of a cationic liposome-based delivery system: enhancing stability and sustained immunological activity in pre-clinical development. Eur. J. Pharm. Biopharm. **76**(3), 404–412 (2010)
66. Nordly, P., Agger, E.M., Andersen, P., Nielsen, H.M., Foged, C.: Incorporation of the TLR4 agonist monophosphoryl lipid a into the bilayer of DDA/TDB liposomes: physico-chemical characterization and induction of CD8(+) T-cell responses in vivo. Pharm. Res. **28**(3), 553–562 (2011)
67. Henriksen-Lacey, M., et al.: Liposomes based on dimethyldioctadecylammonium promote a depot effect and enhance immunogenicity of soluble antigen. J. Control. Release **142**, 180–186 (2010)
68. Gasparini, R., Lai, P.: Utility of virosomal adjuvated influenza vaccines: a review of the literature. J. Prev. Med. Hyg. **51**, 1–6 (2010)
69. Patel, G.B., Zhou, H., KuoLee, R., Chen, W.: Archaeosomes as adjuvants for combination vaccines. J. Liposome Res. **14**, 191–202 (2004)
70. Sharma, S., Mukkur, T.K., Benson, H.A., Chen, Y.: Pharmaceutical aspects of intranasal delivery of vaccines using particulate systems. J. Pharm. Sci. **98**, 812–843 (2009)
71. Henriksen-Lacey, M., et al.: Liposomal cationic charge and antigen adsorption are important properties for the efficient deposition of antigen at the injection site and ability of the vaccine to induce a CMI response. J. Control. Release **145**, 102–108 (2010)
72. Zhong, Z., et al.: A novel liposomal vaccine improves humoral immunity and prevents tumor pulmonary metastasis in mice. Int. J. Pharm. **399**, 156–162 (2010)
73. Pearse, M.J., Drane, D.: ISCOMATRIX adjuvant: a potent inducer of humoral and cellular immune responses. Vaccine **22**, 2391–2395 (2004)
74. Pearse, M.J., Drane, D.: ISCOMATRIX adjuvant for antigen delivery. Adv. Drug Deliv. Rev. **57**, 465–474 (2005)
75. Drane, D., Pearse, M.J.: Immunopotentiators in modern vaccines, pp. 191–215. Elsevier Academic Press, Massachusetts, USA (2006)
76. Rimmelzwaan, G.F., Baars, M., van Amerongen, G., van Beek, R., Osterhaus, A.D.: A single dose of an ISCOM influenza vaccine induces long-lasting protective immunity against homologous challenge infection but fails to protect Cynomolgus macaques against distant drift variants of influenza A (H3N2) viruses. Vaccine **20**, 158–163 (2001)
77. Souza, M., Costantini, V., Azevedo, M.S., Saif, L.J.: A human norovirus-like particle vaccine adjuvanted with ISCOM or mLT induces cytokine and antibody responses and protection to the homologous GII.4 human norovirus in a gnotobiotic pig disease model. Vaccine **25**, 8448–8459 (2007)
78. Rimmelzwaan, G.F., Claas, E.C., van Amerongen, G., de Jong, J.C., Osterhaus, A.D.: ISCOM vaccine induced protection against a lethal challenge with a human H5N1 influenza virus. Vaccine **17**, 1355–1358 (1999)
79. Sjolander, S., Drane, D., Davis, R., Beezum, L., Pearse, M., Cox, J.: Intranasal immunisation with influenza-ISCOM induces strong mucosal as well as systemic antibody and cytotoxic T-lymphocyte responses. Vaccine **19**, 4072–4080 (2001)
80. Madhun, A.S., Haaheim, L.R., Nilsen, M.V., Cox, R.J.: Intramuscular Matrix-M-adjuvanted virosomal H5N1 vaccine induces high frequencies of multifunctional Th1 CD4+ cells and strong antibody responses in mice. Vaccine **27**, 7367–7376 (2009)
81. Sanders, M.T., Deliyannis, G., Pearse, M.J., McNamara, M.K., Brown, L.E.: Single dose intranasal immunization with ISCOMATRIX vaccines to elicit antibody-mediated clearance of influenza virus requires delivery to the lower respiratory tract. Vaccine **27**, 2475–2482 (2009)
82. Skene, C.D., Doidge, C., Sutton, P.: Evaluation of ISCOMATRIX and ISCOM vaccines for immunisation against Helicobacter pylori. Vaccine **26**, 3880–3884 (2008)
83. Pinitkiatisakul, S., Friedman, M., Wikman, M., Mattsson, J.G., Lovgren-Bengtsson, K., Stahl, S., Lunden, A.: Immunogenicity and protective effect against murine cerebral neosporosis of recombinant NcSRS2 in different iscom formulations. Vaccine **25**, 3658–3668 (2007)
84. McBurney, W.T., et al.: In vivo activity of cationic immune stimulating complexes (PLUSCOMs). Vaccine **26**, 4549–4556 (2008)
85. Boyle, J., et al.: The utility of ISCOMATRIX adjuvant for dose reduction of antigen for vaccines requiring antibody responses. Vaccine **25**, 2541–2544 (2007)
86. Plummer, E.M., Manchester, M: Viral nanoparticles and virus-like particles: platforms for contemporary vaccine design. *WIREs* Nanomed. Nanobiotechnol. **3**, 174–196 (2011). doi:10.1002/wnan.119
87. Noad, R., Roy, P.: Virus-like particles as immunogens. Trends Microbiol. **11**, 438–444 (2003)

88. Campo, M.S., Roden, R.B.: Papillomavirus prophylactic vaccines: established successes, new approaches. J. Virol. **84**, 1214–1220 (2010)
89. Ludwig, C., Wagner, R.: Virus-like particles-universal molecular toolboxes. Curr. Opin. Biotechnol. **18**, 537–545 (2007)
90. Buonaguro, L., Tornesello, M.L., Buonaguro, F.M.: Virus-like particles as particulate vaccines. Curr. HIV Res. **8**, 299–309 (2010)
91. Grgacic, E.V., Anderson, D.A.: Virus-like particles: passport to immune recognition. Methods **40**, 60–65 (2006)
92. Schiller, J.T., Castellsague, X., Villa, L.L., Hildesheim, A.: An update of prophylactic human papillomavirus L1 virus-like particle vaccine clinical trial results. Vaccine **26**(Suppl 10), K53–K61 (2008)
93. Pearton, M., et al.: Influenza virus-like particles coated onto microneedles can elicit stimulatory effects on Langerhans cells in human skin. Vaccine **28**, 6104–6113 (2010)
94. Akahata, W., et al.: A virus-like particle vaccine for epidemic Chikungunya virus protects nonhuman primates against infection. Nat. Med. **16**, 334–338 (2010)
95. Quan, F.S., Huang, C., Compans, R.W., Kang, S.M.: Virus-like particle vaccine induces protective immunity against homologous and heterologous strains of influenza virus. J. Virol. **81**, 3514–3524 (2007)
96. Krammer, F., et al.: Influenza virus-like particles as an antigen-carrier platform for the ESAT-6 epitope of Mycobacterium tuberculosis. J. Virol. Methods **167**, 17–22 (2010)
97. Song, J.M., et al.: Protective immunity against H5N1 influenza virus by a single dose vaccination with virus-like particles. Virology **405**, 165–175 (2010)
98. Kang, S.M., et al.: Induction of long-term protective immune responses by influenza H5N1 virus-like particles. PLoS One **4**, e4667 (2009)
99. Muratori, C., Bona, R., Federico, M.: Lentivirus-based virus-like particles as a new protein delivery tool. Methods Mol. Biol. **614**, 111–124 (2010)
100. Torchilin, V.P.: Structure and design of polymeric surfactant-based drug delivery systems. J. Control. Release **73**, 137–172 (2001)
101. Torchilin, V.P.: Micellar nanocarriers: pharmaceutical perspectives. Pharm. Res. **24**, 1–16 (2007)
102. O'Reilly, R.K.: Spherical polymer micelles: nanosized reaction vessels? Philos. Transact. A Math. Phys. Eng. Sci. **365**, 2863–2878 (2007)
103. Morein, B., Sharp, M., Sundquist, B., Simons, K.: Protein subunit vaccines of parainfluenza type 3 virus: immunogenic effect in lambs and mice. J. Gen. Virol. **64**(Pt 7), 1557–1569 (1983)
104. Prabakaran, M., et al.: Reverse micelle-encapsulated recombinant baculovirus as an oral vaccine against H5N1 infection in mice. Antiviral Res. **86**, 180–187 (2010)
105. Bharali, D.J., Khalil, M., Gurbuz, M., Simone, T.M., Mousa, S.A.: Nanoparticles and cancer therapy: a concise review with emphasis on dendrimers. Int. J. Nanomedicine **4**, 1–7 (2009)
106. Patri, A.K., Majoros, I.J., Baker, J.R.: Dendritic polymer macromolecular carriers for drug delivery. Curr. Opin. Chem. Biol. **6**, 466–471 (2002)
107. Klajnert, B., Bryszewska, M.: Dendrimers: properties and applications. Acta Biochim. Pol. **48**, 199–208 (2001)
108. Baker, J. R., Jr.: Dendrimer-based nanoparticles for cancer therapy. Hematology. Am. Soc. Hematol. Educ. Program. **1**, 708–719 (2009)
109. Moreno, C.A., et al.: Preclinical evaluation of a synthetic Plasmodium falciparum MAP malaria vaccine in Aotus monkeys and mice. Vaccine **18**, 89–99 (1999)
110. Nardin, E.H., et al.: A totally synthetic polyoxime malaria vaccine containing Plasmodium falciparum B cell and universal T cell epitopes elicits immune responses in volunteers of diverse HLA types. J. Immunol. **166**, 481–489 (2001)
111. Nardin, E.H., et al.: Synthetic malaria peptide vaccine elicits high levels of antibodies in vaccinees of defined HLA genotypes. J. Infect. Dis. **182**, 1486–1496 (2000)
112. Shampur, M., Padinjarenmattathil, U., Desai, A., Narayanaswamy, J.: Development and immunogenicity of a novel polyetherimine (PETIM) dendrimer based nanoformulated DNA rabies vaccine. Int. J. Infect. Dis. **14**(Suppl 1), E453 (2010). Elsevier Science
113. Dutta, T., Garg, M., Jain, N.K.: Poly(propyleneimine) dendrimer and dendrosome mediated genetic immunization against hepatitis B. Vaccine **26**, 3389–3394 (2008)
114. Skwarczynski, M., Zaman, M., Urbani, C.N., Lin, I.C., Jia, Z., Batzloff, M.R., Good, M.F., Monteiro, M.J., Toth, I.: Polyacrylate dendrimer nanoparticles: a self-adjuvanting vaccine delivery system. Angew. Chem. Int. Ed. Engl. **49**, 5742–5745 (2010)
115. in het Panhuis, M.: Vaccine delivery by carbon nanotubes. Chem. Biol. **10**, 897–898 (2003)
116. Kendall, M.: Engineering of needle-free physical methods to target epidermal cells for DNA vaccination. Vaccine **24**, 4651–4656 (2006). doi:10.1016/j.vaccine.2005.08.066. S0264-410X(05)00841-8 [pii]
117. Kersten, G., Hirschberg, H.: Antigen delivery systems. Expert Rev. Vaccines **3**, 453–462 (2004). doi:10.1586/14760584.3.4.453. ERV030423 [pii]
118. Nandedkar, T.D.: Nanovaccines: recent developments in vaccination. J. Biosci. **34**, 995–1003 (2009)

# Vaccine Delivery Systems: Roles, Challenges and Recent Advances

## 46

Aditya Pattani, Prem N. Gupta, Rhonda M. Curran, and R. Karl Malcolm

## Contents

| | | |
|---|---|---|
| 46.1 | Introduction | 744 |
| 46.2 | Roles of Vaccine Delivery Systems | 744 |
| 46.3 | Important Considerations and Challenges for the Design of Vaccine Delivery Systems | 745 |
| 46.3.1 | The Conformation of an Antigen Needs to Be Protected | 745 |
| 46.3.2 | The Vaccine Delivery Rate Should Be Adequately Evaluated | 745 |
| 46.3.3 | The Adjuvant Component Is Often as Important as the Antigen Itself | 745 |
| 46.3.4 | Dosage Form Design Must Aim to Eliminate the Need for Trained Personnel for Administration of the Vaccine | 745 |
| 46.4 | Concepts in Designing Delivery Systems for Vaccines | 745 |
| 46.4.1 | Particulate Antigens Are More Potent Compared to Soluble Antigens | 745 |
| 46.4.2 | Particulate Delivery Systems Have Inherent Adjuvant Action | 746 |
| 46.5 | Designing Antigen Delivery Systems | 746 |
| 46.5.1 | Transport of Antigen Across a Biological Barrier | 747 |
| 46.5.2 | Biodistribution of the Antigen Delivery System and Their APC-Targeting Potential | 747 |
| 46.5.3 | Antigen Stability in the Delivery System | 747 |
| 46.5.4 | Concomitant Delivery of Antigen and Co-stimulatory Molecules | 747 |
| 46.5.5 | Immunomodulating Activities of Polymers | 747 |
| 46.5.6 | Antigen Dose and Structure Affect the Type of Immunity Induced | 748 |
| 46.6 | Advanced Vaccine Delivery Systems | 748 |
| 46.6.1 | Liposomes | 748 |
| 46.6.2 | Polymeric Microparticles/Nanoparticles | 748 |
| 46.6.3 | Virosomes | 749 |
| 46.6.4 | Immunostimulatory Complexes (ISCOMS) | 749 |
| 46.6.5 | Emulsions | 749 |
| 46.6.6 | Mucoadhesive Polymer/Gel | 749 |
| Conclusions | | 750 |
| References | | 750 |

A. Pattani, PhD
Department of Research and Development,
Kairav Chemofarbe Industries Ltd & NanoXpert Technologies, Mumbai, India

P.N. Gupta, PhD (✉)
Formulation & Drug Delivery Division, Indian Institute of Integrative Medicine, Jammu, India
e-mail: pngupta10@gmail.com

R.M. Curran, PhD
Institute of Nursing and Health Research,
University of Ulster,
Jordanstown, Northern Ireland, UK

R.K. Malcolm, PhD
School of Pharmacy,
Queen's University of Belfast, Belfast, UK

### Abstract

The overwhelming majority of vaccine antigens are biological macromolecules, such as proteins and polysaccharides, typically with a molecular weight greater than 10,000. As such, they need to be delivered to the body in the correct conformation in order to elicit the desired immune response and to effectively target the immune cells. Currently, most vaccines are administered parenterally via the intradermal, subcutaneous or intramuscular route, the choice largely dependent on whether the antigen is in the adsorbed or nonadsorbed state. However, these routes have major drawbacks, including pain associated with the use of needles, the potential for needle contamination, practicalities of needle disposal and the need for a primary healthcare worker. There is now a particular focus on the development of mucosal vaccines, designed for direct application to mucosal surfaces such as those present in the mouth, nose, vagina and rectum.

Often, simple antigen solutions are immunologically ineffective when delivered by these mucosal routes, owing to difficulties associated with mucosal retention and uptake. A diverse range of formulation strategies, including microspheres, liposomes, nanoparticles and virus-like particles, are now being actively investigated. In addition to the design and selection of the antigen candidate, the choice and preparation of the antigen delivery system is crucial to achieve the end goal of vaccination. In this chapter, an overview of the role and considerations for the design of vaccine delivery systems is presented, with particular focus on the challenges and recent advances in the field of colloidal and nano-sized delivery systems.

susceptible to inactivation, loss of conformation and poor absorption by the peroral route. As a result, most vaccines are administered parenterally by needle injection. These problems are further compounded by the fact that antigens are extremely prone to physical, chemical and conformational degradation resulting in vaccine inefficacy [2]. Current vaccination requires the need for trained medical personnel for the administration of parenteral vaccines and for the disposal of used needles and syringes [3]. It is likely that induction of strong, antigen-specific mucosal immune responses will play a critical role in developing vaccines for diseases such as HIV/AIDS. Achieving this goal may necessitate antigen dosing at mucosal sites, with particular measures to overcome the challenges associated with local degradation [4, 5] and retention (e.g. in the nose and vagina). To this end, there is substantial scope globally for the development of accessible and affordable vaccine delivery strategies that maintain or enhance the delivery and efficacy of a given vaccine.

## 46.2 Roles of Vaccine Delivery Systems

A wide range of advanced drug delivery systems have been developed to overcome the various problems and obstacles associated with more conventional drug delivery methods. Table 46.1 provides an overview of the main formulation approaches adopted for the efficient delivery of drug molecules.

## 46.1 Introduction

In the current age of molecularly defined vaccines, the necessary focus on ensuring product safety has inevitably impinged upon clinical efficacy [1]. Most antigen candidates are biological macromolecules and therefore highly

**Table 46.1** Roles of drug delivery systems

| Role of delivery system | Examples |
|---|---|
| Protection against chemical and physical degradation, improved shelf life and elimination of cold chain | Freeze-dried delivery systems/spray-dried vaccines [6] |
| Protection against (chemical/enzymatic) degradation at mucosal sites | Microparticles/nanoparticles [7] |
| Improved efficacy by immune system modulation | Liposomes [8, 9], nanoparticles [10, 11] |
| Improved efficiency through sustained release | Poly(lactide-co-glycolide) microparticles [12] |
| Obviation of the need for trained medical personnel | Microneedles [13] |
| Remove the need for special disposal of needles/syringes | Dissolving microneedles [14, 15] |
| Improved mucosal retention | Vaginal gels [16], nasal gels [17], vaginal rods [18, 19] |

## 46.3 Important Considerations and Challenges for the Design of Vaccine Delivery Systems

### 46.3.1 The Conformation of an Antigen Needs to Be Protected

The preparation process of vaccine delivery systems or constituent excipients of vaccine delivery formulations may cause physical or chemical changes to the vaccine that result in loss of or altered activity [2]. The final activity of the dosage form must be adequately assessed by various immunological and in vivo techniques. The ultimate aim should be stabilisation to such an extent that the need for cold-chain transport is eliminated.

### 46.3.2 The Vaccine Delivery Rate Should Be Adequately Evaluated

Many delivery systems are designed to deliver the vaccine over a prolonged period of time. However, it is known that prolonged or repeated administration of some vaccine candidates can lead to the development of tolerance [20, 21]. This aspect should be carefully evaluated.

### 46.3.3 The Adjuvant Component Is Often as Important as the Antigen Itself

Adjuvants are useful, and sometimes essential, in potentiating an immune response. Thus, adjuvant formulation, dose and release rate should be given equal consideration and may need to be correlated to that of the accompanying antigens.

### 46.3.4 Dosage Form Design Must Aim to Eliminate the Need for Trained Personnel for Administration of the Vaccine

Ideally a vaccine intended for quick and effective mass immunisation should be able to be administered without the need for trained personnel. This would facilitate global and equitable access to the vaccine.

## 46.4 Concepts in Designing Delivery Systems for Vaccines

Clinical experience favours use of subunit vaccines as a safer alternative to traditional organism-based vaccines, despite their often impaired immunogenicity. Use of vaccine delivery systems may overcome this compromise.

### 46.4.1 Particulate Antigens Are More Potent Compared to Soluble Antigens

One advantage of particulate vaccines over soluble antigens arises from their facilitated uptake by antigen-presenting cells (APCs). Antigens associated with particles mimic the particulate nature of pathogens. For example, particulate vaccines are typically a few hundred nanometres to a few microns in size, dimensions comparable to those of common pathogens against which the immune system has evolved to react, and promoting efficient uptake by APCs. Internalisation of particulate vaccines into phagosomes through the mechanism of phagocytosis has important consequences since phagosomes are known to be competent organelles for antigen cross-presentation [22]. This makes particulate vaccines attractive for inducing cellular immune responses and in contrast to soluble antigens which are preferentially presented by the MHC class-II pathway and only poorly cross-presented. Other attractive features of particulate antigens include (i) the possibility to deliver relatively large quantities of particle-associated antigen inside the APCs, (ii) a prolonged intracellular [23] or extracellular [24] release leading to prolonged antigen presentation compared with soluble antigen, and (iii) concomitant delivery of antigen and immunostimulatory components to the same APC [25].

**Table 46.2** Delivery systems with inherent adjuvanticity that have been tested in humans

| Delivery systems | Composition (mechanism of action) | Disease (antigen) |
|---|---|---|
| Liposomes | One or several bilayers of phospholipids | Influenza (monovalent split) |
| Liposomes + MTP-PE | Liposomes that incorporate the synthetic lipid MTP-PE | HIV (gp120) |
| Virosomes | Liposomes that incorporate in their membrane viral fusion proteins. | HAV, influenza, DT, TT |
| ISCOMS | Micellar assemblies made of saponin Quil-A, cholesterol and phospholipids | Influenza (trivalent split), HPV16 (E6/E7), Helicobacter pylori |
| PLGA microparticles | Particles made of homo- and copolymers or lactic and glycolic acids. | TT, HIV |
| MF59® | Squalene/water emulsion stabilised with Span85 and Polysorbate 80 | Influenza (trivalent split), HBV, HSV-2 (gB + gD), HIV-1(gp120), CMV(gB) |
| SBAS-2 | Squalene/water emulsion that incorporate MPL® and QS21® | Malaria (RTS, S), HIV-1 (gp 120) |
| SBAS-4 | Alum gel with MPL® | HBV (HBsAg), HSV |
| Incomplete Freund adjuvant | Water/Drakeol emulsion stabilised with mannide monooleate | HIV-1, Melanoma (gp100) |
| Montanide ISA720 | Emulsion with a metabolizable oil | Malaria (MSP1, MSP2) |
| Detox® | Squalene/water emulsion that incorporate MPL® and CWS | Malaria (R32NS18), Melanoma cell lysates |

Abbreviations: *CMV* cytomegalovirus, *CWS* cell wall skeleton from Mycobacterium phlei, *DT* diphtheria toxoid, *HAV* hepatitis A virus, *HBV* hepatitis B virus, *HPV* human papilloma virus, *HSV* herpes simplex virus, *MPL* monophosphoryl lipid, *MTP-PE* muramyl tripeptide dipalmitoyl phosphatidyl ethanolamine, *PLGA* poly-(D,L)-lactide-co-glycolic acid, *TT* tetanus toxoid

### 46.4.2 Particulate Delivery Systems Have Inherent Adjuvant Action

There are inherent safety concerns associated with traditional vaccines based upon killed, live-attenuated microorganisms or attenuated toxins. For some pathogens (i.e. tuberculosis, HIV), these types of vaccines are not available [26]. Nowadays, there is increasing interest in vaccines based on proteins, peptides or antigen-expressing DNA or RNA. Safer than the use of whole microorganisms, they are, however, poorly immunogenic and require the use of adjuvants [27]. Examples of particulate delivery systems with inherent adjuvanticity that have been tested in humans are summarised in Table 46.2. Adjuvants can enhance specific immune response against the co-administered antigen by two major mechanisms [27]. (i) The particulate delivery system increases the uptake of antigen by APC by either they are directly engulfed by APC or they form a depot of antigen that prolongs exposure thus increasing the chance of the antigen to be uptaken by APC. (ii) The particulate delivery system acts as an immunopotentiator (e.g. cytokines) type of adjuvant directly activating innate immune cells.

## 46.5 Designing Antigen Delivery Systems

Pharmaceutical vaccine formulations that are presently being tested in various experimental and clinical models are generally particulate in nature and obtained by aggregation/cross-linking of antigen [28] or by adsorption or precipitation of antigen on aluminium salts. Alternatively, the antigen can be chemically attached to a pre-formed particular carrier [29] or chemically or physically distributed in or on particles in a more (in liposomes and virus-like particles) or less (in polymeric microspheres) organised way [29, 30]. Typically, any type of particle facilitates the recognition of the vaccine by professional APC as well as the uptake of the vaccine into these cells. Prior to APC uptake, the formulation itself may influence the properties of the recruited

**Table 46.3** Various factors affecting uptake of polymeric particles

| | |
|---|---|
| Particle size | Animal species used for evaluation |
| Particle hydrophobicity | Age of the animal |
| Dose of particle (antigen dose) | Fed state of the animal |
| Administration vehicle | Mucosal layer characteristics |
| Polymer composition | Use of targeting agent on the particles |
| Effect of additives | Method for the quantitation for the extent of uptake |
| Particle surface charge | |

phagocytic cells. The design of the vaccine delivery system may also influence other critical factors, as discussed below.

### 46.5.1 Transport of Antigen Across a Biological Barrier

Vaccine delivery systems should improve antigen passage through relevant biological barriers, such as the intestinal and nasal mucosa. Based on observations over the past few decades, it has been noted that smaller particle sizes generally enhance the ability of particles to transport antigens across the intestinal barrier [31]. The ideal size for a mucosal vaccine carrier would be within the 50–500 nm range, although other factors that affect particle uptake are summarised in Table 46.3 [7].

### 46.5.2 Biodistribution of the Antigen Delivery System and Their APC-Targeting Potential

The size of the polymeric particles influences their distribution after subcutaneous, intradermal or intramuscular administration. When administered by the intramuscular or subcutaneous route, particles with size 20–100 nm can penetrate the extracellular matrix and enter directly into the lymphatic vessels. Once in the lymph, the particles travel to the lymphatic nodes where they are captured by the dense population of immune cells, mostly by dendritic cells, and generate effective immune responses [32]. In addition to the critical role of the particle size, it has been observed that the uptake of nanoparticles by macrophages and dendritic cells can be strongly enhanced if the particles have a cationic surface [33].

### 46.5.3 Antigen Stability in the Delivery System

Vaccine delivery systems may offer the advantage of antigen protection from harsh physiological conditions. However, formulation development conditions are of critical importance since the use of organic solvents, extreme temperatures or high energy inputs can also degrade or aggregate the antigen. Also, the materials used in the fabrication of the delivery systems or their degradation products can also enhance protein deterioration [34].

### 46.5.4 Concomitant Delivery of Antigen and Co-stimulatory Molecules

If the vaccine by itself is not capable of stimulating the expression of molecules necessary for T-cell activation, the combined delivery of antigen and co-stimulation factors within the same pharmaceutical formulation may improve the priming of lymphocytes. This concept has been demonstrated with recombinant viral vectors and liposomes, which provided co-stimulatory molecules such as those of the B7 family [35].

### 46.5.5 Immunomodulating Activities of Polymers

Some polymers can be used as effective protein carriers, and the development of vaccine delivery systems based on liposomes, microspheres, nanoparticles or water-soluble synthetic polymers has received considerable attention, as they can be tailored to meet the specific physical, chemical and immunogenic requirements of a particular antigen [36]. Antigen delivery systems which assist the acidification of endosomes also promote the MHC class-II presentation of the antigen they are carrying. This might be one mechanism by which nano- and microparticles of poly(lactide-co-glycolide) function. On the other hand, pharmaceutical formulations, which contain basic polymers or excipients (e.g. chitosan, ethylene imine, collagen, anionic lipids and surfactants), may prevent antigen presentation by the MHC class-II pathway. It has been shown that chitosan promotes Th1 cytokine responses and MHC class-I antigen presentation [37].

### 46.5.6 Antigen Dose and Structure Affect the Type of Immunity Induced

The amount and amino acid sequence of the antigen that initiates the response also influence the differentiation of CD4 T cells into distinct effector subsets, with high and low density of peptide on the surface of APCs stimulating Th1 or Th2 cell responses, respectively [38]. Hence, when the stability of the antigen is compromised, as may occur in poly(lactide-co-glycolide) microspheres [39], this might also have consequences for the Th1/Th2 skewing of the immune response, since the relative accessibility of different epitopes may have changed. Moreover, peptides that interact strongly with the T-cell receptor tend to stimulate Th1-like responses, whereas peptides that bind weakly tend to stimulate Th2-like responses [38].

## 46.6 Advanced Vaccine Delivery Systems

### 46.6.1 Liposomes

Liposomes consist of one or more phospholipid bilayers enclosing an aqueous phase. Antigens can be encapsulated within the aqueous compartment, linked to the liposomal surface or embedded in the lipid bilayer, all of which can protect the antigen from the surroundings. Variations in size, composition and physicochemical characteristics can render liposomes a versatile platform for antigen delivery. Immunostimulatory properties of liposomes are supposed to arise from (i) their capacity to associate and release antigens over a prolonged time and (ii) their preferential internalisation by APCs. Thus, this higher immunoavailability of the antigen might promote the maturation and antigen presentation by APCs [40]. In liquid form, liposomes have limited stability and tend to aggregate and fuse together, an issue that can be solved to some extent through lyophilization. Among the innumerable types of liposomal vaccine formulations studied over the past few decades, cationic liposomes appear to be particularly immunogenic. For example, liposomes made of dimethyldioctadecylammonium and the immune-modulating glycolipid trehalose dibehenate efficiently promoted the cell-mediated and the humoral immune responses and are currently being evaluated in a Phase I study for the management of tuberculosis [41].

### 46.6.2 Polymeric Microparticles/ Nanoparticles

Microparticles were first used as delivery systems for entrapped antigens in the early 1990s [42]. Owing to a long history in medical applications, the biodegradable poly(D,L-lactide) and poly(D,L-lactic-co-glycolic acid) are probably the most studied materials for parenteral and mucosal antigen delivery [43]. Long-lasting immunity can be induced by the parenteral administration of microparticles made with different ratios of polymers of different molecular weights that hydrolyse over a wide range of times. Long-lasting immunity can also be induced by using particles of mixed sizes, especially larger particles that avoid uptake by macrophages and hence break down at a slower rate [44]. In this manner, one injection of a vaccine can result in long-lasting immunity obviating the need to boost. Besides the encouraging immunological performances of poly(lactide-co-glycolide) (PLGA) based particles, the ensuring stability of encapsulated protein antigens has been found to be an important issue. Indeed, some protein antigens tend to aggregate or degrade upon entrapment into PLGA or during release from the matrix. The exposure of antigen to organic solvent and acidic microenvironment generated during polymer hydrolysis may lead to antigen inactivation. These problems have been partially solved by optimised manufacturing methods or addition of stabilising agents such as $Mg(OH)_2$, other proteins, surfactants or sugars [45, 46].

### 46.6.3 Virosomes

Virosomes are liposomes containing functional viral membrane proteins and have a particle size that resembles that of the viruses. As such, virosomes represent reconstituted empty influenza virus envelopes where the viral proteins confer immunostimulatory properties [47]. Thus, when they are formulated to carry heterologous antigens, virosomes can be considered as delivery systems with intrinsic adjuvant activity. To date, virosomes have been approved in Europe for systemic vaccination against hepatitis A and influenza, and they have also been formulated for intranasal immunisation [48].

### 46.6.4 Immunostimulatory Complexes (ISCOMS)

ISCOMS are characterised by a cagelike structure that incorporates the antigen. They consist of lipids and Quil-A, the active component of the saponin derived from the plant *Quillaja saponaria* that has adjuvant activity [49]. Hydrophobic antigens can be embedded or anchored directly into the lipidic colloidal domains, whereas hydrophilic antigens require modification for efficient entrapment [50]. An interesting feature of ISCOMs is their good stability under varying conditions. ISCOMs have been evaluated in clinical trials in humans, and as well as being well tolerated, they induced strong humoral and cytotoxic T lymphocyte (CTL) responses even at very low antigen doses [51]. Despite their potential and good performance in clinical trials, ISCOM-based vaccines have only approved for veterinary use.

### 46.6.5 Emulsions

The investigation of emulsions as vaccine adjuvants started with Freund's complete adjuvant (a water-in-oil emulsion of a mineral oil, paraffin and killed mycobacteria) [52]. Although capable of generating high antibody titres, this emulsion led to strong adverse reactions which hampered its clinical use. A less toxic version, the Freund's incomplete adjuvant which lacks the mycobacterial component, is still applied in veterinary medicine but has been prohibited for use in human vaccines because of severe adverse events. In the 1990s, a squalene oil-in-water emulsion (MF59) was developed and generated good antibody titres, demonstrating good tolerability and general safety [53]. MF59 is currently approved in Europe for influenza vaccines.

### 46.6.6 Mucoadhesive Polymer/Gel

Some mucoadhesives have adjuvant properties when injected (e.g. sodium alginate), and others (e.g. sodium carboxymethyl cellulose) have been selected for their aqueous viscosity-enhancing properties when used as depot agents in experimental formulations for parenteral vaccines. Recently, mucoadhesive gels have received considerable attention for vaginal vaccine delivery. From a formulation perspective, inducing effective antigen-specific immune responses by cervicovaginal instillation of buffer solution containing solubilized antigen is far from ideal owing to the potential for leakage at the administration site, rapid enzymatic degradation of the antigen, the influence of the menstrual cycle and inadequate exposure of antigen to the mucosal associated lymphoid tissue. In order to improve the efficacy of vaginal vaccine delivery, various mucoadhesive delivery systems, including hydroxyethyl cellulose-based rheologically structured gel vehicles [16] and lyophilised solid dosage formulations [18, 54], have been investigated.

**Fig. 46.1** Scheme for the development of lyophilised dosage form for vaginal vaccine administration

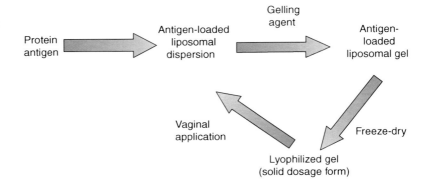

Recently, our group reported the development of liposomal gel formulations, and novel lyophilised variants, comprising HIV-1 envelope glycoprotein, CN54gp140, encapsulated within neutral, positively charged or negatively charged liposomes [19]. Scheme for the development of lyophilised dosage form for vaginal vaccine administration is shown in Fig. 46.1. The CN54gp140 liposomes were evaluated for mean vesicle diameter, polydispersity, morphology, zeta potential and antigen encapsulation efficiency before being incorporated into hydroxyethyl cellulose (HEC) aqueous gel and subsequently lyophilised to produce a rod-shaped solid dosage form for practical vaginal application. The lyophilised liposome–HEC rods were evaluated for moisture content and redispersibility in simulated vaginal fluid. Since these rods are designed to revert to gel form following intravaginal application, mucoadhesive, mechanical (compressibility and hardness) and rheological properties of the reformed gels were evaluated. The liposomes exhibited good encapsulation efficiency and the gels demonstrated suitable mucoadhesive strength. The freeze-dried liposome–HEC formulations represent a novel formulation strategy that could offer potential as stable and practical dosage form.

## Conclusions

Safety concerns associated with traditional vaccine strategies have placed constraints on emerging vaccines that can render them less effective. Integral to overcoming this are vaccine delivery systems, acting not just as simple carriers of vaccines, but as adjuvants and targeting agents offering additional benefits such as needle-free delivery and stabilisation that facilitate wider access. There is huge scope for the development of vaccine delivery systems with uncharted potential to positively impact future global immunisation strategies.

## References

1. Bachmann, M.F., Jennings, G.T.: Vaccine delivery: a matter of size, geometry, kinetics and molecular patterns. Nat. Rev. Immunol. **10**, 787–796 (2010)
2. Pattani, A., et al.: Molecular investigations into vaginal immunization with HIV gp41 antigenic construct H4A in a quick release solid dosage form. Vaccine **30**, 2778–2785 (2012)
3. Donnelly, R.F., et al.: Design, optimization and characterisation of polymeric microneedle arrays prepared by a novel laser-based micromoulding technique. Pharm. Res. **28**, 41–57 (2011)
4. Duerr, A.: Update on mucosal HIV vaccine vectors. Curr. Opin. HIV AIDS **5**, 397–403 (2010)
5. Neutra, M.R., Kozlowski, P.A.: Mucosal vaccines: the promise and the challenge. Nat. Rev. Immunol. **6**, 148–158 (2006)
6. Wong, Y., et al.: Drying a tuberculosis vaccine without freezing. Proc. Natl. Acad. Sci. **104**, 2591–2595 (2007)
7. Vyas, S.P., Gupta, P.N.: Implication of nanoparticles/microparticles in mucosal vaccine delivery. Expert Rev. Vaccines **6**, 401–418 (2007)
8. Perrie, Y., Frederik, P.M., Gregoriadis, G.: Liposome-mediated dna vaccination: the effect of vesicle composition. Vaccine **19**, 3301–3310 (2001)
9. Yan, W., Chen, W., Huang, L.: Mechanism of adjuvant activity of cationic liposome: phosphorylation of a MAP kinase, ERK and induction of chemokines. Mol. Immunol. **44**, 3672–3681 (2007)

10. Arias, M.A., et al.: Carnauba wax nanoparticles enhance strong systemic and mucosal cellular and humoral immune responses to HIV-gp140 antigen. Vaccine **29**, 1258–1269 (2011)
11. Borges, O., et al.: Immune response by nasal delivery of hepatitis B surface antigen and co-delivery of a CPG ODN in alginate coated chitosan nanoparticles. Eur. J. Pharm. Biopharm. **69**, 405–416 (2008)
12. O'Hagan, D.T., Singh, M., Gupta, R.K.: Poly(lactide-co-glycolide) microparticles for the development of single-dose controlled-release vaccines. Adv. Drug Deliv. Rev. **32**, 225–246 (1998)
13. Prausnitz, M.R., Mikszta, J.A., Cormier, M., Andrianov, A.K.: Microneedle-based vaccines. Curr. Top. Microbiol. Immunol. **333**, 369–393 (2009)
14. Sullivan, S.P., et al.: Dissolving polymer microneedle patches for influenza vaccination. Nat. Med. **16**, 915–920 (2010)
15. Donnelly, R. F.: Microneedle-mediated intradermal delivery. In: Donnelly, R. F., Thakur, R. R. S., Morrow, D. I. J., Woolfson, A. D. Microneedle-mediated transdermal and intradermal drug delivery. Blackwell, Chichester, West Sussex (2011)
16. Curran, R.M., et al.: Vaginal delivery of the recombinant HIV-1 clade-C trimeric gp140 envelope protein CN54gp140 within novel rheologically structured vehicles elicits specific immune responses. Vaccine **27**, 6791–6798 (2009)
17. Tiwari, S., et al.: Liposome in situ gelling system: novel carrier based vaccine adjuvant for intranasal delivery of recombinant protein vaccine. Procedia. Vaccinol. **1**, 148–163 (2009)
18. Donnelly, L., et al.: Intravaginal immunization using the recombinant HIV-1 clade-C trimeric envelope glycoprotein CN54gp140 formulated within lyophilized solid dosage forms. Vaccine **29**, 4512–4520 (2011)
19. Gupta, P.N., et al.: Development of liposome gel based formulations for intravaginal delivery of the recombinant HIV-1 envelope protein CN54gp140. Eur. J. Pharm. Sci. **46**, 315–322 (2012)
20. Black, C.A., et al.: Vaginal mucosa serves as an inductive site for tolerance. J. Immunol. **165**, 5077–5083 (2000)
21. Mestecky, J., Moldoveanu, Z., Elson, C.O.: Immune response versus mucosal tolerance to mucosally administered antigens. Vaccine **23**, 1800–1803 (2005)
22. Howland, S.W., Wittrup, K.D.: Antigen release kinetics in the phagosome are critical to cross-presentation efficiency1. J. Immunol. **180**, 1576 (2008)
23. Shen, H., et al.: Enhanced and prolonged cross-presentation following endosomal escape of exogenous antigens encapsulated in biodegradable nanoparticles. Immunology **117**, 78–88 (2006)
24. Singh, M., et al.: Controlled release microparticles as a single dose hepatitis B vaccine: evaluation of immunogenicity in mice. Vaccine **15**, 475–481 (1997)
25. Singh, M., et al.: Cationic microparticles are an effective delivery system for immune stimulatory CPG DNA. Pharm. Res. **18**, 1476–1479 (2001)
26. O'Hagan, D.T., Rappuoli, R.: The safety of vaccines. Drug Discov. Today **9**, 846–854 (2004)
27. O'Hagan, D.T., Valiante, N.M.: Recent advances in the discovery and delivery of vaccine adjuvants. Nat. Rev. Drug Discov. **2**, 727–735 (2003)
28. Watanabe, M., Nagai, M., Funaishi, K., Endoh, M.: Efficacy of chemically cross-linked antigens for acellular pertussis vaccine. Vaccine **19**, 1199–1203 (2000)
29. Jegerlehner, A., et al.: A molecular assembly system that renders antigens of choice highly repetitive for induction of protective B cell responses. Vaccine **20**, 3104–3112 (2002)
30. Kersten, G.F.A., Crommelin, D.J.A.: Liposomes and ISCOMS. Vaccine **21**, 915–920 (2003)
31. Gómez, S., et al.: Gantrez® AN nanoparticles as an adjuvant for oral immunotherapy with allergens. Vaccine **25**, 5263–5271 (2007)
32. Cubas, R., et al.: Virus-like particle (VLP) lymphatic trafficking and immune response generation after immunization by different routes. J. Immunother. **32**, 118–128 (2009)
33. Foged, C., Brodin, B., Frokjaer, S., Sundblad, A.: Particle size and surface charge affect particle uptake by human dendritic cells in an in vitro model. Int. J. Pharm. **298**, 315–322 (2005)
34. Blanco, M.D., Alonso, M.J.: Development and characterization of protein-loaded poly (lactide-co-glycolide) nanospheres. Eur. J. Pharm. Biopharm. **43**, 287–294 (1997)
35. Zajac, P., et al.: Enhanced generation of cytotoxic T lymphocytes using recombinant vaccinia virus expressing human tumor-associated antigens and B7 co-stimulatory molecules. Cancer Res. **58**, 4567–4571 (1998)
36. Ríhová, B.: Immunomodulating activities of soluble synthetic polymer-bound drugs. Adv. Drug Deliv. Rev. **54**, 653–674 (2002)
37. Strong, P., Clark, H., Reid, K.: Intranasal application of chitin microparticles down-regulates symptoms of allergic hypersensitivity to *dermatophagoides pteronyssinus* and *aspergillus fumigatus* in murine models of allergy. Clin. Exp. Allergy **32**, 1794–1800 (2002)
38. Rogers, P.R., Croft, M.: Peptide dose, affinity, and time of differentiation can contribute to the Th1/Th2 cytokine balance. J. Immunol. **163**, 1205–1213 (1999)
39. Johansen, P., Merkle, H.P., Gander, B.: Physicochemical and antigenic properties of tetanus and diphtheria toxoids and steps towards improved stability. Biochim. Biophys. Acta **1425**, 425–436 (1998)
40. Hart, B.A., et al.: Liposome-mediated peptide loading of MHC-DR molecules in vivo. FEBS Lett. **409**, 91–95 (1997)
41. Van Dissel, J.T., et al.: Ag85B–ESAT-6 adjuvanted with IC31® promotes strong and long-lived mycobacterium tuberculosis specific T cell responses in naïve human volunteers. Vaccine **28**, 3571–3581 (2010)
42. O'Hagan, D.T.E.A.: Biodegradable microparticles as controlled release antigen delivery systems. Immunology **73**, 239–242 (1991)
43. Putney, S.D., Burke, P.A.: Improving protein therapeutics with sustained-release formulations. Nat. Biotechnol. **16**, 153–157 (1998)

44. Eldridge, J.H., et al.: Biodegradable microspheres as a vaccine delivery system. Mol. Immunol. **28**, 287–294 (1991)
45. Audran, R., Men, Y., Johansen, P., Gander, B., Corradin, G.: Enhanced immunogenicity of microencapsulated tetanus toxoid with stabilizing agents. Pharm. Res. **15**, 1111–1116 (1998)
46. Gupta, P.N., Khatri, K., Goyal, A.K., Mishra, N., Vyas, S.P.: M-cell targeted biodegradable plga nanoparticles for oral immunization against hepatitis B. J. Drug Target. **15**, 701–713 (2007)
47. Moser, C., Metcalfe, I.C., Viret, J.F.: Virosomal adjuvanted antigen delivery systems. Expert Rev. Vaccines **2**, 189–196 (2003)
48. Durrer, P., et al.: Mucosal antibody response induced with a nasal virosome-based influenza vaccine. Vaccine **21**, 4328–4334 (2003)
49. Barr, I.G., Sjölander, A., Cox, J.C.: ISCOMs and other saponin based adjuvants. Adv. Drug Deliv. Rev. **32**, 247–271 (1998)
50. Lenarczyk, A., et al.: ISCOM® based vaccines for cancer immunotherapy. Vaccine **22**, 963–974 (2004)
51. Ennis, F.A., et al.: Augmentation of human influenza A virus-specific cytotoxic t lymphocyte memory by influenza vaccine and adjuvanted carriers (ISCOMs). Virology **259**, 256–261 (1999)
52. Casals, J., Freund, J.: Sensitization and antibody formation in monkeys injected with tubercle bacilli in paraffin oil. J. Immunol. **36**, 399–404 (1939)
53. Ott, G., Barchfeld, G.L., Nest, G.V.: Enhancement of humoral response against human influenza vaccine with the simple submicron oil/water emulsion adjuvant MF59. Vaccine **13**, 1557–1562 (1995)
54. Pattani, A., et al.: Characterisation of protein stability in rod-insert vaginal rings. Int. J. Pharm. **430**, 89–97 (2012)

# APC-Targeted (DNA) Vaccine Delivery Platforms: Nanoparticle Aided

Pirouz Daftarian, Paolo Serafini, Victor Perez, and Vance Lemmon

## Contents

47.1 Dendritic-Cell (DC) Vaccines and Immunotherapies: From Ex Vivo Loading to In Vivo Targeting .................... 754

47.2 Gene-Based or DNA-Based Vaccines ....... 754

47.3 Improving DNA-Based Vaccines by Prime Boost .......................................... 755

47.4 Next Generation of Vaccines: Targeted Delivery ...................................... 756

47.5 Vaccine: Physical Delivery of to the APC Location ................................. 756

47.6 Manipulating the Spatial-Temporal Immune Response by the Use of Implantable Microdevices ..................... 757

47.7 Targeting APC via Functionalized Nanoparticles ............................................. 757

47.8 Improving DNA-Based Vaccines by Nanocarriers ........................................... 758

47.9 Nanoparticle Platform: PADRE-Derivatized Dendrimer (PDD) ......................................................... 758

47.10 APC Recognition Moiety in PDD ............. 759

47.11 PDD is an APC Opsonized Platforms that Is Immunoenhancing In Vivo ............ 761

47.12 PPD-Mediated TRP2 Vaccination Leads to the Rejection of Established Melanoma .................................................... 761

Conclusion ............................................................. 762

References ............................................................. 763

P. Daftarian, PhD (✉)
Department of Ophthalmology, University of Miami, Miami, FL, USA

Department of Biochemistry & Molecular Biology, University of Miami, Miami, FL, USA
e-mail: PDaftarian@med.miami.edu

P. Serafini, PhD
Department of Microbiology and Immunology, Miller School of Medicine, University of Miami, Miami, FL, USA

V. Perez, PhD
Department of Ophthalmology, University of Miami, Miami, FL, USA

Department of Microbiology and Immunology, University of Miami, Miller School of Medicine, Miami, FL, USA

V. Lemmon, PhD
Miami Project to Cure Paralysis, University of Miami, Miami, FL, USA

### Abstract

Only a small fraction of any administered drug, including a vaccine, reaches its intended target tissue or cells. Homing vaccines to antigen-presenting cells (APC), where they can do their magic, has been the focus of attention for several decades. Since they are equipped with proper co-stimulatory signals, only professional APC are able to correctly process antigens and stimulate T and B cells to mount specific immune responses. With advances in the fields of modern immunology and nanotechnology, the new vaccine delivery platforms are emerging. APC-targeted vaccine delivery is required to elicit protective immunity, to reduce manufacturing costs, to minimize unanticipated effects of vaccines caused by off-target effects (thus reducing immunosuppressive mechanisms and toxicity), and to dramatically increase immunization efficacy. Indeed, in vivo targeting of APCs may represent the best hope for the generation of strategies leading to a personalized immunization without the need of expensive and complex ex vivo manipulation of patients' PBMCs. Here, we will review and compare novel approaches on nanoparticle-based-targeted vaccine platforms, evaluating their molecular mechanisms of action, their translation ability, their efficacy, and their cost.

## 47.1 Dendritic-Cell (DC) Vaccines and Immunotherapies: From Ex Vivo Loading to In Vivo Targeting

*DC Vaccines.* DCs are the most powerful APC and have higher rates of antigen uptake and antigen processing. They too can migrate to lymph nodes and present or cross-present the antigens in the context of self MHC combined by the appropriate co-stimulatory signals resulting in expansion of specific T cells and B cells. Interestingly, DCs can form gap junctions in lymph nodes, a process that promotes cross-priming [1, 2]. DC vaccines are commonly derived from patient-derived monocytes cultured in interleukin-4 and granulocyte macrophage colony-stimulating factor. Such DCs are then loaded with antigen proteins or peptides. The final product, antigen-loaded DC, is then injected back to the patient [2]. DCs have an extra regulation component for the control of the MHC class II locus, which leads to high levels of CIITA transcription (a prerequisite of MHC class II expression) resulting in higher levels of MHC class II transcription [3]. However, the antigen-processing ability of DC is positively correlated with its maturity, and the half-life of MHC class II molecules increases dramatically as the DC matures [3]. The use of antigen-loaded autologous DCs as vaccine has brought hope in many fields, in particular for cancer therapeutic vaccines, due to the FDA approval of the first ever human prostate cancer therapeutic vaccine, which was shown to be safe and capable of inducing tumor antigen-specific immune responses in a substantial part of the vaccinated patients. The DC-based patient-specific prostate cancer vaccine is a breakthrough; however, it has faced marketing challenges partly due to its high cost ~ $93 K and a complex procedure; in UK, for example, only 3 % of oncologists said that it is likely that they use this vaccine [1].

Other attempts to target DCs are aggressively ongoing. CD205 is expressed at high levels mostly in mature dendritic cells in mice, although low levels can be found in lymphocytes (B and T) and granulocytes [4]. Interestingly, early work delivering model antigens using anti-dec205 monoclonal antibody demonstrated that tolerance and T-cell depletion instead of tumor immunity was induced [5]. However, if an adjuvant (i.e., anti-CD40 antibody) is provided, a strong humoral and cellular-based immune response is generated, and it is able to cure 70 % of mice previously challenged with a metastatic melanoma [6]. Thus, despite the fact that in human DC205 is less restricted to the DCs, these experiments indicate that targeting APCs together with an adequate immunoadjuvant can be the key for a successful immunotherapy.

Other target is mannose receptors that are expressed mainly by monocytes, macrophage immature dendritic cells, However, expression has been reported also in different epithelial cell subsets [7]. Coupling an antigen to mannose can enhance MHC class I and II antigen presentation and promote a sustained humoral and cell-mediated immunity [8]. The same seems to hold true in human since in a randomized phase III clinical trial using oxidized mannan coupled to the breast cancer associated antigen Muc1, all the treated patients were tumor-free at 5 years while in the 27 % of the patients treated with the placebo mammary carcinoma recurred [9].

## 47.2 Gene-Based or DNA-Based Vaccines

One promising alternative to traditional vaccine strategies is the use of naked plasmid DNA as a vaccine to prime the immune system. DNA-based vaccination provides a variety of practical benefits for large-scale production that are not as easily achievable with other forms of vaccines including recombinant proteins or whole tumor cells [10–13]. Jon Wolff in 1990 reported that injection of DNA plasmids resulted in a response in mice. Soon after, in 1993, Margaret Liu demonstrated that DNA vaccination was able to elicit both humoral and cellular responses.

Human trials using DNA-based vaccinations started as early as 1996. DNA-based or gene-based vaccination uses a DNA sequence that harbors an intended antigen. Mammalian-based plasmids include a desired DNA sequence of at least one antigen and a mammalian-based promoter. Such a plasmid is injected into a host and uses the cell machinery to translate the protein that the DNA plasmid codes for. DNA vaccines are capable of utilizing both endogenous and exogenous pathways of antigen presentation. Thus, gene-based vaccinations are in nature superior to protein vaccines that primarily utilize the exogenous pathway. Furthermore, another problem that protein-based vaccines face is that the harsh processes of protein production make preservation of the native form of the protein in a vaccine challenging, while DNA-based vaccines result in making the host expressed naturally produced proteins [14].

DNA vaccinations also have been shown to elicit durable immune responses which are critical in immunizations [15]. Moreover, DNA vaccines have been shown to mount cross-neutralizing antibodies responses that are important in pathogens that have antigenic variations and different subtypes. Gene-based vaccination creates a scenario similar to a natural infection except there will be no clinical manifestation as there is no infectious agent involved. Furthermore, a major challenge facing vaccinologists designing vaccines is to identify the proper epitopes or antigen. Gene-based vaccination provides the opportunity to design and make constructs containing genes of multiple candidate antigens, deleting immunosuppressive elements, incorporation of immunoenhancing genes or sequences, which can greatly improve vaccine design [16]. This is important, in particular, for cancer vaccines that are not induced by a virus. One can screen via vaccinating with plasmids encoding for various candidate proteins to determine the best protective antigens [16].

## 47.3 Improving DNA-Based Vaccines by Prime Boost

However, DNA-based vaccination is not perfect and has its own challenges including poor immunogenicity. While numerous clinical trials have proven the safety of this cost effective vaccination strategy, they have also revealed its limitations, since the immune response elicited is insufficient to clear established tumors or infections. For example, the negatively charged cell membrane of mammalian cells ensures that negatively charged nucleic acids do not readily penetrate them. Such cell membranes act as a robust barrier to prevent the transport of large molecules such as DNA, contributing in the low efficacy of DNA-based vaccination. The low in vivo transfection efficiency, the absence of preferential targeting of professional antigen-presenting cells, and the modest intrinsic adjuvant activity of DNA vaccines are thought to be the main reasons for the marginal effect observed in the clinic and in therapeutic murine models. An accepted assumption is that the immunogenic potency of DNA-based vaccines can be significantly increased if the delivery of DNA to professional antigen-presenting cells (APCs) is maximized and if additional immunologic help in the form of adjuvants and/or cytokines is provided [17, 18].

The main challenge of the DNA-based vaccination is that the concentration of the expression of the native antigen is low and even much lower in APC, since the delivery is not targeted. One method to overcome this problem is to use gene-based vaccination as a booster immunization of a protein-based immunization [15]. The scientific explanation is that a booster immunization requires modest amounts of the antigen and thus a combination of vaccination strategies; the conventional method, the use of the protein antigens, followed by a gene-based vaccination should expand T-cell clonotypes that recognize the native form of the protein endogenously and exogenously made by the host. Indeed, Nabel and his colleagues showed that DNA booster immunization post-protein vaccination induced high-magnitude antibody responses against the conserved hemagglutinin stem epitope, and these antibodies were able to neutralize diverse strains, in clinical trials [15]. There are various methods to enhance the vaccination efficacy of DNA vaccines including in vivo electroporation [14] or the use of nanocarriers [19]. In vivo electroporation is a powerful method to enhance DNA vaccination resulting in local DNA uptake, endogenous expression of immunologically relevant components, and recruitment of some T cells [16, 20]. Indeed, there are multiple ongoing clinical trials using various in vivo electroporators showing the safety and the efficacy of such vaccines [16].

## 47.4 Next Generation of Vaccines: Targeted Delivery

Despite the success of DNA vaccines in preclinical nontherapeutic models, although demonstrating safety, they have had only modest therapeutic results in clinical trials and in stringent cancer murine models. The absence of efficient in vivo transfection and any targeted delivery of DNA to the APC are thought to be the main reasons for this failure. As mentioned above, APCs (including macrophages, B cells, and DCs) are crucial for priming a protective immune response against foreign pathogen and tumors [21]. Thus, efficient transfection is considered the main avenue to achieve the promise of DNA vaccines and fully release their potential therapeutic efficacy. To this aim, different strategies have been adopted and can be divided in three main categories: (1) delivering the vaccine physically where an important concentration of APCs is present (i.e., targeting skin Langerhans APCs via dermal electroporation, via transcutaneous release, or via needleless jet-based syringe), (2) manipulating the spatial-temporal immune response by the use of implantable microdevices, and (3) targeting APC via functionalized nanoparticles.

## 47.5 Vaccine: Physical Delivery of to the APC Location

It is now clear that the skin is not only a physical barrier that separates the external environment from the internal organs but, because it is the physical interface of the first interaction with many external pathogens, it is an active immune organ [22, 23]. Professional antigen-presenting cells of the skin comprise epidermal Langerhans cells (CD207/langerin$^+$), dermal langerin, and dermal langerin$^+$ dendritic cells (DCs). Langerhans cells in human skin can induce cytotoxic T lymphocytes, and, in mice, these cells were shown to be the key for tumor antigen cross-presentation and the generation of a protective immune response in melanoma-bearing mice [24]. Because of its accessibility and the elevated concentration of professional APCs, the dermis has become an important site of immunization. The delivery of the vaccine to this anatomical site is thus used as a method for physically facilitating the uptake of the DNA vaccines into professional APCs.

*Dermal* in vivo electroporation has become a gold standard method for DNA immunization. Dermal administration of the vaccine is followed by a microneedle-mediated electroporation of the dermis that maximizes the DNA entry into cells, inducing a danger signal, promoting the maturation of the Langerhans cells, and allowing display of the native form of protein to the immune system. Despite the optimal results that are obtained in naive mice, the situation is different during inflammation in a tumor bearer. Indeed, in the dermis, beside immunogenic APCs, other populations of cells (including tolerogenic APCs) can present, and thus, the "physical" dermal delivery may not assure an adequate protective immunity. Indeed, there are challenges for achieving an optimized reproducible process for eliciting strong immune responses. Eliciting strong immune responses demands generation of high numbers of primed B, T, and in particular CD4+ T helper cells in lymphoid organs. Due to reproducibility challenges, and since in vivo electroporation targets all cells (e.g., muscle cells) indiscriminately, this process includes optimized parameters for in vivo electroporation and the assistance of additional strategies to enhance T-cell arm of the immune system. Furthermore, the use of a normal syringe for intradermal injection requires significant expertise, and the difficulties associated with this technique can explain the interlaboratory variations observed. Nevertheless, the construction of microdevices containing microneedles, tattooing-like vaccine instruments, and needleless syringes [25] is facilitating the use of dermal vaccines with the specific idea to target the place where Langerhans cells reside.

## 47.6 Manipulating the Spatial-Temporal Immune Response by the Use of Implantable Microdevices

An elegant idea that stems from the old concept of isolating DC precursors, inducing their differentiation, loading them with antigens, and promoting their maturation has been proposed by Mooney's group. Instead of trying to selectively transfect the APCs in vivo, the authors re-elaborate the concept of generation, pulsing and inducing the maturation of DC, the same concept used for DC-based vaccines, using the FDA-approved polylactide-co-glycolide (PLG)-based device [26]. This synthetic polymer-based matrix can orchestrate spatially and temporally the release of cytokines, antigens, and danger signal in vivo [26] Basically, the authors envision a method to transfer a whole GMP laboratory into an implantable microdevice.

Briefly, highly porous "sponges" were fabricated from PLG using a gas-foaming process. GM-CSF, CpG oligonucleotides, and tumor antigens were incorporated into scaffolds during the foaming process to allow for their localized and sustained release once implanted in vivo. The authors elegantly showed that CD11b+DC precursors were attracted to the device by the GM-CSF release, whereas CpG-ODN molecules promote the recruitment of CD11c+PDCA1+plasmacytoid DC. When both GM-CSF and CpG-ODN were co-released, an increase number of plasmacytoid DC was observed while no differences in the myeloid DC recruitment was observed. The addition of a tumor lysate to the scaffold not only was able to promote antigen uptake but also, most likely by the release of endogenous immunogenic activators (i.e., DNA, heat-shock protein), promotes the in situ generation of IL-12-producing CD8+DC [26].

In vivo activated and pulsed DC subsets readily migrate to the draining lymph nodes promoting the generation of a strong CTL immunity capable of promoting tumor regression in mice bearing 9-day-old B16 melanoma tumors. It is important to note that this is an extremely stringent therapeutic setting, and no other immune therapies have ever been shown to have a similar therapeutic effect under this condition. Thus, the possibilities offered by the PLG-based devices open a unique possibility for the fine regulation in vivo of the immune response not only by a targeted pulsing of the desired dendritic cell subsets but also by the possibility to regulate the microenvironment during DC recruitment, antigen uptake, and differentiation. Although it is conceivable to adapt this methodology with the use of nucleic acids vaccines, difficulties may arise in protecting DNA and/or RNAs from endogenous nucleases.

## 47.7 Targeting APC via Functionalized Nanoparticles

Nanoparticles are attractive as controlled, targeted nanocarriers for drugs or vaccines because they can diffuse through tissues and penetrate cell membranes transporting their cargos inside the cells. There are two major components in the generation of a nanoplatform for delivering vaccines including DNA and RNA to the antigen-presenting cells: targeting molecules and a loading/encapsulating surface. Examples of nanoparticles lacking targeting molecules exist; their specificity relies in the size of the nanomolecule, the method of administration, and poorly characterized chemical properties. We will focus the rest of this chapter in those nanoparticles in which the targeting of the antigen-presenting cells has been designed to take advantage of an understanding of the immune system and of ligand/receptor interactions conserved in animal models and in humans. Despite the importance of the other nanoplatforms, the lack of information of their specificity in animal models makes it difficult to use them and translate findings into the clinic.

With advances in the understanding of the receptors specific for the different subsets of antigen-presenting cells, the list of ligands used for the specific delivery of vaccines has been growing [30, 31, 34–39]. However, Mannose receptors, DC205, Fc receptors (I, II, and III), CD11c–CD18 integrins, and MHC class II molecules are the most commonly used ligands for APC-specific targeting. Antigens, DNA-harboring antigens, or nanovehicles are decorated with targeting moieties (peptides, monoclonal antibodies, or aptamers) to home into antigen-presenting cells. Examples of targeting moieties used for this purpose are CDllb, CD169, mannose receptor, Dec-205, CDllc, CD21/CD35, CX3CR1, fractalkine, IgG FC receptor and other Fc receptor. Others have used pH-dependent targeting of DCs [27–32]. Even some reasoned that if the nanoparticle at right size (<500 nm) is used, there is no need for targeting DCs as these cells are present in lymph nodes and are highly phagocytic; one therefore needs to only target lymph nodes without the use of a targeting ligand [5, 33–35].

## 47.8 Improving DNA-Based Vaccines by Nanocarriers

Non-covalent complexation of dendrimer-based nanocarriers with DNA or biotherapeutics has been the center of extensive research [36, 37]. To target DCs, Cruz and his colleagues described a nanoparticle composed of poly(d,l-lactide-co-glycolide)-coated superparamagnetic iron oxide and coupled with antibodies recognizing the DC-specific receptor DC-SIG. They traced the nanoparticle since it was also labeled with fluorescent and could show antigen delivery into endolysosomal compartments within 24 h. Since these platforms contain superparamagnetic iron oxide, they need to be coated to become biocompatible and are therefore coated with dextran, poly(lactide-co-glycolide), or other polymers [39].

Liposomes are phospholipid bilayer vesicles and have been used for drug and vaccine delivery for a long time. Liposome is coated with mannose to deliver vaccines with enhance delivery or recruitment of DCs and APC [40–42]. Interestingly, Geng and his colleagues showed that a drug used for high blood pressure, amiloride, was able to enhance the DNA delivery into cells, in vivo, resulting in elicitation of strong immune responses to the encoded antigens [41].

We have reported a platform that can make complex with DNA and escort it onto APC in host. The platform uses a ligand that binds MHC class II-expressing cells that include predominantly professional APC (Fig. 47.1). We will next discuss this platform [37].

## 47.9 Nanoparticle Platform: PADRE-Derivatized Dendrimer (PDD)

DNA is a negative polyelectrolyte, and, thus, the transport of genetic material into cells is hampered by the high negative charge of DNA segments, which prevents its passage across the hydrophobic barriers constituted by cell and nuclear membranes. A general approach to overcome this problem is the use of polycationic agents to complex the DNA and gives rise to condensed nanoparticles with an overall charge close to zero, which can cross membranes more easily [43, 44]. Cationic chitosan and dendrimers, for example, have been the focus of numerous vaccine studies [45, 46]. Dendrimers are available in a variety of sizes (usually up to sixth generation) and with surfaces terminated in alcohol, carboxylic acid, or amine groups.

Polyamidoamine (PAMAM) dendrimers have been extensively used as vehicles for gene delivery, since their surface amine groups are readily protonated at physiological pH, leading to large positive charges on their surfaces, which drive the condensation with DNA plasmids containing the required oligonucleotide sequences. PAMAM

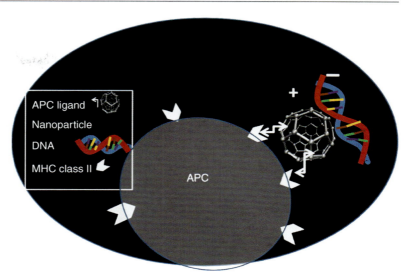

**Fig. 47.1** APC targeting Nanoparticle Peptide-Dervatized-Dendrimer (PDD) with immunopotentiating effects

dendrimers, which were developed by Tomalia, have been extensively investigated for DNA transfection [37, 47]. It is known that extensive synthetic elaboration and derivatization of the amine terminal groups with hydrophobic amino acids decrease the dendrimer transfection efficiency [48], which results from the pronounced loss of surface positive charges and its weakening effect on the electrostatic interactions with DNA [37, 43, 44, 47]. PAMAM dendrimers are highly branched macromolecules spanning from a central core and containing a series of layers, structurally and synthetically distinct. Each layer added to the structure of a dendrimer is referred to as a "generation," in a clear reference to the stepwise growth of the macromolecule. Generation-5 (G5) PAMAM dendrimers, in particular, are ideal for the delivery of DNA (negatively charged) into the cells (also negatively charged) as demonstrated in Fig. 47.2.

## 47.10 APC Recognition Moiety in PDD

The major histocompatibility complex (MHC) genomic region is present in all APC. MHC is an essential component of the immune system that plays important roles in immune responses to pathogens, tumor antigens, as well as in autoimmunity. The proteins encoded by the MHC genes are expressed on the surface of cells that present both self-antigens, from the cell itself, and non-self-antigens, fragments from pathogens or tumor cells, to various T cells enabling them to (i) provide help for initiation immune responses, (ii) help T cells to kill invading pathogens/tumor cells/cells infected with pathogens, and (iii) coordinate the maturation of antibodies against non-self-antigens. Class II MHC molecules are expressed largely on dendritic cells, B lymphocytes, monocytes, and macrophages. Class II MHC molecules are recognized by helper T lymphocytes and induce proliferation of helper T lymphocytes and amplification of the immune response to the particular immunogenic peptide that is displayed.

Class I MHC molecules are expressed on almost all nucleated cells and are recognized by cytotoxic T lymphocytes (CTLs), which then destroy the antigen-bearing cells. CTLs are particularly important in tumor rejection and in fighting viral infections. A Pan D-binding peptide, PADRE, is capable of binding selected MHC molecules. For instance, PADRE was shown to have a high affinity to the six selected DRB1 subtypes [49] and to antigen-presenting cells of PBMC [50]. In fact, PADRE binds to MHC class II of human and mouse [51]. Thus far, PADRE has been employed for its ability to bind

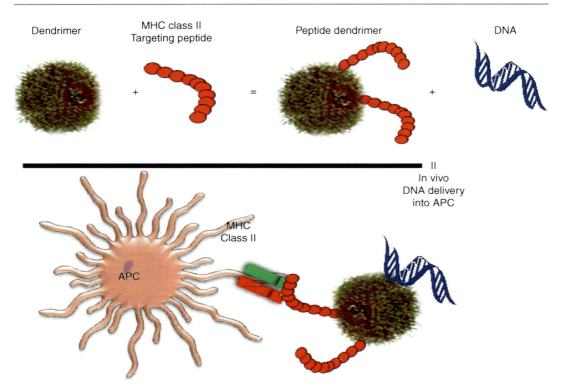

**Fig. 47.2** Peptide-dendrimer platform for in vivo gene delivery: MHC class II-targeting peptides are covalently linked to the dendrimer (PDD). PDD complexed with DNA efficiently targets the vaccines into professional MHC class II + APCs in vivo

APCs and activate CD4 helper T cells and for the first time to target APCs (via its binding to their surface MHC) [37].

Others and we have worked on universal T helper epitopes, for example, a computer-generated Pan DR T helper epitope called PADRE, as an adjuvant for vaccines [52–60]. PADRE, a nonnatural epitope, binds to majority of MHC class II molecules present on antigen-presenting cells such as B cells, T cells, monocytes, macrophages, dendritic cells, and Langerhans cells. We postulated that in order for such universal epitopes to act as a CD4 helper, they first have to bind MHC class or HLA-DR of most alleles, and this capability may be used as a cell-targeting element in a vehicle. Since MHC class II is mainly expressed on professional APC, they must bind with APC. We then conceived that by coupling a universal T helper epitope to a nanovehicle, we may be able to make an APC-targeting nanocarrier.

We selected polyamidoamine (PAMAM) dendrimers for they have been extensively studies with terms of their synthesis, chemical characterizations including charge, molecular weight, number of amines, methods of surface fictionalization, toxicity, and biocompatibility. Therefore, we developed APC-targeting nanocarriers that consist of a generation-5 dendrimer coupled with Pan DR T helper epitopes or PADRE (PADRE dendrimer hereafter referred to as PDD), a T helper epitope of influenza virus hemagglutinin (HA) molecule (PDD2), or one that uses a random control peptide.

Feasibility of translation of a vaccine to human clinical studies and logistics of delivering vaccines globally creates some limitations and challenges for some of the scientifically acceptable platforms. For example, generation of humanized monoclonal antibodies coupled to a nanoparticle that carries a DNA as well as an adjuvant may have a difficult regulatory path, an expensive

**Table 47.1** What a DNA-carrying APC-targeting vaccine platform offers

| Limitations of conventional vaccines | Targeted APC nanocarrier |
|---|---|
| Provide humoral response, but induce poor T-cell responses  Reason: insufficient amount of vaccine at APC | APC targeting induces humoral response and strong T-cell response |
| Poor immunogenicity of many antigens  Reason: weak T-cell response | Activates cell-mediated response, including T helper cells |
| Poor uptake of vaccines by APCs  Reason: protein antigens, DNA, and RNA fail to cross cell membranes (i.e., both are negative charge, no transport mechanisms) | PDD induces uptake via endocytosis |
| Immuno-avoidance mechanisms of pathogens and cancer antigens  Reason: absence of strong T-cell response and off-targeting suppressor mechanisms | Activates cell-mediated response, including T helper cells |
| High cost  Reason: proteins are costly to make significantly less stable than DNA | Can transport DNA vaccines, DNA is stable, with low cost of goods |

scale-up process, and some safety concerns. Furthermore, of the major factors that dictate the success or failure of a nanoparticle product are the Chemistry, Manufacturing, and Controls (CMC) and safety including immunogenicity of the platform. Hence, we have selected core polymer and cell recognition moieties that should have relatively simple CMC and acceptable safety parameters.

## 47.11 PDD is an APC Opsonized Platforms that Is Immunoenhancing In Vivo

As a feasible approach we have attempted to tailor a nanocarrier that has cell-targeting moiety that in addition to targeting APC, it also has intrinsic adjuvant activity. DNA may become adjuvanted and protected from degradation when incorporated in nanoparticles that may have additional advantages [61–67]. Dendrimers show an RNA/DNA-protecting ability as well as increasing the transfection efficiency [68, 69]; however, in vivo they demonstrated only modest success. However, an optimized PADRE-dendrimer nanocarrier, PDD platform, should show active and selective delivery of DNA to APCs in vivo. In addition the intrinsic immune-enhancing characteristics of PDD synergize with the vaccine effects (Table 47.1).

This platform is based on the conjugation of generation 5 poly(amidoamine) (G5-PAMAM) dendrimers, (a DNA-loading surface), with MHC class II-targeting peptides that serve to selectively deliver the dendrimers to APCs while enhancing their immune stimulatory potency. In vitro, DNA conjugated with this platform efficiently transfect murine and human APC. When DNA-peptide-dendrimer complexes are administered subcutaneously in rodents, they preferentially transfect dendritic cells (DC) in the draining lymph nodes, promotes the generation of high affinity T cells, and results in the rejection of established tumors

## 47.12 PPD-Mediated TRP2 Vaccination Leads to the Rejection of Established Melanoma

If targeting DNA vaccine to MHC class II + APC is important, it should show advantage in treating mice bearing tumors, and treatment in a prophylactic setting is not really convincing. B16 melanoma therapeutic mouse setting where tumor is established is known as an aggressive model since tolerogenic mechanisms are elicited, and the delivering the antigen to the professional APC might become the key for successful therapy [70]. PPD nanoparticles were therefore tested in the stringent therapeutic murine tumor melanoma model B16 (Fig. 47.3). The B16 tumor-associated antigen TRP2 is a weak antigen, and

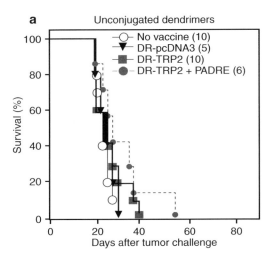 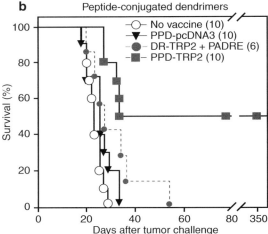

**Fig. 47.3** PPD promotes the rejection of established B16 melanoma tumors. C57Bl/6-bearing mice were injected received as vaccine cDNA3-TRP2 (*gray square*) or pcDNA3 (as control, *black triangle*) subcutaneously distributed in four anatomical sites (10 μg/site), on days 2 and 9 post-tumor implantation. Immunization was performed using either unmodified DR (**a**) or with PPD (**b**). As additional controls, mice received PBS (*white circle*) or unmodified dendrimer loaded with pcDNA3-TRP2 and mixed with the same dose of PADRE peptide theoretically linked to the PPD (DR-TRP2+PADRE; *gray diamond*). Log-Rank $p<0.001$, Holm-Sidak: PPD-TRP2 vs PPD-pcDNA3, $p=0.001$; PPD-TRP2 vs DR-TRP2+PADRE, $p=0.03$; DRTRP2+ PADRE vs no vaccine, $p=0.04$ (This figure was adopted with permission)

only modest results are obtained when TRP2 is used to treat established melanomas [71] unless the antigen vaccination is performed using powerful recombinant vaccinia virus [72] or combinatorial treatment [71, 73]. Coherent with previous results, immunization against TRP2 performed 2 and 9 days after tumor injection using peptide-unconjugated, control dendrimers failed to induce protection (Fig. 47.3a).

However, when the same complexes were mixed with PADRE, a modest statistically significant ($p=0.045$) increment of survival was observed compared to the empty vector-vaccinated mice (Fig. 47.2). Since in this case PADRE was not conjugated but only mixed with the dendrimer, this effect can be attributed to the adjuvant effect of PADRE. When vaccination against TRP2 was performed with PPD (Fig. 47.3b, PPD-TRP2 group), a statistical and clinically significant prolongation of survival was observed, with 50 % of mice remaining tumor-free (PPD-TRP2 vs PPD-pcDNA3: $p<0.005$) for 350 days, the duration of the experiment. These data indicate that PADRE needs to be linked to the dendrimer, as it was optimized with an average of two to three peptides per dendrimer, in order to promote a strong therapeutic effect. These data suggest that the specific targeting to the professional APCs and the adjuvant effect provided by the universal peptide are likely to be both needed for a successful vaccination.

### Conclusion

Fortunately, in the recent years, many novel vaccine platforms have been developed and described. Those that can target lymphoid organs, result in recruitment of more APC/DCs in injection site, can induce and enhance DC migration to draining lymph nodes, or can carry co-stimulating molecules or cytokines. PDD, a DNA-carrying nanocarrier that targets APC, was described that uses a cell recognition element that has intrinsic immunopotentiating capability [17]. Table 47.1 summarizes the advantages of such vaccines to traditional vaccines. Many tumor or pathogen-associated antigens have poor immunogenicity in nature, which is part of their immune-avoidance strategy. A vaccine that targets professional APC while enhancing T helper cells should correct this shortcoming of vaccines. Such technologies may be further improved so they can be formulated in a single dose and even orally.

Furthermore, nanocarrier vaccine platforms may also be used in combination of other vaccine technologies including protein-based vaccines or DNA vaccines delivered by in vivo electroporation [74]. PADRE-PAMAM dendrimer (PDD) nanovaccine delivery may fulfill some of the major challenges facing robust protective vaccines: it provides high transfection efficiency, proper targeting to the APCs, and an adjuvant effect. The platform is based on the synthesis, in a unique nanoparticle, of two technologies already independently tested with modest results (Table 47.1). The universal peptide PADRE binds/activates cells bearing any of the 15 major forms of the HLA-DR and has been widely used as immune adjuvant [75–77].

Its adjuvant (although modest) effect has been confirmed in this report. In striking contrast, the physical conjugation of the PAMAM dendrimer to the MHC class II-targeting peptide dramatically increases the immunogenicity of the platform and the efficacy of DNA vaccines, overcoming the limitations of the current platform for DNA vaccination. Considering that vaccine doses are dramatically lower than those used in drugs or antibiotics and since targeted nanovaccines results in further lowered vaccine doses, it is expected that many of nanoparticle-based vaccine delivery platforms reach into human as good candidate immunization in human diseases, in particular, the DNA-based nanovaccines.

## References

1. Payne, H., Bahl, A., Mason, M., Troup, J., De Bono, J.: Optimizing the care of patients with advanced prostate cancer in the UK: current challenges and future opportunities. BJU Int. **110**, 658–667 (2012). doi:10.1111/j.1464-410X.2011.10886.x
2. Lesterhuis, W.J., et al.: Route of administration modulates the induction of dendritic cell vaccine-induced antigen-specific T cells in advanced melanoma patients. Clin. Cancer Res. **17**, 5725–5735 (2011). doi:10.1158/1078-0432.CCR-11-1261. 1078-0432.CCR-11-1261 [pii]
3. Neefjes, J., Jongsma, M.L., Paul, P., Bakke, O.: Towards a systems understanding of MHC class I and MHC class II antigen presentation. Nat. Rev. Immunol. **11**, 823–836 (2011). doi:10.1038/nri3084. nri3084 [pii]
4. Witmer-Pack, M.D., Swiggard, W.J., Mirza, A., Inaba, K., Steinman, R.M.: Tissue distribution of the DEC-205 protein that is detected by the monoclonal antibody NLDC-145. II. Expression in situ in lymphoid and nonlymphoid tissues. Cell. Immunol. **163**, 157–162 (1995). doi:10.1006/cimm.1995.1110. S0008-8749(85)71110-0 [pii]
5. Bonifaz, L., et al.: Efficient targeting of protein antigen to the dendritic cell receptor DEC-205 in the steady state leads to antigen presentation on major histocompatibility complex class I products and peripheral CD8+ T cell tolerance. J. Exp. Med. **196**, 1627–1638 (2002)
6. Mahnke, K., et al.: Targeting of antigens to activated dendritic cells in vivo cures metastatic melanoma in mice. Cancer Res. **65**, 7007–7012 (2005). doi:10.1158/0008-5472.CAN-05-0938. 65/15/7007 [pii]
7. Irache, J.M., Salman, H.H., Gamazo, C., Espuelas, S.: Mannose-targeted systems for the delivery of therapeutics. Expert Opin. Drug Deliv. **5**, 703–724 (2008). doi:10.1517/17425247.5.6.703
8. Tacken, P.J., de Vries, I.J., Torensma, R., Figdor, C.G.: Dendritic-cell immunotherapy: from ex vivo loading to in vivo targeting. Nat. Rev. Immunol. **7**, 790–802 (2007). doi:10.1038/nri2173. nri2173 [pii]
9. Apostolopoulos, V., et al.: Pilot phase III immunotherapy study in early-stage breast cancer patients using oxidized mannan-MUC1 [ISRCTN71711835]. Breast Cancer Res. **8**, R27 (2006). doi:10.1186/bcr1505. bcr1505 [pii]
10. Widera, G., et al.: Increased DNA vaccine delivery and immunogenicity by electroporation in vivo. J. Immunol. **164**, 4635–4640 (2000). ji_v164n9p4635 [pii]
11. Cappelletti, M., et al.: Gene electro-transfer improves transduction by modifying the fate of intramuscular DNA. J. Gene Med. **5**, 324–332 (2003). doi:10.1002/jgm.352
12. Mennuni, C., et al.: Efficient induction of T-cell responses to carcinoembryonic antigen by a heterologous prime-boost regimen using DNA and adenovirus vectors carrying a codon usage optimized cDNA. Int. J. Cancer **117**, 444–455 (2005). doi:10.1002/ijc.21188
13. Rice, J., et al.: DNA fusion gene vaccination mobilizes effective anti-leukemic cytotoxic T lymphocytes from a tolerized repertoire. Eur. J. Immunol. **38**, 2118–2130 (2008). doi:10.1002/eji.200838213
14. Daftarian, P., et al.: In vivo electroporation and non-protein based screening assays to identify antibodies

against native protein conformations. Hybridoma (Larchmt) **30**, 409–418 (2011). doi:10.1089/hyb.2010.0120
15. Ledgerwood, J.E., et al.: DNA priming and influenza vaccine immunogenicity: two phase 1 open label randomised clinical trials. Lancet Infect. Dis. **11**, 916–924 (2011). doi:10.1016/S1473-3099(11)70240-7. S1473-3099(11)70240-7 [pii]
16. Rice, J., Ottensmeier, C.H., Stevenson, F.K.: DNA vaccines: precision tools for activating effective immunity against cancer. Nat. Rev. Cancer **8**, 108–120 (2008). doi:10.1038/nrc2326. nrc2326 [pii]
17. Liu, M.A.: Immunologic basis of vaccine vectors. Immunity **33**(4), 504–515 (2010). doi:10.1016/j.immuni.2010.10.004. S1074-7613(10)00364-X [pii]
18. Leroux-Roels, G.: Unmet needs in modern vaccinology: adjuvants to improve the immune response. Vaccine **28**(Suppl 3), C25–C36 (2010). doi:10.1016/j.vaccine.2010.07.021. S0264-410X(10)01004-2 [pii]
19. Daftarian, P., et al.: Peptide-conjugated PAMAM dendrimer as a universal platform for antigen presenting cell targeting and effective DNA-based vaccinations. Cancer Res. (2011). doi:10.1158/0008-5472.CAN-11-1766. 0008–5472.CAN-11-1766 [pii]
20. Ahlen, G., et al.: In vivo electroporation enhances the immunogenicity of hepatitis C virus nonstructural 3/4A DNA by increased local DNA uptake, protein expression, inflammation, and infiltration of CD3+ T cells. J. Immunol. **179**, 4741–4753 (2007). 179/7/4741 [pii]
21. Steinman, R.M., Mellman, I.: Immunotherapy: bewitched, bothered, and bewildered no more. Science **305**, 197–200 (2004). 10.1126/science.1099688 305/5681/197 [pii]
22. Nakajima, S., et al.: Langerhans cells are critical in epicutaneous sensitization with protein antigen via thymic stromal lymphopoietin receptor signaling. J. Allergy Clin. Immunol. **129**, 1048–1055 e1046 (2012). doi:10.1016/j.jaci.2012.01.063. S0091-6749(12)00191-1 [pii]
23. Egawa, G., Kabashima, K.: Skin as a peripheral lymphoid organ: revisiting the concept of skin-associated lymphoid tissues. J. Invest. Dermatol. **131**, 2178–2185 (2011). doi:10.1038/jid.2011.198. jid2011198 [pii]
24. Stoitzner, P., et al.: Tumor immunotherapy by epicutaneous immunization requires langerhans cells. J. Immunol. **180**, 1991–1998 (2008). 180/3/1991 [pii]
25. Kis, E.E., Winter, G., Myschik, J.: Devices for intradermal vaccination. Vaccine **30**, 523–538 (2012). doi:10.1016/j. S0264-410X(11)01787-7 [pii]
26. Ali, O.A., Emerich, D., Dranoff, G., Mooney, D.J.: In situ regulation of DC subsets and T cells mediates tumor regression in mice. Sci. Transl. Med. **1**, 8ra19 (2009). doi:10.1126/scitranslmed.3000359. 1/8/8ra19 [pii]
27. Karumuthil-Melethil, S., et al.: Dendritic cell-directed CTLA-4 engagement during pancreatic beta cell antigen presentation delays type 1 diabetes. J. Immunol. **184**, 6695–6708 (2010). doi:10.4049/jimmunol.0903130. jimmunol.0903130 [pii]
28. Kwon, Y.J., Standley, S.M., Goodwin, A.P., Gillies, E.R., Frechet, J.M.: Directed antigen presentation using polymeric microparticulate carriers degradable at lysosomal pH for controlled immune responses. Mol. Pharm. **2**, 83–91 (2005). doi:10.1021/mp0498953
29. Malcherek, G., et al.: MHC class II-associated invariant chain peptide replacement by T cell epitopes: engineered invariant chain as a vehicle for directed and enhanced MHC class II antigen processing and presentation. Eur. J. Immunol. **28**, 1524–1533 (1998)
30. Matsuo, H., et al.: Engineered hepatitis B virus surface antigen L protein particles for in vivo active targeting of splenic dendritic cells. Int. J. Nanomedicine **7**, 3341–3350 (2012). doi:10.2147/IJN.S32813. ijn-7-3341 [pii]
31. Caminschi, I., Maraskovsky, E., Heath, W.R.: Targeting dendritic cells in vivo for cancer therapy. Front. Immunol. **3**, 13 (2012). doi:10.3389/fimmu.2012.00013
32. Kwon, Y.J., James, E., Shastri, N., Frechet, J.M.: In vivo targeting of dendritic cells for activation of cellular immunity using vaccine carriers based on pH-responsive microparticles. Proc. Natl. Acad. Sci. U.S.A. **102**, 18264–18268 (2005). doi:10.1073/pnas.0509541102. 0509541102 [pii]
33. Faraasen, S., et al.: Ligand-specific targeting of microspheres to phagocytes by surface modification with poly(L-lysine)-grafted poly(ethylene glycol) conjugate. Pharm. Res. **20**, 237–246 (2003)
34. Reddy, S.T., Rehor, A., Schmoekel, H.G., Hubbell, J.A., Swartz, M.A.: In vivo targeting of dendritic cells in lymph nodes with poly(propylene sulfide) nanoparticles. J. Control. Release **112**, 26–34 (2006)
35. Bonifaz, L.C., et al.: In vivo targeting of antigens to maturing dendritic cells via the DEC-205 receptor improves T cell vaccination. J. Exp. Med. **199**, 815–824 (2004)
36. Crampton, H.L., Simanek, E.E.: Dendrimers as drug delivery vehicles: non-covalent interactions of bioactive compounds with dendrimers. Polym. Int. **56**, 489–496 (2007). doi:10.1002/pi.2230
37. Daftarian, P., et al.: Peptide-conjugated PAMAM dendrimer as a universal DNA vaccine platform to target antigen-presenting cells. Cancer Res. **71**, 7452–7462 (2011). doi:10.1158/0008-5472.CAN-11-1766. 0008–5472.CAN-11-1766 [pii]
38. Cruz, L.J., et al.: Multimodal imaging of nanovaccine carriers targeted to human dendritic cells. Mol. Pharm. **8**, 520–531 (2011). doi:10.1021/mp100356k
39. Cho, N.H., et al.: A multifunctional core-shell nanoparticle for dendritic cell-based cancer immunotherapy. Nat. Nanotechnol. **6**, 675–682 (2011)
40. Vyas, S.P., Goyal, A.K., Khatri, K.: Mannosylated liposomes for targeted vaccines delivery. Methods Mol. Biol. **605**, 177–188 (2010). doi:10.1007/978-1-60327-360-2_12
41. Geng, S., et al.: Amiloride enhances antigen specific CTL by facilitating HBV DNA vaccine entry into cells. PLoS One **7**, e33015 (2012). doi:10.1371/journal.pone.0033015. PONE-D-11-20345 [pii]
42. Watson, D.S., Endsley, A.N., Huang, L.: Design considerations for liposomal vaccines: influence of

formulation parameters on antibody and cell-mediated immune responses to liposome associated antigens. Vaccine 30, 2256–2272 (2012)
43. Kesharwani, P., Gajbhiye, V., Tekade, R.K., Jain, N.K.: Evaluation of dendrimer safety and efficacy through cell line studies. Curr. Drug Targets 12, 1478–1497 (2011). BSP/CDT/E-Pub/00269 [pii]
44. Tekade, R.K., Dutta, T., Gajbhiye, V., Jain, N.K.: Exploring dendrimer towards dual drug delivery: pH responsive simultaneous drug-release kinetics. J. Microencapsul. 26, 287–296 (2009). doi:10.1080/02652040802312572. 902427985 [pii]
45. van der Lubben, I.M., et al.: Chitosan microparticles for mucosal vaccination against diphtheria: oral and nasal efficacy studies in mice. Vaccine 21, 1400–1408 (2003). S0264410X02006862 [pii]
46. van der Lubben, I.M., Verhoef, J.C., Borchard, G., Junginger, H.E.: Chitosan for mucosal vaccination. Adv. Drug Deliv. Rev. 52, 139–144 (2001). S0169-409X(01)00197-1 [pii]
47. Wang, W., Kaifer, A.E.: Electrochemical switching and size selection in cucurbit[8]uril-mediated dendrimer self-assembly. Angew. Chem. Int. Ed. Engl. 45, 7042–7046 (2006). doi:10.1002/anie.200602220
48. Kojima, C., Regino, C., Umeda, Y., Kobayashi, H., Kono, K.: Influence of dendrimer generation and polyethylene glycol length on the biodistribution of PEGylated dendrimers. Int. J. Pharm. 383, 293–296 (2010). doi:10.1016/j.ijpharm.2009.09.015. S0378-5173(09)00628-0 [pii]
49. Alexander, J., et al.: Development of high potency universal DR-restricted helper epitopes by modification of high affinity DR-blocking peptides. Immunity 1, 751–761 (1994)
50. Neumann, F., et al.: Identification of an antigenic peptide derived from the cancer-testis antigen NY-ESO-1 binding to a broad range of HLA-DR subtypes. Cancer Immunol. Immunother. 53, 589–599 (2004). doi:10.1007/s00262-003-0492-6
51. Kim, D., et al.: Role of IL-2 secreted by PADRE-specific CD4+ T cells in enhancing E7-specific CD8+ T-cell immune responses. Gene Ther. 15, 677–687 (2008)
52. Belot, F., Guerreiro, C., Baleux, F., Mulard, L.A.: Synthesis of two linear PADRE conjugates bearing a deca- or pentadecasaccharide B epitope as potential synthetic vaccines against Shigella flexneri serotype 2a infection. Chemistry 11, 1625–1635 (2005)
53. Alexander, J., et al.: Linear PADRE T helper epitope and carbohydrate B cell epitope conjugates induce specific high titer IgG antibody responses. J. Immunol. 164, 1625–1633 (2000)
54. Decroix, N., Pamonsinlapatham, P., Quan, C.P., Bouvet, J.-P.: Impairment by mucosal adjuvants and cross-reactivity with variant peptides of the mucosal immunity induced by injection of the fusion peptide PADRE-ELDKWA. Clin. Diagn. Lab. Immunol. 10, 1103–1108 (2003)
55. Wei, J., et al.: Dendritic cells expressing a combined PADRE/MUC4-derived polyepitope DNA vaccine induce multiple cytotoxic T-cell responses. Cancer Biother. Radiopharm. 23, 121–128 (2008)
56. Daftarian, P., et al.: Immunization with Th-CTL fusion peptide and cytosine-phosphate-guanine DNA in transgenic HLA-A2 mice induces recognition of HIV-infected T cells and clears vaccinia virus challenge. J. Immunol. 171, 4028–4039 (2003)
57. Daftarian, P., et al.: Eradication of established HPV 16-expressing tumors by a single administration of a vaccine composed of a liposome-encapsulated CTL-T helper fusion peptide in a water-in-oil emulsion. Vaccine 24, 5235–5244 (2006)
58. Daftarian, P., et al.: Novel conjugates of epitope fusion peptides with CpG-ODN display enhanced immunogenicity and HIV recognition. Vaccine 23, 3453–3468 (2005). doi:10.1016/j.vaccine.2005.01.093. S0264-410X(05)00134-9 [pii]
59. Daftarian, P., et al.: Two distinct pathways of immunomodulation improve potency of p53 immunization in rejecting established tumors. Cancer Res. 64, 5407–5414 (2004). doi:10.1158/0008-5472.CAN-04-0169. 64/15/5407 [pii]
60. Daftarian, P.M., et al.: Rejection of large HPV-16 expressing tumors in aged mice by a single immunization of VacciMax encapsulated CTL/T helper peptides. J. Transl. Med. 5, 26 (2007). doi:10.1186/1479-5876-5-26. 1479-5876-5-26 [pii]
61. Myhr, A.I., Myskja, B.K.: Precaution or integrated responsibility approach to nanovaccines in fish farming? A critical appraisal of the UNESCO precautionary principle. Nanoethics 5, 73–86 (2011). 10.1007/s11569-011-0112-4 112 [pii]
62. Clemente-Casares, X., Tsai, S., Yang, Y., Santamaria, P.: Peptide-MHC-based nanovaccines for the treatment of autoimmunity: a "one size fits all" approach? J. Mol. Med. (Berl) 89, 733–742 (2011). doi:10.1007/s00109-011-0757-z
63. Skwarczynski, M., Toth, I.: Peptide-based subunit nanovaccines. Curr. Drug Deliv. 8, 282–289 (2011). BSP/CDD/E-Pub/00085 [pii]
64. Yang, J., et al.: Preparation and antitumor effects of nanovaccines with MAGE-3 peptides in transplanted gastric cancer in mice. Chin. J. Cancer 29, 359–364 (2010). 1000-467X201004359 [pii]
65. Nandedkar, T.D.: Nanovaccines: recent developments in vaccination. J. Biosci. 34, 995–1003 (2009)
66. Salvador-Morales, C., Zhang, L., Langer, R., Farokhzad, O.C.: Immunocompatibility properties of lipid-polymer hybrid nanoparticles with heterogeneous surface functional groups. Biomaterials 30, 2231–2240 (2009)
67. Danesh-Bahreini, M.A., et al.: Nanovaccine for leishmaniasis: preparation of chitosan nanoparticles containing Leishmania superoxide dismutase and evaluation of its immunogenicity in BALB/c mice. Int. J. Nanomedicine 6, 835–842 (2011). doi:10.2147/IJN.S16805. ijn-6-835 [pii]
68. Pietersz, G.A., Tang, C.K., Apostolopoulos, V.: Structure and design of polycationic carriers for gene delivery. Mini Rev. Med. Chem. 6, 1285–1298 (2006)

69. Tekade, R.K., Kumar, P.V., Jain, N.K.: Dendrimers in oncology: an expanding horizon. Chem. Rev. **109**, 49–87 (2009)
70. Hung, C.F., Yang, M., Wu, T.C.: Modifying professional antigen-presenting cells to enhance DNA vaccine potency. Methods Mol. Med. **127**, 199–220 (2006)
71. Jerome, V., Graser, A., Muller, R., Kontermann, R.E., Konur, A.: Cytotoxic T lymphocytes responding to low dose TRP2 antigen are induced against B16 melanoma by liposome-encapsulated TRP2 peptide and CpG DNA adjuvant. J. Immunother. **29**, 294–305 (2006)
72. Bronte, V., et al.: Genetic vaccination with "self" tyrosinase-related protein 2 causes melanoma eradication but not vitiligo. Cancer Res. **60**, 253–258 (2000)
73. Cohen, A.D., et al.: Agonist anti-GITR antibody enhances vaccine-induced CD8(+) T-cell responses and tumor immunity. Cancer Res. **66**, 4904–4912 (2006)
74. Frelin, L., et al.: Electroporation: a promising method for the nonviral delivery of DNA vaccines in humans? Drug News Perspect. **23**, 647–653 (2010)
75. Alexander, J., et al.: The optimization of helper T lymphocyte (HTL) function in vaccine development. Immunol. Res. **18**, 79–92 (1998). doi:10.1007/BF02788751
76. Wu, C.Y., Monie, A., Pang, X., Hung, C.F., Wu, T.C.: Improving therapeutic HPV peptide-based vaccine potency by enhancing CD4+ T help and dendritic cell activation. J. Biomed. Sci. **17**, 88 (2010)
77. Kim, D., et al.: Enhancement of CD4+ T-cell help reverses the doxorubicin-induced suppression of antigen-specific immune responses in vaccinated mice. Gene Ther. **15**, 1176–1183 (2008). gt200879 [pii]10.1038/gt.2008.79

# Lactic Acid Bacteria Vector Vaccines

## 48

### Maria Gomes-Solecki

## Contents

| | | |
|---|---|---|
| 48.1 | **Plague and the Pathogen** | 768 |
| 48.2 | **The Enzootic Cycle and Transmission** | 768 |
| 48.3 | **Diagnostic and Therapy** | 770 |
| 48.3.1 | Clinical Disease | 770 |
| 48.3.2 | Laboratory Diagnosis | 770 |
| 48.3.3 | Classic Therapy | 771 |
| 48.4 | **Vaccine** | 771 |
| 48.4.1 | Preclinical Development, Safety, and Efficacy | 775 |
| 48.4.2 | Clinical Development | 775 |
| 48.5 | **Strengths and Weaknesses** | 776 |
| **References** | | 776 |

M. Gomes-Solecki, DVM
Department of Microbiology,
Immunology and Biochemistry,
University of Tennessee Health Science Center,
Memphis, TN, USA
e-mail: mgomesso@uthsc.edu

### Abstract

Vaccines are currently being developed based on the new concept of designing antigens that can prompt the innate immune system to trigger adaptive immunity and to characterize the T cells that are needed for the desired response. To develop protective immune responses against mucosal pathogens, the delivery route and adjuvants for vaccination are important. The host, however, strives to maintain mucosal homeostasis by responding to mucosal antigens with tolerance. This induction of mucosal immunity through vaccination is a rather difficult task. However, potent mucosal adjuvants, vectors, and other special delivery systems can be used. There is a great need to develop effective mucosal delivery systems that avoid degradation and promote uptake of the antigen in the gastrointestinal tract and stimulate adaptive immune responses, rather than the tolerogenic immune responses seen in studies done with feeding soluble antigens.

Lactic acid bacteria have Generally Recognized As Safe (GRAS) status and have been developed in the past decade as potent adjuvants for mucosal delivery of vaccine antigens. Both *Lactococcus lactis* and *Lactobacillus spp.* have been used. In this chapter I will review the development of a platform technology based in *Lactobacillus plantarum* to deliver prophylactic molecules orally and will provide an example for plague.

## 48.1 Plague and the Pathogen

*Yersinia pestis*, an aerobic, nonmotile, Gram-negative bacillus belonging to the family *Enterobacteriaceae*, is transmitted to humans via flea bite or via aerosol droplet causing bubonic or pneumonic plague, respectively [1, 2].

Most human plague cases present as one of three primary forms—bubonic, septicemic, or pneumonic. Secondary plague septicemia, pneumonia, and meningitis are the most common complications. In the United States an average of seven cases have been reported each year. Case fatalities for untreated bubonic plague range from 40 to 60 %, while untreated septicemic and pneumonic forms of the disease are invariably fatal [3, 4]. In most instances, fatal cases involve patients who do not seek treatment soon enough after becoming sick or are incorrectly diagnosed when they do see a physician. Death usually results from an overwhelming septic shock [5, 6].

The *Y. pestis* genome has been completely sequenced. The genes found in *Y. pestis* are isosequential to alleles found in *Y. pseudotuberculosis* [7]. Thus, it appears that *Y. pestis* evolved 1,500–20,000 years ago from *Y. pseudotuberculosis* and that it has undergone large-scale genetic flux [8]. The genome is 4.63 megabases in size in addition to its three virulence plasmids (10, 70, and 100 Kb) [7, 9, 10]. The 10 kb plasmid, encoding plasminogen activator protease (Pla) [11], and the 100 kb plasmid, encoding F1 capsular antigen, are unique to *Y. pestis* [11–13]. However, all three pathogenic *Yersinia* share the 70 Kb plasmid which encodes genes for important virulence factors required for pathogenicity: the type III secretion system, LcrV, and a series of *Yersinia* outer surface proteins (Yops) [6, 9, 10, 14].

The pathogenicity of *Y. pestis* results from its impressive ability to overcome the defenses of the mammalian host and to overwhelm it with massive growth. The type III secretion system is a mechanism for intracellular targeting of virulence factors shared by several bacterial pathogens [15] which enables the secretion and injection of virulence effectors by a bacterium-host cell contact-dependent mechanism. These proteins have to promote cellular cytotoxicity, inhibit inflammatory cell chemotaxis, induce the production of immunomodulatory cytokines (e.g., IL-10), and prevent formation of protective granulomas [16]. In mice infected with *Y. pestis*, significant levels of IFN gamma and TNF alfa arise only just before death. In contrast, infection with avirulent strains of *Y. pestis* induces the prompt and marked synthesis of these cytokines [16].

## 48.2 The Enzootic Cycle and Transmission

Plague is enzootic in rodents in Africa, Asia, South America, and North America [17]. *Y. pestis* is transmitted from host to host by fleas via blood feeding, through consumption or handling of infectious host tissues, or through inhalation of infectious materials. *Y. pestis* infects an astonishingly broad range of mammals and uses rats, squirrels, mice, prairie dogs, marmots, or gerbils as reservoirs and several arthropod vectors for transmission [2, 4, 6, 18]. This zoonotic infection persists for long periods of time at low levels of prevalence in enzootic cycles that involve partially resistant rodents (enzootic or maintenance hosts) and cause little host mortality. These long periods are interrupted by occasional outbursts or epizootics (i.e., spreading die-offs) among these hosts or epidemics, when the incidence among humans increases [19]. See Fig. 48.1.

Humans acquire this zoonotic infection via an atypical bite from animal fleas, sometimes prompted by an animal's death from plague, after which the flea seeks a new source of blood. The incubation period from fleabite to symptomatic disease is 2–10 days [20]. Humans can be viewed as playing no role in the maintenance of plague in nature because rodent populations and their fleas suffice and because humans are poor transmitters of short-lived outbreaks of pneumonic plague. Most infected fleas come from the domestic black rat *Rattus rattus* or the brown sewer rat *Rattus norvegicus*. The most common and efficient flea vector is *Xenopsylla cheopis*, but many other flea species can transmit plague. The oriental rat flea *X. cheopis* is more susceptible than are other fleas to having the proventriculus of its digestive tract blocked by a blood meal containing *Y. pestis*. Blocked fleas are unable to clear their midguts of infected blood, leading them to

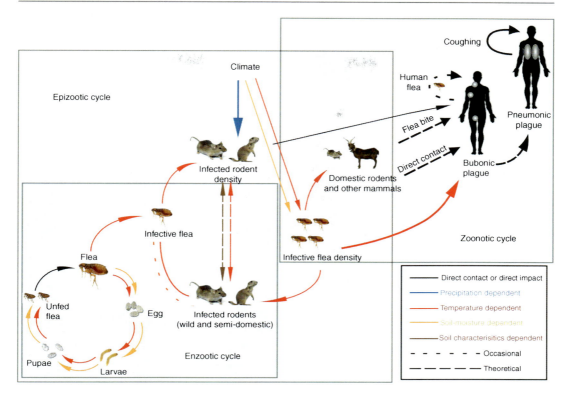

**Fig. 48.1** Schematic of the plague cycle with small mammals as hosts and fleas as vectors. *Arrows* represent connections affected by climate with a color coding depending on the most influential climate variable on this link (i.e., precipitation, temperatures, and other variables indirectly depending on them such as soil characteristics and soil moisture). *Grey rectangles* somewhat arbitrarily delimit epizootic, enzootic, and zoonotic cycles. Note that despite their location at the far end of the cycle, humans often provide the only available information on plague dynamics [19]

bite repeatedly and to regurgitate bacteria into the skin of its next host [17, 21].

*Y. pestis* cells spread from the site of the infected flea bite to the regional lymph nodes, grow to high numbers causing the formation of a bubo, and spill into the bloodstream where bacteria are removed in the liver and spleen. Growth continues in these organs, spreads to others, and causes septicemia. Fleas feeding on septicemic animals complete the infection cycle. In humans, bubonic plague can develop into an infection of the lung (secondary pneumonic plague) that can lead to aerosol transmission (primary pneumonic plague) [2, 6, 22]. In this case, the organism replicates in massive numbers in the lung tissue, and hemorrhagic lesions and immune cell infiltrates destroy the airways and precipitate a rapidly contagious and lethal course of infection followed by further aerosol dissemination of the pathogen [23]. Aerosol transmission can result from naturally occurring pneumonic plague as well as from biological warfare. Multiple antibiotic-resistant strains of *Y. pestis* occur naturally and they can be easily bioengineered [2]. Thus, plague is a Category A bioterrorism agent in need for novel strategies for its prevention.

People of all ages and both sexes are susceptible to the disease. Distributions by age and sex were not given by the World Health Organization [24], but most recent cases occur in children, with a slight preponderance among males [17]. Human plague in all countries where the disease is endemic shows seasonal variation. The peak season corresponds to the timing of epizootics with dying off of susceptible rodents. Seasonality can be correlated with the increase in fertility of rodent fleas, increase in rodent populations, and greater proximity of humans to infected animals [17].

## 48.3 Diagnostic and Therapy

### 48.3.1 Clinical Disease

Bubonic plague is the classic form of the disease. Patients usually develop symptoms of fever, headache, chills, and swollen, extremely tender lymph nodes (buboes) within 2–6 days of contact with the organism either by fleabite or by exposure of open wounds to infected materials. In addition, gastrointestinal complaints such as nausea, vomiting, and diarrhea are common [25]. Skin lesions infrequently develop at the initial site of an infection. Soreness in the affected lymph nodes will sometimes precede swelling [26], and any of the lymph node areas can be involved, depending upon the site of the initial infection. Buboes are typically found in the inguinal and femoral regions but also occur in other nodes [5]. Bacteremia or secondary plague septicemia is frequently seen in patients with bubonic plague [6, 27].

Primary septicemic plague is generally defined as occurring in a patient with positive blood cultures but no palpable lymphadenopathy. Clinically, plague septicemia resembles septicemias caused by other Gram-negative bacteria. Patients are febrile, and most have chills, headache, malaise, and gastrointestinal disturbances. There is some evidence that patients with septicemic plague have a higher incidence of abdominal pain than do bubonic plague patients [25]. The mortality rate for people with septicemic plague is fairly high probably because the antibiotics generally used to treat undifferentiated sepsis are not effective against *Y. pestis* [3, 6, 25, 26].

Primary pneumonic plague is a rare but deadly form of the disease that is spread via respiratory droplets through close contact (2–5 ft) with an infected individual. It progresses rapidly from a febrile flu-like illness to an overwhelming pneumonia with coughing and the production of bloody sputum. The incubation period for primary pneumonic plague is between 1 and 3 days. In general, patients who develop secondary plague pneumonia have a high fatality rate [6].

### 48.3.2 Laboratory Diagnosis

The laboratory diagnosis of plague is based on bacteriological and/or serological evidence [26]. Samples for analysis can include blood, bubo aspirates, sputum, cerebrospinal fluid in patients with plague meningitis, and scrapings from skin lesions, if present. Staining techniques such as the Gram, Giemsa, Wright, or Wayson stain can provide supportive but not presumptive or confirmatory evidence of a plague infection [26]. The mainstay of rapid, bedside diagnosis of bubonic plague is examination of the bubo aspirate under the microscope after Gram or Wayson staining [17]. A diagnosis of plague is confirmed by bacteriological culture to isolate the organism. *Y. pestis* grows on most routine laboratory culture media in 2 days. The colonies are opaque and smooth with irregular edges that have a "hammered metal" appearance when magnified [6, 26] Although not a rapid diagnostic technique, a serological response is often used retrospectively to confirm cases of plague. Alternative methods for diagnosing plague have been developed, including enzyme-linked immunosorbent assays for detection of F1 [17], antigen capture assays (i.e., anti-F1 dipstick assay), and classic and real-time PCR tests that use structural genes for F1 antigen, plasminogen activator, murine toxin, and 16S ribosomal RNA [17, 28] (Table 48.1).

**Table 48.1** Diagnostic methods for detection of *Yersinia pestis* in clinical specimens

| Method | Sensitivity, specificity and rapidity |
|---|---|
| Culture of bubo aspirate, blood or sputum specimen | Highly sensitive if patient is untreated, highly specific, takes 2–3 days for identification |
| Gram or wayson stain of bubo aspirate or sputum specimen | Moderately sensitive, moderately specific, rapid within minutes |
| Immunofluorescent antibody applied to bubo aspirate or sputum specimen | Moderately sensitive, highly specific, rapid within minutes |
| ELISA for F1 antigen in bubo aspirate | Highly sensitive, highly specific, rapid within hours |
| Dipstick for F1 antigen in bubo aspirate | Highly sensitive, highly specific, rapid within minutes |
| PCR for F1 gene in bubo aspirate | Moderately sensitive, highly specific, rapid within hours |

**Table 48.2** Antimicrobial agents for treatment of plague

| Drug | Clinical experience |
|---|---|
| Gentamicin | Monotherapy shown to be effective in the past decade in the United States and Tanzania after streptomycin theraphy was discontinued |
| Streptomycin | Drug of choice from 1948 to ~40 years later, when its use was discontinued in most countries; still used in Madagascar in combination with trimethoprim-sulfamethoxazole |
| Doxycycline or tetracycline | Effective alternative to gentamicin when oral theraphy is preferred; used as a prophylaxis for pneumonic disease |
| Trimethoprim-sulfamethoxazole | Used in combination with streptomycin in Madagascar; recommended as prophylaxis for pneumonic disease |
| Chloramphenicol | Effective but rarely used due to bone marrow toxic effects |
| Cephalosporins and other ß-lactams | Not recommended but effective in experimental animal infection |
| Ciprofloxacin and other fluoroquinolones | Not recommended but effective in experimental animal infection |

### 48.3.3 Classic Therapy

All patients suspected of having bubonic plague should be placed in isolation until 2 days after starting antibiotic treatment to prevent the potential spread of the disease should the patient develop secondary plague pneumonia [6]. The antibiotics and regimes used to treat *Y. pestis* infections and as prophylactic measures are listed in Table 48.2.

Streptomycin has been used to treat plague for over 45 years [29] and still remains the drug of choice. In countries where streptomycin is not available, gentamicin is an effective substitute [30]. Because streptomycin is bacteriolytic, it should be administered with care to prevent the development of endotoxic shock. Due to its toxicity, patients are not usually maintained on streptomycin for the full 10-day treatment regimen but are gradually switched to one of the other antibiotics, usually tetracycline. The tetracyclines are also commonly used for prophylactic therapy, while chloramphenicol is recommended for the treatment of plague meningitis [31]. While *Y. pestis* is susceptible to penicillin in vitro, this antibiotic is considered ineffective against human disease [3]. A randomized comparison of gentamicin and doxycycline in Tanzania, however, indicated that doxycycline was equally effective, without any nephrotoxicity [32]. Thus, doxycycline can be considered an alternative drug of choice. Only one patient was reported to have received successful treatment with ciprofloxacin [17, 33], but neither the fluoroquinolones nor the b-lactams have been subjected to testing in humans. Antibiotic-resistant strains are rare and are not increasing in frequency [6].

### 48.4 Vaccine

Current epidemiological records suggest 4,000 human plague cases annually worldwide [34]. However, due to possible aerosol transmission and fulminant virulence of pneumonic plague, *Y. pestis* has been categorized a Class A bioterrorism agent. Unlike anthrax, *Y. pestis* does not form spores and does not survive outside the body. For this reason, no one has succeeded in developing an effective bioweapon using aerosolized bacteria. In addition, the ability of pneumonic plague to propagate an epidemic is severely restricted by the requirement for close contact with a dying patient, usually on the last day of the patient's life [35]. Thus, the danger of using this organism as a bio weapon has been greatly exaggerated [17, 36]. During World War II Japanese forces released plague-infected fleas from aircraft over Chinese cities [2] although it is doubtful that these crude strategies of attack would be attempted nowadays. However, it is widely accepted that *Y. pestis* can be easily genetically manipulated to create strains with specific engineered traits, such as resistance to the antibiotics used to treat the disease. Thus, efforts are under way to develop new subunit vaccines that will protect against plague pneumonia. Pneumonic plague has an intracellular phase that clamors for a cell-mediated immune response to convey protection [37] which is a major challenge for the development of such vaccines.

The administration of appropriate medical molecules via mucosal routes offers several important advantages over systemic delivery such as reduction of secondary effects, easy administration, and the possibility to modulate both systemic and mucosal immune responses [38].

A major technological challenge is to develop delivery vectors that can survive transit through the gastrointestinal tract while shielding therapeutic or immunogenic molecules from low pH, bile, proteolytic enzymes, antimicrobial peptides, and intestinal peristalsis [39].

Lactic acid bacteria (LAB) are a group of Gram-positive, nonpathogenic, non-sporulating bacteria that include species of Lactobacillus, Lactococcus, Leuconostoc, Pediococcus, and Streptococcus. They have limited biosynthetic abilities and require preformed amino acids, B vitamins, purines, pyrimidines, and a sugar as a carbon and energy source. These nutritional requirements restrict their habitats to those in which the required compounds are abundant. Thus, these highly specialized bacteria occupy a range of niches including milk, plant surfaces, the oral cavity, the gastrointestinal tract, and the vagina of vertebrates [40].

LAB have been consumed for centuries by humans in fermented foods and have an extraordinary safety profile. These intrinsic advantages turn LAB into excellent delivery vectors of novel preventive and therapeutic molecules for humans. A number of studies of oral vaccines generated from genetically engineered pathogenic or commensal bacteria have been reported [41–47]. LAB represent an attractive alternative to the use of other mucosal delivery systems, such as liposomes, microparticles, and attenuated pathogens [40, 48].

Generating live, attenuated vaccines involves a process in which an infectious agent is altered so that it becomes harmless while retaining its ability to interact with the host and stimulate a protective immune response [49]. Live, attenuated pathogenic bacteria, such as derivatives of Mycobacterium, Salmonella, and Bordetella spp., are the most popular live delivery vectors used currently. They are particularly well adapted to interact with mucosal surfaces as they have specialized machinery to initiate the infection process. The major disadvantages of live vaccines include inadequate attenuation and the potential to revert to virulence. These bacteria can reacquire their pathogenic potential and, thus, are not entirely safe for human use, especially by children and by immunosuppressed individuals [48]. Lactic acid bacteria-based vaccines act as live, attenuated vaccines but without the safety concern. LAB have a Generally Recognized As Safe (GRAS) status and, thus, are not likely to cause harm.

Interest in the use of LAB as delivery vehicles stems from a large body of immunological research which shows that a delivery system is needed to avoid degradation and promote uptake of the antigen in the gastrointestinal tract and stimulate adaptive immune responses, rather than the tolerogenic immune responses that are seen in feeding studies with soluble antigens [40, 50, 51].

The production of a desired antigen by LAB can occur in three different cellular locations: (i) intracellular, which allows the protein to escape harsh external environmental conditions (such as gastric juices in the stomach) but requires cellular lysis for protein release and delivery; (ii) extracellular, which allows the release of the protein into the external medium, resulting in direct interaction with the environment (food product or the digestive tract); and (iii) cell wall anchored, which combines the advantages of the other two locations (i.e., interaction between the cell wall-anchored protein and the environment, in addition to protection from proteolytic degradation). In this context, several studies have compared the production of different antigens in LAB, using all three locations and evaluated the subsequent immunological impact [40, 48]. These studies demonstrated that the highest immune response was obtained with cell wall-anchored antigens exposed on the surface of LAB. Therefore, most of the recent LAB vaccination studies have selected surface exposure of the antigen of interest, rather than intra- or extracellular production [48].

Dendritic cells (DCs) play a central role in bridging the innate immune system with the adaptive immune system [52–55]. DCs are found throughout the body and are especially common at mucosal surfaces. With only a single layer of epithelial cells separating the external from the internal world amid the constant need for particle exchange, intestinal dendritic cells (DC) play a key role in maintaining intestinal homeostasis as well as governing protective immune responses against invading pathogens [56]. To avoid activation of self-reactive T cells and to limit unnecessary responses, such as those against commensal flora, DCs can imprint tolerance onto T cells [57]. See Fig. 48.2.

Immature type DCs are enriched underneath the epithelium of mucosal inductive sites and are poised to capture antigens. They extend protrusions between epithelial cells, enabling direct sampling of luminal antigens [58]. Through upregulation of MHC and costimulatory molecules, matured DCs convert into highly efficient antigen-presenting cells [56]. Successful antigen presentation to CD4+ T cells requires recognition of cognate peptide in the context of MHC class II molecules, whereas epitopes presented on MHC class I molecules stimulate Ag-specific CD8+ T cells [56]. When antigen uptake occurs, these DCs change their phenotype by expressing higher levels costimulatory molecules and move to T-cell areas of inductive sites for antigen presentation. Thus, DCs and their derived cytokines play key roles in the induction of antigen-specific effector Th cell responses. In this regard, targeting mucosal DCs is an effective strategy to induce mucosal and systemic immune responses [55, 56].

The ability of some LAB to persist in the gastrointestinal tract may be critically important in the effectiveness of LAB-based vaccines. A comparison of a persisting LAB strain, *L. plantarum*,

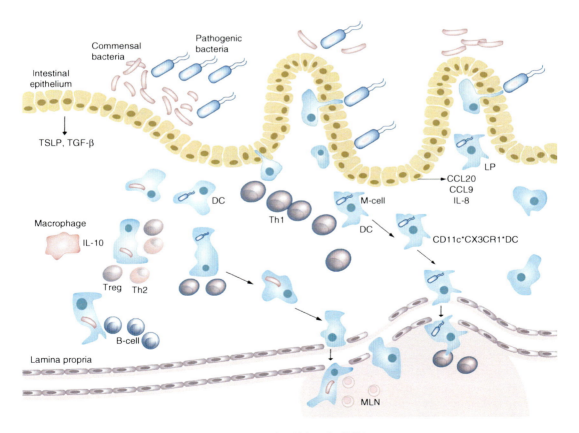

**Fig. 48.2** Sampling bacteria and their products by gut dendritic cells (DCs)

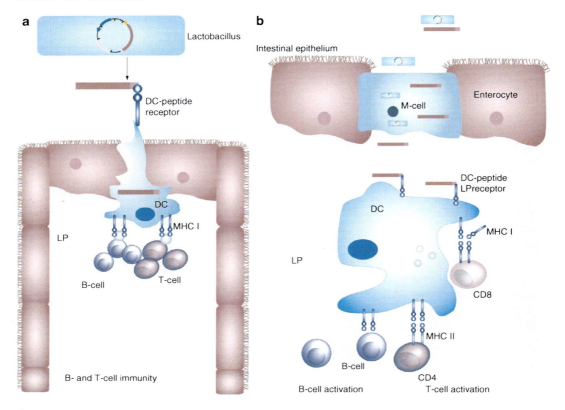

**Fig. 48.3** Delivery of immunogenic subunits to gut dendritic cells (DCs) by probiotic lactobacilli. (**a**) Direct sampling of luminal antigens by protruding DCs in the Lamina Propria (LP). (**b**) Activation of DCs following delivery of lactobacilli through M cells

with a nonpersisting LAB strain, *L. lactis*, identified *L. plantarum* to be more effective at eliciting antigen-specific immunity suggesting that persistence promoted immunogenicity [59]. Furthermore, it has been shown that particular Lactobacillus species induced critical inflammatory cytokines and induced activation and maturation of dendritic cells [60]. It has also been shown that immature DCs efficiently capture Lactobacillus species and these bacteria activated human DCs resulting in the production of proinflammatory cytokines like IL12 and increased proliferation of CD4+ and CD8+ cells and skewed the T-cell response toward a Th1 pathway believed to be involved in effective clearance of microbial pathogens [61, 62]. See Fig. 48.3. Thus, of the lactobacilli strains previously used for vaccine delivery, we chose to study *Lactobacillus plantarum* due to its ability to deliver the expressed antigen, to its ability to persist in the gut [59], and to its ability to act as a potent adjuvant [63].

Evidence suggests that the peptidoglycan layer of some LAB promotes natural immunoadjuvanticity [64, 65, 66] and antigen localization on the cell wall makes it more accessible to the immune system as compared to intracellular or secreted proteins [67]. Leader peptides mark proteins for translocation across the cytoplasmic membrane, and lipid modification is of major importance both for anchoring exported proteins to the membrane and for protein function [68]. It has been shown that lipidation at the first amino acid of the mature *Borrelia burgdorferi* OspA protein is essential to induce an immune response via TLR2 [69, 70]. We examined the influence of posttranslational processing in the localization of OspA lipoprotein on the cell envelope of *L. plantarum*. We discovered that the leader peptide of OspA targets the protein to the cell envelope of *Lactobacillus* and that the Cys [17] is recognized by the *L. plantarum* cell wall-sorting machinery that lipidates and anchors the protein to the cell envelope. The end result is a delivery system that exerts a potent adjuvant effect [71].

Our rationale was to design an oral delivery system based on a commensal, completely safe bacterium (*L. plantarum*) known to skew immune responses toward protective Th1. We used the leader sequence of *B. burgdorferi* OspA which targets the cloned recombinant protein to the cell wall of the bacteria generating a stealthy carrier in which the antigen is not only shielded from gastrointestinal insult but further induces immune responses via TLR2 thereby adding adjuvanticity to the system. Our goal was to generate a platform technology to engineer mucosal vaccines against multiple pathogens. Oral vaccines based on this system should only trip immune responses skewed toward Th1 after the recombinant *L. plantarum* is been up taken by dendritic cells thereby breaking the natural immunological tolerance of the gut.

Using the Lyme disease mouse model, we demonstrated that mice orally immunized with *Lactobacillus plantarum* expressing OspA developed a systemic IgG and mucosal IgA humoral immune response that protected them against tick challenge with *B. burgdorferi* [72]. In addition, human dendritic cells stimulated with this vaccine produced cytokines that polarize T cells to a Th1-type cellular response. Furthermore, a T84 human epithelial cell line did not produce the proinflammatory chemokine IL8 in response to vaccine stimulation, suggesting that this vaccine will not induce local inflammatory responses in the human gastrointestinal tract [71].

Expanding on the application of our technology, we developed an oral vaccine against plague based on LcrV-expressing *L. plantarum*. The low calcium response V (LcrV) is a secreted virulence factor, and parenteral immunization with recombinant protein protects mice from subcutaneous [73] and aerosol challenge with virulent *Y. pestis* [74]. Thus, LcrV is one of the proven vaccine candidates against *Yersinia pestis*. Mice immunized via oral gavage with LcrV-expressing lactobacilli developed systemic (serological) IgG and mucosal (gut, bronchoalveolar, and vaginal) IgA antibody responses specific to LcrV. We observed that human dendritic cells stimulated with lactobacilli expressing LcrV produced cytokines that polarize T cells to a potentially protective Th1 cellular immune response and that the T84 epithelial human cell line did not produce the proinflammatory chemokine IL8 in response to vaccine stimulation. These results replicate the immune response we observed for our Lyme disease vaccine, which was protective. It should be underlined that challenge studies were not performed with the Lactobacillus/LcrV vaccine and therefore the vaccine efficacy was not tested against *Y. pestis*. In addition to *L. plantarum* [75], the commensal, nonpathogenic bacterium *Lactococcus lactis* has been used to deliver LcrV with some success [2, 76, 77].

### 48.4.1 Preclinical Development, Safety, and Efficacy

Previously tested killed whole-cell preparations or live, attenuated plague vaccines are currently not favored in the United States because of safety and efficacy concerns, but live, rationally attenuated strains of *Y. pestis* have been shown in animal models to provide strong protection against both bubonic and pneumonic plague. Modern live *Y. pestis* vaccines should elicit humoral and cellular immune responses against a variety of relevant antigens, providing stronger protection against weaponized *Y. pestis* than vaccines based on only one or two antigens [2, 78, 79].

Recent efforts to create a safe and effective plague vaccine have focused on the development of recombinant subunit vaccines that elicit antibodies against two well-characterized *Y. pestis* antigens, the F1 capsule and the virulence protein LcrV [80, 81, 82]. While there has been some controversy surrounding the efficacy of subunit vaccines in some nonhuman primates [83], several candidates are currently moving toward licensure.

### 48.4.2 Clinical Development

Earlier vaccines to prevent flea-borne plague have been used for more than half a century for people in endemic areas, including 11 million US military personnel deployed to Vietnam in the 1960s and 1970s, but the formalin-killed whole-cell plague vaccine, which did not protect against pneumonic disease, was discontinued by its US manufacturer in 1998. The live, attenuated vaccine EV76 that lacks *pgm* genes has been used for a long time in Europe and other countries but is not commercially available [17].

Subunit vaccines based on rF1 and rV antigens are the most promising prospects and have passed through Phase I and II clinical trials and are well into the licensing process. Although direct determination of efficacy is not possible due to ethical considerations, human immune responses to subunit plague vaccine have shown good correlation with macaque and mouse immune responses [84]. The rV10 vaccine (truncated recombinant LcrV protein) is currently undergoing US Food and Drug Administration pre-Investigational New Drug authorization review for a future phase I trial [1]. Currently, the recombinant F1V (rF1V) being developed at DynPort Vaccine Company is in a Phase 2b clinical trial (www.clinicaltrials.gov) [2].

## 48.5 Strengths and Weaknesses

Currently, there are few mucosal vaccines in use, due to the poor efficiency of new strategies. A common theme that has emerged is that the natural immunity that develops following infection with mucosal transmitted diseases provides minimal protection against secondary challenges with a heterologous pathogen. Thus, a vaccine may need to elicit a different type of immune response altogether to achieve protection in the immunized hosts [85].

However, the ineffective protection afforded by traditional injected vaccines against a large number of pathogens (HIV, malaria, tuberculosis, etc.) has led to active research and development of alternative routes of immunization, such as the mucosal route. The development of safe and effective mucosal adjuvants remains a particular priority for non-replicating mucosal vaccines. These would offer the potential to induce vaccine-specific responses at the mucosal portals of pathogen entry. So, it is necessary to develop nontoxic and effective adjuvants which could permit the use of low immunogenic non-replicating antigens when administered by mucosal routes. Moreover, the use of an adjuvant competent for mucosal administration could allow targeting of mucosal immune cells and generate the desired immune response [85].

A comparison of the advantages and limitations of various methods of mucosal immunization makes it evident that there is no one superior method but there is a drawback that they share: the fact that there is a lack of control in terms of dose that is delivered (as opposed to systemic administered) to each individual is an important problem during the vaccines studies. The choice of any given route of mucosal vaccination and the selection of appropriate adjuvants and formulations will affect vaccine design, process, and manufacturing issues [85]. The major strengths of oral vaccination are the possibility for needle-free administration and the wide distribution in resource-poor countries where electricity for refrigeration of sophisticated drugs and vaccines is absent.

Much work remains to be done, but current research continues to clarify the concepts and provide the tools that are needed to exploit the full potential of mucosal vaccines. However, this research is done on animal models, and it is very important to keep in mind that often these models cannot be extrapolated to humans due to the very significant differences in anatomy, physiology, and immunogenicity [85].

## References

1. Quenee, L.E., Schneewind, O.: Plague vaccines and the molecular basis of immunity against Yersinia pestis. Hum. Vaccin. **5**, 817–823 (2009)
2. Sun, W., Roland, K.L., Curtiss 3rd, R.: Developing live vaccines against plague. J. Infect. Dev. Ctries. **5**, 614–627 (2011)
3. Crook, L.D., Tempest, B.: Plague. A clinical review of 27 cases. Arch. Intern. Med. **152**, 1253–1256 (1992)
4. Craven, R.B., Maupin, G.O., Beard, M.L., Quan, T.J., Barnes, A.M.: Reported cases of human plague infections in the United States, 1970–1991. J. Med. Entomol. **30**, 758–761 (1993)

5. Butler, T.: The black death past and present. 1. Plague in the 1980s. Trans. R. Soc. Trop. Med. Hyg. **83**, 458–460 (1989)
6. Perry, R.D., Fetherston, J.D.: Yersinia pestis–etiologic agent of plague. Clin. Microbiol. Rev. **10**, 35–66 (1997)
7. Parkhill, J., et al.: Genome sequence of Yersinia pestis, the causative agent of plague. Nature **413**, 523–527 (2001). doi:10.1038/35097083
8. Achtman, M., et al.: Yersinia pestis, the cause of plague, is a recently emerged clone of Yersinia pseudotuberculosis. Proc. Natl. Acad. Sci. U.S.A. **96**, 14043–14048 (1999)
9. Cornelis, G.R.: Yersinia type III secretion: send in the effectors. J. Cell Biol. **158**, 401–408 (2002). doi:10.1083/jcb.200205077
10. Fields, K.A., Straley, S.C.: LcrV of Yersinia pestis enters infected eukaryotic cells by a virulence plasmid-independent mechanism. Infect. Immun. **67**, 4801–4813 (1999)
11. Galvan, E.M., Lasaro, M.A., Schifferli, D.M.: Capsular antigen fraction 1 and Pla modulate the susceptibility of Yersinia pestis to pulmonary antimicrobial peptides such as cathelicidin. Infect. Immun. **76**, 1456–1464 (2008). doi:10.1128/IAI.01197-07
12. Miller, J., et al.: Macromolecular organisation of recombinant Yersinia pestis F1 antigen and the effect of structure on immunogenicity. FEMS Immunol. Med. Microbiol. **21**, 213–221 (1998)
13. Li, B., Yang, R.: Interaction between Yersinia pestis and the host immune system. Infect. Immun. **76**, 1804–1811 (2008). doi:10.1128/IAI.01517-07
14. Hamad, M.A., Nilles, M.L.: Roles of YopN, LcrG and LcrV in controlling Yops secretion by Yersinia pestis. Adv. Exp. Med. Biol. **603**, 225–234 (2007). doi:10.1007/978-0-387-72124-8_20
15. Hueck, C.J.: Type III protein secretion systems in bacterial pathogens of animals and plants. Microbiol. Mol. Biol. Rev. **62**, 379–433 (1998)
16. Lathem, W.W., Crosby, S.D., Miller, V.L., Goldman, W.E.: Progression of primary pneumonic plague: a mouse model of infection, pathology, and bacterial transcriptional activity. Proc. Natl. Acad. Sci. U.S.A. **102**, 17786–17791 (2005). doi:10.1073/pnas.0506840102
17. Butler, T.: Plague into the 21st century. Clin. Infect. Dis. **49**, 736–742 (2009). doi:10.1086/604718
18. Butler, T.: Plague and Other Yersinia Infections. Plenum Press, New York (1983)
19. Ben-Ari, T., et al.: Plague and climate: scales matter. PLoS Pathog. **7**, e1002160 (2011). doi:10.1371/journal.ppat.1002160
20. Gage, K.L., et al.: Cases of cat-associated human plague in the Western US, 1977–1998. Clin Infect. Dis. **30**, 893–900 (2000). doi:10.1086/313804
21. Jarrett, C.O., et al.: Transmission of Yersinia pestis from an infectious biofilm in the flea vector. J. Infect. Dis. **190**, 783–792 (2004). doi:10.1086/422695
22. Starnbach, M.N., Straley, S.C.: Yersinia: Strategies that Thwart Immune Defenses, pp. 71–92. Lippincott Williams and Wilkins, Philadelphia (2000)
23. Meyer, K.F.: Pneumonic plague. Bacteriol. Rev. **25**, 249–261 (1961)
24. World Health Organization: Human plague in 2002 and 2004. Wkly. Epidemiol. Rec. **79**, 301–308 (2004)
25. Hull, H.F., Montes, J.M., Mann, J.M.: Septicemic plague in New Mexico. J. Infect. Dis. **155**, 113–118 (1987)
26. Poland, J.D., Barnes, A.M.: Plague in CRC Handbook Series in Zoonoses. Section A. Bacterial, Rickettsial, Chlamydial and Mycotic Diseases, vol. I, pp. 515–559. CRC Press, Boca Raton (1979)
27. Gage, K.L., Lance, S.E., Dennis, D.T., Montenieri, J.A.: Human plague in the United States: a review of cases from 1988–1992 with comments on the likelihood of increased plague activity. Border Epidemiol. Bull. **19**, 1–10 (1992)
28. Cavanaugh, D.C.: K F Meyer's work on plague. J. Infect. Dis. **129**(Suppl), S10–S12 (1974)
29. Meyer, K.F.: Modern therapy of plague. J. Am. Med. Assoc. **144**, 982–985 (1950)
30. Boulanger, L.L., et al.: Gentamicin and tetracyclines for the treatment of human plague: review of 75 cases in new Mexico, 1985–1999. Clin. Infect. Dis. **38**, 663–669 (2004). doi:10.1086/381545
31. Becker, T.M., et al.: Plague meningitis–a retrospective analysis of cases reported in the United States, 1970–1979. West. J. Med. **147**, 554–557 (1987)
32. Mwengee, W., et al.: Treatment of plague with gentamicin or doxycycline in a randomized clinical trial in Tanzania. Clin. Infect. Dis. **42**, 614–621 (2006). doi:10.1086/500137
33. Kuberski, T., Robinson, L., Schurgin, A.: A case of plague successfully treated with ciprofloxacin and sympathetic blockade for treatment of gangrene. Clin. Infect. Dis. **36**, 521–523 (2003). doi:10.1086/367570
34. Stenseth, N.C., et al.: Plague: past, present, and future. PLoS Med. **5**, e3 (2008). doi:10.1371/journal.pmed.0050003
35. Kool, J.L.: Risk of person-to-person transmission of pneumonic plague. Clin. Infect. Dis. **40**, 1166–1172 (2005). doi:10.1086/428617
36. Prentice, M.B., Rahalison, L.: Plague. Lancet **369**, 1196–1207 (2007). doi:10.1016/S0140-6736(07)60566-2
37. Parent, M.A., et al.: Cell-mediated protection against pulmonary Yersinia pestis infection. Infect. Immun. **73**, 7304–7310 (2005). doi:10.1128/IAI.73.11.7304-7310.2005
38. Bermudez-Humaran, L.G.: Lactococcus lactis as a live vector for mucosal delivery of therapeutic proteins. Hum. Vaccin. **5**, 264–267 (2009)
39. Amdekar, S., Dwivedi, D., Roy, P., Kushwah, S., Singh, V.: Probiotics: multifarious oral vaccine against infectious traumas. FEMS Immunol. Med. Microbiol. **58**, 299–306 (2010). doi:10.1111/j.1574-695X.2009.00630.x
40. Wells, J.M., Mercenier, A.: Mucosal delivery of therapeutic and prophylactic molecules using lactic acid bacteria. Nat. Rev. Microbiol. **6**, 349–362 (2008). doi:10.1038/nrmicro1840

41. Anderson, R., Dougan, G., Roberts, M.: Delivery of the Pertactin/P.69 polypeptide of Bordetella pertussis using an attenuated Salmonella typhimurium vaccine strain: expression levels and immune response. Vaccine **14**, 1384–1390 (1996)
42. Ascon, M.A., Hone, D.M., Walters, N., Pascual, D.W.: Oral immunization with a Salmonella typhimurium vaccine vector expressing recombinant enterotoxigenic Escherichia coli K99 fimbriae elicits elevated antibody titers for protective immunity. Infect. Immun. **66**, 5470–5476 (1998)
43. Peters, C., Peng, X., Douven, D., Pan, Z.K., Paterson, Y.: The induction of HIV Gag-specific CD8+ T cells in the spleen and gut-associated lymphoid tissue by parenteral or mucosal immunization with recombinant Listeria monocytogenes HIV Gag. J. Immunol. **170**, 5176–5187 (2003)
44. Lee, J.S., et al.: Mucosal immunization with surface-displayed severe acute respiratory syndrome coronavirus spike protein on Lactobacillus casei induces neutralizing antibodies in mice. J. Virol. **80**, 4079–4087 (2006). doi:10.1128/JVI.80.8.4079-4087.2006
45. Wu, C.M., Chung, T.C.: Mice protected by oral immunization with Lactobacillus reuteri secreting fusion protein of Escherichia coli enterotoxin subunit protein. FEMS Immunol. Med. Microbiol. **50**, 354–365 (2007). doi:10.1111/j.1574-695X.2007.00255.x
46. Kajikawa, A., Satoh, E., Leer, R.J., Yamamoto, S., Igimi, S.: Intragastric immunization with recombinant Lactobacillus casei expressing flagellar antigen confers antibody-independent protective immunity against Salmonella enterica serovar Enteritidis. Vaccine **25**, 3599–3605 (2007). doi:10.1016/j.vaccine.2007.01.055
47. Daniel, C., et al.: Protection against Yersinia pseudotuberculosis infection conferred by a Lactococcus lactis mucosal delivery vector secreting LcrV. Vaccine **27**, 1141–1144 (2009). doi:10.1016/j.vaccine.2008.12.022
48. Bermudez-Humaran, L.G., Kharrat, P., Chatel, J.M., Langella, P.: Lactococci and lactobacilli as mucosal delivery vectors for therapeutic proteins and DNA vaccines. Microb. Cell Fact. **10**(Suppl 1), S4 (2011). doi:10.1186/1475-2859-10-S1-S4
49. Badgett, M.R., Auer, A., Carmichael, L.E., Parrish, C.R., Bull, J.J.: Evolutionary dynamics of viral attenuation. J. Virol. **76**, 10524–10529 (2002)
50. Lavelle, E.C., O'Hagan, D.T.: Delivery systems and adjuvants for oral vaccines. Expert Opin. Drug Deliv. **3**, 747–762 (2006). doi:10.1517/17425247.3.6.747
51. Neutra, M.R., Kozlowski, P.A.: Mucosal vaccines: the promise and the challenge. Nat. Rev. Immunol. **6**, 148–158 (2006). doi:10.1038/nri1777
52. Steinman, R.M.: The dendritic cell system and its role in immunogenicity. Annu. Rev. Immunol. **9**, 271–296 (1991). doi:10.1146/annurev.iy.09.040191.001415
53. Banchereau, J., Steinman, R.M.: Dendritic cells and the control of immunity. Nature **392**, 245–252 (1998). doi:10.1038/32588
54. Pulendran, B., Banchereau, J., Maraskovsky, E., Maliszewski, C.: Modulating the immune response with dendritic cells and their growth factors. Trends Immunol. **22**, 41–47 (2001)
55. Fujkuyama, Y., et al.: Novel vaccine development strategies for inducing mucosal immunity. Expert Rev. Vaccines **11**, 367–379 (2012). doi:10.1586/erv.11.196
56. Bedoui, S., et al.: Different bacterial pathogens, different strategies, yet the aim is the same: evasion of intestinal dendritic cell recognition. J. Immunol. **184**, 2237–2242 (2010). doi:10.4049/jimmunol.0902871
57. Artis, D.: Epithelial-cell recognition of commensal bacteria and maintenance of immune homeostasis in the gut. Nat. Rev. Immunol. **8**, 411–420 (2008). doi:10.1038/nri2316
58. Rescigno, M., et al.: Dendritic cells express tight junction proteins and penetrate gut epithelial monolayers to sample bacteria. Nat. Immunol. **2**, 361–367 (2001). doi:10.1038/86373
59. Grangette, C., et al.: Protection against tetanus toxin after intragastric administration of two recombinant lactic acid bacteria: impact of strain viability and in vivo persistence. Vaccine **20**, 3304–3309 (2002)
60. Christensen, H.R., Frokiaer, H., Pestka, J.J.: Lactobacilli differentially modulate expression of cytokines and maturation surface markers in murine dendritic cells. J. Immunol. **168**, 171–178 (2002)
61. Kalina, W.V., Mohamadzadeh, M.: Lactobacilli as natural enhancer of cellular immune response. Discov. Med. **5**, 199–203 (2005)
62. Mohamadzadeh, M., et al.: Lactobacilli activate human dendritic cells that skew T cells toward T helper 1 polarization. Proc. Natl. Acad. Sci. U.S.A. **102**, 2880–2885 (2005). doi:10.1073/pnas.0500098102
63. Mohamadzadeh, M., Duong, T., Hoover, T., Klaenhammer, T.R.: Targeting mucosal dendritic cells with microbial antigens from probiotic lactic acid bacteria. Expert Rev. Vaccines **7**, 163–174 (2008). doi:10.1586/14760584.7.2.163
64. Perdigon, G., Alvarez, S., Pesce de Ruiz Holgado, A.: Immunoadjuvant activity of oral Lactobacillus casei: influence of dose on the secretory immune response and protective capacity in intestinal infections. J. Dairy Res. **58**, 485–496 (1991)
65. Pouwels, P.H., Leer, R.J., Boersma, W.J.: The potential of Lactobacillus as a carrier for oral immunization: development and preliminary characterization of vector systems for targeted delivery of antigens. J. Biotechnol. **44**, 183–192 (1996). doi:10.1016/0168-1656(95)00140-9
66. Maassen, C.B., et al.: Instruments for oral disease-intervention strategies: recombinant Lactobacillus

casei expressing tetanus toxin fragment C for vaccination or myelin proteins for oral tolerance induction in multiple sclerosis. Vaccine **17**, 2117–2128 (1999)
67. Bermudez-Humaran, L.G., et al.: Controlled intra- or extracellular production of staphylococcal nuclease and ovine omega interferon in Lactococcus lactis. FEMS Microbiol. Lett. **224**, 307–313 (2003)
68. Navarre, W.W., Schneewind, O.: Surface proteins of gram-positive bacteria and mechanisms of their targeting to the cell wall envelope. Microbiol Mol Biol Rev **63**, 174–229 (1999)
69. Weis, J.J., Ma, Y., Erdile, L.F.: Biological activities of native and recombinant Borrelia burgdorferi outer surface protein A: dependence on lipid modification. Infect. Immun. **62**, 4632–4636 (1994)
70. Sellati, T.J., et al.: Treponema pallidum and Borrelia burgdorferi lipoproteins and synthetic lipopeptides activate monocytic cells via a CD14-dependent pathway distinct from that used by lipopolysaccharide. J. Immunol. **160**, 5455–5464 (1998)
71. del Rio, B., Seegers, J.F., Gomes-Solecki, M.: Immune response to Lactobacillus plantarum expressing Borrelia burgdorferi OspA is modulated by the lipid modification of the antigen. PLoS One **5**, e11199 (2010). doi:10.1371/journal.pone.0011199
72. del Rio, B., et al.: Oral immunization with recombinant lactobacillus plantarum induces a protective immune response in mice with Lyme disease. Clin. Vaccine. Immunol. **15**, 1429–1435 (2008). doi:10.1128/CVI.00169-08
73. Leary, S.E., et al.: Active immunization with recombinant V antigen from Yersinia pestis protects mice against plague. Infect. Immun. **63**, 2854–2858 (1995)
74. Alpar, H.O., Eyles, J.E., Williamson, E.D., Somavarapu, S.: Intranasal vaccination against plague, tetanus and diphtheria. Adv. Drug Deliv. Rev. **51**, 173–201 (2001)
75. del Rio, B., et al.: Platform technology to deliver prophylactic molecules orally: an example using the Class A select agent Yersinia pestis. Vaccine **28**, 6714–6722 (2010). doi:10.1016/j.vaccine.2010.07.084
76. Foligne, B., et al.: Prevention and treatment of colitis with Lactococcus lactis secreting the immunomodulatory Yersinia LcrV protein. Gastroenterology **133**, 862–874 (2007). doi:10.1053/j.gastro.2007.06.018
77. Ramirez, K., et al.: Neonatal mucosal immunization with a non-living, non-genetically modified Lactococcus lactis vaccine carrier induces systemic and local Th1-type immunity and protects against lethal bacterial infection. Mucosal Immunol. **3**, 159–171 (2010). doi:10.1038/mi.2009.131
78. Welkos, S., et al.: Determination of the virulence of the pigmentation-deficient and pigmentation-/plasminogen activator-deficient strains of Yersinia pestis in non-human primate and mouse models of pneumonic plague. Vaccine **20**, 2206–2214 (2002)
79. Smiley, S.T.: Current challenges in the development of vaccines for pneumonic plague. Expert Rev. Vaccines **7**, 209–221 (2008). doi:10.1586/14760584.7.2.209
80. Powell, B.S., et al.: Design and testing for a non-tagged F1-V fusion protein as vaccine antigen against bubonic and pneumonic plague. Biotechnol. Prog. **21**, 1490–1510 (2005). doi:10.1021/bp050098r
81. Alvarez, M.L., et al.: Plant-made subunit vaccine against pneumonic and bubonic plague is orally immunogenic in mice. Vaccine **24**, 2477–2490 (2006). doi:10.1016/j.vaccine.2005.12.057
82. Cornelius, C.A., et al.: Immunization with recombinant V10 protects cynomolgus macaques from lethal pneumonic plague. Infect. Immun. **76**, 5588–5597 (2008). doi:10.1128/IAI.00699-08
83. Smiley, S.T.: Immune defense against pneumonic plague. Immunol. Rev. **225**, 256–271 (2008). doi:10.1111/j.1600-065X.2008.00674.x
84. Williamson, E.D., et al.: Human immune response to a plague vaccine comprising recombinant F1 and V antigens. Infect. Immun. **73**, 3598–3608 (2005). doi:10.1128/IAI.73.6.3598-3608.2005
85. Pavot, V., Rochereau, N., Genin, C., Verrier, B., Paul, S.: New insights in mucosal vaccine development. Vaccine **30**, 142–154 (2012). doi:10.1016/j.vaccine.2011.11.003

# Electroporation-Based Gene Transfer

## 49

Mattia Ronchetti, Michela Battista, Claudio Bertacchini, and Ruggero Cadossi

## Contents

| | | |
|---|---|---|
| 49.1 | **Principles of Cell Membrane Electroporation** | 782 |
| 49.2 | **Electrochemotherapy** | 782 |
| 49.3 | **Gene Electrotransfer** | 782 |
| 49.4 | **DNA Electrotransfer Principles** | 783 |
| 49.5 | **Gene Electrotransfer to Target Tissues** | 783 |
| 49.6 | **Electroporation Devices** | 784 |
| 49.7 | **Clinical Applications** | 787 |
| 49.7.1 | Gene Therapy | 787 |
| 49.7.2 | DNA Vaccination | 787 |
| **References** | | 788 |

M. Ronchetti, BSc (✉) • M. Battista, PhD
C. Bertacchini, MSc • R. Cadossi, MD
Laboratory of Clinical Biophysics,
IGEA S.p.A., Carpi, Italy
e-mail: m.ronchetti@igeamedical.com

## Abstract

When an external electric field, under specific pulse conditions, is applied to cells, in suspension or in biological tissues, the permeability of cell membranes is transiently increased. This physical method, termed electroporation, can be used to introduce poorly or non permeant molecules into the cell. The combination of electroporation and chemotherapy is termed electrochemotherapy. Electrochemotherapy enhances local cytotoxicity of hydrophilic drugs that do not easily pass the cell membrane, e.g. Bleomycin. This combination treatment has proven effective in local control of metastatic tumour nodules to the skin, independently of the histotype. Delivery of genetic materials into the target tissues or cells by means of electric pulses is referred to as electroporation based gene transfer. Gene expression level and kinetic patterns after in vivo electroporation mediated delivery can be optimized and adapted for different purposes by employing different applicator configuration, electrical parameters, and target tissues of delivery. The current chapter discusses present knowledge of nonviral gene delivery, the mechanism of DNA electrotransfer, and clinical applications focusing on delivery to skeletal muscle and skin. An overview of the equipment, tissue electroporation device and electrodes, currently available for clinical use of electroporation based gene transfer is provided.

## 49.1 Principles of Cell Membrane Electroporation

Application of an external electric field to a single cell, cell suspensions or biological tissue generates a change in the cell transmembrane potential. At the cell membrane level, applying electric fields above the threshold value of the transmembrane potential (~1.5 V) [1] results in changes to the membrane structure that render the membrane permeable to otherwise non-permeant molecules [2], a phenomenon termed electroporation or electropermeabilisation. By modulating applied electric field parameters within a definite window, it is possible to temporarily permeabilise the membrane, allowing the cell to return to its natural state and thus preserving its viability: this process has been historically named reversible electroporation [3] and has been used in combination with chemotherapeutic agents, a procedure termed electrochemotherapy, or as a delivery method for nucleic acids, gene electrotransfer [4].

The application of electric fields of higher intensity and/or for longer times can cause a permanent disruption of cell membrane permeability, leading to cell death. This process is commonly referred to as irreversible electroporation and represents a novel soft tissue ablation modality currently under clinical investigation [3].

## 49.2 Electrochemotherapy

Electrochemotherapy is the local potentiation, by means of local reversible electroporation of tumour tissues, of the antitumour activity of non-permeant (i.e. bleomycin) or poorly permeant (i.e. cisplatin) drugs already possessing intrinsic cytotoxic activity [5]. The first clinical studies on electrochemotherapy date to the early 1990s [6] and reported effectiveness in local disease control on head and neck squamous cell carcinoma nodules. Following the first heterogenous clinical trials, reviewed elsewhere by Sersa G [7] up to 2006, a prospective, multicenter, international clinical trial, the ESOPE (European Standard Operating Procedures of Electrochemotherapy) study, was conducted to evaluate the efficacy and safety of electrochemotherapy on cutaneous and subcutaneous tumour nodules showing an objective response rate of 85 % (73.7 % complete response rate) on treated tumour nodules [8]. Within the study, treatment protocol was unified through the definition of electrochemotherapy standard operating procedures [9]. Recently, a systematic review and formal meta-analysis of all relevant published literature further confirmed the effectiveness of electrochemotherapy along with potential predictors of tumour response to electrochemotherapy with respect to various treatment conditions [10].

## 49.3 Gene Electrotransfer

Gene therapy is a promising field of medicine in which genes are introduced into the body to treat diseases especially those caused by genetic anomalies or deficiencies. The main obstacle for gene therapy has been the safe, effective and reliable delivery of genes to target tissue. DNA electrotransfer has proved to be an efficient and safe method for delivery of naked DNA in non-viral gene therapy approaches. Following the pioneering in vitro studies in 1982 that demonstrated that DNA could be introduced into living cells by means of electric pulses [4, 5], several studies on the use of electroporation for delivery of molecules to eukaryotic cells and in various tissues have been reported [11]. Over the years, the electroporation devices have been developed, and electric pulse generators and applicators have been created to deliver nucleic acids under different condition and in different tissues. Modern generators allow controlling amplitude, pulse length and number while investigating the effect of the various electric parameters on transfection and expression levels as well as optimisation of gene electrotransfer protocols [12].

The employment of naked DNA eliminates the limitations and the concerns linked to the use of viruses like coding sequence length in the case of the adenovirus [13] and formation of insertional mutations during their integration into the host genome in the case of the retrovirus or lentivirus [14]. In addition, naked DNA is safe and simple to manipulate and generate, and it is entirely constituted by double-stranded DNA with no associated proteins, whereas adenovirus proteins can induce immunological responses and prevent the possibility to re-administer the viral vectors [15]; moreover naked DNA delivered via gene electrotransfer offers the possibility of multiple administrations.

Other physical approaches for nonviral gene therapy include direct injection [16]; plasmid liposome complexes [17], still one of the most common techniques for gene delivery into cells; the biolistic approach for DNA transfer to superficial tissues, like the skin, it employs a device called gene gun that propels plasmid-coated gold microparticles through the cell membrane into the cytoplasm and the nucleus, bypassing the endosomal compartment where DNA can be damaged. Limitations of this method include the low efficiency of the metal particles in reaching the entire tissue due to the low penetration of the particles, the deposition of metal particles into the body with the potential for long-term consequences [18, 19]. Sonoporation applies ultrasound to increase the permeability of cell membrane to macromolecules including plasmid DNA. Indeed, enhancement of gene expression was observed by irradiating ultrasonic wave to the tissue after injection of DNA [20]. In mice, hydrodynamic injection gene transfer is a technique achieved by means of the rapid intravenous injection of a high volume of solution in just a few seconds. Because of the high volume injected, the liquid accumulates in the inferior vena cava, and the injected DNA is captured by the internal organs that are perfused by it [16, 21], mainly the liver. However, because of the very large amount of liquid injected, this approach seems to have limited clinical translation potential.

## 49.4 DNA Electrotransfer Principles

Because of its ease of application, safety and its efficiency, DNA electrotransfer has rapidly expanded and advanced to an efficient methodology for nonviral gene delivery. Several experiments showed that in DNA electrotransfer, intake of exogenous DNA by the cell is controlled by different factors. The major role of high-voltage, short-duration electric pulses is membrane permeabilisation. Subsequently, the DNA is moved by electrophoretic forces towards, as well as across, the permeabilised cell membrane by lower-voltage, longer-duration pulses and enters the cell cytosol [22, 23]. Moreover, it has been showed that when applying the electric field, the presence of DNA facilitates pore formation through direct interaction with the membrane; therefore, DNA must be injected before the electric pulse delivery [22].

The role of electric pulses in DNA electrotransfer has been studied using a combination of pulses, and three ways to deliver pulses for DNA transfer have been identified: a. delivery of solely short, high-amplitude pulses (e.g. six pulses, 100 μs and 1.4 kV cm$^{-1}$) [24], resulting in a reasonable efficacy and a low mortality; b. long, low-amplitude pulse delivery (e.g. eight pulses, 20 ms, 200 kV cm$^{-1}$) [25] (the longer pulses have a better electrophoretic effect, and therefore transfection rates can be increased); and c. short, high-amplitude pulse followed by long, low-amplitude pulses [26]. This pulse combination is based on the concept that the high-amplitude pulse induces permeabilisation, while the following long-duration, low-voltage pulses can drive the DNA across the destabilised membrane. Finally, further experiments demonstrated the importance of the lag between high-voltage and low-voltage pulses: the shorter the lag between HV and LV, the higher the transfection efficiency obtained [21].

## 49.5 Gene Electrotransfer to Target Tissues

Gene transfer into muscle has been studied extensively since the observation of Wolff et al. [27] showed that plasmid DNA can be taken up and expressed in mouse skeletal muscle cells. Skeletal muscles are an attractive target for gene transfer due to abundant presence, easy accessibility, high vascularisation and prolonged preservation and expression of introduced genes following mitosis. [28, 29] Nevertheless, the highly variable efficiency of transfection and the unpredictability of the resulting protein synthesis [25] pose important drawbacks of this method. To overcome these limits, several in vivo studies have been focused on the use of electroporation for gene transfer into skeletal muscle. DNA is injected intramuscularly in relatively high volumes of an isotonic plasmid solution followed by the application of electrical pulses by means of electrodes that are placed around the injection site. Non-invasive, plate electrodes are used for superficial muscles, while invasive needle electrodes are employed to treat deeper tissues.

To ensure a homogenous electric field distribution and a uniform transfection between the electrodes, thus facilitating expression of the plasmid vector, the positioning of the electrodes must be parallel to the muscle fibres [28]. Other factors that are critical for gene expression efficiency include the amount of the injected plasmid, promoter system and area of transfection [30]. DNA electrotransfer to muscle tissue has been proved to be highly efficient, and steady gene expression with transgene production has been detected for more than a year [25] overcoming high variability in the level of protein expression.

Following the muscle, gene electrotransfer has been successfully applied to several other tissues such as the cornea, testis, lung, liver, kidney, bladder tumour and skin [31]. Above all, skin retains several important features that make it an attractive target tissue for gene therapy: it is easy to assess for treatment and for histological and clinical results evaluation, and it contains antigen-presenting cells (Langerhans cells, dendritic cells), which are part of the immune system and make the skin a proper target for DNA vaccination studies [31, 32].

In order to electroporate the skin, electric pulses must be delivered using appropriate electrodes (non-invasive plate and patch electrodes or invasive array, circular or pairwise formation needle electrodes). Several pulse combinations have been tested; the optimal pulse conditions depend on the type of skin transfected – size, thickness and age of the skin must be taken into consideration [33, 34] – type of electrode and type of DNA [35]. A very extensive number of electrical protocols have been evaluated resolving that the most efficient protocols for gene transfer are a combination of high-voltage (HV) and low-voltage (LV) pulses (1 HV 1,000 V/cm, 100 μs + 1 LV 100 V/cm, 400 ms) [36], several short HV pulses (6 HV 1,750 V/CM, 100 μs) [37] and several long LV pulses (8 LV 100 V/cm, 150 ms) [38].

Gene electrotransfer to the skin has been employed to achieve the expression of several clinically relevant plasmids. They include local growth factors such as vascular endothelial growth factor (VEGF), keratinocyte growth factor (KGF) and viral targets for the vaccination towards infection diseases as HIV, hepatitis B, smallpox and malaria [31]. Moreover, the expression of IL-12, for the treatment of melanoma, has been also investigated following delivery in conjunction with electroporation [39, 40]. Preclinical studies employing gene electrotransfer to the muscle and to the skin have been summarised in Table 49.1.

Gene transfection to muscle can give long-term expression lasting up to or beyond 1 year, whereas the duration of expression in the skin is approximately 3–4 weeks [41]. However, in cases where there is no need for long-term expression of transfected plasmid and continuous production of proteins or antigens is not necessary, or desired, the skin represents an ideal target organ for gene electrotransfer.

## 49.6 Electroporation Devices

Electroporation technology is based on pulse generators that use different applicator electrodes, e.g. matrix of needles or plates to deliver suitable electric pulses to the target tissues. The complete system, pulse generators and applicators, is classified as a medical device. Different systems can have different approaches to the pulse generation. When this technology is developed for use in humans, for either small molecules delivery or gene transfection, specific requirements in terms of safety, reliability and efficacy must be met. Currently, there are

**Table 49.1** Summary of preclinical studies describing in vivo gene electrotransfer to the muscle and skin tissues

| Electrode type | | Animal models | DNA plasmid | Voltage range (V/cm) | Duration range | Pulse number range | References |
|---|---|---|---|---|---|---|---|
| Invasive | Skin | Pig, mouse, rat | GFP, Luc, hepatitis B, Hif-1α, KGF, VEGF, PSA, CEA, survivin | 1,125–1,800 HV | 50–100 μs | 2–18 | [37, 59–63] |
| | | | | 50–400 LV | 10–100 ms | 6–10 | [63–66] |
| | | | | 1,000–10,125 + | 50–100 μs + | 1–2 + | [35, 63, 67–69] |
| | | | | 80–275 HV+LV | 275–400 ms | 1–8 | |
| | Muscle | Mouse, sheep, rat, dog | IL-5, GFP, EPO, GHRH, JEV | 100–250 | 50 ms | 5–8 | [89–93] |
| Non-invasive | Skin | Pig, mouse, rat, rabbit | NeoR, LacZ, GFP, OVA, hepatitis B, IL-12, EPO | 400–1,750 (HV) | 100–300 μs | 1–8 | [37, 39, 70–72] |
| | | | | 12–800 (LV) | 2–400 ms | 1–12 | [33, 34, 39, 71–77] |
| | | | | 700–1,000 | 700–1,000 μs | 1 | [35, 36, 41, 72, 78–81] |
| | | | | + | + | + | |
| | | | | 80–200 (HV+LV) | 80–200 ms | 1 | |
| | | | | 1,000 HV | 100 μs | 8 | [84] |
| | Muscle | Mouse, rabbit, rat | GFP, Luc, IL-4, IL-1Ra, VEGF, EPO | 50–600 LV | 100 μs–20 ms | 2–8 | [82, 83, 85–88] |
| | | | | 100–2,000 + | 100 μs + | 1 + | [36, 82, 84] |
| | | | | 80 V/cm HV+LV | 400 ms | 1 | |

Invasive needle electrodes and non-invasive plate electrodes have been employed. Needle electrodes have been used to target the skin and deeper lying tissues such as muscles in larger animals, while plate electrode has been used to treat superficial tissues and rodent muscles. For simplicity, electrical protocols applied are reported as overall ranges
*Abbreviations: V/cm* volt per cm, *HV* high voltage, *LV* low voltage

**Fig. 49.1** Cliniporator™ for electroporation

several electroporators available on the market for preclinical and lab testing; however, only few models are certified [42] and bear a CE mark for human clinical use in the European Union. One such example is the Cliniporator™ developed and marketed by IGEA S.p.A., Italy (Fig. 49.1).

It is possible to divide the electroporators considering the characteristics of the pulse and the technology used to generate it. There are several techniques to generate the pulse, all of them having usually a square or an exponential shape. Very short pulses having a length shorter than 1 μs require specific circuits and up to now are used for electroporation research only. Short pulses, having a length between 1 and 1000 μs, and longer pulses, having a length of several milliseconds, are usually generated with different methods: by a direct capacitor discharge, by the pulse transformer, by high-voltage analog generator or by high-voltage square wave generation [43, 44]. Basic characteristic of an electroporator to be used for gene electrotransfer is to deliver combinations of high-voltage (HV) pulses and low-voltage (LV) pulses. This usually requires two different circuitries to ensure an appropriate precision and good control of the pulse shape.

Additionally, sophisticated devices, such as the Cliniporator, can measure in real-time voltage and current on the load, providing the user with immediate feedback of the treatment being performed. Real-time measurement of applied pulses is also an ideal way to monitor performances of the device, detecting malfunctions early. Since the high-voltage pulse amplitude has a range between one hundred volts and few thousand volts [45], regulations and safety standards [42] require a very good isolation of the high-voltage circuit to ensure the safety of the patients, of the operator and of the device. Another very important safety issue is determined by the variability of working environment conditions (e.g. operating theatre or clinician office) generally unknown beforehand. Energy delivered is influenced by tissue or cell impedance and can vary from point to point and from tissue to tissue. Thus, since high currents can damage cells, tissues and even the device itself, the current is the critical parameter to be controlled.

Finally considering the structure of the device, we see that in some cases the user interface controls directly the power part but in other cases the power part is controlled by an independent electronic circuit and the user interface simply sends the treatment parameters to the power part. This different approach could have an impact about the reliability and the safety in case of failure of the device [43]. Consequently, the design of a device for electroporation has to keep under close control energy delivered to the patient. Design and implementation should focus on a sturdy and fast pulse generator with a high reactivity in case of failures, using redundancies to improve its safety and providing a very user-friendly interface to minimise the risk of user error [42, 43, 45].

## 49.7 Clinical Applications

### 49.7.1 Gene Therapy

Preclinical investigations of gene transfer cover a broad range of studies including the treatment of genetic muscle disease such as Duchenne's dystrophy [46], genetic vaccination [47] and systemic delivery of secretory therapeutic proteins as erythropoietin (EPO) [48], hematopoietic agents such as factor VIII [49], anticancer agents like interferons (IFN-α) [50] and antiangiogenic factors as metargidin [51]. Ongoing clinical trials [52] using the muscle as target tissue for gene transfer are summarised in Table 49.2, while studies employing derma as target tissue are summarised in Table 49.3.

### 49.7.2 DNA Vaccination

Electroporation-mediated DNA vaccination represents one promising use of in vivo gene transfer by electroporation. DNA vaccines have many advantages. DNA vectors are easily produced and manipulated, rapidly tested

**Table 49.2** Ongoing clinical trials using intramuscular (IM) gene electrotransfer

| Pathology | Subjects | Primary outcome | Clinical Trial.gov ID |
|---|---|---|---|
| Prostate cancer | | A phase IIII trial of DNA vaccine with a PSMA27/pDom fusion gene given through intramuscular injection in HLA A2+ patients with prostate carcinomas with or without electroporation | UK-112 |
| Papillomavirus infections (completed) | 24 | Safety and tolerability of escalating doses of VGX-3100, administered by IM injection with EP to adult female subjects postsurgical or ablative treatment of grade 2 or 3 CIN as adjuvant treatment | NCTOO685412 |
| Chronic hepatitis C infections | 12 | Safety and tolerability of electroporation-mediated IM delivery of CHRONVAC-C® in chronically HCV-infected, treatment-naive patients with low viral load | NCT00563 173 |
| Malignant melanoma (completed) | 25 | Safety and feasibility of electroporation-mediated intramuscular delivery of a mouse tyrosinase plasmid DNA vaccine in patients with stage IIB, IIC III or IV melanoma | NCT00471133 |
| HIV infections (completed) | 40 | Safety of an intramuscular prime and boost injection of the ADVAX DNA-based HIV vaccine via TriGrid™ electroporation at three dosing levels | NCT00545987 |
| Healthy adults | 24 | Assessment of the tolerability of the MedPulser DDS device | NCTOO72 146 |
| Cervical intraepithelial neoplasia | 348 | Number of participants with histopathological regression of cervical lesions to CIN 1 or less as a measure of efficacy | NCT01304524 |
| Malignant melanoma | 30 | Safety and tolerability of an investigational immunotherapy, SCIB1, in patients with melanoma whose cancer has spread from the initial tumour (i.e. stage III or stage IV melanoma) | NCT01138410 |
| Leukaemia | 184 | Phase II study of WT1 immunity via DNA fusion gene vaccination in haematological malignancies by intramuscular injection followed by intramuscular electroporation | NCT01334060 |
| HPV-related head and neck cancer | 21 | Safety and feasibility of administration of pNGVL4a-CRT/E7(Detox) DNA vaccine using the intramuscular TriGrid™ Delivery System in combination with cyclophosphamide in HPV-16-associated head and neck cancer | NCT01493154 |
| HIV-1 infection | 12 | Safety, tolerability and immunogenicity of PENNVAX™-B (Gag, Pol, Env)+electroporation in HIV-1-infected adult participants | NCT01082692 |
| Chronic hepatitis C | 32 | Early viral kinetics – second-phase slope of viral decline | NCT01335711 |

**Table 49.3** Ongoing clinical trials using intradermal (ID) electrotransfer

| Pathology | Subjects | Primary outcome | Clinical Trial.gov ID |
|---|---|---|---|
| Prostate cancer | 18 | Assess the feasibility and safety of escalating doses of pVAXrcPSAv53l DNA vaccine, administered intradermally in combination with electroporation in patients with relapse of prostate cancer | NCT00859729 |
| Colorectal cancer | 20 | To evaluate the safety and immunogenicity of a DNA immunisation approach where tetwtCEA DNA will be administered in combination with electroporation | NCT01064375 |
| H1 and H5 influenza virus | 100 | Safety and tolerability of nine different formulations of multiple combination of H1 and H5 HA plasmid administered ID followed by electroporation in healthy adult subjects | NCT01405885 |
| Human influenza | 50 | Safety and tolerability of a DNA-based influenza vaccine composed of a combination of two different H1 HA plasmids administered ID followed by electroporation in healthy elderly adult subjects | NCT01587131 |

and isolated and easily stored and transported. DNA vaccines can promote cellular as well as humoral immune response [53], and if necessary, they can contain several antigen epitope [54]. Electroporation high transfection efficacy increases the immune response compared to injection of naked plasmid alone which is further enhanced by activating antigen-presenting cells (APCs) through danger signals and local inflammation occurring after the delivery of electric pulses and by recruiting immune B and T cells to the site of DNA administration. Moreover, direct transfection of APCs may be important for T cell priming upon DNA and augmented immune response in electroporation-mediated DNA transfer to skin [55, 56]. An additional promising approach using prime-boost protocols which combine adenovirus vector administration and DNA gene electrotransfer has been proposed and positively demonstrated higher levels of immune responses to antigen and increased survival in canine patients affected by B cell lymphoma [57].

DNA vaccine tolerability in humans has been demonstrated in healthy volunteers showing no anti-DNA antibody and no integration of pDNA into host chromosome detection following electroporation-mediated delivery of DNA to the muscle [58] (Tables 49.2 and 49.3). Several electrotransfer DNA vaccine trials for cancer and three phase I clinical studies using DNA vaccine against infectious agents (HIV, cervical intraepithelial neoplasia and HCV) in association with electroporation are currently ongoing [52].

# References

1. Kotnik, T., Bobanovic, F., Miklavcic, D.: Sensitivity of transmembrane voltage induced by applied electric fields – a theoretical analysis. Bioelectrochem. Bioenerg. **43**, 285–291 (1997)
2. Miklavcic, D., Semrov, D., Mekid, H., Mir, L.M.: A validated model of in vivo electric field distribution for electrochemotherapy and for DNA electrotransfer for gene therapy. Biochim. Biophys. Acta **1523**, 73–83 (2000)
3. Rubinsky, B.: Irreversible electroporation in medicine. Technol. Cancer Res. Treat. **6**, 255–260 (2007)
4. Neumann, E., Schaefer-Ridder, M., Wang, Y., Hofschneider, P.H.: Gene transfer into mouse lyoma cells by electroporation in high electric fields. EMBO J. **1**, 841–845 (1982)
5. Silve, A., Mir, L.M.: Cell electropermeabilization and cellular uptake of small molecules: the electrochemotherapy concept. In: Kee, S.T., Gehl, J., Lee, E.W. (eds.) Clinical Aspects of Electroporation 1, pp. 69–82. Springer, New York (2011)
6. Mir, L.M., et al.: Electrochemotherapy, a new antitumour treatment: first clinical trial. C. R. Acad. Sci. III **313**, 613–618 (1991)
7. Sersa, G.: The state of the art of electrochemotherapy before ESOPE study. Advantages and clinical use. Eur. J. Cancer. Suppl. **4**, 52–59 (2006)
8. Marty, M., et al.: Electrochemotherapy – an easy, highly effective and safe treatment of cutaneous and subcutaneous metastases. Results of ESOPE (European Standard Operating Procedures of Electrochemotherapy) study. Eur. J. Cancer. **4**, 3–13 (2006)

9. Mir, L.M., et al.: Standard operating procedures of the electrochemotherapy: instructions for the use of bleomycin or cisplatin administered either systemically or locally and electric pulses delivered by the Cliniporator by means of invasive or noninvasive electrodes. Eur. J. Cancer. Suppl. **4**, 14–25 (2006)
10. Mali, B., Jarm, T., Snoj, M., Sersa, G., Miklavcic, D.: Antitumor effectiveness of electrochemotherapy: a systematic review and meta-analysis. Eur. J. Surg. Oncol. **39**, 4–16 (2012). http://dx.doi.org/10.1016/j.ejso.2012.08.016
11. Mir, L.M., Moller, P.H., André, F., Gehl, J.: Electric pulse-mediated gene delivery to various animal tissues. Adv. Genet. **54**, 83–114 (2005)
12. Durieux, A.C., Bonnefoy, R., Manissolle, C., Freyssenet, D.: High-efficiency gene electrotransfer into skeletal muscle: description and physiological applicability of a new pulse generator. Biochem. Biophys. Res. Commun. **296**(2), 443–450 (2002)
13. Hacein-Bey-Abina, S., et al.: LMO2-associated clonal T cell proliferation in two patients after gene therapy for SCID-X1. Science **302**(5644), 415–419 (2003)
14. Hacein-Bey-Abina, S., Le Deist, F., Carlier, F., Bouneaud, C., Hue, C., De Villartay, J.P., Thrasher, A.J., Wulffraat, N., Sorensen, R., Dupuis-Girod, S., Fischer, A., Davies, E.G., Kuis, W., Leiva, L., Cavazzana-Calvo, M.: Sustained correction of X-linked severe combined immunodeficiency by ex vivo gene therapy. N. Engl. J. Med. **346**(16), 1185–1193 (2002)
15. Couzin, J., Kaiser, J.: Gene therapy. As Gelsinger case ends, gene therapy suffers another blow. Science **307**(5712), 1028 (2005)
16. Budker, V., et al.: Hypothesis: naked plasmid DNA is taken up by cells in vivo by a receptor-mediated process. J. Gene Med. **2**, 76–88 (2000)
17. Dauty, E., Remy, J.S., Blessing, T., Behr, J.P.: Dimerizable cationic detergents with a low cmc condense plasmid DNA into nanometric particles and transfect cells in culture. J. Am. Chem. Soc. **123**, 9227–9234 (2001)
18. Lin, M.T., Pulkkinen, L., Uitto, J., Yoon, K.: The gene gun: current application in cutaneous gene therapy. Int. J. Dermatol. **39**, 161–170 (2000)
19. Davidson, J.M., Krieg, T., Eming, S.A.: Particle-mediated gene therapy of wounds. Wound Repair Regen. **8**, 452–459 (2000)
20. Newman, C.M., Lawrie, A., Brisken, A.F., Cumberland, D.C.: Ultrasound gene therapy: on the road from concept to reality. Echocardiography **18**, 339–347 (2001)
21. Liu, F., Huang, L.: Improving plasmid DNA-mediated liver gene transfer by prolonging its retention in the hepatic vasculature. J. Gene Med. **3**, 569–576 (2001)
22. Sukharev, S.I., Klenchin, V.A., Serov, S.M., Chernomordik, L.V., Chizmadzhev, YuA.: Electroporation and electrophoretic DNA transfer into cells. The effect of DNA interaction with electropores. Biophys. J. **63**(5), 1320–1327 (1992)
23. Klenchin, V.A., Sukharev, S.I., Serov, S.M., Chernomordik, L.V., Chizmadzhev, YuA.: Electrically induced DNA uptake by cells is a fast process involving DNA electrophoresis. Biophys. J. **60**(4), 804–811 (1991)
24. Heller, R., Jaroszeski, M., Atkin, A., et al.: In vivo gene electroinjection and expression in rat liver. FEBS Lett. **389**, 225–228 (1996)
25. Mir, L.M., Bureau, M.F., Gehl, J., Rangara, R., Rouy, D., Caillaud, J.M., Delaere, P., Branellec, D., Schwartz, B., Scherman, D.: High-efficiency gene transfer into skeletal muscle mediated by electric pulses. Proc. Natl. Acad. Sci. U.S.A. **96**(8), 4262–4267 (1999)
26. Bureau, M.F., Gehl, J., Deleuze, V., Mir, L.M., Scherman, D.: Importance of association between permeabilization and electrophoretic forces for intramuscular DNA electrotransfer. Biochim. Biophys. Acta **1474**, 353–359 (2000)
27. Wolff, J.A., et al.: Direct gene transfer into mouse muscle in vivo. Science **247**, 1465–1468 (1990)
28. André, F., Mir, L.M.: DNA electrotransfer: its principles and an updated review of its therapeutic applications. Gene Ther. **11**(1), S33–S42 (2004)
29. Gehl, J.: Electroporation for drug and gene delivery in the clinic: doctors go electric. Methods Mol. Biol. **423**, 351–359 (2008)
30. Hojman, P., Gissel, H., Gehl, J.: Sensitive and precise regulation of haemoglobin after gene transfer of erythropoietin to muscle tissue using electroporation. Gene Ther. **14**(12), 950–959 (2007)
31. Gothelf, A., Gehl, J.: Gene electrotransfer to skin; review of existing literature and clinical perspectives. Curr. Gene Ther. **10**(4), 287–299 (2010)
32. Kutzler, M.A., Weiner, D.B.: DNA vaccines: ready for prime time? Nat. Rev. Genet. **9**, 776–788 (2008)
33. Zhang, L., Li, L., Hoffmann, G.A., Hoffman, R.M.: Depth-targeted efficient gene delivery and expression in the skin by pulsed electric fields: an approach to gene therapy of skin aging and other diseases. Biochem. Biophys. Res. Commun. **220**(3), 633–636 (1996)
34. Chesnoy, S., Huang, L.: Enhanced cutaneous gene delivery following intradermal injection of naked DNA in a high ionic strength solution. Mol. Ther. **5**(1), 57–62 (2002)
35. Gothelf, A., Mahmood, F., Dagnaes-Hansen, F., Gehl, J.: Efficacy of transgene expression in porcine skin as a function of electrode choice. Bioelechemistry **82**(2), 95–102 (2011)
36. Andre, F.M., Gehl, J., Sersa, G., et al.: Efficiency of High- and Low-Voltage Pulse Combinations for Gene Electrotransfer in Muscle, Liver, Tumor, and Skin. Hum. Gene Ther. **19**(11), 1261–1272 (2008)
37. Drabick, J.J., Glasspool-Malone, J., King, A., Malone, R.W.: Cutaneous transfection and immune responses to intradermal nucleic acid vaccination are significantly enhanced by in vivo electropermeabilization. Mol. Ther. **3**(2), 249–255 (2001)
38. Heller, L.C., Jaroszeski, M.J., Coppola, D., Heller, R.: Comparison of electrically mediated and liposome-complexed plasmid DNA delivery to the skin. Genet. Vaccines Ther. **6**, 16 (2008)

39. Heller, R., Schultz, J., Lucas, M.L., et al.: Intradermal delivery of interleukin-12 plasmid DNA by in vivo electroporation. DNA Cell Biol. **20**(1), 21–26 (2001)
40. Daud, A.I., DeConti, R.C., Andrews, S., et al.: Phase I trial of interleukin-12 plasmid electroporation in patients with metastatic melanoma. J. Clin. Oncol. **26**(36), 5896–5903 (2008)
41. Gothelf, A., Eriksen, J., Hojman, P., Gehl, J.: Duration and level of transgene expression after gene electrotransfer to skin in mice. Gene Ther. **17**(7), 839–845 (2010)
42. European Standard EN60601-1: Medical Electrical Equipment – Part 1: General Requirements for Basic Safety and Essential Performance, 3rd edn. British Standards Institution, London (2007)
43. Bertacchini, C., et al.: Design of an irreversible electroporation system for clinical use. Technol Cancer Res Treat **6**(4), 313–320 (2007)
44. Rebersek, M., Miklavcic, D.: Advantages and disadvantages of different concepts of electroporation pulse generation. ATKAFF **52**, 12–19 (2011)
45. Puc, M., Rebersek, S., Miklavcic, D.: Requirements for a clinical electrochemotherapy device – electroporator. Radiol. Oncol. **31**, 368–373 (1997)
46. Murakami, T., Nishi, T., Kimura, E., Goto, T., Maeda, Y., Ushio, Y., Uchino, M., Sunada, Y.: Full-length dystrophin cDNA transfer into skeletal muscle of adult mdx mice by electroporation. Muscle Nerve **27**(2), 237–241 (2003)
47. Rosati, M., et al.: Increased immune responses in rhesus macaques by DNA vaccination combined with electroporation. Vaccine **26**(40), 5223–5229 (2008)
48. Payen, E., Bettan, M., Rouyer-Fessard, P., Beuzard, Y., Scherman, D.: Improvement of mouse beta-thalassemia by electrotransfer of erythropoietin cDNA. Exp. Hematol. **29**(3), 295–300 (2001)
49. Long, Y.C., Jaichandran, S., Ho, L.P., Tien, S.L., Tan, S.Y., Kon, O.L.: FVIII gene delivery by muscle electroporation corrects murine hemophilia A. J. Gene Med. **7**(4), 494–505 (2005)
50. Zhang, G.H., et al.: Gene expression and antitumor effect following im electroporation delivery of human interferon alpha 2 gene. Acta Pharmacol. Sin. **24**(9), 891–896 (2003)
51. Trochon-Joseph, V., et al.: Evidence of antiangiogenic and antimetastatic activities of the recombinant disintegrin domain of metargidin. Cancer Res. **64**, 2062–2069 (2004)
52. ClinicalTrial.gov. US. National Institute of Health. www.clinicaltrials.gov (2012)
53. Tuting, T., Storkus, W.J., Falo Jr., L.D.: DNA immunization targeting the skin: molecular control of adaptive immunity. J. Invest. Dermatol. **111**(2), 183–188 (1998)
54. Medi, B.M., Hoselton, S., Marepalli, R.B., Singh, J.: Skin targeted DN vaccine delivery using electroporation in rabbits. I: efficacy. Int. J. Pharm. **294**(1–2), 53–63 (2005)
55. Hirao, L.A., Wu, L., Khan, A.S., Satishchandran, A., Draghia-Akli, R., Weiner, D.B.: Intradermal/subcutaneous immunization by electroporation improves plasmid vaccine delivery and potency in pigs and rhesus macaques. Vaccine **26**(3), 440–448 (2008)
56. Liu, M.A.: DNA vaccines: a review. J. Intern. Med. **253**(4), 402–410 (2003)
57. Peruzzi, D., et al.: A vaccine targeting telomerase enhances survival of dogs affected by B-cell lymphoma. Mol. Ther. **18**, 1559–1567 (2010)
58. Rune, K., Torunn, E.T., Dag, K., Jacob, M.: Clinical evaluation of pain and muscle damage induced by electroporation of skeletal muscle in humans abstract from American Society of Gene Therapy 7th annual meeting. June 2–6, 2004 Minneapolis, Minnesota, USA. Mol. Ther. **9**(Supp 1), S1–S435 (2004)
59. Glasspool-Malone, J., Drabick, J.J., Somiari, S., Malone, R.W.: Efficient nonviral cutaneous transfection. Mol. Ther. **2**, 140–146 (2000)
60. Byrnes, C.K., et al.: Electroporation enhances transfection efficiency in murine cutaneous wounds. Wound Repair Regen. **12**, 397–403 (2004)
61. Marti, G., et al.: Electroporative transfection with KGF-1 DNA improves wound healing in a diabetic mouse model. Gene Ther. **11**, 1780–1785 (2004)
62. Lin, M.P., et al.: Delivery of plasmid DNA expression vector for keratinocyte growth factor-1 using electroporation to improve cutaneous wound healing in a septic rat model. Wound Repair Regen. **14**, 618–624 (2006)
63. Roos, A.K., et al.: Enhancement of cellular immune response to a prostate cancer DNA vaccine by intradermal electroporation. Mol. Ther. **13**, 320–327 (2006)
64. Kang, J.H., Toita, R., Niidome, T., Katayama, Y.: Effective delivery of DNA into tumor cells and tissues by electroporation of polymer- DNA complex. Cancer Lett. **265**, 281–288 (2008)
65. Liu, L., et al.: Age-dependent impairment of HIF-1alpha expression in diabetic mice: correction with electroporation-facilitated gene therapy increases wound healing, angiogenesis, and circulating angiogenic cells. J. Cell. Physiol. **217**, 319–327 (2008)
66. Ferraro, B., Cruz, Y.L., Coppola, D., Heller, R.: Intradermal delivery of plasmid VEGF(165) by electroporation promotes wound healing. Mol. Ther. **17**, 651–657 (2009)
67. Brave, A., et al.: Late administration of plasmid DNA by intradermal electroporation efficiently boosts DNA-primed T and B cell responses to carcinoembryonic antigen. Vaccine **27**, 3692–3696 (2009)
68. Lladser, A., et al.: Intradermal DNA electroporation induces survivin-specific CTLs, suppresses angiogenesis and confers protection against mouse melanoma. Cancer Immunol. Immunother. **59**, 81–92 (2009)
69. Roos, A.K., Eriksson, F., Walters, D.C., Pisa, P., King, A.D.: Optimization of skin electroporation in mice to increase tolerability of DNA vaccine delivery to patients. Mol. Ther. **17**, 1637–1642 (2009)

70. Titomirov, A.V., Sukharev, S., Kistanova, E.: In vivo electroporation and stable transformation of skin cells of newborn mice by plasmid DNA. Biochim. Biophys. Acta **1088**, 131–134 (1991)
71. Lucas, M.L., Jaroszeski, M.J., Gilbert, R., Heller, R.: In vivo electroporation using an exponentially enhanced pulse: a new waveform. DNA Cell Biol. **20**, 183–188 (2001)
72. Pavselj, N., Preat, V.: DNA electrotransfer into the skin using a combination of one high- and one low-voltage pulse. J. Control. Release **106**, 407–415 (2005)
73. Maruyama, H., et al.: Skin-targeted gene transfer using in vivo electroporation. Gene Ther. **8**, 1808–1812 (2001)
74. Zhang, L., Nolan, E., Kreitschitz, S., Rabussay, D.P.: Enhanced delivery of naked DNA to the skin by non-invasive in vivo electroporation. Biochim. Biophys. Acta **1572**, 1–9 (2002)
75. Lee, P.Y., Chesnoy, S., Huang, L.: Electroporatic delivery of TGFbeta1 gene works synergistically with electric therapy to enhance diabetic wound healing in db/db mice. J. Invest. Dermatol. **123**, 791–798 (2004)
76. Thanaketpaisarn, O., Nishikawa, M., Yamashita, F., Hashida, M.: Tissue- specific characteristics of in vivo electric gene: transfer by tissue and intravenous injection of plasmid DNA. Pharm. Res. **22**, 883–891 (2005)
77. Heller, L.C., et al.: Optimization of cutaneous electrically mediated plasmid DNA delivery using novel electrode. Gene Ther. **14**, 275–280 (2007)
78. Vandermeulen, G., et al.: Optimisation of intradermal DNA electrotransfer for immunisation. J. Control. Release **124**, 81–87 (2007)
79. Vandermeulen, G., et al.: Skin-specific promoters for genetic immunisation by DNA electroporation. Vaccine **27**, 4272–4277 (2009)
80. Vandermeulen, G., et al.: Effect of tape stripping and adjuvants on immune response after intradermal DNA electroporation. Pharm. Res. **26**, 1745–1751 (2009)
81. Gothelf, A., Hojman, P., Gehl, J.: Therapeutic levels of erythropoietin (EPO) achieved after gene electrotransfer to skin in mice. Gene Ther. (2010). doi:10.1038/gt.2010.46
82. Ho, S.H., et al.: Protection against collagen-induced arthritis by electrotransfer of an expression plasmid for the interleukin 4. Biochem. Biophys. Res. Commun. **321**, 759–766 (2004)
83. Cukjati, D., Batiuskaite, D., André, F., Miklavcic, D., Mir, L.M.: Real time electroporation control for accurate and safe in vivo non-viral gene therapy. Bioelectrochemistry **70**, 501–507 (2007)
84. Hojman, P., et al.: Physiological effects of high- and low-voltage pulse combinations for gene electrotransfer in muscle. Hum. Gene Ther. **19**, 1249–1260 (2008)
85. Hojman, P., Zibert, J.R., Gissel, H., Eriksen, J., Gehl, J.: Gene expression profiles in skeletal muscle after gene electrotransfer. BMC Mol. Biol. **8**, 56 (2007)
86. Jeong, J.G., et al.: Electrotransfer of human IL-1Ra into skeletal muscles reduces the incidence of murine collagen-induced arthritis. J. Gene Med. **6**, 1125–1133 (2004)
87. Abruzzese, R.V., et al.: Ligand-dependent regulation of vascular endothelial growth factor and erythropoietin expression by a plasmid-based autoinducible gene-switch system. Mol. Ther. **2**, 276–287 (2000)
88. Bettan, M., et al.: High level protein secretion into blood circulation after electric pulse-mediated gene transfer into skeletal muscle. Mol. Ther. **2**, 204–210 (2000)
89. Aihara, H., Miyazaki, J.: Gene transfer into muscle by electroporation in vivo. Nat. Biotechnol. **16**, 867–870 (1998)
90. Scheerlinck, J.P., et al.: In vivo electroporation improves immune responses to DNA vaccination in sheep. Vaccine **22**, 1820–1825 (2004)
91. Terada, Y., et al.: Efficient and ligand-dependent regulated erythropoietin production by naked dna injection and in vivo electroporation. Am. J. Kidney Dis. **38**, S50–S53 (2001)
92. Tone, C.M., Cardoza, D.M., Carpenter, R.H., Draghia-Akli, R.: Long-term effects of plasmid-mediated growth hormone releasing hormone in dogs. Cancer Gene Ther. **11**, 389–396 (2004)
93. Wu, C.J., Lee, S.C., Huang, H.W., Tao, M.H.: In vivo electroporation of skeletal muscles increases the efficacy of Japanese encephalitis virus DNA vaccine. Vaccine **22**, 1457–1464 (2004)

# Why Does an I.M. Immunization Work?

## 50

Emanuela Bartoccioni

## Contents

| | | |
|---|---|---|
| 50.1 | Introduction | 793 |
| 50.2 | The Skeletal Muscle | 794 |
| 50.3 | The Inflammatory Response into the Muscle: The Trigger | 794 |
| 50.4 | The Adaptive Immune Response into the Muscle: Who Does What? | 795 |
| 50.5 | Dendritic Cells as Professional APC | 796 |
| 50.6 | Muscle Cells as Nonprofessional APC | 798 |
| 50.7 | Tolerance: The Other Side of the Coin | 800 |
| 50.8 | Is There a Recipe for a Perfect DNA Vaccine? | 800 |
| Conclusions | | 801 |
| References | | 801 |

### Abstract

The skeletal muscle has been long viewed as a site of election for immunization due to the sustained stay of the antigens in this tissue. Besides a passive function during immune priming, however, recent studies have shed light on an active role of muscle cells (fibers, satellite cells, and stromal cells) in inflammatory and immune response. In this context, the chemical nature of the antigen (into tissues and cells?) and the choice of adjuvants contribute to determine the fate of the antigen itself and the resulting immune response. In particular, this applies to modern vaccine approaches that involve the use of nucleic acid to locally synthesize the protein antigen. Based on these novel concepts, the possibility of manipulating therapeutically the mechanisms of peripheral tolerance at the level of muscle tissue will be discussed in the light of the recent literature.

## 50.1 Introduction

Immunization consists in the induction of immunological memory to protect against subsequent natural infection by a pathogen, through the specific response of T and B cells and the production of neutralizing antibodies directed against the foreign antigen. Traditionally, the protein antigen is mixed with adjuvants and injected into the muscle. The adjuvants are necessary to trigger

E. Bartoccioni, PhD
Department of Laboratory Medicine,
General Pathology Institute,
Università Cattolica S. Cuore, Rome, Italy
e-mail: ebartoc@rm.unicatt.it

innate immune mechanisms that are the initial events that dictate the outcome of the adaptive immune response [1, 2].

The immune response always begins with an inflammatory reaction that subsequently evolves into an antigen-specific response, through the involvement of monocytes/macrophages, dendritic cells (DC), and B and T cells. In the early 1990s, new gene therapy technologies, in which non-replicating bacterial plasmids encoding proteins were injected intramuscularly, were set up. This novel approach revealed that proteins endogenously expressed in the muscle in vivo from recombinant plasmids stimulated the immune response. High levels of protein expression can in fact be achieved in muscle tissue, and the direct i.m. transfer of recombinant DNA presents several advantages, such as specific immune response to the protein antigen coupled with an adjuvant-like action of the plasmid, low costs, and great safety [3, 4].

## 50.2 The Skeletal Muscle

The skeletal muscle is the most abundant tissue of the human body. It consists of myofibers, a class of large syncytial multinuclear cells, which are terminally differentiated permanent cells. Mature myofibers have their origin from the fusion of mononuclear satellite stem cells, which can proliferate to regenerate and repair the damaged tissue: following injury to a muscle fiber, satellite cells become activated myoblasts, proliferate, and fuse into primitive multinuclear cells, which then differentiate into mature muscle fibers [5, 6]. These activated satellite cells may be cultured and expanded in vitro, and the resulting myoblasts may fuse in myotubes, the in vitro counterpart of differentiated adult muscle cells. Several kinds of resident macrophages are also present within muscle tissue, and recently also monocyte-derived and conventional DCs residing in the skeletal muscle before and after i.m. immunization have been well characterized in mice [7].

## 50.3 The Inflammatory Response into the Muscle: The Trigger

During the innate response, the inflammatory stimulus is the trigger that initiates a complex network of cell reactions, and its physical, chemical, or biological nature determines the quality of the consequent response. Actually, while stimuli of any nature cause an innate response, only biological agents may induce its transition to adaptive response. Microbial proteins, lipids, carbohydrates, and nucleic acids contain common structural patterns collectively called PAMPs (pathogen-associated molecular patterns) that bind to innate immune receptors (toll-like receptor, TLR) present on cell surface or endosomal membrane of responding cells [8]. The TLR family includes more than ten members, with different ligand specificities and differential expression among cell types [9].

While PAMPs are naturally present in biological agents, the immunization process needs that adjuvants with similar function are added to purified protein antigens to obtain full adaptive response; they are necessary to activate DC, which then become able to take up incoming antigens and carry them to draining lymph node (LN). Further, they may be used to influence the magnitude and the type of the specific response to produce the most effective immunity for each specific pathogen. In addition to conventional adjuvants (such as aluminum salts or MF59), it is now taking hold the use of natural as well as synthetic ligands for well-defined TLR, and a number of these are now in clinical or late preclinical stages [1].

Moreover, in the last few years, evidence has been accumulating that muscle cells can respond to several kinds of such molecules and actively participate to several steps of the innate and adaptive immune responses, by expressing membrane molecules and cytokines. Stimulation of TLR2 and TLR4, in particular, induces the mRNA of the CCL2 and CXCL1 chemokines in

**Table 50.1** The subcellular localization of several toll-like receptors (TLRs), their natural ligands, their synthetic agonist, and the presence in muscle are shown

|      | Subcellular localization | Natural ligand(s) | Synthetic agonist(s) | Expression in muscle cells | References |
|------|--------------------------|-------------------|----------------------|----------------------------|------------|
| TLR2 | Plasma membrane | Lipoproteins/lipopeptides Peptidoglycan | BPPcysMPEG Pam3CSK4 | Yes (not in myoblasts) | [12–14] |
| TLR3 | Endosomes | dsRNA | Poly-IC, poly-ICLC, poly-IC$_{12}$-U | Yes | [15] |
| TLR4 | Plasma membrane | LPS | MPL | Yes | [14, 16] |
| TLR5 | Plasma membrane | Flagellin | Flagellin-antigen fusion proteins | Yes (not in myoblasts) | [14] |
| TLR7/TLR8 | Endosomes | ssRNA | Imiquimod, resiquimod | ? | [17] |
| TLR9 | Endosomes | Bacterial/viral DNA | CpG oligodeoxynucleotides | Yes | [14] |

C2C12 mouse myoblasts [10]. Similarly, the stimulation of TLR3 in cultured human myoblasts activates the transcription factor NFκB and elicits IL-8 secretion, while the binding of unmethylated CpG dinucleotides to TLR9 triggers the release of IL-12, TNFα, and IFNα, a critical step in cellular immune response [11]. Characteristics of the main TLRs and their expression in muscle are summarized in Table 50.1.

Thus, in the context of i.m. immunization, the injury caused by the needle is by itself the first trigger for inflammation, but it is only in the presence of foreign biological molecules able to bind to TLRs that the adaptive response may begin. When DNA vaccination is used, bacterial DNA from the plasmid backbone, which includes sequence motifs like the CpG dinucleotide, may directly activate bystander cells through TLR9.

Finally, DNA vaccine technology has suffered until a short time ago from an inadequate immunogenicity; however, the recent development of new gene delivery techniques such as the electrotransfer, which allows a greater delivery and expression of the transgene upon the application of an electric field, has significantly enhanced the immune response, possibly thanks to the damage and inflammation induced in muscle tissue by the electric field [4, 18].

## 50.4 The Adaptive Immune Response into the Muscle: Who Does What?

There are two kinds of adaptive responses, humoral immunity and cell-mediated immunity, brought about by different components of the immune system and directed against different types of microbes.

Extracellular microbes (such as bacteria and viruses before their entry into cells) and their toxins (poisonous molecules they produce against the host) are neutralized by antibodies secreted by B cells, while intracellular microbes (such as intracellular bacteria and viruses) require a more complex defense strategy. In fact, intracellular microbes survive and proliferate within the host cells and phagocytes, where they are inaccessible to circulating antibodies; thus, cell-mediated immunity ($T_{H1}$ and $T_{CTL}$ cells) becomes necessary to destroy microbes residing in phagocytes or to kill infected cells in order to eliminate the reservoirs of infection.

Vaccination with protein antigens mostly induces humoral response. Instead, DNA vaccination is able to stimulate both kinds of response because injected plasmid DNA may induce both extracellular and/or intracellular antigen expression (Figs. 50.1 and 50.2). Injected plasmid DNA may enter ("transfect" in jargon) muscle cells as

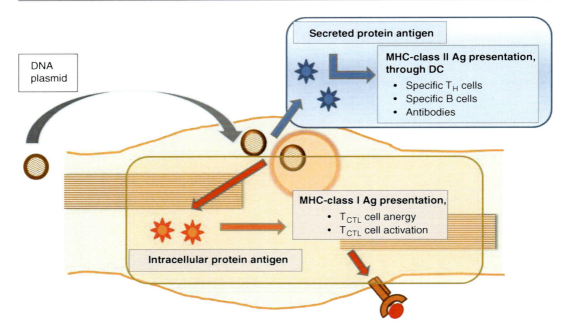

**Fig. 50.1** Expression of plasmid DNA-encoded antigens in muscle fibers. Different potential outcomes for DNA-driven antigen expression in muscle cells are depicted. Secretory proteins (i.e., proteins harboring a signal peptide) are released in the extracellular space, where they behave as soluble antigens; thus, they can be taken up by DC and presented through the MHC II route, leading to CD4+ T cell activation and antibody production by B cells. Non-secreted proteins (*red pathway*) enter the MHC I route, and their surface exposure can elicit either CD8+ T cell anergy or activation, based on the concurrent expression of costimulatory molecules as induced by an inflammatory microenvironment. Note that presentation of secretory proteins through the MHC I route can also occur (not shown)

well as resident professional antigen-presenting cells (APCs), like macrophages and DC. Whichever the producing cells, the protein antigen, if secreted, may induce the normal humoral response (like an extracellular microbe). In particular, transfected APC may present the antigen to $T_{CTL}$ cell through two distinct routes: either directly, through MHC class I molecules as any other endogenous protein, or by the so-called cross-priming mechanism, whereby muscle cell fragments containing the protein would be taken up by professional APC and the antigen presented – through MHC class I molecules – to $T_{CTL}$ cells (Fig. 50.3) [19]. Another form of presentation recently described may be mediated by the direct transfer of MHC I-peptide complexes from bystander cells to the DC surface ("cross-dressing") [20]. Moreover, captured antigen will also be presented to T helper ($T_H$) lymphocytes through the conventional MHC class II-dependent route. These professional APCs are DCs that migrate to draining lymph nodes, to bring about the adaptive immune process.

But let's take a closer look to how DCs work in adaptive immunity.

## 50.5 Dendritic Cells as Professional APC

Dendritic cells, when activated by foreign antigens via the TLR, start an adaptive immune response to these antigens. The kind of immune response initiated depends on the type of DC and also on the particular innate immune signals received. DCs regulate CD4+ T helper ($T_H$) cell differentiation through cytokines and membrane B7 molecules. There is a family of these molecules, with different characteristics. Both B7-1 and B7-2 activate or suppress CD4+ T cells, depending on which ligand is expressed on CD4+ T cell: the presence of CD28 on T cells is able to

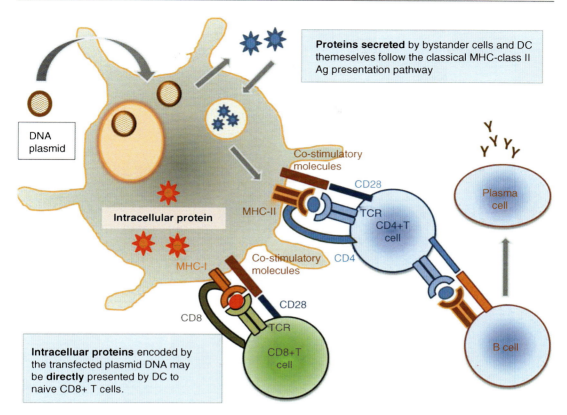

**Fig. 50.2** Central role of professional APCs in DNA vaccination. APC (dendritic cells) are also targeted by injected DNA and express DNA-encoded antigens. Proteins secreted by DC and by surrounding cells (muscle fibers, stromal cells) are taken up and exposed for presentation to CD4+ T cells through the MHC II route. Intracellular protein antigens ("pseudoinfection") are presented through the MHC I route to naive CD8+ T lymphocytes. Successful antigen-specific priming of both CD4+ and CD8+ T cells requires the presence of costimulatory molecules (B7-1, B7-2) expressed at the surface of professional APC and co-engaged with MHC molecules by T cell receptor coreceptors (TCR and CD28, respectively). This complex and highly specialized multimolecular contact between APC and T lymphocytes has been named, by analogy with connections between neuronal cells, "immunological synapse." The interaction between B and T cells is also depicted. B cells express MHC II and costimulatory molecules and can present secreted antigens (once recognized and captured through the BCR) to T cells. Contact with T helper lymphocytes fully primes B cells for antibody production (T cell-dependent response)

transduce activating signals, while CTLA-4 induces T cell suppression [21]. The inducible costimulator ligand (ICOSL, also known as B7-h, B7-H2) stimulates not only the effector or memory T cell responses but also the generation of regulatory T (Treg) cells [22, 23].

It is well established that $T_H$ cells may differentiate into $T_{H1}$ cells, required for the generation of memory CD8+ T cells, in response to intracellular microbes, or into $T_{H2}$ cells in response to extracellular microbes. This different polarization is a function of the different set of cytokines produced by DC that in turn reflects the different TLR-mediated innate signal received from the pathogen.

Immature DCs are present into the tissues under steady-state conditions, while TLR agonists stimulate their maturation and promote their migration to draining lymph nodes.

Immature and mature DCs are characterized by different sets of membrane molecules, like MHC class II and B7 family molecules and chemokine receptors. There are two major DC subsets: myeloid (or conventional or classical) DC and plasmacytoid DC, which are not completely overlapping in mouse and man [24]. Recently,

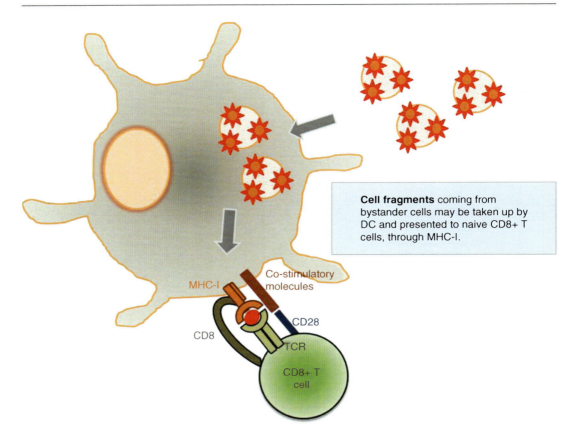

**Fig. 50.3** Cross presentation by dendritic cells. DCs can prime CD8+ cells, other than with endogenous antigens (including those encoded by transfected DNA) and also with exogenous antigens taken up as part of cell fragments or microvesicles or from surrounding cells. This modality of antigen presentation, called cross presentation, involves the intracellular processing of the antigen

human CD141+ DCs have been identified as the human counterpart of mouse CD8+ DCs; they are particularly interesting as being able to capture exogenous antigens and present them through MHC class I molecules (cross presentation) to CD8+ $T_{CTL}$ cells [25].

A recent interesting study on the characterization of dendritic cells present within the mouse muscle during i.m. immunization has shown that large numbers of resident interstitial DC were able to capture Ag in the muscle, migrate to draining LN, and efficiently activate naive T cells (Fig. 50.4), while DC arising from blood monocytes achieved these abilities largely only after the addition of LPS to sterile alum-adjuvanted antigen [7].

## 50.6 Muscle Cells as Nonprofessional APC

Even if transfected DCs play a dominant role in DNA vaccine-mediated immunity, they account for only a part of the total induced response, with local nonprofessional APC having a significant role in adaptive immune response.

Muscle cells are able to actively participate in the induction of immunity and to behave as nonprofessional APC. Muscle cells express receptors for cytokines and PAMPs that enable them to respond to an inflammatory milieu, by secreting cytokines and chemokines and expressing adhesion molecules [26]. In particular, some membrane protein necessary to the APC function, like

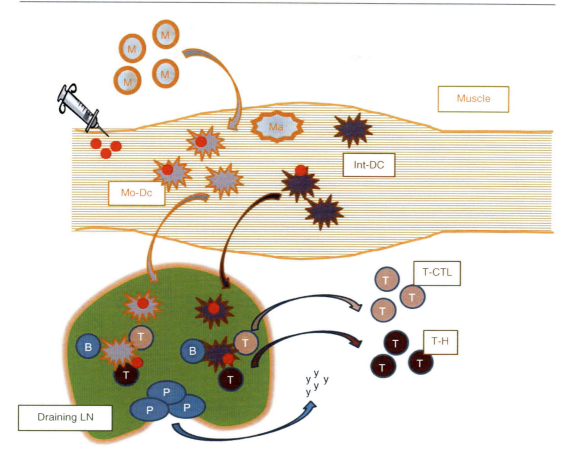

**Fig. 50.4** Different classes of DCs involved in muscle response to injected antigens. Muscle contains interstitial/resident dendritic cells (Int-DC) that monitor the immunological status of the tissue and transport injected antigens to lymph nodes (draining LN) in order to activate specific B and T cell response. Of note, mobilization of resident DC requires also the presence of inflammatory stimuli (LPS, adjuvants? See text); another class of DCs differentiates from monocytes (M) that migrate to muscle in the context of inflammatory response (Mo-DC), capture the antigen, and transport it to LN to fulfill APC duties. These different classes of muscle DC have just begun to be molecularly characterized. B cell-derived plasma cells (P) secreting antibodies (Y) are depicted. Macrophages (Ma) may derive from monocytes or be resident

MHC class I and class II molecules, and costimulatory molecules have been observed in several experimental systems [27–29]. Importantly, following plasmid transfection, muscle cells express several molecules necessary to antigen presentation such as MHC class I and costimulatory molecules like BB-1 and B7-1, gaining the ability to present immunogenic DNA-encoded antigens to $T_{CTL}$ cells [30].

The idea that muscle fibers act as APC faces a number of important problems.

First, muscle cells expressing new antigen proteins should appear to the immune system like infected "different" cells and be destroyed by the same cytotoxic-specific T cells they induce; however, muscle fibers are large syncytial postmitotic cells that are unlikely to undergo immune killing by lymphocytes, as it is also suggested by evidence that antigens transfected by DNA electrotransfer extensively diffuse in muscle tissue and are still expressed after 9 months, thus confirming long-term survival of producing cells [18].

Second, although antigen production by muscle cells seems to be durable, the APC function conferred by DNA transfection is expected to be transient in nature due to the extinction of the inflammatory activity that accompanies DNA transfer (see above). Nevertheless, a variety of model systems have shown that both memory T and B cells can persist in the absence of the antigen, ensuring the effectiveness of vaccination even in the absence of active antigen presentation [31–34].

Third, antigen-presenting muscle cells may induce tolerance instead of immunity.

## 50.7 Tolerance: The Other Side of the Coin

It thus appears that the skeletal muscle may be easily manipulated to express many immunologically relevant molecules and this makes it a promising target for DNA vaccination. However, the immune system is able to induce tolerance at the level of this tissue, as appears from the rarity of autoimmune diseases directed against the muscle. Understanding the mechanisms involved in the induction and maintenance of immune tolerance, as well as anergy (a state of unresponsiveness to antigen), is essential to efficiently address i.m. vaccination.

In general, in noninflammatory conditions the expression of neoantigens could induce tolerance instead of immunity. This may be the case in which the antigen is expressed in transgenic animals.

Very recently, it has been shown that immunological tolerance to muscle autoantigens involves peripheral deletion of autoreactive CD8+ $T_{CTL}$ cells [35]. In this model, transgenic mice expressed a membrane-bound form of ovalbumin (OVA) neoantigen specifically in the skeletal muscle (SM-OVA). In these animals, there were no signs of autoimmunity, but the immunization with OVA protein induced CD4+-dependent response and anti-OVA IgG antibody production, indicating that OVA-specific CD4+ cells were present in OVA mice and were not intrinsically tolerant to OVA, but simply ignored the antigen unless presented in the appropriate inflammatory context.

Conversely, when immunization with vesicular stomatitis virus encoding OVA (VSV-OVA) was used to elicit an OVA-specific $T_{CTL}$ cell response, no cytotoxic activity was observed; moreover, adoptively transferred OVA-specific CD8+ $T_{CTL}$ cells were selectively deleted from the periphery of SM-OVA mice. This process that is not related to T regulatory cells probably involves both the partial T cell activation in absence of costimulatory molecules at the surface of APC and the inadequacy of cytokine presence in noninflammatory conditions. Interestingly, it appears that when the antigen is permanently expressed by the myofibers, the peripheral tolerance is stronger than immune response.

Muscle-induced tolerance thus represents a biologically relevant phenomenon with respect to DNA vaccination, since in vivo transfection mimics under several aspects the intracellular presence of the antigen in transgenic mice.

## 50.8 Is There a Recipe for a Perfect DNA Vaccine?

Based on the above considerations, the success of DNA vaccination could depend by several factors:

1. The plasmid DNA should transfect not only muscle cells, which constitute a large volume of tissue useful for the initial response, but also DCs that, in an inflammatory milieu, start the complex machinery to reach a whole effector and memory response.
2. The plasmid backbone should contain unmethylated CpG motifs, which per se behave as adjuvants and bind intracellular TLR9, polarizing the response of $T_H$ toward $T_{H1}$ cells.
3. Adjuvants should also be used, in order to guarantee the inflammatory trigger necessary to activate the DCs.
4. The amount of expressed antigen, which correlates with the level of the immune response, should be sufficiently high and the expression long enough to establish immune memory. Once memory is established, presence of the antigen is probably no longer necessary (see above).

## Conclusions

Novel knowledge on muscle immunobiology clearly indicates that this tissue, far from acting as a simple mechanical support, actively contributes to the establishment of vaccine-induced immunity. This is especially true when DNA-based vaccines are concerned due to the remarkable protein synthetic and secretory capacity of muscle fibers, coupled with a previously unsuspected ability to cross talk with inflammatory cells and cytokines. In keeping, DNA-based intramuscular vaccination has raised important expectations as an avenue to circumvent the pitfalls of conventional vaccine strategies especially against intracellular pathogens, and the first clinical trials in this direction are on the way.

Another largely overlooked possibility is to exploit muscle immunological properties in order to manipulate immune tolerance. Initial experimental evidence obtained in transgenic animals, together with clinical experience of the rarity of muscle-directed autoimmunity, stands in support of this idea.

Are we going to use muscle to induce tolerance against DNA-encoded proteins? If so, how are we going to avoid what will in this case represent an unwanted immune response? Is muscle the way against autoimmunity?

To those who don't have the crystal ball nor believe in immunological dogmas, the future is probably bringing many surprises. Stay tuned, and keep your muscles fit.

## References

1. Coffman, R.L., Sher, A., Seder, R.A.: Vaccine adjuvants: putting innate immunity to work. Immunity **33**, 492–503 (2010)
2. Desmet, C.J., Ishii, K.J.: Nucleic acid sensing at the interface between innate and adaptive immunity in vaccination. Nat. Rev. Immunol. **12**, 479–491 (2012)
3. Rice, J., Ottensmeier, C.H., Stevenson, F.K.: DNA vaccines: precision tools for activating effective immunity against cancer. Nat. Rev. Cancer **8**, 108–120 (2008)
4. Rochard, A., Scherman, D., Bigey, P.: Genetic immunization with plasmid DNA mediated by electrotransfer. Hum. Gene Ther. **22**, 789–798 (2011)
5. Wiendl, H., Hohlfeld, R., Kieseier, B.C.: Immunobiology of muscle: advances in understanding an immunological microenvironment. Trends Immunol. **26**, 373–380 (2005)
6. Bischoff, R., Franzini-Armstrong, C.: Satellite and stem cells in muscle regeneration. In: Engel, A., Franzini-Armstrong, C. (eds.) Myology, pp. 66–86. McGraw-Hill, New York (2004)
7. Langlet, C., Tamoutounour, S., Henri, S., Luche, H., Ardouin, L., Grégoire, C., Malissen, B., Guilliams, M.: CD64 expression distinguishes monocyte-derived and conventional dendritic cells and reveals their distinct role during intramuscular immunization. J. Immunol. **188**, 1751–1760 (2012)
8. Janeway Jr., C.A.: Approaching the asymptote? Evolution and revolution in immunology. Cold Spring Harb. Symp. Quant. Biol. **54**, 1–13 (1989)
9. Zarember, K.A., Godowski, P.J.: Tissue expression of human Toll-like receptors and differential regulation of Toll-like receptor mRNAs in leukocytes in response to microbes, their products, and cytokines. J. Immunol. **168**, 554–561 (2002)
10. Gurunathan, S., Wu, C.Y., Freidag, B.L., Seder, R.A.: DNA vaccines: a key for inducing long-term cellular immunity. Curr. Opin. Immunol. **12**, 442–447 (2000)
11. Krieg, A.M.: CpG motifs in bacterial DNA and their immune effects. Annu. Rev. Immunol. **20**, 709–760 (2002)
12. Lombardi, V., Van Overtvelt, L., Horiot, S., Moussu, H., Chabre, H., Louise, A., Balazuc, A.M., Mascarell, L., Moingeon, P.: Toll-like receptor 2 agonist Pam3CSK4 enhances the induction of antigen-specific tolerance via the sublingual route. Clin. Exp. Allergy **38**, 1819–1829 (2008)
13. Prajeeth, C.K., Jirmo, A.C., Krishnaswamy, J.K., Ebensen, T., Guzman, C.A., Weiss, S., Constabel, H., Schmidt, R.E., Behrens, G.M.: The synthetic TLR2 agonist BPPcysMPEG leads to efficient cross-priming against co-administered and linked antigens. Eur. J. Immunol. **40**, 1272–1283 (2010)
14. Boyd, J.H., Divangahi, M., Yahiaoui, L., Gvozdic, D., Qureshi, S., Petrof, B.J.: Toll-like receptors differentially regulate CC and CXC chemokines in skeletal muscle via NF-kappaB and calcineurin. Infect. Immun. **74**, 6829–6838 (2006)
15. Schreiner, B., Voss, J., Wischhusen, J., Dombrowski, Y., Steinle, A., Lochmuller, H., Dalakas, M., Melms, A., Wiendl, H.: Expression of toll-like receptors by human muscle cells in vitro and in vivo: TLR3 is highly expressed in inflammatory and HIV myopathies, mediates IL-8 release and up-regulation of NKG2D-ligands. FASEB J. **20**, 118–120 (2006)
16. Casella, C.R., Mitchell, T.C.: Putting endotoxin to work for us: monophosphoryl lipid A as a safe and effective vaccine adjuvant. Cell. Mol. Life Sci. **65**, 3231–3240 (2008)
17. Gorden, K.B., Gorski, K.S., Gibson, S.J., Kedl, R.M., Kieper, W.C., Qiu, X., Tomai, M.A., Alkan, S.S., Vasilakos, J.P.: Synthetic TLR agonists reveal

functional differences between human TLR7 and TLR8. J. Immunol. **174**, 1259–1268 (2005)
18. Mir, L.M., Bureau, M.F., Gehl, J., Rangara, R., Rouy, D., Caillaud, J.M., Delaere, P., Branellec, D., Schwartz, B., Scherman, D.: High-efficiency gene transfer into skeletal muscle mediated by electric pulses. Proc. Natl. Acad. Sci. U.S.A. **96**, 4262–4267 (1999)
19. Joffre, O.P., Segura, E., Savina, A., Amigorena, S.: Cross-presentation by dendritic cells. Nat. Rev. Immunol. **12**, 557–569 (2012)
20. Wakim, L.M., Bevan, M.J.: Cross-dressed dendritic cells drive memory CD8+ T-cell activation after viral infection. Nature **471**, 629–632 (2011)
21. Krummel, M.F., Allison, J.P.: CD28 and CTLA-4 have opposing effects on the response of T cells to stimulation. J. Exp. Med. **182**, 459–465 (1995)
22. Sharpe, A.H., Freeman, G.J.: The B7-CD28 superfamily. Nat. Rev. Immunol. **2**, 116–126 (2002)
23. Ito, T., Yang, M., Wang, Y.H., Lande, R., Gregorio, J., Perng, O.A., Qin, X.F., Liu, Y.J., Gilliet, M.: Plasmacytoid dendritic cells prime IL-10-producing T regulatory cells by inducible costimulator ligand. J. Exp. Med. **204**, 105–115 (2007)
24. Palucka, K., Banchereau, J., Mellman, I.: Designing vaccines based on biology of human dendritic cell subsets. Immunity **33**, 464–478 (2010)
25. Haniffa, M., Shin, A., Bigley, V., McGovern, N., Teo, P., See, P., Wasan, P.S., Wang, X.N., Malinarich, F., Malleret, B., Larbi, A., Tan, P., Zhao, H., Poidinger, M., Pagan, S., Cookson, S., Dickinson, R., Dimmick, I., Jarrett, R.F., Renia, L., Tam, J., Song, C., Connolly, J., Chan, J.K., Gehring, A., Bertoletti, A., Collin, M., Ginhoux, F.: Human tissues contain CD141hi crosspresenting dendritic cells with functional homology to mouse CD103+ nonlymphoid dendritic cells. Immunity **37**, 60–73 (2012)
26. Marino, M., Scuderi, F., Provenzano, C., Bartoccioni, E.: Skeletal muscle cells: from local inflammatory response to active immunity. Gene Ther. **18**, 109–116 (2010)
27. Goebels, N., Michaelis, D., Wekerle, H., Hohlfeld, R.: Human myoblasts as antigen presenting cells. J. Immunol. **149**, 661–667 (1992)
28. Behrens, L., Kerschensteiner, M., Misgeld, T., Goebels, N., Wekerle, H., Hohlfeld, R.: Human muscle cells express a functional costimulatory molecule distinct from B7.1 (CD80) and B7.2 (CD86) in vitro and in inflammatory lesions. J. Immunol. **161**, 5943–5951 (1998)
29. Curnow, J., Corlett, L., Willcox, N., Vincent, A.: Presentation by myoblasts of an epitope from endogenous acetylcholine receptor indicates a potential role in the spreading of the immune response. J. Neuroimmunol. **115**, 127–134 (2001)
30. Shirota, H., Petrenko, L., Hong, C., Klinman, D.M.: Potential of transfected muscle cells to contribute to DNA vaccine immunogenicity. J. Immunol. **179**, 329–336 (2007)
31. Hou, S., Hyland, L., Ryan, K.W., Portner, A., Doherty, P.C.: Virus-specific CD8+ T-cell memory determined by clonal burst size. Nature **369**, 652–654 (1994)
32. Lau, L.L., Jamieson, B.D., Somasundaram, T., Ahmed, R.: Cytotoxic T-cell memory without antigen. Nature **369**, 648–652 (1994)
33. Sallusto, F., Lanzavecchia, A., Araki, K., Ahmed, R.: From vaccines to memory and back. Immunity **33**, 451–463 (2010)
34. Zielinski, C.E., Corti, D., Mele, F., Pinto, D., Lanzavecchia, A., Sallusto, F.: Dissecting the human immunologic memory for pathogens. Immunol. Rev. **240**, 40–51 (2011)
35. Franck, E., Bonneau, C., Jean, L., Henry, J.P., Lacoume, Y., Salvetti, A., Boyer, O., Adriouch, S.: Immunological tolerance to muscle autoantigens involves peripheral deletion of autoreactive CD8+ T cells. PLoS One **7**, e36444 (2012)

# Part VIII
# Patenting, Manufacturing, Registration

## Overview of Part VIII

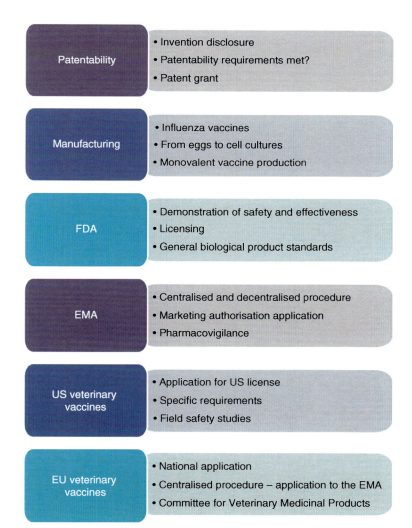

The way from a feasibility study to a product launch is time- and cost-intensive. A successful new vaccine is the culmination of many carefully planned different steps in laboratory under GLP, patent and marketing department, and, last but not least, finally the application for marketing authorization. The proof of concept demonstrates that the candidate vaccines work in principle. Successful clinical studies according to international GCP guidelines are the basis for a license. The GMP controlled manufacturing process, including periodic facility inspections, is part of the licensure. Vaccine safety is controlled post marketing by pharmacovigilance.

*Patentability.* Both the USA and China dominate first (earliest priority) and second (a subsequent family member) patent filings claiming active ingredients of vaccines against infectious diseases, with Europe and Australia rising to dominance for second patent filings in the vaccine field. Claims reciting a medical indication using a known or new vaccine composition must be inventive and sufficiently disclosed. The EPC limits the ability of an applicant to claim aspects of a vaccine identified as a method of treatment, surgery, or diagnosis in view of various public policies underlying the patentability exclusion under Article 53(c) EPC 2000.

*Manufacturing.* The development and the production of influenza vaccines are based on a complex manufacturing process starting with the selection and development of optimal candidate vaccine viruses, and it requires various dynamic interactions with regulatory authorities and health-care officials. Most influenza vaccine production is based on classical egg-based technology, a technology that has been used to produce seasonal vaccine for more than 30 years. Novel cell-culture technologies can offer various advantages over egg-based manufacturing methods and most likely will supplement the current egg-based technology.

*FDA regulations for human vaccines.* Vaccine licensure is based on a demonstration of safety and effectiveness, as well as the ability of the license holder to manufacture the product in a consistent manner within the defined and agreed upon specifications. After licensure, monitoring of the vaccine and production activities, including periodic facility inspections, must continue as long as the manufacturer holds a license for the product.

*EMA regulations for human vaccines.* Although the European pharmaceutical legislation does not provide a formal definition, vaccines are typically considered medicinal products containing one or more immunogenic antigens intended for prophylaxis against infectious disease. Medicinal products containing one or more immunogenic antigens for the treatment of disease, e.g., chronic HIV infection, chronic hepatitis B or C infection, cancer, or Alzheimer's disease, are typically referred to as therapeutic vaccines or active immunotherapy.

*Vet vaccines in the USA.* The Center for Veterinary Biologics (CVB), United States Department of Agriculture (USDA), located in Ames, Iowa, has regulatory jurisdiction over veterinary biologics in the USA for the diagnosis, prevention, and treatment of animal diseases. All veterinary vaccines sold within the USA must have either a US Veterinary Biological Product License, produced within a licensed establishment, or a US Veterinary Biological Permit.

*Vet vaccines in the EU.* Veterinary vaccines must be authorized by the relevant competent authorities in the EU. An application for a Marketing Authorization (MA) must be submitted, and this should include a dossier that demonstrates the quality of the vaccine and the safety and efficacy in the target species. It should be noted that safety must also be demonstrated for the user, the environment, and the consumer.

# Patentability of Vaccines: A Practical Perspective

## 51

Stacey J. Farmer and Martin Grund

## Contents

| | | |
|---|---|---|
| 51.1 | Introductory Remarks | 808 |
| 51.2 | Patentable Subject Matter Including Exclusions and Exceptions | 810 |
| 51.3 | Principles of Novelty and State of the Art | 814 |
| 51.4 | Claims Reciting a Medical Indication Using a Known or New Vaccine Composition Must Be Inventive and Sufficiently Disclosed | 815 |
| 51.5 | A Few Words on Post-Patent Grant Procedures | 817 |
| 51.6 | Concluding Remarks | 819 |
| References | | 820 |

S.J. Farmer, PhD (✉) • M. Grund, PhD
Grund Intellectual Property Group, Munich, Germany
e-mail: farmer@grundipg.com

### Abstract

This chapter provides a non-exhaustive survey of current structures governing patent granting and defense mechanisms, patentability approaches, and notable legal trends relating to vaccine-based innovations. Developing a strategically meaningful global patent portfolio entails a balanced approach between a thoughtful strategy that safeguards an actual or contemplated market position such that a reliable revenue stream can be maintained, with a careful selection of inventions for patenting given the constantly evolving worldwide patent systems and landscapes. A robust patent strategy typically begins by building upon a core technology, with an eye to capturing secondary subject-matter springing from such first inventions. For vaccine developments, initial patent filings might include a variety of components affording a new structural or therapeutic function, including antigens (any of nucleic acid sequences, proteins, with accompanying expression technologies), adjuvants, and excipients, delivery platforms, administration regimes, and target patient groups, either alone or in combination. Secondary follow-on patents may extend exclusivity for these foundational technologies by claiming improvements and/or further advances of existing technical teachings. It is our hope that this chapter will enable the reader to identify and understand the impact of developing new patent rights as an important business tool, in addition to identifying potential opportunities and uncovering niche markets for developing and/or exploiting vaccine technologies, as well as recognizing existing patent rights that might present a barrier to planned or ongoing research initiatives.

This chapter addresses patentability schemes, trends, and landmark judicial holdings that potentially impact vaccine-based innovations. Our analysis will primarily focus on the European and US systems in terms of what constitutes patentable subject matter and claiming strategies commonly employed to protect vaccine-oriented inventions. It is our hope that this analysis will be useful not only when carrying out activities leading up to the acquisition of intellectual property rights but also for enabling the reader to identify and understand the impact of already existing patent rights that might present a barrier to planned or ongoing research initiatives, to recognize when licensing of exiting rights might be desirable, and to identify potential areas where new opportunities or niche markets for vaccine technologies might be pursued.

## 51.1 Introductory Remarks

The biotechnology industry relies on a robust patent system, at least for gaining access to the financial reward and investment required for funding the ever-skyrocketing cost of research and development, especially for the long and costly regulatory process of clinical trials that all vaccine-related products must withstand. In addition, companies and organizations rely on these intangible assets for inspiring innovation opportunities and partnerships, or as objects for sale, whether outright or via licensing arrangements.

It is indisputable that current vaccination programs effectively prevent millions of deaths each year worldwide, by immunologically guarding susceptible individuals from the threat or occurrence of vaccine-preventable diseases. In 2012, the World Intellectual Property Organization (WIPO) published the results of a massive undertaking by the French institution "France Innovation Scientifique & Transfert" (FIST) that dissected and analyzed, in very fine detail, the historical global patenting profile for vaccine-oriented inventions. The report entitled "Patent Landscape Report on Vaccines for Selected Infectious Diseases" [1] ("Report") uncovers a provocative worldwide pattern of patenting and innovation activity encompassing a wide variety of technologies in the vaccine space and also provides a deeper analysis of these activities in, e.g., Brazil, India, and China. A read of the Report reveals that both the USA and China dominate first (earliest priority) and second (a subsequent family member) patent filings claiming active ingredients of vaccines against infectious diseases, with Europe and Australia rising to dominance for second patent filings in the vaccine field [2]. This vigorous pattern of filing activities suggests that, for entities working in the vaccine area, rewarding innovative efforts is imperative, given the hefty costs associated with research, regulatory requirements, and commercialization of these technologies. This is why a harmonized reliable approach to patent protection, preferably on a global level, is of great consequence.

There are an abundance of European Union (EU) Directives, associated European Community (EC) Regulations, other Guidelines, and explanatory notes that attempt to both harmonize and regulate biotechnology activities, such as the use of live recombinant viral vector vaccines, attenuated and not, for the prevention and treatment of various infectious diseases in both the human and veterinary contexts. These legal authorities also govern the quality and nonclinical and clinical aspects of bringing vaccines from bench to market [3]. In addition to vaccines based on viral vectors, there exist other types of vaccines, including bacterial vaccines, DNA-based vaccines [4] including plasmid DNA vaccines, recombinant protein vaccines, subunit and toxoid vaccines, combined vaccines, synthetic vaccines, and also "associations" of immunological veterinary medicinal products (IVMPs), any of which may feature suspensions of killed or attenuated microorganisms or products or derivatives from microorganisms. Supplementary to these vaccine modalities are other substances that might be coadministered such as antigens, antibodies, adjuvants, and other excipients. All of these biological substances, in addition to their method of manufacture, nucleic acid sequences, delivery platforms, cell lines and cultures used for expression, therapeutic targets and clinical indications, related peptides and proteins such as antibodies, and improvements on any of the foregoing, besides being subject to EU law, may also form part of an inventive disclosure leading to a patent property right.

One EU Directive, in particular Directive 98/44/EC, provides a unified legal framework for protecting biotechnology inventions throughout the European Union and has been successfully implemented into national laws of all EU member states, as well as the European Patent Convention (EPC), thereby making it applicable to biotechnological inventions filed before the European Patent Office (EPO). At present, the primary mode of securing patent protection in the European market, especially for biotechnology innovations such as vaccines, is by way of the European patent ("EP Patent"), which is granted by the EPO. Once an EP Patent has been granted, it is regarded as a bundle of rights that may be currently "validated" in up to 38 individual countries which are signatories of the EPC ("EPC Member States"), upon the completion of various formalities before the relevant national patent authority.

An alternative patenting option that would offer unitary patent protection in 25 of the EU member states, except Italy and Spain, is still evolving. This new regime is envisioned to provide a unitary patent document for an invention captured in a single application, which will be substantively evaluated under existing procedures before the EPO and subject to a new "Unified Patent Court," which will exercise exclusive jurisdiction over civil litigation related to infringement and validity [5]. In addition to these administrative and legal advantages that will be afforded to individual and company inventors alike, significant financial benefit may also be realized in view of the dramatically reduced translation costs currently associated with EP Patent validation procedures. Although the initial agreements have been signed, they must still be ratified by a certain number of countries under a particular protocol before the unitary patent system can become legally effective.

Over 30 years have passed since the first EPC entered into force ("EPC 1973"), and since then, the patent business has been booming on a global basis. Noting that the well-established EPC 1973 could benefit from an update, in view of the proliferation of technologies such as biotechnology, the enactment of various international treaties, and the recent upsurge in the number of EPC Member States, the Administrative Council of the European Patent Organization initiated a major effort to revise the EPC. The aim of the revision was to modernize the European patent system while maintaining the proven foundational principles of substantive and procedural patent law that were enshrined in the EPC 1973.

One major factor inspiring this recognition for change was the need to harmonize European patent law with various legislative initiatives such as new and revised EU Directives and other international agreements that significantly impact intellectual property rights on the global stage, including the TRIPS Agreement (Trade-Related Aspects of Intellectual Property Rights, resulting from the Uruguay Round Table Agreement of 1994) and the Patent Law Treaty, which was signed by the EPO in 2001. Another ongoing situation inspiring the EPC update is the ever-increasing number of EPC Contracting States (38 at present), which demands both simplicity and flexibility in the procedures leading up to the European patent grant.

The revised version of the EPC 1973 ("EPC 2000") was achieved by a delegation of the Contracting States, WIPO, and other parties participating in a Diplomatic Conference taking place at the Munich-based European Patent Office ("EPO") headquarters in November 2000. It entered into force on December 13, 2007. One

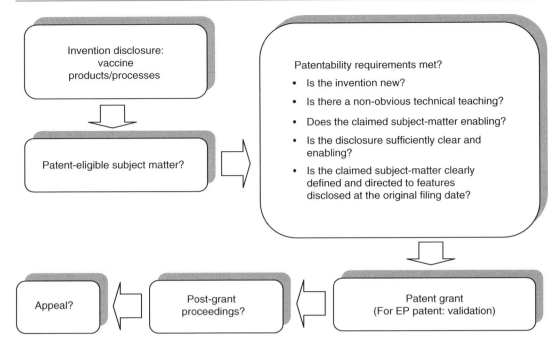

**Fig. 51.1** Life cycle of an invention

major outcome of the EPC revision involved the transfer of the detailed aspects of EPO procedures from the Articles (which can be amended only through a Diplomatic Conference consisting of all Contracting States) to the Implementing Regulations ("Rules") (which can be easily amended simply by a decision of the EPO Administrative Council). Over the last few years, this particular change has allowed the EPO to readily adapt EPO practice to reflect changes in patenting laws and policies developed at the EU level, most notably the recent shifts in biotechnology-oriented provisions such as the patentability of human embryonic stem cells, and the scope of protection afforded to biotechnology inventions involving genetic material under EU Directive 98/44/EC [6].

In this chapter, we will consider EPC provisions, both substantive and procedural in nature, and especially their impact on the patenting of vaccine-based innovations. A top-level schematic depicting a patenting life cycle is shown in Fig. 51.1. We will also discuss certain patent claiming strategies commonly to the biotechnology industry when building a patent portfolio. In tandem with this analysis, we will briefly address parallel provisions and case law under the US system insofar that meaningful practical differences exist, giving special attention to the new patent law regime effective March 16, 2013, which will at least significantly impact how novelty of an invention is evaluated before the US Patent and Trademark Office ("US PTO").

## 51.2 Patentable Subject Matter Including Exclusions and Exceptions

A threshold question, which must be answered positively before any substantive patentability analysis gets underway, is whether an invention is eligible for patent protection in the first place. This initial inquiry is carried out in Europe under the authority of Article 52(1) EPC 2000, which states: "European patents shall be granted for any inventions, in all fields of technology, provided that they are new, involve an inventive step and are susceptible of industrial application." This wording aligns with the language of section 27.1, first sentence, of the TRIPS Agreement, thus making clear that European patents are available for all inventions that yield a technical teaching [7]. Biotechnology inventions are defined in the EPC as "inventions which concern a product consisting of or containing biological material or a process by means of which biological material is produced, processed or used" and should be interpreted under EU Directive 98/44/EC as a supplementary measure [8].

However, there remain several notable exclusions and exceptions from patentability that periodically impact biotechnology innovations, which are embodied in Articles 52 and 53 EPC, respectively. For example, Article 52(2) EPC precludes patents for inventions encompassing discoveries and scientific theories [9]. Article 53(c) EPC 2000 reflects the EPO's philosophical stance in which methods of treatment, including methods using vaccines, are regarded under European practice as a "patentability exception," [10] thus prohibiting the granting of a European patent for any method construed as encompassing a treatment, surgical steps, or diagnostic features that are, or even could be, fairly practiced on the living animal and human body. Hence, any vaccine-based claims deemed to fall under this exception will be refused. Interestingly, and different from the US practice, instead of excluding these methods based on a legal fiction that these inventions lack industrial applicability (under Article 57 EPC) [11], since they can readily constitute technical inventions within the plain meaning of Article 52(1) EPC, the EPC casts such methods as patentability exceptions in view of public health and related policy concerns, e.g., that treatment of a patient should not be impeded merely because a physician is forced to negotiate a patent license in order to pursue a particular course of therapy. Significantly, Article 53(c) EPC 2000 does not apply to products, in particular substances or compositions, for use in any of these excepted methods, a reward of compromise to inventions caught by this provision.

The issue of patentable subject matter, in particular for biotechnology innovations, has been at the forefront of judicial inquiries in a number of jurisdictions, most recently in Australia, where the Federal Court of Australia ruled [12] that nucleic acid (both deoxyribonucleic acid/DNA and ribonucleic acid/RNA) that has been isolated from the naturally existing cellular environment, as a product of human intervention, is fit to be patentable subject matter as a "manner of manufacture" under the Australian Patents Act [13]. The key to this decision was the subject nucleic acid composition claimed in Myriad Genetic Inc.'s disputed patent [14], namely, the isolated polynucleotides comprising all or a portion of the BRCA1 locus or of a mutated BRCA1 locus of the breast cancer gene BRCA1, was found to have been artificially purged from the cellular environment and thus "isolated." In addition, the disclosed in vitro results using this isolated nucleic acid composition were shown to be useful for the claimed identifying and/or diagnosing an individual's susceptibility to breast and/or ovarian cancer.

In parallel proceedings in the USA, the question of whether Myriad's "isolated" genes as described and claimed in counterpart US patents [15] constitute patentable subject matter under US patent law [16] was decided by the US Supreme Court ("USSC") on June 13, 2013 [17]. Significantly, this ruling overturned the finding rendered by the appellate court, the US Court of Appeals for the Federal Circuit (CAFC), by holding that composition claims directed to "isolated" DNA molecules are not patent-eligible products of nature because they represent a naturally occurring composition of matter [18]. The scientific and legal basis supporting the USSC's denial of patent protection to these isolated, but natural forms of DNA lies in the observations that Myriad did not create or alter the genetic information encoded by the BRCA 1/BRCA2 genes, nor did they change the structure of the DNA itself. Despite the acknowledgement that Myriad indeed characterized an "important and useful gene," simply "separating that gene from its surrounding genetic material is not an act of invention" and "does not by itself satisfy the §101 inquiry [19]." The full impact of the USSC's pronouncement is currently unclear, as tens of thousands of issued US patents are potentially invalid insofar as their claims encompass isolated, naturally occurring nucleic acid sequences within the meaning of this decision. Importantly, the USSC did admit that synthetic, artificially created sequences, e.g. cDNA sequences, might qualify as patent-eligible subject-matter since they are synthetic, exon-only creations that do not exist in nature [20].

However, this landmark holding did not directly address the CAFC's careful determination that claims to methods merely "comparing" or "analyzing" DNA sequences are not patent-eligible because these types of claims recite no transformative steps and instead relate to unpatentable abstract, mental steps [21]. This aspect of the lower court's decision therefore remains intact, and thus the applicable standard for claims reciting these types of analytical steps.

Following the Australian decision, one must wonder how an "isolated" nucleic acid will be construed in matters of patent infringement in Europe. The UK High Court, in the highly contentious Monsanto case [22], narrowly interpreted the term "isolated" as recited by a claim directed to an "isolated DNA sequence" as meaning a fragment physically separated from other molecular components in the cellular milieu. Consequently, it found that Monsanto's patent, which described and claimed a gene encoding an enzyme conferring resistance to a particular herbicide, was not infringed by the importation of soybean meal created from soybeans harboring this gene. The Court opined that the meal did not contain the isolated gene sequence and also did not react with certain antibodies in the manner required by Monsanto's patent claims. The European Courts of Justice concurred with this holding, noting that such an interpretation was consistent with Article 9 of EU Directive 98/44/EC [23].

While the UK Monsanto decision is not codified in Australian, US, or EPO law per se, it does underscore the need to give particular attention to the drafting of claims encompassing "isolated" nucleic acids. It may be wise to incorporate enough detailed definitional language at least to satisfy the "manner of manufacture" requirement, or "composition of matter" terms as the case might be, to avoid triggering an unduly narrow construction of this term before a particular court. Moreover, in the context of vaccines, it might be useful to include claims directed to nucleic acid compositions as they appear in the actual (envisioned) commercial product or to any new intermediate substances isolatable during the course of vaccine manufacture.

The impact of the foregoing decisions, and judicially pending matters, regarding the patentability of isolated nucleic acid sequences cannot be exaggerated, as vaccine patents often describe and claim embodiments relating to isolated nucleic sequences in some respect, e.g., sequences of the microorganisms or other biological components forming the vaccine, with related expression and delivery vectors. Although Rule 29(2) EPC 2000 positively acknowledges that, e.g., human gene sequences are a priori patentable, any disclosed sequences must first pass muster as patent-eligible subject matter by not merely cast as a bare "discovery" that is otherwise excluded by Article 52(2)(a) EPC [24]. Rather, the sequence must be mechanically "isolated" (or otherwise "artificial" for US patent rights) from its naturally occurring cellular milieu and also be new, relate to a useful technical teaching material to the described inventive concept, and be sufficiently disclosed and claimed with particular clarity.

Under US practice, patent-eligible subject matter is governed by Title 35 of the United States Code (U.S.C.) at §101, which defines the standard of patent "utility," i.e., the disclosed and claimed invention defines a specific, substantial, and credible utility [25]. Under European practice, this is tantamount to the requirement that the claimed subject matter "is industrially applicable" and that the invention provides a technical contribution to the art [26]. Vaccine compositions (and methods for their manufacture) do not commonly trigger utility issues, since vaccines usually provide some kind of immunological effect against viruses or other microorganisms amounting to a credible utility as understood by a person of ordinary skill in the art. However, a disclosed utility that requires further research to fully characterize, or at least reasonably confirm, a "real world" immunogenic, therapeutic effect might fail 35 U.S.C. §101. Care should therefore be taken during the drafting process when disclosing and claiming subject matter relating to the prevention and treatment of an unspecified disease or condition and/or when using uncharacterized or partially characterized proteins, therapeutic proteins or antibodies, DNA fragments encoding a full open reading frame (ORF), or receptor targets that these physiological entities are preferably structurally, or at least functionally, defined in a manner that directly links them to a tangible therapeutic effect.

The Federal Circuit has considered the question of patent-eligible subject matter under 35 U.S.C. § 101 on a further topic that appears highly relevant to vaccine innovations insofar as they disclose and claim dosage and administration regimes [27]. The patents at issue related to a schedule of infant immunization for infectious diseases designed to reduce the risk of chronic immune-mediated disorders. Two out of the three disputed patents claimed methods of screening and comparing information on immunization schedules with the occurrence of chronic disease for identifying a lower risk schedule and then immunizing accordingly by administering the vaccine using said schedule. The claims of a third disputed patent recited screening and comparing steps, but did not include the subsequent immunization step.

The lower first instance court denied patentability for all three patents as being based on some abstract idea that a relationship exists between the infant immunization schedule for infectious diseases and the later occurrence of the specified chronic disorders. However, relying on the Supreme Court's guidance in the *Bilski* decision [28], the intermediate CAFC instead concluded that the "exclusions from patent eligibility should be applied narrowly." [29] The CAFC also held that the presence of a mental step is not itself fatal to 35 U.S.C. § 101 eligibility, and even though it had serious doubts about the substantive patentability of the claims, two of the three patents at least described patent-eligible subject matter because they recited a physical step of "immunizing" on a determined schedule, which was held to be a specific, tangible application sufficient to meet the requirements of 35 U.S.C. § 101. Whether these inventions also met the remaining substantive patentability requirements was a matter to be resolved under the relevant provisions of US patent law. The third patent, however, was held invalid for claiming patent-ineligible subject matter, i.e., a purely mental, abstract process of collecting and comparing immunization data for establishing immunization schedules optimized for treating particular clinical conditions, which was barred under 35 U.S.C. § 101.

This US jurisprudence regarding dosing regimes seems to comport with European practice, as set forth by the EPO Enlarged Board of Appeals decision G 2/08, which first recognized the validity of "administration patents." In this case, the disputed patent claimed the use of known nicotinic acid or a compound metabolized to nicotinic acid for treating hyperlipidemia using a particular oral administration schedule. The Enlarged Board held that "where it is already known to use a medicament to treat an illness, Article 54(5) EPC does not exclude that this medicament be patented for use in a different treatment by therapy of the same illness." [30] Yet, the Board cautioned that patentability was "subject to compliance with the other provisions of the EPC, in particular novelty and inventive step," [31] which might prove to be an insurmountable hurdle with such so-called administration claims.

This proved evident in the EPO Technical Boards of Appeal case T 1760/08, where the Board stated that the patent specification failed to disclose any showing that the claimed compound at the claimed dosage imparted any unexpected improvement over any other described dosage in terms of providing a superior bioavailability [32]. Because one of the described working examples indicated that the administration of the claimed compound at an increasing stepwise dosage (i.e., administered in a range of 250, 500, or 1,000 mg daily) showed better results with a rising amount of the compound, the selection of the 500 mg dose for inclusion in the claims was held to be merely an arbitrary choice from the disclosed alternatives. Accordingly, although the dosage regime itself was regarded as patent-eligible matter, because no superior technical teaching resulted from this selection, the patent failed for lacking an inventive step over the prior art.

The preceding demonstrates that the patent application drafter must be vigilant when preparing and claiming embodiments directed to vaccine technologies. While the threshold of utility would appear to be readily satisfied by a vaccine composition having a demonstrated immunogenic effect, obtaining sufficiently broad protection for this type of invention commonly involves claiming aspects external to the vaccine itself, such as nucleic acid sequences of the derivative microorganism and related expression vectors, and dosing regimens useful for achieving the immunoprotective result. Each of these embodiments must initially successfully satisfy the utility requirement as set forth in the laws governing the various global patent systems, a feat that can be deceptively complex.

## 51.3 Principles of Novelty and State of the Art

In European practice, Article 54(1) EPC specifies that an invention is considered new if it does not form part of the state of the art. The EPC, like most other patent law systems worldwide, adopts the principle of absolute novelty for determining what constitutes prior art, as enshrined by Article 54(2) EPC: "The state of the art shall be held to comprise everything made available to the public by means of a written or oral description by use, or in any other way before the date of filing of the European patent application." [33] But for a very narrow exception in cases where the invention was disclosed either at a specified public convention or inadvertently due to some abuse [34], there is no grace period recognized for an invention disclosure that is deemed to be "publicly available" prior to its official filing date (or priority date as applicable) [35]. Therefore, a potential patent applicant must zealously safeguard its invention from any such acts, including any public, nonconfidential use of the invention that brings the inventive concept to the attention of the public, until the invention is safely filed at the patent office. This novelty approach is common to virtually all patent systems worldwide, including, but certainly not limited to, Japan, China, Australia, Canada, India, and Brazil.

The aforementioned "first-to-file" doctrine has been at least partially adopted by the US patent system under the changes implemented by the Leahy-Smith America Invents Act ("AIA"), which represents a historic reform of US patent law and the most comprehensive since the last major revision in 1952. The new first-inventor-to-file regime provides that a claimed invention will not be found novel if it was already patented, described in a printed publication, in public use, on sale, or otherwise available to the public, anywhere in the world, before the effective filing date of the claimed invention [36]. However, an important exception to this prior art definition was preserved in the new law: The US PTO will not regard a public disclosure (or other publicly available use) as forming part of the state of the art if this disclosure was made by or from the inventor or joint inventor, or by another who obtained the subject matter disclosed directly or indirectly from the inventor, if such disclosure was made 1 year or less before the effective filing date of a claimed invention [37]. Moreover, the AIA reverses the historic unequal treatment of US provisional applications and foreign priority documents regarding their prior art status by eliminating the *Hilmer* doctrine [38].

The new "first-to-file" prior art principles apply to fresh patent applications filed on the effective date of the new US law; [39] however, the benefits of the former "first-to-invent" prior art doctrine remaining in force for older applications can be irreversibly lost in favor of the new structure should the applicant amend a claim that lacks support in the application as originally filed. The possible consequence of this situation is that a prior publication of the invention that was otherwise excepted from the prior art under the old law becomes a potentially novelty-destroying disclosure to the application under the new approach to defining the state of the art.

Parties to a joint research agreement, an arrangement that would apparently be common in the vaccine development industry, should be aware that any public disclosure made in an application or patent *by another* shall not be prior art if the subject matter and the claimed invention were commonly owned, or subject to an obligation of assignment to the same person, not later than the effective filing date of the claimed invention [40]. This means that the claimed invention must have resulted from activities defined by the scope of the joint research agreement, that the invention was made by/on behalf of at least one party to a joint research agreement in effect on/before the effective filing date of the claimed invention, and, importantly, that the application documents disclose the parties to this agreement.

Returning to European definition of "prior art," a somewhat complicated extension of the "first-to-file" principle lies in the notion of "elder European rights" which was introduced to preclude double patenting and is defined by Article 54(3) EPC 2000: [41] "Additionally, the content of the European patent application as filed, of which the dates of filing are prior to the date referred to in (2) and which were published on or after that date, shall be considered as comprised in the state of the art [42]. " Hence, a European (or a Euro-PCT) patent application having an earlier priority date, disclosing the same subject matter, and validly published after the filing date of a later European application will serve as a bar to novelty, but shall not be used for the purposes of evaluating inventive step. Therefore, "obvious" equivalents to any claimed features in the earlier application are not considered to form part of the state of the art for the later application.

However, for pending European applications (including Euro-PCT applications) and European patents granted before the EPC 2000 entered into force, the Article 54(3) EPC 2000 provisions do not apply. Rather, the "elder" European right only imparts this prior art effect for designated states that overlap (or "collide") with the later application if the designation fees were validly paid in view of Article 54(4) and Rule 23a EPC 1973. The typical result of such a situation was different sets of claims for different designated states in view of different documents that make up the state of the art. To complicate matters further, if a designated state in the earlier application is withdrawn subsequent to the publication date, but the designation fees were validly paid, the prior art effect still applies for that particular EPC Member State.

Illustrating the practical effect of Article 54(3) EPC 1973 v. 2000: Under former EP practice, Applicant A validly pays designation fees only for DE, FR, GB for European application EP-A. Upon publication, EP-A has a prior art effect (novelty only) for later-filed European patent applications only in DE, FR, and GB insofar as these states are jointly designated and designation fees are validly paid. However, under the EPC 2000, Applicant B validly pays designation fees for IT and SE for later-filed European application EP-B. In this example, upon publication, EP-A will have a prior art effect (novelty) against EP-B in all EPC Contracting States designated by EP-B, even if EP-B does not designate the states of DE, FR, and GB at all. Therefore, under the EPC 2000, the prior art definition for "elder" European rights has been greatly simplified.

## 51.4 Claims Reciting a Medical Indication Using a Known or New Vaccine Composition Must Be Inventive and Sufficiently Disclosed

The EPC limits the ability of an applicant to claim aspects of a vaccine identified as a method of treatment, surgery, or diagnosis in view of various public policies underlying the patentability exclusion under Article 53(c) EPC 2000. Exemplary embodiments commonly found in vaccine-based invention disclosures include methods relating to diagnosis of disease, and/or other means of detection, immunomodulation, immunostimulation, treatment, prophylaxis therapies, monitoring, and evaluation, and also to vaccine administration techniques that might amount to surgical steps. Offering a compromise for innovations made in such medical fields, the EPC did carve out a narrow exception, embodied in Articles 54(4) and 54(5) EPC 2000, allowing applicants to claim known substances or compositions for a first or further medical use in these excluded methods, respectively.

Specifically, Article 54(4) EPC 2000 recites: "Paragraphs (2) and (3) shall not exclude the patentability of any substance or composition, comprised in the state of the art, for use in a method referred to in Article 53(c) EPC 2000, provided that its use for any such method referred to in that paragraph is not comprised in the state of the art." Hence, this paragraph provides the requisite authority for the "first medical indication claim," which can be broadly worded to recite a contemplated therapeutic purpose, for example, *"Composition X as a medicament"* or *"A vaccine comprising composition X."* This format is thus useful for an inventor who characterizes a new or known "composition X" for the first time in a medical context, such as a therapeutic use as a vaccine. Should a prior art disclosure reasonably describe that the same composition X is useful as a vaccine, this would amount to a novelty-destroying disclosure unless the patent application discloses that composition X is used in a new and specific therapeutic context that is unknown in the prior art, i.e., a second or further medical indication.

Once a first medical use of a known compound has been described, any newly discovered second or further medical use of a known substance or composition might be patent eligible under Article 54(5) EPC 2000, which recites: "Paragraphs 2 & 3 shall also not exclude the patentability of any substance or composition referred to in paragraph 4 for any specific use in a method referred to in Article 53(c) EPC, provided that such use is not comprised in the state of the art." This paragraph thus aims to provide a clear basis for the formerly known Swiss-type "second medical indication claim." Because Article 54(5) specifies that the composition is "for use" in a particular application, claims directed to new uses for a known vaccine should contain this language to avoid lack of clarity and resulting lack of novelty objections. Such a claim in the vaccine context may be simply worded: *"Vaccine X for use in treating/preventing disease Y"* or *"Composition comprising vaccine X for use in a specific therapeutic application."* In these exemplary claims, "vaccine composition X" is already known as a medicament, but not for the specific, purpose-related use as stated in the claim.

The EPO might grant a broad claim for a first medical use even where only one specific medical use of the substance has been meaningfully disclosed in the patent description [43]. However, we offer a cautionary note as to the extent of experimental evidence that could be required to validly support certain types of vaccine claims, at least in order to meet the rigorous demands set forth under Article 83 EPC. This provision stipulates that an invention be disclosed "in a manner sufficiently clear and complete for it to be carried out by a person skilled in the art." For claims directed to a first or a second medical use under Article 54(4) or 54(5), respectively, it is not only necessary that the skilled person is enabled for making the claimed compounds on the basis of the written technical disclosure in the application, but the application must also provide direct and sufficient evidence, considering common general knowledge in the technical field, that the claimed therapeutic effect is attained for Article 83 EPC to be met since said effect represents a functional technical feature of the claim [44].

For example, a claim reciting, e.g., *"A protein ... or compound derived from bacterium A for use in a therapy that elicits an immune response against said protein (or compound)"* would require the patent description to present credible evidence or other guidance that said protein or compound is mechanistically linked to the induction of a humoral or cellular immune response. However, a claim that functionally defines the known compound or substance, e.g., *"Bacterium A for use in a vaccine therapy to prevent disease Y"* must meet a somewhat more demanding burden of proof to sufficiently evidence that the described invention provides an enabling disclosure beyond mere theory, e.g., that bacteria A confers an identifiable therapeutic efficacy or a verifiable immunoprotective effect in preventing disease Y following a challenge. These principles are in place to ensure that the skilled person reading the patent disclosure is put in the position of carrying out the invention, with a reasonable expectation of success, without encountering any undue burden, and without having to engage in an experimental research program in order to practice the invention as it is claimed.

It is apparent, under the prevailing EPO standards, that to convincingly establish and link a general immunoprotective effect with a physiological target following the administration of a vaccine composition can be a daunting task. For example, in the EPO Technical Boards of Appeal decision T 187/93, the Board refused a patent application because the claims generalized a disclosed technical teaching of the immunoprotective effects of a particular herpes simplex virus (HSV) type 1 or type 2 glycoprotein D protein to all membrane-bound viral proteins in general, including all herpes viral membrane proteins and all herpes simplex viral membrane proteins. This broadening of the application's fair technical teaching, which was otherwise limited to experimental evidence showing an immunoprotective effect derived from glycoprotein D of HSV type 1 or 2, was found to be excessive and unfounded. The application thus failed Article 83 EPC because the notional skilled person was not in a position to carry out the invention as broadly claimed due to a lack of an enabling teaching in the application as filed. In addition, the Board noted that applicant's arguments of entitlement to the broad claim scope were incompatible with those successfully made in support of inventive step per Article 56 EPC, which focused more narrowly on methods directed to the membrane glycoprotein D of herpes simplex type 1 or type 2 and the induced immunoprotection in an immunized subject against an in vivo challenge by these viral strains [45].

In EPO decision T 219/01, a patent disclosed and claimed, as a first medical use, "an HIV vaccine comprising gp120" in addition to "gp120 for eliciting a protective immune response against HIV." Unfortunately, although the patent did provide credible technical data evidencing vaccination in chimpanzees, a post-published AIDSVAX clinical study involving the claimed vaccine showed that the vaccine utterly failed to confer protection in human subjects. Based on this evidence, which was submitted to the EPO by an opponent during a post-grant opposition procedure, the patent was revoked for failing to provide a sufficient, enabling disclosure under Article 83 EPC as the claims did encompass the use of the vaccine in humans, which was no longer regarded as technically plausible [46].

If the description of a patent specification provides no more than a vague indication of a possible medical use for a chemical compound yet to be identified, later submitted more detailed post-published evidence cannot be used to remedy the fundamental insufficiency of disclosure of such subject matter at the filing date. The claim format sanctioned by Article 54(5) EPC requires that attaining the claimed therapeutic effect, i.e., the claimed specific "use" of the known compound, is a functional technical feature of the claim. As a consequence, under Article 83 EPC, unless this is already known to the skilled person at the priority date, the application must disclose the suitability of the product to be manufactured for the claimed therapeutic application. In EPO decision T 609/02, one of the granted patent claims under attack recited a use of a compound *"for use against over-expression of steroid hormone-responsive or steroid hormone-like compound responsive gene(s)."* However, the application failed to provide any data showing which steroid hormone-responsive or steroid-like compound was actually contemplated by the claims or how the overexpression of these genes could be prevented. An important aspect of this holding was that a simple statement made in a patent specification, i.e., that compound X may be used to treat disease Y, is simply not enough to ensure sufficiency of disclosure for a claimed pharmaceutical. Instead, for a new medical use, the written description is obligated to provide at least some meaningful information and/or evidence, e.g., experimental testing that demonstrates the claimed compound has a direct effect on a metabolic mechanism specifically linked to the described clinical state [47].

## 51.5 A Few Words on Post-Patent Grant Procedures

Following the allowance of a European patent, several post-grant procedures may be initiated that may wholly invalidate, or at least partially restrict, the scope of the granted patent right. One option at the EPO, particularly common in the biotechnology field, is a centralized inter partes opposition procedure, which involves a third-party challenge to the validity of an EP Patent that must be filed within 9 months after the publication of the patent grant [48]. The scope of the challenge, set forth under Article 100 EPC, allows an opponent to contest the patent on the grounds that the subject matter of the claims is not new, inventive, or industrially applicable, is excluded from patentability, contains added matter in relation to the application as filed, and/or is insufficiently disclosed. We have already touched on many of these issues in this chapter. To prove the case, the opponent must file a detailed brief setting forth at least one of the above grounds, with supplemental evidence in the form of scientific publications, patent documents, declarations, and other expert testimonies being filed as deemed appropriate. The patent proprietor may respond to the notice and file supporting evidence to refute the opponent's allegations as desired. The opposition procedure may result in a patent being fully revoked, being amended with restricted claims, or fully untouched. Subsequent to the opposition period, an EP Patent may only be challenged in an EPC Member State where the patent is validated and in force [49]. Should an adversely affected party to the proceedings decide to lodge an appeal against the Opposition Division's decision, the EPO has detailed mechanisms in place to guide this procedure at the second instance [50].

The central limitation procedure, introduced in the revised EPC 2000 as Articles 105a–c, offers a patent proprietor the possibility for ex parte limitation or full revocation of a granted European patent having effect "ab initio" in all EPC Member States where the patent has been validated [51]. This centralized procedure beneficially affords the patentee not only with an opportunity to amend the substance of a potentially invalid European patent but also a chance to avoid often lengthy and costly proceedings before each of the relevant national courts. Notable aspects of this procedure are that it is initiated exclusively by the patentee (although third parties may present observations); [52] the limitation request is prescribed for the claims only with no explicit provision for amending the description or drawings (which should already support the amended claims); however, such amendments may be possible upon petition. Moreover, EPO limitation proceedings do not take precedence over existing national proceedings (revocation proceedings in particular), which may be stayed or suspended pending the outcome of the final EPO decision if the national court so decides. The scope of the EPO's inquiry under Article 105b EPC 2000, and as implemented by the Rules [53], is restricted to whether the patentee's amendment(s) in fact limits the claims (and are not purely cosmetic), whether the amendment meets the clarity and conciseness requirements under Article 84 EPC, and whether the amendment meets the stringent requirements against added subject matter per Article 123 (2) and (3) EPC (no added matter extending the subject matter beyond the content of the application as filed or of the at grant, respectively).

If the Examining Division deems that the patentee has not complied with the above requirements, a further opportunity may be given but is discretionary since the patentee is certainly free to file an additional limitation request in the future. Interestingly, the limitation procedure does not oblige the Examining Division to substantively examine whether the amended claims actually avoid the prior art or if they satisfy the EPC patentability requirements pursuant to Articles 52–57 EPC. Once the limitation request has been allowed, the patentee must provide a translation of the limited claims in all three official EPO languages (and into the official language of each relevant EPC Contracting State as applicable) in addition to submitting payment of a printing fee. Of importance, in case the EPO rejects the limitation or revocation request, no option to appeal this decision is possible.

Turning to US practice, the recent AIA introduces and/or modifies a variety of post-patent grant procedures, which will undoubtedly open up new and strategically important options for dealing with a US patent following its allowance. For example, the AIA establishes a new "post-grant review" ("PGR") and "inter partes review" [54] ("IPR"), which can be strategically employed to challenge and potentially destroy a competitor's US patent.

In greater detail, the PGR is an "opposition-style" mechanism akin to EPO procedures, where a non-patentee third party may allege the invalidity of one or more patent claims of a US patent based on a broad range of grounds [55] similar to those set forth under Article 100 EPC, compared to former USPTO reexamination procedures. Unlike in European opposition proceedings, a US patent can be attacked as lacking clarity or definiteness, which is not allowed under European practice (i.e., under Article 84 EPC). However, like EPO oppositions, a PGR petition must be filed not later than 9 months after the date of patent grant, or issue date of a US reissue patent, as the case may be.

The threshold requirement for admitting a PGR action requires that the information presented in the opponent's petition, if not rebutted, demonstrates that it is "more likely than not" that at least one of the US patent claims challenged in the petition is not patentable [56]. These requirements appear to correspond to the European patent procedure, which requires "a statement of the extent to which the European patent is opposed and of the grounds on which the opposition is based, as well as an indication of the facts and evidence presented in support of these grounds." [57] The PGR proceedings are a matter of public record and are conducted by the newly created Patent Trial and Appeal Board ("PTAB"), consisting of a three-person Administrative Patent Judge panel, which differs from European oppositions where the Opposition Division consists of three examiners, one of which is usually the primary Examiner who conducted the grant proceedings. Although the PGR proceedings can be completely carried out in writing, each party has the right to oral proceedings as desired; corresponding provisions exist in the EPC [58]. Apparently, the PTAB is more limited with respect to the permitted scope of discovery compared to the EPO, which is at least theoretically not restricted by the facts, evidence, and arguments provided by the parties per Article 114 (1) EPC. Moreover, the Opposition Division may invoke its discretionary power to examine grounds of opposition not even raised by any opponent to the proceedings [59]. PGR actions are terminated either by a PTAB decision or by a settlement between the parties, but this decision is ultimately at the discretion of the US PTO.

The post-grant inter partes review ("IPR") option is intended to eventually replace the existing inter partes reexamination procedure and allows a third-party non-patent owner to challenge a US patent after the later of (1) 9 months from the issue or reissue date of a US patent (i.e., after the PGR period has expired); (2) if a PGR has been instituted, then from the date of its termination; or (3) within 1 year after petitioner is served with a complaint where patentee alleges infringement of the patent. The IPR procedure, which is conducted before the PTAB, has no related counterpart in the EPC. An IPR action can be initiated throughout the life of the US patent. In contrast to the PGR, an IPR action can initiated only based on the grounds of novelty and lack of inventive step, using only patents and printed publications.

## 51.6 Concluding Remarks

The foregoing represents a non-exhaustive overview of existing structures governing patent granting and defense procedures and certain patentability approaches that are suitable not only for vaccine-based innovations but for biological inventions generally. A strong patent position typically begins by building upon a core technology, with an eye to capturing secondary subject matter that frequently emanates from these first inventions. For vaccine innovations, initial patent filings might focus on cell cloning, expression, and therapeutic function of isolated vaccine sequences (which may or may not be deposited with an officially sanctioned institution), while secondary follow-on patents may extend exclusivity for the foundational technologies by claiming improvements and/or further developed or related technical teachings such as purification methods, drug delivery methods and systems, pharmaceutical compositions, and dosing regimens and additional clinical indications, any of which could be fairly subject to an extension of the monopoly period in the EU by application for a supplementary protection certificate (SPC) where regulatory approval is required. As is evident from our foregoing analysis, developing an optimized global patent portfolio entails a balanced approach between a prudent strategy that best safeguards a desired market position such that a reliable funding pipeline can be maintained, with a careful selection of inventions for patenting in the face of constantly evolving patent systems and landscapes.

## References

1. The Report can be viewed here: http://www.wipo.int/freepublications/en/patents/946/wipo_pub_946_3.pdf (Last visited 28 Feb 2013)
2. *See* Report, Appendix 6 at Figures 1, 2
3. EU/EC legislation regulating vaccine testing and authorization in Europe include (each website visited December 10, 2012): EU Directive 2001/83/EC – on medicinal products for human use http://ec.europa.eu/enterprise/pharmaceuticals/eudralex/vol-1/dir_2001_83_cons/dir2001_83_cons_20081230_en.pdf, EU Directive 2001/20/EC – Clinical trials directive http://europa.eu/eur-lex/pri/en/oj/dat/2001/l_121/l_12120010501en00340044.pdf
4. Concept paper on guidance for DNA vaccines, EMEA/CHMP/308136/2007 Committee for the Medicinal Products for Human Use (CHMP), 15 March 2012
5. For information about the Unitary Patent, including EU documents and current status, *see*: http://ec.europa.eu/internal_market/indprop/patent/index_en.htm (Last visited 1 Mar 2013)
6. *See Brüstle vs. Greenpeace*, Case C-34/10 (18 October 2011) and *Monsanto Technology LLC v Cefetra BV and Others*, Case C-428/08 (6 July 2010), respectively
7. *Basic Proposal for the Revision of the EPC, Preparatory Documents*: CA/PL 6/99; CA/PL PV 9, points 24–27; CA/PL PV 14, points 143–156; CA/100/00, pages 37–40; CA/124/00, points 12–16; CA/125/00, points 45–73; MR/2/00, pages 43–44; MR/8/00; MR/15/00; MR/16/00; MR/24/00, pages 69–71)
8. *See* Rule 26 EPC 2000, which specifies additional patent-eligible inventions in the life sciences
9. *See* Article 52(2)(a) EPC 2000
10. *See* Article 53(c) EPC 2000
11. Article 57 EPC demands that inventions be industrially applicable, i.e., capable of being exploited in a commercial context. Since methods relating to treatment, diagnosis, and surgery can be readily adapted to a wide range of industries, they are indeed "industrially applicable" but are simply excepted from patentability for reasons relating to public policy
12. *Cancer Voices Australia v Myriad Genetics Inc* [2013] NSD643/2010, Federal Court of Australia (Sydney), decision dated 15 February 2013
13. *See* Part 3, Division 1, Section 18, Patents Act 1990, Act No. 83 of 1990 as amended, taking into account amendments up to Act No. 35 of 2012. This holding is also consistent with European practice; *see* Rule 27(a) EPC
14. The patent in suit is Australian Pat. No. 686004, with a priority date of 12 August 1994, granted with 30 different claims. Only the validity of claims 1–3 was at issue in the proceedings
15. At issue were claims 1, 2, 5, 6, and 7 of U. S. Patent 5,747,282; claim 1 of U. S. Patent 5,693,473; and claims 1, 6, and 7 of U. S. Patent 5,837,492
16. 35 U.S.C. § 101 provides: "Whoever invents or discovers any new and useful process, machine, manufacture, or composition of matter, or any new and useful improvement thereof, may obtain a patent therefore, subject to the conditions and requirements of this title"
17. The full text of the decision (Case No. 12-398) can be viewed here: http://www.supremecourt.gov/opinions/12pdf/12-398_1b7d.pdf. (Last viewed 29 July 2013)
18. *See Association for Molecular Pathology (AMP) and ACLU v. USPTO and Myriad Genetics* 689 F.3d, 1303 (Fed. Cir. 2012)
19. *See* Slip Opinion, No. 12-398 at 2; 569 U. S. __ (2013) at 12
20. *See* Slip Opinion No. 12-398 at 3; 569 U. S. __ (2013) at 16–17
21. *See Mayo Collaborative Servs. v. Prometheus Labs., Inc.*, 132 S. Ct. 1289 (Fed. Cir. 2012)
22. *See Monsanto Technology LLC v Cargill International SA & Anor* [2007] EWHC 2257 (Pat) (10 October 2007)
23. *See Monsanto Technology LLC v Cefetra BV and Others*, Case C-428/08 (6 July 2010)
24. Rule 29(2) EPC 2000 states: "An element isolated from the human body or otherwise produced by means of a technical process, including the sequence or partial sequence of a gene, may constitute a patentable invention, even if the structure of that element is identical to that of a natural element"
25. 35 U.S.C. §101 states: "Whoever invents or discovers any new and useful process, machine, manufacture, or composition of matter, or any new and useful improvement thereof, may obtain a patent therefore, subject to the conditions and requirements of this title"
26. These requirements are set forth in Article 57 and Article 56 EPC, respectively
27. *Classen Immunotherapies, Inc., v. Biogen Idec*, 2006-1634 -1649 (Fed. Cir. August 31, 2011)
28. *Bilski v. Kappos,* 561 US __, 130 S. Ct. 3218, 177 L. Ed. 2d 792 (2010)
29. *Id.* at 18–19
30. *See* EPO Enlarged Boards of Appeals decision G 2/08 at reason 6.1
31. *Id.* at reason 5.10.9
32. *See* EPO Technical Boards of Appeal decision T 1760/08 at reason 3.3
33. Both Article 54(1) and Article 54(2) EPC 2000 retain the same wording as their counterpart provisions in the EPC 1973 so no material changes to practice resulted when the EPC 2000 took effect
34. *See* Article 55 EPC

35. Unlike the US patent system, the EPC does not recognize a grace period prior to filing a patent application, whereby the inventor may disclose his invention without that disclosure forming part of the prior art. However, Article 55 EPC specifies that certain "non-prejudicial disclosures" that protect an applicant from a novelty-destroying disclosure of the invention before the filing date caused either by "evident abuse" (e.g., violation of a confidentiality agreement or theft of the invention) or where the invention was displayed at an EPO-certified international exhibition. To qualify for Article 55 EPC protection, the disclosure must have occurred *no earlier than 6 months prior* to the application's filing (not priority) date
36. *See* new 35 U.S.C. § 102(a)(1), which largely reflects the prior art definitions stated in former 35 U.S.C. 102(a) and 35 U.S.C. 102(b). Furthermore, US patents, US Patent Application Publications, and PCT International Application Publications that were effectively filed by a different applicant will be considered as part of the state of the art under new 35 U.S.C. § 102(a)(2)
37. *See* new 35 U.S.C. § 102(b)(1),(2)
38. *In re Hilmer* 359 F.2 859 (C.C.P.A. 1966) held that a patent application's foreign filing date could be used as shield itself from later-filed or published cited prior art at the US PTO, but could not be used as affirmative prior art. This situation frequently disadvantaged foreign inventors who filed a first application outside the USA and then relied on Article 4 of the Paris Convention in filing a later US application. Under these circumstances, the foreign application could not be used as prior art against another US application because it was not "filed in the United States." Therefore, the foreign priority application date received unequal treatment by the US PTO, since no offensive benefit could be gained from the earlier disclosure.
39. These changes under the AIA come into force on March 16, 2013
40. *See* new 35 U.S.C. § 102(b)(2)(C)
41. Article 54(3) EPC 2000 differs from its predecessor by removing the language that the publication occur "under Article 93" and is regarded by the EPO as a "minor editorial amendment" (Preparatory documents: CA/PL 17/99; CA/PL PV 10, points 19–21; CA/PV 14, point 6; CA/100/00, pages 43–44; MR/6/00, pages 3–4; MR/2/00, pages 47–48; MR/24/00, page 71)
42. *Basic Proposal for the Revision of the EPC, Preparatory Documents*: CA/PL 17/99; CA/PL PV 10, points 19–21; CA/PV 14, point 6; CA/100/00, pages 43–44: Article 54(3) EPC 2000 partly simplifies Article 54(3) EPC 1973 practice since most European applications routinely designate all Contracting States. Therefore, the complicated prior art searches demanded by Article 54(4) EPC 1973 appeared to confer an advantage to only a limited number of applicants
43. *See* Rule 42(e) EPC 2000
44. An elegant discussion on this point can be found in EPO Technical Boards of Appeal decision T 1496/08, e.g., at reason 14
45. *See* EPO Technical Boards of Appeal decision T 187/93, e.g., at reasons 19–35
46. *See* EPO Technical Boards of Appeal decision T 219/01, reasons 4 and 5.2, but *compare*: in decision T 716/08, the Board held that a therapeutic effect was found where an application disclosed that the mere presence of antibodies against the 48 kDa ISAV protein in the rabbit serum used for screening the lambda bacteriophage cDNA library was evidence that the 48 kDa ISAV protein was antigenic and, consequently, could also be a useful constituent of a subunit vaccine, *see* reasons at 17–24 and 47
47. *See* EPO Technical Boards of Appeal decision T 609/02, e.g., at reason 9
48. *See* Articles 99–105 EPC, which govern EPO opposition procedures
49. *See* Article 138 EPC for particular provisions that must be applied to these procedures
50. *See* Articles 106–111 EPC, which govern EPO appeals procedures
51. Article 68 EPC 2000: When the decision to revoke or limit the European patent is published per Article 105b(3) EPC 2000, the effects of this decision apply ab initio or retroactively from the outset via Article 64 (rights conferred at patent grant) and Article 67 (rights conferred after publication of the European patent application)
52. Observations may be submitted by third parties in an official EPO language per Article 115 EPC 2000
53. For example, Rule 95(2) EPC 2000 governs the scope of examination and admissibility of the limitation request, i.e., patentee's compliance with the substantive requirements for limitation of the European patent
54. These new procedures are found in the AIA at SEC. 6 in 35 U.S.C. §§ 311–319 and 35 U.S.C. §§ 321–329, respectively
55. U.S.C. 282(b), any patentability ground specified in 35 U.S.C. at Part II; 35 U.S.C. 282(c), failure to comply with any requirement of 35 U.S.C. §§112, 251 such as written description, definiteness, and clarity
56. 35 U.S.C. §324(a),(b)
57. Rule 76(2)(c) EPC
58. Article 116 EPC
59. Rule 81(1) EPC

# Influenza Cell-Culture Vaccine Production

## 52

Markus Hilleringmann, Björn Jobst, and Barbara C. Baudner

## Contents

| | | |
|---|---|---|
| 52.1 | **Introduction** | 824 |
| 52.1.1 | Influenza Disease | 824 |
| 52.1.2 | Influenza Viruses and Pandemics | 824 |
| 52.1.3 | Influenza Drugs | 825 |
| 52.1.4 | Influenza Vaccines | 826 |
| 52.2 | **World Health Organization (WHO) and Influenza Surveillance** | 826 |
| 52.2.1 | Influenza Surveillance | 826 |
| 52.2.2 | The WHO Global Influenza Surveillance Network (GISN) | 826 |
| 52.2.3 | Influenza Vaccine Recommendations | 826 |
| 52.2.4 | Seed Strains | 827 |
| 52.2.5 | Vaccine Potency/Reference Reagents | 828 |
| 52.3 | **Influenza Vaccine Production** | 828 |
| 52.4 | **Cell-Culture-Based Vaccine Production** | 829 |
| 52.4.1 | Cell Propagation | 830 |
| 52.4.2 | Virus Production | 830 |
| 52.4.3 | Virus Purification | 830 |
| 52.4.4 | Virus Inactivation and Splitting and Subunit Extraction | 831 |
| 52.4.5 | Polishing | 831 |
| 52.4.6 | Vaccine Characterization and Quality Tests (Bulk Release) | 831 |
| 52.4.7 | Formulation (Mixing and Filling) | 832 |
| 52.4.8 | Final Quality Control and Lot Release | 832 |
| 52.5 | **Influenza Vaccine Release** | 832 |
| 52.5.1 | Clinical Trials and Regulatory Agency Review | 832 |
| 52.6 | **Improved Influenza Vaccines and Future Trends** | 832 |
| **References** | | 835 |

M. Hilleringmann, PhD (✉)
Department of Applied Sciences and Mechatronics,
FG Protein Biochemistry and Cellular Microbiology,
University of Applied Sciences Munich,
Munich, Germany
email: markus.hilleringmann@hm.edu

B. Jobst, PhD
Manufacturing Science and Technology (MS&T),
Novartis Vaccines and Diagnostics GmbH,
Marburg, Germany

B.C. Baudner, PhD
Vaccine Research, Novartis Vaccines and Diagnostics srl,
Siena, Italy

### Abstract

Influenza vaccination is currently the principal means of reducing or counteracting influenza mortality and morbidity burden in the community. Since the early development of monovalent killed-virus vaccine formulations in the 1940s, different principal strategies were followed by vaccine manufactures resulting in a variety of influenza vaccines (e.g., inactivated whole-virus vaccines, live attenuated vaccines, detergent or solvent "split" vaccines, subunit vaccines, and adjuvanted vaccines). Actually two main production processes, the classical egg-based technology and more recently cell-culture-based operations, can be distinguished. In addition different routes of immunization allow the generation of intramuscular-, intradermal-, and intranasal-influenza vaccines.

The development and the production of influenza vaccines is based on a complex manufacturing process starting with the selection and development of optimal candidate vaccine viruses, and it requires various dynamic interactions with regulatory authorities and health-care officials. Planning for vaccine supplies and use as well as provision of other related health-care resources are essential components of a comprehensive seasonal and pandemic influenza response. Rapid spread of influenza viruses during seasonal epidemics and occasional pandemics tightly frames the whole process if vaccine is to be manufactured and delivered on time. There is a continuous effort to develop new and safe influenza vaccines and improve reagents for strain-specific potency testing to face complex influenza-related challenges better.

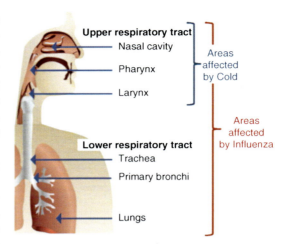

**Fig. 52.1** Areas affected by cold and influenza

## 52.1 Introduction

### 52.1.1 Influenza Disease

Influenza has a viral origin and often results in an acute respiratory illness affecting mainly the nose, throat, bronchi, and, occasionally, lungs (Fig. 52.1). The infection is characterized by sudden onset of high fever, aching muscles, headache and severe malaise, nonproductive cough, sore throat, and rhinitis. At the early stage of influenza infection, it might be difficult to distinguish it from the common cold, although influenza could be recognized by a high fever with a sudden onset and extreme fatigue [1].

The virus is transmitted easily from person to person via droplets and small particles produced when infected people cough or sneeze; therefore influenza tends to spread rapidly in seasonal epidemics occurring yearly during autumn and winter [2].

Most people recover from fever and other symptoms within a week without requiring medical attention. However, mainly among high-risk groups (the very young, elderly, or chronically ill), influenza can result in hospitalizations and deaths. Worldwide, annual epidemics cause about three to five million cases of severe illness and about 250,000–500,000 deaths [3].

### 52.1.2 Influenza Viruses and Pandemics

There are three types of influenza viruses: A, B (both seasonal), and C, with type A and B influenza cases occurring much more frequently than type C influenza. Type A influenza viruses are further typed into subtypes according to different kinds and combinations of virus surface glycoproteins – hemagglutinin (HA or H, respectively) and neuraminidase (NA or N, respectively) (Fig. 52.2) [3, 4]. Among many subtypes of influenza A viruses, currently influenza A(H1N1) and A(H3N2) subtypes are circulating among humans. Influenza viruses are well adapted to human populations and have the capacity to evade the immune system through mechanisms of mutations (antigenic drift) and swapping of surface protein genes between distantly related virus strains (antigenic shift) [5] (Fig. 52.2). When a major change occurs, the risk of a human pandemic arises [6].

Three influenza pandemics occurred during the twentieth century, the most serious being the Spanish influenza. The last pandemic of the past century occurred in 1968, and the responsible

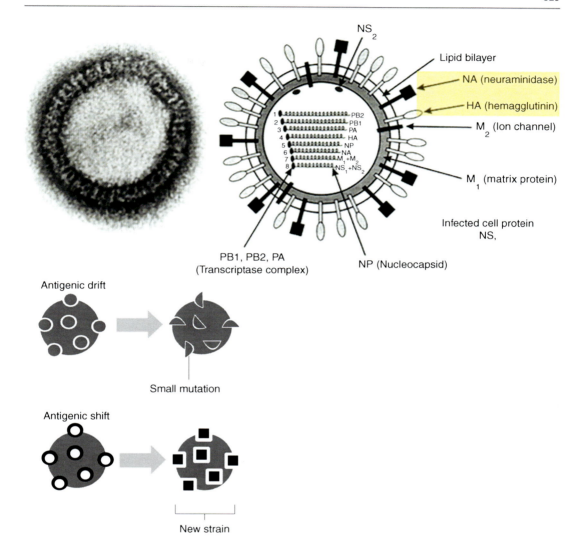

**Fig. 52.2** Influenza virus; antigenic drift and shift. Antigenic drift – The gradual alteration by point mutations of the hemagglutinin (HA) and neuraminidase (NA) proteins within a type or subtype. Antigenic drift results in the inability of antibodies to previous strains to neutralize the mutant virus. Antigenic drift occurs in both influenza A and B viruses and causes periodic epidemics. Antigenic shift – The appearance in the human population of an influenza A virus that has exchanged the genes encoding its HA protein or NA protein with another influenza virus, often a virus that infects animals rather than humans. Antigenic shift is responsible for worldwide pandemics

virus infected an estimated one to three million people throughout the world. The first pandemic of the present century occurred in 2009 and was caused by a H1N1 strain (A/California/07/09).

In 1997, a novel avian influenza virus, H5N1, first infected humans in China. Since its emergence, the H5N1 virus has spread from Asia to Europe and Africa, resulting in the infection of millions of poultry and wild birds. Several hundred human cases and deaths have been reported by the WHO [7].

### 52.1.3 Influenza Drugs

Antiviral drugs such as adamantanes (amantadine and remantadine) and the neuraminidase inhibitors zanamivir and oseltamivir (Tamiflu®)

[8, 9] have been used to treat influenza; however, some influenza viruses develop resistance to antiviral medicines, limiting the effectiveness of treatment. WHO constantly monitors antiviral susceptibility in the circulating influenza viruses [3].

### 52.1.4 Influenza Vaccines

According to the WHO, vaccination against influenza is the most effective way to prevent infection and severe outcomes caused by influenza viruses. Vaccination is especially important for people at higher risk of serious influenza complications and for people who live with or take care of high-risk individuals.

Various influenza vaccines have been available for many years [10] (Fig. 52.3). Seasonal influenza vaccines are usually trivalent, each dose containing 15 µg of each of two influenza A subtypes (e.g., H1N1 and H3N2) and 15 µg of one influenza B strain. There are currently three predominant types of inactivated influenza vaccine: whole-virus, split, and subunit vaccines [11, 12] (Fig. 52.3).

Among healthy adults, influenza vaccine can prevent 70–90 % of influenza-specific illness. Among the elderly, the vaccine reduces severe illnesses and complications by up to 60 % and deaths by 80 % [3].

## 52.2 World Health Organization (WHO) and Influenza Surveillance

### 52.2.1 Influenza Surveillance

As a result of genetic mutation, influenza viruses evolve frequently, and new strains quickly replace the older ones. Thus, vaccines formulated for 1 year may be ineffective in the following year, and continuous global monitoring and frequent reformulation of influenza vaccines is needed. For this reason, the WHO coordinates an international surveillance system to monitor the epidemiology of influenza viruses (Fig. 52.4) [13]. The surveillance system is undertaken year-round and allows for the detailed analysis of circulating influenza viruses isolated from both humans and animals, especially birds (e.g., H5N1) and pigs, and is able to detect newly evolved antigenic variants of the influenza A (H3N2 and H1N1) and B strains to which human populations are likely to be susceptible [14].

### 52.2.2 The WHO Global Influenza Surveillance Network (GISN)

The GISN monitors which influenza viruses are circulating in humans around the world throughout the year [13]. GISN comprises:
- 5 WHO Collaborating Centers (Atlanta, Beijing, London, Melbourne, Tokyo)
- 136 National Influenza Centers in 106 countries
- 11 H5 Reference Laboratories
- 4 Essential Regulatory Laboratories

The major technical roles of GISN are to:
- Monitor human influenza disease burden
- Monitor antigenic drift and other changes (such as antiviral drug resistance) in seasonal influenza viruses
- Obtain suitable virus isolates for updating of influenza vaccines
- Detect and obtain isolates of new influenza viruses infecting humans, especially those with pandemic potential

### 52.2.3 Influenza Vaccine Recommendations

Since 1973, twice a year the WHO convenes technical consultations to recommend which dominant circulating strains should be included in the vaccine – for the northern hemisphere in February and for the southern hemisphere in September. These recommendations are based on information provided by the WHO Global Influenza Surveillance Network (GISN), now the WHO Global Influenza Surveillance and Response System (GISRS) (Fig. 52.4). Recently, influenza A(H5N1), A(H9N2), and other subtypes of influenza viruses have also been taken into consideration by GISRS for pandemic preparedness purposes [14].

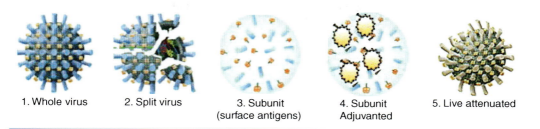

1. Whole virus vaccines - consisting of complete viruses which have been 'killed' or inactivated, so that they are not infectious but retain their strain - specific antigenic properties.
2. Split virus vaccines-consisting of inactivated virus particles disrupted by detergent treatment. These vaccines contain both surface and internal antigens.
3. Subunit or surface antigen vaccines - consisting essentially of purified hemagglutinin (HA) and neuraminidase (NA) from which other virus components have been removed.
4. Adjuvanted Subunit - consisting essentially of purified surface antigens (HA and NA) only plus an adjuvant added.
5. Live attenuated (cold - adapted) virus vaccines consisting of weakened (non - pathogenic) whole virus .In which the live virus in the vaccine can only multiply in the cooler nasal passages and which are administered intranasally

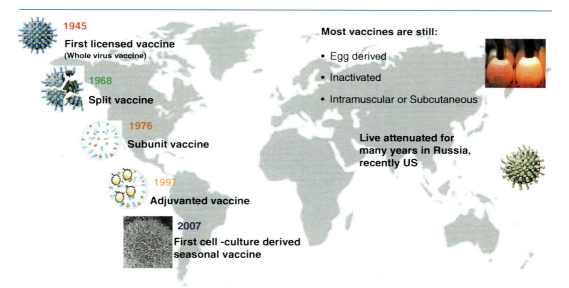

**Fig. 52.3** Influenza vaccines

## 52.2.4 Seed Strains

Once the WHO recommends which dominant circulating strains should be included in the vaccine, WHO Collaborating Centers start to generate and analyze seed strains for vaccine production [14]. Since the 1970s, this has been mainly done by genetic reassortment for influenza A strains. Embryonated eggs are coinfected with the field strain selected for the vaccine and an A/PR8/34 (or similar) donor strain that is known to give good yields on eggs [15]. High-growth progeny virus is analyzed to confirm the presence of surface glycoproteins from the field

**Fig. 52.4** WHO Global Influenza Surveillance and Response System (GISRS)

strain. A second technology is reverse genetics, a patented technology, available to attenuate highly pathogenic viruses and reassort the attenuated HA and NA with backbone virus [16]. Once developed, these candidate reassortants are characterized for their antigenic and genetic properties before being released to interested institutions on request (Fig. 52.4).

### 52.2.5 Vaccine Potency/Reference Reagents

Antigen standards and sheep antisera are subsequently developed and standardized by Essential Regulatory Laboratories (ERLs), in collaboration with vaccine manufacturers. Strain-specific reagents are made available to manufacturers worldwide on request. The antigen standard and sheep antisera are needed for single radial immunodiffusion (SRID) testing in order to quantify the produced antigen in the influenza vaccine bulks and release the vaccines [17].

## 52.3 Influenza Vaccine Production

The manufacture of influenza vaccine has several unique aspects that render the production process challenging:
(a) Influenza vaccine composition must match actual global WHO epidemiological surveillance data. Therefore, an updated vaccine formulation is developed each time.
(b) Yearly licensure has to be obtained for every amendment.
(c) There are very tight time slots and short windows of opportunity to adopt processes and respond to changes. The production process is one step of many, involving various institutions.
(d) Seasonal and pandemic influenza vaccines are interrelated with a major impact on production capacities.

Therefore, manufacturers of influenza vaccines are continuously optimizing existing processes and developing new and novel methods of preparing seasonal influenza vaccines, as well as pandemic vaccine candidates. Special efforts are made to increase production capacities and introduce more automated processes, designing more flexible and time-saving manufacturing methods.

Most influenza vaccine production is based on classical egg-based technology, a technology that has been used to produce seasonal vaccine for more than 30 years [18, 19]. Novel cell-culture technologies can offer various advantages over egg-based manufacturing methods and most likely will supplement the current egg-based technology (Fig. 52.5) [20, 21].

Independent of which production process is used, either egg-based or cell-culture-based, the downstream production process could produce a whole-virus vaccine, a split vaccine, or a subunit vaccine from the viral harvest. All manufacturing processes must go through extensive testing and validation before licensure, and the production of

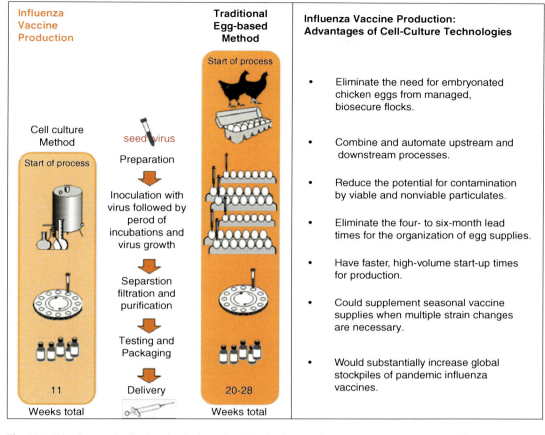

**Fig. 52.5** Vaccine production by classical egg-based technology and novel cell-culture processes: Timeline comparison and advantages of cell-culture-based manufacturing methods

each batch is tightly controlled. In the current chapter we will focus on recent cell-culture influenza vaccine production, describing the procedure for a subunit influenza vaccine.

## 52.4 Cell-Culture-Based Vaccine Production

Vaccine seed strains are distributed to the vaccine producers by WHO Influenza reference centers in order to evaluate their suitability for vaccine production. Factors such as yield when grown in cell culture (or in eggs), antigenic stability, inactivation, and purification process are evaluated. The experiences with the vaccine candidate strains at the manufacturer from all participating companies are considered for the final endorsement of the vaccine strains by the European Medicines Agency (EMA) and the United States Food and Drug Administration (FDA). The meetings with the national authorities are typically held 1 month after the initial WHO decision on composition.

Different cell lines can be used to produce the influenza virus such as VERO [22], a kidney cell from the African green monkey, and Madin-Darby canine kidney (MDCK) cells [23], which originated in 1958 from the kidney of a healthy dog. Alternative expression systems for recombinant influenza proteins, like insect cells, are under study and evaluation. A MDCK cell line was optimized for the production of influenza vaccine and tested with numerous virus variants. It has proven to be particularly suitable for the production of influenza vaccines as it grows in suspension. Thus it requires no surface to proliferate, simplifying industrial production considerably.

Cell-culture influenza vaccine production resulting in a subunit influenza vaccine can be divided into the following steps (1.-8.; see Figs. 52.6, 52.7, 52.8, 52.9, 52.10, 52.11, and 52.12). A summarized scheme is shown in Fig. 52.13.

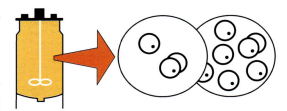

**Fig. 52.6** Cell propagation

## 52.4.1 Cell Propagation

Preparation of a cell line for propagation starts with the thawing of the cell line "seed" lot (e.g., MDCK). For a "First-pass" cell line propagation begins with the small-scale pre-culture propagation of seed cells. The cells are stored in liquid nitrogen at −196 °C. For the production, minimum amounts – a so-called cryo-tube containing 1 ml with approximately ten million cells – are thawed and proliferated in three steps from a 10-l-volume to 100 l and then to 1,000 l. At each stage the cells receive the optimum conditions for growth with respect to temperature, oxygen, pH value, and nutrient supplies. The proliferation of the cells in fermenters (stainless steel tanks) is constantly monitored with the aid of a computer system that automatically checks all data and accurately records each step. Cell proliferation takes place in a contained fermenter system in so-called clean rooms, resulting in maximum safety and purity for employees, environment, patient, and product. Approximately 3 weeks after the removal of the cells from the nitrogen refrigeration, sufficient cells have grown in the 1,000-l-fermenter. In contrast, it can take up to 6 months to organize the egg supplies, flock setting, etc., for initial inoculation in classical egg-based technology (Fig. 52.6).

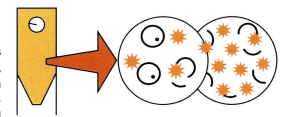

**Fig. 52.7** Virus production

**Fig. 52.8** Virus purification

amounts for downstream processing are produced. During the course of this process, the cells die off and viruses are released in the medium/supernatant (Fig. 52.7).

## 52.4.3 Virus Purification

The first step of a long series of purification procedures is so-called separation, by which the virus suspension is separated from the cell residue. In a subsequent chromatography step, the virus is captured and separated from the medium solution. The next step, ultrafiltration, further concentrates the intermediate product. This occurs in an automated and closed system, whereas harvesting of an egg-based virus is largely a manual process that requires extracting infected cells and fluids and then collecting the virus (Fig. 52.8).

## 52.4.2 Virus Production

Upon achieving a certain predetermined density, the cells are transferred to a 2,500-l fermenter via a closed pipe system. The influenza virus is added to the tank to start infecting the mammalian cells. Virus replication in the host cell requires several days (approximately 3 days) until sufficient virus

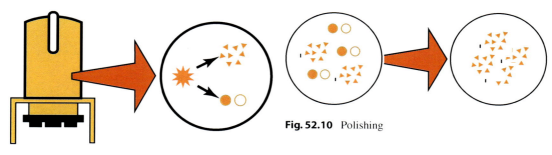

**Fig. 52.9** Virus inactivation and splitting and subunit extraction

**Fig. 52.10** Polishing

## 52.4.4 Virus Inactivation and Splitting and Subunit Extraction

After the purification the virus is inactivated (rendered unable to cause infection) via a chemical reaction process. Most manufacturers use either beta-propiolactone or formaldehyde for this purpose. Virus splitting/extraction of surface antigens (HA and NA) with detergent follows, because only fractions of specific surface proteins are required for the subsequent subunit influenza vaccine (Fig. 52.9).

## 52.4.5 Polishing

Splitting of the virus is followed by ultracentrifugation for removal of undesired viral core structures that have no role in eliciting neutralizing antibodies. This purification technology is basically similar to the egg-based vaccine process, and the resulting purified subunit vaccine is very similar in composition to egg-based vaccines. After further purification and concentration steps, approximately 10 l of antigen concentrate of a virus strain are left of the original 2,500 l coming from infection. Since the seasonal influenza vaccines contain three (or more recently four) viral strains (H1N1, H3N2, and one or two B strains), the production process must be performed three to four times – once for every strain (influenza vaccine monobulk) in order to generate a trivalent or quadrivalent vaccine (Fig. 52.10).

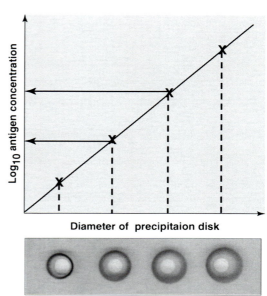

**Fig. 52.11** SRID Assay

## 52.4.6 Vaccine Characterization and Quality Tests (Bulk Release)

Subsequently the three individual bulks are analyzed for their HA antigen content by a single radial immunodiffusion (SRID) assay (Fig. 52.11) [24, 25]. SRID assay is a technique used in immunology to determine the quantity of an antigen by measuring the diameters of circles of precipitin complexes surrounding samples of the antigen that mark the boundary between the antigen and an antibody suspended in a medium, such as an agar gel. Measurement of the diameters of the precipitin discs of the unknown solutions allows an estimation of the antigen concentration to be made by simple

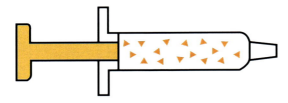

**Fig. 52.12** Mixing and filling

interpolation to diameters of the precipitin discs of standard antigen solutions with known concentration.

The reagents need to be specific for the strains in the vaccine and take approximately 3 months to be produced and distributed (see above: *Vaccine Potency/Reference Reagents*). This can delay the release of the vaccine. Additional testing for process-related impurities, sterility, and residual infectious virus as well as stability studies is also required for each manufactured bulk and for the release of material (Fig. 52.11) [26].

### 52.4.7 Formulation (Mixing and Filling)

The three individual influenza antigen monobulks are subsequently diluted into standard doses using sterile phosphate-buffered salt solutions in a concentration similar to bodily fluids. Final vaccine is then sterile filtered and added to bottles, prefilled syringes, or in nasal spray containers (Fig. 52.12).

### 52.4.8 Final Quality Control and Lot Release

Each trivalent batch or "lot" (like the individual monobulks) undergoes rigorous quality control testing to ensure it meets regulatory agency standards for safety, efficacy, and stability prior to vaccine release [25]. At this point, the development phase of an influenza cell-culture subunit vaccine is complete. All that remains is to complete the licensing process (Fig. 52.13).

## 52.5 Influenza Vaccine Release

### 52.5.1 Clinical Trials and Regulatory Agency Review

Vaccines are classified as biological products and regulated by both the United States Food and Drug Administration (FDA) and the European Medicines Agency (EMA). In Europe clinical trials are requested to demonstrate the safety and immunogenicity of each new influenza vaccine formulation. This is not the case in the United States or the southern hemisphere.

Vaccine lots are also carefully labeled in case problems are identified after the vaccine has been distributed. Vaccine cannot be delivered until they are formally released by the regulatory agencies [26, 27] (Fig. 52.14).

All in all, it takes about 16 weeks from the extraction of the cells until a finished seasonal influenza vaccine is packaged and ready for delivery. As mentioned above, external factors such as availability of strain-specific reagents can influence the timeline. The complex process of a seasonal influenza vaccine campaign including various dynamic interactions of public health institutions/laboratories and vaccine manufactures is depicted schematically in Fig. 52.15.

## 52.6 Improved Influenza Vaccines and Future Trends

In order to provide safe and effective influenza vaccines on time, their production requires an accurate coordination of a highly complex process involving a wide range of expertise in public health institutions and laboratories and vaccine manufacturers. Influenza vaccine composition must match actual global WHO epidemiological surveillance data, which reflect the continuing evolution of influenza viruses. Therefore, an updated vaccine formulation is developed on an almost annual basis. Current seasonal influenza vaccine production for the northern and southern hemispheres and process adjustments take place within two compressed production cycles that are

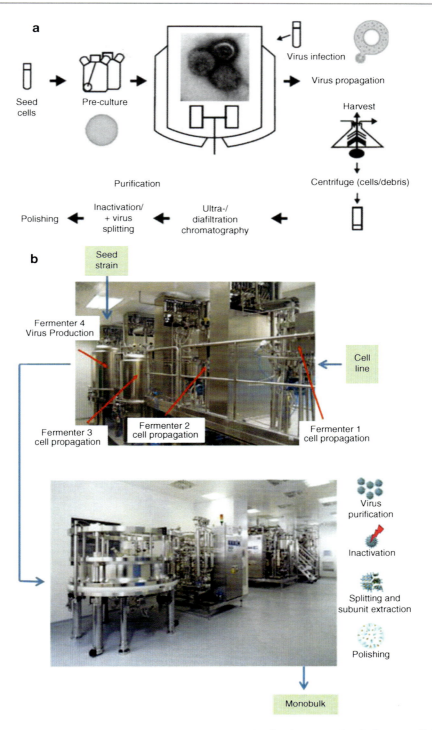

**Fig. 52.13** Influenza vaccine monobulk generation. The production of a monovalent bulk containing one virus strain takes approximately 6 weeks from the thawing of the cells to the finished antigen concentrate. For trivalent vaccine three rounds of monobulk manufacturing are needed. After the monovalent bulks are available, it takes another 10 weeks for mixing, filling, packaging, and regulatory approval before the final vaccine product can be shipped ((**a**): schematic workflow; (**b**): process showing examples of manufacturing equipment)

restricted to about 6 months each. These time constraints make new technical and clinical development programs very challenging.

Especially during influenza pandemics, vaccine production faces a race against the time.

**Clinical trial (in Europe)**

**Regulatory agency review and release**

**Final vaccine**

**Fig. 52.14** Influenza vaccine release

Although during the 2009 H1N1 influenza pandemic, first vaccines were available approximately 3 months after the declaration of the pandemic, quantities of the vaccine were limited, and only a minority could be vaccinated in time – before the pandemic waves had peaked. More flexible platforms that enable more rapid vaccine production and facilitate regulatory processes are highly desirable, and both manufacturers and governments together with research institutions have encouraged a wave of innovation to develop novel methods in influenza vaccine manufacturing [28], as recently reviewed in detail by Dormitzer et al. [29]. Approaches range from research or early development projects to already licensed products. However, new procedures for influenza vaccines can also be expected to lead to higher complexity of vaccine production and standardization, raising new safety and regulatory challenges that have to be addressed.

Historically, most efforts have been put on improving already licensed egg-based vaccines and their manufacturing, focusing on increased production capacities and the automation of the production processes [19]. The introduction of cell-culture-based influenza production brings several advantages over egg-based vaccine production, with a high potential to better face future influenza-related challenges [20, 21, 30]. Generally, improvements in seed generation procedures and development of vaccine potency/reference reagents [31, 32] will speed influenza vaccine production. The introduction of new

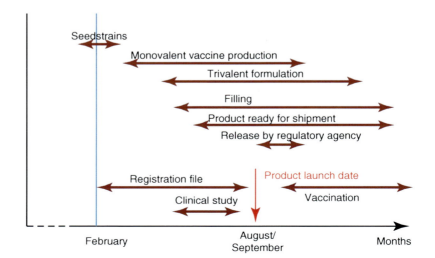

**Fig. 52.15** "Cycle" of a seasonal vaccination campaign for the northern hemisphere starting from influenza vaccine seed strains

assay strategies to estimate the HA content could further accelerate product release [33–35]. Additional new technologies for influenza vaccines are being explored, including recombinant strategies (to generate recombinant subunit-based influenza vaccines [36–38] or recombinant viruslike particle (VLP)-based influenza vaccines) [39, 40] and approaches for the production of "live attenuated influenza vaccines (LAIV)" [41–43] and "live vectored influenza vaccines" [44, 45].

The efficacy of influenza vaccines as well as vaccine manufacturing capacity can be further enhanced by the addition of adjuvants to the vaccine formulation [46]. Adjuvants are substances included in vaccines to improve antibody production and the immune response of the recipient or to decrease the amount of antigen (dose size) required in the vaccine. The latter is the most effective way to increase global vaccine manufacturing capacity. Several adjuvants have demonstrated the capacity to increase antibody levels to those considered protective, using lower antigen doses. Importantly, the broader use of the squalene-based emulsion MF59 during the 2009 H1N1 global pandemic allowed its safety profile to be further validated in a large and diverse populations of humans, including young children [47, 48] and pregnant women [49].

To improve coverage, efforts are undertaken to generate "higher-valent" vaccine formulations, e.g., quadrivalent influenza vaccines (QIV) containing four individual influenza antigen monobulks [50, 51]. The ultimate goal would be the generation of a "universal influenza vaccine," which need not be altered as influenza strains change. One approach to a universal vaccines is the inclusion of suitably presented, conserved antigens resulting in an effective, long-lasting vaccine formulation to confer broad coverage against diverse strains [39, 50–55].

Novel delivery approaches (e.g., oral, transdermal, nasal), different from traditional intramuscular and subcutaneous administration, could simplify vaccine application, at the same time increasing safety and patient compliance [43, 56–58].

In summary, technical and scientific innovations will further promote safe and innovative vaccine regimens that could significantly broaden and extend protection against influenza. The increased demand for effective influenza vaccines throughout the world will challenge not only the technical capacity of vaccine companies, but also their financial strength and strategic vision. Importantly these challenges must be aligned with efforts of public health officials to effectively implement these vaccines to control influenza.

**Acknowledgements** The authors are grateful to Heidi Trusheim and Karsten Kattmann for critical reading of the manuscript.

# References

1. Eccles, R.: Understanding the symptoms of the common cold and influenza. Lancet Infect. Dis. **5**, 718–25 (2005). doi:10.1016/S1473-3099(05)70270-X
2. Health topics influenza – WHO World Health Organization, Geneva, Switzerland. http://www.who.int/topics/influenza/en/ (2013)
3. Influenza (Seasonal) Fact sheet N°211 – WHO World Health Organization, Geneva, Switzerland. http://www.who.int/mediacentre/factsheets/fs211/en/ (2009)
4. Seasonal influenza (flu)/influenza (flu) viruses – CDC Centers for Disease Control and Prevention, Atlanta, USA. http://www.cdc.gov/flu/about/viruses/index.htm
5. Webster, R.G., Laver, W.G.: Antigenic variation in influenza virus. Biology and chemistry. Prog. Med. Virol. **13**, 271–338 (1971)
6. Hay, A.J., Gregory, V., Douglas, A.R., Lin, Y.P.: The evolution of human influenza viruses. Philos. Trans. R. Soc. Lond. B Biol. Sci. **356**, 1861–70 (2001). doi:10.1098/rstb.2001.0999
7. Gasparini, R., Amicizia, D., Lai, P.L., Panatto, D.: Aflunov(®): a prepandemic influenza vaccine. Expert Rev. Vaccines **11**, 145–57 (2012). doi:10.1586/erv.11.170
8. Beigel, J., Bray, M.: Current and future antiviral therapy of severe seasonal and avian influenza. Antiviral Res. **78**, 91–102 (2008). doi:10.1016/j.antiviral.2008.01.003
9. Jefferson, T., et al.: Neuraminidase inhibitors for preventing and treating influenza in healthy adults and children. Cochrane Database Syst. Rev. **18**, 1 (2012). doi:10.1002/14651858.CD008965.pub3
10. Oxford, J., Gilbert, A., Lambkin-Williams, R.: Influenza vaccines have a short but illustrious history of dedicated science enabling the rapid global production of A/Swine (H1N1) vaccine in the current pandemic. In: Del Giudice, G., Rappuoli, R. (eds.) Influenza Vaccines for the Future, pp. 115–147. Birkhauser Inc, Basel (2011)
11. Verma, R., Khanna, P., Chawla, S.: Influenza vaccine: an effective preventive vaccine for developing countries. Hum. Vaccin. Immunother. **8**, 675–8 (2012). doi:10.4161/hv.19516

12. Ellebedy, A.H., Webby, R.J.: Influenza vaccines. Vaccine **27**(Suppl 4), D65–D68 (2009). doi:10.1016/j.vaccine.2009.08.038
13. WHO Global Influenza Surveillance Network (GISN) Surveillance and Vaccine Development – WHO Collaborating Centre for Reference and Research on Influenza (VIDRL), North Melbourne, Australia. http://www.influenzacentre.org/centre_GISN.htm
14. Gerdil, C.: The annual production cycle for influenza vaccine. Vaccine **21**, 1776–1779 (2003). pii: S0264410X03000719
15. Burnet, F.M.: Growth of influenza virus in the allantoic cavity of the chick embryo. Aust. J. Exp. Biol. Med. Sci. **19**, 291–295 (1941)
16. Hoffmann, E., Neumann, G., Kawaoka, Y., Hobom, G., Webster, R.G.: A DNA transfection system for generation of influenza A virus from eight plasmids. Proc. Natl. Acad. Sci. U.S.A. **97**, 6108–6113 (2000). doi:10.1073/pnas.100133697
17. Expert Committee on Biological Standardization Geneva, 17 to 21 October 2011 Proposed Generic Protocol for the Calibration of Seasonal/Pandemic Influenza Antigen Working Reagents by WHO Essential Regulatory Laboratories (WHO/BS/2011.2183). http://www.who.int/biologicals/expert_committee/BS2011.2183_Flu_vax_ERL_calibration_protocol.pdf
18. Hickling J., D'Hondt E. A review of production technologies for influenza virus vaccines, and their suitability for deployment in developing countries for influenza pandemic preparedness – WHO World Health Organization Initiative for Vaccine Research Geneva Switzerland Date: 20 December 2006. www.who.int/entity/vaccine_research/diseases/influenza/Flu_vacc_manuf_tech_report.pdf
19. Matthews, J.T.: Egg-based production of influenza vaccine: 30 years of commercial experience. The Bridge **36**, 17–24 (2006)
20. Rappuoli, R.: Cell-culture-based vaccine production: technological options. The Bridge **36**, 25–30 (2006)
21. Dormitzer, P.R.: Cell culture-derived influenza vaccines. In: Del Giudice, G., Rappuoli, R. (eds.) Influenza Vaccines for the Future, pp. 293–312. Birkhauser Inc, Basel (2011)
22. Kistner, O., Barrett, P.N., Mundt, W., Reiter, M., Schober-Bendixen, S., Dorner, F.: Development of a mammalian cell (Vero) derived candidate influenza virus vaccine. Vaccine **16**, 960–968 (1998). pii: S0264-410X(97)00301-0
23. Palache, A.M., Brands, R., van Scharrenburg, G.: Immunogenicity and reactogenicity of influenza subunit vaccines produced in MDCK cells or fertilised chicken eggs. J. Infect. Dis. **176**(Suppl 1), S20–S23 (1997)
24. Schild, G.C., Wood, J.M., Newman, R.W.: A single radial-immunodiffusion technique for the assay of influenza hemagglutinin antigen. WHO Bull. **52**, 223–231 (1975)
25. Wood, J.M., Schild, G.C., Newman, R.W., Seagroatt, V.: Application of an improved single radial-immunodiffusion technique for the assay of influenza hemagglutinin antigen content of whole virus and subunit vaccines. Dev. Biol. Stand. **39**, 193–200 (1977)
26. Recommendations for the production and control of influenza vaccine (inactivated) © World Health Organization WHO Technical Report Series, No. 927, Annex 3 (2005), http://www.who.int/vaccine_research/diseases/influenza/TRS_927_ANNEX_3_Influenza_2005.pdf
27. Wood, J.M., Levandowski, R.A.: The influenza vaccine licensing process. Vaccine **21**, 1786–1788 (2003). pii: S0264410X03000732
28. Sambhara, S., Rappuoli, R.: Improving influenza vaccines. Expert Rev. Vaccines **11**, 871–872 (2012). doi:10.1586/erv.12.79
29. Dormitzer, P.R., Tsai, T.F., Del Giudice, G.: New technologies for influenza vaccines. Hum. Vaccin. Immunother. **8**, 45–58 (2012). doi:10.4161/hv.8.1.18859
30. Montomoli, E., et al.: Cell culture-derived influenza vaccines from Vero cells: a new horizon for vaccine production. Expert Rev. Vaccines **11**, 587–94 (2012). doi:10.1586/erv.12.24
31. Strecker, T., et al.: Exploring synergies between academia and vaccine manufacturers: a pilot study on how to rapidly produce vaccines to combat emerging pathogens. Clin. Chem. Lab. Med. **50**, 1275–9 (2012). doi:10.1515/cclm-2011-0650
32. WHO Writing Group, Ampofo W. K. et al.: Improving influenza vaccine virus selection: report of a WHO informal consultation held at WHO headquarters, Geneva, Switzerland, 14–16 June 2010. Influenza Other Respi. Viruses. **6**, 142–152 (2012). doi: 10.1111/j.1750-2659.2011.00277.x
33. Kapteyn, J.C., et al.: HPLC-based quantification of haemagglutinin in the production of egg and MDCK cell-derived influenza virus seasonal and pandemic vaccines. Vaccine **27**, 1468–77 (2009). doi:10.1016/j.vaccine.2008.11.113
34. Lorbetskie, B., et al.: Optimization and qualification of a quantitative reversed-phase HPLC method for hemagglutinin in influenza preparations and its comparative evaluation with biochemical assays. Vaccine **29**, 3377–89 (2011). doi:10.1016/j.vaccine.2011.02.090
35. Williams, T.L., et al.: Quantification of influenza virus hemagglutinins in complex mixtures using isotope dilution tandem mass spectrometry. Vaccine **26**, 2510–20 (2008). doi:10.1016/j.vaccine.2008.03.014
36. Cox, M.M.: Recombinant protein vaccines produced in insect cells. Vaccine **30**, 1759–66 (2012). doi:10.1016/j.vaccine.2012.01.016
37. Baxter, R., et al.: Evaluation of the safety, reactogenicity and immunogenicity of FluBlok® trivalent recombinant baculovirus-expressed hemagglutinin influenza vaccine administered intramuscularly to healthy adults 50–64 years of age. Vaccine **29**, 2272–8 (2011). doi:10.1016/j.vaccine.2011.01.039
38. Song, L., et al.: Efficacious recombinant influenza vaccines produced by high yield bacterial expression: a solution to global pandemic and seasonal needs. PLoS One **3**, e2257 (2008). doi:10.1371/journal.pone.0002257

39. Kang, S.M., Kim, M.C., Compans, R.W.: Virus-like particles as universal influenza vaccines. Expert Rev. Vaccines **11**, 995–1007 (2012). doi:10.1586/erv.12.70
40. Haynes, J.R.: Influenza virus-like particle vaccines. Expert Rev. Vaccines **8**, 435–45 (2009). doi:10.1586/erv.09.8
41. Gasparini, R., Amicizia, D., Lai, P.L., Panatto, D.: Live attenuated influenza vaccine–a review. J. Prev. Med. Hyg. **52**, 95–101 (2011)
42. Monto, A.S., et al.: Comparative efficacy of inactivated and live attenuated influenza vaccines. N. Engl. J. Med. **361**, 1260–7 (2009). doi:10.1056/NEJMoa0808652
43. Carter, N.J., Curran, M.P.: Live attenuated influenza vaccine (FluMist®; Fluenz™): a review of its use in the prevention of seasonal influenza in children and adults. Drugs **71**, 1591–622 (2011). doi:10.2165/11206860-000000000-00000
44. Kopecky-Bromberg, S.A., Palese, P.: Recombinant vectors as influenza vaccines. Curr. Top. Microbiol. Immunol. **333**, 243–67 (2009). doi:10.1007/978-3-540-92165-3_13
45. Lambe, T.: Novel viral vectored vaccines for the prevention of influenza. Mol. Med. **18**, 1153–60 (2012). doi:10.2119/molmed.2012.00147
46. O'Hagan, D.T., Tsai, T., Reed, S.: Emulsion-based adjuvants for improved influenza vaccines. In: Del Giudice, G., Rappuoli, R. (eds.) Influenza Vaccines for the Future, pp. 327–357. Birkhauser Inc, Basel (2011)
47. Vesikari, T., Pellegrini, M., Karvonen, A., Groth, N., Borkowski, A., et al.: Enhanced immunogenicity of seasonal influenza vaccines in young children using MF59 adjuvant. Pediatr. Infect. Dis. J. **28**, 563–571 (2009). doi:10.1097/INF.0b013e31819d6394
48. Vesikari, T., Knuf, M., Wutzler, P., Karvonen, A., Kieninger-Baum, D., et al.: Oil-in-water emulsion adjuvant with influenza vaccine in young children. N. Engl. J. Med. **365**, 1406–16 (2011). doi:10.1056/NEJMoa1010331
49. Heikkinen, T., Young, J., van Beek, E., Franke, H., Verstraeten, T., et al.: Safety of MF59-adjuvanted A/H1N1 influenza vaccine in pregnancy: a comparative cohort study. Am. J. Obstet. Gynecol. **207**, 177.e1–8 (2012). doi:10.1016/j.ajog.2012.07.007
50. Ambrose, C.S., Levin, M.J.: The rationale for quadrivalent influenza vaccines. Hum. Vaccin. Immunother. **8**, 81–88 (2012). doi:10.4161/hv.8.1.17623
51. Barr, I.G., Jelley, L.L.: The coming era of quadrivalent human influenza vaccines: who will benefit? Drugs **72**, 2177–85 (2012). doi:10.2165/11641110-000000000-00000
52. Shaw, A.R.: Universal influenza vaccine: the holy grail? Expert Rev. Vaccines **11**, 923–927 (2012). doi:10.1586/erv.12.73
53. Du, L., Zhou, Y., Jiang, S.: Research and development of universal influenza vaccines. Microbes Infect. **12**, 280–6 (2010). doi:10.1016/j.micinf.2010.01.001
54. Kang, S.M., Song, J.M., Compans, R.W.: Novel vaccines against influenza viruses. Virus Res. **162**, 31–38 (2011). doi:10.1016/j.virusres.2011.09.037
55. Rudolph, W., Ben Yedidia, T.: A universal influenza vaccine: where are we in the pursuit of this "Holy Grail"? Hum. Vaccin. **7**, 10–11 (2011). pii: 14925
56. Belshe, R.B., Newman, F.K., Cannon, J., Duane, C., Treanor, J., et al.: Serum antibody responses after intradermal vaccination against influenza. N. Engl. J. Med. **351**, 2286–94 (2004). doi:10.1056/NEJMoa043555
57. Kenney, R.T., Frech, S.A., Muenz, L.R., Villar, C.P., Glenn, G.M.: Dose sparing with intradermal injection of influenza vaccine. N. Engl. J. Med. **351**, 2295–301 (2004). doi:10.1056/NEJMoa043540
58. Ansaldi, F., Durando, P., Icardi, G.: Intradermal influenza vaccine and new devices: a promising chance for vaccine improvement. Expert Opin. Biol. Ther. **11**, 415–27 (2011). doi:10.1517/14712598.2011.557658

# United States Food and Drug Administration: Regulation of Vaccines

## 53

Valerie Marshall

## Contents

| | | |
|---|---|---|
| 53.1 | **General Requirements** | 839 |
| 53.2 | **Overview of the Regulatory Process** | 841 |
| 53.2.1 | Pre-investigational Stage (Pre-IND) | 841 |
| 53.2.2 | IND Stage | 841 |
| 53.2.3 | Biologics License Application (BLA) | 842 |
| 53.2.4 | Post-marketing | 843 |
| 53.3 | **Preclinical Guidelines** | 843 |
| 53.4 | **Clinical Guidelines** | 843 |
| **References** | | 844 |

### Abstract

Vaccine development is a complex process guided by regulatory requirements that are designed to ensure the licensure of safe and effective products. Vaccines are subject to rigorous regulatory oversight throughout their life cycle including scientific and clinical assessments. This chapter will focus on the United States regulatory process for development and licensure of preventive vaccines for infectious disease indications.

## 53.1 General Requirements

The legal framework for the regulation of vaccines is primarily derived from Section 351 of the Public Health Service (PHS) Act (42 USC 262) [1] and from certain sections of the U.S. Food, Drug, and Cosmetic Act (FD&C Act) (21 USC §321) [2]. Vaccines meet the criteria of both a drug and a biological product since the Food, Drug, and Cosmetic Act defines drugs as "articles intended for use in the diagnosis, cure, mitigation, treatment, or prevention of disease" [2].

The PHS Act and the FD&C Act are implemented through the Code of Federal Regulations (CFR), which contains the general rules published in the *Federal Register* by agencies of the federal government. The federal regulations that apply specifically to licensure of vaccines and other biologicals are Title 21 CFR 600 through 680 [3]. Important regulations and legislation

V. Marshall, MPH
United States Public Health
Service Commissioned Corps,
Food and Drug Administration (FDA),
Center for Biologics Evaluation and Research,
Office of Vaccines Research and Review,
Rockville, MD, USA
e-mail: valerie.marshall@fda.hhs.gov

**Table 53.1** US legislation and regulations applicable to the development, manufacture, and licensure of vaccines

| | |
|---|---|
| Public Health Service Act (42 USC 262–63) Section 351 | |
| Food, Drug, and Cosmetics Act (21 USC 301–392) | |
| Prescription Drug User Fee Act, 1992, 1997, 2002, 2007, and 2012 | |
| Food and Drug Administration Amendments Act, 2007 | |
| Food and Drug Administration Safety and Innovation Act, 2012 | |
| Title 21 of the Code of Federal Regulations | |
| Part 25 | Environmental impact considerations |
| Part 50 | Protection of human subjects |
| Part 58 | Good laboratory practice for nonclinical laboratory studies |
| Part 201 | Labeling |
| Part 202 | Prescription drug advertising |
| Parts 210–211 | Good manufacturing practices, cGMP |
| Part 312 | Investigational New Drug (IND) application |
| Part 601 | Licensing |
| Part 600 and 610 | General Biological Product Standards |

**Table 53.2** Select FDA guidance document related to product testing and manufacture

| |
|---|
| Guidance for Industry: Characterization and Qualification of Cell Substrates and Other Biological Materials Used in the Production of Viral Vaccines for Infectious Disease Indications, February, 2010 |
| Guidance for Industry: CGMP for Phase 1 Investigational Drugs, July 2008 |
| Guidance for Industry: Considerations for Plasmid DNA Vaccines for Infectious Disease Indications, November 2007 |
| Guidance for Industry: Considerations for Developmental Toxicity Studies for Preventive and Therapeutic Vaccines for Infectious Disease Indications, February 2006 |
| Guidance for Industry: Analytical Procedures and Methods Validation, August 2000 |
| Guidance for Industry for the Evaluation of Combination Vaccines for Preventable Diseases: Production, Testing and Clinical Studies, April 1997 |

applicable to the development, manufacture, and licensure of vaccines are delineated in Table 53.1.

Vaccine licensure is based on a demonstration of safety and effectiveness, as well as the ability of the license holder to manufacture the product in a consistent manner within the defined and agreed upon specifications. Section 351 of the PHS Act (42 USC 262) states that a biologics license application can be approved based on a demonstration that "…(a) the biological product that is the subject of the application is safe, pure and potent; and (b) the facility in which the biological product is manufactured, processed, packed or held meets standards designed to assure that the biological product continues to be safe, pure, and potent." Title 21 of the CFR, Part 600, provides the following definitions for safety, purity, and potency:

*Safety* is defined as the "relative freedom from harmful effect to persons affected, directly or indirectly, by a product when prudently administered, taking into consideration the character of the product in relation to the condition of the recipient at the time."

*Purity* is defined as the "relative freedom from extraneous matter in the finished product, whether or not harmful to the recipient or deleterious to the product."

*Potency* is defined as the "specific ability or capacity of the product, as indicated by appropriate laboratory tests or by adequately controlled clinical data obtained through administration of the product in the manner intended, to effect a given result."

Considering the diversity of novel vaccine products, applying these criteria requires careful consideration of the product characteristics, the methods of manufacture, the target population, and the indication.

The FDA periodically publishes guidance documents that describe the interpretation of a regulation and/or the agency's current thinking related to specific aspects of the manufacture and preclinical and clinical evaluation of drugs and biological products. In contrast to regulations, guidance documents are not binding; thus, the manufacturer may choose alternative approaches other than those described in pertinent guidance in order to comply with laws and regulations. Refer to Tables 53.2 and 53.3 for select FDA guidance documents on product testing and manufacturing and clinical development of preventive vaccines for infectious disease indications.

Important legislation applicable to vaccine development has evolved over time, in part, to keep pace with technological and scientific advances in the pharmaceutical and biomedical

**Table 53.3** Select FDA guidance document related to clinical studies

| |
|---|
| Draft Guidance for Industry: Determining the Extent of Safety Data Collection Needed in Late Stage Premarket and Postapproval Clinical Investigations, February 2012 |
| Draft Guidance for Industry Non-Inferiority Clinical Trials, March 2010 |
| Guidance for Industry: Clinical Data Needed to Support the Licensure of Pandemic Influenza Vaccines, May 2007 |
| Guidance for Industry: Clinical Data Needed to Support the Licensure of Seasonal Inactivated Influenza Vaccines, May 2007 |
| Guidance for Industry: Toxicity Grading Scale for Healthy Adult and Adolescent Volunteers Enrolled in Preventive Vaccine Clinical Trials, September 2007 |

industries. The Prescription Drug User Fee Act (PDUFA), first enacted in 1992, granted the FDA authority to collect user fees from manufacturers to expedite the review of drug and biological applications and establish post-marketing drug-safety activities in accordance with performance goals [4]. The legislation was reauthorized in 1997 (PDUFA II) [5], 2002 (PDUFA III) [6], 2007 (PDUFA IV) [7], and 2012 (PDUFA V) [8].

The Food and Drug Administration Amendments Act (FDAAA) of 2007 added significant reform to the regulation of drugs and biologicals and provided the FDA with additional funding and new authorities [7]. For example, FDAAA added many new provisions to the Food, Drug, and Cosmetic Act by which FDA can, provided certain conditions are met, require post-marketing studies and clinical trials for prescription drug and biological products at the time of approval or after approval including requiring risk evaluation and mitigation strategies (REMS) and safety-related labeling changes [9].

The Food and Drug Administration Safety and Innovation Act ("FDASIA") of 2012 further revised existing law including reauthorizing and amending several drug and medical device provisions, establishing new user fee statutes for biosimilar biologics and generic drugs as well as reauthorizing two programs that encourage pediatric drug development [10]. FDASIA also provides FDA with new authority concerning drug shortages, among other things.

## 53.2 Overview of the Regulatory Process

The Center for Biologics Evaluation and Research (CBER) of the U.S. Food and Drug Administration (FDA) is the federal agency responsible for regulating vaccines and other biological products. The review of vaccine applications occurs among CBER's Office of Vaccines Research and Review, Office of Compliance and Biologics Quality, and Office of Biostatistics and Epidemiology. CBER provides regulatory guidance through four major stages of vaccine development: pre-investigational new drug application (pre-IND) stage, investigational new drug application (IND) stage, licensing (biologics license application (BLA)), and post-marketing (Fig. 53.1). Regulatory oversight increases at each stage of development.

CBER established the Managed Review Process to provide a systematic and effective approach to the regulatory review of all regulatory submissions to ensure the safety and effectiveness of approved biological products. A multidisciplinary review team, comprised of a regulatory project manager, clinical/medical officers, product reviewers, statisticians, pharmacology/toxicology reviewers, and other scientific experts with various backgrounds, reviews vaccine applications and other regulatory submissions in accordance with FDA regulations and mandated time lines.

### 53.2.1 Pre-investigational Stage (Pre-IND)

The pre-investigational new drug (pre-IND) application stage consists primarily of laboratory testing, the development of the manufacturing process, and refinement of the methodologies by which the vaccine candidate can be characterized. During this phase, candidate vaccines are evaluated in preclinical tests in vivo and in vitro to determine their suitability for entry into human trials.

### 53.2.2 IND Stage

If a sponsor wants to initiate a clinical investigation with a vaccine candidate, an Investigational

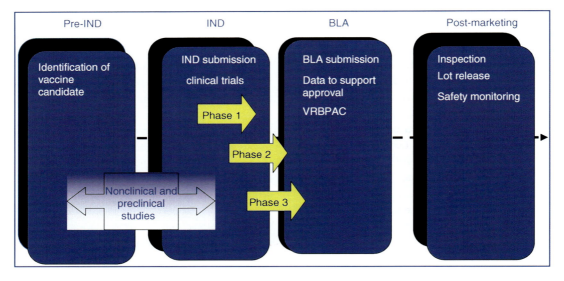

**Fig. 53.1** Stages of product development and FDA oversight through vaccine development

New Drug Application (IND) must be submitted. Title 21 CFR 312 describes the content of an original IND submission and the regulatory requirements for conduct of clinical trials under the IND regulations [3]. Briefly, the IND submission includes a description of raw materials, the method of manufacture and quality testing, preclinical animal safety data, the proposed Phase I clinical protocol, and the qualifications of the clinical investigator. The clinical immunogenicity, safety, and efficacy of a vaccine are evaluated in the various phases of IND studies as defined in 21 CFR 312.21 [11]. Review of the IND submission allows the FDA to monitor the safety of clinical trial subjects and ensure that the study design permits a thorough evaluation of the vaccine's effectiveness and safety. Amendments to the IND over the course of product development include more definitive clinical trials and product quality testing as well as scale up of the manufacturing methods.

### 53.2.3 Biologics License Application (BLA)

The licensing stage follows the IND stage when IND studies are completed and data demonstrating the safety and effectiveness of the product for a specific use have been collected and analyzed. The sponsor may submit a biologics license application (BLA) to manufacture and distribute the product commercially. The regulations that pertain to submission of a BLA can be found in 21 CFR 600 through 680 [12]. The BLA contains the data derived from nonclinical and clinical studies that demonstrate that a vaccine meets prescribed requirements for safety, purity, and potency. Other mandatory elements of a BLA submission include a description of the manufacturing process that demonstrates control and consistency of the production process, stability data, and a proposed package insert and labeling.

FDA may request that manufacturers present the data supporting the safety and efficacy of the vaccine to the Vaccines and Related Biological Products Advisory Committee (VRBPAC) prior to completion of the BLA review by CBER. The VRBPAC is a standing FDA advisory committee composed of scientific experts and clinicians who comment on the adequacy of the data to support safety and efficacy in the target population and may also consider unique complex clinical, manufacturing, and testing issues. The committee's recommendations are strongly considered in the CBER's decision to license a vaccine.

## 53.2.4 Post-marketing

After licensure, monitoring of the vaccine and production activities, including periodic facility inspections, must continue as long as the manufacturer holds a license for the product. Manufacturing establishments are inspected to determine whether licensed products are manufactured and tested as described in the license application and in accordance with applicable regulations. In addition, all licensed vaccines must undergo appropriate lot-release testing as mandated by Title 21 CFR 610 [13].

The FDA continues its oversight of licensed products through post-marketing surveillance. Manufacturers are required to provide ongoing reports of the safety of licensed vaccines. The Vaccine Adverse Event Reporting System (VAERS) accepts reports of any adverse event that may be associated with US-licensed vaccines from health-care providers, manufacturers, and the public. The Centers for Disease Control and Prevention (CDC) and the FDA monitor VAERS for adverse reactions of concern or for trends that may be attributed to a vaccine. In addition to VAERS, FDA is developing enhanced post-marketing surveillance systems to track the safety of drugs and biological products, including vaccines. In May 2008, FDA launched the Sentinel Initiative aimed to develop and implement a safety surveillance system that will complement existing systems already in place to monitor adverse events linked to the use of drugs and biological products. The Post-Licensure Rapid Immunization Safety Monitoring (PRISM) is a new national system for active vaccine safety surveillance within the Sentinel Initiative [14].

After a vaccine is approved, the vaccine manufacturer may conduct post-marketing studies to further evaluate the safety of the vaccine, which may allow identification of rare adverse reactions not detected during pre-licensure studies. In addition, Title IX of FDAAA 2007 gave the FDA increased authority to require certain post-marketing studies and clinical trials for drugs and biological products, and, under certain circumstances, safety-related labeling changes as well as risk evaluation and mitigation strategies [9].

## 53.3 Preclinical Guidelines

Preclinical testing is a prerequisite to move a candidate vaccine from the laboratory to the clinic and includes all aspects of testing prior to introducing a candidate product into humans such as product characterization, proof of concept/immunogenicity studies, and safety testing in animals. In addition to demonstrating the preclinical safety and biological activity of a candidate vaccine, preclinical data should support the proposed clinical formulation. Detailed information on the source and quality of starting materials and the manufacturing processes is also necessary. Specific requirements for preclinical studies are product specific and dependent on the type of vaccine (i.e., recombinant, live, attenuated), the manufacturing process, and the mechanism of action.

Preclinical studies are aimed at defining the in vitro and in vivo characteristics of candidate vaccines including safety and immunogenicity evaluations in animal models. Preclinical studies in animals are valuable tools to identify possible risks to the humans and help to plan protocols for subsequent clinical studies. The US FDA approach to nonclinical safety testing of preventive vaccines is summarized in the guidance document entitled "WHO guidelines on nonclinical evaluation of Vaccines," published by the WHO [15]. This document provides basic principles and approaches to nonclinical safety evaluation of vaccines that are based on a case-by-case approach and allow flexibility for testing requirements.

It is recommended that manufacturers discuss the preclinical and early clinical testing approaches and requirements with CBER during the Pre-IND Meeting. Dialogue with CBER will help clarify approaches to designing preclinical and clinical studies to support proceeding to and conduct of clinical trials.

## 53.4 Clinical Guidelines

Vaccines are extensively evaluated in the specific population to which they will be administered to generate safety and efficacy data that can be used as a basis for approval.

Clinical studies are governed by regulations in 21 CFR 312 [3]. Randomized, double-blind, controlled trials are the gold standard for demonstrating the efficacy of candidate vaccines.

Clinical development occurs in three stages, commonly referred to as Phase I, II, and III studies. Phase 1 studies are designed to evaluate the safety and immunogenicity of a candidate vaccine in a small number of healthy subjects. Phase 2 studies can enroll several hundred subjects, to further evaluate the safety and immune response to the vaccine and provide preliminary estimates on rates of common adverse events and often laboratory abnormalities. Phase III studies are large-scale trials that provide the critical documentation of the vaccine's safety and effectiveness needed to support licensure. Phase III studies are usually conducted in populations that represent the target population for the product. The clinical data obtained during Phase III studies should be generated with a product that has been manufactured by a defined process using product specifications intended for marketing in the USA.

Well-defined endpoints should be chosen prior to the start of a clinical study to lend scientific and statistical credibility to results. Clinical disease efficacy endpoints provide the greatest scientific rigor for evaluating vaccines. The appropriate endpoint(s) with respect to efficacy is dependent upon the characteristics of the infectious disease and the candidate vaccine and usually relates to clinically significant disease morbidity or mortality [16].

Vaccines are an important public health intervention for protecting people and communities from the mortality and morbidity associated with many infectious diseases. The FDA provides regulatory oversight throughout the complex vaccine development and licensure process. After licensure, vaccine safety is continually monitored through lot-release testing, inspections, and product surveillance. The FDA ensures the safety, effectiveness, and availability of licensed vaccines through its comprehensive review mechanisms.

## References

1. Public Health Service Act. July 1, 1944, Chap. 373, Title III, Sec. 351, 58 Stat. 702, currently codified at 42 United States Code, Sec. 262
2. Federal Food, Drug and Cosmetic Act. 21 United States Code, Sec. 321 (1938)
3. Code of Federal Regulations, Washington, DC, Office of the Federal Register, National Archives & Records Administration. Title 21, Part 312 (2013)
4. Prescription Drug User Fee Act, Public Law No. 102–571 (1992)
5. Food and Drug Administration Modernization Act of 1997, Public Law No. 105–115 (1997)
6. Public Health Security and Bioterrorism Preparedness and Response Act of 2002, Public Law No. 107–188 (2002)
7. Food and Drug Administration Amendments Act of 2007, Public Law No. 110–85 (2007)
8. Food and Drug Administration Safety and Innovation Act (FDASIA), Public Law No. 112–144 (2012)
9. Gruber, M.F.: The review process for vaccines for preventive and therapeutic infectious disease indications regulated by the US FDA: impact of the FDA Amendments Act 2007. Expert Rev. Vaccines **10**(7), 1011–1019 (2011)
10. Regulatory Information. Food and Drug Administration Safety and Innovation Act (FDASIA). 13 July 2013. Food and Drug Administration. http://www.fda.gov/RegulatoryInformation/Legislation/FederalFoodDrugandCosmeticActFDCAct/SignificantAmendmentstotheFDCAct/FDASIA/ucm20027187.htm. Accessed 25 July 2013
11. Code of Federal Regulations, Washington, DC, Office of the Federal Register, National Archives & Records Administration. Title 21 CFR Part 312.21 (2013)
12. Code of Federal Regulations, Washington, DC, Office of the Federal Register, National Archives & Records Administration. Title 21 CFR Parts 600 through 680 (2013)
13. Code of Federal Regulations, Washington, DC, Office of the Federal Register, National Archives & Records Administration. Title 21 CFR Part 610 (2013)
14. Nguyen, M., Ball, R., Midthun, K., Lieu, T.A.: The Food and Drug Administration's Post-Licensure Rapid Immunization Safety Monitoring program: strengthening the federal vaccine safety enterprise. Pharmacoepidemiol. Drug Saf. **21**(Suppl 1), 291–297 (2012)
15. World Health Organization, WHO Guidelines on Nonclinical Evaluation of Vaccines. http://www.who.int/biologicals/publications/nonclinical_evaluation_vaccines_nov_2003.pdf. Accessed on 25 July 2013
16. Hudgens, M., Gilbert, P.G., Gulf, S.G.: Endpoints in vaccine trials. Stat. Methods Med. Res. **13**(2), 89–114 (2004)

# Vaccines: EU Regulatory Requirements

## 54

Bettina Klug, Patrick Celis, Robin Ruepp, and James S. Robertson

## Contents

| | | |
|---|---|---|
| 54.1 | Regulatory Process and Definitions | 845 |
| 54.2 | Assistance to SMEs Developing Vaccines | 847 |
| 54.3 | Structure and Content of the Marketing Authorisation Application | 847 |
| 54.4 | Requirements for the MAA | 848 |
| 54.4.1 | Quality Requirements | 848 |
| 54.4.2 | Non-clinical Guidance | 848 |
| 54.4.3 | Clinical Guidelines | 849 |
| References | | 849 |

B. Klug, MD (✉)
Paul-Ehrlich-Institut, Langen, Germany
e-mail: bettina.klug@pei.de

P. Celis, PhD • R. Ruepp, PhD
European Medicines Agency (EMA),
Westferry Circus, Canary Wharf, London, UK

J.S. Robertson, PhD
National Institute for Biological Standards and Control,
Blanche Lane, South Mimms,
Potters Bar, Hertfordshire, UK

### Abstract

The European pharmaceutical legislation provides a comprehensive framework for the marketing authorisation of vaccines; depending on the nature of the product, three different routes of application are possible. For innovative in particular recombinant vaccines, the centralised marketing authorisation procedure is mandatory.

This book chapter provides an overview on the regulatory process and the requirements for a marketing authorisation application for prophylactic and therapeutic vaccines in Europe. The chapter highlights the most relevant guidelines for the quality, non-clinical and clinical development of vaccines.

## 54.1 Regulatory Process and Definitions

Innovative vaccines, and in particular recombinant vaccines (recombinant protein-based vaccines and recombinant viral-vectored vaccines), must be evaluated and approved in the European Union (EU) via the centralised procedure. Other novel vaccines can also be approved centrally if

---

The views expressed in this article are the personal views of the author(s) and may not be understood or quoted as being made on behalf of or reflection of the position of the European Medicines Agency or one of its committees or working parties, the Paul-Ehrlich Institute or the National Institute for Biological Standards and Control.

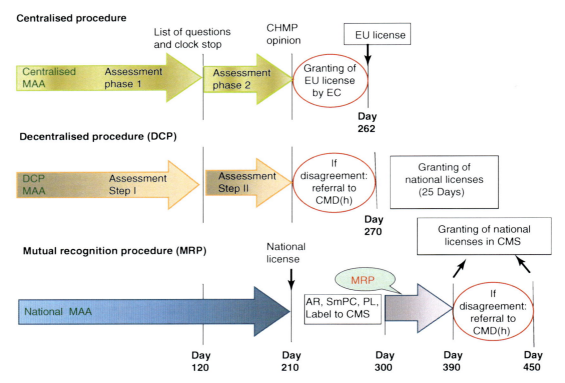

**Fig. 54.1** Description of the three routes of marketing authorisation procedures for medicinal products in the European Union. Only the main procedural steps are depicted. Referral and re-examination procedures are not included in this figure. For more information, consult the Notice to Applicants [1, 2]. Not all procedures are open for all medicines (e.g. recombinant vaccines and ATMPs have to follow the centralised procedure). For the centralised procedure (on the basis of a positive opinion from the CHMP), the European Commission will grant a marketing authorisation (licence) that is valid in the entire EU; the DCP and the MRP result in licences in more than one EU member state (from 2 to 28). *Abbreviations*: *CHMP* Committee for Medicinal Products for Human Use, *MAA* Marketing Authorisation Application, *EC* European Commission, *EU* European Union; *CMD(h)* Co-ordination Group for Mutual Recognition and Decentralised Procedures – Human, *AR* Assessment Report, *SmPC* Summary of Product Characteristics, *PL* Package Leaflet, *CMS* Concerned Member State

justified by the applicant (eligibility to the centralised procedure under the 'optional scope' as outlined in Article 3 of Regulation (EC) No 726/2004). The centralised evaluation of a Marketing Authorisation Application (MAA) is undertaken by the European Medicines Agency's (EMA) Scientific Committees and takes up to 210 days, after which the Committee for Medicinal Products for Human Use (CHMP) adopts an opinion recommending or rejecting the authorisation. In the case of a positive opinion, it then falls upon the European Commission (EC) to grant the marketing authorisation, which will be valid in all EU member states (more information on the centralised procedure can be found in the Notice to Applicants Volume 2A [1]). For the authorisation of traditional non-recombinant vaccines in the EU, the developer can submit the MAA to one or more national competent authorities for medicines for review (for more information on the mutual recognition and decentralised procedure, please refer to Notice to Applicants Volume 2A [2]). The different routes of marketing authorisation procedures for medicinal products in the European Union are described in Fig. 54.1.

Although the European pharmaceutical legislation does not provide a formal definition, vaccines are typically considered medicinal products containing one or more immunogenic antigens intended for prophylaxis against infectious disease. Medicinal products containing one or more immunogenic antigens for the treatment of disease, e.g. chronic HIV infection, chronic hepatitis B or C infection, cancer or Alzheimer's disease, are typically referred to as therapeutic vaccines or active immunotherapy. The same scientific principles for their product development as for prophylactic vaccines against infectious diseases apply.

Vaccines against infectious diseases that are based on viral (or other) vectors or on DNA plasmids are specifically excluded from the definition of a Gene Therapy Medicinal Product (GTMP) [3], whereas the same vector or plasmid incorporating a tumour antigen (a so-called 'cancer vaccine') is considered a GTMP in the EU and therefore falls within the definition of, and corresponding regulatory framework for, an Advanced Therapy Medicinal Product (ATMP). Cell-based cancer immunotherapy products (e.g. sipuleucel-T licensed in the USA) or cell-based vaccines against infectious diseases (e.g. dendritic cells loaded with peptide fragments from a human pathogen) are also ATMPs (specifically somatic cell therapy medicinal products). For ATMPs, a specific regulatory framework has been established [4] aimed at facilitating the development and licensing of these novel products via the Committee for Advanced Therapies (CAT). For all MAAs for ATMPs, the CAT prepares a draft opinion, before the CHMP adopts its final opinion.

## 54.2 Assistance to SMEs Developing Vaccines

Novel prophylactic and therapeutic vaccines are often developed by small companies or academic spin offs. Within the EU, several incentives have been set up to help steer such developers through the maze of regulatory and scientific requirements. Most notably is the establishment at the EMA of the Office for Micro-, Small- and Medium-sized Enterprises (SME office) and also at some national authorities for medicines. Companies with an SME status (assigned by the EMA) can benefit from incentives that focus on the main financial and administrative hurdles for SMEs in the pre-marketing authorisation procedures. For example, they can benefit from regulatory assistance by the SME office and obtain fee reductions for scientific advice and other scientific services, whilst the fee for an MAA and associated inspections can be deferred. In addition, for ATMPs, application for certification of the quality/non-clinical data can be made prior to actual MAA submission.

Another important tool for early interactions with developers of novel vaccines and ATMPs is briefing meetings with the EMA's Innovation Task Force (ITF). The ITF meetings allow informal exchange and dialogue between developers and regulators on procedural and scientific issues. Early interaction can also take place in the form of meetings with the secretariat of the Scientific Advice Working Party (SAWP) (in preparation of a scientific advice request) or of the Committee for Orphan Medicinal Products (COMP) (in preparation of an application for orphan drug status) [5].

## 54.3 Structure and Content of the Marketing Authorisation Application

The requirements for the structure and content of the MAA are laid down in the Common Technical Document (CTD) and provide for a harmonised structure and format for MAAs in Europe, Japan and the USA [6].

In Europe, the legal provision for implementation of the CTD format are set out in Annex I to Directive 2001/83/EC, laying down the technical requirements for the quality, non-clinical and clinical documentation for all categories of medicinal products specified in this Annex (including vaccines) [7]. For medicinal products that present specific features, such as vaccines and ATMPs, these requirements need to be adapted. The provision of tailored requirements not only addresses the highly specific nature of these products but also allows a flexible approach to account for the rapid technological and scientific developments in the biotechnological field. The high level technical requirements for the content of the MAA provided in Annex I are further substantiated by regulatory and scientific guidelines specific to various aspects of product development.

As for all medicinal products, vaccine and ATMP developers are required to submit a Paediatric Investigation Plan (PIP) to the EMA's

Paediatric Committee (PDCO) for review and agreement. The PIP describes planned clinical studies in children, including the proposed timing of the studies and formulation adaptations to make it suitable for children. The Paediatric Regulations (EC) No. 1901/2006 and No. 1902/2006 [8, 9] require that the PIP is submitted to the EMA as early as possible (ideally soon after the phase I clinical trial) and compliance with the agreed PIP is checked at the time of MAA submission.

The MAA must also include a detailed Risk Management Plan (RMP) that defines a set of pharmacovigilance activities and interventions that identify, characterise, prevent or minimise risks relating to the medicinal product, including the assessment of the effectiveness of those interventions. New pharmacovigilance legislation came into operation in July 2012 [10, 11], and new provisions for Periodic Safety Update Reports (PSURs), RMPs, safety signals and Post-Authorisation Safety Studies (PASS) were introduced. In addition, literature monitoring and several tools for product safety reviews at EU level are part of this legislation. A Pharmacovigilance Risk Assessment Committee (PRAC) has been established at the EMA, and as one of its tasks, the PRAC assesses the RMP. Further guidance on the new pharmacovigilance legislation is published in the form of Good Pharmacovigilance Practicess (GVP) modules [12]. The GVP modules cover medicines authorised centrally via the EMA as well as medicines authorised at national level. The full set of 16 final modules is scheduled to be available by early 2013.

An MAA is also required to contain an evaluation of the potential environmental risks posed by medicinal products in the form of an Environmental Risk Assessment (ERA) [13].

## 54.4 Requirements for the MAA

The nature of an antigen(s) is highly variable, involving complex viral or microbial particles, synthetic peptides, recombinant proteins, virus-like particles, immune-modulating antibodies, gene therapy or cell-based products. The development of vaccines, typically antigens intended for prophylactic use against infectious diseases, is addressed in a variety of guidelines on vaccines.

The development of therapeutic cancer vaccines is addressed in the Guideline on the evaluation of anticancer medicinal products [14].

Adjuvanted vaccines have been licensed in the EU for some time, and there is considerable interest in developing new adjuvants for both existing and novel vaccines. Indeed the area is quite complex and the nature and mode of action of novel adjuvants is quite broad. EU guidance on the development and regulatory approval for an adjuvant is available [15, 16]. An important aspect in developing a novel adjuvant is to show that it does enhance the immune response to the antigen with associated clinical benefit. The safety of novel adjuvants is also an important factor.

### 54.4.1 Quality Requirements

The quality section of an MAA provides a thorough characterisation of the medicinal product (the vaccine), a detailed description of the manufacturing process starting from the establishment and characterisation of seed and cell banks through to the final formulation of the vaccine, a description of all raw materials and components used in the manufacturing process and a description and validation of all quality control tests applied during the manufacturing process and on the product (vaccine) itself. The quality section should also address the consistency of vaccine production and the stability of the vaccine and describe an appropriate and validated potency assay for the vaccine.

There is no generic EU guideline addressing the quality requirements of vaccines; however, the requirements are not dissimilar from that for most biological medicinal products. Guidance is available for some specific vaccines including smallpox vaccine [17], influenza vaccine [18–20], recombinant viral-vectored vaccines [21] and DNA vaccines [22]. Guidance for both influenza vaccines and DNA vaccines is currently being revised [23, 24].

### 54.4.2 Non-clinical Guidance

For vaccines protecting against infectious diseases, in general the aspects covered by the Note for Guidance on preclinical pharmacological and toxicological testing of vaccines [25] should be followed. Of note, not all aspects of a classical non-clinical development programme need to be covered for these vaccines, e.g. pharmacokinetics is generally not required for vaccines; however, non-clinical immunogenicity or challenge studies may be considered relevant depending on the vaccine. More specific non-clinical guidance is available for vaccines containing adjuvants [15], for smallpox vaccines [17] and for live recombinant viral-vectored vaccines [21].

In contrast to prophylactic vaccines, there is only limited non-clinical guidance for 'therapeutic vaccines'; however, information from some of the above guidelines may be valuable and applicable based on the characteristics of the particular type of vaccine concerned. Specific guidance on cell-based medicinal products [26] or anticancer pharmaceuticals [27] may be useful for vaccines in this category.

### 54.4.3 Clinical Guidelines

The EU Guideline on clinical evaluation of vaccines [28] provides a comprehensive account of the requirements for the clinical development of prophylactic vaccines for infectious diseases. It is supplemented by clinical section of the Guideline on quality, non-clinical and clinical aspects of live recombinant viral-vectored vaccines [21].

In the development of any new vaccine, adequate data on immunogenicity should be assembled during the clinical development programme. This includes characterisation of the immune response, investigation of an appropriate dose and schedule, persistence of immunity and consideration on booster doses. Additionally, for vectored vaccines, the determination and characterisation of the pre-existing immunity to the vector should be addressed.

Pharmacokinetic studies are usually not required for vaccines. However, such studies might be applicable when new delivery systems are employed or when the vaccine contains novel adjuvants or excipients.

Demonstration of protective efficacy may not be necessary and or feasible for all vaccines since this will be influenced by the prevalence and characteristics of the target infectious disease. If clinically validated correlates of protection have been established, immunogenicity studies may be considered sufficient, such as for hepatitis B, tetanus or diphtheria vaccines.

Vaccine effectiveness reflects direct (vaccine induced) and indirect (population related) protection during routine use. Thus it may be possible and highly desirable to assess vaccine effectiveness during the post-authorisation period.

So far only limited guidance for the clinical development of therapeutic vaccines is available. However, the evaluation of the immunogenicity of therapeutic vaccines should follow the same principles as for any other vaccine [28]. For demonstrating the efficacy of a therapeutic vaccine, the specific guideline for the particular condition should be consulted (if available).

The Guideline on the evaluation of anticancer medicinal products addresses the clinical development of therapeutic anticancer vaccines [14]. The choice of target population as well as a discussion of the outcome measures takes the particularity of a vaccine into account. Induction of an effective immune and clinical response may need more time to develop compared to classical cytotoxic compounds leading to disease progression prior to the onset of beneficial biological activities or clinical effects.

## References

1. Notice to Applicants Volume 2A – procedures for marketing authorisation. Chapter 4: centralised procedure. http://ec.europa.eu/health/files/eudralex/vol--2/a/chap4rev200604_en.pdf
2. Notice to Applicants Volume 2A – procedures for marketing authorisation. Chapter 2: mutual recognition. http://ec.europa.eu/health/files/eudralex/vol-2/a/vol2a_chap2_2007-02_en.pdf
3. Directive 2009/120/EC amending Directive 2001/83/EC of the European Parliament and of the Council on the Community code relating to medicinal products for human use as regards advanced therapy medicinal products. http://ec.europa.eu/health/files/eudralex/vol-1/dir_2009_120/dir_2009_120_en.pdf
4. Regulation (EC) No 1394/2007 of the European Parliament and of the Council of 13 November 2007 on advanced therapy medicinal products and amending Directive 2001/83/EC and Regulation (EC). http://ec.europa.eu/health/files/eudralex/vol-1/reg_2007_1394/reg_2007_1394_en.pdf

5. Klug, B., et al.: Regulatory structures for gene therapy medicinal products in the European Union. Methods Enzymol. **507**, 337–354 (2012)
6. Notice to Applicants Volume 2 B – presentation and format of the dossier – Common Technical Document (CTD). http://ec.europa.eu/health/files/eudralex/vol-2/b/update_200805/ctd_05-2008_en.pdf
7. Annex 1 to directive 2001/83/EC – on medicinal products for human use. http://ec.europa.eu/health/files/eudralex/vol-1/dir_2001_83_cons2009/2001_83_cons2009_en.pdf
8. Regulation (EC) No 1901/2006 of the European Parliament and of the Council of 12 December 2006 on medicinal products for paediatric use and amending Regulation (EEC) No 1768/92, Directive 2001/20/EC, Directive 2001/83/EC and Regulation (EC) No 726/2004. http://ec.europa.eu/health/files/eudralex/vol-1/reg_2006_1901/reg_2006_1901_en.pdf
9. Regulation (EC) No 1902/2006 of the European Parliament and of the Council of 20 December 2006 amending Regulation 1901/2006 on medicinal products for paediatric use. http://ec.europa.eu/health/files/eudralex/vol-1/reg_2006_1902/reg_2006_1902_en.pdf
10. Commission Regulation (EU) No 1235/2010 of the European Parliament and of the Council of 15 December 2010 amending, as regards pharmacovigilance of medicinal products for human use, Regulation (EC) No 726/2004 laying down Community procedures for the authorisation and supervision of medicinal products for human and veterinary use and establishing a European Medicines Agency, and Regulation (EC) No 1394/2007 on advanced therapy medicinal products. http://ec.europa.eu/health/files/eudralex/vol-1/reg_2010_1235/reg_2010_1235_en.pdf
11. Directive 2010/84/EU of the European Parliament and of the Council of 15 December 2010 amending, as regards pharmacovigilance, Directive 2001/83/EC on the Community code relating to medicinal products for human use. http://ec.europa.eu/health/files/eudralex/vol-1/dir_2010_84/dir_2010_84_en.pdf and corrigendum. http://ec.europa.eu/health/files/eudralex/vol-1/dir_2010_84_cor/dir_2010_84_cor_en.pdf
12. Good pharmacovigilance practices. http://www.ema.europa.eu/ema/index.jsp?curl=pages/regulation/document_listing/document_listing_000345.jsp&mid=WC0b01ac058058f32c
13. Guideline on the environmental risk assessment of medicinal products for human use. EMEA/CHMP/SWP/4447/00 corr. 1. http://www.ema.europa.eu/docs/en_GB/document_library/Scientific_guideline/2009/10/WC500003978.pdf
14. Guideline on the evaluation of anticancer medicinal products in man. CPMP/EWP/205/95 Rev. 4. http://www.ema.europa.eu/docs/en_GB/document_library/Scientific_guideline/2013/01/WC500137128.pdf
15. Guideline on adjuvants in vaccines for human use. EMEA/CHMP/VEG/134716/2004. http://www.ema.europa.eu/docs/en_GB/document_library/Scientific_guideline/2009/09/WC500003809.pdf
16. Explanatory note on immunomodulators for the guideline on adjuvants in vaccines for human use. CHMP/VWP/244894/2006. http://www.ema.europa.eu/docs/en_GB/document_library/Scientific_guideline/2009/09/WC500003810.pdf
17. Note for guidance on the development of vaccinia virus-based vaccines against smallpox. CPMP/1100/02. http://www.ema.europa.eu/docs/en_GB/document_library/Scientific_guideline/2009/09/WC500003900.pdf
18. Note for guidance on the harmonisation of requirements for influenza vaccines. CPMP/BWP/214/96. http://www.ema.europa.eu/docs/en_GB/document_library/Scientific_guideline/2009/09/WC500003945.pdf
19. Note for guidance on cell-culture-inactivated influenza vaccines. Annex to the note for guidance on the harmonisation of requirements for influenza vaccines. CPMP/BWP/2490/00. http://www.ema.europa.eu/docs/en_GB/document_library/Scientific_guideline/2009/09/WC500003877.pdf
20. Points to consider on the development of live attenuated influenza vaccines. EMEA/CPMP/BWP/1765/99. http://www.ema.europa.eu/docs/en_GB/document_library/Scientific_guideline/2009/09/WC500003899.pdf
21. Guideline on quality, non-clinical and clinical aspects of live recombinant viral vectored vaccines. EMA/CHMP/VWP/141697/2009. http://www.ema.europa.eu/docs/en_GB/document_library/Scientific_guideline/2010/08/WC500095721.pdf
22. Note for guidance on the quality, preclinical and clinical aspects of gene transfer medicinal products. CPMP/BWP/3088/99. http://www.ema.europa.eu/docs/en_GB/document_library/Scientific_guideline/2009/10/WC500003987.pdf
23. Concept paper on the revision of guidelines for influenza vaccines. EMA/CHMP/VWP/734330/2011. http://www.ema.europa.eu/docs/en_GB/document_library/Scientific_guideline/2011/10/WC500115612.pdf
24. Concept paper on guidance for DNA vaccines. EMEA/CHMP/308136/2007. http://www.ema.europa.eu/docs/en_GB/document_library/Scientific_guideline/2012/03/WC500124898.pdf
25. Note for guidance on pre-clinical safety evaluation of biotechnology-derived pharmaceuticals (ICH S6). CPMP/SWP/465/95. http://www.ema.europa.eu/docs/en_GB/document_library/Scientific_guideline/2009/10/WC500004004.pdf
26. Guideline on human cell-based medicinal products. EMEA/CHMP/410869/2006. http://www.ema.europa.eu/docs/en_GB/document_library/Scientific_guideline/2009/09/WC500003894.pdf
27. Note for guidance on non-clinical evaluation for anticancer pharmaceuticals (ICH S9). EMEA/CHMP/ICH/646107/2008. http://www.ema.europa.eu/ema/index.jsp?curl=pages/regulation/general/general_content_000400.jsp&mid=WC0b01ac0580029570
28. Guideline on clinical evaluation of new vaccines. EMEA/CHMP/VWP/164653/2005. http://www.ema.europa.eu/docs/en_GB/document_library/Scientific_guideline/2009/09/WC500003870.pdf

# Licensing and Permitting of Veterinary Vaccines in the USA: US Regulatory Requirements

## 55

Louise M. Henderson and AdaMae Lewis

## Contents

| | | |
|---|---|---|
| 55.1 | Overview of the Regulatory Process | 852 |
| 55.2 | General Requirements | 853 |
| 55.3 | Specific Requirements | 853 |
| 55.4 | Pre-clinical Studies | 854 |
| 55.5 | Clinical Studies | 855 |
| References | | 857 |

### Abstract

The Center for Veterinary Biologics (CVB), United States Department of Agriculture (USDA), located in Ames, Iowa, has regulatory jurisdiction over veterinary biologics in the USA for the diagnosis, prevention, and treatment of animal diseases. Regulatory jurisdiction is established by the Virus-Serum-Toxin Act (VSTA) [1] of 1913 (amended in 1985) which requires that all veterinary biologics available within the USA are pure, safe, potent, and effective (not worthless, dangerous, contaminated, or harmful). Regulations can be found in Title 9 of the Code of Federal Regulations (9 CFR), Subchapter E, Parts 101–122 [2]. Veterinary Services Memoranda (VS Memo) that are available online provide detailed requirements with more specific guidance. The Web site provides links to the relevant guidance documents [3]. It is important to note that the USDA has regulatory authority only over those veterinary vaccines specifically for the prevention and treatment of animal diseases; the Food and Drug Administration (FDA) has regulatory jurisdiction over veterinary vaccines for other purposes (e.g., vaccines targeting some hormones or pain receptors). This chapter provides an overview of the regulations and guidance for licensing (products produced in the USA) and permitting (products produced outside the USA but sold within the USA) those vaccines under USDA regulatory jurisdiction.

L.M. Henderson, PhD (✉)
Henderson Consulting, LLC,
Consultants for Veterinary Biologics,
LLC, Ames, IA, USA
e-mail: lmhenderson@consultantsforveterinarybiologics.com

A. Lewis, PhD
Lewis Biologics, Inc.,
Consultants for Veterinary Biologics,
LLC, Ames, IA, USA

## 55.1 Overview of the Regulatory Process

The Center for Veterinary Biologics (CVB), United States Department of Agriculture (USDA), located in Ames, Iowa,[1] has regulatory jurisdiction over veterinary biologics in the USA for the diagnosis, prevention, and treatment of animal diseases. Regulatory jurisdiction is established by the Virus-Serum-Toxin Act (VSTA) [1] of 1913 (amended in 1985) which requires that all veterinary biologics available within the USA are pure, safe, potent, and effective (not worthless, dangerous, contaminated, or harmful). Regulations can be found in Title 9 of the Code of Federal Regulations (9 CFR), Subchapter E, Parts 101–122 [2]. Veterinary Services Memoranda (VS Memo) that are available online provide detailed requirements with more specific guidance. The Web site provides links to the relevant guidance documents [3]. It is important to note that the USDA has regulatory authority only over those veterinary vaccines specifically for the prevention and treatment of animal diseases; the Food and Drug Administration (FDA) has regulatory jurisdiction over veterinary vaccines for other purposes (e.g., vaccines targeting some hormones or pain receptors). This chapter will focus on those vaccines under USDA regulatory jurisdiction.

The CVB is comprised of two functional units: (1) Policy, Evaluation, and Licensing (PEL) is responsible for developing regulatory policies, evaluating applications for licensure and permitting, evaluating all data supporting License and Permit applications, assessing risk of proposed products prior to licensure, awarding US Veterinary Biologics Establishment Licenses, awarding US Veterinary Biological Product Licenses (for products produced in the USA), awarding US Veterinary Biologic Permits (for veterinary biologics imported into the USA), laboratory testing at the CVB Laboratory, and production of some standard reagents for testing of licensed biologics; (2) Inspection and Compliance (IC) is responsible for setting and enforcing standards for facilities, record keeping, review of manufacturer personnel qualifications, and quality control as well as for inspections of manufacturing production and testing sites and approval of serials of products prior to release for sale.

All veterinary vaccines sold within the USA must have either a US Veterinary Biological Product License and be produced within a licensed establishment or a US Veterinary Biological Permit [4]. All labels must be approved before use to ensure that claims provided on vaccines are consistent with the scientific data provided to, and approved by, the CVB and are not false or misleading [5, 6]. In general, the requirements for production are very similar regardless of country of manufacture [7, 8]. For importation, holders of permits must have a person in the USA in a licensed facility responsible for receipt and quarantine of imported vaccine prior to release for sale. US licensed vaccines can be exported from the USA provided they meet the importing countries' requirements. In addition vaccines not licensed by the USDA can be exported under the Food and Drug Administration (FDA) Export Reform and Enhancement Act of 1996; these unlicensed products must meet the requirements of the importing company, conform to GMP requirements or other international standards, and not have the US Biologics Establishment License on the label [9, 10].

Biotechnology-derived vaccines [11] are subject to the same regulations as conventionally derived (live, modified-live, killed, and subunit) vaccines, although the characterization of the Master Seed includes evaluation at the molecular level for identity and purity; the only additional relevant regulation is the National Environmental Policy Act (NEPA) [12] which requires live (replication-competent) genetically engineered vaccines to undergo a thorough risk assessment prior to release in the environment.

Applicants do not submit complete dossiers; rather, the licensing/permitting process is highly interactive. Applicants are assigned a Staff Officer (Reviewer) in the PEL who is the primary contact for the applicant. The Reviewer communicates

---

[1] Center for Veterinary Biologics, 1920 Dayton Avenue, Ames, Iowa 50010 phone: 515-337-6100

with the applicant throughout the process, providing feedback on each submission and approving proposed protocols prior to the applicant initiating required studies in host animals. Prior to obtaining either a license or permit, the CVB must approve production and distribution of experimental vaccines [13, 14]. Before studies are performed in US animals, the CVB will review the data previously submitted in order to ensure there is not undue risk in performing the intended studies in host animals; therefore, preliminary safety and stability data must provide reasonable assurance of product safety under the intended uses. As applicants progress through the licensing/permitting process, a Biologics Specialist (Inspector) will be assigned to work with the firm to ensure that facilities, record keeping systems, and personnel meet requirements. This provides a means for identifying issues early in the process, allowing mitigation of problems as early as possible.

The CVB does not regulate by GMP per se; the regulatory oversight, however, does include oversight of the same issues and concerns. Manufacturers must adhere to detailed requirements set for each product that are provided in the Outline of Production for that vaccine written by the manufacturer and approved by the CVB. This provides regulatory oversight of all ingredients, seeds, processes, conditions, manufacture, storage, and testing of the vaccine. The processes must be controlled within the allowed ranges of time, temperature, pH, test results, etc., and be overseen by qualified personnel. Before each serial (a single lot of vaccine, produced and tested as a homogenous lot) is released, CVB review of test results is required. This provides confidence that each serial has been produced according to the established requirements in an approved manufacturing facility by competent personnel and meets the established standards. Results of testing each serial for purity, potency, and safety are submitted to the CVB; if acceptable, CVB-IC will release the serial (allow the manufacturer to sell the serial) [15]. The CVB Laboratory may test any serial prior to release, during dating should adverse events be reported, and at end of dating for that serial.

## 55.2 General Requirements

*Facilities.* In order to ensure that veterinary vaccines in the USA have been produced as required, there are very specific requirements the manufacturing facility must meet [16]. Blueprints, plot plans, and legends must be submitted to the CVB [17]. Facilities documents are reviewed and manufacturing sites are inspected to ensure that all standards are met [18]. Air handling and workflow are evaluated to ensure that vaccines are produced under conditions that minimize risk of contamination. All live agents must be approved before they can be brought into production facilities. Certain agents of foreign animal diseases (FADs) are prohibited. Ingredients must meet specific requirements to minimize risk of introducing contamination during production. Quality control records are inspected for compliance with requirements and facilities are inspected for compliance with regulations [19].

*Manufacturing Methods.* The CVB does not specify manufacturing methods; rather, the manufacturer provides a detailed Outline of Production that is reviewed and approved by the CVB [20, 21]. The Outline of Production details the ingredients used in manufacturing vaccines, the methods used, the storage conditions, and the testing requirements. All serials must be tested for purity, safety, and potency prior to release [22]. In-process procedures must be validated adequately (i.e., equipment must be calibrated, processes must be shown to be adequate to meet the intended objective) [23]. Revisions to the Outline of Production must be approved before manufacturer's revise production. A highly controlled manufacturing process (as detailed in the Outline of Production) using ingredients that meet specified standards in a facility that is subject to inspections then provides confidence that the variation between serials is minimized and serials should be pure, safe, potent, and efficacious.

## 55.3 Specific Requirements

*Master Seeds and Cells.* One of the key components of the regulatory system in the USA is the establishment and characterization of Master

Seeds and/or Master Cells for use in manufacturing veterinary vaccines [24]. Vaccines are produced using approved Master Seeds and/or Master Cells. Although not required, almost all licensed/permitted vaccines are manufactured from approved Master Seeds and Master Cells. A Master Seed is the bacterial or viral disease agent. A Master Cell is established for growth of viruses and other obligate intracellular microorganisms. The Master Seeds and Master Cells are a single lot of culture which has been highly characterized for purity and identity by the manufacturer with confirmatory purity and identity testing by the CVB Laboratory. When using primary cells, the manufacture must test each lot of primary cells for purity and identity.

For genetically engineered vaccines, additional detailed information must be submitted that genetically and phenotypically characterizes the Master Seed including a description of the source of all genetic material used to produce the recombinant Master Seed, the processes used to produce it, and the results of testing done during and after development of the Master Seed. The information must be provided in a Summary Information Format (SIF) [25]. The SIFs are specific for the intended use of the Master Seed. Category I recombinant Master Seeds are used in products that are replication incompetent. This includes all those used for inactivated (killed) vaccines, subunit vaccines, suicide vectors that fail to replicate in the target host animal, and monoclonal antibodies (for therapeutic, prophylactic, or diagnostic use). Category II recombinant Master Seeds are live gene-deleted microorganisms with no foreign nucleic acid insertions. Category III recombinant Master Seeds are used in live vectored vaccines and provide sufficient data to allow a thorough risk assessment prior to authorization for field safety studies.

*Ingredients.* All ingredients used in the manufacture of licensed veterinary vaccines in the USA must meet quality standards and must not introduce risk [26]. Ingredients of animal origin must be shown to be from acceptable sources that have little or no risk of bovine spongiform encephalopathy (BSE). [27]

*Sterility and Safety.* All serials of vaccine must be shown to be pure prior to release [28–30]. In the case of inactivated vaccines, the 9 CFR specifies the required sterility tests [31]. For live vaccines, manufacturers must demonstrate lack of contamination. All serials must also be tested for safety in small numbers of animals [32–38]. For some vaccines, lab animals may be injected with vaccine and observed for adverse events. In other cases, host animals must be injected with the serial and observed for adverse events.

## 55.4 Pre-clinical Studies

*Potency Assay.* All serials must be tested for potency. Potency assays are intended as a correlate to efficacy; therefore, potency assays must correlate the immunogenic (antigenic) content in each serial of vaccine to host-animal efficacy studies. For many diseases, a specific potency assay is required; these can be found in the 9 CFR [39]. For modified-live vaccines, enumeration of the viral or bacterial count may be sufficient; with genetically modified-live vaccines, stability of the inserted antigen expression must be demonstrated as well.

For inactivated vaccines, the CVB allows host-animal or lab-animal potency assays (vaccination serology or vaccination challenge), but manufacturers are encouraged to develop in vitro assays quantifying immunogenic components of vaccines and correlate the assay to host-animal efficacy [40]. The use of monoclonal antibodies recognizing major immunogens has been used extensively to develop these assays. Manufacturers must develop and optimize the assay and demonstrate a dose response in the assay. The assay must be shown to be sensitive, specific, and rugged. Use of purified, highly characterized standards and product-like references (e.g., the efficacy serial) in each assay to monitor assay performance over time is encouraged. Manufacturers must monitor the potency assay and demonstrate stability and consistency. They are encouraged to submit potency assay validation prior to the onset of efficacy studies so that the potency assay can be carefully evaluated by the CVB prior to the host-animal efficacy, at which time the potency assay will be correlated to efficacy. With careful planning and well-characterized standards, manufacturers are able to maintain valid references for use in a well-developed assay, providing a means to demonstrate potency with confidence while reducing animal testing in the future. The CVB participates in the International Cooperation on Harmonisation of Technical Requirements for Registration of Veterinary Medicinal Products (VICH) efforts to harmonize potency assays intended to reduce animal testing which began in 1996 in cooperation with the European Union and Japan [41].

*Back-Passage Studies.* Modified-live viral vaccines must be shown to remain attenuated through five serial back passages through the host animal. This is intended to ensure that attenuation is a result of significant molecular mutation that is not easily reversed during passage through the host animal leading to a virulent strain arising from the use of the modified-live vaccine [42].

*Stability Studies.* Vaccines must be shown to be stable using real-time studies in which vaccine is stored according to label recommendations. Once established, stability need not be demonstrated for each serial.

## 55.5 Clinical Studies

*Efficacy Studies Including Correlation to Potency Assay.* Efficacy of each vaccine must be demonstrated in host animals using vaccine at the minimum effective dose (as demonstrated by the manufacturer). For some vaccines, efficacy study requirements are specified in the 9 CFR, which may provide detailed specifications of the required efficacy study [43]. For many vaccines, no specific efficacy study is specified. Each vaccine must be shown to be efficacious in each host species at the minimum age for which it is recommended. In some cases, pregnant animals must be used; in other cases, neonates or young animals must be used. These requirements are based on the recommended use of the vaccine. Studies must be randomized, blinded, and scientifically rigorous. Efficacy studies require challenge with the disease agent using randomly assigned vaccinates and control animals vaccinated with placebo vaccine (controls). The challenge is expected to be sufficient to develop significant clinical signs in controls. Differences between vaccinates and controls must be statistically and clinically significant [44]. If no challenge model is known, field studies may be allowed to demonstrate efficacy in large numbers of animals exposed under natural conditions. The outcome of the efficacy study will determine the claim allowed. Additional claims will require additional studies. Specific efficacy study claims and care of animals during studies are addressed in guidance available online [45, 46]. The CVB recommends adherence to VICH Guidelines [47].

Potency assays must correlate to host-animal efficacy studies, preferably by testing of the efficacy serial at the time the efficacy study is initiated. Usually, the efficacy serial is used as a reference each time a serial is tested; the serial under test is compared to the efficacy serial and must be shown to have greater antigenic content than the efficacy serial. The reference must be replaced when deterioration is detected or at expiration. It is important to note that vaccines must be at or above the efficacy serial at expiration. For modified-live vaccines, this requires that the bacterial or viral count at serial release is higher than the efficacy serial at release.

*Duration of Immunity Studies.* For some diseases in which there is little known about the duration of immunity induced by vaccination, manufacturers must establish the length of time vaccinated animals are protected. This usually requires an additional host-animal vaccination-challenge study unless sufficient data exists to allow serologic data to demonstrate immunity.

*Field Safety Studies.* Following efficacy studies, manufacturers must demonstrate safety in US animals using a minimum of two serials that meet all Outline of Production requirements (including potency above that of efficacy) [48].

```
Licensing a Vererinary Biologic in the U.S. (9 CFR 101 - 122)
                        │
Application for US Veterinary Biological Product License (9
CFR 102; VS Memo 800.50; APHIS Form 2003)
                        │
Outline of Production (submitted With APHIS Form 2015)(9 CFR 114.8-
114.9; VS Memo 800.206)detailing manufacturing and testing; process
validation data; revised throughout licensure
                        │
Master Seed and Cell reports with purity and identity test results (9 CFR
101.7; VS Memo 800.109); confirmatory identity and purity testing by CVB
Lab. If genetically engineered, includes SIF (VS Memo 800.205)
                        │
Backpassage study (9 CFR 102.5 and 104.5; VS Memo 800.201) for
live products
                        │
Laboratory validation of potency assay (9 CFR 113.8; VS Memo 800.112)
                        │
Detailed protocol for efficacy study in host animal
                        │
Permission to ship experimental product for efficacy sudy (9 CFR 103.3)
                        │
Efficacy vaccination-challenge study using serial at
highest passage in host animals at youngest age
recommended for licensure: include duration of immunity
if needed (9 CFR 113; VS Memo 800.202, 800.200)
                        │
Approval of efficacy study establishes minimum          Correlation of
Potency, minimum age animal, vaccination                potency assay to
recommendations, claim allowed                          efficacy
                        │
Preparation of 3 consecutive prelicense serials
                        │
Safety studies in lab animals, safety studies in host: studies of adjuvant safety
                        │
Protocol for field safety studies, request to ship experimental product for field
safety studies. CVB performs risk assessment prior to authorization
(includeds public notification if product is live genetically engineered); risk
mitigation prior to anthorization to perform field safety studies
                        │
Field safety studies in large numbers of host animals in 3
geographically distinct regions of U.S. using at least 2 serials
                        │
Submit labels (9 CFR 112: VS Memo 800.54)
                        │
**License awarded**
                        │
Approval of new claims and changes       Stability studies
to manufacture or testing
```

Application for US Veterinary Biologics Establishment License (9 CFR 102: VS Memo 800.101 APHIS Form 2001)

Articles of Incorporation (9CFR 102.3), Water quality statement, (9 CFR 108.11) Qualifications of Personnel (9 CFR 102.4, 114.7: VS Memo 800.63: APHIS Form 2007)

Facility blueprints. plot plans. legends (9 CFR 108.2-108.5 VS Memo 800.78)

Announced CVB Inspection prior to Establishment License (VS Memo 800.91)

**Establishment License awarded same time as first Product License**

Unannounced post-license inspections for compliance

Release of each serial after review of test results

This requires vaccinating large numbers of animals (often hundreds) in at least three geographically distinct areas of the USA. Field safety tests are intended to be performed in different husbandry settings or in animals with different endemic disease conditions. Animals are observed for adverse events for several weeks following vaccination. For food animals, residue clearance must also be determined to support withdrawal recommendations.

*Discussion.* This is a very broad overview of the regulatory process for veterinary vaccines in the USA. Much more detailed information is available at the CVB Web site [49]. There are a large number of very specific requirements that have not been mentioned in this discussion. Regulations tend to be general in nature, with more specific guidelines available in Veterinary Services Memorandum. Requirements are based on ensuring the purity, safety, potency, and efficacy of veterinary vaccines sold within or from the USA. Requirements are adapted as new issues are recognized, e.g., outbreaks of disease or emergence of new animal diseases. Conditional licenses can be approved for emergency situations for vaccines prior to completion of efficacy or potency testing. Exceptions may be granted to regulations when scientifically justified or when emergency situations arise that threaten the health of animals or of humans exposed to animal diseases. Training is available every spring. Manufacturers and foreign regulatory officials are encouraged to attend to learn more about the process and meet CVB personnel [50].

## References

1. Virus-Serum-Toxin Act, US Code of Federal Regulations 21, Parts 151–159. http://www.aphis.usda.gov/animal_health/vet_biologics/vb_regs_and_guidance.shtml (2012)
2. Code of Federal Regulations, Title 9 (9 CFR). http://www.aphis.usda.gov/animal_health/vet_biologics/vb_regs_and_guidance.shtml (2012)
3. CVB Web Site. http://www.aphis.usda.gov/animal_health/vet_biologics/vb_regs_and_guidance.shtml
4. 9 CFR 102, Licenses for biological products. http://www.aphis.usda.gov/animal_health/vet_biologics/vb_cfr.shtml (2012)
5. 9 CFR 112, Packaging and Labeling. http://www.aphis.usda.gov/animal_health/vet_biologics/vb_cfr.shtml (2012)
6. Veterinary Services Memorandum 800.54, Guidelines for Preparation and review of Labeling Materials (February 17, 1986). http://www.aphis.usda.gov/animal_health/vet_biologics/vb_vs_memos.shtml (2012)
7. Veterinary Services Memorandum 800.50, Basic License Requirements and Guidelines for Submission of Materials in Support of Licensure (February 9, 2011). http://www.aphis.usda.gov/animal_health/vet_biologics/vb_vs_memos.shtml (2012)
8. Veterinary Services Memorandum 800.101, U.S. Veterinary Biological Product Permits for Distribution and Sale (2012). http://www.aphis.usda.gov/animal_health/vet_biologics/vb_vs_memos.shtml (2012)
9. 21 CFR, Section 382. http://www.fda.gov/ (2012)
10. Veterinary Services Memorandum 800.94, FDA Export Reform and Enhancement Act of 1996, http://www.aphis.usda.gov/animal_health/vet_biologics/vb_vs_memos.shtml (2012)
11. Veterinary Services Memorandum 800.205, General Licensing Considerations: Biotechnology-derived Veterinary Biologics Categories I, II, III (May 28, 2003). http://www.aphis.usda.gov/animal_health/vet_biologics/vb_vs_memos.shtml (2012)
12. National Environmental Policy Act. http://ceq.hss.doe.gov/ (1969)
13. 9 CFR 103, Experimental production, distribution, and evaluation of biological products prior to licensure. http://www.aphis.usda.gov/animal_health/vet_biologics/vb_regs_and_guidance.shtml (2012)
14. Veterinary Services Memorandum 800.67, Shipment of Experimental Veterinary Biological Products (November 16, 2011) http://www.aphis.usda.gov/animal_health/vet_biologics/vb_vs_memos.shtml (2012)
15. Veterinary Services Memorandum 800.53, Release of Biological Products (April 2, 2001). http://www.aphis.usda.gov/animal_health/vet_biologics/vb_vs_memos.shtml (2012)
16. 9 CFR 108, Facilities requirements for licensed establishments. http://www.aphis.usda.gov/animal_health/vet_biologics/vb_regs_and_guidance.shtml (2012)
17. Veterinary Services Memorandum 800.78, Preparation and Submission of Facilities Documents (November 11, 2010). http://www.aphis.usda.gov/animal_health/vet_biologics/vb_vs_memos.shtml (2012)
18. Veterinary Services Memorandum 800.91, Categories of Inspection for Licensed Veterinary Biologics Establishments (May 13, 1999). http://www.aphis.usda.gov/animal_health/vet_biologics/vb_vs_memos.shtml (2012)
19. 9 CFR 115, Inspections. http://www.aphis.usda.gov/animal_health/vet_biologics/vb_regs_and_guidance.shtml (2012)
20. 9 CFR 114, Production requirements for biological products. http://www.aphis.usda.gov/animal_health/vet_biologics/vb_regs_and_guidance.shtml (2012)

21. Veterinary Services Memorandum 800.206, General Licensing Considerations: Preparing Outlines of Production for Vaccines, Bacterins, Antigens, and Toxoids (April 13, 2012). http://www.aphis.usda.gov/animal_health/vet_biologics/vb_vs_memos.shtml (2012)
22. 9 CFR 113.6, Animal and Plant Health Inspection Service Testing. http://www.aphis.usda.gov/animal_health/vet_biologics/vb_cfr.shtml (2012)
23. 9 CFR 109, Sterilization and pasteurization at licensed establishments. http://www.aphis.usda.gov/animal_health/vet_biologics/vb_regs_and_guidance.shtml (2012)
24. 9 CFR 101.7, Seed organisms. http://www.aphis.usda.gov/animal_health/vet_biologics/vb_regs_and_guidance.shtml (2012)
25. Risk analysis for veterinary biologics. http://www.aphis.usda.gov/animal_health/vet_biologics/vb_sifs.shtml (2011)
26. 9 CFR 113.50 through 113.55, Ingredient requirements. http://www.aphis.usda.gov/animal_health/vet_biologics/vb_regs_and_guidance.shtml (2012)
27. Veterinary Services Memorandum 800.51, Additives in animal biological products (November 7, 2007). http://www.aphis.usda.gov/animal_health/vet_biologics/vb_vs_memos.shtml (2012)
28. 9 CFR 113.27, Detection of extraneous viable bacteria and fungi except in live vaccine. http://www.aphis.usda.gov/animal_health/vet_biologics/vb_regs_and_guidance.shtml (2012)
29. 9 CFR 113,27, Detection of extraneous viable bacteria and fungi in live vaccines. http://www.aphis.usda.gov/animal_health/vet_biologics/vb_regs_and_guidance.shtml (2012)
30. 9 CFR 113.28, Detection of mycoplasma contamination. http://www.aphis.usda.gov/animal_health/vet_biologics/vb_regs_and_guidance.shtml (2012)
31. 9 CFR sterility tests. http://www.aphis.usda.gov/animal_health/vet_biologics/vb_regs_and_guidance.shtml (2012)
32. 9 CFR 113.33, Mouse safety tests. http://www.aphis.usda.gov/animal_health/vet_biologics/vb_regs_and_guidance.shtml (2012)
33. 113.38, Guinea pig safety test. http://www.aphis.usda.gov/animal_health/vet_biologics/vb_regs_and_guidance.shtml (2012)
34. 9 CFR 113.39, Cat safety tests. http://www.aphis.usda.gov/animal_health/vet_biologics/vb_regs_and_guidance.shtml (2012)
35. 9 CFR 113.40, Dog safety tests. http://www.aphis.usda.gov/animal_health/vet_biologics/vb_regs_and_guidance.shtml (2012)
36. 9 CFR 113.41, Calf safety test. http://www.aphis.usda.gov/animal_health/vet_biologics/vb_regs_and_guidance.shtml (2012)
37. 9 CFR 113.44, Swine safety test. http://www.aphis.usda.gov/animal_health/vet_biologics/vb_regs_and_guidance.shtml (2012)
38. 9 CFR 113.45, Sheep safety test. http://www.aphis.usda.gov/animal_health/vet_biologics/vb_regs_and_guidance.shtml (2012)
39. 9 CFR 113.8, in vitro tests for serial release. http://www.aphis.usda.gov/animal_health/vet_biologics/vb_regs_and_guidance.shtml (2012)
40. Veterinary Services Memorandum 800.112, guidelines for validation of in vitro potency assays (August 29, 2011). http://www.aphis.usda.gov/animal_health/vet_biologics/vb_vs_memos.shtml (2012)
41. International Cooperation on Harmonisation of Technical Requirements for Registration of Veterinary Medicinal Products. http://www.vichsec.org/ (1996)
42. Veterinary Services Memorandum 800.201, general licensing considerations: backpassage studies (June 25, 2008). http://www.aphis.usda.gov/animal_health/vet_biologics/vb_vs_memos.shtml (2012)
43. 9 CFR 113, Standard requirements. http://www.aphis.usda.gov/animal_health/vet_biologics/vb_regs_and_guidance.shtml (2012)
44. Veterinary services memorandum 800.202, general licensing considerations: efficacy studies (June 14, 2002). http://www.aphis.usda.gov/animal_health/vet_biologics/vb_vs_memos.shtml (2012)
45. Veterinary Services Memorandum 800.200, general licensing considerations: study practices and documentation (June 14, 2002). http://www.aphis.usda.gov/animal_health/vet_biologics/vb_vs_memos.shtml (2012)
46. Veterinary Services Memorandum 800.301, good clinical practices (July 26, 2001). http://www.aphis.usda.gov/animal_health/vet_biologics/vb_vs_memos.shtml (2012)
47. Veterinary Services Memorandum 800.207, general licensing considerations: Target Animal Safety (TAS) studies prior to product licensure – VICH guideline 44 (July 6, 2010) http://www.aphis.usda.gov/animal_health/vet_biologics/vb_vs_memos.shtml (2012)
48. Veterinary Services Memorandum 800.204, general licensing considerations: field safety studies (March 16, 2007). http://www.aphis.usda.gov/animal_health/vet_biologics/vb_vs_memos.shtml (2012)
49. Center for Veterinary Biologics. http://www.aphis.usda.gov/animal_health/vet_biologics/vb_regs_and_guidance.shtml (2012)
50. Institute for International Cooperation in Animal Biologics, Veterinary Biologics Training Program. http://www.cfsph.iastate.edu/IICAB/ (2012)

# Veterinary Vaccines: EU Regulatory Requirements

## Rhona Banks

## Contents

56.1 Overview of the Regulatory Process............ 859
56.2 General Requirements................................. 861
56.3 Specific Requirements.................................. 862
56.4 Preclinical Guidelines.................................. 862
56.5 Clinical Guidelines...................................... 863
Conclusions........................................................ 863
References......................................................... 864

### Abstract

Veterinary vaccines must be authorised by the relevant competent authorities in the EU. An application for a Marketing Authorisation (MA) must be submitted and this should include a dossier that demonstrates the quality of the vaccine and the safety and efficacy in the target species. It should be noted that safety must also be demonstrated for the user, the environment and the consumer.

## 56.1 Overview of the Regulatory Process

Veterinary vaccines must be authorised by the relevant competent authorities in the EU. An application for a Marketing Authorisation (MA) must be submitted, and this should include a dossier that demonstrates the quality of the vaccine and the safety and efficacy in the target species. It should be noted that safety must also be demonstrated for the user, the environment and the consumer.

All vaccines must be registered in line with Directive 2001/82/EC [1] as amended by Directive 2004/28/EC [2] and Directive 2009/09 [3] and there is a Consolidated Directive available [4]. These describe the studies and information which will be needed to support the claims for the product.

There are a number of different routes for applying for a Marketing Authorisation and Notice to Applicants Volume 6A gives an overview [5].

R. Banks, MIBiol, PhD
Department of Biologicals Regulatory Affairs,
Triveritas Ltd., Bank Barn,
How Mill, Brampton, UK
e-mail: rhona.banks@triveritas.com

**Table 56.1** Outline of centralised procedure (210 assessment days). Summarised from Notice to Applicants [6]

| Day | Action |
|---|---|
| xx | Submit your dossier to EMA, Rapporteur and Co-Rapporteur – see recommended submission dates on EMA website |
| xx | EMA send the applicant an acknowledgement of receipt of the dossier and, within 10 working days following receipt, will complete its validation |
| xx | At the end of the validation process the EMA starts the procedure<br>Send requested parts of the dossier to all CVMP members within 1 month of the start of the procedure |
| 1 | Start of the procedure – Timetable prepared by EMA and sent to Applicant |
| 70 | Rapporteur's Assessment Report sent to the Co-rapporteur, CVMP members and EMA Secretariat |
| 85 | Co-rapporteur's critique of the Rapporteur's Assessment Report sent to Rapporteur, CVMP members and EMA Secretariat. These reports are sent to the Applicant by the EMA Secretariat (making it clear that they do not yet represent the position of the CVMP.) |
| 100 | Rapporteur, Co-rapporteur, other CVMP members and EMA receive comments from Members of the CVMP |
| 115 | Draft list of questions from Rapporteur and Co-Rapporteur (with overall conclusions and overview of scientific data) sent to CVMP members + EMA |
| 120 | CVMP adopts list of questions, overall conclusions and review of the scientific data and this is sent to the Applicant by the EMA. By Day 120, adoption by CVMP of request for GMP inspection, if necessary<br>*Clock stop*<br>Applicant has 3 months to respond but may request an additional 3 months if a justification is given |
| 121 | Submission of the responses, including revised SPC, labelling and package leaflet text in English – Target dates for the submission of the responses are published on the EMA web-site<br>Restart of the clock – Project Manager sends revised timetable in consultation with Rapporteur and Co-Rapporteur (usually as follows) |
| 160 | Joint response Assessment Report from Rapporteur and Co-Rapporteur sent to CVMP members and EMA and Applicant (note this does not yet represent position of the CVMP). Where applicable, GMP inspection to be done |
| 170 | Deadline for comments from CVMP Members to be sent to Rapporteur and Co-Rapporteur, EMA and other CVMP members |
| 180 | CVMP discussion of the draft Opinion (SPC, labelling and package leaflet) and decision if an oral explanation by the Applicant is needed. For any oral explanation, clock is stopped so Applicant can prepare this |
| 181 | Restart of the clock and oral explanation (if needed). Project Manager sends updated SPC and product literature in English to the Applicant |
| 210 | Adoption of CVMP Opinion + CVMP Assessment Report |
| 211 | Transmission to Applicant of CVMP Opinion + CVMP Assessment Report |
| 211–237 | Translations of SPC and product literature and administrative procedures (there is also a specific timetable for the different stages) |

- National – application to one Member State. This licence can then form the basis for mutual recognition in other Member States. The original country will act as the Reference Member State.
- Mutual recognition – several Member States can be included in a specific procedure to recognise a national licence of a Reference Member State.
- Decentralised procedure – application to several Member States at the same time, selecting one as the Reference Member State.
- Centralised procedure – application to the European Medicines Agency for a single evaluation by the Committee for Veterinary Medicinal Products (CVMP). If the CVMP gives a positive opinion, then the EU Commission makes the final decision and grants an authorisation for all EU member states. An outline of the 210-day procedure is given in Table 56.1.

All veterinary vaccines for the EU must be produced in facilities that are in compliance with Good Manufacturing Practice (GMP), and these are subject to regular inspection by EU authorities [7].

Medicinal products which are developed using DNA recombination techniques to make genetically modified organisms (GMO) must be authorised through the centralised procedure [8] and these types of vaccine are also subject to additional requirements as specified in Directive 2001/18/EC [9] discussed below.

Medicinal products that are developed using other biotechnological methods which are considered by the EMA to be a significant innovation may also be authorised via the centralised procedure. It is very likely that novel vaccines including DNA/RNA and proteins will be eligible for application by the centralised procedure, and so this route is the main focus of this chapter. There is a very helpful question and answer on the EMA webpage for innovative products [10].

This regulatory system may sound complex but there is some help along the way as the European Medicines Agency (EMA) wants to encourage applicants with innovative products to engage with them as early as possible. In addition an application can be made to the Scientific Advice Working Party (SAWP) on specific questions [11]. Note that the applicant should state what their opinion is on a particular issue and provide supporting data for each question. The SAWP will consider the questions and give an answer with their rationale – so they may or may not agree with the position of the applicant.

## 56.2 General Requirements

Marketing Authorisation Holders (MAH) must be established in the European Economic Area (EEA) so this is the first point to address prior to applying for a Marketing Authorisation. The application for an MA should consist of a complete dossier which will cover all the relevant aspects of quality, safety and efficacy for the product. Thus the whole assessment of the product is done at one time; there is no phased approval.

A Summary of Product Characteristics (SPC) must be included in Part 1 of the dossier, and this document includes a brief description of the product and presentations and all the proposed claims, indications, methods of use and safety warnings. The dossier should include experimental trials, field trials or bibliography which will support all of the statements on the SPC.

There is an SPC template on the EMA website which should be used, and there is also an SPC Guideline for Immunologicals [12] which gives an indication of what items should be included for each section as well as some standard wording to cover certain situations.

Dossiers must also include a Detailed Description of the Pharmacovigilance System (DDPS) and there is an EMA guideline [13] which should be followed. Once an MA is granted the product will be subject to post-marketing monitoring under the pharmacovigilance system and there is a defined time and method for reporting Periodic Safety Updates.

There is concern in the EU about the availability of authorised veterinary medicinal products particularly for minor uses/minor species (MUMS). The EMA has developed a guideline specifically on data requirements for immunological MUMS products [14], and this outlines where some reductions in studies could be acceptable. There is a list of infectious agents causing particular diseases in various animal species which are not fully managed by current products. An application should be made to the EMA to have a proposed product agreed as having MUMS status. The CVMP will consider the supporting documentation and grant MUMS status and may also award financial incentives for products which are greatly needed.

The Commission has introduced a regulation [15] which is aimed to promote innovation and the development of new medicines by micro, small- and medium-sized enterprises (SMEs) that are developing medicines for human or veterinary use. The SME Office at the EMA offers assistance to SMEs, including practical help and a number of financial incentives. There is an extensive user guide available on the EMA website and full details of how to apply for SME status are given [16].

Electronic submissions of dossiers are encouraged and there is a question and answer document [17] on the EMA website.

## 56.3 Specific Requirements

General and some specific requirements for quality, safety and efficacy of immunological veterinary medicinal products (IVMPs) are included in Directive 2009/09 [3], and this also effectively gives the format in which the dossier should be presented. The European Pharmacopoeia (Ph Eur) has both general and specific monographs which are applicable for vaccines. The general monograph 'Vaccines for Veterinary use 0062' [18] should be complied with (note that this monograph has been updated for the beginning of 2013) and in addition there are monographs on safety [19] and efficacy of vaccines [20]. Uniquely there are some specific monographs for particular types of veterinary vaccines and these include summary descriptions of how safety and efficacy should be demonstrated against a particular disease (e.g. monograph 2072 on Pasteurella vaccine (inactivated) for sheep [21]). Note that there are many specific Ph Eur monographs for veterinary vaccines so it is important to consider these when developing a new IVMP for Europe.

There is a new EMA guideline [22] which replaces several of the general and specific guidelines for immunological veterinary medicinal products that were in Notice to Applicants Volume 7 – the latest IVMP guidelines are displayed on the EMA website [23]. This new guideline does outline important items on quality, safety and efficacy which may not be entirely obvious from the Directive and Ph Eur monographs. However, the new guideline does not always contain as much detailed information as the older ones so the applicant needs to carefully consider how to develop their product.

Vaccines containing or consisting of genetically modified organisms (GMOs) are subject to additional provisions in Directive 2001/18/EC [9]. This established a step-by-step approval process so that an assessment of risks to human health and the environment must be undertaken before any GMO can be released into the environment or placed on the market. For experimental release, e.g. for a field trial, a notification must be submitted to the competent authorities of the Member State where it is intended to release the GMO. The notification must include a technical dossier which gives the necessary information so that the environmental risk can be assessed. Note that in most cases the regulatory authorities who grant national licences and also give approval for field trials are not the same as the competent authorities who approve the experimental release of GMOs. It is thus necessary to be in contact with more than one authority in a particular Member State in order to get the permission both to release the GMO and to actually run a field trial.

When submitting an application for a GMO vaccine a copy of the written consent of the competent authority for the deliberate release into the environment must be included. A complete technical dossier on the GMO as detailed in Annex III A of Directive 2001/28/EC [9] and an extensive environmental risk assessment must be provided. This information should be included in a separate volume of the dossier so that it can be provided to the competent authorities for GMOs. The EMA has a specific standard procedure to manage consultations with the national competent authorities for GMOs during the normal assessment of the whole dossier [24].

## 56.4 Preclinical Guidelines

There are many scientific guidelines for IVMPs on the EMA website [25] and these cover quality, safety and efficacy issues. There are two VICH guidelines (applicable in the EU, USA and Japan) on the testing for reversion to virulence of live vaccine strains [26] and testing methods to demonstrate safety of live and inactivated vaccines [27]. It is usually expected that laboratory studies on safety will be carried out in line with Good Laboratory Practice (GLP) and laboratory studies on efficacy should be carried out in line with Good Clinical Practice. In any case the studies should be properly planned with a protocol and fully written up in a study report. All data on a particular product should be included in the dossier whether or not it gives positive findings.

Many veterinary vaccines may contain more than one active ingredient (considered as a combination vaccine). In addition it may be possible to administer two or more separately licensed IVMPs at the same time (considered as an association). There is a draft guideline currently being prepared on how these types of vaccine should be evaluated [28].

One issue which may affect all types of veterinary vaccine is whether maternally derived antibody (MDA) interferes with the development of an immune response and thus reduces efficacy. It is necessary to either show that MDA levels have waned by the time vaccination is recommended or demonstrate that MDA does not interfere with efficacy against a virulent challenge [29].

For veterinary vaccines to be used in food-producing animals it is necessary to consider whether there are any pharmacologically active substances included for which a maximum residue limit (MRL) is required under Article 1 of Regulation (EC) No 470/2009 [30]. Any adjuvants or excipients included in the IVMP should be considered, and there is an EMA guideline [31] which includes substances that do not fall within the scope of the regulation and another EMA guideline [32] which indicates data that would be needed to add a substance to the acceptable list.

## 56.5 Clinical Guidelines

Laboratory trials should usually be supported with field trials using batches of vaccine prepared according to the manufacturing process in the MA application. As a commercial-type batch should be used, both safety and efficacy can be studied in the same field trial [3]. These trials should be carried out in line with Good Clinical Practice, and a predefined statistical evaluation of the results should be done. Ideally field trials should be carried out in the EU at two different sites (preferably in different countries although this is not really specified anywhere). Some Ph Eur monographs also include a summary of the design of field trials, e.g. Neonatal piglet colibacillosis vaccine (inactivated) 0962 [33].

There is a recent guideline on the design of studies to evaluate the safety and efficacy of fish vaccines [34]. This includes information on both laboratory and field trials.

### Conclusions

There is considerable complexity in the EU regulatory requirements for IVMPs, but the information is all publicly available and a lot of help is provided on the EMA website for centralised applications. For other routes of application there is also useful information provided on the heads of medicines agency veterinary website [35].

It is worth applying for scientific advice from the EMA when developing a novel vaccine as this may have an impact on what studies are needed to support an MA application. The advice is not binding on the CVMP but it is almost always followed by them, so if the applicant does follow the scientific advice, there is a good chance of success.

One relatively recent initiative is the possibility to include a benefit-risk assessment in Part 1 of the dossier. This should provide a scientific evaluation of the positive therapeutic effects of the vaccine in relation to any risks that may be associated with the quality and safety of the product for animals or humans [36]. A benefit-risk assessment has always been part of the evaluation carried out by regulators, but there is increasing awareness that there is an underlying risk if there is no prophylactic or therapeutic treatment for a particular disease.

This chapter can only give a broad idea of the regulations, directives, guidelines, recommendations and general experience which is relevant for the EU regulatory process for veterinary vaccines. There is already a remarkable variety of veterinary vaccines to cover many different target species, and the hope is that this can be extended even further with new technologies. With each veterinary vaccine, there are often unique issues to address so it is well worth discussing your project with EU regulators at an early stage taking advantage of the EMA's scientific advice procedure and also considering whether a MUMS approach is relevant.

## References

1. Directive 2001/82/EC of the European Parliament and of the Council of 6 November 2001 on the Community Code relating to veterinary medicinal products. http://eur-lex.europa.eu/LexUriServ/LexUriServ.do?uri=OJ:L:2001:311:0001:0066:en:pdf (2001)
2. Directive 2004/28/EC of the European Parliament and of the Council of 31 March 2004 amending Directive 2001/82/EC on the Community Code relating to veterinary medicinal products. http://ec.europa.eu/health/files/eudralex/vol-5/dir_2004_28/dir_2004_28_en.pdf (2004)
3. Commission Directive 2009/9/EC of 10 February 2009 amending Directive 2001/82/EC of the European Parliament and of the Council on the Community code relating to medicinal products for veterinary use. Official Journal L 44, 14/2/2009. p. 10–61. http://ec.europa.eu/health/files/eudralex/vol-5/dir_2009_9/dir_2009_9_en.pdf (2009)
4. Consolidated Directive 2001/82/EC as amended by Directive 2004/28/EC of the European Parliament and of the Council of 31 March 2004, Directive 2009/9/EC of 10 February 2009, Regulation (EC) No 470/2009 of the European Parliament and of the Council of 6 May 2009, Directive 2009/53/EC of the European Parliament and of the Council of 18 June 2009 and Regulation (EC) No 596/2009 of the European Parliament and of the Council of 18 June 2009 http://ec.europa.eu/health/files/eudralex/vol-5/dir_2001_82_cons2009/dir_2001_82_cons2009_en.pdf (2009)
5. EudraLex – Volume 6 notice to applicants and regulatory guidelines for medicinal products for veterinary use. Volume 6A procedures for marketing authorisation. http://ec.europa.eu/health/files/eudralex/vol-6/a/vol6a_chap1_2007-01_en.pdf (2007)
6. EudraLex – Volume 6 notice to applicants and regulatory guidelines for medicinal products for veterinary use. Volume 6A procedures for marketing authorisation chapter 4 centralised procedure. http://ec.europa.eu/health/files/eudralex/vol-6/a/vol6a_chap4_2006_05_en.pdf (2006)
7. Commission Directive 91/412/EEC of 23 July 1991 laying down the principles and guidelines of good manufacturing practice for veterinary medicinal products http://ec.europa.eu/health/files/eudralex/vol-5/dir_1991_412/dir_1991_412_en.pdf (1991)
8. Regulation (EC) No 726/2004 of the European Parliament and of the Council of 31 March 2004 laying down Community procedures for the authorisation and supervision of medicinal products for human and veterinary use and establishing a European Medicines Agency. http://eur-lex.europa.eu/LexUriServ/LexUriServ.do?uri=OJ:L:2004:136:0001:0033:EN:PDF (2004)
9. Directive 2001/18/EC of the European Parliament and of the Council of 12 March 2001 on the deliberate release into the environment of genetically modified organisms and repealing Council Directive 90/220/EEC. http://ec.europa.eu/health/files/eudralex/vol-1/dir_2001_18/dir_2001_18_en.pdf (2001)
10. European Medicines Agency: Veterinary presubmission Q & A for innovative products. http://www.emea.europa.eu/ema/index.jsp?curl=pages/regulation/general/general_content_000171.jsp&mid=WC0b01ac058002d9ab (2012)
11. European Medicines Agency: EMA/CVMP/172329/2004-Rev.3. Guidance for companies requesting scientific advice. http://www.emea.europa.eu/docs/en_GB/document_library/Regulatory_and_procedural_guideline/2009/10/WC500004147.pdf (2012)
12. EudraLex – Volume 6 notice to applicants and regulatory guidelines for medicinal products for veterinary use. Volume 6C regulatory guidelines. Summary of the product characteristics SPC – immunologicals. http://ec.europa.eu/health/files/eudralex/vol-6/c/spc_immunologicals_rev3_08-06-2007_en.pdf (2007)
13. European Medicines Agency: EMA/1531641/2010. Pre-submission instruction on the detailed description of the pharmacovigilance system of a marketing authorisation holder; to be submitted with a marketing authorisation application for a veterinary medicinal product http://www.ema.europa.eu/docs/en_GB/document_library/Regulatory_and_procedural_guideline/2012/03/WC500123503.pdf (2012)
14. European Medicines Agency: EMA/CVMP/IWP/123243/2006-Rev.2. Guideline on Data requirements for Immunological veterinary medicinal products intended for minor use or minor species/limited markets. http://www.ema.europa.eu/docs/en_GB/document_library/Scientific_guideline/2010/04/WC500089628.pdf (2010)
15. Regulation (EC) No 2049/2005 of 15 December 2005 laying down, pursuant to Regulation (EC) No 726/2004 of the European Parliament and of the Council, rules regarding the payment of fees to, and the receipt of administrative assistance from, the European Medicines Agency by micro, small and medium-sized enterprises. http://eur-lex.europa.eu/LexUriServ/LexUriServ.do?uri=OJ:L:2005:329:0004:0007:EN:PDF (2005)
16. European Medicines Agency: EMA/204919/2010 user guide for micro, small and medium-sized enterprises (SMEs) http://www.ema.europa.eu/docs/en_GB/document_library/Regulatory_and_procedural_guideline/2009/10/WC500004134.pdf (2010)
17. European Medicines Agency: EMA/613295/2011-Rev.1. Electronic submission of veterinary dossiers questions and answers. http://www.emea.europa.eu/docs/en_GB/document_library/Other/2012/05/WC500127591.pdf (2012)
18. European Pharmacopoeia 7th Ed. Vaccines for veterinary use. Monograph 04/2013: 0062. European Pharmacopoeia, Strasbourg, France. (2013)
19. European Pharmacopoeia 7th Ed. Evaluation of safety of veterinary vaccines and immunosera Monograph 04/2013:50206. European Pharmacopoeia, Strasbourg, France. (2013)

20. European Pharmacopoeia 7th Ed. Evaluation of efficacy of veterinary vaccines and immunosera. Monograph 04/2008:50207. (2013)
21. European Pharmacopoeia 7th Ed. Pasteurella vaccine (inactivated) for sheep. Monograph 04/2013:2072. (2013)
22. European Medicines Agency: EMA/CVMP/IWP/206555/2010. Guideline on requirements for the production and control of immunological veterinary medicinal products. http://www.ema.europa.eu/docs/en_GB/document_library/Scientific_guideline/2012/06/WC500128997.pdf (2012)
23. European Medicines Agency: Immunologicals guidelines. http://www.ema.europa.eu/ema/index.jsp?curl=pages/regulation/general/general_content_000194.jsp&mid=WC0b01ac058002dd33 (2012)
24. European Medicines Agency. SOP/V/4012. Evaluation of veterinary medicinal products containing or consisting of Genetically Modified Organisms. http://www.ema.europa.eu/docs/en_GB/document_library/Standard_Operating_Procedure_-_SOP/2009/09/WC500003079.pdf (2012)
25. European Medicines Agency: Immunologicals guidelines. http://www.emea.europa.eu/ema/index.jsp?curl=pages/regulation/general/general_content_000194.jsp&mid=WC0b01ac058002dd33 (2012)
26. VICH Topic GL41. EMEA/CVMP/VICH/1052/2004. Guideline on target animal safety: examination of live veterinary vaccines in target animals for absence of reversion to virulence. http://www.ema.europa.eu/docs/en_GB/document_library/Scientific_guideline/2009/10/WC500004552.pdf (2007)
27. VICH Topic GL44 Step 7. EMEA/CVMP/VICH/359665/2005. Guideline on target animal safety for veterinary live and inactivated vaccines. http://www.ema.europa.eu/docs/en_GB/document_library/Scientific_guideline/2009/10/WC500004553.pdf (2008)
28. European Medicines Agency: EMA/CVMP/IWP/594618/2010 draft. Guideline on the requirements for combined vaccines and associations of immunological veterinary medicinal products (IVMPs). http://www.ema.europa.eu/docs/en_GB/document_library/Scientific_guideline/2011/11/WC500118227.pdf (2011)
29. European Medicines Agency:. EMA/CVMP/IWP/439467/2007 Reflection paper on the demonstration of a possible impact of maternally derived antibodies on vaccine efficacy in young animals. http://www.ema.europa.eu/docs/en_GB/document_library/Scientific_guideline/2010/03/WC500076626.pdf (2010)
30. Regulation (EC) No 470/2009 of the European Parliament and of the Council of 6 May 2009 laying down Community procedures for the establishment of residue limits of pharmacologically active substances in foodstuffs of animal origin, repealing Council Regulation (EEC) No 2377/90 and amending Directive 2001/82/EC of the European Parliament and of the Council and Regulation (EC) No 726/2004 of the European Parliament and of the Council. http://ec.europa.eu/health/files/eudralex/vol-5/reg_2009-470/reg_470_2009_en.pdf (2009)
31. European Medicines Agency: EMA/CVMP/519714/2009-Rev.12 Substances considered as not falling within the scope of Regulation (EC) No 470/2009, with regards to residues of veterinary medicinal products in foodstuffs of animal origin. http://www.ema.europa.eu/docs/en_GB/document_library/Regulatory_and_procedural_guideline/2009/10/WC500004958.pdf (2012)
32. European Medicines Agency: EMA/CVMP/516817/2009 Guideline on data to be provided in support of a request to include a substance in the list of substances considered as not falling within the scope of Regulation (EC) No 470/2009. http://www.ema.europa.eu/docs/en_GB/document_library/Scientific_guideline/2010/11/WC500099149.pdf (2010)
33. European Pharmacopoeia 7th Ed. Neonatal piglet colibacillosis vaccine (inactivated). Monograph 04/2013:0962. (2013)
34. European Medicines Agency: EMA/CVMP/IWP/314550/2010 Guideline on the design of studies to evaluate the safety and efficacy of fish vaccines. http://www.ema.europa.eu/docs/en_GB/document_library/Scientific_guideline/2011/11/WC500118226.pdf (2011)
35. Heads of Medicine Agency: Veterinary medicines. http://www.hma.eu/veterinary.html (2012)
36. European Medicines Agency: EMEA/CVMP/248499/2007 recommendation on the evaluation of the benefit-risk balance of veterinary medicinal products. http://www.ema.europa.eu/docs/en_GB/document_library/Other/2009/10/WC500005264.pdf (2009)

# Index

**A**

Adjuvants, 464, 484, 519, 527, 537, 551, 562, 584, 603, 623, 633, 649, 682, 703, 717, 728, 745, 754, 774, 787, 793, 809, 827, 848, 863
*Agrobacterium tumefaciens*, 552
Agroinfiltration, 554, 556, 557
Alginate, 592, 593, 749
Allergen, 489, 503, 528, 544, 614, 718
Allergen-specific immunotherapy, 495–499, 505, 506, 625
Allergic sensitization, 490–493, 497
Aluminum, 484, 518, 520, 528, 530, 624, 633–639, 660, 661, 720, 794
Anergy, 509, 510, 796, 800
Angiotensin vaccine, 455, 469
Antigen presenting cells, 479, 484, 490, 491, 497, 507, 510, 520, 529, 543, 544, 552, 562, 567, 568, 581, 583, 587, 625, 635, 637, 649, 659, 660, 674, 680, 683, 690, 721, 729, 745, 755–760, 773, 784, 788, 796
Antigen specific immunotherapy (ASI), 478, 481
Antirenin antibodies, 452
Atherosclerosis, 450–459, 722
Atopy, 490, 492, 493
Autoantibodies, 478, 480–486
Autoantigens, 450, 478, 481–483, 485, 487, 800

**B**

Bacille Calmette-Guerin (BCG), 487, 497, 588, 604, 703, 704
Bacterial toxins, 518, 530, 537–547, 728
BALT. *See* Bronchus-associated lymphoid tissue (BALT)
Basophils, 490, 492, 494, 495, 504–506
Benefit-risk assessment, 863
Biodefense vaccines, 720
Biodistribution, 596, 659, 660, 747
Biofactories, 552–555
Biomarker, 459, 597, 678, 686, 689, 690, 722
Blueprint, 678–679, 853
B16 melanoma, 757, 761, 762
Body mass index, 464, 465, 467, 474
Bronchoalveolar lavage fluid (BALF), 513
Bronchus-associated lymphoid tissue (BALT), 581, 583, 584, 604

**C**

Carbon nanotubes, 729, 730, 735–737
Cell culture process, 829
Centralised procedure, 845, 846, 860, 861
Chitosan, 518, 584, 589, 592–593, 614–616, 623–629, 652, 662, 703, 721, 729, 732, 748, 758
Coat and poke, 701–703
Committee for Veterinary Medicinal Products (CVMP), 860, 861, 863
Common mucosal immune system (CIMS), 580, 604
Common Technical Document (CTD), 847
Controlled release, 591, 646, 649
CpG oligonucleotides (CpG), 458, 521, 530, 542, 544, 567, 568, 573, 609, 611, 661, 662, 703, 722, 729, 757, 795, 800
Cross presentation, 587, 610, 612, 637, 661, 662, 730, 745, 746, 798
Cross-reactivity, 565, 684–685
Cytotoxic T lymphocyte (CTL), 530, 533, 583, 591, 626, 637, 674, 680, 687, 691, 707, 732–734, 749, 756, 757, 759

**D**

Damage-associated molecular patterns (DAMP), 609, 637, 639
Databases, 679–681, 684, 690, 691
Dendrimers, 518, 644, 645, 653, 654, 657, 658, 675, 729, 730, 735, 736, 758–763
Dendritic cells, 456, 485, 491, 507, 521, 529, 543, 552, 567, 604, 626, 635, 660, 680, 698, 718, 728, 747, 754, 773, 784, 794, 847
Dry powders, 498, 518, 584, 586, 588, 593, 594, 717–725

**E**

Electrochemotherapy, 782
Electroporation, 675, 707, 708, 756, 757, 763, 781–788
EMA. *See* European Medicines Agency (EMA)
Emulsion, 518, 520, 530, 611, 634, 644–646, 650, 651, 732, 746, 749, 835
Endotoxin, 518, 529, 538, 544, 558, 611, 689
Energy homeostasis, 464, 466–467, 473, 474
Eosinophils, 490, 495, 504, 512, 513, 635–637

Epitope, 458, 472, 485, 491, 496, 505–508, 510–513, 546, 558, 565, 566, 569, 652, 661, 663, 674, 680, 684–688, 691, 736, 748, 755, 756, 760, 773, 788
European Medicines Agency (EMA), 465, 657, 805, 829, 832, 845–848, 860–863
European Patent Office (EPO), 809
Exotoxin, 518, 538, 540, 541, 544, 545, 547, 608–610

## F
FDA. *See* Food and Drug Administration (FDA)
Field safety studies, 854, 855
First to file, 814, 815
Food and Drug Administration (FDA), 465, 466, 545, 564, 567, 611, 649, 657, 701, 710, 718, 719, 728, 754, 757, 776, 804, 829, 832, 839–844, 852
Food intake, 450, 464, 466–470, 472–474

## G
GAD. *See* Glutamic acid decarboxylase (GAD)
GALT. *See* Gut-associated lymphoid tissue (GALT)
Generally Recognized As Safe (GRAS), 649, 675, 772
Genetically modified organisms (GMO), 861, 862
Genetic prediction, 479
Gene vaccine, 497
Ghrelin, 450, 463–474
Glutamic acid decarboxylase (GAD), 450, 478, 480–484, 487
GMP. *See* Good manufacturing practice (GMP)
Gold, 530, 596, 652, 662, 663, 704, 756, 783, 844
Good manufacturing practice (GMP), 513, 514, 729, 757, 804, 852, 853, 860
GRAS. *See* Generally Recognized As Safe (GRAS)
Gut-associated lymphoid tissue (GALT), 450, 506–513, 581, 604–607, 610, 612

## H
Heat shock proteins, 458, 478, 484–485, 487, 518, 551–559, 728, 757
House dust mite, 450, 491, 492, 503–514, 614
Hygiene hypothesis, 492, 493
Hyperresponsiveness, 498, 512, 513, 626
Hypertension, 450–459, 464, 466, 469, 574
paradox, 452, 459
Hypo-allergens, 495–496
Hypothalamus, 450, 464, 466–469, 471, 473

## I
IL 10, 479, 484, 485, 491, 495, 506, 507, 509, 510, 608, 609, 626, 768
Immune-stimulating complexes (ISCOMs), 589, 591, 612, 648, 650, 674, 728–730, 733–734, 746, 749
Immunocomplex, 473
Immunoconjugate, 469–473

Immunotoxins, 537–547
Inflammasome, 573, 609, 626, 637–639, 690, 708
Influenza Surveillance Network, 826–828
Influenza virus, 532, 557, 562, 568, 593, 703, 704, 706, 734, 749, 760, 788, 824–827, 829, 830, 832, 835
Insulin, 469, 471, 472, 478, 480–487, 589, 591, 593–595
ISCOMs. *See* Immune-stimulating complexes (ISCOMs)

## J
Japanese cedar, 450, 503, 504, 506, 507, 510, 512, 514

## L
Lactic acid bacteria, 675, 767–776
*Lactobacillus* spp., 675
LADA. *See* Latent autoimmune diabetes in adults (LADA)
Langerhans cells, 498, 581, 703, 756, 757, 760, 784
Laser vaccine adjuvant, 518, 520–521
Latent autoimmune diabetes in adults (LADA), 478, 484, 485, 487
Leukotrienes, 490
Licensure, 775, 804, 805, 828, 839–841, 843, 844, 852
Lipid A, 450, 520, 528–532, 567, 573, 611, 614, 628, 661, 719, 721, 722, 728, 733
Lipid metabolism targets, 457
Lipopolysaccharide, 491, 527–533, 538, 609, 611, 637, 703
Lipoproteins, 456–458, 465, 563–565, 661, 774, 795
Liposomes, 530, 566–571, 573, 574, 589–591, 611, 612, 647, 649, 650, 652, 659, 661, 728–733, 744, 746–750, 758, 772, 783
Lyme diseases, 561–574, 775
Lymph node, 479, 490, 498, 499, 508, 520, 521, 581, 583, 585, 604, 605, 607, 636, 638, 680, 729, 754, 757, 758, 761, 762, 769, 770, 794, 796, 797, 799

## M
Major histocompatibility complex (MHC), 479, 481, 491, 506, 508, 510, 522, 523, 538, 539, 544, 556, 558, 587, 610, 612, 636, 637, 675, 680, 681, 683, 684, 687, 688, 691, 724, 730, 733, 745, 748, 754, 758–761, 763, 773, 796–799
MALT. *See* Mucosa-associated lymphoid tissue (MALT)
Manufacturing, 513, 587, 590, 624, 629, 634, 720, 737, 749, 761, 776, 804, 828, 829, 833–835, 840–843, 848, 852–854, 860, 863
MAP kinase, 638, 639
Marketing authorization, 804, 805
Mast cell, 490, 491, 494–496, 498, 504–506, 511, 635, 680
Master Seeds, 852–854
M cells, 507, 508, 581–583, 605, 607, 609–611, 718, 730, 731
Metallochelating, 566, 568, 569, 571, 573

# Index

MF59, 518, 530, 634, 728, 729, 746, 749, 794, 835
Microdevices, 756, 757
Microneedles, 521, 525, 674, 697–711, 735, 744, 750, 756, 757
Microvesicles, 798
Modified vaccinia virus Ankara, 497
Monobulk, 831–833, 835
Monophosphoryl lipid A, 520, 530, 567, 573, 611, 614, 628, 719, 721, 722, 728, 733
Mucosa-associated lymphoid tissue (MALT), 581, 604, 607, 718
Multiepitope vaccine, 558
Muramyl dipeptide, 538, 540, 541, 544, 566, 567, 573, 584, 690

## N
NALT. *See* Nasopharynx-associated lymphoid tissue (NALT)
Nanocarriers, 580, 586, 591, 597, 646, 647, 651, 662, 729–731, 733–735, 737, 756–758, 760–763
Nanoliposome, 518, 561–574
Nanomedicines, 643–663
Nanoparticles, 566, 586, 612, 625, 644, 703, 719, 728, 744, 753
Nasopharynx-associated lymphoid tissue (NALT), 581, 583, 604, 718, 719
Neuropeptides, 466–469
Non-professional APC, 675, 798–800

## O
Obesity, 450, 456, 463–474
Oral
 adjuvant, 518, 603–616
 tolerance, 458, 503–514, 544, 730
Oxidized-LDL, 456

## P
PADRE, 675, 758–763
PAMAM dendrimers, 675, 730, 758–761, 763
PAMP. *See* Pathogen-associated molecular patterns (PAMP)
Particulate vaccines, 729–730, 733–735, 745
Patentability, 804, 807–820
Pathogen-associated molecular patterns (PAMP), 573, 609–612, 616, 661, 675, 794, 798
Pattern-recognising receptors (PRR), 573, 606, 607, 609, 611, 613, 616, 626, 639, 680, 683
Permeabilisation, 783
Peyer's patches, 506, 508, 581, 583, 604, 607, 612, 613, 730, 737
Pharmacovigilance, 804, 848, 861
Photothermal effects, 523
Plague, 768–771, 775
Plant, 506, 507, 512–514, 518, 545, 551–559, 611, 675, 686, 749, 772
Poke and patch, 701–703, 706

Polishing, 831
Polymer, 518, 568, 584, 589–595, 613–615, 624, 625, 643–663, 701, 705, 707, 731, 732, 735–737, 747–750, 757, 758, 761
Polymer architecture, 653, 654
Polypeptides, 506, 518, 542, 543, 555–558, 637, 655–657, 699
Postgenomic era, 677–691
Potency assay, 848, 854, 855
Powders, 498, 518, 584, 586, 588, 593, 594, 625, 629, 674, 717–725
Prime-boost, 755–756, 788
Proinflammatory, 458, 529, 556, 573, 597, 637, 639, 721, 722, 774, 775
Proteome, 555, 679–680, 683, 684, 686, 689
Pulmonary immunization, 580, 582, 584, 585, 590, 591, 593, 595–597

## Q
Quality control, 627, 660, 832, 848, 852, 853

## R
Registration, 566, 804, 805, 855
Regulatory T cells, 450, 479, 481, 483, 484, 487, 491, 505, 509, 510, 607, 626, 660
Renin-angiotensin system, 450, 452, 453, 459
Reverse vaccinology, 562, 678, 681, 683, 685, 686, 688, 689, 691
Rice seed-based vaccine, 514

## S
Self-adjuvanted immunogen, 551–559
Shiga toxin, 538, 540, 541, 545
Skeletal muscle, 783, 784, 794, 800
Skin
 anatomy, 699
 patch, 498
Stratum corneum, 498, 521, 568, 698–701, 706–711
Sub-cellular localization, 685–688, 691, 795
Sublingual immunotherapy, 495, 505, 506
Subunit vaccine, 528, 562, 564, 565, 588, 616, 633, 678, 681, 683, 689, 690, 702, 721, 745, 771, 775, 776, 821, 826, 828, 831, 832, 852, 854
Superantigen-associated shock, 724

## T
T cell receptor, 479, 506, 510, 539, 544, 680, 681, 683, 687, 748, 797
Th1/Th2 response, 530
TLR. *See* Toll-like receptors (TLR)
Tolerogens, 482, 487, 505, 506, 510, 604, 605, 608, 609, 611, 616, 626, 756, 761, 772
Toll-like receptors (TLR), 484, 491, 520, 529, 538, 539, 544, 557, 573, 609, 614–616, 626, 637, 661, 662, 682, 683, 722, 728, 729, 794–797

Transcriptome, 679–680, 683–686
Transferosomes, 568
Transmembrane potential, 675, 782
Type 1 diabetes mellitus, 478
Type I allergy, 450, 489–499
Tyrosine phosphatase-like protein, 480

**U**

US Patent and Trademark Office (US PTO), 810, 814, 819

**V**

Vaccine Adverse Events Reportin System, 843
Vaccine
    design, 557, 562, 674, 677–691, 755, 776
    licensure, 804, 840
    potency, 828, 832, 834
Validation, 513, 689, 809, 828, 840, 848, 854, 860
Vector vaccine, 497, 767–776, 808

Veterinary medicinal products, 808, 855, 860–862
Veterinary vaccine, 565, 567, 634, 805, 851–857, 859–863
Virosomes, 566, 568, 733, 746, 749
Virus-like particles, 464, 469, 497, 554, 663, 703, 744, 746, 848
Viscosan, 625
Vitamin A, 509, 610, 682

**W**

World Health Organization (WHO), 449, 769, 826–828, 832, 843
World Intellectual Property Organization (WIPO), 808, 809

**Z**

Zinc, 480, 538, 539, 596–597, 610

Printed by Publishers' Graphics LLC